STUDENT'S SOLUTIONS MANUAL

BEVERLY FUSFIELD

COLLEGE ALGEBRA
ELEVENTH EDITION

Margaret L. Lial
American River College
John Hornsby
University of New Orleans
David I. Schneider
University of Maryland
Callie J. Daniels
St. Charles Community College

PEARSON

Boston   Columbus   Indianapolis   New York   San Francisco   Upper Saddle River
Amsterdam   Cape Town   Dubai   London   Madrid   Milan   Munich   Paris   Montreal   Toronto
Delhi   Mexico City   Sao Paulo   Sydney   Hong Kong   Seoul   Singapore   Taipei   Tokyo

The author and publisher of this book have used their best efforts in preparing this book. These efforts include the development, research, and testing of the theories and programs to determine their effectiveness. The author and publisher make no warranty of any kind, expressed or implied, with regard to these programs or the documentation contained in this book. The author and publisher shall not be liable in any event for incidental or consequential damages in connection with, or arising out of, the furnishing, performance, or use of these programs.

Reproduced by Pearson from electronic files supplied by the author.

Copyright © 2013, 2009, 2005 Pearson Education, Inc.
Publishing as Pearson, 75 Arlington Street, Boston, MA 02116.

All rights reserved. No part of this publication may be reproduced, stored in a retrieval system, or transmitted, in any form or by any means, electronic, mechanical, photocopying, recording, or otherwise, without the prior written permission of the publisher. Printed in the United States of America.

ISBN-13: 978-0-321-79138-2
ISBN-10: 0-321-79138-X

1 2 3 4 5 6 EB 16 15 14 13 12

www.pearsonhighered.com

PEARSON

CONTENTS

3 Polynomial and Rational Functions

4 Inverse, Exponential, and Logarithmic Functions

5 Systems and Matrices

6   Analytic Geometry

7   Further Topics in Algebra

Chapter R

REVIEW OF BASIC CONCEPTS

Section R.1   Sets

1. The set {4, 5, 6, ..., 15} has a limited number of elements, so it is a finite set. Yes, 10 is an element of the set.

3. The set $\left\{1, \frac{1}{2}, \frac{1}{4}, \frac{1}{8}, ...\right\}$ has an unlimited number of elements, so it is an infinite set. No, 10 is not an element of the set.

5. The set $\{x \mid x$ is a natural number larger than 11$\}$, which can also be written as {11, 12, 13, 14, ...}, has an unlimited number of elements, so it is an infinite set. No, 10 is not an element of the set.

7. There are an infinite number of fractions between 1 and 2, so $\{x \mid x$ is a fraction between 1 and 2$\}$ is an infinite set. No, 10 is not an element of the set.

9. The elements of the set {12, 13, 14, ..., 20} are all the natural numbers from 12 to 20 inclusive. There are 9 elements in the set, {12, 13, 14, 15, 16, 17, 18, 19, 20}.

11. Each element of the set $\left\{1, \frac{1}{2}, \frac{1}{4}, ..., \frac{1}{32}\right\}$ after the first is found by multiplying the preceding number by $\frac{1}{2}$. There are 6 elements in the set, $\left\{1, \frac{1}{2}, \frac{1}{4}, \frac{1}{8}, \frac{1}{16}, \frac{1}{32}\right\}$.

13. To find the elements of the set {17, 22, 27, ..., 47}, start with 17 and add 5 to find the next number. There are 7 elements in the set, {17, 22, 27, 32, 37, 42, 47}.

15. When you list all elements in the set {all natural numbers greater than 7 and less than 15}, you obtain {8, 9, 10, 11, 12, 13, 14}.

17. 6 is an element of the set {3, 4, 5, 6}, so we write $6 \in \{3, 4, 5, 6\}$.

19. −4 is not an element of {4, 6, 8, 10}, so we write $-4 \notin \{4, 6, 8, 10\}$.

21. 0 is an element of {2, 0, 3, 4}, so we write $0 \in \{2, 0, 3, 4\}$.

23. {3} is a subset of {2, 3, 4, 5}, not an element of {2, 3, 4, 5}, so we write $\{3\} \notin \{2, 3, 4, 5\}$.

25. {0} is a subset of {0, 1, 2, 5}, not an element of {0, 1, 2, 5}, so we write $\{0\} \notin \{0, 1, 2, 5\}$.

27. 0 is not an element of $\varnothing$, since the empty set contains no elements. Thus, $0 \notin \varnothing$.

29. $3 \in \{2, 5, 6, 8\}$
Since 3 is not one of the elements in {2, 5, 6, 8}, the statement is false.

31. $1 \in \{3, 4, 5, 11, 1\}$
Since 1 is one of the elements of {3, 4, 5, 11, 1}, the statement is true.

33. $9 \notin \{2, 1, 5, 8\}$
Since 9 is not one of the elements of {2, 1, 5, 8}, the statement is true.

35. $\{2, 5, 8, 9\} = \{2, 5, 9, 8\}$
This statement is true because both sets contain exactly the same four elements.

37. $\{5, 8, 9\} = \{5, 8, 9, 0\}$
These two sets are not equal because {5, 8, 9, 0} contains the element 0, which is not an element of {5, 8, 9}. Therefore, the statement is false.

39. $\{x \mid x$ is a natural number less than 3$\} = \{1, 2\}$
Since 1 and 2 are the only natural numbers less than 3, this statement is true.

For Exercises 41–51, A = {2, 4, 6, 8, 10, 12}, B = {2, 4, 8, 10}, C = {4, 10, 12}, D = {2, 10}, and U = {2, 4, 6, 8, 10, 12, 14}.

41. $A \subseteq U$
This statement says "A is a subset of U." Since every element of A is also an element of U, the statement is true.

43. $D \subseteq B$
Since both elements of D, 2 and 10, are also elements of B, D is a subset of B. The statement is true.

45. $A \subseteq B$
Set A contains two elements, 6 and 12, that are not elements of B. Thus, A is not a subset of B. The statement is false.

47. $\varnothing \subseteq A$
The empty set is a subset of every set, so the statement is true.

Copyright © 2013 Pearson Education, Inc.

49. $\{4, 8, 10\} \subseteq B$

Since 4, 8, and 10 are all elements of B, $\{4, 8, 10\}$ is a subset of B. The statement is true.

51. $B \subseteq D$

Since B contains two elements, 4 and 8, that are not elements of D, B is not a subset of D. The statement is false.

53. Every element of $\{2, 4, 6\}$ is also an element of $\{3, 2, 5, 4, 6\}$, so $\{2, 4, 6\}$ is a subset of $\{3, 2, 5, 4, 6\}$.

We write $\{2, 4, 6\} \subseteq \{3, 2, 5, 4, 6\}$.

55. Since 0 is an element of $\{0, 1, 2\}$, but is not an element of $\{1, 2, 3, 4, 5\}$, $\{0, 1, 2\}$ is not a subset of $\{1, 2, 3, 4, 5\}$. We write $\{0, 1, 2\} \not\subseteq \{1, 2, 3, 4, 5\}$.

57. The empty set is a subset of every set, so $\varnothing \subseteq \{1, 4, 6, 8\}$.

59. $\{5, 7, 9, 19\} \cap \{7, 9, 11, 15\} = \{7, 9\}$

Since 7 and 9 are the only elements belonging to both sets, the statement is true.

61. $\{2, 1, 7\} \cup \{1, 5, 9\} = \{1\}$

$\{2, 1, 7\} \cup \{1, 5, 9\} = \{1, 2, 5, 7, 9\}$, while $\{2, 1, 7\} \cap \{1, 5, 9\} = \{1\}$. Therefore, the statement is false.

63. $\{3, 2, 5, 9\} \cap \{2, 7, 8, 10\} = \{2\}$

Since 2 is the only element belonging to both sets, the statement is true.

65. $\{3, 5, 9, 10\} \cap \varnothing = \{3, 5, 9, 10\}$

In order to belong to the intersection of two sets, an element must belong to both sets. Since the empty set contains no elements, $\{3, 5, 9, 10\} \cap \varnothing = \varnothing$, so the statement is false.

67. $\{1, 2, 4\} \cup \{1, 2, 4\} = \{1, 2, 4\}$

Since the two sets are equal, their union contains the same elements, namely 1, 2, and 4. Thus, the statement is true.

69. $\varnothing \cup \varnothing = \varnothing$

Since the empty set contains no elements, the statement is true.

For Exercises 71–93,
$U = \{0, 1, 2, 3, 4, 5, 6, 7, 8, 9, 10, 11, 12, 13\}$,
$M = \{0, 2, 4, 6, 8\}$, $N = \{1, 3, 5, 7, 9, 11, 13\}$,
$Q = \{0, 2, 4, 6, 8, 10, 12\}$, and $R = \{0, 1, 2, 3, 4\}$.

71. $M \cap R$

The only elements belonging to both M and R are 0, 2, and 4, so $M \cap R = \{0, 2, 4\}$.

73. $M \cup N$

The union of two sets contains all elements that belong to either set or to both sets.
$M \cup N = \{0, 1, 2, 3, 4, 5, 6, 7, 8, 9, 11, 13\}$

75. $M \cap N$

There are no elements which belong to both M and N, so $M \cap N = \varnothing$. M and N are disjoint sets.

77. $N \cup R = \{0, 1, 2, 3, 4, 5, 7, 9, 11, 13\}$

79. N'

The set N' is the complement of set N, which means the set of all elements in the universal set U that do not belong to N.
$N' = Q$ or $\{0, 2, 4, 6, 8, 10, 12\}$

81. $M' \cap Q$

First form M', the complement of M. M' contains all elements of U that are not elements of M. Thus, $M' = \{1, 3, 5, 7, 9, 10, 11, 12, 13\}$. Now form the intersection of M' and Q. Thus, we have $M' \cap Q = \{10, 12\}$.

83. $\varnothing \cap R$

Since the empty set contains no elements, there are no elements belonging to both $\varnothing$ and R. Thus, $\varnothing$ and R are disjoint sets, and $\varnothing \cap R = \varnothing$.

85. $N \cup \varnothing$

Since $\varnothing$ contains no elements, the only elements belonging to N or $\varnothing$ are the elements of N. Thus, $\varnothing$ and N are disjoint sets, and $N \cup \varnothing = N$ or $\{1, 3, 5, 7, 9, 11, 13\}$.

87. $(M \cap N) \cup R$

First form the intersection of M and N. Since M and N have no common elements (they are disjoint), $M \cap N = \varnothing$. Thus, $(M \cap N) \cup R = \varnothing \cup R$. Now, since $\varnothing$ contains no elements, the only elements belonging to R or $\varnothing$ are the elements of R. Thus, $\varnothing$ and R are disjoint sets, and $\varnothing \cup R = R$ or $\{0, 1, 2, 3, 4\}$.

89. $(Q \cap M) \cup R$

First form the intersection of Q and M. We have $Q \cap M = \{0, 2, 4, 6, 8\} = M$. Now form the union of this set with R. We have $(Q \cap M) \cup R = M \cup R = \{0, 1, 2, 3, 4, 6, 8\}$.

Copyright © 2013 Pearson Education, Inc.

91. $(M' \cup Q) \cap R$

First, find M', the complement of M. We have $M' = \{1, 3, 5, 7, 9, 10, 11, 12, 13\}$. Next, form the union of M' and Q. We have $M' \cup Q = \{0, 1, 2, 3, 4, 5, 6, 7, 8, 9, 10, 11, 12, 13\} = U$. Thus, we have $(M' \cup Q) \cap R = U \cap R = R$ or $\{0, 1, 2, 3, 4\}$.

93. $Q' \cap (N' \cap U)$

First, find Q', the complement of Q. We have $Q' = \{1, 3, 5, 7, 9, 11, 13\} = N$. Now find N', the complement of N. We have $N' = \{0, 2, 4, 6, 8, 10, 12\} = Q$. Next, form the intersection of N' and U. We have $N' \cap U = Q \cap U = Q$ Finally, we have $Q' \cap (N' \cap U) = Q' \cap Q = \varnothing$ Since the intersection of Q' and $(N' \cap U)$ is $\varnothing$, Q' and $(N' \cap U)$ are disjoint sets.

Section R.2 Real Numbers and Their Properties

1. (a) 0 is a whole number. Therefore, it is also an integer, a rational number, and a real number. 0 belongs to B, C, D, F.

(b) 34 is a natural number. Therefore, it is also a whole number, an integer, a rational number, and a real number. 34 belongs to A, B, C, D, F.

(c) $-\frac{9}{4}$ is a rational number and a real number. $-\frac{9}{4}$ belongs to D, F.

(d) $\sqrt{36} = 6$ is a natural number. Therefore, it is also a whole number, an integer, a rational number, and a real number. $\sqrt{36}$ belongs to A, B, C, D, F.

(e) $\sqrt{13}$ is an irrational number and a real number. $\sqrt{13}$ belongs to E, F.

(f) $2.16 = \frac{216}{100} = \frac{54}{25}$ is a rational number and a real number. 2.16 belongs to D, F.

3. False. Positive integers are whole numbers, but negative integers are not.

5. False. No irrational numbers are integers.

7. True. Every natural number is a whole number.

9. True. Some rational numbers are whole numbers.

For Exercises 11–15,
$A = \left\{ -6, -\frac{12}{4}, -\frac{5}{8}, -\sqrt{3}, 0, \frac{1}{4}, 1, 2\pi, 3, \sqrt{12} \right\}$.

11. 1 and 3 are natural numbers.

13. -6, $-\frac{12}{4}$ (or -3), 0, 1, and 3 are integers.

15. $-\sqrt{3}$, 2π and $\sqrt{12}$ are irrational numbers.

17. $-2^4 = -(2 \cdot 2 \cdot 2 \cdot 2) = -16$

19. $(-2)^4 = (-2) \cdot (-2) \cdot (-2) \cdot (-2) = 16$

21. $(-3)^5 = (-3) \cdot (-3) \cdot (-3) \cdot (-3) \cdot (-3) = -243$

23. $-2 \cdot 3^4 = -2 \cdot (3 \cdot 3 \cdot 3 \cdot 3) = -2 \cdot 81 = -162$

25. $-2 \cdot 5 + 12 \div 3 = -10 + 12 \div 3$
$= -10 + 4 = -6$

27. $-4(9 - 8) + (-7)(2)^3 = -4(1) + (-7)(2)^3$
$= -4(1) + (-7) \cdot 8$
$= -4 + (-7) \cdot 8$
$= -4 + (-56) = -60$

29. $(4 - 2^3)(-2 + \sqrt{25}) = (4 - 8)(-2 + 5)$
$= (-4)(3) = -12$

31. $\left(-\frac{2}{9} - \frac{1}{4} \right) - \left[-\frac{5}{18} - \left(-\frac{1}{2} \right) \right]$
$= \left(-\frac{8}{36} - \frac{9}{36} \right) - \left(-\frac{5}{18} + \frac{9}{18} \right)$
$= \left(-\frac{17}{36} \right) - \left(\frac{4}{18} \right) = -\frac{17}{36} - \frac{8}{36} = -\frac{25}{36}$

33. $\dfrac{-8 + (-4)(-6) \div 12}{4 - (-3)} = \dfrac{-8 + 24 \div 12}{4 + 3}$
$= \dfrac{-8 + 2}{7}$
$= \dfrac{-6}{7} = -\dfrac{6}{7}$

For Exercises 35–47, $p = -4$, $q = 8$, and $r = -10$.

35. $-p^2 - 7q + r^2 = (-4)^2 - 7 \cdot 8 + (-10)^2$
$= (-4)^2 - 7 \cdot 8 + 100$
$= -16 - 7 \cdot 8 + 100$
$= -16 - 56 + 100$
$= -72 + 100 = 28$

Copyright © 2013 Pearson Education, Inc.

37. $\dfrac{q+r}{q+p} = \dfrac{8+(-10)}{8+(-4)} = \dfrac{-2}{4} = -\dfrac{1}{2}$

39. $\dfrac{3q}{r} - \dfrac{5}{p} = \dfrac{3\cdot 8}{-10} - \dfrac{5}{-4} = \dfrac{24}{-10} - \dfrac{5}{-4} = -\dfrac{12}{5} - \dfrac{5}{-4}$

$\quad = -\dfrac{12}{5} + \dfrac{5}{4} = -\dfrac{48}{20} + \dfrac{25}{20} = -\dfrac{23}{20}$

41. $\dfrac{5r}{2p-3r} = \dfrac{5(-10)}{2(-4)-3(-10)} = \dfrac{-50}{-8+30}$

$\quad = \dfrac{-50}{22} = -\dfrac{25}{11}$

43. $\dfrac{\frac{q}{2} - \frac{r}{3}}{\frac{3p}{4} + \frac{q}{8}} = \dfrac{\frac{8}{2} - \frac{-10}{3}}{\frac{3(-4)}{4} + \frac{8}{8}} = \dfrac{4 + \frac{10}{3}}{-3 + 1} = \dfrac{\frac{22}{3}}{-2} = -\dfrac{11}{3}$

45. $\dfrac{-(p+2)^2 - 3r}{2-q} = \dfrac{-(-4+2)^2 - 3(-10)}{2-8}$

$\quad = \dfrac{-(-2)^2 - 3(-10)}{-6}$

$\quad = \dfrac{-4 - 3(-10)}{-6} = \dfrac{-4 - (-30)}{-6}$

$\quad = \dfrac{-4+30}{-6} = \dfrac{26}{-6} = -\dfrac{13}{3}$

47. $\dfrac{3p + 3(4+p)^3}{r+8} = \dfrac{3(-4) + 3(4+(-4))^3}{(-10)+8}$

$\quad = \dfrac{-12 + 3(0)^3}{-2} = \dfrac{-12}{-2} = 6$

49. distributive   **51.** inverse

53. identity   **55.** commutative

57. associative   **59.** closure

61. No; in general $a - b \ne b - a$. Examples will vary, i.e. if $a = 15$ and $b = 0$, then $a - b = 15 - 0 = 15$, but $b - a = 0 - 15 = -15$.

63. $\dfrac{10}{11}(22z) = \left(\dfrac{10}{11}\cdot 22\right)z = 20z$

65. $(m+5) + 6 = m + (5+6) = m + 11$

67. $\dfrac{3}{8}\left(\dfrac{16}{9}y + \dfrac{32}{27}z - \dfrac{40}{9}\right)$

$\quad = \dfrac{3}{8}\left(\dfrac{16}{9}y\right) + \dfrac{3}{8}\left(\dfrac{32}{27}z\right) - \dfrac{3}{8}\left(\dfrac{40}{9}\right)$

$\quad = \left(\dfrac{3}{8}\cdot\dfrac{16}{9}\right)y + \left(\dfrac{3}{8}\cdot\dfrac{32}{27}\right)z - \dfrac{5}{3}$

$\quad = \dfrac{2}{3}y + \dfrac{4}{9}z - \dfrac{5}{3}$

69. $8p - 14p = (8-14)p = -6p$

71. $-4(z-y) = -4z - (-4y) = -4z + 4y$

73. The process in your head should be the following:

$72\cdot 17 + 28\cdot 17 = (72+28)(17)$

$\qquad\qquad\qquad\;\; = (100)(17) = 1700$

75. The process in your head should be the following:

$123\dfrac{5}{8}\cdot 1\dfrac{1}{2} - 23\dfrac{5}{8}\cdot 1\dfrac{1}{2} = \left(123\dfrac{5}{8} - 23\dfrac{5}{8}\right)\left(1\dfrac{1}{2}\right)$

$\qquad\qquad\qquad\qquad = (100)\left(1\dfrac{1}{2}\right) = 150$

77. This statement is false since $|6-8| = |-2| = 2$ and $|6| - |8| = 6 - 8 = -2$. A corrected statement would be $|6-8| \ne |6| - |8|$ or $|6-8| = |8| - |6|$

79. This statement is true since $|-5|\cdot|6| = 5\cdot 6 = 30$ and $|-5\cdot 6| = |-30| = 30$.

81. This statement is false. For example if you let $a = 2$ and $b = 6$ then $|2-6| = |-4| = 4$ and $|a| - |b| = |2| - |6| = 2 - 6 = -4$. A corrected statement is $|a-b| = |b| - |a|$, if $b > a > 0$.

83. $|-10| = 10$ **85.** $-\left|\dfrac{4}{7}\right| = -\dfrac{4}{7}$

87. $-|-8| = -8$

89. $P_d = |P - 125| = |116 - 125| = |-9| = 9$

The P_d value for a woman whose actual systolic pressure is 116 and whose normal value should be 125 is 9.

Copyright © 2013 Pearson Education, Inc.

For Exercises 91–97, $x = -4$ and $y = 2$.

91. $|3x - 2y| = |3(-4) - 2(2)| = |-12 - 4| = |-16| = 16$

93. $|-3x + 4y| = |-3(-4) + 4(2)| = |12 + 8| = |20| = 20$

95. $\dfrac{2|y| - 3|x|}{|xy|} = \dfrac{2|2| - 3|-4|}{|-4(2)|} = \dfrac{2(2) - 3(4)}{|-8|} = \dfrac{4 - 12}{8} = \dfrac{-8}{8} = -1$

97. $\dfrac{|-8y + x|}{-|x|} = \dfrac{|-8(2) + (-4)|}{-|-4|} = \dfrac{|-16 + (-4)|}{-|-4|} = \dfrac{|-20|}{-4} = \dfrac{20}{-4} = -5$

99. Property 2   **101.** Property 3   **103.** Property 1

105. $d(P,Q) = |-1 - (-4)| = |-1 + 4| = |3| = 3$ or $d(P,Q) = |-4 - (-1)| = |-4 + 1| = |-3| = 3$

107. $d(Q,R) = |8 - (-1)| = |8 + 1| = |9| = 9$ or $d(Q,R) = |-1 - 8| = |-9| = 9$

109. $xy > 0$ if x and y have the same sign.

111. $\dfrac{x}{y} < 0$ if x and y have different signs.

113. Since x^3 has the same sign as x, $\dfrac{x^3}{y} > 0$ if x and y have the same sign.

115. Since $|-16 - (+1)| = |-17| = 17$ and $|1 - (-16)| = |17| = 17$, the number of strokes between their scores is 17.

117. $\text{BAC} = 48 \times 3.2 \times 0.075 \div 190 - 2 \times 0.015 \approx 0.031$

119. Exercise 117: $\text{BAC} = 48 \times 3.2 \times 0.075 \div 215 - 2 \times 0.015 \approx 0.024$
Exercise 118: $\text{BAC} = 36 \times 4.0 \times 0.075 \div 160 - 3 \times 0.015 = 0.023$
The increased weight results in lower BACs.

For exercises 121–131, we use the formula $R = \dfrac{\left(250 \cdot \dfrac{C}{A}\right) + \left(1000 \cdot \dfrac{T}{A}\right) + \left(12.5 \cdot \dfrac{Y}{A}\right) + 6.25 - \left(1250 \cdot \dfrac{I}{A}\right)}{3}$, where

$0 \le \dfrac{C}{A} \le 0.775$, $0 \le \dfrac{T}{A} \le 0.11875$, $0 \le \dfrac{Y}{A} \le 12.5$, and $0 \le \dfrac{I}{A} \le 0.095$.

121. First, find the value of the numerator.

$\left(250 \cdot \dfrac{C}{A}\right) + \left(1000 \cdot \dfrac{T}{A}\right) + \left(12.5 \cdot \dfrac{Y}{A}\right) + 6.25 - \left(1250 \cdot \dfrac{I}{A}\right)$

$= \left(250 \cdot \dfrac{324}{492}\right) + \left(1000 \cdot \dfrac{36}{492}\right) + \left(12.5 \cdot \dfrac{3900}{492}\right) + 6.25 - \left(1250 \cdot \dfrac{1}{492}\right) \approx 332.98$

$R \approx \dfrac{332.98}{3} \approx 111.0$

123. First, find the value of the numerator.

$\left(250 \cdot \dfrac{C}{A}\right) + \left(1000 \cdot \dfrac{T}{A}\right) + \left(12.5 \cdot \dfrac{Y}{A}\right) + 6.25 - \left(1250 \cdot \dfrac{I}{A}\right)$

$= \left(250 \cdot \dfrac{312}{475}\right) + \left(1000 \cdot \dfrac{28}{475}\right) + \left(12.5 \cdot \dfrac{3922}{475}\right) + 6.25 - \left(1250 \cdot \dfrac{11}{475}\right) \approx 303.67$

$R \approx \dfrac{303.67}{3} = 101.2$

Copyright © 2013 Pearson Education, Inc.

125. First, find the value of the numerator.

$$\left(250 \cdot \frac{C}{A}\right)+\left(1000 \cdot \frac{T}{A}\right)+\left(12.5 \cdot \frac{Y}{A}\right)+6.25-\left(1250 \cdot \frac{I}{A}\right)$$

$$=\left(250 \cdot \frac{240}{389}\right)+\left(1000 \cdot \frac{17}{389}\right)+\left(12.5 \cdot \frac{3200}{389}\right)+6.25-\left(1250 \cdot \frac{5}{389}\right) \approx 290.95$$

$$R \approx \frac{290.95}{3} \approx 97.0$$

127. First, find the value of the numerator.

$$\left(250 \cdot \frac{C}{A}\right)+\left(1000 \cdot \frac{T}{A}\right)+\left(12.5 \cdot \frac{Y}{A}\right)+6.25-\left(1250 \cdot \frac{I}{A}\right)$$

$$=\left(250 \cdot \frac{306}{489}\right)+\left(1000 \cdot \frac{25}{489}\right)+\left(12.5 \cdot \frac{3622}{489}\right)+6.25-\left(1250 \cdot \frac{10}{489}\right) \approx 280.84$$

$$R \approx \frac{280.84}{3} \approx 93.6$$

129. First, find the value of the numerator.

$$\left(250 \cdot \frac{C}{A}\right)+\left(1000 \cdot \frac{T}{A}\right)+\left(12.5 \cdot \frac{Y}{A}\right)+6.25-\left(1250 \cdot \frac{I}{A}\right)$$

$$=\left(250 \cdot \frac{365}{574}\right)+\left(1000 \cdot \frac{24}{574}\right)+\left(12.5 \cdot \frac{4370}{574}\right)+6.25-\left(1250 \cdot \frac{12}{574}\right) \approx 276.07$$

$$R \approx \frac{276.07}{3} \approx 92.0$$

131. First, find the value of the numerator.

$$\left(250 \cdot \frac{C}{A}\right)+\left(1000 \cdot \frac{T}{A}\right)+\left(12.5 \cdot \frac{Y}{A}\right)+6.25-\left(1250 \cdot \frac{I}{A}\right)$$

$$=\left(250 \cdot \frac{357}{571}\right)+\left(1000 \cdot \frac{28}{571}\right)+\left(12.5 \cdot \frac{3705}{571}\right)+6.25-\left(1250 \cdot \frac{9}{571}\right) \approx 273.0$$

$$R \approx \frac{273.0}{3} \approx 91.0$$

133. First, find the value of the numerator.

$$\left(250 \cdot \frac{C}{A}\right)+\left(1000 \cdot \frac{T}{A}\right)+\left(12.5 \cdot \frac{Y}{A}\right)+6.25-\left(1250 \cdot \frac{I}{A}\right)$$

$$=\left(250 \cdot \frac{336}{497}\right)+\left(1000 \cdot \frac{49}{497}\right)+\left(12.5 \cdot \frac{4557}{497}\right)+6.25-\left(1250 \cdot \frac{10}{497}\right) \approx 363.32$$

$$R \approx \frac{363.32}{3} \approx 121.1$$

Section R.3   Polynomials

1. $\left(-4x^5\right)\left(4x^2\right)=(-4 \cdot 4)\left(x^5 x^2\right)$
$$=-16x^{5+2}=-16x^7$$

3. $n^6 \cdot n^4 \cdot n = n^{6+4+1}=n^{11}$

5. $9^3 \cdot 9^5 = 9^{3+5}=9^8$

7. $\left(-3m^4\right)\left(6m^2\right)\left(-4m^5\right)$
$$=\left[(-3)(6)(-4)\right]\left(m^4 m^2 m^5\right)$$
$$=72m^{4+2+5}=72m^{11}$$

9. $\left(5x^2 y\right)\left(-3x^3 y^4\right)=\left[5(-3)\right]\left(x^2 x^3\right)\left(y y^4\right)$
$$=-15x^{2+3}y^{1+4}=-15x^5 y^5$$

Copyright © 2013 Pearson Education, Inc.

11. $(2^2)^5 = 2^{2 \cdot 5} = 2^{10}$

13. $\left(-6x^2\right)^3 = (-6)^3 \left(x^2\right)^3 = (-6)^3 x^6 = -216x^6$

15. $-(4m^3 n^0)^2 = -\left[4^2 (m^3)^2 (n^0)^2\right]$
$$= -4^2 m^{3 \cdot 2} n^{0 \cdot 2}$$
$$= -4^2 m^6 n^0 = -\left(4^2\right) m^6 \cdot 1$$
$$= -4^2 m^6 = -16 m^6$$

17. $\left(\dfrac{r^8}{s^2}\right)^3 = \dfrac{(r^8)^3}{(s^2)^3} = \dfrac{r^{8 \cdot 3}}{s^{2 \cdot 3}} = \dfrac{r^{24}}{s^6}$

19. $\left(\dfrac{-4m^2}{tp^2}\right)^4 = \dfrac{(-4)^4 (m^2)^4}{t^4 \left(p^2\right)^4} = \dfrac{(-4)^4 m^8}{t^4 p^8} = \dfrac{256 m^8}{t^4 p^8}$

21. $-\left(\dfrac{x^3 y^5}{z}\right)^0 = -1$

23. **(a)** $6^0 = 1$; B   **(b)** $-6^0 = -1$; C

   (c) $(-6)^0 = 1$; B   **(d)** $-(-6)^0 = -1$; C

25. Answers will vary.
$$x^2 + x^2 = 2x^2$$

27. $-5x^{11}$ is a polynomial. It is a monomial since it has one term. It has degree 11.

29. $6x + 3x^4$ is a polynomial. It has degree 4 since the term $3x^4$ has the greatest degree. It is a binomial since there are two terms.

31. $-7z^5 - 2z^3 + 1$ is a polynomial. It has degree 5 since the term $-7z^5$ has the greatest degree. It is a trinomial since there are three terms.

33. $15a^2 b^3 + 12a^3 b^8 - 13b^5 + 12b^6$ is a polynomial. It has degree 11 since the term $12a^3 b^8$ has the greatest degree. It is not a monomial, binomial, or trinomial.

35. $\dfrac{3}{8} x^5 - \dfrac{1}{x^2} + 9$ is not a polynomial since one term has a variable in the denominator.

37. 5 is a polynomial of degree 0. It is a monomial.

39. $\left(5x^2 - 4x + 7\right) + \left(-4x^2 + 3x - 5\right)$
$$= \left[5 + (-4)\right] x^2 + (-4 + 3) x + \left[7 + (-5)\right]$$
$$= 1 \cdot x^2 + (-1)x + 2 = x^2 - x + 2$$

41. $2\left(12y^2 - 8y + 6\right) - 4\left(3y^2 - 4y + 2\right)$
$$= 2\left(12y^2\right) - 2(8y) + 2(6) - 4\left(3y^2\right)$$
$$\qquad\qquad - 4(-4y) - 4 \cdot 2$$
$$= 24y^2 - 16y + 12 - 12y^2 + 16y - 8$$
$$= 12y^2 + 4$$

43. $\left(6m^4 - 3m^2 + m\right) - \left(2m^3 + 5m^2 + 4m\right) + \left(m^2 - m\right)$
$$= 6m^4 - 3m^2 + m - 2m^3 - 5m^2 - 4m + m^2 - m$$
$$= 6m^4 - 2m^3 + (-3 - 5 + 1)m^2 + (1 - 4 - 1)m$$
$$= 6m^4 - 2m^3 + (-7)m^2 + (-4)m$$
$$= 6m^4 - 2m^3 - 7m^2 - 4m$$

45. $(4r - 1)(7r + 2) = 4r(7r) + 4r(2) - 1(7r) - 1(2)$
$$= 28r^2 + 8r - 7r - 2$$
$$= 28r^2 + r - 2$$

47. $x^2 \left(3x - \dfrac{2}{3}\right)\left(5x + \dfrac{1}{3}\right)$
$$= x^2 \left[\left(3x - \dfrac{2}{3}\right)\left(5x + \dfrac{1}{3}\right)\right]$$
$$= x^2 \left[(3x)(5x) + (3x)\left(\dfrac{1}{3}\right) - \dfrac{2}{3}(5x) - \dfrac{2}{3}\left(\dfrac{1}{3}\right)\right]$$
$$= x^2 \left(15x^2 + x - \dfrac{10}{3}x - \dfrac{2}{9}\right)$$
$$= x^2 \left(15x^2 + \dfrac{3}{3}x - \dfrac{10}{3}x - \dfrac{2}{9}\right)$$
$$= x^2 \left(15x^2 - \dfrac{7}{3}x - \dfrac{2}{9}\right) = 15x^4 - \dfrac{7}{3}x^3 - \dfrac{2}{9}x^2$$

49. $4x^2 \left(3x^3 + 2x^2 - 5x + 1\right)$
$$= 4x^2 \left(3x^3\right) + 4x^2 \left(2x^2\right) - 4x^2 (5x) + 4x^2 \cdot 1$$
$$= 12x^5 + 8x^4 - 20x^3 + 4x^2$$

Copyright © 2013 Pearson Education, Inc.

51. $(2z-1)(-z^2+3z-4)$

$= (2z-1)(-z^2)+(2z-1)(3z)-(2z-1)(4)$

$= 2z(-z^2)-1(-z^2)+2z(3z)-1(3z)$
$\quad - (2z)(4)-(-1)(4)$

$= -2z^3+z^2+6z^2-3z-8z-(-4)$

$= -2z^3+7z^2-11z+4$

We may also multiply vertically.

$$
\begin{array}{r}
-z^2+3z-4 \\
2z-1 \\
\hline
z^2-3z+4 \quad \leftarrow -1(-z^2+3z-4) \\
-2z^3+6z^2-8z \quad \quad \leftarrow 2z(-z^2+3z-4) \\
\hline
-2z^3+7z^2-11z+4
\end{array}
$$

53. $(m-n+k)(m+2n-3k)$

$= (m-n+k)(m)+(m-n+k)(2n)$
$\quad - (m-n+k)(3k)$

$= m^2-mn+km+2mn-2n^2$
$\quad +2kn-3km+3kn-3k^2$

$= m^2+mn-2n^2-2km+5kn-3k^2$

We may also multiply vertically.

$$
\begin{array}{r}
m-n+k \\
m+2n-3k \\
\hline
-3km+3kn-3k^2 \\
2mn-2n^2 \quad\quad +2kn \\
m^2-mn \quad\quad +km \\
\hline
m^2+mn-2n^2-2km+5kn-3k^2
\end{array}
$$

55. $(2x+3)(2x-3)(4x^2-9)$

Note that $(2x+3)(2x-3)$ is a special product, the difference of the sum and difference of squares.

$(2x+3)(2x-3)(4x^2-9)=(4x^2-9)(4x^2-9)$

Note that $(4x^2-9)(4x^2-9)$ is a special

product, the square of a binomial.

$(4x^2-9)(4x^2-9)=16x^4-72x^2+81$

Thus,

$(2x+3)(2x-3)(4x^2-9)=16x^4-72x^2+81.$

57. $(x+1)(x+1)(x-1)(x-1)$

$= [(x+1)(x+1)][(x-1)(x-1)]$

$= (x^2+2x+1)(x^2-2x+1)$

$= x^2(x^2-2x+1)+2x(x^2-2x+1)$
$\quad\quad +1(x^2-2x+1)$

$= x^4-2x^3+x^2+2x^3-4x^2+2x+x^2-2x+1$

$= x^4-2x^2+1$

Alternatively,

$(x+1)(x+1)(x-1)(x-1)$

$= (x+1)(x-1)(x+1)(x-1)$

$= (x^2-1)(x^2-1)$

$= x^4-2x^2+1$

59. $(2m+3)(2m-3)=(2m)^2-3^2$
$\quad\quad\quad\quad\quad\quad = 4m^2-9$

61. $(4x^2-5y)(4x^2+5y)=(4x^2)^2-(5y)^2$
$\quad\quad\quad\quad\quad\quad\quad = 16x^4-25y^2$

63. $(4m+2n)^2=(4m)^2+2(4m)(2n)+(2n)^2$
$\quad\quad\quad\quad\quad = 16m^2+16mn+4n^2$

65. $(5r-3t^2)^2=(5r)^2-2(5r)(3t^2)+(3t^2)^2$
$\quad\quad\quad\quad\quad = 25r^2-30rt^2+9t^4$

67. $\left[(2p-3)+q\right]^2$

$= (2p-3)^2+2(2p-3)(q)+q^2$

$= (2p)^2-2(2p)(3)+(3)^2+4pq-6q+q^2$

$= 4p^2-12p+9+4pq-6q+q^2$

69. $[(3q+5)-p][(3q+5)+p]$

$= (3q+5)^2-p^2$

$= [(3q)^2+2(3q)(5)+5^2]-p^2$

$= 9q^2+30q+25-p^2$

71. $[(3a+b)-1]^2=(3a+b)^2-2(3a+b)(1)+1^2$

$\quad\quad\quad = (9a^2+6ab+b^2)-2(3a+b)+1$

$\quad\quad\quad = 9a^2+6ab+b^2-6a-2b+1$

73. $(y+2)^3=(y+2)^2(y+2)$

$= (y^2+4y+4)(y+2)$

$= y^3+4y^2+4y+2y^2+8y+8$

$= y^3+6y^2+12y+8$

Copyright © 2013 Pearson Education, Inc.

75. $(q-2)^4 = (q-2)^2(q-2)^2$

$= (q^2 - 4q + 4)(q^2 - 4q + 4)$

$= q^4 - 4q^3 + 4q^2 - 4q^3 + 16q^2$

$\quad - 16q + 4q^2 - 16q + 16$

$= q^4 - 8q^3 + 24q^2 - 32q + 16$

77. $\left(p^3 - 4p^2 + p\right) - \left(3p^2 + 2p + 7\right)$

$= p^3 - 4p^2 + p - 3p^2 - 2p - 7$

$= p^3 - 7p^2 - p - 7$

79. $(7m + 2n)(7m - 2n) = (7m)^2 - (2n)^2$

$= 49m^2 - 4n^2$

81. $-3(4q^2 - 3q + 2) + 2(-q^2 + q - 4)$

$= -12q^2 + 9q - 6 - 2q^2 + 2q - 8$

$= -14q^2 + 11q - 14$

83. $p(4p - 6) + 2(3p - 8) = 4p^2 - 6p + 6p - 16$

$= 4p^2 - 16$

85. $-y(y^2 - 4) + 6y^2(2y - 3)$

$= -y^3 + 4y + 12y^3 - 18y^2$

$= 11y^3 - 18y^2 + 4y$

87.

$$-2x^2 \overline{\smash{\big)}\, -4x^7 - 14x^6 + 10x^4 - 14x^2} \quad \begin{array}{l} 2x^5 + 7x^4 - 5x^2 + 7 \end{array}$$

$\underline{-4x^7}$

$\quad\quad -14x^6$

$\quad\quad \underline{-14x^6}$

$\quad\quad\quad\quad 10x^4$

$\quad\quad\quad\quad \underline{10x^4}$

$\quad\quad\quad\quad\quad\quad -14x^2$

$\quad\quad\quad\quad\quad\quad \underline{-14x^2}$

$\quad\quad\quad\quad\quad\quad\quad\quad 0$

$\dfrac{-4x^7 - 14x^6 + 10x^4 - 14x^2}{-2x^2} = -2x^5 + 7x^4 - 5x^2 + 7$

89.

$$x - 2 \overline{\smash{\big)}\, 4x^3 - 3x^2 + 0x + 1} \quad \begin{array}{l} 4x^2 + 5x + 10 \end{array}$$

$\underline{4x^3 - 8x^2}$

$\quad\quad 5x^2 + 0x$

$\quad\quad \underline{5x^2 - 10x}$

$\quad\quad\quad\quad 10x + 1$

$\quad\quad\quad\quad \underline{10x - 20}$

$\quad\quad\quad\quad\quad\quad 21$

$\dfrac{4x^3 - 3x^2 + 1}{x - 2} = 4x^2 + 5x + 10 + \dfrac{21}{x - 2}$

91.

$$3m + 2 \overline{\smash{\big)}\, 6m^3 + 7m^2 - 4m + 2} \quad \begin{array}{l} 2m^2 + m - 2 \end{array}$$

$\underline{6m^3 + 4m^2}$

$\quad\quad 3m^2 - 4m$

$\quad\quad \underline{3m^2 + 2m}$

$\quad\quad\quad\quad -6m + 2$

$\quad\quad\quad\quad \underline{-6m - 4}$

$\quad\quad\quad\quad\quad\quad 6$

$\dfrac{6m^3 + 7m^2 - 4m + 2}{3m + 2} = 2m^2 + m - 2 + \dfrac{6}{3m + 2}$

93.

$$x^2 + 0x + 3 \overline{\smash{\big)}\, x^4 + 0x^3 + 5x^2 + 5x + 27} \quad \begin{array}{l} x^2 \quad\quad + 2 \end{array}$$

$\underline{x^4 + 0x^3 + 3x^2}$

$\quad\quad\quad 2x^2 + 5x + 27$

$\quad\quad\quad \underline{2x^2 + 0x + 6}$

$\quad\quad\quad\quad\quad\quad 5x + 21$

$\dfrac{x^4 + 5x^2 + 5x + 27}{x^2 + 3} = x^2 + 2 + \dfrac{5x + 21}{x^2 + 3}$

95. $99 \times 101 = (100 - 1)(100 + 1) = 100^2 - 1^2$

$= 10,000 - 1 = 9999$

97. $102^2 = (100 + 2)^2 = 100^2 + 2(100)(2) + 2^2$

$= 10,000 + 400 + 4 = 10,404$

99. **(a)** The area of the largest square is

$s^2 = (x + y)^2$.

(b) The areas of the two squares are

x^2 and y^2. The area of each rectangle is

xy. Therefore, the area of the largest

square can be written as $x^2 + 2xy + y^2$.

(c) Answers will vary. The total area must

equal the sum of the four parts.

(d) It reinforces the special product for

squaring a binomial:

$(x + y)^2 = x^2 + 2xy + y^2$.

101. **(a)** The volume is

$V = \dfrac{1}{3}h(a^2 + ab + b^2)$

$= \dfrac{1}{3}(200)(314^2 + 314 \times 756 + 756^2)$

$\approx 60,501,000\,\text{ft}^3$

(b) The shape becomes a rectangular box

with a square base. Its volume is given by

length × width × height or b^2h.

Copyright © 2013 Pearson Education, Inc.

(c) If we let $a = b$, then

$$V = \frac{1}{3}h(a^2 + ab + b^2) \text{ becomes}$$

$$V = \frac{1}{3}h(b^2 + bb + b^2), \text{ which simplifies}$$

to $V = hb^2$. Yes, the Egyptian formula gives the same result.

103. $x = 1940$

$0.001147(1940)^2 - 4.5905(1940) + 4595 \approx 6.3$

The formula is exact.

105. $x = 1990$

$0.001147(1990)^2 - 4.5905(1990) + 4595 \approx 2.1$

The formula is 0.3 low.

107. $(0.25^3)(400^3) = [(0.25)(400)]^3$

$$= 100^3 = 1,000,000$$

109. $\dfrac{4.2^5}{2.1^5} = \left(\dfrac{4.2}{2.1}\right)^5 = 2^5 = 32$

Section R.4   Factoring Polynomials

1. The greatest common factor is 12.

$12m + 60 = 12(m) + 12(5) = 12(m + 5)$

3. The greatest common factor is $8k$.

$8k^3 + 24k = 8k(k^2) + 8k(3) = 8k(k^2 + 3)$

5. The greatest common factor is xy.

$xy - 5xy^2 = xy \cdot 1 - xy(5y) = xy(1 - 5y)$

7. The greatest common factor is $-2p^2q^4$.

$-4p^3q^4 - 2p^2q^5$

$= (-2p^2q^4)(2p) + (-2p^2q^4)(q)$

$= -2p^2q^4(2p + q)$

9. The greatest common factor is $4k^2m^3$.

$4k^2m^3 + 8k^4m^3 - 12k^2m^4$

$= (4k^2m^3)(1) + (4k^2m^3)(2k^2) - (4k^2m^3)(3m)$

$= 4k^2m^3(1 + 2k^2 - 3m)$

11. The greatest common factor is $2(a + b)$.

$2(a + b) + 4m(a + b)$

$= [2(a + b)](1) + [2(a + b)](2m)$

$= 2(a + b)(1 + 2m)$

13. $(5r - 6)(r + 3) - (2r - 1)(r + 3)$

$= (r + 3)[(5r - 6) - (2r - 1)]$

$= (r + 3)[5r - 6 - 2r + 1]$

$= (r + 3)(3r - 5)$

15. $2(m - 1) - 3(m - 1)^2 + 2(m - 1)^3$

$= (m - 1)[2 - 3(m - 1) + 2(m - 1)^2]$

$= (m - 1)[2 - 3m + 3 + 2(m^2 - 2m + 1)]$

$= (m - 1)(2 - 3m + 3 + 2m^2 - 4m + 2)$

$= (m - 1)(2m^2 - 7m + 7)$

17. The completely factored form of $4x^2y^5 - 8xy^3$

is $4xy^3(xy^2 - 2)$.

19. $6st + 9t - 10s - 15 = (6st + 9t) - (10s + 15)$

$= 3t(2s + 3) - 5(2s + 3)$

$= (2s + 3)(3t - 5)$

21. $2m^4 + 6 - am^4 - 3a$

$= (2m^4 + 6) - (am^4 + 3a)$

$= 2(m^4 + 3) - a(m^4 + 3) = (m^4 + 3)(2 - a)$

23. $p^2q^2 - 10 - 2q^2 + 5p^2$

$= p^2q^2 - 2q^2 + 5p^2 - 10$

$= q^2(p^2 - 2) + 5(p^2 - 2)$

$= (p^2 - 2)(q^2 + 5)$

25. The positive factors of 6 could be 2 and 3, or 1 and 6. Since the middle term is negative, we know the factors of 4 must both be negative. As factors of 4, we could have −1 and −4, or −2 and −2. Try different combinations of these factors until the correct one is found.

$6a^2 - 11a + 4 = (2a - 1)(3a - 4)$

27. The positive factors of 3 are 1 and 3. Since the middle term is positive, we know the factors of 8 must both be positive. As factors of 8, we could have 1 and 8, or 2 and 4. Try different combinations of these factors until the correct one is found.

$3m^2 + 14m + 8 = (3m + 2)(m + 4)$

29. The positive factors of 15 are 1 and 15, or 3 and 5. Since the middle term is positive, we know the factors of 8 must both be positive. As factors of 8, we could have 1 and 8, or 2 and 4. Trying different combinations of these factors we find that $15p^2 + 24p + 8$ is prime.

Copyright © 2013 Pearson Education, Inc.

31. Factor out the greatest common factor, $2a$:

$12a^3 + 10a^2 - 42a = 2a(6a^2 + 5a - 21)$. Now

factor the trinomial by trial and error:

$6a^2 + 5a - 21 = (3a + 7)(2a - 3)$. Thus,

$12a^3 + 10a^2 - 42a = 2a(3a + 7)(2a - 3)$.

33. The positive factors of 6 could be 2 and 3, or 1 and 6. As factors of –6, we could have –1 and 6, –6 and 1, –2 and 3, or –3 and 2. Try different combinations of these factors until the correct one is found.

$6k^2 + 5kp - 6p^2 = (2k + 3p)(3k - 2p)$

35. The positive factors of 5 can only be 1 and 5. As factors of –6, we could have –1 and 6, –6 and 1, –2 and 3, or –3 and 2. Try different combinations of these factors until the correct one is found.

$5a^2 - 7ab - 6b^2 = (5a + 3b)(a - 2b)$

37. The positive factors of 12 could be 4 and 3, 2 and 6, or 1 and 12. The factors of –1 are 1 and –1. Try different combination of these factors until the correct one is found.

$12x^2 - xy - y^2 = (4x + y)(3x - y)$

39. Factor out the greatest common factor, $2a^2$:

$24a^4 + 10a^3b - 4a^2b^2 = 2a^2\left(12a^2 + 5ab - 2b^2\right)$

Now factor the trinomial by trial and error:

$12a^2 + 5ab - 2b^2 = (4a - b)(3a + 2b)$

Thus,

$24a^4 + 10a^3b - 4a^2b^2 = 2a^2\left(12a^2 + 5ab - 2b^2\right)$

$= 2a^2(4a - b)(3a + 2b)$

41. $9m^2 - 12m + 4 = (3m)^2 - 12m + 2^2$

$= (3m)^2 - 2(3m)(2) + 2^2$

$= (3m - 2)^2$

43. $32a^2 + 48ab + 18b^2$

$= 2(16a^2 + 24ab + 9b^2)$

$= 2[(4a)^2 + 24ab + (3b)^2]$

$= 2[(4a)^2 + 2(4a)(3b) + (3b)^2] = 2(4a + 3b)^2$

45. $4x^2y^2 + 28xy + 49 = (2xy)^2 + 28xy + 7^2$

$= (2xy)^2 + 2(2xy)(7) + 7^2$

$= (2xy + 7)^2$

47. $(a - 3b)^2 - 6(a - 3b) + 9$

$= (a - 3b)^2 - 6(a - 3b) + 3^2$

$= (a - 3b)^2 - 2(a - 3b)(3) + 3^2$

$= [(a - 3b) - 3]^2 = (a - 3b - 3)^2$

49. (a) Since $(x + 5y)^2 = x^2 + 10xy + 25y^2$, a matches B.

(b) Since $(x - 5y)^2 = x^2 - 10xy + 25y^2$, b matches C.

(c) Since $(x + 5y)(x - 5y) = x^2 - 25y^2$, c matches A.

(d) Since $(5y + x)(5y - x) = 25y^2 - x^2$, d matches D.

51. $9a^2 - 16 = (3a)^2 - 4^2$

$= (3a + 4)(3a - 4)$

53. $x^4 - 16 = \left(x^2 + 4\right)\left(x^2 - 4\right)$

$= \left(x^2 + 4\right)(x + 2)(x - 2)$

55. $25s^4 - 9t^2 = (5s^2)^2 - (3t)^2$

$= (5s^2 + 3t)(5s^2 - 3t)$

57. $(a + b)^2 - 16 = (a + b)^2 - 4^2$

$= [(a + b) + 4][(a + b) - 4]$

$= (a + b + 4)(a + b - 4)$

59. $p^4 - 625 = (p^2)^2 - 25^2 = (p^2 + 25)(p^2 - 25)$

$= (p^2 + 25)(p^2 - 5^2)$

$= (p^2 + 25)(p + 5)(p - 5)$

Note that $p^2 + 25$ is a prime factor.

61. $8 - a^3 = 2^3 - a^3 = (2 - a)(2^2 + 2 \cdot a + a^2)$

$= (2 - a)(4 + 2a + a^2)$

63. $125x^3 - 27 = (5x)^3 - 3^3$

$= (5x - 3)\left[(5x)^2 + 5x \cdot 3 + 3^2\right]$

$= (5x - 3)(25x^2 + 15x + 9)$

65. $27y^9 + 125z^6$

$= (3y^3)^3 + (5z^2)^3$

$= (3y^3 + 5z^2)\left[(3y^3)^2 - (3y^3)(5z^2) + (5z^2)^2\right]$

$= (3y^3 + 5z^2)(9y^6 - 15y^3z^2 + 25z^4)$

Copyright © 2013 Pearson Education, Inc.

67. $(r+6)^3 - 216$
$= (r+6)^3 - 6^3$
$= \left[(r+6)-6\right]\left[(r+6)^2 + (r+6)(6) + 6^2\right]$
$= \left[(r+6)-6\right]\left[r^2 + 12r + 36 + (r+6)(6) + 6^2\right]$
$= \left[r+6-6\right]\left[r^2 + 12r + 36 + 6r + 36 + 36\right]$
$= r(r^2 + 18r + 108)$

69. $27 - (m+2n)^3$
$= 3^3 - (m+2n)^3$
$= \left[3-(m+2n)\right]\cdot\left[3^2 + (3)(m+2n) + (m+2n)^2\right]$
$= \left[3-(m+2n)\right]\cdot$
$\qquad \left[3^2 + (3)(m+2n) + m^2 + 4mn + 4n^2\right]$
$= (3-m-2n)(9 + 3m + 6n + m^2 + 4mn + 4n^2)$

71. The correct complete factorization of $x^4 - 1$ is choice B: $(x^2+1)(x+1)(x-1)$. Choice A is not a complete factorization, since $x^2 - 1$ can be factored as $(x+1)(x-1)$. The other choices are not correct factorizations of $x^4 - 1$.

73. $x^6 - 1 = (x^3)^2 - 1^2$
$\qquad = (x^3+1)(x^3-1)$ or $(x^3-1)(x^3+1)$
Use the patterns for the difference of cubes and sum of cubes to factor further. Since
$x^3 - 1 = (x-1)(x^2+x+1)$ and
$x^3 + 1 = (x+1)(x^2-x+1)$,
we obtain the factorization
$x^6 - 1 = (x^3-1)(x^3+1)$
$\qquad = (x-1)(x^2+x+1)(x+1)(x^2-x+1)$

75. From Exercise 73, we have
$x^6 - 1 = (x-1)(x^2+x+1)(x+1)(x^2-x+1)$.
From Exercise 74, we have
$x^6 - 1 = (x-1)(x+1)(x^4+x^2+1)$.
Comparing these answers, we see that
$x^4 + x^2 + 1 = (x^2-x+1)(x^2+x+1)$.

77. The answer in Exercise 75 and the final line in Exercise 76 are the same.

79. Let $x = 3k - 1$. This substitution gives
$7(3k-1)^2 + 26(3k-1) - 8 = 7x^2 + 26x - 8$
$\qquad = (7x-2)(x+4)$
Replacing x with $3k - 1$ gives
$7(3k-1)^2 + 26(3k-1) - 8$
$= [7(3k-1)-2][(3k-1)+4]$
$= (21k - 7 - 2)(3k - 1 + 4)$
$= (21k-9)(3k+3) = 3(7k-3)(3)(k+1)$
$= 9(7k-3)(k+1)$.

81. Let $x = a - 4$. This substitution gives
$9(a-4)^2 + 30(a-4) + 25$
$= 9x^2 + 30x + 25$
$= (3x)^2 + 2(3x)(5) + 5^2 = (3x+5)^2$
Replacing x by $a - 4$ gives
$9(a-4)^2 + 30(a-4) + 25$
$= [3(a-4)+5]^2$
$= (3a - 12 + 5)^2 = (3a-7)^2$

83. $m^4 - 3m^2 - 10$
Let $x = m^2$, so that $x^2 = (m^2)^2 = m^4$.
$x^2 - 3x - 10 = (x-5)(x+2)$.
Replacing x with m^2 gives
$m^4 - 3m^2 - 10 = (m^2-5)(m^2+2)$.

85. $4b^2 + 4bc + c^2 - 16$
$= (4b^2 + 4bc + c^2) - 16$
$= \left[(2b)^2 + 2(2b)(c) + c^2\right] - 16$
$= (2b+c)^2 - 4^2 = [(2b+c)+4][(2b+c)-4]$
$= (2b+c+4)(2b+c-4)$

87. $x^2 + xy - 5x - 5y = (x^2+xy) - (5x+5y)$
$\qquad = x(x+y) - 5(x+y)$
$\qquad = (x+y)(x-5)$

89. $p^4(m-2n) + q(m-2n) = (m-2n)(p^4+q)$

91. $4z^2 + 28z + 49 = (2z)^2 + 2(2z)(7) + 7^2$
$\qquad = (2z+7)^2$

93. $1000x^3 + 343y^3$
$= (10x)^3 + (7y)^3$
$= (10x+7y)\left[(10x)^2 - (10x)(7y) + (7y)^2\right]$
$= (10x+7y)(100x^2 - 70xy + 49y^2)$

Copyright © 2013 Pearson Education, Inc.

95. $125m^6 - 216 = (5m^2)^3 - 6^3$
$$= (5m^2 - 6)\left[(5m^2)^2 + 5m^2(6) + 6^2\right]$$
$$= (5m^2 - 6)(25m^4 + 30m^2 + 36)$$

97. $64 + (3x + 2)^3$
$$= 4^3 + (3x + 2)^3$$
$$= \left[4 + (3x + 2)\right]\left[4^2 - (4)(3x + 2) + (3x + 2)^2\right]$$
$$= \left[4 + (3x + 2)\right]\left[\begin{matrix}4^2 - (4)(3x + 2)\\ +9x^2 + 12x + 4\end{matrix}\right]$$
$$= (4 + 3x + 2)(16 - 12x - 8 + 9x^2 + 12x + 4)$$
$$= (3x + 6)(9x^2 + 12)$$
$$= 3(x + 2)(3)(3x^2 + 4) = 9(x + 2)(3x^2 + 4)$$

99. $(x + y)^3 - (x - y)^3$
Let $a = x + y$ and let $b = x - y$. Then we have
$$(x + y)^3 - (x - y)^3 = a^3 - b^3$$
$$= (a - b)(a^2 + ab + b^2)$$
Replacing a and b gives
$$(a - b)(a^2 - ab + b^2)$$
$$= \left[(x + y) - (x - y)\right]$$
$$\cdot \left[(x + y)^2 + (x + y)(x - y) + (x - y)^2\right]$$
$$= 2y\left[\begin{matrix}x^2 + 2xy + y^2 + (x^2 - y^2)\\ + x^2 - 2xy + y^2\end{matrix}\right]$$
$$= 2y(3x^2 + y^2)$$

101. $144z^2 + 121$
The sum of squares cannot be factored.
$144z^2 + 121$ is prime.

103. $(x + y)^2 - (x - y)^2$
$$= [(x + y) + (x - y)][(x + y) - (x - y)]$$
$$= (x + y + x - y)(x + y - x + y)$$
$$= (2x)(2y) = 4xy$$

105. Answers will vary.

107. $49x^2 - \dfrac{1}{25} = \left(7x - \dfrac{1}{5}\right)\left(7x + \dfrac{1}{5}\right)$

109. $\dfrac{25}{9}x^4 - 9y^2 = \left(\dfrac{5}{3}x^2 + 3y\right)\left(\dfrac{5}{3}x^2 - 3y\right)$

111. $4z^2 + bz + 81 = (2z)^2 + bz + 9^2$ will be a
perfect trinomial if $bz = \pm 2(2z)(9) \Rightarrow$
$bz = \pm 36z \Rightarrow b = \pm 36$.
If $b = 36$, $4z^2 + 36z + 81 = (2z + 9)^2$.
If $b = -36$, $4z^2 - 36z + 81 = (2z - 9)^2$.

113. $100r^2 - 60r + c = (10r)^2 - \underbrace{2(10r)(3)}_{60r} + c$
will be a perfect trinomial if $c = 3^2 = 9$.
If $c = 9$, $100r^2 - 60r + 9 = (10r - 3)^2$.

Section R.5   Rational Expressions

1. In the rational expression $\dfrac{x + 3}{x - 6}$, the solution
to the equation $x - 6 = 0$ is excluded from the domain.
$x - 6 = 0 \Rightarrow x = 6$
The domain is $\{x \mid x \neq 6\}$.

3. In the rational expression $\dfrac{3x + 7}{(4x + 2)(x - 1)}$, the
solution to the equation
$(4x + 2)(x - 1) = 0$ is excluded from the domain.
$(4x + 2)(x - 1) = 0$
$4x + 2 = 0$ or $x - 1 = 0$
$4x = -2$ $x = 1$
$x = -\frac{1}{2}$
The domain is $\left\{x \left| x \neq -\dfrac{1}{2}, 1\right.\right\}$.

5. In the rational expression $\dfrac{12}{x^2 + 5x + 6}$, the
solution to the equation $x^2 + 5x + 6 = 0$ is
excluded from the domain.
$x^2 + 5x + 6 = 0$
$(x + 3)(x + 2) = 0$
$x + 3 = 0$ or $x + 2 = 0$
$x = -3$ $x = -2$
The domain is $\{x \mid x \neq -3, -2\}$.

7. In the rational expression $\dfrac{x^2 - 1}{x + 1}$, the solution
to the equation $x + 1$ is excluded from the
domain: $x + 1 = 0 \Rightarrow x = -1$
The domain is $\{x \mid x \neq -1\}$.

Copyright © 2013 Pearson Education, Inc.

9. In the rational expression $\dfrac{x^3-1}{x-1}$, the solution to the equation $x-1$ is excluded from the domain: $x-1=0 \Rightarrow x=1$

The domain is $\{x \mid x \neq 1\}$.

11. $\dfrac{8x^2+16x}{4x^2} = \dfrac{8x(x+2)}{4x^2} = \dfrac{2\cdot 4x(x+2)}{x\cdot 4x}$

$= \dfrac{2(x+2)}{x} = \dfrac{2x+4}{x}$

13. $\dfrac{3(3-t)}{(t+5)(t-3)} = \dfrac{3(3-t)(-1)}{(t+5)(t-3)(-1)} = \dfrac{-3}{t+5}$

15. $\dfrac{8k+16}{9k+18} = \dfrac{8(k+2)}{9(k+2)} = \dfrac{8}{9}$

17. $\dfrac{m^2-4m+4}{m^2+m-6} = \dfrac{(m-2)(m-2)}{(m-2)(m+3)} = \dfrac{m-2}{m+3}$

19. $\dfrac{8m^2+6m-9}{16m^2-9} = \dfrac{(2m+3)(4m-3)}{(4m+3)(4m-3)} = \dfrac{2m+3}{4m+3}$

21. $\dfrac{x^3+64}{x+4} = \dfrac{(x+4)(x^2-4x+16)}{x+4}$

$= x^2-4x+16$

23. $\dfrac{15p^3}{9p^2} \div \dfrac{6p}{10p^2} = \dfrac{15p^3}{9p^2} \cdot \dfrac{10p^2}{6p} = \dfrac{150p^5}{54p^3}$

$= \dfrac{25\cdot 6p^5}{9\cdot 6p^3} = \dfrac{25p^2}{9}$

25. $\dfrac{2k+8}{6} \div \dfrac{3k+12}{2} = \dfrac{2k+8}{6} \cdot \dfrac{2}{3k+12}$

$= \dfrac{2(k+4)(2)}{6(3)(k+4)} = \dfrac{2}{9}$

27. $\dfrac{x^2+x}{5} \cdot \dfrac{25}{xy+y} = \dfrac{x(x+1)}{5} \cdot \dfrac{25}{y(x+1)}$

$= \dfrac{25x(x+1)}{5y(x+1)} = \dfrac{5x}{y}$

29. $\dfrac{4a+12}{2a-10} \div \dfrac{a^2-9}{a^2-a-20} = \dfrac{4a+12}{2a-10} \cdot \dfrac{a^2-a-20}{a^2-9}$

$= \dfrac{4(a+3)}{2(a-5)} \cdot \dfrac{(a-5)(a+4)}{(a+3)(a-3)}$

$= \dfrac{2(a+4)}{a-3} = \dfrac{2a+8}{a-3}$

31. $\dfrac{p^2-p-12}{p^2-2p-15} \cdot \dfrac{p^2-9p+20}{p^2-8p+16}$

$= \dfrac{(p-4)(p+3)}{(p-5)(p+3)} \cdot \dfrac{(p-5)(p-4)}{(p-4)(p-4)} = 1$

33. $\dfrac{m^2+3m+2}{m^2+5m+4} \div \dfrac{m^2+5m+6}{m^2+10m+24}$

$= \dfrac{m^2+3m+2}{m^2+5m+4} \cdot \dfrac{m^2+10m+24}{m^2+5m+6}$

$= \dfrac{(m+2)(m+1)}{(m+4)(m+1)} \cdot \dfrac{(m+6)(m+4)}{(m+3)(m+2)} = \dfrac{m+6}{m+3}$

35. $\dfrac{x^3+y^3}{x^3-y^3} \cdot \dfrac{x^2-y^2}{x^2+2xy+y^2}$

$= \dfrac{(x+y)(x^2-xy+y^2)}{(x-y)(x^2+xy+y^2)} \cdot \dfrac{(x+y)(x-y)}{(x+y)(x+y)}$

$= \dfrac{x^2-xy+y^2}{x^2+xy+y^2}$

37. $\dfrac{xz-xw+2yz-2yw}{z^2-w^2} \cdot \dfrac{4z+4w+xz+wx}{16-x^2}$

$= \dfrac{x(z-w)+2y(z-w)}{(z+w)(z-w)} \cdot \dfrac{4(z+w)+x(z+w)}{(4+x)(4-x)}$

$= \dfrac{(z-w)(x+2y)}{(z+w)(z-w)} \cdot \dfrac{(z+w)(4+x)}{(4+x)(4-x)} = \dfrac{x+2y}{4-x}$

39. Expressions (B) and (C) are both equivalent to -1, since the numerator and denominator are additive inverses.

B. $\dfrac{-x-4}{x+4} = \dfrac{-1(x+4)}{x+4} = -1$

C. $\dfrac{x-4}{4-x} = \dfrac{-1(4-x)}{4-x} = -1$

41. $\dfrac{3}{2k} + \dfrac{5}{3k} = \dfrac{3\cdot 3}{2k\cdot 3} + \dfrac{5\cdot 2}{3k\cdot 2} = \dfrac{9}{6k} + \dfrac{10}{6k} = \dfrac{19}{6k}$

43. $\dfrac{1}{6m} + \dfrac{2}{5m} + \dfrac{4}{m} = \dfrac{1\cdot 5}{6m\cdot 5} + \dfrac{2\cdot 6}{5m\cdot 6} + \dfrac{4\cdot 30}{m\cdot 30}$

$= \dfrac{5}{30m} + \dfrac{12}{30m} + \dfrac{120}{30m} = \dfrac{137}{30m}$

45. $\dfrac{1}{a} - \dfrac{b}{a^2} = \dfrac{1\cdot a}{a\cdot a} - \dfrac{b}{a^2} = \dfrac{a}{a^2} - \dfrac{b}{a^2} = \dfrac{a-b}{a^2}$

47. $\dfrac{5}{12x^2y} - \dfrac{11}{6xy} = \dfrac{5}{12x^2y} - \dfrac{11\cdot 2x}{6xy\cdot 2x}$

$= \dfrac{5}{12x^2y} - \dfrac{22x}{12x^2y} = \dfrac{5-22x}{12x^2y}$

Copyright © 2013 Pearson Education, Inc.

49. $\dfrac{17y+3}{9y+7}-\dfrac{-10y-18}{9y+7}=\dfrac{(17y+3)-(-10y-18)}{9y+7}=\dfrac{17y+3+10y+18}{9y+7}$

$$=\dfrac{27y+21}{9y+7}=\dfrac{3(9y+7)}{9y+7}=3$$

51. $\dfrac{1}{x+z}+\dfrac{1}{x-z}=\dfrac{1\cdot(x-z)}{(x+z)(x-z)}+\dfrac{1\cdot(x+z)}{(x-z)(x+z)}=\dfrac{(x-z)+(x+z)}{(x-z)(x+z)}=\dfrac{x-z+x+z}{(x+z)(x-z)}=\dfrac{2x}{(x+z)(x-z)}$

53. Since $a-2=(-1)(2-a)$ we have

$$\dfrac{3}{a-2}-\dfrac{1}{2-a}=\dfrac{3}{a-2}-\dfrac{1(-1)}{(2-a)(-1)}=\dfrac{3}{a-2}-\dfrac{-1}{a-2}=\dfrac{3-(-1)}{a-2}=\dfrac{4}{a-2}$$

We may also use $2-a$ as the common denominator.

$$\dfrac{3}{a-2}-\dfrac{1}{2-a}=\dfrac{3(-1)}{(a-2)(-1)}-\dfrac{1}{2-a}=\dfrac{-3}{2-a}-\dfrac{1}{2-a}=\dfrac{-4}{2-a}$$

The two results, $\dfrac{4}{a-2}$ and $\dfrac{-4}{2-a}$, are equivalent rational expressions.

55. Since $2x-y=(-1)(y-2x)$

$$\dfrac{x+y}{2x-y}-\dfrac{2x}{y-2x}=\dfrac{x+y}{2x-y}-\dfrac{2x(-1)}{(y-2x)(-1)}=\dfrac{x+y}{2x-y}-\dfrac{-2x}{2x-y}=\dfrac{x+y-(-2x)}{2x-y}=\dfrac{x+y+2x}{2x-y}=\dfrac{3x+y}{2x-y}$$

We may also use $y-2x$ as the common denominator. In this case, our result will be $\dfrac{-3x-y}{y-2x}$. The two results

are equivalent rational expressions.

57. $\dfrac{4}{x+1}+\dfrac{1}{x^2-x+1}-\dfrac{12}{x^3+1}=\dfrac{4}{x+1}+\dfrac{1}{x^2-x+1}-\dfrac{12}{(x+1)(x^2-x+1)}$

$$=\dfrac{4(x^2-x+1)}{(x+1)(x^2-x+1)}+\dfrac{1(x+1)}{(x+1)(x^2-x+1)}-\dfrac{12}{(x+1)(x^2-x+1)}$$

$$=\dfrac{4(x^2-x+1)+(x+1)-12}{(x+1)(x^2-x+1)}=\dfrac{4x^2-4x+4+x+1-12}{(x+1)(x^2-x+1)}$$

$$=\dfrac{4x^2-3x-7}{(x+1)(x^2-x+1)}=\dfrac{(4x-7)(x+1)}{(x+1)(x^2-x+1)}=\dfrac{4x-7}{x^2-x+1}$$

59. $\dfrac{3x}{x^2+x-12}-\dfrac{x}{x^2-16}=\dfrac{3x}{(x-3)(x+4)}-\dfrac{x}{(x-4)(x+4)}=\dfrac{3x(x-4)}{(x-3)(x+4)(x-4)}-\dfrac{x(x-3)}{(x-4)(x+4)(x-3)}$

$$=\dfrac{3x(x-4)-x(x-3)}{(x-3)(x+4)(x-4)}=\dfrac{3x^2-12x-x^2+3x}{(x-3)(x+4)(x-4)}=\dfrac{2x^2-9x}{(x-3)(x+4)(x-4)}$$

61. $\dfrac{1+\frac{1}{x}}{1-\frac{1}{x}}=\dfrac{x\left(1+\frac{1}{x}\right)}{x\left(1-\frac{1}{x}\right)}=\dfrac{x\cdot1+x\left(\frac{1}{x}\right)}{x\cdot1-x\left(\frac{1}{x}\right)}=\dfrac{x+1}{x-1}$

63. $\dfrac{\frac{1}{x+1}-\frac{1}{x}}{\frac{1}{x}}=\dfrac{x(x+1)\left(\frac{1}{x+1}-\frac{1}{x}\right)}{x(x+1)\left(\frac{1}{x}\right)}=\dfrac{x(x+1)\left(\frac{1}{x+1}\right)-x(x+1)\left(\frac{1}{x}\right)}{x(x+1)\left(\frac{1}{x}\right)}=\dfrac{x-(x+1)}{x+1}=\dfrac{x-x-1}{x+1}=\dfrac{-1}{x+1}$

Copyright © 2013 Pearson Education, Inc.

65.
$$\frac{1+\frac{1}{1-b}}{1-\frac{1}{1+b}} = \frac{(1-b)(1+b)\left(1+\frac{1}{1-b}\right)}{(1-b)(1+b)\left(1-\frac{1}{1+b}\right)} = \frac{(1-b)(1+b)(1) + (1-b)(1+b)\left(\frac{1}{1-b}\right)}{(1-b)(1+b)(1) - (1-b)(1+b)\left(\frac{1}{1+b}\right)} = \frac{(1-b)(1+b) + (1+b)}{(1-b)(1+b) - (1-b)}$$

$$= \frac{(1+b)[(1-b)+1]}{(1-b)[(1+b)-1]} = \frac{(1+b)(2-b)}{(1-b)b} \text{ or } \frac{(2-b)(1+b)}{b(1-b)}$$

67.
$$\frac{\frac{1}{a^3+b^3}}{\frac{1}{a^2+2ab+b^2}} = \frac{\frac{1}{(a+b)\left(a^2-ab+b^2\right)}}{\frac{1}{(a+b)^2}} = \frac{\frac{1}{(a+b)\left(a^2-ab+b^2\right)}(a+b)^2\left(a^2-ab+b^2\right)}{\frac{1}{(a+b)^2}(a+b)^2\left(a^2-ab+b^2\right)} = \frac{a+b}{a^2-ab+b^2}$$

69.
$$\frac{m-\frac{1}{m^2-4}}{\frac{1}{m+2}} = \frac{m-\frac{1}{(m+2)(m-2)}}{\frac{1}{m+2}} = \frac{(m+2)(m-2)\left(m-\frac{1}{(m+2)(m-2)}\right)}{(m+2)(m-2)\left(\frac{1}{m+2}\right)} = \frac{(m+2)(m-2)(m) - (m+2)(m-2)\left(\frac{1}{(m+2)(m-2)}\right)}{(m+2)(m-2)\left(\frac{1}{m+2}\right)}$$

$$= \frac{m(m+2)(m-2)-1}{m-2} = \frac{m(m^2-4)-1}{m-2} = \frac{m^3-4m-1}{m-2}$$

71.
$$\frac{\frac{3}{p^2-16}+p}{\frac{1}{p-4}} = \frac{\frac{3}{(p+4)(p-4)}+p}{\frac{1}{p-4}} = \frac{(p+4)(p-4)\left[\frac{3}{(p+4)(p-4)}+p\right]}{(p+4)(p-4)\left[\frac{1}{p-4}\right]} = \frac{(p+4)(p-4)\left(\frac{3}{(p+4)(p-4)}\right)+(p+4)(p-4)p}{p+4}$$

$$= \frac{3+(p^2-16)p}{p+4} = \frac{3+p^3-16p}{p+4} = \frac{p^3-16p+3}{p+4}$$

73.
$$\frac{\frac{y+3}{y}-\frac{4}{y-1}}{\frac{y}{y-1}+\frac{1}{y}} = \frac{y(y-1)\left(\frac{y+3}{y}-\frac{4}{y-1}\right)}{y(y-1)\left(\frac{y}{y-1}+\frac{1}{y}\right)} = \frac{y(y-1)\left(\frac{y+3}{y}\right)-y(y-1)\left(\frac{4}{y-1}\right)}{y(y-1)\left(\frac{y}{y-1}\right)+y(y-1)\left(\frac{1}{y}\right)} = \frac{(y-1)(y+3)-4y}{y^2+y-1}$$

$$= \frac{y^2+2y-3-4y}{y^2+y-1} = \frac{y^2-2y-3}{y^2+y-1}$$

75.
$$\frac{\frac{1}{x+h}-\frac{1}{x}}{h} = \frac{x(x+h)\left(\frac{1}{x+h}-\frac{1}{x}\right)}{x(x+h)(h)} = \frac{x(x+h)\left(\frac{1}{x+h}\right)-x(x+h)\left(\frac{1}{x}\right)}{x(x+h)(h)}$$

$$= \frac{x-(x+h)}{x(x+h)(h)} = \frac{x-x-h}{x(x+h)(h)} = \frac{-h}{x(x+h)(h)} = \frac{-1}{x(x+h)}$$

77.
$$\frac{\frac{1}{(x+h)^2+9}-\frac{1}{x^2+9}}{h} = \frac{\left[(x+h)^2+9\right]\left(x^2+9\right)\left[\frac{1}{(x+h)^2+9}-\frac{1}{x^2+9}\right]}{\left[(x+h)^2+9\right]\left(x^2+9\right)h}$$

$$= \frac{\left[(x+h)^2+9\right]\left(x^2+9\right)\left[\frac{1}{(x+h)^2+9}\right]-\left[(x+h)^2+9\right]\left(x^2+9\right)\left[\frac{1}{x^2+9}\right]}{\left[(x+h)^2+9\right]\left(x^2+9\right)h}$$

$$= \frac{\left(x^2+9\right)-\left[(x+h)^2+9\right]}{\left[(x+h)^2+9\right]\left(x^2+9\right)h} = \frac{x^2+9-\left[x^2+2xh+h^2+9\right]}{\left[(x+h)^2+9\right]\left(x^2+9\right)h}$$

$$= \frac{x^2+9-x^2-2xh-h^2-9}{\left[(x+h)^2+9\right]\left(x^2+9\right)h} = \frac{-2xh-h^2}{\left[(x+h)^2+9\right]\left(x^2+9\right)h}$$

$$= \frac{h(-2x-h)}{\left[(x+h)^2+9\right]\left(x^2+9\right)h} = \frac{-2x-h}{\left[(x+h)^2+9\right]\left(x^2+9\right)}$$

Copyright © 2013 Pearson Education, Inc.

79. Altitude of 7000 feet, $x = 7$ (thousand)
The distance from the origin is
$$\frac{7-7}{0.639(7)+1.75} = 0,$$
which represents 0 miles.

80. Altitude of 1200 feet, $x = 1.2$ (thousand)
The distance from the origin is
$$\frac{7-1.2}{0.639(1.2)+1.75} \approx 2.305,$$
which represents about 2305 miles.

81. $y = \dfrac{6.7x}{100-x}$; let $x = 75$ (75%)
$$y = \frac{6.7(75)}{100-75} = \frac{502.5}{25} = 20.1$$
The cost of removing 75% of the pollutant is 20.1 thousand dollars.

82. $y = \dfrac{6.7x}{100-x}$; let $x = 95$ (95%).
$$y = \frac{6.7(95)}{100-95} = \frac{636.5}{5} = 127.3$$
The cost of removing 95% of the pollutant is 127.3 thousand dollars.

Section R.6   Rational Exponents

1. (a) $4^{-2} = \dfrac{1}{4^2} = \dfrac{1}{16}$; B

(b) $-4^{-2} = -\left(4^{-2}\right) = -\dfrac{1}{4^2} = -\dfrac{1}{16}$; D

(c) $(-4)^{-2} = \dfrac{1}{(-4)^2} = \dfrac{1}{16}$; B

(d) $-(-4)^{-2} = -\dfrac{1}{(-4)^2} = -\dfrac{1}{16}$; D

3. $(-4)^{-3} = \dfrac{1}{(-4)^3} = \dfrac{1}{-64} = -\dfrac{1}{64}$

5. $-(-5)^{-4} = -\dfrac{1}{(-5)^4} = -\dfrac{1}{625}$

7. $\left(\dfrac{1}{3}\right)^{-2} = \dfrac{1}{\left(\frac{1}{3}\right)^2} = \dfrac{1}{\frac{1}{9}} = 9$, also
$$\left(\frac{1}{3}\right)^{-2} = \frac{1}{\left(\frac{1}{3}\right)^2} = \left(\frac{3}{1}\right)^2 = 3^2 = 9$$

9. $(4x)^{-2} = \dfrac{1}{(4x)^2} = \dfrac{1}{16x^2}$

11. $4x^{-2} = 4 \cdot x^{-2} = 4 \cdot \dfrac{1}{x^2} = \dfrac{4}{x^2}$

13. $-a^{-3} = -\dfrac{1}{a^3}$

15. $\dfrac{4^8}{4^6} = 4^{8-6} = 4^2 = 16$

17. $\dfrac{x^{12}}{x^8} = x^{12-8} = x^4$

19. $\dfrac{r^7}{r^{10}} = r^{7-10} = r^{-3} = \dfrac{1}{r^3}$

21. $\dfrac{6^4}{6^{-2}} = 6^{4-(-2)} = 6^6$

Because 6^6 is a relatively large number, it is generally acceptable not to simplify it to be 46,656.

23. $\dfrac{4r^{-3}}{6r^{-6}} = \dfrac{4}{6} \cdot \dfrac{r^{-3}}{r^{-6}} = \dfrac{2}{3} r^{-3-(-6)} = \dfrac{2}{3} r^3 = \dfrac{2r^3}{3}$

25. $\dfrac{16m^{-5}n^4}{12m^2 n^{-3}} = \dfrac{16}{12} \cdot \dfrac{m^{-5}}{m^2} \cdot \dfrac{n^4}{n^{-3}} = \dfrac{4}{3} m^{-5-2} n^{4-(-3)}$
$$= \frac{4}{3} m^{-7} n^7 = \frac{4n^7}{3m^7}$$

27. $-4r^{-2}\left(r^4\right)^2 = -4r^{-2}r^8 = -4r^{-2+8} = -4r^6$

29. $\left(5a^{-1}\right)^4 \left(a^2\right)^{-3} = 5^4 a^{-4} a^{-6} = 5^4 a^{-4+(-6)}$
$$= 5^4 a^{-10} = \frac{5^4}{a^{10}}$$

Because 5^4 is a relatively large number, it is generally acceptable not to simplify to $\dfrac{625}{a^{10}}$.

31. $\dfrac{\left(p^{-2}\right)^0}{5p^{-4}} = \dfrac{p^0}{5p^{-4}} = \dfrac{1}{5} \cdot \dfrac{p^0}{p^{-4}}$
$$= \frac{1}{5} p^{0-(-4)} = \frac{1}{5} p^4 = \frac{p^4}{5}$$

Copyright © 2013 Pearson Education, Inc.

33. $\dfrac{(3pq)q^2}{6p^2q^4} = \dfrac{3pq^{1+2}}{6p^2q^4} = \dfrac{3pq^3}{6p^2q^4}$

$\quad = \dfrac{3}{6} \cdot \dfrac{p}{p^2} \cdot \dfrac{q^3}{q^4} = \dfrac{1}{2} p^{1-2}q^{3-4}$

$\quad = \dfrac{1}{2} p^{-1}q^{-1} = \dfrac{1}{2pq}$

35. $\dfrac{4a^5\left(a^{-1}\right)^3}{\left(a^{-2}\right)^{-2}} = \dfrac{4a^5 a^{-3}}{a^4} = \dfrac{4a^{5+(-3)}}{a^4}$

$\quad = \dfrac{4a^2}{a^4} = 4a^{2-4} = 4a^{-2} = \dfrac{4}{a^2}$

37. $\dfrac{(5x)^{-2}\left(5x^3\right)^{-3}}{\left(5^{-2}x^{-3}\right)^3} = \dfrac{5^{-2}x^{-2} \cdot 5^{-3}x^{-9}}{5^{-6}x^{-9}} = \dfrac{5^{-5}x^{-11}}{5^{-6}x^{-9}}$

$\quad = 5^{-5-(-6)} x^{-11-(-9)} = 5x^{-2} = \dfrac{5}{x^2}$

39. $169^{1/2} = 13$, because $13^2 = 169$.

41. $16^{1/4} = 2$, because $2^4 = 16$.

43. $\left(-\dfrac{64}{27}\right)^{1/3} = -\dfrac{4}{3}$, because $\left(-\dfrac{4}{3}\right)^3 = -\dfrac{64}{27}$.

45. $(-4)^{1/2}$ is not a real number, because no real number, when squared, will yield a negative quantity.

47. (a) $\left(\dfrac{4}{9}\right)^{3/2} = \left[\left(\dfrac{4}{9}\right)^{1/2}\right]^3 = \left(\dfrac{2}{3}\right)^3 = \dfrac{8}{27}$; E

(b) $\left(\dfrac{4}{9}\right)^{-3/2} = \left(\dfrac{9}{4}\right)^{3/2} = \left[\left(\dfrac{9}{4}\right)^{1/2}\right]^3$

$\quad = \left(\dfrac{3}{2}\right)^3 = \dfrac{27}{8}$; G

(c) $-\left(\dfrac{9}{4}\right)^{3/2} = -\left[\left(\dfrac{9}{4}\right)^{1/2}\right]^3$

$\quad = -\left(\dfrac{3}{2}\right)^3 = -\dfrac{27}{8}$; F

(d) $-\left(\dfrac{4}{9}\right)^{-3/2} = -\left(\dfrac{9}{4}\right)^{3/2} = -\left[\left(\dfrac{9}{4}\right)^{1/2}\right]^3$

$\quad = -\left(\dfrac{3}{2}\right)^3 = -\dfrac{27}{8}$; F

49. $8^{2/3} = \left(8^{1/3}\right)^2 = 2^2 = 4$

51. $100^{3/2} = \left(100^{1/2}\right)^3 = 10^3 = 1000$

53. $-81^{3/4} = -\left(81^{1/4}\right)^3 = -3^3 = -27$

55. $\left(\dfrac{27}{64}\right)^{-4/3} = \left(\dfrac{64}{27}\right)^{4/3} = \left[\left(\dfrac{64}{27}\right)^{1/3}\right]^4$

$\quad = \left(\dfrac{4}{3}\right)^4 = \dfrac{256}{81}$

57. $3^{1/2} \cdot 3^{3/2} = 3^{1/2+3/2} = 3^{4/2} = 3^2 = 9$

59. $\dfrac{64^{5/3}}{64^{4/3}} = 64^{5/3-4/3} = 64^{1/3} = 4$

61. $y^{7/3} \cdot y^{-4/3} = y^{7/3+(-4/3)} = y^{3/3} = y^1 = y$

63. $\dfrac{k^{1/3}}{k^{2/3} \cdot k^{-1}} = \dfrac{k^{1/3}}{k^{2/3+(-1)}} = \dfrac{k^{1/3}}{k^{2/3+(-3/3)}}$

$\quad = \dfrac{k^{1/3}}{k^{-1/3}} = k^{1/3-(-1/3)} = k^{2/3}$

65. $\dfrac{\left(x^{1/4}y^{2/5}\right)^{20}}{x^2} = \dfrac{x^5 y^8}{x^2} = x^{5-2}y^8 = x^3 y^8$

67. $\dfrac{\left(x^{2/3}\right)^2}{\left(x^2\right)^{7/3}} = \dfrac{x^{4/3}}{x^{14/3}} = x^{4/3-14/3} = x^{-10/3} = \dfrac{1}{x^{10/3}}$

69. $\left(\dfrac{16m^3}{n}\right)^{1/4}\left(\dfrac{9n^{-1}}{m^2}\right)^{1/2} = \dfrac{16^{1/4}m^{3/4}}{n^{1/4}} \cdot \dfrac{9^{1/2}n^{-1/2}}{m}$

$\quad = \dfrac{2m^{3/4}}{n^{1/4}} \cdot \dfrac{3n^{-1/2}}{m}$

$\quad = (2\cdot3)\dfrac{m^{3/4}}{m} \cdot \dfrac{n^{-1/2}}{n^{1/4}}$

$\quad = 6m^{(3/4)-1}n^{(-1/2)-(1/4)}$

$\quad = 6m^{(3/4)-(4/4)}n^{(-2/4)-(1/4)}$

$\quad = 6m^{-1/4}n^{-3/4} = \dfrac{6}{m^{1/4}n^{3/4}}$

Copyright © 2013 Pearson Education, Inc.

71. $\dfrac{p^{1/5}p^{7/10}p^{1/2}}{\left(p^3\right)^{-1/5}} = \dfrac{p^{1/5+7/10+1/2}}{p^{-3/5}}$

$\qquad = \dfrac{p^{2/10+7/10+5/10}}{p^{-6/10}} = \dfrac{p^{14/10}}{p^{-6/10}}$

$\qquad = p^{(14/10)-(-6/10)}$

$\qquad = p^{20/10} = p^2$

73. (a) Let $w = 25$

$\qquad t = 31,293w^{-1.5} = 31,293\cdot(25)^{-3/2}$

$\qquad = 31,293\cdot(25^{1/2})^{-3} = 31,293\cdot 5^{-3}$

$\qquad = \dfrac{31,293}{125} \approx 250$ sec

(b) To double the weight, replace w with $2w$

$\quad$ to get $\dfrac{31,293}{(2w)^{1.5}} = \dfrac{31,293}{2^{3/2}w^{1.5}} = \dfrac{1}{2^{3/2}}(t)$; so

$\quad$ the holding time changes by a factor of

$\qquad \dfrac{1}{2^{3/2}} \approx 0.3536$.

75. $y^{5/8}(y^{3/8} - 10y^{11/8}) = y^{5/8}y^{3/8} - 10y^{5/8}y^{11/8}$

$\qquad = y^{5/8+3/8} - 10y^{5/8+11/8}$

$\qquad = y - 10y^2$

77. $-4k(k^{7/3} - 6k^{1/3}) = -4k^1k^{7/3} + 24k^1k^{1/3}$

$\qquad = -4k^{10/3} + 24k^{4/3}$

79. $(x + x^{1/2})(x - x^{1/2})$ has the form

$\quad (a+b)(a-b) = a^2 - b^2$.

$\quad (x + x^{1/2})(x - x^{1/2}) = x^2 - (x^{1/2})^2 = x^2 - x$

81. $\left(r^{1/2} - r^{-1/2}\right)^2$

$\quad = \left(r^{1/2}\right)^2 - 2\left(r^{1/2}\right)\left(r^{-1/2}\right) + \left(r^{-1/2}\right)^2$

$\quad = r - 2r^{(1/2)+(-1/2)} + r^{-1} = r - 2r^0 + r^{-1}$

$\quad = r - 2\cdot 1 + r^{-1} = r - 2 + r^{-1} = r - 2 + \dfrac{1}{r}$

83. Factor $4k^{-1} + k^{-2}$, using the common factor

$\quad k^{-2}$: $4k^{-1} + k^{-2} = k^{-2}(4k + 1)$

85. Factor $4t^{-2} + 8t^{-4}$, using the common factor

$\quad 4t^{-4}$: $4t^{-2} + 8t^{-4} = 4t^{-4}(t^2 + 2)$

87. Factor $9z^{-1/2} + 2z^{1/2}$, using the common

$\quad$ factor $z^{-1/2}$:

$\quad 9z^{-1/2} + 2z^{1/2} = z^{-1/2}(9 + 2z)$

89. Factor $p^{-3/4} - 2p^{-7/4}$, using the common

$\quad$ factor $p^{-7/4}$:

$\quad p^{-3/4} - 2p^{-7/4} = p^{-7/4}(p^{4/4} - 2)$

$\qquad = p^{-7/4}(p - 2)$

91. Factor $-4a^{-2/5} + 16a^{-7/5}$, using the common

$\quad$ factor $4a^{-7/5}$:

$\quad -4a^{-2/5} + 16a^{-7/5} = 4a^{-7/5}(-a + 4)$

93. Factor $(p+4)^{-3/2} + (p+4)^{-1/2} + (p+4)^{1/2}$,

$\quad$ using the common factor $(p+4)^{-3/2}$.

$\quad (p+4)^{-3/2} + (p+4)^{-1/2} + (p+4)^{1/2}$

$\quad = (p+4)^{-3/2}\cdot[1 + (p+4) + (p+4)^2]$

$\quad = (p+4)^{-3/2}\cdot(1 + p + 4 + p^2 + 8p + 16)$

$\quad = (p+4)^{-3/2}(p^2 + 9p + 21)$

95. Factor

$\quad 2(3x+1)^{-3/2} + 4(3x+1)^{-1/2} + 6(3x+1)^{1/2}$

$\quad$ using the common factor $2(3x+1)^{-3/2}$:

$\quad 2(3x+1)^{-3/2} + 4(3x+1)^{-1/2} + 6(3x+1)^{1/2}$

$\quad = 2(3x+1)^{-3/2}\left[1 + 2(3x+1) + 3(3x+1)^2\right]$

$\quad = 2(3x+1)^{-3/2}\left[1 + (6x+2) + 3(9x^2 + 6x + 1)\right]$

$\quad = 2(3x+1)^{-3/2}\left[1 + 6x + 2 + 27x^2 + 18x + 3\right]$

$\quad = 2(3x+1)^{-3/2}\left(27x^2 + 24x + 6\right)$

$\quad = 2(3x+1)^{-3/2}\cdot 3\left(9x^2 + 8x + 2\right)$

$\quad = 6(3x+1)^{-3/2}\left(9x^2 + 8x + 2\right)$

97. Factor

$\quad 4x(2x+3)^{-5/9} + 6x^2(2x+3)^{4/9} - 8x^3(2x+3)^{13/9}$

$\quad$ using the common factor $2x(2x+3)^{-5/9}$:

$\quad 4x(2x+3)^{-5/9} + 6x^2(2x+3)^{4/9} - 8x^3(2x+3)^{13/9}$

$\quad = 2x(2x+3)^{-5/9}\left[2 + 3x(2x+3) - 4x^2(2x+3)^2\right]$

$\quad = 2x(2x+3)^{-5/9}$

$\qquad \cdot\left[2 + (6x^2 + 9x) - 4x^2(4x^2 + 12x + 9)\right]$

$\quad = 2x(2x+3)^{-5/9}$

$\qquad \cdot\left[2 + 6x^2 + 9x - 16x^4 - 48x^3 - 36x^2\right]$

$\quad = 2x(2x+3)^{-5/9}\left[-16x^4 - 48x^3 - 30x^2 + 9x + 2\right]$

Copyright © 2013 Pearson Education, Inc.

99. $\dfrac{a^{-1}+b^{-1}}{(ab)^{-1}} = \dfrac{\frac{1}{a}+\frac{1}{b}}{\frac{1}{ab}} = \dfrac{\frac{1\cdot b}{a\cdot b}+\frac{1\cdot a}{b\cdot a}}{\frac{1}{ab}}$

$= \dfrac{\frac{b+a}{ab}}{\frac{1}{ab}} = \dfrac{b+a}{ab}\cdot\dfrac{ab}{1} = b+a$

101. $\dfrac{r^{-1}+q^{-1}}{r^{-1}-q^{-1}}\cdot\dfrac{r-q}{r+q} = \dfrac{\frac{1}{r}+\frac{1}{q}}{\frac{1}{r}-\frac{1}{q}}\cdot\dfrac{r-q}{r+q}$

$= \dfrac{rq\left(\frac{1}{r}+\frac{1}{q}\right)}{rq\left(\frac{1}{r}-\frac{1}{q}\right)}\cdot\dfrac{r-q}{r+q}$

$= \dfrac{q+r}{q-r}\cdot\dfrac{r-q}{r+q} = \dfrac{r-q}{q-r}$

$= \dfrac{r-q}{-1(r-q)} = -1$

103. $\dfrac{x-9y^{-1}}{\left(x-3y^{-1}\right)\left(x+3y^{-1}\right)} = \dfrac{x-\dfrac{9}{y}}{\left(x-\dfrac{3}{y}\right)\left(x+\dfrac{3}{y}\right)}$

$= \dfrac{x-\dfrac{9}{y}}{x^2-\dfrac{9}{y^2}} = \dfrac{y^2\left(x-\dfrac{9}{y}\right)}{y^2\left(x^2-\dfrac{9}{y^2}\right)}$

$= \dfrac{y^2x-9y}{y^2x^2-9}$ or $\dfrac{y(xy-9)}{x^2y^2-9}$

105. $\dfrac{(x^2+1)^4(2x)-x^2(4)(x^2+1)^3(2x)}{(x^2+1)^8}$

$= \dfrac{(2x)(x^2+1)^3\left[(x^2+1)-4x^2\right]}{(x^2+1)^8}$

$= \dfrac{(2x)\left[(x^2+1)-4x^2\right]}{(x^2+1)^5} = \dfrac{(2x)(1-3x^2)}{(x^2+1)^5}$

107. $\dfrac{4(x^2-1)^3+8x(x^2-1)^4}{16(x^2-1)^3}$

$= \dfrac{4(x^2-1)^3\left[1+2x(x^2-1)\right]}{16(x^2-1)^3}$

$= \dfrac{1+2x(x^2-1)}{4} = \dfrac{1+2x^3-2x}{4}$

109. $\dfrac{2(2x-3)^{1/3}-(x-1)(2x-3)^{-2/3}}{(2x-3)^{2/3}}$

$= \dfrac{(2x-3)^{1/3}\left[2-(x-1)(2x-3)^{-1}\right]}{(2x-3)^{2/3}}$

$= \dfrac{2-(x-1)(2x-3)^{-1}}{(2x-3)^{1/3}} = \dfrac{2-\frac{x-1}{2x-3}}{(2x-3)^{1/3}}$

$= \dfrac{2(2x-3)-(x-1)}{(2x-3)^{4/3}} = \dfrac{3x-5}{(2x-3)^{4/3}}$

111. Let x = length of side of cube. Then $3x$ = length of side of bigger cube (side tripled). x^3 is the volume of the cube, and $(3x)^3 = 3^3x^3 = 27x^3$ is the volume of the bigger cube. The volume will change by a factor of 27.

113. $0.2^{2/3}\cdot 40^{2/3} = (0.2\cdot 40)^{2/3} = (8^{1/3})^2 = 2^2 = 4$

115. $\dfrac{2^{2/3}}{2000^{2/3}} = \left(\dfrac{2}{2000}\right)^{2/3}$

$= \left(\dfrac{1}{1000}\right)^{2/3}\left[\left(\dfrac{1}{1000}\right)^{1/3}\right]^2$

$= \left(\dfrac{1}{10}\right)^2 = \dfrac{1}{100}$

Section R.7   Radical Expressions

1. $\sqrt[3]{125} = 125^{1/3} = 5$ **3.** $\sqrt[4]{81} = 81^{1/4} = 3$

5. $\sqrt[3]{-125} = (-125)^{1/3} = -5$

7. $\sqrt[4]{-81} = (-81)^{1/4}$

This expression is not a real number.

9. $\sqrt[5]{32} = 32^{1/5} = 2$

11. $-\sqrt[5]{-32} = -(-32)^{1/5} = -(-2) = 2$

13. (a) F; $(-3x)^{1/3} = \sqrt[3]{-3x}$

(b) H; $(-3x)^{-1/3} = \dfrac{1}{(-3x)^{1/3}} = \dfrac{1}{\sqrt[3]{-3x}}$

(c) G; $(3x)^{1/3} = \sqrt[3]{3x}$

(d) C; $(3x)^{-1/3} = \dfrac{1}{\sqrt[3]{3x}}$

15. $m^{2/3} = \sqrt[3]{m^2}$ or $\left(\sqrt[3]{m}\right)^2$

17. $(2m+p)^{2/3} = \sqrt[3]{(2m+p)^2}$ or $\left(\sqrt[3]{2m+p}\right)^2$

19. $\sqrt[5]{k^2} = k^{2/5}$

Copyright © 2013 Pearson Education, Inc.

21. $-3\sqrt{5p^3} = -3(5p^3)^{1/2} = -3 \cdot 5^{1/2} p^{3/2}$

23. A

25. It is true for all $x \ge 0$.

27. $\sqrt[4]{x^4} = |x|$

29. $\sqrt{25k^4m^2} = \sqrt{\left(5k^2m\right)^2} = |5k^2m| = 5k^2|m|$

31. $\sqrt{\left(4x-y\right)^2} = |4x-y|$

33. $\sqrt[3]{81} = \sqrt[3]{27 \cdot 3} = \sqrt[3]{27} \cdot \sqrt[3]{3} = 3\sqrt[3]{3}$

35. $-\sqrt[4]{32} = -\sqrt[4]{16 \cdot 2} = -\sqrt[4]{16} \cdot \sqrt[4]{2} = -2\sqrt[4]{2}$

37. $\sqrt{14} \cdot \sqrt{3pqr} = \sqrt{14 \cdot 3pqr} = \sqrt{42pqr}$

39. $\sqrt[3]{7x} \cdot \sqrt[3]{2y} = \sqrt[3]{7x \cdot 2y} = \sqrt[3]{14xy}$

41. $-\sqrt{\dfrac{9}{25}} = -\dfrac{\sqrt{9}}{\sqrt{25}} = -\dfrac{3}{5}$

43. $-\sqrt[3]{\dfrac{5}{8}} = -\dfrac{\sqrt[3]{5}}{\sqrt[3]{8}} = -\dfrac{\sqrt[3]{5}}{2}$

45. $\sqrt[4]{\dfrac{m}{n^4}} = \dfrac{\sqrt[4]{m}}{\sqrt[4]{n^4}} = \dfrac{\sqrt[4]{m}}{n}$

47. $3\sqrt[5]{-3125} = 3\sqrt[5]{\left(-5\right)^5} = 3(-5) = -15$

49. $\sqrt[3]{16(-2)^4(2)^8} = \sqrt[3]{2^4 \cdot (-2)^4 2^8} = \sqrt[3]{2^4 \cdot 2^4 \cdot 2^8}$
$= \sqrt[3]{2^{15} \cdot 2} = \sqrt[3]{2^{15}} \cdot \sqrt[3]{2}$
$= 2^5 \cdot \sqrt[3]{2} = 32\sqrt[3]{2}$

51. $\sqrt{8x^5z^8} = \sqrt{2 \cdot 4 \cdot x^4 \cdot x \cdot z^8}$
$= \sqrt{4x^4z^8} \cdot \sqrt{2x} = 2x^2z^4\sqrt{2x}$

53. $\sqrt[4]{x^4 + y^4}$ cannot be simplified further.

55. $\sqrt{\dfrac{2}{3x}} = \dfrac{\sqrt{2}}{\sqrt{3x}} = \dfrac{\sqrt{2}}{\sqrt{3x}} \cdot \dfrac{\sqrt{3x}}{\sqrt{3x}} = \dfrac{\sqrt{2 \cdot 3x}}{\sqrt{9x^2}} = \dfrac{\sqrt{6x}}{3x}$

57. $\sqrt{\dfrac{x^5y^3}{z^2}} = \dfrac{\sqrt{x^5y^3}}{\sqrt{z^2}} = \dfrac{\sqrt{x^4xy^2y}}{z}$
$= \dfrac{\sqrt{x^4y^2} \cdot \sqrt{xy}}{z} = \dfrac{x^2y\sqrt{xy}}{z}$

59. $\sqrt[3]{\dfrac{8}{x^4}} = \dfrac{\sqrt[3]{8}}{\sqrt[3]{x^4}} = \dfrac{2}{\sqrt[3]{x^3 \cdot x}} = \dfrac{2}{x\sqrt[3]{x}}$
$= \dfrac{2}{x\sqrt[3]{x}} \cdot \dfrac{\sqrt[3]{x^2}}{\sqrt[3]{x^2}} = \dfrac{2\sqrt[3]{x^2}}{x \cdot x} = \dfrac{2\sqrt[3]{x^2}}{x^2}$

61. $\sqrt[4]{\dfrac{g^3h^5}{9r^6}} = \dfrac{\sqrt[4]{g^3h^5}}{\sqrt[4]{9r^6}} = \dfrac{\sqrt[4]{h^4\left(g^3h\right)}}{\sqrt[4]{r^4\left(9r^2\right)}} = \dfrac{h\sqrt[4]{g^3h}}{r\sqrt[4]{9r^2}}$
$= \dfrac{h\sqrt[4]{g^3h}}{r\sqrt[4]{3^2r^2}} \cdot \dfrac{\sqrt[4]{3^2r^2}}{\sqrt[4]{3^2r^2}} = \dfrac{h\sqrt[4]{3^2g^3hr^2}}{r\sqrt[4]{3^4r^4}}$
$= \dfrac{h\sqrt[4]{9g^3hr^2}}{3r^2}$

63. $\sqrt[8]{3^4} = 3^{4/8} = 3^{1/2} = \sqrt{3}$

65. $\sqrt[3]{\sqrt{4}} = \sqrt[3]{4^{1/2}} = (4^{1/2})^{1/3} = 4^{1/6}$
$= (2^2)^{1/6} = 2^{2/6} = 2^{1/3} = \sqrt[3]{2}$

67. $\sqrt[4]{\sqrt[3]{2}} = \sqrt[4]{2^{1/3}} = (2^{1/3})^{1/4} = 2^{1/12} = \sqrt[12]{2}$

69. $8\sqrt{2x} - \sqrt{8x} + \sqrt{72x}$
$= 8\sqrt{2x} - \sqrt{4 \cdot 2x} + \sqrt{36 \cdot 2x}$
$= 8\sqrt{2x} - 2\sqrt{2x} + 6\sqrt{2x} = 12\sqrt{2x}$

71. $2\sqrt[3]{3} + 4\sqrt[3]{24} - \sqrt[3]{81}$
$= 2\sqrt[3]{3} + 4\sqrt[3]{8 \cdot 3} - \sqrt[3]{27 \cdot 3}$
$= 2\sqrt[3]{3} + 4(2)\sqrt[3]{3} - 3\sqrt[3]{3}$
$= 2\sqrt[3]{3} + 8\sqrt[3]{3} - 3\sqrt[3]{3} = 7\sqrt[3]{3}$

73. $\sqrt[4]{81x^6y^3} - \sqrt[4]{16x^{10}y^3}$
$= \sqrt[4]{\left(81x^4\right)x^2y^3} - \sqrt[4]{\left(16x^8\right)x^2y^3}$
$= 3x\sqrt[4]{x^2y^3} - 2x^2\sqrt[4]{x^2y^3}$

75. Cannot be simplified further

77. This product has the pattern
$(a+b)(a-b) = a^2 - b^2$, the difference of squares.
$\left(\sqrt{2}+3\right)\left(\sqrt{2}-3\right) = \left(\sqrt{2}\right)^2 - 3^2 = 2 - 9 = -7$

79. This product has the pattern
$(a-b)(a^2 + ab + b^2) = a^3 - b^3$, the difference of cubes.
$\left(\sqrt[3]{11}-1\right)\left(\sqrt[3]{11^2} + \sqrt[3]{11} + 1\right) = \left(\sqrt[3]{11}\right)^3 - 1^3$
$= 11 - 1 = 10.$

Copyright © 2013 Pearson Education, Inc.

81. This product has the pattern
$(a+b)^2 = a^2 + 2ab + b^2$, the square of a binomial.

$$\left(\sqrt{3} + \sqrt{8}\right)^2 = \left(\sqrt{3}\right)^2 + 2\left(\sqrt{3}\right)\left(\sqrt{8}\right) + \left(\sqrt{8}\right)^2$$
$$= 3 + 2\sqrt{24} + 8 = 11 + 2\sqrt{4 \cdot 6}$$
$$= 11 + 2(2)\sqrt{6} = 11 + 4\sqrt{6}$$

83. This product can be found by using the FOIL method.

$$\left(3\sqrt{2} + \sqrt{3}\right)\left(2\sqrt{3} - \sqrt{2}\right)$$
$$= 3\sqrt{2}\left(2\sqrt{3}\right) - 3\sqrt{2}\left(\sqrt{2}\right) + \sqrt{3}\left(2\sqrt{3}\right) - \sqrt{3}\sqrt{2}$$
$$= 6\sqrt{6} - 3 \cdot 2 + 2 \cdot 3 - \sqrt{6} = 6\sqrt{6} - 6 + 6 - \sqrt{6}$$
$$= 5\sqrt{6}$$

85.
$$\frac{\sqrt[3]{mn} \cdot \sqrt[3]{m^2}}{\sqrt[3]{n^2}} = \sqrt[3]{\frac{mnm^2}{n^2}} = \sqrt[3]{\frac{m^3}{n}}$$
$$= \frac{\sqrt[3]{m^3}}{\sqrt[3]{n}} \cdot \frac{\sqrt[3]{n^2}}{\sqrt[3]{n^2}} = \frac{m\sqrt[3]{n^2}}{n}$$

87.
$$\sqrt[3]{\frac{2}{x^6}} - \sqrt[3]{\frac{5}{x^9}} = \frac{\sqrt[3]{2}}{\sqrt[3]{x^6}} - \frac{\sqrt[3]{5}}{\sqrt[3]{x^9}} = \frac{\sqrt[3]{2}}{x^2} - \frac{\sqrt[3]{5}}{x^3}$$
$$= \frac{\sqrt[3]{2} \cdot x}{x^2 \cdot x} - \frac{\sqrt[3]{5}}{x^3} = \frac{x\sqrt[3]{2}}{x^3} - \frac{\sqrt[3]{5}}{x^3}$$
$$= \frac{x\sqrt[3]{2} - \sqrt[3]{5}}{x^3}$$

89.
$$\frac{1}{\sqrt{2}} + \frac{3}{\sqrt{8}} + \frac{1}{\sqrt{32}} = \frac{1}{\sqrt{2}} + \frac{3}{\sqrt{4 \cdot 2}} + \frac{1}{\sqrt{16 \cdot 2}}$$
$$= \frac{1}{\sqrt{2}} + \frac{3}{2\sqrt{2}} + \frac{1}{4\sqrt{2}}$$
$$= \frac{4 \cdot 1}{4\sqrt{2}} + \frac{3 \cdot 2}{2 \cdot 2\sqrt{2}} + \frac{1}{4\sqrt{2}}$$
$$= \frac{4}{4\sqrt{2}} + \frac{6}{4\sqrt{2}} + \frac{1}{4\sqrt{2}}$$
$$= \frac{4 + 6 + 1}{4\sqrt{2}} = \frac{11}{4\sqrt{2}}$$
$$= \frac{11\sqrt{2}}{4\sqrt{2} \cdot \sqrt{2}} = \frac{11\sqrt{2}}{4 \cdot 2} = \frac{11\sqrt{2}}{8}$$

91.
$$\frac{-4}{\sqrt[3]{3}} + \frac{1}{\sqrt[3]{24}} - \frac{2}{\sqrt[3]{81}}$$
$$= \frac{-4}{\sqrt[3]{3}} + \frac{1}{\sqrt[3]{8 \cdot 3}} - \frac{2}{\sqrt[3]{27 \cdot 3}} = \frac{-4}{\sqrt[3]{3}} + \frac{1}{2\sqrt[3]{3}} - \frac{2}{3\sqrt[3]{3}}$$
$$= \frac{-4 \cdot 6}{\sqrt[3]{3} \cdot 6} + \frac{1 \cdot 3}{2\sqrt[3]{3} \cdot 3} - \frac{2 \cdot 2}{3\sqrt[3]{3} \cdot 2}$$
$$= \frac{-24}{6\sqrt[3]{3}} + \frac{3}{6\sqrt[3]{3}} - \frac{4}{6\sqrt[3]{3}} = \frac{-24 + 3 - 4}{6\sqrt[3]{3}} = \frac{-25}{6\sqrt[3]{3}}$$
$$= \frac{-25}{6\sqrt[3]{3}} \cdot \frac{\sqrt[3]{3^2}}{\sqrt[3]{3^2}} = \frac{-25\sqrt[3]{9}}{6 \cdot 3} = \frac{-25\sqrt[3]{9}}{18}$$

93.
$$\frac{\sqrt{3}}{\sqrt{5} + \sqrt{3}} = \frac{\sqrt{3}}{\sqrt{5} + \sqrt{3}} \cdot \frac{\sqrt{5} - \sqrt{3}}{\sqrt{5} - \sqrt{3}}$$
$$= \frac{\sqrt{3}\left(\sqrt{5} - \sqrt{3}\right)}{\left(\sqrt{5}\right)^2 - \left(\sqrt{3}\right)^2}$$
$$= \frac{\sqrt{3}\sqrt{5} - \sqrt{3}\sqrt{3}}{5 - 3} = \frac{\sqrt{15} - 3}{2}$$

95.
$$\frac{\sqrt{7} - 1}{2\sqrt{7} + 4\sqrt{2}} = \frac{\sqrt{7} - 1}{2\sqrt{7} + 4\sqrt{2}} \cdot \frac{2\sqrt{7} - 4\sqrt{2}}{2\sqrt{7} - 4\sqrt{2}}$$
$$= \frac{\left(\sqrt{7} - 1\right)\left(2\sqrt{7} - 4\sqrt{2}\right)}{\left(2\sqrt{7} + 4\sqrt{2}\right)\left(2\sqrt{7} - 4\sqrt{2}\right)}$$
$$= \frac{\sqrt{7} \cdot 2\sqrt{7} - \sqrt{7} \cdot 4\sqrt{2} - 1 \cdot 2\sqrt{7} + 1 \cdot 4\sqrt{2}}{\left(2\sqrt{7}\right)^2 - \left(4\sqrt{2}\right)^2}$$
$$= \frac{2 \cdot 7 - 4\sqrt{14} - 2\sqrt{7} + 4\sqrt{2}}{4 \cdot 7 - 16 \cdot 2}$$
$$= \frac{14 - 4\sqrt{14} - 2\sqrt{7} + 4\sqrt{2}}{28 - 32}$$
$$= \frac{14 - 4\sqrt{14} - 2\sqrt{7} + 4\sqrt{2}}{-4}$$
$$= \frac{-2\left(-7 + 2\sqrt{14} + \sqrt{7} - 2\sqrt{2}\right)}{-4}$$
$$= \frac{-7 + 2\sqrt{14} + \sqrt{7} - 2\sqrt{2}}{2}$$

97.
$$\frac{p - 4}{\sqrt{p} + 2} = \frac{p - 4}{\sqrt{p} + 2} \cdot \frac{\sqrt{p} - 2}{\sqrt{p} - 2} = \frac{(p - 4)\left(\sqrt{p} - 2\right)}{\left(\sqrt{p}\right)^2 - 2^2}$$
$$= \frac{(p - 4)\left(\sqrt{p} - 2\right)}{p - 4} = \sqrt{p} - 2$$

Copyright © 2013 Pearson Education, Inc.

99.
$$\frac{3m}{2+\sqrt{m+n}} = \frac{3m}{2+\sqrt{m+n}} \cdot \frac{2-\sqrt{m+n}}{2-\sqrt{m+n}}$$
$$= \frac{3m\left(2-\sqrt{m+n}\right)}{2^2 - \left(\sqrt{m+n}\right)^2}$$
$$= \frac{3m\left(2-\sqrt{m+n}\right)}{4-(m+n)}$$
$$= \frac{3m\left(2-\sqrt{m+n}\right)}{4-m-n}$$

101.
$$\frac{5\sqrt{x}}{2\sqrt{x}+\sqrt{y}} = \frac{5\sqrt{x}\left(2\sqrt{x}-\sqrt{y}\right)}{\left(2\sqrt{x}+\sqrt{y}\right)\left(2\sqrt{x}-\sqrt{y}\right)}$$
$$= \frac{5\sqrt{x}\left(2\sqrt{x}-\sqrt{y}\right)}{4x-y}$$

103. $S = 15.18\sqrt[9]{n} = 15.18\sqrt[9]{4} \approx 17.7$
The speed of the boat with a four-man crew is approx. 17.7 ft/sec

105. Windchill temperature
$$= 35.74 + 0.6215T - 35.75V^{0.16} + 0.4275TV^{0.16}$$
Windchill temperature
$$= 35.74 + 0.6215(10) - 35.75\left(30^{0.16}\right)$$
$$+ 0.4275(10)\left(30^{0.16}\right) \approx -12.3°$$

107. $\sqrt[4]{8} \cdot \sqrt[4]{2} = \sqrt[4]{8\cdot 2} = \sqrt[4]{16} = 2$

109. $\dfrac{\sqrt[5]{320}}{\sqrt[5]{10}} = \sqrt[5]{\dfrac{320}{10}} = \sqrt[5]{32} = 2$

111. $\dfrac{\sqrt[3]{15}}{\sqrt[3]{5}} \cdot \sqrt[3]{9} = \sqrt[3]{3} \cdot \sqrt[3]{9} = \sqrt[3]{27} = 3$

113. $\dfrac{355}{113} = 3.1415929\ldots$ and $\pi \approx 3.1415926\ldots$, so it gives six decimal places of accuracy.

115. $\dfrac{3927}{1250} = 3.1416$ and $\pi \approx 3.1415926\ldots$, so it first differs in the fourth decimal place.

Chapter R   Review Exercises

1. The elements of the set {6, 8, 10, ..., 20} are the even numbers from 6 to 20 inclusive. The elements in the set are {6, 8, 10, 12, 14, 16, 18, 20}.

3. True. The set of negative integers = {..., −4, −3, −2, −1}, while the set of whole numbers = {0, 1, 2, 3, ...}. The two sets do not intersect, and so they are disjoint.

5. True

7. False. The two sets are not equal because they do not have the same elements.

9. True   **11.** True

13. $A' = \{2,6,9,10\}$ **15.** $B \cap E = \varnothing$

17. $D \cap \varnothing = \varnothing$

19. $(C \cap D) \cup B = \{1,3\} \cup B = \{1,2,3,4,6,8\}$

21. $\varnothing' = \{1, 2, 3, 4, 5, 6, 7, 8, 9, 10\}$ or U

23. $-12, -6, -\sqrt{4}$ (or −2), 0, and 6 are integers.

25. $\dfrac{4\pi}{5}$ is an irrational number and a real number.

27. 0 is a whole number, an integer, a rational number, and a real number.

29. Answers will vary. Sample answer: The reciprocal of a product is the product of the reciprocals.

31. Answers will vary. Sample answer: A product raised to a power is the product of each factor raised to the power.

33. Answers will vary. Sample answer: A quotient raised to a power is the quotient of the dividend raised to the power and the divisor raised to the power.

35. commutative   **37.** associative

39. identity

41. In a sample of 5000 students,
43% + 16% + 12% or 71% are expected to be over 19, or 5000(0.71) = 3550 students.

43. $(-4-1)(-3-5) - 2^3 = (-5)(-8) - 8$
$$= 40 - 8 = 32$$

45. $\left(-\dfrac{5}{9} - \dfrac{2}{3}\right) - \dfrac{5}{6} = \left(-\dfrac{5}{9} - \dfrac{6}{9}\right) - \dfrac{5}{6}$
$$= \dfrac{-11}{9} - \dfrac{5}{6} = -\dfrac{22}{18} - \dfrac{15}{18} = -\dfrac{37}{18}$$

Copyright © 2013 Pearson Education, Inc.

47. $\dfrac{6(-4)-3^2(-2)^3}{-5[-2-(-6)]} = \dfrac{6(-4)-9(-8)}{-5[-2+6]}$

$= \dfrac{6(-4)-9(-8)}{-5(4)}$

$= \dfrac{-24-(-72)}{-20} = \dfrac{-24+72}{-20}$

$= \dfrac{48}{-20} = -\dfrac{12}{5}$

49. Let $a = -1$, $b = -2$, $c = 4$.

$-c(2a-5b) = -4[2(-1)-5(-2)]$

$\qquad = -4\left[-2-(-10)\right] = -4(-2+10)$

$\qquad = -4(8) = -32$

51. Let $a = -1$, $b = -2$, $c = 4$.

$\dfrac{9a+2b}{a+b+c} = \dfrac{9(-1)+2(-2)}{-1+(-2)+4}$

$\qquad = \dfrac{-9+(-4)}{-3+4} = \dfrac{-13}{1} = -13$

53. $(3q^3-9q^2+6)+(4q^3-8q+3)$

$= 3q^3-9q^2+6+4q^3-8q+3$

$= 7q^3-9q^2-8q+9$

55. $(8y-7)(2y+7) = 16y^2+56y-14y-49$

$\qquad\qquad\qquad\quad = 16y^2+42y-49$

57. $(3k-5m)^2 = (3k)^2-2(3k)(5m)+(5m)^2$

$\qquad\qquad\quad = 9k^2-30km+25m^2$

59. $\dfrac{30m^3-9m^2+22m+5}{5m+1}$

$\begin{array}{r}
6m^2-3m\ +5 \\
5m+1\overline{)30m^3-9m^2+22m+5} \\
\underline{30m^3+6m^2} \\
-15m^2+22m \\
\underline{-15m^2-\ \ 3m} \\
25m+5 \\
\underline{25m+5} \\
0
\end{array}$

Thus,

$\dfrac{30m^3-9m^2+22m+5}{5m+1} = 6m^2-3m+5.$

61. $\dfrac{3b^3-8b^2+12b-30}{b^2+4}$

Insert the missing term in the divisor with a 0 coefficient.

$\begin{array}{r}
3b\ -8 \\
b^2+0b+4\overline{)3b^3-8b^2+12b-30} \\
\underline{3b^3+0b^2+12b} \\
-8b^2+0b-30 \\
\underline{-8b^2+0b-32} \\
2
\end{array}$

Thus,

$\dfrac{3b^3-8b^2+12b-30}{b^2+4} = 3b-8+\dfrac{2}{b^2+4}.$

63. $3(z-4)^2+9(z-4)^3 = 3(z-4)^2[1+3(z-4)]$

$\qquad\qquad\qquad\qquad = 3(z-4)^2(1+3z-12)$

$\qquad\qquad\qquad\qquad = 3(z-4)^2(3z-11)$

65. $z^2-6zk-16k^2 = (z-8k)(z+2k)$

67. $48a^8-12a^7b-90a^6b^2$

$= 6a^6(8a^2-2ab-15b^2)$

$= 6a^6(4a+5b)(2a-3b)$

69. $49m^8-9n^2 = (7m^4)^2-(3n)^2$

$\qquad\qquad\quad = (7m^4+3n)(7m^4-3n)$

71. $6(3r-1)^2+(3r-1)-35$

Let $x = 3r-1$. With this substitution,

$6(3r-1)^2+(3r-1)-35$ becomes

$6x^2+x-35$.

Factor the trinomial by trial and error.

$6x^2+x-35 = (3x-7)(2x+5)$

Replacing x with $3r-1$ gives

$[3(3r-1)-7][2(3r-1)+5]$

$= (9r-3-7)(6r-2+5) = (9r-10)(6r+3)$

$= 3(9r-10)(2r+1).$

73. $xy+2x-y-2 = (xy+2x)-(y+2)$

$\qquad\qquad\qquad = x(y+2)-(y+2)$

$\qquad\qquad\qquad = (y+2)(x-1)$

75. $(3x-4)^2+(x-5)(2)(3x-4)(3)$

$= (3x-4)[(3x-4)+(x-5)(2)(3)]$

$= (3x-4)[3x-4+6x-30]$

$= (3x-4)(9x-34)$

77. $\dfrac{k^2+k}{8k^3} \cdot \dfrac{4}{k^2-1} = \dfrac{k(k+1)(4)}{8k^3(k+1)(k-1)}$

$\qquad\qquad\qquad = \dfrac{4k}{8k^3(k-1)} = \dfrac{1}{2k^2(k-1)}$

Copyright © 2013 Pearson Education, Inc.

79. $\dfrac{x^2+x-2}{x^2+5x+6} \div \dfrac{x^2+3x-4}{x^2+4x+3}$

$= \dfrac{x^2+x-2}{x^2+5x+6} \cdot \dfrac{x^2+4x+3}{x^2+3x-4}$

$= \dfrac{(x+2)(x-1)}{(x+3)(x+2)} \cdot \dfrac{(x+3)(x+1)}{(x+4)(x-1)} = \dfrac{x+1}{x+4}$

81. $\dfrac{p^2-36q^2}{p^2-12pq+36q^2} \cdot \dfrac{p^2-5pq-6q^2}{p^2+2pq+q^2}$

$= \dfrac{(p+6q)(p-6q)}{(p-6q)^2} \cdot \dfrac{(p-6q)(p+q)}{(p+q)^2}$

$= \dfrac{p+6q}{p+q}$

83. $\dfrac{m}{4-m} + \dfrac{3m}{m-4} = \dfrac{m(-1)}{(4-m)(-1)} + \dfrac{3m}{m-4}$

$= \dfrac{-m}{m-4} + \dfrac{3m}{m-4} = \dfrac{2m}{m-4}$

We may also use $4-m$ as the common denominator. In this case, the result will be $\dfrac{-2m}{4-m}$. The two results are equivalent rational expressions.

85. $\dfrac{p^{-1}+q^{-1}}{1-(pq)^{-1}} = \dfrac{\frac{1}{p}+\frac{1}{q}}{1-\frac{1}{pq}} = \dfrac{pq\left(\frac{1}{p}+\frac{1}{q}\right)}{pq\left(1-\frac{1}{pq}\right)}$

$= \dfrac{pq\left(\frac{1}{p}\right)+pq\left(\frac{1}{q}\right)}{pq(1)-pq\left(\frac{1}{pq}\right)} = \dfrac{q+p}{pq-1}$

87. $\left(-\dfrac{5}{4}\right)^{-2} = \left(-\dfrac{4}{5}\right)^{2} = \dfrac{16}{25}$

89. $(5z^3)(-2z^5) = -10z^{3+5} = -10z^8$

91. $(-6p^5w^4m^{12})^0 = 1$ Definition of a^0

93. $\dfrac{-8y^7p^{-2}}{y^4p^3} = -8y^{7-(-4)}p^{(-2)-(-3)} = -8y^{11}p$

95. $\dfrac{(p+q)^4(p+q)^{-3}}{(p+q)^6} = (p+q)^{4+(-3)-6}$

$= (p+q)^{-5} = \dfrac{1}{(p+q)^5}$

97. $(7r^{1/2})(2r^{3/4})(-r^{1/6}) = -14r^{1/2+3/4+1/6}$

$= -14r^{17/12}$

99. $\dfrac{y^{5/3}\cdot y^{-2}}{y^{-5/6}} = \dfrac{y^{5/3+(-2)}}{y^{-5/6}} = \dfrac{y^{5/3+(-6/3)}}{y^{-5/6}} = \dfrac{y^{-1/3}}{y^{-5/6}}$

$= y^{-1/3-(-5/6)} = y^{-2/6+5/6}$

$= y^{3/6} = y^{1/2}$

101. $\dfrac{\left(p^{15}q^{12}\right)^{-4/3}}{\left(p^{24}q^{16}\right)^{-3/4}} = \dfrac{p^{15(-4/3)}q^{12(-4/3)}}{p^{24(-3/4)}q^{16(-3/4)}} = \dfrac{p^{-20}q^{-16}}{p^{-18}q^{-12}}$

$= p^{-20-(-18)}q^{-16-(-12)}$

$= p^{-2}q^{-4} = \dfrac{1}{p^2q^4}$

103. $\sqrt{200} = \sqrt{100\cdot2} = \sqrt{100}\cdot\sqrt{2} = 10\sqrt{2}$

105. $\sqrt[4]{1250} = \sqrt[4]{625\cdot2} = \sqrt[4]{625}\cdot\sqrt[4]{2} = 5\sqrt[4]{2}$

107. $-\sqrt[3]{\dfrac{2}{5p^2}} = -\dfrac{\sqrt[3]{2}}{\sqrt[3]{5p^2}} = -\dfrac{\sqrt[3]{2}}{\sqrt[3]{5p^2}}\cdot\dfrac{\sqrt[3]{25p}}{\sqrt[3]{25p}}$

$= -\dfrac{\sqrt[3]{2\cdot25p}}{\sqrt[3]{125p^3}} = -\dfrac{\sqrt[3]{50p}}{5p}$

109. $\sqrt[4]{\sqrt[3]{m}} = \left(\sqrt[3]{m}\right)^{1/4} = (m^{1/3})^{1/4} = m^{1/3\cdot1/4}$

$= m^{1/12} = \sqrt[12]{m}$

111. This product has the pattern $(a+b)(a^2-ab+b^2) = a^3+b^3$, the sum of two cubes.

$\left(\sqrt[3]{2}+4\right)\left(\sqrt[3]{2^2}-4\sqrt[3]{2}+16\right)$

$= \left(\sqrt[3]{2}+4\right)\left[\left(\sqrt[3]{2}\right)^2 - \sqrt[3]{2}\cdot4+4^2\right]$

$= \left(\sqrt[3]{2}\right)^3 + 4^3 = 2+64 = 66$

113. $\sqrt{18m^3} - 3m\sqrt{32m} + 5\sqrt{m^3}$

$= \sqrt{9m^2\cdot2m} - 3m\sqrt{16\cdot2m} + 5\sqrt{m^2m}$

$= 3m\sqrt{2m} - 4\cdot3m\sqrt{2m} + 5m\sqrt{m}$

$= 3m\sqrt{2m} - 12m\sqrt{2m} + 5m\sqrt{m}$

$= -9m\sqrt{2m} + 5m\sqrt{m}$ or $m\left(-9\sqrt{2m}+5\sqrt{m}\right)$

115. $\dfrac{6}{3-\sqrt{2}} = \dfrac{6}{3-\sqrt{2}}\cdot\dfrac{3+\sqrt{2}}{3+\sqrt{2}}$

$= \dfrac{6\left(3+\sqrt{2}\right)}{9-2} = \dfrac{6\left(3+\sqrt{2}\right)}{7}$

117. $x(x^2+5) = x\cdot x^2 + x\cdot5 = x^3+5x$

119. $(m^2)^3 = m^{2\cdot3} = m^6$

Copyright © 2013 Pearson Education, Inc.

121. $\dfrac{\left(\frac{a}{b}\right)}{2} = \left(\dfrac{a}{b}\right) \cdot \left(\dfrac{1}{2}\right) = \dfrac{a}{2b}$

123. $\dfrac{1}{(-2)^3} = \dfrac{1}{-8} = (-2)^{-3}$ or

$\dfrac{1}{(-2)^3} = \dfrac{1}{-8} = -\dfrac{1}{8} = -2^{-3}$

125. $\left(\dfrac{8}{7} + \dfrac{a}{b}\right)^{-1} = \left(\dfrac{8b}{7b} + \dfrac{7a}{7b}\right)^{-1}$

$= \left(\dfrac{8b + 7a}{7b}\right)^{-1} = \dfrac{7b}{8b + 7a}$

Chapter R   Test

1. False. $B' = \{2,4,6,7,8\}$

2. True

3. False. $(B \cap C) \cup D = \{1\} \cup D = \{1,4\}$

4. True

5. **(a)** $-13, -\dfrac{12}{4}$ (or -3), 0, and $\sqrt{49}$ (or 7) are integers.

   (b) $-13, -\dfrac{12}{4}$ (or -3), 0, $\dfrac{3}{5}$, 5.9 (or $\frac{59}{10}$), and $\sqrt{49}$ (or 7) are rational numbers.

   (c) All numbers in the set are real numbers.

6. Let $x = -2, y = -4, z = 5$.

$\left|\dfrac{x^2 + 2yz}{3(x+z)}\right| = \left|\dfrac{(-2)^2 + 2(-4)(5)}{3(-2+5)}\right|$

$= \left|\dfrac{4 + (-40)}{3(3)}\right| = \left|\dfrac{-36}{9}\right| = |-4| = 4$

7. **(a)** associative property

   (b) commutative

   (c) distributive

   (d) inverse

For exercise 8, we use the formula

$R = \dfrac{\left(250 \cdot \frac{C}{A}\right) + \left(1000 \cdot \frac{T}{A}\right) + \left(12.5 \cdot \frac{Y}{A}\right) + 6.25 - \left(1250 \cdot \frac{I}{A}\right)}{3}$

from section R.2.

8. $A = 514, C = 363, Y = 4388, T = 34, I = 11$
First, find the value of the numerator.

$\left(250 \cdot \dfrac{C}{A}\right) + \left(1000 \cdot \dfrac{T}{A}\right) + \left(12.5 \cdot \dfrac{Y}{A}\right)$
$\qquad + 6.25 - \left(1250 \cdot \dfrac{I}{A}\right)$

$= \left(250 \cdot \dfrac{363}{514}\right) + \left(1000 \cdot \dfrac{34}{514}\right) + \left(12.5 \cdot \dfrac{4388}{514}\right)$
$\qquad + 6.25 - \left(1250 \cdot \dfrac{11}{514}\right)$

≈ 328.92

$R \approx \dfrac{328.92}{3} \approx 109.6$

Drew Brees' quarterback rating was about 109.6.

9. $(x^2 - 3x + 2) - (x - 4x^2) + 3x(2x+1)$
$= x^2 - 3x + 2 - x + 4x^2 + 6x^2 + 3x$
$= 11x^2 - x + 2$

10. $(6r - 5)^2 = (6r)^2 - 2(6r)(5) + 5^2$
$= 36r^2 - 60r + 25$

11. $(t + 2)(3t^2 - t + 4)$

$\quad\quad 3t^2\ -t + 4$
$\quad\quad\quad\quad\quad t + 2$
$\quad\quad\overline{6t^2 - 2t + 8}$
$\quad 3t^3 - t^2 + 4t$
$\quad\overline{3t^3 + 5t^2 + 2t + 8}$

12. $\dfrac{2x^3 - 11x^2 + 28}{x - 5}$

$\quad\quad\quad\quad 2x^2\ -x\ -5$
$x - 5\overline{)2x^3 - 11x^2 + 0x + 28}$
$\quad\quad\underline{2x^3 - 10x^2}$
$\quad\quad\quad\quad -x^2 + 0x$
$\quad\quad\quad\quad \underline{-x^2 + 5x}$
$\quad\quad\quad\quad\quad\quad -5x + 28$
$\quad\quad\quad\quad\quad\quad \underline{-5x + 25}$
$\quad\quad\quad\quad\quad\quad\quad\quad 3$

Thus,

$\dfrac{2x^3 - 11x^2 + 28}{x - 5} = 2x^2 - x - 5 + \dfrac{3}{x - 5}.$

Copyright © 2013 Pearson Education, Inc.

13. For the year 2000, $x = 5$.
Adjusted poverty threshold
$$\approx 4.872x^2 + 170.2x + 7787$$
$$= 4.872 \cdot 5^2 + 170.2 \cdot 5 + 7787$$
$$= 121.8 + 851 + 7787 = 8759.8 \approx \$8760$$

14. For the year 2008, $x = 13$
Adjusted poverty threshold
$$\approx 4.872x^2 + 170.2x + 7787$$
$$= 4.872 \cdot 13^2 + 170.2 \cdot 13 + 7787$$
$$= 823.368 + 2212.6 + 7787 = 10,822.968$$
$$\approx \$10,823$$

15. $6x^2 - 17x + 7 = (3x - 7)(2x - 1)$

16. $x^4 - 16 = (x^2)^2 - 4^2$
$$= (x^2 + 4)(x^2 - 4)$$
$$= (x^2 + 4)(x^2 - 2^2)$$
$$= (x^2 + 4)(x + 2)(x - 2)$$

17. $24m^3 - 14m^2 - 24m = 2m(12m^2 - 7m - 12)$
$$= 2m(4m + 3)(3m - 4)$$

18. $x^3 y^2 - 9x^3 - 8y^2 + 72$
$$= (x^3 y^2 - 9x^3) - (8y^2 - 72)$$
$$= x^3(y^2 - 9) - 8(y^2 - 9) = (x^3 - 8)(y^2 - 9)$$
$$= (x^3 - 2^3)(y^2 - 3^2)$$
$$= (x - 2)(x^2 + 2x + 4)(y + 3)(y - 3)$$

19. $(a - b)^2 + 2(a - b) = (a - b)(a - b + 2)$

20. $1 - 27x^6 = (1 - 3x^2)(1 + 3x^2 + 9x^4)$

21.
$$\frac{5x^2 - 9x - 2}{30x^3 + 6x^2} \div \frac{x^4 - 3x^2 - 4}{2x^8 + 6x^7 + 4x^6}$$
$$= \frac{5x^2 - 9x - 2}{30x^3 + 6x^2} \cdot \frac{2x^8 + 6x^7 + 4x^6}{x^4 - 3x^2 - 4}$$
$$= \frac{(5x + 1)(x - 2)}{6x^2(5x + 1)} \cdot \frac{2x^6(x^2 + 3x + 2)}{(x^2 - 4)(x^2 + 1)}$$
$$= \frac{(5x + 1)(x - 2)}{6x^2(5x + 1)} \cdot \frac{2x^6(x + 2)(x + 1)}{(x + 2)(x - 2)(x^2 + 1)}$$
$$= \frac{2x^6(x + 1)}{6x^2(x^2 + 1)} = \frac{x^4(x + 1)}{3(x^2 + 1)}$$

22.
$$\frac{x}{x^2 + 3x + 2} + \frac{2x}{2x^2 - x - 3}$$
$$= \frac{x}{(x + 2)(x + 1)} + \frac{2x}{(2x - 3)(x + 1)}$$
The least common denominator is
$(x + 2)(x + 1)(2x - 3)$.
$$= \frac{x(2x - 3)}{(x + 2)(x + 1)(2x - 3)} + \frac{2x(x + 2)}{(2x - 3)(x + 1)(x + 2)}$$
$$= \frac{2x^2 - 3x}{(x + 2)(x + 1)(2x - 3)} + \frac{2x^2 + 4x}{(x + 2)(x + 1)(2x - 3)}$$
$$= \frac{2x^2 - 3x + 2x^2 + 4x}{(x + 2)(x + 1)(2x - 3)} = \frac{4x^2 + x}{(x + 2)(x + 1)(2x - 3)}$$
$$= \frac{x(4x + 1)}{(x + 2)(x + 1)(2x - 3)}$$

23.
$$\frac{a + b}{2a - 3} - \frac{a - b}{3 - 2a} = \frac{a + b}{2a - 3} - \frac{(a - b)(-1)}{(3 - 2a)(-1)}$$
$$= \frac{a + b}{2a - 3} + \frac{a - b}{2a - 3} = \frac{2a}{2a - 3}$$
If $3 - 2a$ is used as the common denominator,
the result will be $\dfrac{-2a}{3 - 2a}$. The rational
expressions $\dfrac{2a}{2a - 3}$ and $\dfrac{-2a}{3 - 2a}$ are
equivalent.

24.
$$\frac{y - 2}{y - \frac{4}{y}} = \frac{y(y - 2)}{y\left(y - \frac{4}{y}\right)} = \frac{y^2 - 2y}{y^2 - 4}$$
$$= \frac{y(y - 2)}{(y + 2)(y - 2)} = \frac{y}{y + 2}$$

25. $\sqrt{18x^5 y^8} = \sqrt{(9x^4 y^8)(2x)}$
$$= \sqrt{9x^4 y^8} \cdot \sqrt{2x} = 3x^2 y^4 \sqrt{2x}$$

26. $\sqrt{32x} + \sqrt{2x} - \sqrt{18x}$
$$= \sqrt{16 \cdot 2x} + \sqrt{2x} - \sqrt{9 \cdot 2x}$$
$$= 4\sqrt{2x} + \sqrt{2x} - 3\sqrt{2x} = 2\sqrt{2x}$$

27. $\left(\sqrt{x} - \sqrt{y}\right)\left(\sqrt{x} + \sqrt{y}\right) = \left(\sqrt{x}\right)^2 - \left(\sqrt{y}\right)^2$
$$= x - y$$

28.
$$\frac{14}{\sqrt{11} - \sqrt{7}} = \frac{14}{\sqrt{11} - \sqrt{7}} \cdot \frac{\sqrt{11} + \sqrt{7}}{\sqrt{11} + \sqrt{7}}$$
$$= \frac{14\left(\sqrt{11} + \sqrt{7}\right)}{11 - 7} = \frac{14\left(\sqrt{11} + \sqrt{7}\right)}{4}$$
$$= \frac{7\left(\sqrt{11} + \sqrt{7}\right)}{2}$$

Copyright © 2013 Pearson Education, Inc.

29. $\left(\dfrac{x^{-2}y^{-1/3}}{x^{-5/3}y^{-2/3}}\right)^3 = \dfrac{x^{-6}y^{-1}}{x^{-5}y^{-2}}$

$= x^{-6-(-5)}y^{-1-(-2)} = x^{-1}y = \dfrac{y}{x}$

30. $\left(-\dfrac{64}{27}\right)^{-2/3} = \left(-\dfrac{27}{64}\right)^{2/3} = \left[\left(-\dfrac{27}{64}\right)^{1/3}\right]^2$

$= \left(-\dfrac{3}{4}\right)^2 = \dfrac{9}{16}$

31. False. For all real numbers x, $\sqrt{x^2} = |x|$.

32. Let $L = 3.5$.

$t = 2\pi\sqrt{\dfrac{L}{32}} = 2\pi\sqrt{\dfrac{3.5}{32}} \approx 2.1$

The period of a pendulum 3.5 ft long is approximately 2.1 seconds.

Copyright © 2013 Pearson Education, Inc.

Chapter 1

EQUATIONS AND INEQUALITIES

Section 1.1   Linear Equations

1. Solve the equation $2x + 5 = x - 3$
$$2x + 5 = x - 3$$
$$2x + 5 - x = x - 3 - x$$
$$x + 5 = -3$$
$$x + 5 - 5 = -3 - 5$$
$$x = -8$$
Moreover, replacing x with -8 in $2x + 5 = x - 3$ yields a true statement. Therefore, the given statement is true.

3. The equations $x^2 = 4$ and $x + 2 = 4$ are not equivalent. The first has a solution set $\{-2, 2\}$ while the solution set of the second is $\{2\}$. Since the equations do not have the same solution set, they are not equivalent. The given statement is false.

5. Answers will vary.

7. B cannot be written in the form $ax + b = 0$. A can be written as $15x - 7 = 0$ or $15x + (-7) = 0$, C can be written as $2x = 0$ or $2x + 0 = 0$, and D can be written as $-0.04x - 0.4 = 0$ or $-0.04x + (-0.4) = 0$.

9. $5x + 4 = 3x - 4$
$$2x + 4 = -4$$
$$2x = -8 \Rightarrow x = -4$$
Solution set: $\{-4\}$

11. $6(3x - 1) = 8 - (10x - 14)$
$$18x - 6 = 8 - 10x + 14$$
$$18x - 6 = 22 - 10x$$
$$28x - 6 = 22$$
$$28x = 28 \Rightarrow x = 1$$
Solution set: $\{1\}$

13. $\dfrac{5}{6}x - 2x + \dfrac{4}{3} = \dfrac{5}{3}$
$$6 \cdot \left[\dfrac{5}{6}x - 2x + \dfrac{4}{3} \right] = 6 \cdot \dfrac{5}{3}$$
$$5x - 12x + 8 = 10$$
$$-7x + 8 = 10$$
$$-7x = 2 \Rightarrow x = -\dfrac{2}{7}$$
Solution set: $\left\{ -\dfrac{2}{7} \right\}$

15. $3x + 5 - 5(x + 1) = 6x + 7$
$$3x + 5 - 5x - 5 = 6x + 7$$
$$-2x = 6x + 7$$
$$-8x = 7 \Rightarrow x = \dfrac{7}{-8} = -\dfrac{7}{8}$$
Solution set: $\left\{ -\dfrac{7}{8} \right\}$

17. $2[x - (4 + 2x) + 3] = 2x + 2$
$$2(x - 4 - 2x + 3) = 2x + 2$$
$$2(-x - 1) = 2x + 2$$
$$-2x - 2 = 2x + 2$$
$$-2 = 4x + 2$$
$$-4 = 4x \Rightarrow -1 = x$$
Solution set: $\{-1\}$

19. $\dfrac{1}{14}(3x - 2) = \dfrac{x + 10}{10}$
$$70 \cdot \left[\dfrac{1}{14}(3x - 2) \right] = 70 \cdot \left[\dfrac{x + 10}{10} \right]$$
$$5(3x - 2) = 7(x + 10)$$
$$15x - 10 = 7x + 70$$
$$8x - 10 = 70$$
$$8x = 80 \Rightarrow x = 10$$
Solution set: $\{10\}$

21. $0.2x - 0.5 = 0.1x + 7$
$$10(0.2x - 0.5) = 10(0.1x + 7)$$
$$2x - 5 = x + 70$$
$$x - 5 = 70 \Rightarrow x = 75$$
Solution set: $\{75\}$

23. $-4(2x - 6) + 8x = 5x + 24 + x$
$$-8x + 24 + 8x = 6x + 24$$
$$24 = 6x + 24$$
$$0 = 6x \Rightarrow 0 = x$$
Solution set: $\{0\}$

25. $0.5x + \dfrac{4}{3}x = x + 10$
$$\dfrac{1}{2}x + \dfrac{4}{3}x = x + 10$$
$$6\left(\dfrac{1}{2}x + \dfrac{4}{3}x \right) = 6(x + 10)$$
$$3x + 8x = 6x + 60$$
$$11x = 6x + 60$$
$$5x = 60 \Rightarrow x = 12$$
Solution set: $\{12\}$

Copyright © 2013 Pearson Education, Inc.

27.
$$0.08x + 0.06(x+12) = 7.72$$
$$100\left[0.08x + 0.06(x+12)\right] = 100 \cdot 7.72$$
$$8x + 6(x+12) = 772$$
$$8x + 6x + 72 = 772$$
$$14x + 72 = 772$$
$$14x = 700 \Rightarrow x = 50$$

Solution set: $\{50\}$

29.
$$4(2x+7) = 2x + 22 + 3(2x+2)$$
$$8x + 28 = 2x + 22 + 6x + 6$$
$$8x + 28 = 8x + 28$$
$$28 = 28 \Rightarrow 0 = 0$$

identity; $\{$all real numbers$\}$

31.
$$2(x-8) = 3x - 16$$
$$2x - 16 = 3x - 16$$
$$-16 = x - 16 \Rightarrow 0 = x$$

conditional equation; $\{0\}$

33.
$$0.3(x+2) - 0.5(x+2) = -0.2x - 0.4$$
$$10\left[0.3(x+2) - 0.5(x+2)\right] = 10\left[-0.2x - 0.4\right]$$
$$3(x+2) - 5(x+2) = -2x - 4$$
$$3x + 6 - 5x - 10 = -2x - 4$$
$$-2x - 4 = -2x - 4$$
$$0 = 0$$

identity; $\{$all real numbers$\}$

35.
$$4(x+7) = 2(x+12) + 2(x+1)$$
$$4x + 28 = 2x + 24 + 2x + 2$$
$$4x + 28 = 4x + 26$$
$$28 = 26$$

contradiction; $\varnothing$

37. Answers will vary. In solving an equation, you cannot multiply (or divide) both sides of an equation by zero. This is essentially what happened when the student divided both sides of the equation by x. To solve the equation, the student should isolate the variable term, which leads to the solution set {0}.
$$5x = 4x$$
$$5x - 4x = 4x - 4x$$
$$x = 0$$

39.
$$V = lwh$$
$$\frac{V}{wh} = \frac{lwh}{wh}$$
$$l = \frac{V}{wh}$$

41.
$$P = a + b + c$$
$$P - a - b = c$$
$$c = P - a - b$$

43.
$$A = \frac{1}{2}h(B+b)$$
$$2A = 2\left[\frac{1}{2}h(B+b)\right]$$
$$2A = h(B+b)$$
$$2A = Bh + bh$$
$$2A - bh = Bh$$
$$\frac{2A - bh}{h} = \frac{Bh}{h}$$
$$B = \frac{2A - bh}{h} = \frac{2A}{h} - b$$

45.
$$S = 2\pi rh + 2\pi r^2$$
$$S - 2\pi r^2 = 2\pi rh$$
$$\frac{S - 2\pi r^2}{2\pi r} = \frac{2\pi rh}{2\pi r}$$
$$h = \frac{S - 2\pi r^2}{2\pi r} = \frac{S}{2\pi r} - r$$

47.
$$S = 2lw + 2wh + 2hl$$
$$S - 2lw = 2wh + 2hl$$
$$S - 2lw = (2w + 2l)h$$
$$\frac{S - 2lw}{2w + 2l} = \frac{(2w + 2l)h}{2w + 2l}$$
$$h = \frac{S - 2lw}{2w + 2l}$$

49.
$$2(x - a) + b = 3x + a$$
$$2x - 2a + b = 3x + a$$
$$-3a + b = x$$
$$x = -3a + b$$

51.
$$ax + b = 3(x - a)$$
$$ax + b = 3x - 3a$$
$$3a + b = 3x - ax$$
$$3a + b = (3 - a)x$$
$$\frac{3a + b}{3 - a} = x$$
$$x = \frac{3a + b}{3 - a}$$

53.
$$\frac{x}{a - 1} = ax + 3$$
$$(a - 1)\left[\frac{x}{a - 1}\right] = (a - 1)(ax + 3)$$
$$x = a^2x + 3a - ax - 3$$
$$3 - 3a = a^2x - ax - x$$
$$3 - 3a = (a^2 - a - 1)x$$
$$\frac{3 - 3a}{a^2 - a - 1} = x$$
$$x = \frac{3 - 3a}{a^2 - a - 1}$$

Copyright © 2013 Pearson Education, Inc.

55. $a^2x + 3x = 2a^2$
$$\left(a^2 + 3\right)x = 2a^2$$
$$x = \frac{2a^2}{a^2 + 3}$$

57. $3x = \left(2x - 1\right)\left(m + 4\right)$
$$3x = 2xm + 8x - m - 4$$
$$m + 4 = 2xm + 5x$$
$$m + 4 = \left(2m + 5\right)x$$
$$\frac{m + 4}{2m + 5} = x$$
$$x = \frac{m + 4}{2m + 5}$$

59. (a) Here, $r = 0.04$, $P = 3150$, and
$$t = \frac{6}{12} = \frac{1}{2} \text{ (year)}.$$
$$I = Prt = 3150\left(0.04\right)\left(\frac{1}{2}\right) = \$63$$
The interest is \$63.

(b) The amount Miguel must pay Julio at the end of the six months is
$\$3150 + \$63 = \$3213$.

61. $F = \frac{9}{5}C + 32$
$$F = \frac{9}{5} \cdot 40 + 32 = 72 + 32 = 104$$
Therefore, $40°C = 104°F$.

63. $C = \frac{5}{9}\left(F - 32\right)$
$$C = \frac{5}{9}\left(50 - 32\right) = \frac{5}{9} \cdot 18 = 10$$
Therefore, $50°F = 10°C$.

65. $C = \frac{5}{9}\left(F - 32\right)$
$$C = \frac{5}{9}\left(100 - 32\right) = \frac{5}{9} \cdot 68 \approx 37.8$$
Therefore, $100°F \approx 37.8°C$.

67. $C = \frac{5}{9}\left(F - 32\right)$
$$C = \frac{5}{9}\left(867 - 32\right) = \frac{5}{9} \cdot 835 \approx 463.9$$
Therefore, $865°F \approx 463.9°C$.

69. $C = \frac{5}{9}\left(F - 32\right)$
$$C = \frac{5}{9}\left(7 - 32\right) = \frac{5}{9} \cdot \left(-25\right) \approx -13.9$$
Therefore, $7°F \approx -13.9°C$.

Section 1.2 Applications and Modeling with Linear Equations

1. 15 minutes is $\frac{1}{4}$ of an hour, so multiply 80 mph by $\frac{1}{4}$ to get a distance of 20 mi.

3. 2% is 0.02, so multiply \$500 by 0.02 and by 4 yrs to get interest of \$40

5. Concentration A, 24%, cannot possibly be the concentration of the mixture because it is less than both the original concentrations.

7. D

9. In the formula $P = 2l + 2w$, let
$P = 294$ and $w = 57$.
$$294 = 2l + 2 \cdot 57$$
$$294 = 2l + 114$$
$$180 = 2l \Rightarrow 90 = l$$
The length is 90 cm.

11. Let x = the length of the shortest side.
Then $2x$ = the length of each of the longer sides.
The perimeter of a triangle is the sum of the measures of the three sides.
$$x + 2x + 2x = 30 \Rightarrow 5x = 30 \Rightarrow x = 6$$
The length of the shortest side is 6 cm.

13. Let x = the length of the shortest side.
Then $2x - 200$ = the length of the longest side and the length of the middle side is
$\left(2x - 200\right) - 200 = 2x - 400$.
The perimeter of a triangle is the sum of the measures of the three sides.
$$x + \left(2x - 200\right) + \left(2x - 400\right) = 2400$$
$$x + 2x - 200 + 2x - 400 = 2400$$
$$5x - 600 = 2400$$
$$5x = 3000 \Rightarrow x = 600$$
The length of the shortest side is 600 ft. The middle side is $2 \cdot 600 - 400 = 1200 - 400$
$= 800$ ft. The longest side is $2 \cdot 600 - 200$
$= 1200 - 200 = 1000$ ft.

Copyright © 2013 Pearson Education, Inc.

15. Let l = the length of the book.
Then $l - 0.42$ = the width of the book.
Use the formula for the perimeter of a rectangle.
$$P = 2l + 2w$$
$$5.96 = 2l + 2(l - 0.42)$$
$$5.96 = 2l + 2l - 0.84$$
$$5.96 = 4l - 0.84$$
$$6.8 = 4l \Rightarrow 1.7 = l$$
The length of the book is 1.7 cm, and the width of the book is $1.7 - 0.42 = 1.28$ cm.

17. Let h = the height of box.
Use the formula for the surface area of a rectangular box.
$$S = 2lw + 2wh + 2hl$$
$$496 = 2 \cdot 18 \cdot 8 + 2 \cdot 8h + 2h \cdot 18$$
$$496 = 288 + 16h + 36h$$
$$496 = 288 + 52h$$
$$208 = 52h \Rightarrow 4 = h$$
The height of the box is 4 ft.

19. Let x = the time (in hours) spent on the way to the business appointment.

	r	t	d
Morning	50	x	$50x$
Afternoon	40	$x + \frac{1}{4}$	$40\left(x + \frac{1}{4}\right)$

The distance on the way to the business appointment is the same as the return trip, so
$$50x = 40\left(x + \tfrac{1}{4}\right)$$
$$50x = 40x + 10$$
$$10x = 10 \Rightarrow x = 1$$
Since she drove 1 hr, her distance traveled would be $50 \cdot 1 = 50$ mi.

21. Let x = David's speed (in mph) on bike.
Then $x + 4.5$ = David's speed (in mph) driving.

	r	t	d
Car	$x + 4.5$	20 min = $\frac{1}{3}$ hr	$\frac{1}{3}(x + 4.5)$
Bike	x	45 min = $\frac{3}{4}$ hr	$\frac{3}{4}x$

The distance by bike and car are the same, so
$$\tfrac{1}{3}(x + 4.5) = \tfrac{3}{4}x$$
$$12\left[\tfrac{1}{3}(x + 4.5)\right] = 12\left[\tfrac{3}{4}x\right]$$
$$4(x + 4.5) = 9x$$
$$4x + 18 = 9x$$
$$18 = 5x \Rightarrow \frac{18}{5} = x$$

Since his rate is $\frac{18}{5}$ (or 3.6) mph, David travels
$$\tfrac{3}{4}\left(\tfrac{18}{5}\right) = \tfrac{27}{10} = 2.7 \text{ mi} \text{ to work.}$$

23. Let x = time (in hours) it takes for Russ and Janet to be 1.5 mi apart.

	r	t	d
Mary	7	x	$7x$
Janet	5	x	$5x$

Since Mary's rate is faster than Janet's, she travels farther than Janet in the same amount of time. To have the difference between Mary and Janet to be 1.5 mi, solve the following equation.
$$7x - 5x = 1.5 \Rightarrow 2x = 1.5 \Rightarrow x = 0.75$$
It will take 0.75 hr = 45 min for Mary and Janet to be 1.5 mi apart.

25. We need to determine how many meters are in 26 miles.
$$26 \text{ mi} \cdot \frac{5,280 \text{ ft}}{1 \text{ mi}} \cdot \frac{1 \text{ m}}{3.281 \text{ ft}} \approx 41,840.9 \text{ m}$$
Usain Bolt's rate in the 100-m dash would be
$$r = \frac{d}{t} = \frac{100}{9.69} \text{ meters per second.}$$
Thus, the time it would take for Usain to run the 26-mi marathon would be
$$t = \frac{d}{r} = \frac{41,840.9}{\frac{100}{9.69}} = 41,840.9 \cdot \frac{9.69}{100} \approx 4054 \text{ sec.}$$
Since there are 60 seconds in one minute and $60 \cdot 60 = 3,600$ seconds in one hour,
$$4054 \text{ sec} = 1 \cdot 3600 + 7 \cdot 60 + 34 \text{ sec}$$
or 1 hr, 7 min, 34 sec. This is about $\frac{1}{2}$ the world record time.

27. Let x = speed (in km/hr) of Callie's boat. When Callie is traveling upstream, the current slows her down, so we subtract the speed of the current from the speed of the boat. When she is traveling downstream, the current speeds her up, so we add the speed of the current to the speed of the boat.

	r	t	d
Upstream	$x - 5$	20 min = $\frac{1}{3}$ hr	$\frac{1}{3}(x - 5)$
Downstream	$x + 5$	15 min = $\frac{1}{4}$ hr	$\frac{1}{4}(x + 5)$

Since the distance upstream and downstream are the same, we must solve the following equation.

(continued on next page)

Copyright © 2013 Pearson Education, Inc.

(*continued*)

$$\tfrac{1}{3}(x-5)=\tfrac{1}{4}(x+5)$$
$$12\left[\tfrac{1}{3}(x-5)\right]=12\left[\tfrac{1}{4}(x+5)\right]$$
$$4(x-5)=3(x+5)$$
$$4x-20=3x+15$$
$$x-20=15\Rightarrow x=35$$

The speed of Callie's boat is 35 km per hour.

29. Let x = the amount of 5% acid solution (in gallons).

Strength	Gallons of Solution	Gallons of Pure Acid
5%	x	$0.05x$
10%	5	$0.10\cdot 5=0.5$
7%	$x+5$	$0.07(x+5)$

The number of gallons of pure acid in the 5% solution plus the number of gallons of pure acid in the 10% solution must equal the number of gallons of pure acid in the 7% solution.

$$0.05x+0.5=0.07(x+5)$$
$$0.05x+0.5=0.07x+0.35$$
$$0.5=0.02x+0.35$$
$$0.15=0.02x$$
$$\frac{0.15}{0.02}=x\Rightarrow x=7.5=7\tfrac{1}{2}\,\text{gal}$$

$7\tfrac{1}{2}$ gallons of the 5% solution should be added.

31. Let x = the amount of 100% alcohol solution (in liters).

Strength	Liters of Solution	Liters of Pure Alcohol
100%	x	$1x=x$
10%	7	$0.10\cdot 7=0.7$
30%	$x+7$	$0.30(x+7)$

The number of liters of pure alcohol in the 100% solution plus the number of liters of pure alcohol in the 10% solution must equal the number of liters of pure alcohol in the 30% solution.

$$x+0.7=0.30(x+7)$$
$$x+0.7=0.30x+2.1$$
$$0.7x+0.7=2.1$$
$$0.7x=1.4$$
$$x=\frac{1.4}{0.7}=\frac{14}{7}=2\,\text{L}$$

2 L of the 100% solution should be added.

33. Let x = the amount of water (in mL).

Strength	Milliliters of Solution	Milliliters of Salt
6%	8	$0.06(8)=0.48$
0%	x	$0(x)=0$
4%	$8+x$	$0.04(8+x)$

The number of milliliters of salt in the 6% solution plus the number of milliliters of salt in the water (0% solution) must equal the number of milliliters in the 4% solution.

$$0.48+0=0.04(8+x)$$
$$0.48=0.32+0.04x$$
$$0.16=0.04x$$
$$\frac{0.16}{0.04}=x\Rightarrow x=\frac{16}{4}=4\,\text{mL}$$

To reduce the saline concentration to 4%, 4 mL of water should be added.

35. Let x = amount of the short-term note. Then $240,000 - x$ = amount of the long-term note.

Note Amount	Interest Rate	Time (years)	Interest Paid
x	2%	1	$x(0.02)(1)$
$240,000-x$	2.5%	1	$(240,000-x)(0.025)(1)$
			5500

The amount of interest from the 2% note plus the amount of interest from the 2.5% note must equal the total amount of interest.

$$0.02x+0.025(240,000-x)=5500$$
$$0.02x+6000-0.025x=5500$$
$$-0.005x+6000=5500$$
$$-0.005x=-500$$
$$x=100,000$$

The amount of the short-term note is $100,000 and the amount of the long-term note is $240,000 - $100,000 = $140,000.

37. Let x = amount invested at 2.5%. Then $2x$ = amount invested at 3%.

Amount in Account	Interest Rate	Interest
x	2.5%	$0.025x$
$2x$	3%	$0.03(2x)=0.06x$
		850

The amount of interest from the 2.5% account plus the amount of interest from the 3% account must equal the total amount of interest.

(*continued on next page*)

Copyright © 2013 Pearson Education, Inc.

(*continued*)

$$0.025x + 0.03(2x) = 850$$
$$0.025x + 0.06x = 850$$
$$0.085x = 850 \Rightarrow x = \$10,000$$

Janet deposited \$10,000 at 2.5% and 2(\$10,000) = \$20,000 at 3%.

39. 30% of \$200,000 is \$60,000, so after paying her income tax, Linda had \$140,000 left to invest. Let x = amount invested at 1.5%. Then $140,000 - x$ = amount invested at 4%.

Amount Invested	Interest Rate	Interest
x	1.5%	$0.015x$
$140,000 - x$	4%	$0.04(140,000 - x)$
140,000		4350

$$0.015x + 0.04(140,000 - x) = 4350$$
$$0.015x + 5600 - 0.04x = 4350$$
$$-0.025x + 5600 = 4350$$
$$-0.025x = -1250$$
$$x = \$50,000$$

Linda invested \$50,000 at 1.5% and \$140,000 - \$50,000 = \$90,000 at 4%.

41. (a) $y = 100 - 0.02x$
$$y = 100 - 0.02(2400) = 100 - 48 = 52$$
The yearly cost is \$52.

(b) $50 = 100 - 0.02x$
$$-50 = -0.02x$$
$$2500 = x$$
The yearly cost of membership will be \$50 if the club purchases are \$2500.

(c) $0 = 100 - 0.02x$
$$-100 = -0.02x$$
$$5000 = x$$
The yearly cost of membership will be \$0 if the club purchases are \$5000.

43. (a) Let x = the number of hours. Then $F = 100(140)x = 14,000x$.

(b) Since $33 \ \mu g/ft^3$ causes irritation, the room would need $33 \cdot 800 = 26,400 \ \mu g$ to cause irritation.
$$F = 14,000x$$
$$26,400 = 14,000x$$
$$\frac{26,400}{14,000} = x$$
$$x \approx 1.9 \text{ hr}$$
It will take about 1.9 hours for concentrations to reach $33 \ \mu g/ft^3$.

45. (a) In 2014, $x = 5$.
$$y = 0.2309x + 18.35$$
$$y = 0.2309 \cdot 5 + 18.35 = 1.1545 + 18.35$$
$$= 19.5045$$
The projected enrollment for Fall 2014 is approximately 19.5 million.

(b) $y = 0.2309x + 18.35$
$$20 = 0.2309x + 18.35$$
$$1.65 = 0.2309x$$
$$x = \frac{1.65}{0.2309} \approx 7.15$$
Enrollment is projected to reach 20 million during the year 2016.

(c) They are quite close.

(d) The year 1997 is represented by $x = -12$.
$$y = 0.2309x + 18.35$$
$$y = 0.2309(-12) + 18.35$$
$$= -2.7708 + 18.35 \approx 15.6$$
According to the model, the enrollment was approximately 15.6 million

(e) Answers will vary.

Section 1.3   Complex Numbers

1. true

3. true

5. false (Every real number is a complex number.)

7. −4 is real and complex.

9. $13i$ is complex, pure imaginary and nonreal complex.

11. $5 + i$ is complex and nonreal complex.

13. π is real and complex.

15. $\sqrt{-25} = 5i$ is complex, pure imaginary and nonreal complex.

17. $\sqrt{-25} = i\sqrt{25} = 5i$

19. $\sqrt{-10} = i\sqrt{10}$

21. $\sqrt{-288} = i\sqrt{288} = i\sqrt{144 \cdot 2} = 12i\sqrt{2}$

23. $-\sqrt{-18} = -i\sqrt{18} = -i\sqrt{9 \cdot 2} = -3i\sqrt{2}$

25. $\sqrt{-13} \cdot \sqrt{-13} = i\sqrt{13} \cdot i\sqrt{13}$
$$= i^2\left(\sqrt{13}\right)^2 = -1 \cdot 13 = -13$$

27. $\sqrt{-3} \cdot \sqrt{-8} = i\sqrt{3} \cdot i\sqrt{8} = i^2\sqrt{3 \cdot 8}$
$$= -1 \cdot \sqrt{24} = -\sqrt{4 \cdot 6} = -2\sqrt{6}$$

Copyright © 2013 Pearson Education, Inc.

29. $\dfrac{\sqrt{-30}}{\sqrt{-10}} = \dfrac{i\sqrt{30}}{i\sqrt{10}} = \sqrt{\dfrac{30}{10}} = \sqrt{3}$

31. $\dfrac{\sqrt{-24}}{\sqrt{8}} = \dfrac{i\sqrt{24}}{\sqrt{8}} = i\sqrt{\dfrac{24}{8}} = i\sqrt{3}$

33. $\dfrac{\sqrt{-10}}{\sqrt{-40}} = \dfrac{i\sqrt{10}}{i\sqrt{40}} = \sqrt{\dfrac{10}{40}} = \sqrt{\dfrac{1}{4}} = \dfrac{1}{2}$

35. $\dfrac{\sqrt{-6} \cdot \sqrt{-2}}{\sqrt{3}} = \dfrac{i\sqrt{6} \cdot i\sqrt{2}}{\sqrt{3}} = i^2 \sqrt{\dfrac{6 \cdot 2}{3}}$

$\qquad = -1 \cdot \sqrt{\dfrac{12}{3}} = -\sqrt{4} = -2$

37. $\dfrac{-6 - \sqrt{-24}}{2} = \dfrac{-6 - \sqrt{-4 \cdot 6}}{2} = \dfrac{-6 - 2i\sqrt{6}}{2}$

$\qquad = \dfrac{2\left(-3 - i\sqrt{6}\right)}{2} = -3 - i\sqrt{6}$

39. $\dfrac{10 + \sqrt{-200}}{5} = \dfrac{10 + \sqrt{-100 \cdot 2}}{5}$

$\qquad = \dfrac{10 + 10i\sqrt{2}}{5} = \dfrac{5\left(2 + 2i\sqrt{2}\right)}{5}$

$\qquad = 2 + 2i\sqrt{2}$

41. $\dfrac{-3 + \sqrt{-18}}{24} = \dfrac{-3 + \sqrt{-9 \cdot 2}}{24} = \dfrac{-3 + 3i\sqrt{2}}{24}$

$\qquad = \dfrac{3\left(-1 + i\sqrt{2}\right)}{24} = \dfrac{-1 + i\sqrt{2}}{8}$

$\qquad = -\dfrac{1}{8} + \dfrac{\sqrt{2}}{8}i$

43. $(3 + 2i) + (9 - 3i) = (3 + 9) + \left[2 + (-3)\right]i$

$\qquad = 12 + (-1)i = 12 - i$

45. $(-2 + 4i) - (-4 + 4i)$

$\qquad = \left[-2 - (-4)\right] + (4 - 4)i$

$\qquad = 2 + 0i = 2$

47. $(2 - 5i) - (3 + 4i) - (-2 + i)$

$\qquad = \left[2 - 3 - (-2)\right] + (-5 - 4 - 1)i$

$\qquad = 1 + (-10)i = 1 - 10i$

49. $-i\sqrt{2} - 2 - \left(6 - 4i\sqrt{2}\right) - \left(5 - i\sqrt{2}\right)$

$\qquad = (-2 - 6 - 5) + \left[-\sqrt{2} - \left(-4\sqrt{2}\right) - \left(-\sqrt{2}\right)\right]i$

$\qquad = -13 + 4i\sqrt{2}$

51. $(2 + i)(3 - 2i)$

$\qquad = 2(3) + 2(-2i) + i(3) + i(-2i)$

$\qquad = 6 - 4i + 3i - 2i^2 = 6 - i - 2(-1)$

$\qquad = 6 - i + 2 = 8 - i$

53. $(2 + 4i)(-1 + 3i)$

$\qquad = 2(-1) + 2(3i) + 4i(-1) + 4i(3i)$

$\qquad = -2 + 6i - 4i + 12i^2 = -2 + 2i + 12(-1)$

$\qquad = -2 + 2i - 12 = -14 + 2i$

55. $(3 - 2i)^2 = 3^2 - 2(3)(2i) + (2i)^2$

$\qquad = 9 - 12i - 4 = 5 - 12i$

57. $(3 + i)(3 - i) = 3^2 - i^2 = 9 - (-1) = 10$

59. $(-2 - 3i)(-2 + 3i) = (-2)^2 - (3i)^2 = 4 - 9i^2$

$\qquad = 4 - 9(-1) = 13$

61. $\left(\sqrt{6} + i\right)\left(\sqrt{6} - i\right) = \left(\sqrt{6}\right)^2 - i^2$

$\qquad = 6 - (-1) = 6 + 1 = 7$

63. $i(3 - 4i)(3 + 4i) = i\left[(3 - 4i)(3 + 4i)\right]$

$\qquad = i\left[3^2 - (4i)^2\right]$

$\qquad = i\left[9 - 16i^2\right]$

$\qquad = i\left[9 - 16(-1)\right]$

$\qquad = i(9 + 16) = 25i$

65. $3i(2 - i)^2 = 3i\left(2^2 - 2(2i) + i^2\right)$

$\qquad = 3i(4 - 4i - 1) = 3i(3 - 4i)$

$\qquad = 9i - 12i^2 = 9i - 12(-1)$

$\qquad = 12 + 9i$

67. $(2 + i)(2 - i)(4 + 3i) = \left[(2 + i)(2 - i)\right](4 + 3i)$

$\qquad = \left[2^2 - i^2\right](4 + 3i)$

$\qquad = \left[4 - (-1)\right](4 + 3i)$

$\qquad = 5(4 + 3i) = 20 + 15i$

69. $\dfrac{6 + 2i}{1 + 2i} = \dfrac{(6 + 2i)(1 - 2i)}{(1 + 2i)(1 - 2i)}$

$\qquad = \dfrac{6 - 12i + 2i - 4i^2}{1^2 - (2i)^2} = \dfrac{6 - 10i - 4(-1)}{1 - 4i^2}$

$\qquad = \dfrac{6 - 10i + 4}{1 - 4(-1)} = \dfrac{10 - 10i}{1 + 4} = \dfrac{10 - 10i}{5}$

$\qquad = \dfrac{10}{5} - \dfrac{10}{5}i = 2 - 2i$

Copyright © 2013 Pearson Education, Inc.

71. $\dfrac{2-i}{2+i} = \dfrac{(2-i)(2-i)}{(2+i)(2-i)} = \dfrac{2^2 - 2(2i) + i^2}{2^2 - i^2}$

$= \dfrac{4 - 4i + (-1)}{4 - (-1)} = \dfrac{3 - 4i}{5} = \dfrac{3}{5} - \dfrac{4}{5}i$

73. $\dfrac{1-3i}{1+i} = \dfrac{(1-3i)(1-i)}{(1+i)(1-i)} = \dfrac{1 - i - 3i + 3i^2}{1^2 - i^2}$

$= \dfrac{1 - 4i + 3(-1)}{1 - (-1)} = \dfrac{1 - 4i - 3}{2}$

$= \dfrac{-2 - 4i}{2} = \dfrac{-2}{2} - \dfrac{4}{2}i = -1 - 2i$

75. $\dfrac{-5}{i} = \dfrac{-5(-i)}{i(-i)} = \dfrac{5i}{-i^2}$

$= \dfrac{5i}{-(-1)} = \dfrac{5i}{1} = 5i \text{ or } 0 + 5i$

77. $\dfrac{8}{-i} = \dfrac{8 \cdot i}{-i \cdot i} = \dfrac{8i}{-i^2}$

$= \dfrac{8i}{-(-1)} = \dfrac{8i}{1} = 8i \text{ or } 0 + 8i$

79. $\dfrac{2}{3i} = \dfrac{2(-3i)}{3i \cdot (-3i)} = \dfrac{-6i}{-9i^2} = \dfrac{-6i}{-9(-1)}$

$= \dfrac{-6i}{9} = -\dfrac{2}{3}i \text{ or } 0 - \dfrac{2}{3}i$

Note: In the above solution, we multiplied the numerator and denominator by the complex conjugate of $3i$, namely $-3i$. Since there is a reduction in the end, the same results can be achieved by multiplying the numerator and denominator by $-i$.

81. $I = 5 + 7i, Z = 6 + 4i$

$E = IZ$

$E = (5 + 7i)(6 + 4i)$

$= 5(6) + 5(4i) + 7i(6) + 7i(4i)$

$= 30 + 20i + 42i + 28i^2$

$= 30 + 62i - 28$

$= 2 + 62i$

83. $I = 10 + 4i, E = 88 + 128i$

$E = IZ$

$88 + 128i = (10 + 4i)Z$

$Z = \dfrac{88 + 128i}{10 + 4i}$

$Z = \dfrac{(88 + 128i)(10 - 4i)}{(10 + 4i)(10 - 4i)}$

$Z = \dfrac{88(10) + 88(-4i) + 128i(10) + 128i(-4i)}{10^2 - (4i)^2}$

$Z = \dfrac{880 - 352i + 1280i - 512i^2}{100 - (-16)}$

$Z = \dfrac{880 - 352i + 1280i + 512}{116}$

$Z = \dfrac{1392 + 928i}{116} = 12 + 8i$

85. $i^{25} = i^{24} \cdot i = \left(i^4\right)^6 \cdot i = 1^6 \cdot i = i$

87. $i^{22} = i^{20} \cdot i^2 = \left(i^4\right)^5 \cdot (-1) = 1^5 \cdot (-1) = -1$

89. $i^{23} = i^{20} \cdot i^3 = \left(i^4\right)^5 \cdot i^3 = 1^5 \cdot (-i) = -i$

91. $i^{32} = \left(i^4\right)^8 = 1^8 = 1$

93. $i^{-13} = i^{-16} \cdot i^3 = \left(i^4\right)^{-4} \cdot i^3 = 1^{-4} \cdot (-i) = -i$

95. $\dfrac{1}{i^{-11}} = i^{11} = i^8 \cdot i^3 = \left(i^4\right)^2 \cdot i^3 = 1^2 \cdot (-i) = -i$

97. Answers will vary.

99. We need to show that $\left(\dfrac{\sqrt{2}}{2} + \dfrac{\sqrt{2}}{2}i\right)^2 = i$.

$\left(\dfrac{\sqrt{2}}{2} + \dfrac{\sqrt{2}}{2}i\right)^2$

$= \left(\dfrac{\sqrt{2}}{2}\right)^2 + 2 \cdot \dfrac{\sqrt{2}}{2} \cdot \dfrac{\sqrt{2}}{2}i + \left(\dfrac{\sqrt{2}}{2}i\right)^2$

$= \dfrac{2}{4} + 2 \cdot \dfrac{2}{4}i + \dfrac{2}{4}i^2 = \dfrac{1}{2} + i + \dfrac{1}{2}i^2$

$= \dfrac{1}{2} + i + \dfrac{1}{2}(-1) = \dfrac{1}{2} + i - \dfrac{1}{2} = i$

101. If $-2 + i$ is a solution of the equation, then substituting that value for x makes a true statement.

$x^2 + 4x + 5 = 0$

$(-2 + i)^2 + 4(-2 + i) + 5 = 0$

$4 - 4i + i^2 - 8 + 4i + 5 = 0$

$4 - 4i - 1 - 8 + 4i + 5 = 0$

$0 = 0$

Copyright © 2013 Pearson Education, Inc.

Section 1.4   Quadratic Equations

1. $x^2 = 25$
$x = \pm\sqrt{25} = \pm 5$; G

3. $x^2 + 5 = 0$
$x^2 = -5$
$x = \pm\sqrt{-5} = \pm i\sqrt{5}$; C

5. $x^2 = -20$
$x = \pm\sqrt{-20} = \pm 2i\sqrt{5}$; H

7. $x - 5 = 0$
$x = 5$; D

9. D is the only one set up for direct use of the zero-factor property.
$(3x - 1)(x - 7) = 0$
$3x - 1 = 0$ or $x - 7 = 0$
$x = \frac{1}{3}$ or $x = 7$
Solution set: $\left\{\frac{1}{3}, 7\right\}$

11. C is the only one that does not require Step 1 of the method of completing the square.
$x^2 + x = 12$ Note:
$x^2 + x + \frac{1}{4} = 12 + \frac{1}{4}$ $\left[\frac{1}{2} \cdot 1\right]^2 = \left(\frac{1}{2}\right)^2 = \frac{1}{4}$
$\left(x + \frac{1}{2}\right)^2 = \frac{49}{4}$
$x + \frac{1}{2} = \pm\sqrt{\frac{49}{4}}$

$x + \frac{1}{2} = \pm\frac{7}{2}$
$x = -\frac{1}{2} \pm \frac{7}{2}$
$-\frac{1}{2} - \frac{7}{2} = \frac{-8}{2} = -4$ and $-\frac{1}{2} + \frac{7}{2} = \frac{6}{2} = 3$
Solution set: $\{-4, 3\}$

13. $x^2 - 5x + 6 = 0$
$(x - 2)(x - 3) = 0$
$x - 2 = 0 \Rightarrow x = 2$ or $x - 3 = 0 \Rightarrow x = 3$
Solution set: $\{2, 3\}$

15. $5x^2 - 3x - 2 = 0$
$(5x + 2)(x - 1) = 0$
$5x + 2 = 0 \Rightarrow x = -\frac{2}{5}$ or $x - 1 = 0 \Rightarrow x = 1$
Solution set: $\left\{-\frac{2}{5}, 1\right\}$

17. $-4x^2 + x = -3$
$0 = 4x^2 - x - 3$
$0 = (4x + 3)(x - 1)$
$4x + 3 = 0 \Rightarrow x = -\frac{3}{4}$ or $x - 1 = 0 \Rightarrow x = 1$
Solution set: $\left\{-\frac{3}{4}, 1\right\}$

19. $x^2 - 100 = 0$
$(x + 10)(x - 10) = 0$
$x + 10 = 0 \Rightarrow x = -10$ or $x - 10 = 0 \Rightarrow x = 10$
Solution set: $\{-10, 10\}$

21. $4x^2 - 4x + 1 = 0$
$(2x - 1)^2 = 0$
$2x - 1 = 0 \Rightarrow 2x = 1 \Rightarrow x = \frac{1}{2}$
Solution set: $\left\{\frac{1}{2}\right\}$

23. $25x^2 + 30x + 9 = 0$
$(5x + 3)^2 = 0$
$5x + 3 = 0 \Rightarrow 5x = -3 \Rightarrow x = -\frac{3}{5}$
Solution set: $\left\{-\frac{3}{5}\right\}$

25. $x^2 = 16$
$x = \pm\sqrt{16} = \pm 4$
Solution set: $\{\pm 4\}$

27. $27 - x^2 = 0$
$27 = x^2$
$x = \pm\sqrt{27} = \pm 3\sqrt{3}$
Solution set: $\left\{\pm 3\sqrt{3}\right\}$

29. $x^2 = -81$
$x = \pm\sqrt{-81} = \pm 9i$
Solution set: $\{\pm 9i\}$

31. $(3x - 1)^2 = 12$
$3x - 1 = \pm\sqrt{12}$
$3x = 1 \pm 2\sqrt{3}$
$x = \frac{1 \pm 2\sqrt{3}}{3}$
Solution set: $\left\{\frac{1 \pm 2\sqrt{3}}{3}\right\}$

Copyright © 2013 Pearson Education, Inc.

33. $(x+5)^2 = -3$

$x+5 = \pm\sqrt{-3}$

$x+5 = \pm i\sqrt{3}$

$x = -5 \pm i\sqrt{3}$

Solution set: $\left\{-5 \pm i\sqrt{3}\right\}$

35. $(5x-3)^2 = -3$

$5x-3 = \pm\sqrt{-3}$

$5x-3 = \pm i\sqrt{3}$

$5x = 3 \pm i\sqrt{3}$

$x = \dfrac{3 \pm i\sqrt{3}}{5} = \dfrac{3}{5} \pm \dfrac{\sqrt{3}}{5}i$

Solution set: $\left\{\dfrac{3}{5} \pm \dfrac{\sqrt{3}}{5}i\right\}$

37. $x^2 - 4x + 3 = 0$

$x^2 - 4x + 4 = -3 + 4$ Note:

$\left[\frac{1}{2}\cdot(-4)\right]^2 = (-2)^2 = 4$

$(x-2)^2 = 1$

$x-2 = \pm\sqrt{1}$

$x-2 = \pm 1$

$x = 2 \pm 1$

$2 - 1 = 1$ and $2 + 1 = 3$

Solution set: $\{1, 3\}$

39. $2x^2 - x - 28 = 0$

$x^2 - \frac{1}{2}x - 14 = 0$ Multiply by $\frac{1}{2}$.

$x^2 - \frac{1}{2}x + \frac{1}{16} = 14 + \frac{1}{16}$

Note: $\left[\frac{1}{2}\cdot\left(-\frac{1}{2}\right)\right]^2 = \left(-\frac{1}{4}\right)^2 = \frac{1}{16}$

$\left(x - \frac{1}{4}\right)^2 = \frac{225}{16}$

$x - \frac{1}{4} = \pm\sqrt{\frac{225}{16}}$

$x - \frac{1}{4} = \pm\frac{15}{4} \Rightarrow x = \frac{1}{4} \pm \frac{15}{4}$

$\frac{1}{4} - \frac{15}{4} = \frac{-14}{4} = -\frac{7}{2}$ and $\frac{1}{4} + \frac{15}{4} = \frac{16}{4} = 4$

Solution set: $\left\{-\frac{7}{2}, 4\right\}$

41. $x^2 - 2x - 2 = 0$

$x^2 - 2x + 1 = 2 + 1$

Note: $\left[\frac{1}{2}\cdot(-2)\right]^2 = (-1)^2 = 1$

$(x-1)^2 = 3$

$x - 1 = \pm\sqrt{3}$

$x = 1 \pm \sqrt{3}$

Solution set: $\left\{1 \pm \sqrt{3}\right\}$

43. $2x^2 + x = 10$

$x^2 + \frac{1}{2}x = 5$

$x^2 + \frac{1}{2}x + \frac{1}{16} = 5 + \frac{1}{16}$ Note:

$\left[\frac{1}{2}\cdot\frac{1}{2}\right]^2 = \left(\frac{1}{4}\right)^2 = \frac{1}{16}$

$\left(x + \frac{1}{4}\right)^2 = \frac{81}{16}$

$x + \frac{1}{4} = \pm\sqrt{\frac{81}{16}}$

$x + \frac{1}{4} = \pm\frac{9}{4}$

$x = -\frac{1}{4} \pm \frac{9}{4}$

$-\frac{1}{4} - \frac{9}{4} = \frac{-10}{4} = -\frac{5}{2}$ and $-\frac{1}{4} + \frac{9}{4} = \frac{8}{4} = 2$

Solution set: $\left\{-\frac{5}{2}, 2\right\}$

45. $-2x^2 + 4x + 3 = 0$

$x^2 - 2x - \frac{3}{2} = 0$

$x^2 - 2x + 1 = \frac{3}{2} + 1$ Note:

$\left[\frac{1}{2}\cdot(-2)\right]^2 = (-1)^2 = 1$

$(x-1)^2 = \frac{5}{2}$

$x - 1 = \pm\sqrt{\frac{5}{2}} = \pm\frac{\sqrt{10}}{2}$

$x = 1 \pm \frac{\sqrt{10}}{2} = \frac{2 \pm \sqrt{10}}{2}$

Solution set: $\left\{\dfrac{2 \pm \sqrt{10}}{2}\right\}$

47. $-4x^2 + 8x = 7$

$x^2 - 2x = -\frac{7}{4}$

$x^2 - 2x + 1 = -\frac{7}{4} + 1$ Note:

$\left[\frac{1}{2}\cdot(-2)\right]^2 = (-1)^2 = 1$

$(x-1)^2 = \frac{-3}{4}$

$x - 1 = \pm\sqrt{\frac{-3}{4}} = \pm\frac{i\sqrt{3}}{2}$

$x = 1 \pm \frac{\sqrt{3}}{2}i$

Solution set: $\left\{1 \pm \dfrac{\sqrt{3}}{2}i\right\}$

49. Francisco is incorrect because $c = 0$ and the

quadratic formula, $x = \dfrac{-b \pm \sqrt{b^2 - 4ac}}{2a}$, can

be evaluated with $a = 1$, $b = -8$, and $c = 0$.

Copyright © 2013 Pearson Education, Inc.

51. $x^2 - x - 1 = 0$

Let $a = 1, b = -1,$ and $c = -1.$

$x = \dfrac{-b \pm \sqrt{b^2 - 4ac}}{2a}$

$= \dfrac{-(-1) \pm \sqrt{(-1)^2 - 4(1)(-1)}}{2(1)}$

$= \dfrac{1 \pm \sqrt{1 + 4}}{2} = \dfrac{1 \pm \sqrt{5}}{2}$

Solution set: $\left\{ \dfrac{1 \pm \sqrt{5}}{2} \right\}$

53. $x^2 - 6x = -7$

$x^2 - 6x + 7 = 0$

Let $a = 1, b = -6,$ and $c = 7.$

$x = \dfrac{-b \pm \sqrt{b^2 - 4ac}}{2a}$

$= \dfrac{-(-6) \pm \sqrt{(-6)^2 - 4(1)(7)}}{2(1)} = \dfrac{6 \pm \sqrt{36 - 28}}{2}$

$= \dfrac{6 \pm \sqrt{8}}{2} = \dfrac{6 \pm 2\sqrt{2}}{2} = 3 \pm \sqrt{2}$

Solution set: $\left\{ 3 \pm \sqrt{2} \right\}$

55. $x^2 = 2x - 5$

$x^2 - 2x + 5 = 0$

Let $a = 1, b = -2,$ and $c = 5.$

$x = \dfrac{-b \pm \sqrt{b^2 - 4ac}}{2a}$

$= \dfrac{-(-2) \pm \sqrt{(-2)^2 - 4(1)(5)}}{2(1)}$

$= \dfrac{2 \pm \sqrt{4 - 20}}{2} = \dfrac{2 \pm \sqrt{-16}}{2} = \dfrac{2 \pm 4i}{2} = 1 \pm 2i$

Solution set: $\left\{ 1 \pm 2i \right\}$

57. $-4x^2 = -12x + 11$

$0 = 4x^2 - 12x + 11$

Let $a = 4, b = -12,$ and $c = 11.$

$x = \dfrac{-b \pm \sqrt{b^2 - 4ac}}{2a}$

$= \dfrac{-(-12) \pm \sqrt{(-12)^2 - 4(4)(11)}}{2(4)}$

$x = \dfrac{12 \pm \sqrt{144 - 176}}{8} = \dfrac{12 \pm \sqrt{-32}}{8}$

$= \dfrac{12 \pm 4i\sqrt{2}}{8} = \dfrac{12}{8} \pm \dfrac{4\sqrt{2}}{8}i = \dfrac{3}{2} \pm \dfrac{\sqrt{2}}{2}i$

Solution set: $\left\{ \dfrac{3}{2} \pm \dfrac{\sqrt{2}}{2}i \right\}$

59. $\frac{1}{2}x^2 + \frac{1}{4}x - 3 = 0$

$4\left(\frac{1}{2}x^2 + \frac{1}{4}x - 3 \right) = 4 \cdot 0$

$2x^2 + x - 12 = 0$

Let $a = 2, b = 1,$ and $c = -12.$

$x = \dfrac{-b \pm \sqrt{b^2 - 4ac}}{2a} = \dfrac{-1 \pm \sqrt{1^2 - 4(2)(-12)}}{2(2)}$

$= \dfrac{-1 \pm \sqrt{1 + 96}}{4} = \dfrac{-1 \pm \sqrt{97}}{4}$

Solution set: $\left\{ \dfrac{-1 \pm \sqrt{97}}{4} \right\}$

61. $0.2x^2 + 0.4x - 0.3 = 0$

$10\left(0.2x^2 + 0.4x - 0.3 \right) = 10 \cdot 0$

$2x^2 + 4x - 3 = 0$

Let $a = 2, b = 4,$ and $c = -3.$

$x = \dfrac{-b \pm \sqrt{b^2 - 4ac}}{2a}$

$= \dfrac{-4 \pm \sqrt{4^2 - 4(2)(-3)}}{2(2)} = \dfrac{-4 \pm \sqrt{16 + 24}}{4}$

$= \dfrac{-4 \pm \sqrt{40}}{4} = \dfrac{-4 \pm 2\sqrt{10}}{4} = \dfrac{-2 \pm \sqrt{10}}{2}$

Solution set: $\left\{ \dfrac{-2 \pm \sqrt{10}}{2} \right\}$

63. $(4x - 1)(x + 2) = 4x$

$4x^2 + 7x - 2 = 4x \Rightarrow 4x^2 + 3x - 2 = 0$

Let $a = 4, b = 3,$ and $c = -2.$

$x = \dfrac{-b \pm \sqrt{b^2 - 4ac}}{2a} = \dfrac{-3 \pm \sqrt{3^2 - 4(4)(-2)}}{2(4)}$

$= \dfrac{-3 \pm \sqrt{9 + 32}}{8} = \dfrac{-3 \pm \sqrt{41}}{8}$

Solution set: $\left\{ \dfrac{-3 \pm \sqrt{41}}{8} \right\}$

Copyright © 2013 Pearson Education, Inc.

65. $(x-9)(x-1) = -16$

$x^2 - 10x + 9 = -16$

$x^2 - 10x + 25 = 0$

Let $a = 1$, $b = -10$, and $c = 25$.

$x = \dfrac{-(-10) \pm \sqrt{(-10)^2 - (4)(1)(25)}}{2}$

$= \dfrac{10 \pm \sqrt{100 - 100}}{2} = \dfrac{10 \pm 0}{2} = 5$

Solution set: {5}

67.

$x^3 - 8 = 0$

$x^3 - 2^3 = 0$

$(x-2)(x^2 + 2x + 4) = 0$

$x - 2 = 0 \Rightarrow x = 2$ or

$x^2 + 2x + 4 = 0$

$a = 1$, $b = 2$, and $c = 4$

$x = \dfrac{-b \pm \sqrt{b^2 - 4ac}}{2a}$

$= \dfrac{-2 \pm \sqrt{2^2 - 4(1)(4)}}{2(1)}$

$= \dfrac{-2 \pm \sqrt{4 - 16}}{2} = \dfrac{-2 \pm \sqrt{-12}}{2}$

$= \dfrac{-2 \pm 2i\sqrt{3}}{2} = -1 \pm \sqrt{3}i$

Solution set: $\left\{2, -1 \pm \sqrt{3}i\right\}$

69.

$x^3 + 27 = 0$

$x^3 + 3^3 = 0$

$(x+3)(x^2 - 3x + 9) = 0$

$x + 3 = 0 \Rightarrow x = -3$ or

$x^2 - 3x + 9 = 0$

$a = 1$, $b = -3$, and $c = 9$

$x = \dfrac{-b \pm \sqrt{b^2 - 4ac}}{2a}$

$= \dfrac{-(-3) \pm \sqrt{(-3)^2 - 4(1)(9)}}{2(1)} = \dfrac{3 \pm \sqrt{9 - 36}}{2}$

$= \dfrac{3 \pm \sqrt{-27}}{2} = \dfrac{3 \pm 3i\sqrt{3}}{2} = \dfrac{3}{2} \pm \dfrac{3\sqrt{3}}{2}i$

Solution set: $\left\{-3, \frac{3}{2} \pm \frac{3\sqrt{3}}{2}i\right\}$

71. $s = \dfrac{1}{2}gt^2$

$2s = 2\left[\dfrac{1}{2}gt^2\right] \Rightarrow 2s = gt^2 \Rightarrow \dfrac{2s}{g} = \dfrac{gt^2}{g} \Rightarrow$

$t^2 = \dfrac{2s}{g} \Rightarrow t = \pm\sqrt{\dfrac{2s}{g}} = \dfrac{\pm\sqrt{2s}}{\sqrt{g}} \cdot \dfrac{\sqrt{g}}{\sqrt{g}} = \dfrac{\pm\sqrt{2sg}}{g}$

73. $F = \dfrac{kMv^2}{r}$

$rF = r\left[\dfrac{kMv^2}{r}\right] \Rightarrow Fr = kMv^2 \Rightarrow$

$\dfrac{Fr}{kM} = \dfrac{kMv^2}{kM} \Rightarrow v^2 = \dfrac{Fr}{kM} \Rightarrow v = \pm\sqrt{\dfrac{Fr}{kM}} \Rightarrow$

$v = \dfrac{\pm\sqrt{Fr}}{\sqrt{kM}} \cdot \dfrac{\sqrt{kM}}{\sqrt{kM}} = \dfrac{\pm\sqrt{FrkM}}{kM}$

75. $r = r_0 + \dfrac{1}{2}at^2$

$r - r_0 = \dfrac{1}{2}at^2$

$2(r - r_0) = at^2$

$\dfrac{2(r - r_0)}{a} = t^2$

$\pm\sqrt{\dfrac{2(r - r_0)}{a}} = \dfrac{\pm\sqrt{2a(r - r_0)}}{a} = t$

77. $h = -16t^2 + v_0 t + s_0$

$16t^2 - v_0 t + h - s_0 = 0$

$16t^2 - v_0 t + (h - s_0) = 0$ $\qquad a = 16, b = -v_0,$
$\qquad\qquad\qquad\qquad\qquad\qquad c = h - s_0$

$x = \dfrac{-b \pm \sqrt{b^2 - 4ac}}{2a}$

$= \dfrac{-(-v_0) \pm \sqrt{(-v_0)^2 - 4(16)(h - s_0)}}{2(16)}$

$= \dfrac{v_0 \pm \sqrt{v_0^2 - 64(h - s_0)}}{32}$

$= \dfrac{v_0 \pm \sqrt{v_0^2 - 64h + 64s_0}}{32}$

Copyright © 2013 Pearson Education, Inc.

79.
$$4x^2 - 2xy + 3y^2 = 2$$
$$4x^2 - 2xy + 3y^2 - 2 = 0$$

(a) Solve for x in terms of y.
$$4x^2 - (2y)x + (3y^2 - 2) = 0$$
$$a = 4, b = -2y, \text{ and } c = 3y^2 - 2$$
$$x = \frac{-b \pm \sqrt{b^2 - 4ac}}{2a}$$
$$= \frac{-(-2y) \pm \sqrt{(-2y)^2 - 4(4)(3y^2 - 2)}}{2(4)}$$
$$= \frac{2y \pm \sqrt{4y^2 - 16(3y^2 - 2)}}{8}$$
$$= \frac{2y \pm \sqrt{4y^2 - 48y^2 + 32}}{8}$$
$$= \frac{2y \pm \sqrt{32 - 44y^2}}{8} = \frac{2y \pm \sqrt{4(8 - 11y^2)}}{8}$$
$$= \frac{2y \pm 2\sqrt{8 - 11y^2}}{8} = \frac{y \pm \sqrt{8 - 11y^2}}{4}$$

(b) Solve for y in terms of x.
$$3y^2 - (2x)y + (4x^2 - 2) = 0$$
$$a = 3, b = -2x, \text{ and } c = 4x^2 - 2$$
$$y = \frac{-b \pm \sqrt{b^2 - 4ac}}{2a}$$
$$= \frac{-(-2x) \pm \sqrt{(-2x)^2 - 4(3)(4x^2 - 2)}}{2(3)}$$
$$= \frac{2x \pm \sqrt{4x^2 - 12(4x^2 - 2)}}{6}$$
$$= \frac{2x \pm \sqrt{4x^2 - 48x^2 + 24}}{6}$$
$$= \frac{2x \pm \sqrt{24 - 44x^2}}{6} = \frac{2x \pm \sqrt{4(6 - 11x^2)}}{6}$$
$$= \frac{2x \pm 2\sqrt{6 - 11x^2}}{6} = \frac{x \pm \sqrt{6 - 11x^2}}{3}$$

81.
$$2x^2 + 4xy - 3y^2 = 2$$
$$2x^2 + 4xy - 3y^2 - 2 = 0$$

a. Solve for x in terms of y.
$$2x^2 + (4y)x + (-3y^2 - 2) = 0$$
$$a = 2, b = 4y, \text{ and } c = -3y^2 - 2$$

$$x = \frac{-b \pm \sqrt{b^2 - 4ac}}{2a}$$
$$= \frac{-(4y) \pm \sqrt{(4y)^2 - 4(2)(-3y^2 - 2)}}{2(2)}$$
$$= \frac{-4y \pm \sqrt{16y^2 - 8(-3y^2 - 2)}}{4}$$
$$= \frac{-4y \pm \sqrt{16y^2 + 24y^2 + 16}}{4}$$
$$= \frac{-4y \pm \sqrt{40y^2 + 16}}{4}$$
$$= \frac{-4y \pm \sqrt{4(10y^2 + 4)}}{4}$$
$$= \frac{-4y \pm 2\sqrt{10y^2 + 4}}{4}$$
$$= \frac{-2y \pm \sqrt{10y^2 + 4}}{2}$$

b. Solve for y in terms of x.
$$-3y^2 + (4x)y + (2x^2 - 2) = 0$$
$$a = -3, b = 4x, c = 2x^2 - 2$$
$$x = \frac{-b \pm \sqrt{b^2 - 4ac}}{2a}$$
$$= \frac{-(4x) \pm \sqrt{(4x)^2 - 4(-3)(2x^2 - 2)}}{2(-3)}$$
$$= \frac{-4x \pm \sqrt{16x^2 + 12(2x^2 - 2)}}{-6}$$
$$= \frac{-4x \pm \sqrt{16x^2 + 24x^2 - 24}}{-6}$$
$$= \frac{-4x \pm \sqrt{40x^2 - 24}}{-6}$$
$$= \frac{-4x \pm \sqrt{4(10x^2 - 6)}}{-6}$$
$$= \frac{-4x \pm 2\sqrt{10x^2 - 6}}{-6}$$
$$= \frac{2x \pm \sqrt{10x^2 - 6}}{3}$$

83. $x^2 - 8x + 16 = 0$
$$a = 1, b = -8, \text{ and } c = 16$$
$$b^2 - 4ac = (-8)^2 - 4(1)(16) = 64 - 64 = 0$$
one rational solution (a double solution)

Copyright © 2013 Pearson Education, Inc.

85. $3x^2 + 5x + 2 = 0$
$a = 3, b = 5,$ and $c = 2$
$b^2 - 4ac = 5^2 - 4(3)(2) = 25 - 24 = 1 = 1^2$
two distinct rational solutions

87. $4x^2 = -6x + 3$
$4x^2 + 6x - 3 = 0$
$a = 4, b = 6,$ and $c = -3$
$b^2 - 4ac = 6^2 - 4(4)(-3) = 36 + 48 = 84$
two distinct irrational solutions

89. $9x^2 + 11x + 4 = 0$
$a = 9, b = 11,$ and $c = 4$
$b^2 - 4ac = 11^2 - 4(9)(4) = 121 - 144 = -23$
two distinct nonreal complex solutions

91. $8x^2 - 72 = 0$
$a = 8, b = 0,$ and $c = -72$
$b^2 - 4ac = 0^2 - 4(8)(-72) = 2304 = 48^2$
two distinct rational solutions

93. It is not possible for the solution set of a quadratic equation with integer coefficients to consist of a single irrational number. Additional responses will vary.

In exercises 95–97, there are other possible answers.

95. $x = 4$ or $x = 5$
$x - 4 = 0$ or $x - 5 = 0$
$(x-4)(x-5) = 0$
$x^2 - 5x - 4x + 20 = 0$
$x^2 - 9x + 20 = 0$
$a = 1, b = -9,$ and $c = 20$

97. $x = 1 + \sqrt{2}$ or $x = 1 - \sqrt{2}$
$x - (1+\sqrt{2}) = 0$ or $x - (1-\sqrt{2}) = 0$
$[x - (1+\sqrt{2})][x - (1-\sqrt{2})] = 0$
$x^2 - x(1-\sqrt{2}) - x(1+\sqrt{2})$
$\qquad + (1+\sqrt{2})(1-\sqrt{2}) = 0$
$x^2 - x + x\sqrt{2} - x - x\sqrt{2} + [1^2 - (\sqrt{2})^2] = 0$
$x^2 - 2x + (1-2) = 0$
$x^2 - 2x - 1 = 0$
$a = 1, b = -2,$ and $c = -1$

Chapter 1 Quiz
(Sections 1.1–1.4)

1. $3(x-5) + 2 = 1 - (4 + 2x)$
$3x - 15 + 2 = 1 - 4 - 2x$
$3x - 13 = -3 - 2x$
$5x - 13 = -3$
$5x = 10 \Rightarrow x = 2$
Solution set $\{2\}$

3. $ay + 2x = y + 5x$
$ay - 3x = y$
$-3x = y - ay = y(1-a)$
$3x = y(a-1)$
$\dfrac{3x}{a-1} = y$

5. Substitute 1999 for x in the equation:
$y = 0.126(1999) - 246.25 \approx 5.62$
So, the model predicts that the minimum hourly wage for 1999 was $5.62. The difference between the actual minimum wage and the predicted wage is
$5.62 - $5.15 = $0.47.

7. $\dfrac{7-2i}{2+4i} = \dfrac{7-2i}{2+4i} \cdot \dfrac{2-4i}{2-4i} = \dfrac{14 - 28i - 4i + (-8)}{4 - (-16)}$
$= \dfrac{6-32i}{20} = \dfrac{6}{20} - \dfrac{32}{20}i = \dfrac{3}{10} - \dfrac{8}{5}i$

9. $x^2 - 29 = 0 \Rightarrow x^2 = 29 \Rightarrow x = \pm\sqrt{29}$
Solution set: $\{\pm 29\}$

Section 1.5 Applications and Modeling with Quadratic Equations

1. The length of the parking area is $2x + 200$, while the width is x, so the area is $(2x + 200)x$. Set the area equal to 40,000.
$(2x + 200)x = 40,000$, so choice A is the correct choice.

3. Use the Pythagorean theorem with $a = x$, $b = 2x - 2$, and $c = x + 4$.
$x^2 + (2x-2)^2 = (x+4)^2$
The correct choice is D.

5. Let $x =$ the first integer. Then $x + 1 =$ the next consecutive integer.
$x(x+1) = 56 \Rightarrow x^2 + x = 56$
$x^2 + x - 56 = 0 \Rightarrow (x+8)(x-7) = 0$
$x + 8 = 0 \Rightarrow x = -8$ or $x - 7 = 0 \Rightarrow x = 7$
If $x = -8$, then $x + 1 = -7$. If $x = 7$, then $x + 1 = 8$. So the two integers are −8 and −7, or 7 and 8.

Copyright © 2013 Pearson Education, Inc.

7. Let x = the first even integer. Then $x + 2$ = the next consecutive even integer.

$$x(x+2) = 168 \Rightarrow x^2 + 2x = 168$$
$$x^2 + 2x - 168 = 0 \Rightarrow (x+14)(x-12) = 0$$
$$x + 14 = 0 \Rightarrow x = -14 \text{ or}$$
$$x - 12 = 0 \Rightarrow x = 12$$

If $x = -14$, then $x + 2 = -12$. If $x = 12$, then $x + 2 = 14$. so, the two even integers are -14 and -12, or 12 and 14.

9. Let x = the first odd integer. Then $x + 2$ = the next consecutive odd integer.

$$x(x+2) = 63 \Rightarrow x^2 + 2x = 63 \Rightarrow$$
$$x^2 + 2x - 63 = 0 \Rightarrow (x+9)(x-7) = 0$$
$$x + 9 = 0 \Rightarrow x = -9 \text{ or}$$
$$x - 7 = 0 \Rightarrow x = 7$$

If $x = -9$, then $x + 2 = -7$. If $x = 7$, then $x + 2 = 9$. so, the two odd integers are -9 and -7, or 7 and 9.

11. Let x = the first odd integer. Then $x + 2$ = the next consecutive odd integer.

$$x^2 + (x+2)^2 = 202$$
$$x^2 + x^2 + 4x + 4 = 202$$
$$2x^2 + 4x + 4 = 202 \Rightarrow 2x^2 + 4x - 198 = 0$$
$$2(x^2 + 2x - 99) = 0 \Rightarrow x^2 + 2x - 99 = 0$$
$$(x+11)(x-9) = 0$$
$$x + 11 = 0 \Rightarrow x = -11 \text{ or}$$
$$x - 9 = 0 \Rightarrow x = 9$$

If $x = -11$, then $x + 2 = -9$. If $x = 9$, then $x + 2 = 11$. So the two integers are -11 and -9, or 9 and 11.

13. Let x = the first even integer. Then $x + 2$ = the next consecutive even integer.

$$(x+2)^2 - x^2 = 84$$
$$x^2 + 4x + 4 - x^2 = 84 \Rightarrow 4x + 4 = 84 \Rightarrow$$
$$4x = 80 \Rightarrow x = 20$$

If $x = 20$, then $x + 2 = 22$. The two integers are 20 and 22.

15. Let x = the length of one leg, $x + 2$ = the length of the other leg, and $x + 4$ = the length of the hypotenuse. (Remember that the hypotenuse is the longest side in a right triangle.) The Pythagorean theorem gives

$$x^2 + (x+2)^2 = (x+4)^2$$
$$x^2 + x^2 + 4x + 4 = x^2 + 8x + 16$$
$$x^2 - 4x - 12 = 0 \Rightarrow (x-6)(x+2) = 0$$
$$x - 6 = 0 \Rightarrow x = 6 \text{ or}$$
$$x + 2 = 0 \Rightarrow x = -2$$

Length cannot be negative, so reject that solution. If $x = 6$, then $x + 2 = 8$ and $x + 4 = 10$. The sides of the right triangle are 6, 8, and 10.

17. Let x = the length of the side of the smaller square. Then $x + 3$ = the length of the side of the larger square.

$$(x+3)^2 + x^2 = 149$$
$$x^2 + 6x + 9 + x^2 = 149 \Rightarrow 2x^2 + 6x - 140 = 0$$
$$x^2 + 3x - 70 = 0 \Rightarrow (x-7)(x+10) = 0$$
$$x - 7 = 0 \Rightarrow x = 7 \text{ or}$$
$$x + 10 = 0 \Rightarrow x = -10$$

Length cannot be negative, so reject that solution. If $x = 7$, then $x + 3 = 10$. The length of the side of smaller square is 7 in., and the length of the side of the larger square is 10 in.

19. Use the figure and equation A from Exercise 1.

$$x(2x+200) = 40,000$$
$$2x^2 + 200x = 40,000$$
$$2x^2 + 200x - 40,000 = 0$$
$$x^2 + 100x - 20,000 = 0$$
$$(x-100)(x+200) = 0$$
$$x = 100 \text{ or } x = -200$$

The negative solution is not meaningful. If $x = 100$, then $2x + 200 = 400$. The dimensions of the lot are 100 yd by 400 yd.

21. Let x = the width of the strip of floor around the rug.

The dimensions of the carpet are $15 - 2x$ by $12 - 2x$. Since $A = lw$, the equation for the carpet area is $(15 - 2x)(12 - 2x) = 108$. Put this equation in standard form and solve by factoring.

$$(15 - 2x)(12 - 2x) = 108$$
$$180 - 30x - 24x + 4x^2 = 108$$
$$180 - 54x + 4x^2 = 108$$
$$4x^2 - 54x + 72 = 0$$
$$2x^2 - 27x + 36 = 0$$
$$(2x-3)(x-12) = 0$$
$$2x - 3 = 0 \Rightarrow x = \frac{3}{2}$$
$$x - 12 = 0 \Rightarrow x = 12$$

(*continued on next page*)

Copyright © 2013 Pearson Education, Inc.

(continued)

The solutions of the quadratic equation are $\frac{3}{2}$ and 12. We eliminate 12 as meaningless in this problem. If $x = \frac{3}{2}$, then $15 - 2x = 12$ and $12 - 2x = 9$. The dimensions of the carpet are 9 ft by 12 ft.

23. Let $x =$ the width of the metal. The dimensions of the base of the box are $x - 4$ by $x + 6$.

Since the formula for the volume of a box is $V = lwh$, we have

$$(x+6)(x-4)(2) = 832$$
$$(x+6)(x-4) = 416$$
$$x^2 - 4x + 6x - 24 = 416$$
$$x^2 + 2x - 24 = 416$$
$$x^2 + 2x - 440 = 0 \Rightarrow (x+22)(x-20) = 0$$
$$x + 22 = 0 \Rightarrow x = -22 \text{ or}$$
$$x - 20 = 0 \Rightarrow x = 20$$

The negative solution is not meaningful. If $x = 20$, then $x + 10 = 30$. The dimensions of the sheet of metal are 20 in by 30 in.

25. Let $h =$ height and $r =$ radius.
Surface area $= 2\pi rh + 2\pi r^2$
$$8\pi = 2\pi r(3) + 2\pi r^2$$
$$8\pi = 6\pi r + 2\pi r^2$$
$$0 = 2\pi r^2 + 6\pi r - 8\pi$$
$$0 = 2\pi\left(r^2 + 3r - 4\right) \Rightarrow 0 = (r+4)(r-1)$$
$$r + 4 = 0 \Rightarrow r = -4 \text{ or } r - 1 = 0 \Rightarrow r = 1$$
The r represents the radius of a cylinder, so -4 is not reasonable. The radius of the circular top is approximately 1 ft.

27. Let $x =$ length of side of square. Area $= x^2$ and perimeter $= 4x$
$$x^2 = 4x \Rightarrow x^2 - 4x = 0 \Rightarrow x(x-4) = 0 \Rightarrow$$
$$x = 0 \text{ or } x = 4$$
We reject 0 since x must be greater than 0. The side of the square measures 4 units.

29. Let $h =$ height and $r =$ radius.
Area of side $= 2\pi rh$ and Area of circle $= \pi r^2$
Surface area $=$ area of side $+$ area of top $+$ area of bottom
Surface area $= 2\pi rh + \pi r^2 + \pi r^2 = 2\pi rh + 2\pi r^2$
$$371 = 2\pi r(12) + 2\pi r^2$$
$$371 = 24\pi r + 2\pi r^2$$
$$0 = 2\pi r^2 + 24\pi r - 371$$
$$a = 2\pi, b = 24\pi, \text{ and } c = -371$$
$$r = \frac{-b \pm \sqrt{b^2 - 4ac}}{2a} r$$
$$= \frac{-24\pi \pm \sqrt{(24\pi)^2 - 4(2\pi)(-371)}}{2(2\pi)}$$
$$= \frac{-24\pi \pm \sqrt{576\pi^2 + 2968\pi}}{4\pi}$$
$$r \approx -15.75 \text{ or } r \approx 3.75$$
The negative solution is not meaningful. The radius of the circular top is approximately 3.75 cm.

31. Let $h =$ the height of the dock.
Then $2h + 3 =$ the length of the rope from the boat to the top of the dock.
Apply the Pythagorean theorem to the triangle shown in the text.
$$h^2 + 12^2 = (2h+3)^2$$
$$h^2 + 144 = (2h)^2 + 2(6h) + 3^2$$
$$h^2 + 144 = 4h^2 + 12h + 9$$
$$0 = 3h^2 + 12h - 135$$
$$0 = h^2 + 4h - 45 \Rightarrow 0 = (h+9)(h-5)$$
$$h + 9 = 0 \Rightarrow h = -9 \text{ or } h - 5 = 0 \Rightarrow h = 5$$
The negative solution is not meaningful. The height of the dock is 5 ft.

33. Let $r =$ radius of circle and $x =$ length of side of square. The radius is $\frac{1}{2}$ the length of the side of the square. Area $= x^2$
$$800 = x^2 \Rightarrow x = \sqrt{800} = 20\sqrt{2} \Rightarrow$$
$$r = 10\sqrt{2}$$

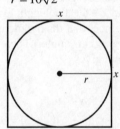

The radius is $10\sqrt{2}$ feet.

Copyright © 2013 Pearson Education, Inc.

35. Let x = length of ladder
Distance from building to ladder = $8 + 2 = 10$.
Distance from ground to window = 13
Apply the Pythagorean theorem.
$$a^2 + b^2 = c^2$$
$$10^2 + 13^2 = x^2 \Rightarrow 100 + 169 = x^2 \Rightarrow$$
$$269 = x^2 \Rightarrow \pm\sqrt{269} = x$$
$$x \approx -16.4 \text{ or } x \approx 16.4$$
The negative solution is not meaningful. The worker will need a 16.4-ft ladder.

37. Let x = length of short leg, $x + 700$ = length of long leg, and $x + 700 + 100$ or $x + 800$ = length of hypotenuse.

Apply the Pythagorean theorem.
$$c^2 = a^2 + b^2$$
$$(x+800)^2 = x^2 + (x+700)^2$$
$$x^2 + 1600x + 640,000$$
$$\qquad = x^2 + x^2 + 1400x + 490,000$$
$$0 = x^2 - 200x - 150,000$$
$$0 = (x+300)(x-500)$$
$$x + 300 = 0 \Rightarrow x = -300 \text{ or}$$
$$x - 500 = 0 \Rightarrow x = 500$$
The negative solution is not meaningful.
500 = length of short leg
500 + 700 = 1200 = length of long leg
1200 + 100 = 1300 = length of hypotenuse
500 + 1200 + 1300 = 3000 = length of walkway. The total length is 3000 yd.

39. (a) $s = -16t^2 + v_0 t$
$$s = -16t^2 + 96t$$
$$80 = -16t^2 + 96t$$
$$16t^2 - 96t + 80 = 0$$
$$a = 16, b = -96 \text{ and } c = 80$$
$$t = \frac{-b \pm \sqrt{b^2 - 4ac}}{2a}$$
$$= \frac{-(-96) \pm \sqrt{(-96)^2 - 4(16)(80)}}{2(16)}$$
$$= \frac{96 \pm \sqrt{9216 - 5120}}{32}$$

$$t = \frac{96 \pm \sqrt{4096}}{32} = \frac{96 \pm 64}{32}$$
$$t = \frac{96 - 64}{32} = 1 \text{ or } t = \frac{96 + 64}{32} = 5$$
The projectile will reach 80 ft at 1 sec and 5 sec.

(b) $s = -16t^2 + 96t$
$$0 = -16t^2 + 96t \Rightarrow 0 = -16t(t-6)$$
$$-16t = 0 \Rightarrow t = 0 \quad \text{or} \quad t - 6 = 0 \Rightarrow t = 6$$
The projectile will return to the ground after 6 sec.

41. (a) $s = -16t^2 + v_0 t$
$$s = -16t^2 + 32t$$
$$80 = -16t^2 + 32t$$
$$16t^2 - 32t + 80 = 0$$
$$a = 16, b = -32 \text{ and } c = 80$$
$$t = \frac{-b \pm \sqrt{b^2 - 4ac}}{2a}$$
$$= \frac{-(-32) \pm \sqrt{(-32)^2 - 4(16)(80)}}{2(16)}$$
$$= \frac{32 \pm \sqrt{1024 - 5120}}{32}$$
$$= \frac{32 \pm \sqrt{-4096}}{32} = \frac{32 \pm 64i}{32}$$
The projectile will not reach 80 ft.

(b) $s = -16t^2 + 32t$
$$0 = -16t^2 + 32t \Rightarrow 0 = -16t(t-2) \Rightarrow$$
$$-16t = 0 \Rightarrow t = 0 \text{ or } t - 2 = 0 \Rightarrow t = 2$$
The projectile will return to the ground after 2 sec.

43. The height of the ball is given by
$$h = -2.7t^2 + 30t + 6.5.$$

(a) When the ball is 12 ft above the moon's surface, $h = 12$. Set $h = 12$ and solve for t.
$$12 = 2.7t^2 + 30t + 6.5$$
$$2.7t^2 - 30t + 5.5 = 0$$
Use the quadratic formula with $a = 2.7$, $b = -30$, and $c = 5.5$.
$$t = \frac{30 \pm \sqrt{900 - 4(2.7)(5.5)}}{2(2.7)} = \frac{30 \pm \sqrt{840.6}}{5.4}$$
$$\frac{30 + \sqrt{840.6}}{5.4} \approx 10.92 \text{ or } \frac{30 - \sqrt{840.6}}{5.4} \approx 0.19$$
Therefore, the ball reaches 12 ft first after 0.19 sec (on the way up), then again after 10.92 sec (on the way down).

Copyright © 2013 Pearson Education, Inc.

(b) When the ball returns to the surface, $h = 0$.

$0 = -2.7t^2 + 30t + 6.5$

Use the quadratic formula with $a = -2.7$, $b = 30$, and $c = 6.5$.

$t = \dfrac{-30 \pm \sqrt{900 - 4(-2.7)(6.5)}}{2(-2.7)}$

$= \dfrac{-30 \pm \sqrt{970.2}}{-5.4}$

$\dfrac{-30 + \sqrt{970.2}}{-5.4} \approx -0.21$ or

$\dfrac{-30 - \sqrt{970.2}}{-5.4} \approx 11.32$

The negative solution is not meaningful. Therefore, the ball returns to the surface after 11.32 sec.

45. (a) Let $x = 50$.

$T = 0.00787(50)^2 - 1.528(50) + 75.89 \approx 19.2$

The exposure time when $x = 50$ ppm is approximately 19.2 hr.

(b) Let $T = 3$ and solve for x.

$3 = 0.00787x^2 - 1.528x + 75.89$

$0.00787x^2 - 1.528x + 72.89 = 0$

Use the quadratic formula with $a = 0.00787$, $b = -1.528$, and $c = 72.89$.

$x = \dfrac{-(-1.528) \pm \sqrt{(-1.528)^2 - 4(0.00787)(72.89)}}{2(0.00787)}$

$= \dfrac{1.528 \pm \sqrt{2.334784 - 2.2945772}}{0.01574}$

$= \dfrac{1.528 \pm \sqrt{0.0402068}}{0.01574}$

$\dfrac{1.528 + \sqrt{0.0402068}}{0.01574} \approx 109.8$ or

$\dfrac{1.528 - \sqrt{0.0402068}}{0.01574} \approx 84.3$

We reject the potential solution 109.8 because it is not in the interval [50, 100]. So, 84.3 ppm carbon monoxide concentration is necessary for a person to reach the 4% to 6% CoHb level in 3 hr.

47. (a) 2010 is represented by $x = 6$. Substitute $x = 6$ into the equation to find y:

$y = 1.493x^2 + 7.279x + 684.4$

$y = 1.493(6)^2 + 7.279(6) + 684.4$

≈ 781.8 million tons

In 2010, emissions were about 781.8 million tons.

(b) Let $y = 700$ and solve for x.

$700 = 1.493x^2 + 7.279x + 684.4 \Rightarrow$

$1.493x^2 + 7.279x - 15.6 = 0$

Use the quadratic formula with $a = 1.493$, $b = 7.279$, and $c = -15.6$.

$x = \dfrac{-7.279 \pm \sqrt{7.279^2 - 4(1.493)(-15.6)}}{2(1.493)}$

$= \dfrac{-7.279 \pm \sqrt{146.147041}}{2.986}$

$\approx \dfrac{-7.279 \pm 12.089}{2.986} \approx 1.6$ or -6.5

We reject the potential solution −6.5 because it is gives a year before 2004. So, the model predicts that the emissions reached 700 million tons about 1.6 years after 2004, which is during 2005.

49. Let $x = 2$ and solve for y.

$y = 318.4x^2 + 1612x + 8596$

$y = 318.4(2)^2 + 1612(2) + 8596$

$= 318.4(4) + 3224 + 8596 \approx 13{,}094$

In 2006, the revenue from Internet publishing and broadcasting was about $13,094 million.

51. For each $20 increase in rent over $300, one unit will remain vacant. Therefore, for x $20 increases, x units will remain vacant. Therefore, the number of rented units will be $80 - x$.

53. $300 + 20x$ is the rent for each apartment, and $80 - x$ is the number of apartments that will be rented at that cost. The revenue generated will then be the product of $80 - x$ and $300 + 20x$, so the correct expression is

$(80 - x)(300 + 20x)$

$= 24{,}000 + 1600x - 300x - 20x^2$

$= 24{,}000 + 1300x - 20x^2$.

55. Let x = number of passengers in excess of 75. Then $225 - 5x$ = the cost per passenger (in dollars) and $75 + x$ = the number of passengers.

(Cost per passenger)(Number of passengers) = Revenue

$(225 - 5x)(75 + x) = 16{,}000$

$16{,}875 + 225x - 375x - 5x^2 = 16{,}000$

$16{,}875 - 150x - 5x^2 = 16{,}000$

$0 = 5x^2 + 150x - 875$

$0 = x^2 + 30x - 175 \Rightarrow 0 = (x + 35)(x - 5)$

(continued on next page)

Copyright © 2013 Pearson Education, Inc.

(continued)

$x+35=0 \Rightarrow x=-35$ or $x-5=0 \Rightarrow x=5$
The negative solution is not meaningful. Since there are 5 passengers in excess of 75, the total number of passengers is 80.

57. Let x = number of weeks the manager should wait. Then $100 + 5x$ =number of pounds and $0.40 - 0.02\,x$ = cost per pound
(Cost per pound)(Number of pounds) = Revenue

$$(0.40-0.02x)(100+5x)=38.40$$
$$40+2x-2x-0.1x^2=38.40$$
$$40-0.1x^2=38.40$$
$$-0.1x^2=-1.6$$
$$-10(-0.1x^2)=-10(-1.6)$$
$$x^2=16 \Rightarrow x=\pm4$$

The negative solution is not meaningful. The farmer should wait 4 weeks to get an average revenue of $38.40 per tree.

Section 1.6   Other Types of Equations and Applications

1. $\dfrac{5}{2x+3}-\dfrac{1}{x-6}=0$

$2x+3\ne0 \Rightarrow x\ne-\frac{3}{2}$ and $x-6\ne0 \Rightarrow x\ne6$.

3. $\dfrac{3}{x-2}+\dfrac{1}{x+1}=\dfrac{3}{x^2-x-2}$

or $\dfrac{3}{x-2}+\dfrac{1}{x+1}=\dfrac{3}{(x-2)(x+1)}$

$x-2\ne0 \Rightarrow x\ne2$ and $x+1\ne0 \Rightarrow x\ne-1$

5. $\dfrac{1}{4x}-\dfrac{2}{x}=3$

$4x\ne0 \Rightarrow x\ne0$

7. $\dfrac{2x+5}{2}-\dfrac{3x}{x-2}=x$

The least common denominator is $2(x-2)$, which is equal to 0 if $x=2$. Therefore, 2 cannot possibly be a solution of this equation.

$$2(x-2)\left[\frac{2x+5}{2}-\frac{3x}{x-2}\right]=2(x-2)(x)$$
$$(x-2)(2x+5)-2(3x)=2x(x-2)$$
$$2x^2+5x-4x-10-6x=2x^2-4x$$
$$-5x-10=-4x \Rightarrow -10=x$$

The restriction $x\ne2$ does not affect the result. Therefore, the solution set is $\{-10\}$.

9. $\dfrac{x}{x-3}=\dfrac{3}{x-3}+3$

The least common denominator is $x-3$, which is equal to 0 if $x=3$. Therefore, 3 cannot possibly be a solution of this equation.

$$(x-3)\left(\frac{x}{x-3}\right)=(x-3)\left[\frac{3}{x-3}+3\right]$$
$$x=3+3(x-3)$$
$$x=3+3x-9$$
$$x=3x-6 \Rightarrow -2x=-6 \Rightarrow x=3$$

The only possible solution is 3. However, the variable is restricted to real numbers except 3. Therefore, the solution set is: $\varnothing$.

11. $\dfrac{-2}{x-3}+\dfrac{3}{x+3}=\dfrac{-12}{x^2-9}$ or

$\dfrac{-2}{x-3}+\dfrac{3}{x+3}=\dfrac{-12}{(x+3)(x-3)}$

The least common denominator is $(x+3)(x-3)$, which is equal to 0 if $x=-3$ or $x=3$.

Therefore, 3and 3 cannot possibly be solutions of this equation.

$$(x+3)(x-3)\left[\frac{-2}{x-3}+\frac{3}{x+3}\right]$$
$$=(x+3)(x-3)\left(\frac{-12}{(x+3)(x-3)}\right)$$
$$-2(x+3)+3(x-3)=-12$$
$$-2x-6+3x-9=-12$$
$$-15+x=-12 \Rightarrow x=3$$

The only possible solution is 3. However, the variable is restricted to real numbers except $-3$and 3. Therefore, the solution set is: $\varnothing$.

13. $\dfrac{4}{x^2+x-6}-\dfrac{1}{x^2-4}=\dfrac{2}{x^2+5x+6}$ or

$\dfrac{4}{(x+3)(x-2)}-\dfrac{1}{(x+2)(x-2)}=\dfrac{2}{(x+2)(x+3)}$

The least common denominator is $(x+3)(x-2)(x+2)$, which is equal to 0 if $x=-3$ or $x=2$ or $x=-2$. Therefore, $-3$and 2 and -2 cannot possibly be solutions of this equation.

$$(x+3)(x-2)(x+2)$$
$$\cdot\left[\frac{4}{(x+3)(x-2)}-\frac{1}{(x+2)(x-2)}\right]$$
$$=(x+3)(x-2)(x+2)\left(\frac{2}{(x+2)(x+3)}\right)$$

(continued)

Copyright © 2013 Pearson Education, Inc.

(*continued*)

$$4(x+2)-1(x+3)=2(x-2)$$
$$4x+8-x-3=2x-4$$
$$3x+5=2x-4\Rightarrow x+5=-4\Rightarrow x=-9$$

The restrictions $x\neq -3$, $x\neq 2$, and $x\neq -2$ do not affect the result. Therefore, the solution set is $\{-9\}$.

15. $\dfrac{2x+1}{x-2}+\dfrac{3}{x}=\dfrac{-6}{x^2-2x}$ or

$$\dfrac{2x+1}{x-2}+\dfrac{3}{x}=\dfrac{-6}{x(x-2)}$$

Multiply each term in the equation by the least common denominator, $x(x-2)$, assuming $x\neq 0,2$.

$$x(x-2)\left[\dfrac{2x+1}{x-2}+\dfrac{3}{x}\right]=x(x-2)\left(\dfrac{-6}{x(x-2)}\right)$$
$$x(2x+1)+3(x-2)=-6$$
$$2x^2+x+3x-6=-6$$
$$2x^2+4x-6=-6$$
$$2x^2+4x=0\Rightarrow 2x(x+2)=0$$
$$2x=0\Rightarrow x=0 \quad\text{or}\quad x+2=0\Rightarrow x=-2$$

Because of the restriction $x\neq 0$, the only valid solution is –2. The solution set is $\{-2\}$.

17. $\dfrac{x}{x-1}-\dfrac{1}{x+1}=\dfrac{2}{x^2-1}$ or

$$\dfrac{x}{x-1}-\dfrac{1}{x+1}=\dfrac{2}{(x+1)(x-1)}$$

Multiply each term in the equation by the least common denominator, $(x+1)(x-1)$, assuming $x\neq\pm 1$.

$$(x+1)(x-1)\left[\dfrac{x}{x-1}-\dfrac{1}{x+1}\right]$$
$$=(x+1)(x-1)\left(\dfrac{2}{(x+1)(x-1)}\right)$$
$$x(x+1)-(x-1)=2\Rightarrow x^2+x-x+1=2$$
$$x^2+1=2\Rightarrow x^2-1=0$$
$$(x+1)(x-1)=0\Rightarrow x+1=0\Rightarrow x=-1\text{ or }$$
$$x-1=0\Rightarrow x=1$$

Because of the restriction $x\neq\pm 1$, the solution set is $\varnothing$.

19. $\dfrac{5}{x^2}-\dfrac{43}{x}=18$

Multiply each term in the equation by the least common denominator, x^2, assuming $x\neq 0$.

$$x^2\left[\dfrac{5}{x^2}-\dfrac{43}{x}\right]=x^2(18)$$
$$5-43x=18x^2\Rightarrow 0=18x^2+43x-5$$
$$0=(2x+5)(9x-1)$$
$$2x+5=0\Rightarrow x=-\tfrac{5}{2}\text{ or }$$
$$9x-1=0\Rightarrow x=\tfrac{1}{9}$$

The restriction $x\neq 0$ does not affect the result. Therefore, the solution set is $\left\{-\tfrac{5}{2},\tfrac{1}{9}\right\}$.

21. $2=\dfrac{3}{2x-1}+\dfrac{-1}{(2x-1)^2}$

Multiply each term in the equation by the least common denominator, $(2x-1)^2$, assuming $x\neq\tfrac{1}{2}$.

$$(2x-1)^2(2)=(2x-1)^2\left[\dfrac{3}{2x-1}+\dfrac{-1}{(2x-1)^2}\right]$$
$$2(4x^2-4x+1)=3(2x-1)-1$$
$$8x^2-8x+2=6x-3-1$$
$$8x^2-8x+2=6x-4\Rightarrow 8x^2-14x+6=0$$
$$2(4x^2-7x+3)=0\Rightarrow 2(4x-3)(x-1)=0$$
$$4x-3=0\Rightarrow x=\tfrac{3}{4}\text{ or }x-1=0\Rightarrow x=1$$

The restriction $x\neq\tfrac{1}{2}$ does not affect the result. Therefore the solution set is $\left\{\tfrac{3}{4},1\right\}$.

23. $\dfrac{2x-5}{x}=\dfrac{x-2}{3}$

Multiply each term in the equation by the least common denominator, $3x$, assuming $x\neq 0$.

$$3x\left(\dfrac{2x-5}{x}\right)=3x\left(\dfrac{x-2}{3}\right)$$
$$3(2x-5)=x(x-2)\Rightarrow 6x-15=x^2-2x\Rightarrow$$
$$0=x^2-8x+15=(x-3)(x-5)$$
$$x-3=0\Rightarrow x=3 \quad\text{or}\quad x-5=0\Rightarrow x=5$$

The restriction $x\neq 0$ does not affect the result. Therefore, the solution set is $\{3, 5\}$.

Copyright © 2013 Pearson Education, Inc.

25. $\dfrac{2x}{x-2} = 5 + \dfrac{4x^2}{x-2}$

Multiply each term in the equation by the least common denominator, $x - 2$, assuming $x \neq 2$.

$$(x-2)\left(\dfrac{2x}{x-2}\right) = (x-2)\left[5 + \dfrac{4x^2}{x-2}\right]$$

$$2x = 5(x-2) + 4x^2$$

$$2x = 5x - 10 + 4x^2$$

$$0 = 4x^2 + 3x - 10$$

$$0 = (x+2)(4x-5)$$

$$x+2 = 0 \Rightarrow x = -2 \quad \text{or} \quad 4x-5 = 0 \Rightarrow x = \tfrac{5}{4}$$

The restriction $x \neq 2$ does not affect the result.

Therefore the solution set is $\left\{-2, \tfrac{5}{4}\right\}$.

27. Let x = the amount of time (in hours) it takes Joe and Sam to paint the house.

	r	t	Part of the Job Accomplished
Joe	$\frac{1}{3}$	x	$\frac{1}{3}x$
Sam	$\frac{1}{5}$	x	$\frac{1}{5}x$

Since Joe and Sam must accomplish 1 job (painting a house), we must solve the following equation.

$$\tfrac{1}{3}x + \tfrac{1}{5}x = 1$$

$$15\left[\tfrac{1}{3}x + \tfrac{1}{5}x\right] = 15 \cdot 1$$

$$5x + 3x = 15 \Rightarrow 8x = 15 \Rightarrow x = \dfrac{15}{8} = 1\tfrac{7}{8}$$

It takes Joe and Sam $1\tfrac{7}{8}$ hr working together to paint the house.

29. Let x = the amount of time (in hours) it takes plant A to produce the pollutant. Then $2x$ = the amount of time (in hours) it takes plant B to produce the pollutant.

	Rate	Time	Part of the Job Accomplished
Pollution from A	$\frac{1}{x}$	26	$\frac{1}{x}(26)$
Pollution from B	$\frac{1}{2x}$	26	$\frac{1}{2x}(26)$

Since plant A and B accomplish 1 job (producing the pollutant), we must solve the following equation.

$$\tfrac{1}{x}(26) + \tfrac{1}{2x}(26) = 1$$

$$\dfrac{26}{x} + \dfrac{13}{x} = 1$$

$$x\left[\dfrac{39}{x}\right] = x \cdot 1 \Rightarrow 39 = x$$

Plant B will take $2 \cdot 39 = 78$ hr to produce the pollutant.

31. Let x = the amount of time (in hours) to fill the pool with both pipes open.

	Rate	Time	Part of the Job Accomplished
Inlet pipe	$\frac{1}{5}$	x	$\frac{1}{5}x$
Outlet pipe	$\frac{1}{8}$	x	$\frac{1}{8}x$

Filling the pool is 1 whole job, but because the outlet pipe empties the pool, its contribution should be subtracted from the contribution of the inlet pipe.

$$\tfrac{1}{5}x - \tfrac{1}{8}x = 1$$

$$40\left[\tfrac{1}{5}x - \tfrac{1}{8}x\right] = 40 \cdot 1 \Rightarrow 8x - 5x = 40 \Rightarrow$$

$$3x = 40 \Rightarrow x = \dfrac{40}{3} = 13\tfrac{1}{3}\,\text{hr}$$

It took $13\tfrac{1}{3}$ hr to fill the pool.

33. Let x = the amount of time (in minutes) to fill the sink with both pipes open.

	Rate	Time	Part of the Job Accomplished
Tap	$\frac{1}{5}$	x	$\frac{1}{5}x$
Drain	$\frac{1}{10}$	x	$\frac{1}{10}x$

Filling the sink is 1 whole job, but because the sink is draining, its contribution should be subtracted from the contribution of the taps.

$$\tfrac{1}{5}x - \tfrac{1}{10}x = 1$$

$$10\left[\tfrac{1}{5}x - \tfrac{1}{10}x\right] = 10 \cdot 1 \Rightarrow 2x - x = 10 \Rightarrow x = 10$$

It will take 10 minutes to fill the sink if Mark forgets to put in the stopper.

Copyright © 2013 Pearson Education, Inc.

35. $x - \sqrt{2x+3} = 0$

$$x = \sqrt{2x+3} \Rightarrow x^2 = \left(\sqrt{2x+3}\right)^2$$

$$x^2 = 2x+3 \Rightarrow x^2 - 2x - 3 = 0 \Rightarrow$$

$$(x+1)(x-3) = 0 \Rightarrow x = -1 \text{ or } x = 3$$

Check $x = -1$.

$$x - \sqrt{2x+3} = 0$$

$$-1 - \sqrt{2(-1)+3} \overset{?}{=} 0$$

$$-1 - \sqrt{-2+3} = 0$$

$$-1 - \sqrt{1} = 0 \Rightarrow -1 - 1 = 0 \Rightarrow -2 = 0$$

This is a false statement. -1 is not a solution.

Check $x = 3$.

$$x - \sqrt{2x+3} = 0$$

$$3 - \sqrt{2(3)+3} \overset{?}{=} 0$$

$$3 - \sqrt{6+3} = 0$$

$$3 - \sqrt{9} = 0 \Rightarrow 3 - 3 = 0 \Rightarrow 0 = 0$$

This is a true statement. 3 is a solution.

Solution set: $\{3\}$

37. $\sqrt{3x+7} = 3x+5$

$$\left(\sqrt{3x+7}\right)^2 = (3x+5)^2$$

$$3x+7 = 9x^2 + 30x + 25$$

$$0 = 9x^2 + 27x + 18$$

$$0 = 9\left(x^2 + 3x + 2\right) = 9(x+2)(x+1)$$

$x = -2$ or $x = -1$

Check $x = -2$.

$$\sqrt{3x+7} = 3x+5$$

$$\sqrt{3(-2)+7} \overset{?}{=} 3(-2)+5$$

$$\sqrt{-6+7} = -6+5$$

$$\sqrt{1} = -1 \Rightarrow 1 = -1$$

This is a false statement. -2 is not a solution.

Check $x = -1$

$$\sqrt{3x+7} = 3x+5$$

$$\sqrt{3(-1)+7} \overset{?}{=} 3(-1)+5$$

$$\sqrt{-3+7} = -3+5$$

$$\sqrt{4} = 2 \Rightarrow 2 = 2$$

This is a true statement. -1 is a solution.

Solution set: $\{-1\}$

39. $\sqrt{4x+5} - 6 = 2x - 11$

$$\sqrt{4x+5} = 2x - 5$$

$$\left(\sqrt{4x+5}\right)^2 = (2x-5)^2$$

$$4x+5 = 4x^2 - 20x + 25$$

$$0 = 4x^2 - 24x + 20$$

$$0 = 4\left(x^2 - 6x + 5\right) = 4(x-1)(x-5)$$

$x = 1$ or $x = 5$

Check $x = 1$.

$$\sqrt{4x+5} - 6 = 2x - 11$$

$$\sqrt{4(1)+5} - 6 \overset{?}{=} 2(1) - 11$$

$$\sqrt{4+5} - 6 = 2 - 11$$

$$\sqrt{9} - 6 = -9$$

$$3 - 6 = -9 \Rightarrow -3 = -9$$

This is a false statement. 1 is not a solution.

Check $x = 5$.

$$\sqrt{4x+5} - 6 = 2x - 11$$

$$\sqrt{4(5)+5} - 6 \overset{?}{=} 2(5) - 11$$

$$\sqrt{20+5} - 6 = 10 - 11$$

$$\sqrt{25} - 6 = -1 \Rightarrow 5 - 6 = -1 \Rightarrow -1 = -1$$

This is a true statement. 5 is a solution.

Solution set: $\{5\}$

41. $\sqrt{4x} - x + 3 = 0$

$$\sqrt{4x} = x - 3$$

$$\left(\sqrt{4x}\right)^2 = (x-3)^2$$

$$4x = x^2 - 6x + 9$$

$$0 = x^2 - 10x + 9 = (x-1)(x-9)$$

$$x = 1 \text{ or } x = 9$$

Check $x = 1$.

$$\sqrt{4x} - x + 3 = 0$$

$$\sqrt{4(1)} - 1 + 3 \overset{?}{=} 0$$

$$\sqrt{4} - 1 + 3 = 0$$

$$2 - 1 + 3 = 0 \Rightarrow 4 = 0$$

This is a false statement. 1 is not a solution.

Check $x = 9$.

$$\sqrt{4x} - x + 3 = 0$$

$$\sqrt{4(9)} - 9 + 3 \overset{?}{=} 0$$

$$\sqrt{36} - 9 + 3 = 0$$

$$6 - 9 + 3 = 0 \Rightarrow 0 = 0$$

This is a true statement. 9 is a solution.

Solution set: $\{9\}$

Copyright © 2013 Pearson Education, Inc.

43. $\sqrt{x} - \sqrt{x-5} = 1$

$\sqrt{x} = 1 + \sqrt{x-5} \Rightarrow \left(\sqrt{x}\right)^2 = \left(1 + \sqrt{x-5}\right)^2$

$x = 1 + 2\sqrt{x-5} + (x-5)$

$x = x + 2\sqrt{x-5} - 4 \Rightarrow 4 = 2\sqrt{x-5}$

$2 = \sqrt{x-5} \Rightarrow 2^2 = \left(\sqrt{x-5}\right)^2$

$4 = x - 5 \Rightarrow 9 = x$

Check $x = 9$.

$\sqrt{x} - \sqrt{x-5} = 1$

$\sqrt{9} - \sqrt{9-5} \overset{?}{=} 1$

$\quad 3 - \sqrt{4} = 1 \Rightarrow 3 - 2 = 1 \Rightarrow 1 = 1$

This is a true statement.

Solution set is: $\{9\}$

45. $\sqrt{x+7} + 3 = \sqrt{x-4}$

$\left(\sqrt{x+7} + 3\right)^2 = \left(\sqrt{x-4}\right)^2$

$(x+7) + 6\sqrt{x+7} + 9 = x - 4$

$x + 6\sqrt{x+7} + 16 = x - 4 \Rightarrow 6\sqrt{x+7} = -20$

$3\sqrt{x+7} = -10 \Rightarrow \left(3\sqrt{x+7}\right)^2 = (-10)^2$

$9(x+7) = 100 \Rightarrow 9x + 63 = 100$

$9x = 37 \Rightarrow x = \dfrac{37}{9}$

Check $x = \frac{37}{9}$.

$\sqrt{x+7} + 3 = \sqrt{x-4}$

$\sqrt{\frac{37}{9} + 7} + 3 \overset{?}{=} \sqrt{\frac{37}{9} - 4}$

$\sqrt{\frac{37}{9} + \frac{63}{9}} + 3 = \sqrt{\frac{37}{9} - \frac{36}{9}}$

$\sqrt{\frac{100}{9}} + 3 = \sqrt{\frac{1}{9}}$

$\frac{10}{3} + 3 = \frac{1}{3} \Rightarrow \frac{10}{3} + \frac{9}{3} = \frac{1}{3} \Rightarrow \frac{19}{3} = \frac{1}{3}$

This is a false statement.

Solution set: $\varnothing$

47. $\sqrt{2x+5} - \sqrt{x+2} = 1$

$\sqrt{2x+5} = \sqrt{x+2} - 1$

$\left(\sqrt{2x+5}\right)^2 = \left(\sqrt{x+2} - 1\right)^2$

$2x + 5 = (x+2) - 2\sqrt{x+2} + 1$

$2x + 5 = x + 3 - 2\sqrt{x+2}$

$x + 2 = -2\sqrt{x+2}$

$(x+2)^2 = \left(-2\sqrt{x+2}\right)^2$

$x^2 + 4x + 4 = 4(x+2)$

$x^2 + 4x + 4 = 4x + 8$

$0 = x^2 - 4$

$0 = (x+2)(x-2) \Rightarrow x = \pm 2$

Check $x = 2$.

$\sqrt{2x+5} - \sqrt{x+2} = 1$

$\sqrt{2(2)+5} - \sqrt{2+2} \overset{?}{=} 1$

$\sqrt{4+5} - \sqrt{4} \overset{?}{=} 1$

$\sqrt{9} - \sqrt{4} \overset{?}{=} 1$

$3 - 2 \overset{?}{=} 1 \Rightarrow 1 = 1$

This is a true statement. 2 is a solution.

Check $x = -2$.

$\sqrt{2x+5} - \sqrt{x+2} = 1$

$\sqrt{2(-2)+5} - \sqrt{-2+2} \overset{?}{=} 1$

$\sqrt{-4+5} - \sqrt{0} \overset{?}{=} 1$

$\sqrt{1} - \sqrt{0} \overset{?}{=} 1$

$1 - 0 \overset{?}{=} 1 \Rightarrow 1 = 1$

This is a true statement. -2 is a solution.

Solution set: $\{\pm 2\}$

49. $\sqrt{3x} = \sqrt{5x+1} - 1$

$\left(\sqrt{3x}\right)^2 = \left(\sqrt{5x+1} - 1\right)^2$

$3x = (5x+1) - 2\sqrt{5x+1} + 1$

$3x = 5x + 2 - 2\sqrt{5x+1}$

$2\sqrt{5x+1} = 2 + 2x \Rightarrow \sqrt{5x+1} = 1 + x$

$\left(\sqrt{5x+1}\right)^2 = (1+x)^2 \Rightarrow 5x+1 = 1 + 2x + x^2$

$0 = x^2 - 3x \Rightarrow 0 = x(x-3) \Rightarrow$

$x = 0 \text{ or } x = 3$

Check $x = 0$.

$\sqrt{3x} = \sqrt{5x+1} - 1$

$\sqrt{3(0)} \overset{?}{=} \sqrt{5(0)+1} - 1$

$\sqrt{0} = \sqrt{0+1} - 1 \Rightarrow 0 = \sqrt{1} - 1$

$0 = 1 - 1 \Rightarrow 0 = 0$

This is a true statement. 0 is a solution.

Check $x = 3$.

$\sqrt{3x} = \sqrt{5x+1} - 1$

$\sqrt{3(3)} \overset{?}{=} \sqrt{5(3)+1} - 1$

$\sqrt{9} = \sqrt{15+1} - 1 \Rightarrow 3 = \sqrt{16} - 1$

$3 = 4 - 1 \Rightarrow 3 = 3$

This is a true statement. 3 is a solution.

Solution set: $\{0, 3\}$

Copyright © 2013 Pearson Education, Inc.

51.
$$\sqrt{x+2} = 1 - \sqrt{3x+7}$$
$$\left(\sqrt{x+2}\right)^2 = (1-\sqrt{3x+7})^2$$
$$x+2 = 1 - 2\sqrt{3x+7} + (3x+7)$$
$$x+2 = 3x+8 - 2\sqrt{3x+7}$$
$$2\sqrt{3x+7} = 2x+6$$
$$2\sqrt{3x+7} = 2(x+3)$$
$$\sqrt{3x+7} = x+3 \Rightarrow \left(\sqrt{3x+7}\right)^2 = (x+3)^2$$
$$3x+7 = x^2+6x+9 \Rightarrow 0 = x^2+3x+2$$
$$0 = (x+2)(x+1)$$
$$x = -2 \text{ or } x = -1$$
Check $x = -2$.
$$\sqrt{x+2} = 1-\sqrt{3x+7}$$
$$\sqrt{-2+2} \overset{?}{=} 1 - \sqrt{3(-2)+7}$$
$$\sqrt{0} = 1 - \sqrt{-6+7}$$
$$0 = 1 - \sqrt{1}$$
$$0 = 1-1 \Rightarrow 0 = 0$$
This is a true statement. −2 is a solution.
Check $x = -1$.
$$\sqrt{x+2} = 1-\sqrt{3x+7}$$
$$\sqrt{-1+2} \overset{?}{=} 1 - \sqrt{3(-1)+7}$$
$$\sqrt{1} = 1 - \sqrt{-3+7}$$
$$1 = 1 - \sqrt{4}$$
$$1 = 1-2 \Rightarrow 1 = -1$$
This is a false statement.
−1 is not a solution.
Solution set: {−2}

53.
$$\sqrt{2\sqrt{7x+2}} = \sqrt{3x+2}$$
$$\left(\sqrt{2\sqrt{7x+2}}\right)^2 = \left(\sqrt{3x+2}\right)^2$$
$$2\sqrt{7x+2} = 3x+2$$
$$\left(2\sqrt{7x+2}\right)^2 = (3x+2)^2$$
$$4(7x+2) = 9x^2+12x+4$$
$$28x+8 = 9x^2+12x+4$$
$$0 = 9x^2-16x-4$$
$$0 = (9x+2)(x-2)$$
$$x = -\tfrac{2}{9} \text{ or } x = 2$$
Check $x = -\tfrac{2}{9}$.
$$\sqrt{2\sqrt{7x+2}} = \sqrt{3x+2}$$
$$\sqrt{2\sqrt{7\left(-\tfrac{2}{9}\right)+2}} \overset{?}{=} \sqrt{3\left(-\tfrac{2}{9}\right)+2}$$
$$\sqrt{2\sqrt{-\tfrac{14}{9}+2}} = \sqrt{-\tfrac{2}{3}+2}$$
$$\sqrt{2\sqrt{-\tfrac{14}{9}+\tfrac{18}{9}}} = \sqrt{-\tfrac{2}{3}+\tfrac{6}{3}}$$
$$\sqrt{2\sqrt{\tfrac{4}{9}}} = \sqrt{\tfrac{4}{3}} \Rightarrow \sqrt{2\left(\tfrac{2}{3}\right)} = \tfrac{\sqrt{4}}{\sqrt{3}}$$

$$\sqrt{\tfrac{4}{3}} = \tfrac{2}{\sqrt{3}} \cdot \tfrac{\sqrt{3}}{\sqrt{3}} \Rightarrow \tfrac{\sqrt{4}}{\sqrt{3}} = \tfrac{2\sqrt{3}}{3}$$
$$\tfrac{2}{\sqrt{3}} \cdot \tfrac{\sqrt{3}}{\sqrt{3}} = \tfrac{2\sqrt{3}}{3} \Rightarrow \tfrac{2\sqrt{3}}{3} = \tfrac{2\sqrt{3}}{3}$$
This is a true statement. $-\tfrac{2}{9}$ is a solution.
Check $x = 2$.
$$\sqrt{2\sqrt{7x+2}} = \sqrt{3x+2}$$
$$\sqrt{2\sqrt{7(2)+2}} \overset{?}{=} \sqrt{3(2)+2}$$
$$\sqrt{2\sqrt{14+2}} = \sqrt{6+2}$$
$$\sqrt{2\sqrt{16}} = \sqrt{8}$$
$$\sqrt{2(4)} = 2\sqrt{2}$$
$$\sqrt{8} = 2\sqrt{2} \Rightarrow 2\sqrt{2} = 2\sqrt{2}$$
This is a true statement. 2 is a solution.
Solution set: $\left\{-\tfrac{2}{9}, 2\right\}$

55.
$$3 - \sqrt{x} = \sqrt{2\sqrt{x}-3}$$
$$\left(3-\sqrt{x}\right)^2 = \left(\sqrt{2\sqrt{x}-3}\right)^2$$
$$9 - 6\sqrt{x} + x = 2\sqrt{x}-3$$
$$12 + x = 8\sqrt{x}$$
$$(12+x)^2 = \left(8\sqrt{x}\right)^2$$
$$144 + 24x + x^2 = 64x$$
$$x^2 - 40x + 144 = 0$$
$$(x-36)(x-4) = 0 \Rightarrow x = 36 \text{ or } x = 4$$
Check $x = 36$.
$$3 - \sqrt{x} = \sqrt{2\sqrt{x}-3}$$
$$3 - \sqrt{36} \overset{?}{=} \sqrt{2\sqrt{36}-3}$$
$$3 - 6 = \sqrt{2(6)-3}$$
$$-3 = \sqrt{12-3} \Rightarrow -3 = \sqrt{9} \Rightarrow -3 = 3$$
This is a false statement. 36 is not a solution.
Check $x = 4$.
$$3 - \sqrt{x} = \sqrt{2\sqrt{x}-3}$$
$$3 - \sqrt{4} \overset{?}{=} \sqrt{2\sqrt{4}-3}$$
$$3 - 2 = \sqrt{2(2)-3}$$
$$1 = \sqrt{4-3} \Rightarrow 1 = \sqrt{1} \Rightarrow 1 = 1$$
This is a true statement. 4 is a solution.
Solution set: $\{4\}$

Copyright © 2013 Pearson Education, Inc.

57. $\sqrt[3]{4x+3} = \sqrt[3]{2x-1}$

$\left(\sqrt[3]{4x+3}\right)^3 = \left(\sqrt[3]{2x-1}\right)^3$

$4x+3 = 2x-1 \Rightarrow 2x = -4 \Rightarrow x = -2$

Check $x = -2$.

$\sqrt[3]{4(-2)+3} = \sqrt[3]{2(-2)-1}$

$\sqrt[3]{-5} \overset{?}{=} \sqrt[3]{-5} \Rightarrow -\sqrt[3]{5} = -\sqrt[3]{5}$

This is a true statement. -2 is a solution.

Solution set: $\{-2\}$

59. $\sqrt[3]{5x^2-6x+2} - \sqrt[3]{x} = 0$

$\sqrt[3]{5x^2-6x+2} = \sqrt[3]{x}$

$\left(\sqrt[3]{5x^2-6x+2}\right)^3 = \left(\sqrt[3]{x}\right)^3$

$5x^2 - 6x + 2 = x$

$5x^2 - 7x + 2 = 0$

$(5x-2)(x-1) = 0 \Rightarrow x = \frac{2}{5}$ or $x = 1$

Check $x = \frac{2}{5}$.

$\sqrt[3]{5x^2-6x+2} - \sqrt[3]{x} = 0$

$\sqrt[3]{5\left(\frac{2}{5}\right)^2 - 6\left(\frac{2}{5}\right) + 2} - \sqrt[3]{\frac{2}{5}} \overset{?}{=} 0$

$\sqrt[3]{5\left(\frac{4}{25}\right) - \frac{12}{5} + 2} - \sqrt[3]{\frac{2}{5}} = 0$

$\sqrt[3]{\frac{4}{5} - \frac{12}{5} + \frac{10}{5}} - \sqrt[3]{\frac{2}{5}} = 0$

$\sqrt[3]{\frac{2}{5}} - \sqrt[3]{\frac{2}{5}} = 0 \Rightarrow 0 = 0$

This is a true statement. $\frac{2}{5}$ is a solution.

Check $x = 1$.

$\sqrt[3]{5x^2-6x+2} - \sqrt[3]{x} = 0$

$\sqrt[3]{5(1)^2 - 6(1) + 2} - \sqrt[3]{1} \overset{?}{=} 0$

$\sqrt[3]{5(1) - 6 + 2} - 1 = 0$

$\sqrt[3]{5 - 6 + 2} - 1 = 0$

$\sqrt[3]{1} - 1 = 0 \Rightarrow 1 - 1 = 0 \Rightarrow 0 = 0$

This is a true statement. 1 is a solution.

Solution set: $\left\{\frac{2}{5}, 1\right\}$

61. $\sqrt[4]{x-15} = 2 \Rightarrow \left(\sqrt[4]{x-15}\right)^4 = 2^4 \Rightarrow$

$x - 15 = 16 \Rightarrow x = 31$

Check $x = 31$.

$\sqrt[4]{x-15} = 2 \Rightarrow \sqrt[4]{31-15} \overset{?}{=} 2$

$\sqrt[4]{16} = 2 \Rightarrow 2 = 2$

This is a true statement.

Solution set: $\{31\}$

63. $\sqrt[4]{x^2+2x} = \sqrt[4]{3} \Rightarrow \left(\sqrt[4]{x^2+2x}\right)^4 = \left(\sqrt[4]{3}\right)^4$

$x^2 + 2x = 3 \Rightarrow x^2 + 2x - 3 = 0$

$(x+3)(x-1) = 0 \Rightarrow x = -3$ or $x = 1$

Check $x = -3$.

$\sqrt[4]{x^2+2x} = \sqrt[4]{3}$

$\sqrt[4]{(-3)^2 + 2(-3)} \overset{?}{=} \sqrt[4]{3}$

$\sqrt[4]{9-6} = \sqrt[4]{3} \Rightarrow \sqrt[4]{3} = \sqrt[4]{3}$

This is a true statement. -3 is a solution.

Check $x = 1$.

$\sqrt[4]{x^2+2x} = \sqrt[4]{3}$

$\sqrt[4]{1^2 + 2(1)} \overset{?}{=} \sqrt[4]{3}$

$\sqrt[4]{1+2} = \sqrt[4]{3} \Rightarrow \sqrt[4]{3} = \sqrt[4]{3}$

This is a true statement. 1 is a solution.

Solution set: $\{-3, 1\}$

65. $x^{3/2} = 125$

$\left(x^{3/2}\right)^{2/3} = 125^{2/3}$

$x = 5^2 = 25$

Check $x = 25$.

$x^{3/2} = 125$

$25^{3/2} \overset{?}{=} 125$

$5 = 5$

This is a true statement. 25 is a solution.

Solution set: $\{25\}$

67. $(x^2+24x)^{1/4} = 3 \Rightarrow \left[\left(x^2+24x\right)^{1/4}\right]^4 = 3^4 \Rightarrow$

$x^2 + 24x = 81 \Rightarrow x^2 + 24x - 81 = 0 \Rightarrow$

$(x+27)(x-3) = 0 \Rightarrow x + 27 = 0 \Rightarrow x = -27$ or

$x - 3 = 0 \Rightarrow x = 3$

Check $x = -27$.

$(x^2+24x)^{1/4} = 3$

$\left[(-27)^2 + 24(-27)\right]^{1/4} \overset{?}{=} 3$

$(729 - 648)^{1/4} = 3$

$81^{1/4} = 3 \Rightarrow 3 = 3$

This is a true statement. -27 is a solution.

Check $x = 3$.

$(x^2+24x)^{1/4} = 3$

$\left[3^2 + 24(3)\right]^{1/4} \overset{?}{=} 3$

$(9 + 72)^{1/4} = 3 \Rightarrow 81^{1/4} = 3 \Rightarrow 3 = 3$

This is a true statement. 3 is a solution.

Solution set: $\{-27, 3\}$

Copyright © 2013 Pearson Education, Inc.

69. $(x-3)^{2/5} = 4$

$\left[(x-3)^{2/5}\right]^{5/2} = \pm 4^{5/2}$

$x - 3 = \pm 32$

$x - 3 = 32 \Rightarrow x = 35$

$x - 3 = -32 \Rightarrow x = -29$

Check $x = 35$.

$(x-3)^{2/5} = 4$

$(35-3)^{2/5} \overset{?}{=} 4$

$32^{2/5} = 4 \Rightarrow 4 = 4$

This is a true statement. 35 is a solution.

Check $x = -29$.

$(x-3)^{2/5} = 4$

$(-29-3)^{2/5} \overset{?}{=} 4$

$(-32)^{2/5} = 4$

$(-2)^2 = 4 \Rightarrow 4 = 4$

This is a true statement. -29 is a solution.

Solution set: $\{-29, 35\}$

71. $(2x+5)^{1/3} - (6x-1)^{1/3} = 0$

$(2x+5)^{1/3} = (6x-1)^{1/3}$

$\left[(2x+5)^{1/3}\right]^3 = \left[(6x-1)^{1/3}\right]^3$

$2x + 5 = 6x - 1$

$-4x = -6$

$x = \frac{3}{2}$

Check $x = \frac{3}{2}$.

$(2x+5)^{1/3} - (6x-1)^{1/3} = 0$

$\left(2 \cdot \frac{3}{2} + 5\right)^{1/3} - \left(6 \cdot \frac{3}{2} - 1\right)^{1/3} \overset{?}{=} 0$

$8^{1/3} - 8^{1/3} = 0 \Rightarrow 0 = 0$

This is a true statement. $\frac{3}{2}$ is a solution.

Solution set: $\left\{\frac{3}{2}\right\}$

73. $(2x-1)^{2/3} = x^{1/3}$

$[(2x-1)^{2/3}]^3 = (x^{1/3})^3$

$(2x-1)^2 = x \Rightarrow 4x^2 - 4x + 1 = x$

$4x^2 - 5x + 1 = 0 \Rightarrow (4x-1)(x-1) = 0 \Rightarrow$

$x = \frac{1}{4} \text{ or } x = 1$

Check $x = \frac{1}{4}$.

$(2x-1)^{2/3} = x^{1/3} \Rightarrow \left[2\left(\frac{1}{4}\right)-1\right]^{2/3} \overset{?}{=} \left(\frac{1}{4}\right)^{1/3}$

$\left[\frac{1}{2}-1\right]^{2/3} = \frac{1}{\sqrt[3]{4}} \cdot \frac{\sqrt[3]{2}}{\sqrt[3]{2}} \Rightarrow \left[-\frac{1}{2}\right]^{2/3} = \frac{\sqrt[3]{2}}{2} \Rightarrow$

$\left[\left(-\frac{1}{2}\right)^2\right]^{1/3} = \frac{\sqrt[3]{2}}{2} \Rightarrow \left(\frac{1}{4}\right)^{1/3} = \frac{\sqrt[3]{2}}{2}$

$\frac{1}{\sqrt[3]{4}} \cdot \frac{\sqrt[3]{2}}{\sqrt[3]{2}} = \frac{\sqrt[3]{2}}{2} \Rightarrow \frac{\sqrt[3]{2}}{2} = \frac{\sqrt[3]{2}}{2}$

This is a true statement. $\frac{1}{4}$ is a solution.

Check $x = 1$.

$(2x-1)^{2/3} = x^{1/3} \Rightarrow \left[2(1)-1\right]^{2/3} \overset{?}{=} (1)^{1/3}$

$[2-1]^{2/3} = 1 \Rightarrow 1^{2/3} = 1 \Rightarrow 1 = 1$

This is a true statement. 1 is a solution.

Solution set: $\left\{\frac{1}{4}, 1\right\}$

75. $x^{2/3} = 2x^{1/3} \Rightarrow \left(x^{2/3}\right)^3 = \left(2x^{1/3}\right)^3 \Rightarrow$

$x^2 = 8x \Rightarrow x^2 - 8x = 0 \Rightarrow x(x-8) = 0 \Rightarrow$

$x = 0 \text{ or } x = 8$

Check $x = 0$.

$x^{2/3} = 2x^{1/3}$

$0^{2/3} \overset{?}{=} 2\left(0^{1/3}\right) \Rightarrow 0 = 2 \cdot 0 \Rightarrow 0 = 0$

This is a true statement. 0 is a solution.

Check $x = 8$.

$x^{2/3} = 2x^{1/3} \Rightarrow 8^{2/3} \overset{?}{=} 2\left(8^{1/3}\right)$

$\left(8^2\right)^{1/3} = 2 \cdot 2 \Rightarrow 64^{1/3} = 4 \Rightarrow 4 = 4$

This is a true statement. 8 is a solution.

Solution set: $\{0, 8\}$

77. $2x^4 - 7x^2 + 5 = 0$

Let $u = x^2$; then $u^2 = x^4$. With this substitution, the equation becomes

$2u^2 - 7u + 5 = 0$.

$2u^2 - 7u + 5 = 0 \Rightarrow (u-1)(2u-5) = 0 \Rightarrow$

$u = 1 \text{ or } u = \frac{5}{2}$

To find x, replace u with x^2.

$x^2 = 1 \Rightarrow x = \pm 1$ or

$x^2 = \frac{5}{2} \Rightarrow x = \pm\sqrt{\frac{5}{2}} = \pm\frac{\sqrt{5}}{\sqrt{2}} \cdot \frac{\sqrt{2}}{\sqrt{2}} = \pm\frac{\sqrt{10}}{2}$

Solution set: $\left\{\pm 1, \pm\frac{\sqrt{10}}{2}\right\}$

Copyright © 2013 Pearson Education, Inc.

79. $x^4 + 2x^2 - 15 = 0$

Let $u = x^2$; then $u^2 = x^4$.

$u^2 + 2u - 15 = 0 \Rightarrow (u - 3)(u + 5) = 0$.

$\qquad u = 3$ or $u = -5$

To find x, replace u with x^2.

$x^2 = 3 \Rightarrow x = \pm\sqrt{3}$ or

$x^2 = -5 \Rightarrow x = \pm\sqrt{-5} = \pm i\sqrt{5}$

Solution set: $\left\{\pm\sqrt{3}, \pm i\sqrt{5}\right\}$

81. $(x-1)^{2/3} + (x-1)^{1/3} - 12 = 0$

Let $u = (x-1)^{1/3}$ then,

$u^2 = \left[(x-1)^{1/3}\right]^2 = (x-1)^{2/3}$.

$u^2 + u - 12 = 0 \Rightarrow (u+4)(u-3) = 0 \Rightarrow$

$u = -4$ or $u = 3$

To find x, replace u with $(x-1)^{1/3}$.

$(x-1)^{1/3} = -4 \Rightarrow \left[(x-1)^{1/3}\right]^3 = (-4)^3 \Rightarrow$

$\qquad x - 1 = -64 \Rightarrow x = -63$ or

$(x-1)^{1/3} = 3 \Rightarrow \left[(x-1)^{1/3}\right]^3 = 3^3 \Rightarrow$

$\qquad x - 1 = 27 \Rightarrow x = 28$

Check $x = -63$.

$(x-1)^{2/3} + (x-1)^{1/3} - 12 = 0$

$(-63-1)^{2/3} + (-63-1)^{1/3} - 12 \overset{?}{=} 0$

$(-64)^{2/3} + (-64)^{1/3} - 12 = 0$

$\left[(-64)^{1/3}\right]^2 - 4 - 12 = 0$

$(-4)^2 - 4 - 12 = 0$

$\qquad 16 - 4 - 12 = 0 \Rightarrow 0 = 0$

This is a true statement. −63 is a solution.

Check $x = 28$.

$(x-1)^{2/3} + (x-1)^{1/3} - 12 = 0$

$(28-1)^{2/3} + (28-1)^{1/3} - 12 \overset{?}{=} 0$

$27^{2/3} + 27^{1/3} - 12 = 0$

$\left[27^{1/3}\right]^2 + 3 - 12 = 0$

$\qquad 3^2 + 3 - 12 = 0$

$\qquad 9 + 3 - 12 = 0 \Rightarrow 0 = 0$

This is a true statement. 28 is a solution.

Solution set: $\{-63, 28\}$

83. $(x+1)^{2/5} - 3(x+1)^{1/5} + 2 = 0$

Let $u = (x+1)^{1/5}$ then,

$u^2 = \left[(x+1)^{1/5}\right]^2 = (x+1)^{2/5}$.

$u^2 - 3u + 2 = 0 \Rightarrow (u-1)(u-2) = 0 \Rightarrow$

$u = 1$ or $u = 2$

To find x, replace u with $(x+1)^{1/5}$.

$(x+1)^{1/5} = 1 \Rightarrow \left[(x+1)^{1/5}\right]^5 = 1^5 \Rightarrow$ or

$\qquad x + 1 = 1 \Rightarrow x = 0$

$(x+1)^{1/5} = 2 \Rightarrow \left[(x+1)^{1/5}\right]^5 = 2^5 \Rightarrow$

$\qquad x + 1 = 32 \Rightarrow x = 31$

Check $x = 0$.

$(x+1)^{2/5} - 3(x+1)^{1/5} + 2 = 0$

$(0+1)^{2/5} - 3(0+1)^{1/5} + 2 \overset{?}{=} 0$

$1^{2/5} - 3(1)^{1/5} + 2 = 0$

$1 - 3(1) + 2 = 0 \Rightarrow 1 - 3 + 2 = 0$

This is a true statement. 0 is a solution.

Check $x = 31$.

$(x+1)^{2/5} - 3(x+1)^{1/5} + 2 = 0$

$(31+1)^{2/5} - 3(31+1)^{1/5} + 2 \overset{?}{=} 0$

$32^{2/5} - 3(32)^{1/5} + 2 = 0$

$\left[(32)^{1/5}\right]^2 - 3(2) + 2 = 0$

$\qquad 2^2 - 6 + 2 = 0$

$\qquad 4 - 6 + 2 = 0 \Rightarrow 0 = 0$

This is a true statement. 31 is a solution.

Solution set: $\{0, 31\}$

85. $6(x+2)^4 - 11(x+2)^2 = -4$

$6(x+2)^4 - 11(x+2)^2 + 4 = 0$

Let $u = (x+2)^2$ then $u^2 = (x+2)^4$.

$6u^2 - 11u + 4 = 0 \Rightarrow (3u - 4)(2u - 1) = 0 \Rightarrow$

$u = \frac{4}{3}$ or $u = \frac{1}{2}$

To find x, replace u with $(x+2)^2$.

$(x+2)^2 = \frac{4}{3} \Rightarrow x + 2 = \pm\sqrt{\frac{4}{3}} = \pm\frac{2\sqrt{3}}{3}$ or

$\qquad x = -2 \pm \frac{2\sqrt{3}}{3} = -\frac{6}{3} \pm \frac{2\sqrt{3}}{3} = \frac{-6 \pm 2\sqrt{3}}{3}$

$(x+2)^2 = \frac{1}{2} \Rightarrow x + 2 = \pm\sqrt{\frac{1}{2}} = \pm\frac{\sqrt{2}}{2}$

$\qquad x = -2 \pm \frac{\sqrt{2}}{2} = -\frac{4}{2} \pm \frac{\sqrt{2}}{2} = \frac{-4 \pm \sqrt{2}}{2}$

Solution set: $\left\{\frac{-6 \pm 2\sqrt{3}}{3}, \frac{-4 \pm \sqrt{2}}{2}\right\}$

Copyright © 2013 Pearson Education, Inc.

87. $10x^{-2} + 33x^{-1} - 7 = 0$

Let $u = x^{-1}$; then $u^2 = x^{-2}$.

$10u^2 + 33u - 7 = 0 \Rightarrow (2u + 7)(5u - 1) = 0$

$u = -\frac{7}{2}$ or $u = \frac{1}{5}$

To find x, replace u with x^{-1}.

$x^{-1} = -\frac{7}{2} \Rightarrow x = -\frac{2}{7}$ or $x^{-1} = \frac{1}{5} \Rightarrow x = 5$

Solution set: $\left\{-\frac{2}{7}, 5\right\}$

89. $x^{-2/3} + x^{-1/3} - 6 = 0$

Let $u = x^{-1/3}$; then $u^2 = \left(x^{-1/3}\right)^2 = x^{-2/3}$.

$u^2 + u - 6 = 0 \Rightarrow (u + 3)(u - 2) = 0$

$u = -3$ or $u = 2$

To find x, replace u with $x^{-1/3}$.

$x^{-1/3} = -3 \Rightarrow \left(x^{-1/3}\right)^{-3} = (-3)^{-3} \Rightarrow$ or

$\quad x = \frac{1}{(-3)^3} \Rightarrow x = -\frac{1}{27}$

$x^{-1/3} = 2 \Rightarrow \left(x^{-1/3}\right)^{-3} = 2^{-3} \Rightarrow$

$\quad x = \frac{1}{2^3} \Rightarrow x = \frac{1}{8}$

Check $x = -\frac{1}{27}$.

$x^{-2/3} + x^{-1/3} - 6 = 0$

$\left(-\frac{1}{27}\right)^{-2/3} + \left(-\frac{1}{27}\right)^{-1/3} - 6 \overset{?}{=} 0$

$(-27)^{2/3} + (-27)^{1/3} - 6 = 0$

$\left[(-27)^{1/3}\right]^2 - 3 - 6 = 0$

$(-3)^2 - 3 - 6 = 0$

$9 - 3 - 6 = 0 \Rightarrow 0 = 0$

This is a true statement.

Check $x = \frac{1}{8}$.

$x^{-2/3} + x^{-1/3} - 6 = 0$

$\left(\frac{1}{8}\right)^{-2/3} + \left(\frac{1}{8}\right)^{-1/3} - 6 \overset{?}{=} 0$

$8^{2/3} + 8^{1/3} - 6 = 0$

$\left(8^{1/3}\right)^2 + 2 - 6 = 0$

$2^2 + 2 - 6 = 0$

$4 + 2 - 6 = 0 \Rightarrow 0 = 0$

This is a true statement.

Solution set: $\left\{-\frac{1}{27}, \frac{1}{8}\right\}$

91. $16x^{-4} - 65x^{-2} + 4 = 0$

Let $u = x^{-2}$; then $u^2 = x^{-4}$. Solve the resulting equation by factoring:

$16u^2 - 65u + 4 = 0 \Rightarrow (u - 4)(16u - 1) = 0 \Rightarrow$

$\quad u = 4$ or $u = \frac{1}{16}$

Find x by replacing u with x^{-2}:

$x^{-2} = 4 \Rightarrow x^2 = \frac{1}{4} \Rightarrow x = \pm\frac{1}{2}$

$x^{-2} = \frac{1}{16} \Rightarrow x^2 = 16 \Rightarrow x = \pm 4$

Check $x = \frac{1}{2}$

$16\left(\frac{1}{2}\right)^{-4} - 65\left(\frac{1}{2}\right)^{-2} + 4 = 0$

$16(2)^4 - 65(2)^2 + 4 \overset{?}{=} 0$

$16(16) - 65(4) + 4 = 0$

$256 - 260 + 4 = 0$

$0 = 0$

This is a true statement, so $\frac{1}{2}$ is a solution.

Check $x = -\frac{1}{2}$

$16\left(-\frac{1}{2}\right)^{-4} - 65\left(-\frac{1}{2}\right)^{-2} + 4 = 0$

$16(-2)^4 - 65(-2)^2 + 4 \overset{?}{=} 0$

$16(16) - 65(4) + 4 = 0$

$256 - 260 + 4 = 0$

$0 = 0$

This is a true statement, so $-\frac{1}{2}$ is a solution.

Check $x = 4$

$16(4)^{-4} - 65(4)^{-2} + 4 = 0$

$16\left(\frac{1}{4}\right)^4 - 65\left(\frac{1}{4}\right)^2 + 4 \overset{?}{=} 0$

$16\left(\frac{1}{256}\right) - 65\left(\frac{1}{16}\right) + 4 = 0$

$\frac{1}{16} - \frac{65}{16} + 4 = 0$

$0 = 0$

This is a true statement, so 4 is a solution.

Check $x = -4$

$16(-4)^{-4} - 65(-4)^{-2} + 4 = 0$

$16\left(-\frac{1}{4}\right)^4 - 65\left(-\frac{1}{4}\right)^2 + 4 \overset{?}{=} 0$

$16\left(\frac{1}{256}\right) - 65\left(\frac{1}{16}\right) + 4 = 0$

$\frac{1}{16} - \frac{65}{16} + 4 = 0$

$0 = 0$

This is a true statement, so -4 is a solution.

Solution set: $\left\{\pm\frac{1}{2}, \pm 4\right\}$

Copyright © 2013 Pearson Education, Inc.

93. $x - \sqrt{x} - 12 = 0$

Let $u = \sqrt{x}$; then $u^2 = x$. Solve the resulting equation by factoring.

$u^2 - u - 12 = 0 \Rightarrow (u-4)(u+3) = 0$

$u = 4$ or $u = -3$

To find x, replace u with $\sqrt{x}$.

$\sqrt{x} = 4 \Rightarrow \left(\sqrt{x}\right)^2 = 4^2 \Rightarrow x = 16$ or

$\sqrt{x} = -3 \Rightarrow \left(\sqrt{x}\right)^2 = (-3)^2 \Rightarrow x = 9$

But $\sqrt{9} \neq -3$

So when $u = -3$, there is no solution for x.

Solution set: $\{16\}$

95. Answers will vary.

97. $d = k\sqrt{h}$ for h

$\dfrac{d}{k} = \sqrt{h} \Rightarrow \dfrac{d^2}{k^2} = h$

So, $h = \dfrac{d^2}{k^2}$.

99. $m^{3/4} + n^{3/4} = 1$ for m

$m^{3/4} = 1 - n^{3/4}$

Raise both sides to the $\frac{4}{3}$ power.

$(m^{3/4})^{4/3} = (1 - n^{3/4})^{4/3}$

$m = (1 - n^{3/4})^{4/3}$

101. $\dfrac{E}{e} = \dfrac{R+r}{r}$ for e

$er\left(\dfrac{E}{e}\right) = er\left(\dfrac{R+r}{r}\right)$

Multiply both sides by er.

$Er = eR + er$

$Er = e(R+r) \Rightarrow \dfrac{Er}{R+r} = e$

So, $e = \dfrac{Er}{R+r}$.

Summary Exercises on Solving Equations

1. $4x - 3 = 2x + 3 \Rightarrow 2x - 3 = 3 \Rightarrow$

$2x = 6 \Rightarrow x = 3$

Solution set: $\{3\}$

3. $x(x+6) = 9 \Rightarrow x^2 + 6x = 9 \Rightarrow x^2 + 6x - 9 = 0$

Solve by completing the square.

$x^2 + 6x + 9 = 9 + 9$

Note: $\left[\frac{1}{2} \cdot 6\right]^2 = 3^2 = 9$

$(x+3)^2 = 18 \Rightarrow x + 3 = \pm\sqrt{18} \Rightarrow$

$x + 3 = \pm 3\sqrt{2} \Rightarrow x = -3 \pm 3\sqrt{2}$

Solve by the quadratic formula.

Let $a = 1, b = 6,$ and $c = -9$.

$x = \dfrac{-b \pm \sqrt{b^2 - 4ac}}{2a}$

$= \dfrac{-6 \pm \sqrt{6^2 - 4(1)(-9)}}{2(1)} = \dfrac{-6 \pm \sqrt{36 + 36}}{2}$

$= \dfrac{-6 \pm \sqrt{72}}{2} = \dfrac{-6 \pm 6\sqrt{2}}{2} = -3 \pm 3\sqrt{2}$

Solution set: $\left\{-3 \pm \sqrt{2}\right\}$

5.
$$\sqrt{x+2} + 5 = \sqrt{x+15}$$
$$\left(\sqrt{x+2} + 5\right)^2 = \left(\sqrt{x+15}\right)^2$$
$$(x+2) + 10\sqrt{x+2} + 25 = x + 15 \Rightarrow$$
$$x + 27 + 10\sqrt{x+2} = x + 15 \Rightarrow$$
$$27 + 10\sqrt{x+2} = 15 \Rightarrow$$
$$10\sqrt{x+2} = -12 \Rightarrow 5\sqrt{x+2} = -6$$
$$\left(5\sqrt{x+2}\right)^2 = (-6)^2 \Rightarrow$$
$$25(x+2) = 36 \Rightarrow 25x + 50 = 36$$
$$25x = -14 \Rightarrow x = -\dfrac{14}{25}$$

Check $x = -\frac{14}{25}$.

$\sqrt{x+2} + 5 = \sqrt{x+15}$

$\sqrt{-\frac{14}{25} + 2} + 5 \stackrel{?}{=} \sqrt{-\frac{14}{25} + 15}$

$\sqrt{-\frac{14}{25} + \frac{50}{25}} + 5 = \sqrt{-\frac{14}{25} + \frac{375}{25}}$

$\sqrt{\frac{36}{25}} + 5 = \sqrt{\frac{361}{25}} \Rightarrow \frac{6}{5} + \frac{25}{5} = \frac{19}{5} \Rightarrow \frac{31}{5} = \frac{19}{5}$

This is a false statement. Solution set: $\varnothing$

7. $\dfrac{3x+4}{3} - \dfrac{2x}{x-3} = x$

The least common denominator is $3(x-3)$, which is equal to 0 if $x = 3$. Therefore, 3 cannot possibly be a solution of this equation.

$3(x-3)\left[\dfrac{3x+4}{3} - \dfrac{2x}{x-3}\right] = 3(x-3)(x)$

$(x-3)(3x+4) - 3(2x) = 3x(x-3)$

$3x^2 + 4x - 9x - 12 - 6x = 3x^2 - 9x$

(continued on next page)

Copyright © 2013 Pearson Education, Inc.

(continued)

$$3x^2 - 11x - 12 = 3x^2 - 9x$$
$$-11x - 12 = -9x$$
$$-12 = 2x \Rightarrow -6 = x$$

The restriction $x \neq 3$ does not affect the result. Therefore, the solution set is $\{-6\}$.

9. $5 - \dfrac{2}{x} + \dfrac{1}{x^2} = 0$

The least common denominator is x^2, which is equal to 0 if $x = 0$. Therefore, 0 cannot possibly be a solution of this equation.

$$x^2 \left[5 - \frac{2}{x} + \frac{1}{x^2} \right] = x^2(0) \Rightarrow 5x^2 - 2x + 1 = 0$$

Solve by completing the square.

$$x^2 - \frac{2}{5}x + \frac{1}{5} = 0 \qquad \text{Multiply by } \tfrac{1}{5}.$$

$$x^2 - \frac{2}{5}x + \frac{1}{25} = -\frac{1}{5} + \frac{1}{25}$$

$$\text{Note:} \left[\frac{1}{2} \cdot \left(-\frac{2}{5} \right) \right]^2 = \left(-\frac{1}{5} \right)^2 = \frac{1}{25}$$

$$\left(x - \frac{1}{5} \right)^2 = -\frac{5}{25} + \frac{1}{25} = \frac{-4}{25}$$

$$x - \frac{1}{5} = \pm \sqrt{\frac{-4}{25}}$$

$$x - \frac{1}{5} = \pm \frac{2}{5}i \Rightarrow x = \frac{1}{5} \pm \frac{2}{5}i$$

Solve by the quadratic formula.
Let $a = 5$, $b = -2$, and $c = 1$.

$$x = \frac{-b \pm \sqrt{b^2 - 4ac}}{2a}$$

$$= \frac{-(-2) \pm \sqrt{(-2)^2 - 4(5)(1)}}{2(5)}$$

$$= \frac{2 \pm \sqrt{4 - 20}}{10} = \frac{2 \pm \sqrt{-16}}{10}$$

$$= \frac{2 \pm 4i}{10} = \frac{2}{10} \pm \frac{4}{10}i = \frac{1}{5} \pm \frac{2}{5}i$$

The restriction $x \neq 0$ does not affect the result. Therefore, the solution set is $\left\{ \frac{1}{5} \pm \frac{2}{5}i \right\}$.

11. $x^{-2/5} - 2x^{-1/5} - 15 = 0$

Let $u = x^{-1/5}$; then $u^2 = \left(x^{-1/5} \right)^2 = x^{-2/5}$.

$$u^2 - 2u - 15 = 0 \Rightarrow (u + 3)(u - 5) = 0 \Rightarrow$$
$$u = -3 \text{ or } u = 5$$

To find x, replace u with $x^{-1/5}$.

$$x^{-1/5} = -3 \Rightarrow \left(x^{-1/5} \right)^{-5} = (-3)^{-5}$$

$$x = \frac{1}{(-3)^5} \Rightarrow x = -\frac{1}{243} \qquad \text{or}$$

$$x^{-1/5} = 5 \Rightarrow \left(x^{-1/5} \right)^{-5} = 5^{-5} \Rightarrow x = \frac{1}{5^5} \Rightarrow$$
$$x = \frac{1}{3125}$$

Check $x = -\frac{1}{243}$.

$$x^{-2/5} - 2x^{-1/5} - 15 = 0$$

$$\left(-\frac{1}{243} \right)^{-2/5} - 2 \left(-\frac{1}{243} \right)^{-1/5} - 15 = 0 ?$$

$$(-243)^{2/5} - 2(-243)^{1/5} - 15 = 0$$

$$\left[(-243)^{1/5} \right]^2 - 2(-3) - 15 = 0$$

$$(-3)^2 + 6 - 15 = 0 \Rightarrow 9 + 6 - 15 = 0 \Rightarrow 0 = 0$$

This is a true statement. $-\frac{1}{243}$ is a solution.

Check $x = \frac{1}{3125}$.

$$x^{-2/5} - 2x^{-1/5} - 15 = 0$$

$$\left(\frac{1}{3125} \right)^{-2/5} - 2 \left(\frac{1}{3125} \right)^{-1/5} - 15 \overset{?}{=} 0$$

$$(3125)^{2/5} - 2(3125)^{1/5} - 15 = 0$$

$$\left[(3125)^{1/5} \right]^2 - 2(5) - 15 = 0$$

$$5^2 - 10 - 15 = 0$$

$$25 - 10 - 15 = 0 \Rightarrow 0 = 0$$

This is a true statement. $\frac{1}{3125}$ is a solution.

Solution set: $\left\{ -\frac{1}{243}, \frac{1}{3125} \right\}$

13. $x^4 - 3x^2 - 4 = 0$

Let $u = x^2$; then $u^2 = x^4$.

$$u^2 - 3u - 4 = 0 \Rightarrow (u + 1)(u - 4) = 0 \Rightarrow$$
$$u = -1 \text{ or } u = 4$$

To find x, replace x with x^2.

$$x^2 = -1 \Rightarrow x = \pm\sqrt{-1} = \pm i \text{ or}$$
$$x^2 = 4 \Rightarrow x = \pm\sqrt{4} = \pm 2$$

Solution set: $\{\pm i, \pm 2\}$

15. $\sqrt[6]{2x+1} = \sqrt[6]{9} \Rightarrow \left(\sqrt[6]{2x+1} \right)^6 = \left(\sqrt[6]{9} \right)^6$

$$2x + 1 = 9 \Rightarrow 2x = 8 \Rightarrow x = 4$$

Check $x = 4$.

$$\sqrt[6]{2x+1} = \sqrt[6]{9} \Rightarrow \sqrt[6]{2(4)+1} \overset{?}{=} \sqrt[6]{9}$$

$$\sqrt[6]{8+1} = \sqrt[6]{9} \Rightarrow \sqrt[6]{9} = \sqrt[6]{9}$$

This is a true statement.
Solution set: $\{4\}$

Copyright © 2013 Pearson Education, Inc.

17. $3\left[2x-(6-2x)+1\right]=5x$

$3(2x-6+2x+1)=5x$

$3(4x-5)=5x$

$12x-15=5x \Rightarrow -15=-7x \Rightarrow$

$\dfrac{-15}{-7}=x \Rightarrow x=\dfrac{15}{7}$

Solution set: $\left\{\dfrac{15}{7}\right\}$

19. $(14-2x)^{2/3}=4$

$[(14-2x)^{2/3}]^3=4^3$

$(14-2x)^2=64$

$196-56x+4x^2=64$

$4x^2-56x+132=0$

$4\left(x^2-14x+33\right)=0$

$4(x-3)(x-11)=0 \Rightarrow x=3 \text{ or } x=11$

Check $x=3$.

$(14-2x)^{2/3}=4$

$\left[14-2(3)\right]^{2/3} \stackrel{?}{=} 4$

$(14-6)^{2/3}=4 \Rightarrow 8^{2/3}=4$

$\left(8^{1/3}\right)^2=4 \Rightarrow 2^2=4 \Rightarrow 4=4$

This is a true statement.

Check $x=11$.

$(14-2x)^{2/3}=4$

$\left[14-2(11)\right]^{2/3} \stackrel{?}{=} 4$

$(14-22)^{2/3}=4 \Rightarrow (-8)^{2/3}=4$

$\left[(-8)^{1/3}\right]^2=4 \Rightarrow (-2)^2=4 \Rightarrow 4=4$

This is a true statement.

Solution set: $\{3,11\}$

21. $\dfrac{3}{x-3}=\dfrac{3}{x-3}$

The least common denominator is $(x-3)$ which is equal to 0 if $x-3$. Therefore, 3 cannot possibly be a solution of this equation.

Solution set: $\{x \mid x \neq 3\}$.

Section 1.7   Inequalities

1. $x<-6$

The interval includes all real numbers less than -6 not including -6. The correct interval notation is $(-\infty,-6)$, so the correct choice is F.

3. $-2<x\leq 6$

The interval includes all real numbers from -2 to 6, not including -2, but including 6. The correct interval notation is $(-2, 6]$, so the correct choice is A.

5. $x \geq -6$

The interval includes all real numbers greater than or equal to -6, so it includes -6. The correct interval notation is $[-6,\infty)$, so the correct choice is I.

7. The interval shown on the number line includes all real numbers between -2 and 6, including -2, but not including 6. The correct interval notation is $[-2, 6)$, so the correct choice is B.

9. The interval shown on the number line includes all real numbers less than -3, not including -3, and greater than 3, not including 3. The correct interval notation is $(-\infty,-3)\cup(3,\infty)$, so the correct choice is E.

11. Answers will vary.

13. $-2x+8\leq 16 \Rightarrow -2x+8-8\leq 16-8 \Rightarrow$

$-2x\leq 8 \Rightarrow \dfrac{-2x}{2}\leq\dfrac{8}{2} \Rightarrow x\geq -4$

Solution set: $[-4, \infty)$

15. $-2x-2\leq 1+x$

$-2x-2+2\leq 1+x+2$

$-2x\leq x+3 \Rightarrow -2x-x\leq 3 \Rightarrow$

$-3x\leq 3 \Rightarrow \dfrac{-3x}{-3}\geq\dfrac{3}{-3} \Rightarrow x\geq -1$

Solution set: $[-1,\infty)$

17. $3(x+5)+1\geq 5+3x$

$3x+10+1\geq 5+3x \Rightarrow 11\geq 5$

The inequality is true when x is any real number.

Solution set: $(-\infty,\infty)$

19. $8x-3x+2<2(x+7)$

$5x+2<2x+14$

$5x+2-2x<2x+14-2x$

$3x+2<14 \Rightarrow 3x+2-2<14-2 \Rightarrow$

$3x<12 \Rightarrow \dfrac{3x}{3}<\dfrac{12}{3} \Rightarrow x<4$

Solution set: $(-\infty,4)$

Copyright © 2013 Pearson Education, Inc.

21.
$$\frac{4x+7}{-3} \le 2x+5$$
$$(-3)\left(\frac{4x+7}{-3}\right) \ge (-3)(2x+5)$$
$$4x+7 \ge -6x-15$$
$$4x+7+6x \ge -6x-15+6x$$
$$10x+7 \ge -15$$
$$10x+7-7 \ge -15-7 \Rightarrow 10x \ge -22 \Rightarrow$$
$$\frac{10x}{10} \ge \frac{-22}{10} \Rightarrow x \ge -\frac{11}{5}$$
Solution set: $\left[-\frac{11}{5}, \infty\right)$

23.
$$\frac{1}{3}x + \frac{2}{5}x - \frac{1}{2}(x+3) \le \frac{1}{10}$$
$$30\left[\frac{1}{3}x + \frac{2}{5}x - \frac{1}{2}(x+3)\right] \le 30\left[\frac{1}{10}\right]$$
$$10x+12x-15(x+3) \le 3$$
$$10x+12x-15x-45 \le 3$$
$$7x-45 \le 3$$
$$7x-45+45 \le 3+45$$
$$7x \le 48$$
$$\frac{7x}{7} \le \frac{48}{7} \Rightarrow x \le \frac{48}{7}$$
Solution set: $\left(-\infty, \frac{48}{7}\right]$

25. $C = 50x + 5000; R = 60x$
The product will at least break even when
$R \ge C$. Set $R \ge C$ and solve for x.
$60x \ge 50x + 5000 \Rightarrow 10x \ge 5000 \Rightarrow x \ge 500$
The break-even point is at $x = 500$.
This product will at least break even if the
number of units of picture frames produced is
in interval $[500, \infty)$.

27. $C = 105x + 900; R = 85x$
The product will at least break even when
$R \ge C$. Set $R \ge C$ and solve for x.
$85x \ge 105x + 900 \Rightarrow -20x \ge 900 \Rightarrow x \le -45$
The product will never break even.

29.
$$-5 < 5+2x < 11$$
$$-5-5 < 5+2x-5 < 11-5$$
$$-10 < 2x < 6$$
$$\frac{-10}{2} < \frac{2x}{2} < \frac{6}{2}$$
$$-5 < x < 3$$
Solution set: $(-5, 3)$

31.
$$10 \le 2x+4 \le 16$$
$$10-4 \le 2x+4-4 \le 16-4$$
$$6 \le 2x \le 12$$
$$\frac{6}{2} \le \frac{2x}{2} \le \frac{12}{2}$$
$$3 \le x \le 6$$
Solution set: [3, 6]

33.
$$-11 > -3x+1 > -17$$
$$-11-1 > -3x+1-1 > -17-1$$
$$-12 > -3x > -18$$
$$\frac{-12}{-3} < \frac{-3x}{-3} < \frac{-18}{-3}$$
$$4 < x < 6$$
Solution set: (4, 6)

35.
$$-4 \le \frac{x+1}{2} \le 5$$
$$2(-4) \le 2\left(\frac{x+1}{2}\right) \le 2(5)$$
$$-8 \le x+1 \le 10$$
$$-8-1 \le x+1-1 \le 10-1$$
$$-9 \le x \le 9$$
Solution set: [−9, 9]

37.
$$-3 \le \frac{x-4}{-5} < 4$$
$$(-5)(-3) \ge (-5)\left(\frac{x-4}{-5}\right) > (-5)(4)$$
$$15 \ge x-4 > -20$$
$$15+4 \ge x-4+4 > -20+4$$
$$19 \ge x > -16$$
$$-16 < x \le 19$$
Solution set: (−16, 19]

39. $x^2 - x - 6 > 0$
Step 1: Find the values of x that satisfy
$x^2 - x - 6 = 0$.
$$x^2 - x - 6 = 0 \Rightarrow (x+2)(x-3) = 0$$
$$x+2 = 0 \Rightarrow x = -2 \quad \text{or} \quad x-3 = 0 \Rightarrow x = 3$$
Step 2: The two numbers divide a number line
into three regions.

Interval A $(-\infty, -2)$	Interval B $(-2, 3)$	Interval C $(3, \infty)$

(continued on next page)

Copyright © 2013 Pearson Education, Inc.

(*continued*)

Step 3: Choose a test value to see if it satisfies the inequality, $x^2 - x - 6 > 0$.

Interval	Test Value	Is $x^2 - x - 6 > 0$ True or False?
A: $(-\infty, -2)$	-3	$(-3)^2 - (-3) - 6 \overset{?}{>} 0$ $6 > 0$ True
B: $(-2, 3)$	0	$0^2 - 0 - 6 \overset{?}{>} 0$ $-6 > 0$ False
C: $(3, \infty)$	4	$4^2 - 4 - 6 \overset{?}{>} 0$ $6 > 0$ True

Solution set: $(-\infty, -2) \cup (3, \infty)$

41. $2x^2 - 9x \le 18$

Step 1: Find the values of x that satisfy the corresponding equation.

$$2x^2 - 9x = 18$$
$$2x^2 - 9x - 18 = 0$$
$$(2x + 3)(x - 6) = 0$$

$2x + 3 = 0 \Rightarrow x = -\tfrac{3}{2}$ or $x - 6 = 0 \Rightarrow x = 6$

Step 2: The two numbers divide a number line into three regions.

Interval A $(-\infty, -\tfrac{3}{2})$ Interval B $(-\tfrac{3}{2}, 6)$ Interval C $(6, \infty)$

$-\tfrac{3}{2}$ 0 6

Step 3: Choose a test value to see if it satisfies the inequality, $2x^2 - 9x \le 18$

Interval	Test Value	Is $2x^2 - 9x \le 18$ True or False?
A: $\left(-\infty, -\tfrac{3}{2}\right)$	-2	$2(-2)^2 - 9(-2) \overset{?}{\le} 18$ $26 \le 18$ False
B: $\left(-\tfrac{3}{2}, 6\right)$	0	$2(0)^2 - 9(0) \overset{?}{\le} 18$ $0 \le 18$ True
C: $(6, \infty)$	7	$2(7)^2 - 9(7) \overset{?}{\le} 18$ $35 \le 18$ False

Solution set: $\left[-\tfrac{3}{2}, 6\right]$

43. $-x^2 - 4x - 6 \le -3$

Step 1: Find the values of x that satisfy the corresponding equation.

$$-x^2 - 4x - 6 = -3$$
$$x^2 + 4x + 3 = 0$$
$$(x + 3)(x + 1) = 0$$

$x + 3 = 0 \Rightarrow x = -3$ or $x + 1 = 0 \Rightarrow x = -1$

Step 2: The two numbers divide a number line into three regions.

Interval A $(-\infty, -3)$ Interval B $(-3, -1)$ Interval C $(-1, \infty)$

-3 -1 0

Step 3: Choose a test value to see if it satisfies the inequality, $-x^2 - 4x - 6 \le -3$

Interval	Test Value	Is $-x^2 - 4x - 6 \le -3$ True or False?
A: $(-\infty, -3)$	-4	$-(-4)^2 - 4(-4) - 6 \overset{?}{\le} -3$ $-6 \le -3$ True
B: $(-3, -1)$	-2	$-(-2)^2 - 4(-2) - 6 \overset{?}{\le} -3$ $-2 \le -3$ False
C: $(-1, \infty)$	0	$-(0)^2 - 4(0) - 6 \overset{?}{\le} -3$ $-6 \le -3$ True

Solution set: $(-\infty, -3] \cup [-1, \infty)$

45. $x(x - 1) \le 6 \Rightarrow x^2 - x \le 6 \Rightarrow x^2 - x - 6 \le 0$

Step 1: Find the values of x that satisfy $x^2 - x - 6 = 0$.

$$x^2 - x - 6 = 0$$
$$(x + 2)(x - 3) = 0$$

$x + 2 = 0 \Rightarrow x = -2$ or $x - 3 = 0 \Rightarrow x = 3$

Step 2: The two numbers divide a number line into three regions.

Interval A $(-\infty, -2)$ Interval B $(-2, 3)$ Interval C $(3, \infty)$

-2 0 3

(*continued on next page*)

Copyright © 2013 Pearson Education, Inc.

(continued)

Step 3: Choose a test value to see if it satisfies the inequality, $x(x-1) \le 6$.

Interval	Test Value	Is $x(x-1) \le 6$ True or False?
A: $(-\infty, -2)$	-3	$-3(-3-1) \overset{?}{\le} 6$ $12 \le 6$ False
B: $(-2, 3)$	0	$0(0-1) \overset{?}{\le} 6$ $0 \le 6$ True
C: $(3, \infty)$	4	$4(4-1) \overset{?}{\le} 6$ $12 \le 6$ False

Solution set: $[-2, 3]$

47. $x^2 \le 9$

Step 1: Find the values of x that satisfy $x^2 \le 9$

$$x^2 = 9$$
$$x^2 - 9 = 0$$
$$(x+3)(x-3) = 0$$
$$x+3 = 0 \Rightarrow x = -3 \quad \text{or} \quad x-3 = 0 \Rightarrow x = 3$$

Step 2: The two numbers divide a number line into three regions.

Interval A $(-\infty, -3)$ Interval B $(-3, 3)$ Interval C $(3, \infty)$
$-3 \quad 0 \quad 3$

Step 3: Choose a test value to see if it satisfies the inequality, $x^2 \le 9$.

Interval	Test Value	Is $x^2 \le 9$ True or False?
A: $(-\infty, -3)$	-4	$(-4)^2 \overset{?}{\le} 9$ $16 \le 9$ False
B: $(-3, 3)$	0	$(0)^2 \overset{?}{\le} 9$ $0 \le 9$ True
C: $(3, \infty)$	4	$(4)^2 \overset{?}{\le} 9$ $16 \le 9$ False

Solution set: $[-3, 3]$

49. $x^2 + 5x + 7 < 0$

Step 1: Find the values of x that satisfy $x^2 + 5x + 7 = 0$.
Use the quadratic formula to solve the equation.
Let $a = 1$, $b = 5$, and $c = 7$.

$$x = \frac{-b \pm \sqrt{b^2 - 4ac}}{2a} = \frac{-5 \pm \sqrt{5^2 - 4(1)(7)}}{2(1)}$$
$$= \frac{-5 \pm \sqrt{25 - 28}}{2} = \frac{-5 \pm \sqrt{-3}}{2}$$
$$= \frac{-5 \pm i\sqrt{3}}{2} = \frac{-5}{2} \pm \frac{\sqrt{3}}{2} i$$

Step 2: The number line is one region, $(-\infty, \infty)$.

Interval A $(-\infty, \infty)$
0

Step 3: Since there are no real values of x that satisfy $x^2 + 5x + 7 = 0$, $x^2 + 5x + 7$ is either always positive or always negative. By substituting an arbitrary value such as $x = 0$, we see that $x^2 + 5x + 7$ will be positive and thus the solution set is $\varnothing$.

Interval	Test Value	Is $x^2 + 5x - 2 < 0$ True or False?
A: $(-\infty, \infty)$	0	$0^2 + 5(0) + 7 \overset{?}{<} 0$ $7 < 0$ False

Solution set: $\varnothing$

51. $x^2 - 2x \le 1 \Rightarrow x^2 - 2x - 1 \le 0$

Step 1: Find the values of x that satisfy $x^2 - 2x - 1 = 0$.
Use the quadratic formula to solve the equation.
Let $a = 1$, $b = -2$, and $c = -1$.

$$x = \frac{-b \pm \sqrt{b^2 - 4ac}}{2a}$$
$$= \frac{-(-2) \pm \sqrt{(-2)^2 - 4(1)(-1)}}{2(1)}$$
$$= \frac{2 \pm \sqrt{4+4}}{2} = \frac{2 \pm \sqrt{8}}{2} = \frac{2 \pm 2\sqrt{2}}{2} = 1 \pm \sqrt{2}$$

$1 - \sqrt{2} \approx -0.4$ or $1 + \sqrt{2} \approx 2.4$

(continued on next page)

Copyright © 2013 Pearson Education, Inc.

(*continued*)

Step 2: The two numbers divide a number line into three regions.

Interval A   Interval B   Interval C
$(-\infty,1-\sqrt{2})$ $(1-\sqrt{2},1+\sqrt{2})$ $(1+\sqrt{2},\infty)$

Step 3: Choose a test value to see if it satisfies the inequality, $x^2-2x\le 1$.

Interval	Test Value	Is $x^2-2x\le 1$ True or False?
A: $\left(-\infty,1-\sqrt{2}\right)$	-1	$(-1)^2-2(-1)\overset{?}{\le}1$ $3\le 1$ False
B: $\left(1-\sqrt{2},1+\sqrt{2}\right)$	0	$0^2-2(0)\overset{?}{\le}1\Rightarrow 0\le 1$ True
C: $\left(1+\sqrt{2},\infty\right)$	3	$3^2-2(3)\overset{?}{\le}1\Rightarrow 3\le 1$ False

Solution set: $\left[1-\sqrt{2},1+\sqrt{2}\right]$

53. A; $(x+3)^2$ is equal to zero when $x=-3$. For any other real number, $(x+3)^2$ is positive.

$(x+3)^2\ge 0$ has solution set $(-\infty,\infty)$.

55. $(3x-4)(x+2)(x+6)=0$

Set each factor to zero and solve.

$3x-4=0\Rightarrow x=\frac{4}{3}$ or $x+2=0\Rightarrow x=-2$ or
$x+6=0\Rightarrow x=-6$

Solution set: $\left\{\frac{4}{3},-2,-6\right\}$

57.

Interval	Test Value	Is $(3x-4)(x+2)(x+6)\le 0$ True or False?
A: $(-\infty,-6)$	-10	$[3(-10)-4][-10+2]$ $\cdot[-10+6]\overset{?}{\le}0$ $-1088\le 0$ True
B: $(-6,-2)$	-4	$[3(-4)-4][-4+2]$ $\cdot[-4+6]\overset{?}{\le}0$ $64\le 0$ False
C: $\left(-2,\frac{4}{3}\right)$	0	$[3(0)-4][0+2][0+6]\overset{?}{\le}0$ $-48\le 0$ True

D: $\left(\frac{4}{3},\infty\right)$	4	$[3(4)-4][4+2][4+6]\overset{?}{\le}0$ $480\le 0$ False

59. $(2x-3)(x+2)(x-3)\ge 0$

Step 1: Solve $(2x-3)(x+2)(x-3)=0$.

Set each factor to zero and solve.

$2x-3=0\Rightarrow x=\frac{3}{2}$ or $x+2=0\Rightarrow x=-2$ or
$x-3=0\Rightarrow x=3$

Solution set: $\left\{-2,\frac{3}{2},3\right\}$

Step 2: Plot the solutions $-2,\frac{3}{2}$, and 3 on a number line.

Step 3: Choose a test value to see if it satisfies the inequality, $(2x-3)(x+2)(x-3)\ge 0$.

Interval	Test Value	Is $(2x-3)(x+2)(x-3)\ge 0$ True or False?
A: $(-\infty,-2)$	-3	$[2(-3)-3][-3+2]$ $\cdot[-3-3]\overset{?}{\ge}0$ $-54\ge 0$ False
B: $\left(-2,\frac{3}{2}\right)$	0	$[2(0)-3][0+2]$ $\cdot[0-3]\overset{?}{\ge}0$ $18\ge 0$ True
C: $\left(\frac{3}{2},3\right)$	2	$[2(2)-3][2+2]$ $\cdot[2-3]\overset{?}{\ge}0$ $-4\ge 0$ False
D: $(3,\infty)$	4	$[2(4)-3][4+2]$ $\cdot[4-3]\overset{?}{\ge}0$ $30\ge 0$ True

Solution set: $\left[-2,\frac{3}{2}\right]\cup[3,\infty)$

Copyright © 2013 Pearson Education, Inc.

61. $4x - x^3 \geq 0$

Step 1: Solve $4x - x^3 = 0$.
$$4x - x^3 = 0 \Rightarrow x(4 - x^2) = 0 \Rightarrow$$
$$x(2 + x)(2 - x) = 0$$

Set each factor to zero and solve.
$x = 0$ or $2 + x = 0 \Rightarrow x = -2$ or
$2 - x = 0 \Rightarrow x = 2$
Solution set: $\{-2, 0, 2\}$

Step 2: The values -2, 0, and 2 divide the number line into four intervals.

Step 3: Choose a test value to see if it satisfies the inequality, $4x - x^3 \geq 0$.

Interval	Test Value	Is $4x - x^3 \geq 0$ True or False?
A: $(-\infty, -2)$	-3	$4(-3) - (-3)^3 \overset{?}{\geq} 0$ $15 \geq 0$ True
B: $(-2, 0)$	-1	$4(-1) - (-1)^3 \overset{?}{\geq} 0$ $-3 \geq 0$ False
C: $(0, 2)$	1	$4(1) - 1^3 \overset{?}{\geq} 0$ $3 \geq 0$ True
D: $(2, \infty)$	3	$4(3) - 3^3 \overset{?}{\geq} 0$ $-15 \geq 0$ False

Solution set: $(-\infty, -2] \cup [0, 2]$

63. $(x + 1)^2 (x - 3) < 0$

Step 1: Solve $(x + 1)^2 (x - 3) = 0$.
Set each distinct factor to zero and solve.
$x + 1 = 0 \Rightarrow x = -1$ or $x - 3 = 0 \Rightarrow x = 3$
Solution set: $\{-1, 3\}$

Step 2: The values -1 and 3 divide the number line into three intervals.

Step 3: Choose a test value to see if it satisfies the inequality, $(x + 1)^2 (x - 3) < 0$.

Interval	Test Value	Is $(x + 1)^2 (x - 3) < 0$ True or False?
A: $(-\infty, -1)$	-2	$(-2 + 1)^2 (-2 - 3) \overset{?}{<} 0$ $-5 < 0$ True
B: $(-1, 3)$	0	$(0 + 1)^2 (0 - 3) \overset{?}{<} 0$ $-3 < 0$ True
C: $(3, \infty)$	4	$(4 + 1)^2 (4 - 3) \overset{?}{<} 0$ $25 < 0$ False

Solution set: $(-\infty, -1) \cup (-1, 3)$

65. $x^3 + 4x^2 - 9x \geq 36$

Step 1: Solve $x^3 + 4x^2 - 9x \geq 36$
$$x^3 + 4x^2 - 9x \geq 36$$
$$x^3 + 4x^2 - 9x - 36 = 0$$
$$x^2(x + 4) - 9(x + 4) = 0$$
$$(x + 4)(x^2 - 9) = 0$$
$$(x + 4)(x + 3)(x - 3) = 0$$

Set each factor to zero and solve.
$x + 4 = 0 \Rightarrow x = -4$ or $x + 3 = 0 \Rightarrow x = -3$ or
$x - 3 = 0 \Rightarrow x = 3$
Solution set: $\{-4, -3, 3\}$

Step 2: The values $-4, -3$, and 3 divide the number line into four intervals.

Step 3: Choose a test value to see if it satisfies the inequality, $x^3 + 4x^2 - 9x \geq 36$

Interval	Test Value	Is $x^3 + 4x^2 - 9x \geq 36$ True or False?
A: $(-\infty, -4)$	-5	$(-5)^3 + 4(-5)^2 - 9(-5) \overset{?}{\geq} 36$ $20 \geq 0$ False
B: $(-4, -3)$	-3.5	$(-3.5)^3 + 4(-3.5)^2$ $-9(-3.5) \overset{?}{\geq} 36$ $37.625 \geq 0$ True

(continued on next page)

Copyright © 2013 Pearson Education, Inc.

(continued)

Interval	Test Value	Is $x^3 + 4x^2 - 9x \geq 36$ True or False?
C: $(-3, 3)$	0	$0^3 + 4(0)^2 - 9(0) \overset{?}{\geq} 36$ $0 \geq 0$ False
D: $(3, \infty)$	4	$4^3 + 4(4)^2 - 9(4) \overset{?}{\geq} 36$ $92 \geq 36$ True

Solution set: $[-4, -3] \cup [3, \infty)$

67. $x^2(x+4)^2 \geq 0$

Step 1: Solve $x^2(x+4)^2 \geq 0$.
Set each distinct factor to zero and solve.
$x = 0$ or $x + 4 = 0 \Rightarrow x = -4$
Solution set: $\{-4, 0\}$
Step 2: The values -4 and 0 divide the
number line into three intervals.

Step 3: Choose a test value to see if it satisfies
the inequality, $x^2(x+4)^2 \geq 0$.

Interval	Test Value	Is $x^2(x+4)^2 \geq 0$ True or False?
A: $(-\infty, -4)$	-5	$(-5)^2(-5+4)^2 \overset{?}{\geq} 0$ $25 \geq 0$ True
B: $(-4, 0)$	-1	$(-1)^2(-1+4)^2 \overset{?}{\geq} 0$ $9 \geq 0$ True
C: $(0, \infty)$	1	$1^2(1+4)^2 \overset{?}{\geq} 0$ $25 \geq 0$ True

Solution set: $(-\infty, \infty)$

69. $\dfrac{x-3}{x+5} \leq 0$

Since one side of the inequality is already 0, we
start with Step 2.
Step 2: Determine the values that will cause
either the numerator or denominator to equal 0.
$x - 3 = 0 \Rightarrow x = 3$ or $x + 5 = 0 \Rightarrow x = -5$
The values -5 and 3 divide the number line
into three regions. Use an open circle on -5
because it makes the denominator equal 0.

Step 3: Choose a test value to see if it satisfies
the inequality, $\dfrac{x-3}{x+5} \leq 0$.

Interval	Test Value	Is $\frac{x-3}{x+5} \leq 0$ True or False?
A: $(-\infty, -5)$	-6	$\frac{-6-3}{-6+5} \overset{?}{\leq} 0$ $9 \leq 0$ False
B: $(-5, 3)$	0	$\frac{0-3}{0+5} \overset{?}{\leq} 0$ $-\frac{3}{5} \leq 0$ True
C: $(3, \infty)$	4	$\frac{4-3}{4+5} \overset{?}{\leq} 0$ $\frac{1}{9} \leq 0$ False

Interval B satisfies the inequality. The endpoint
-5 is not included because it makes the
denominator 0.
Solution set: $(-5, 3]$

71. $\dfrac{1-x}{x+2} < -1$

Step 1: Rewrite the inequality so that 0 is on
one side and there is a single fraction on the
other side.

$$\frac{1-x}{x+2} < -1 \Rightarrow \frac{x-1}{x+2} > 1$$
$$\frac{x-1}{x+2} - 1 > 0 \Rightarrow \frac{x-1}{x+2} - \frac{x+2}{x+2} > 0$$
$$\frac{x-1-(x+2)}{x+2} > 0 \Rightarrow \frac{x-1-x-2}{x+2} > 0$$
$$\frac{-3}{x+2} > 0$$

Step 2: Since the numerator is a constant,
determine the values that will cause
denominator to equal 0.
$x + 2 = 0 \Rightarrow x = -2$
The value -2 divides the number line into two
regions.

(continued on next page)

Copyright © 2013 Pearson Education, Inc.

(*continued*)

Step 3: Choose a test value to see if it satisfies

the inequality, $\dfrac{1-x}{x+2} < -1$

Interval	Test Value	Is $\dfrac{1-x}{x+2} < -1$ True or False?
A: $(-\infty, -2)$	-3	$\dfrac{-1-(-3)}{-3+2} \overset{?}{<} -1$ $-2 < -1$ True
B: $(-2, \infty)$	-1	$\dfrac{1-(-1)}{-1+2} \overset{?}{<} 1$ $2 < -1$ False

Solution set: $(-\infty, -2)$

73. $\dfrac{3}{x-6} \le 2$

Step 1: Rewrite the inequality so that 0 is on one side and there is a single fraction on the other side.

$$\dfrac{3}{x-6} - 2 \le 0 \Rightarrow \dfrac{3}{x-6} - \dfrac{2(x-6)}{x-6} \le 0$$

$$\dfrac{3-2(x-6)}{x-6} \le 0 \Rightarrow \dfrac{3-2x+12}{x-6} \le 0$$

$$\dfrac{15-2x}{x-6} \le 0$$

Step 2: Determine the values that will cause either the numerator or denominator to equal 0.

$15 - 2x = 0 \Rightarrow x = \frac{15}{2}$ or $x - 6 = 0 \Rightarrow x = 6$

The values 6 and $\frac{15}{2}$ divide the number line into three regions. Use an open circle on 6 because it makes the denominator equal 0.

Interval A   Interval B   Interval C
$(-\infty, 6)$ $(6, \frac{15}{2})$ $(\frac{15}{2}, \infty)$

6   7 $\frac{15}{2}$

Step 3: Choose a test value to see if it satisfies

the inequality, $\dfrac{3}{x-6} \le 2$.

Interval	Test Value	Is $\dfrac{3}{x-6} \le 2$ True or False?
A: $(-\infty, 6)$	0	$\dfrac{3}{0-6} \overset{?}{\le} 2$ $-\frac{1}{2} \le 2$ True
B: $(6, \frac{15}{2})$	7	$\dfrac{3}{7-6} \overset{?}{\le} 2$ $3 \le 2$ False

C: $\left(\frac{15}{2}, \infty\right)$	8	$\dfrac{3}{8-6} \overset{?}{\le} 2$ $\frac{3}{2} \le 2$ True

Intervals A and C satisfiy the inequality. The endpoint 6 is not included because it makes the denominator 0.

Solution set: $\left(-\infty, 6\right) \cup \left[\frac{15}{2}, \infty\right)$

75. $\dfrac{-4}{1-x} < 5$

Step 1: Rewrite the inequality to compare a single fraction to 0:

$$\dfrac{-4}{1-x} < 5 \Rightarrow \dfrac{-4}{1-x} - 5 < 0$$

$$\dfrac{-4}{1-x} - \dfrac{5(1-x)}{1-x} < 0 \Rightarrow \dfrac{-4-5(1-x)}{1-x} < 0$$

$$\dfrac{-4-5+5x}{1-x} < 0 \Rightarrow \dfrac{-9+5x}{1-x} < 0$$

Step 2: Determine the values that will cause either the numerator or denominator to equal 0.

$-9 + 5x = 0 \Rightarrow x = \frac{9}{5}$ or $1 - x = 0 \Rightarrow x = 1$

The values 1 and $\frac{9}{5}$ divide the number line into three regions.

Interval A   Interval B   Interval C
$(-\infty, 1)$ $(1, \frac{9}{5})$ $(\frac{9}{5}, \infty)$

0   1 $\frac{9}{5}$ 2

Step 3: Choose a test value to see if it satisfies

the inequality, $\dfrac{-4}{1-x} < 5$

Interval	Test Value	Is $\dfrac{-4}{1-x} < 5$ True or False?
A: $(-\infty, 1)$	0	$\dfrac{-4}{1-0} \overset{?}{<} 5$ $-4 < 5$ True
B: $\left(1, \frac{9}{5}\right)$	$\frac{6}{5}$	$\dfrac{-4}{1-\frac{6}{5}} \overset{?}{<} 5$ $20 < 5$ False
C: $\left(\frac{9}{5}, \infty\right)$	2	$\dfrac{-4}{1-2} \overset{?}{<} 5$ $4 < 5$ True

Solution set: $\left(-\infty, 1\right) \cup \left(\frac{9}{5}, \infty\right)$

Copyright © 2013 Pearson Education, Inc.

77. $\dfrac{10}{3+2x} \le 5$

Step 1: Rewrite the inequality so that 0 is on one side and there is a single fraction on the other side.

$$\dfrac{10}{3+2x} - 5 \le 0 \Rightarrow \dfrac{10}{3+2x} - \dfrac{5(3+2x)}{3+2x} \le 0$$

$$\dfrac{10-5(3+2x)}{3+2x} \le 0 \Rightarrow \dfrac{10-15-10x}{3+2x} \le 0$$

$$\dfrac{-10x-5}{3+2x} \le 0$$

Step 2: Determine the values that will cause either the numerator or denominator to equal 0.

$-10x-5 = 0 \Rightarrow x = -\frac{1}{2}$ or $3+2x = 0 \Rightarrow x = -\frac{3}{2}$

The values $-\frac{3}{2}$ and $-\frac{1}{2}$ divide the number line into three regions. Use an open circle on $-\frac{3}{2}$ because it makes the denominator equal 0.

Step 3: Choose a test value to see if it satisfies the inequality, $\dfrac{10}{3+2x} \le 5$.

Interval	Test Value	Is $\frac{10}{3+2x} \le 5$ True or False?
A: $\left(-\infty, -\frac{3}{2}\right)$	-2	$\frac{10}{3+2(-2)} \overset{?}{\le} 5$ $-10 \le 5$ True
B: $\left(-\frac{3}{2}, -\frac{1}{2}\right)$	-1	$\frac{10}{3+2(-1)} \overset{?}{\le} 5$ $10 \le 5$ False
C: $\left(-\frac{1}{2}, \infty\right)$	0	$\frac{10}{3+2(0)} \overset{?}{\le} 5$ $\frac{10}{3} \le 5$ True

Intervals A and C satisfy the inequality. The endpoint $-\frac{3}{2}$ is not included because it makes the denominator 0.

Solution set: $\left(-\infty, -\frac{3}{2}\right) \cup \left[-\frac{1}{2}, \infty\right)$

79. $\dfrac{7}{x+2} \ge \dfrac{1}{x+2}$

Step 1: Rewrite the inequality so that 0 is on one side and there is a single fraction on the other side.

$$\dfrac{7}{x+2} - \dfrac{1}{x+2} \ge 0 \Rightarrow \dfrac{6}{x+2} \ge 0$$

Step 2: Since the numerator is a constant, determine the value that will cause the denominator to equal 0.

$x+2 = 0 \Rightarrow x = -2$

The value -2 divides the number line into two regions. Use an open circle on -2 because it makes the denominator equal 0.

Step 3: Choose a test value to see if it satisfies the inequality, $\dfrac{7}{x+2} \ge \dfrac{1}{x+2}$.

Interval	Test Value	Is $\frac{7}{x+2} \ge \frac{1}{x+2}$ True or False?
A: $(-\infty, -2)$	-3	$\frac{7}{-3+2} \overset{?}{\ge} \frac{1}{-3+2}$ $-7 \ge -1$ False
B: $(-2, \infty)$	0	$\frac{7}{0+2} \overset{?}{\ge} \frac{1}{0+2}$ $\frac{7}{2} \ge \frac{1}{2}$ True

Interval B satisfies the inequality. The endpoint -2 is not included because it makes the denominator 0.

Solution set: $(-2, \infty)$

81. $\dfrac{3}{2x-1} > \dfrac{-4}{x}$

Step 1: Rewrite the inequality so that 0 is on one side and there is a single fraction on the other side.

$$\dfrac{3}{2x-1} + \dfrac{4}{x} > 0$$

$$\dfrac{3x}{x(2x-1)} + \dfrac{4(2x-1)}{x(2x-1)} > 0$$

$$\dfrac{3x+4(2x-1)}{x(2x-1)} > 0$$

$$\dfrac{3x+8x-4}{x(2x-1)} > 0 \Rightarrow \dfrac{11x-4}{x(2x-1)} > 0$$

Step 2: Determine the values that will cause either the numerator or denominator to equal 0.

$11x-4 = 0 \Rightarrow x = \frac{4}{11}$ or $x = 0$ or

$2x-1 = 0 \Rightarrow x = \frac{1}{2}$

The values 0, $\frac{4}{11}$, and $\frac{1}{2}$ divide the number line into four regions.

(continued on next page)

Copyright © 2013 Pearson Education, Inc.

(continued)

Step 3: Choose a test value to see if it satisfies the inequality, $\dfrac{3}{2x-1} > \dfrac{-4}{x}$.

Interval	Test Value	Is $\dfrac{3}{2x-1} > \dfrac{-4}{x}$ True or False?
A: $(-\infty,0)$	-1	$\dfrac{3}{2(-1)-1} \overset{?}{>} \dfrac{-4}{-1}$ $-1 > 4$ False
B: $\left(0,\frac{4}{11}\right)$	$\frac{1}{11}$	$\dfrac{3}{2\left(\frac{1}{11}\right)-1} \overset{?}{>} \dfrac{-4}{\frac{1}{11}}$ or $-\dfrac{11}{3} \overset{?}{>} -44$ $-3\frac{2}{3} > -44$ True
C: $\left(\frac{4}{11},\frac{1}{2}\right)$	$\frac{9}{22}$	$\dfrac{3}{2\left(\frac{9}{22}\right)-1} \overset{?}{>} \dfrac{-4}{\frac{9}{22}}$ or $-\dfrac{33}{2} \overset{?}{>} -\dfrac{88}{9}$ $-16\frac{1}{2} > -9\frac{7}{9}$ False
D: $\left(\frac{1}{2},\infty\right)$	1	$\dfrac{3}{2(1)-1} \overset{?}{>} \dfrac{-4}{1}$ $3 > -4$ True

Solution set: $\left(0,\frac{4}{11}\right) \cup \left(\frac{1}{2},\infty\right)$

83. $\dfrac{4}{2-x} \ge \dfrac{3}{1-x}$

Step 1: Rewrite the inequality so that 0 is on one side and there is a single fraction on the other side.

$$\dfrac{4}{2-x} \ge \dfrac{3}{1-x}$$
$$\dfrac{4}{2-x} - \dfrac{3}{1-x} \ge 0$$
$$\dfrac{4(1-x)}{(2-x)(1-x)} - \dfrac{3(2-x)}{(1-x)(2-x)} \ge 0$$
$$\dfrac{4(1-x) - 3(2-x)}{(x-2)(1-x)} \ge 0$$
$$\dfrac{4 - 4x - 6 + 3x}{(2-x)(1-x)} \ge 0$$
$$\dfrac{-2 - x}{(2-x)(1-x)} \ge 0$$

Step 2: Determine the values that will cause either the numerator or denominator to equal 0.
$-2 - x = 0 \Rightarrow x = -2$ or $2 - x = 0 \Rightarrow x = 2$ or
$1 - x = 0 \Rightarrow x = 1$
The values -2, 1, and 2 divide the number line into four regions. Use an open circle on 1 and 2 because they make the denominator equal 0.

Interval A $(-\infty,-2)$	Interval B $(-2,1)$	Interval C $(1,2)$	Interval D $(2,\infty)$

(number line with points -2, 0, 1, 2)

Step 3: Choose a test value to see if it satisfies the inequality, $\dfrac{4}{2-x} \ge \dfrac{3}{1-x}$

Interval	Test Value	Is $\dfrac{4}{2-x} \ge \dfrac{3}{1-x}$ True or False?
A: $(-\infty,-2)$	-3	$\dfrac{4}{-2-(-3)} \overset{?}{\ge} \dfrac{3}{-1-(-3)}$ $4 \ge \frac{3}{2}$ True
B: $(-2,1)$	0	$\dfrac{4}{2-0} \overset{?}{\ge} \dfrac{3}{1-0}$ $2 \ge 3$ False
C: $(1,2)$	1.5	$\dfrac{4}{2-1.5} \overset{?}{\ge} \dfrac{3}{1-1.5}$ $8 \ge -6$ True
D: $(2,\infty)$	3	$\dfrac{4}{2-3} \overset{?}{\ge} \dfrac{3}{1-3}$ $-4 \ge -\frac{3}{2}$ False

Intervals A and C satisfy the inequality. The endpoints 1 and 2 are not included because they make the denominator 0.
Solution set: $(-\infty,-2] \cup (1,2)$

85. $\dfrac{x+3}{x-5} \le 1$

Step 1: Rewrite the inequality so that 0 is on one side and there is a single fraction on the other side.

$$\dfrac{x+3}{x-5} - 1 \le 0 \Rightarrow \dfrac{x+3}{x-5} - \dfrac{x-5}{x-5} \le 0$$
$$\dfrac{x+3-(x-5)}{x-5} \le 0 \Rightarrow \dfrac{x+3-x+5}{x-5} \le 0$$
$$\dfrac{8}{x-5} \le 0$$

Step 2: Since the numerator is a constant, determine the value that will cause the denominator to equal 0.
$x - 5 = 0 \Rightarrow x = 5$
The value 5 divides the number line into two regions. Use an open circle on 5 because it makes the denominator equal 0.

(continued on next page)

Copyright © 2013 Pearson Education, Inc.

(*continued*)

Step 3: Choose a test value to see if it satisfies the inequality, $\frac{x+3}{x-5} \le 1$.

Interval	Test Value	Is $\frac{x+3}{x-5} \le 1$ True or False?
A: $(-\infty, 5)$	0	$\frac{0+3}{0-5} \overset{?}{\le} 1$ $-\frac{3}{5} \le 1$ True
B: $(5, \infty)$	6	$\frac{6+3}{6-5} \overset{?}{\le} 1$ $9 \le 1$ False

Interval A satisfies the inequality. The endpoint 5 is not included because it makes the denominator 0.

Solution set: $(-\infty, 5)$

87. $\frac{2x-3}{x^2+1} \ge 0$

Since one side of the inequality is already 0, we start with Step 2.

Step 2: Determine the values that will cause either the numerator or denominator to equal 0.

$2x - 3 = 0$ or $x^2 + 1 = 0$

$x = \frac{3}{2}$ has no real solutions

$\frac{3}{2}$ divides the number line into two intervals.

Step 3: Choose a test value to see if it satisfies the inequality, $\frac{2x-3}{x^2+1} \ge 0$.

Interval	Test Value	Is $\frac{2x-3}{x^2+1} \ge 0$ True or False?
A. $(-\infty, \frac{3}{2})$	0	$\frac{2(0)-3}{0^2+1} \overset{?}{\ge} 0$ $-3 \ge 0$ False
B. $(\frac{3}{2}, \infty)$	2	$\frac{2(2)-3}{2^2+1} \overset{?}{\ge} 0$ $\frac{1}{5} \ge 0$ True

Solution set: $[\frac{3}{2}, \infty)$

89. $\frac{(5-3x)^2}{(2x-5)^3} > 0$

Since one side of the inequality is already 0, we start with Step 2.

Step 2: Determine the values that will cause either the numerator or denominator to equal 0.

$5 - 3x = 0 \Rightarrow x = \frac{5}{3}$ or $2x - 5 = 0 \Rightarrow x = \frac{5}{2}$

The values $\frac{5}{3}$ and $\frac{5}{2}$ divide the number line into three intervals.

Step 3: Choose a test value to see if it satisfies the inequality, $\frac{(5-3x)^2}{(2x-5)^3} > 0$.

Interval	Test Value	Is $\frac{(5-3x)^2}{(2x-5)^3} > 0$ True or False?
A: $(-\infty, \frac{5}{3})$	0	$\frac{(5-3\cdot0)^2}{(2\cdot0-5)^3} \overset{?}{>} 0$ $-\frac{1}{5} > 0$ False
B: $(\frac{5}{3}, \frac{5}{2})$	2	$\frac{(5-3\cdot2)^2}{(2\cdot2-5)^3} \overset{?}{>} 0$ $-1 > 0$ False
C: $(\frac{5}{2}, \infty)$	3	$\frac{(5-3\cdot3)^2}{(2\cdot3-5)^3} \overset{?}{>} 0$ $16 > 0$ True

Solution set: $(\frac{5}{2}, \infty)$

91. $\frac{(2x-3)(3x+8)}{(x-6)^3} \ge 0$

Since one side of the inequality is already 0, we start with Step 2.

Step 2: Determine the values that will cause either the numerator or denominator to equal 0.

$2x - 3 = 0$ or $3x + 8 = 0$ or $x - 6 = 0$

$x = \frac{3}{2}$ or $x = -\frac{8}{3}$ or $x = 6$

The values $-\frac{8}{3}$, $\frac{3}{2}$, and 6 divide the number line into four intervals. Use an open circle on 6 because it makes the denominator equal 0.

(continued on next page)

Copyright © 2013 Pearson Education, Inc.

(continued)

Step 3: Choose a test value to see if it satisfies

the inequality, $\dfrac{(2x-3)(3x+8)}{(x-6)^3} \geq 0$.

Interval	Test Value	Is $\dfrac{(2x-3)(3x+8)}{(x-6)^3} \geq 0$ True or False?
A: $\left(-\infty, -\frac{8}{3}\right)$	-3	$\dfrac{[2(-3)-3][3(-3)+8]}{(-3-6)^3} \overset{?}{\geq} 0$ $-\frac{1}{81} \geq 0$ False
B: $\left(-\frac{8}{3}, \frac{3}{2}\right)$	0	$\dfrac{(2\cdot0-3)(3\cdot0+8)}{(0-6)^3} \overset{?}{\geq} 0$ $\frac{1}{9} \geq 0$ True
C: $\left(\frac{3}{2}, 6\right)$	2	$\dfrac{(2\cdot2-3)(3\cdot2+8)}{(2-6)^3} \overset{?}{\geq} 0$ $-\frac{7}{32} \geq 0$ False
D: $(6, \infty)$	7	$\dfrac{(2\cdot7-3)(3\cdot7+8)}{(7-6)^3} \overset{?}{\geq} 0$ $319 \geq 0$ True

Solution set: $\left[-\frac{8}{3}, \frac{3}{2}\right] \cup (6, \infty)$

93. (a) Let $R = 6.5$ and then solve for x.
$$6.5 = 0.3022x + 4.542$$
$$1.958 = 0.3022x \Rightarrow 6.5 \approx x$$
The model predicts that the receipts reached \$6.5 billion about 6.5 years after 1989 which was in 1995.

(b) Let $R = 10$ and then solve for x.
$$10 = 0.3022x + 4.542$$
$$5.458 = 0.3022x \Rightarrow 18.1 \approx x$$
The model predicts that the receipts reach \$10 billion about 18.1 years after 1989 which was in 2007.

95. $-16t^2 + 220t \geq 624$
Step 1: Find the values of x that satisfy
$$-16t^2 + 220t = 624.$$
$$-16t^2 + 220t = 624 \Rightarrow 0 = 16t^2 - 220t + 624$$
$$0 = 4t^2 - 55t + 156$$
$$0 = (t-4)(4t-39)$$

$t - 4 = 0 \Rightarrow t = 4$ or $4t - 39 = 0 \Rightarrow t = \frac{39}{4} = 9.75$

Step 2: The two numbers divide a number line into three regions, where $t \geq 0$.

Step 3: Choose a test value to see if it satisfies the inequality, $-16t^2 + 220t \geq 624$.

Interval	Test Value	Is $-16t^2 + 220t \geq 624$ True or False?
A: $(0,4)$	1	$-16\cdot1^2 + 220\cdot1 \overset{?}{\geq} 624$ $204 \geq 624$ False
B: $(4, 9.75)$	5	$-16\cdot5^2 + 220\cdot5 \overset{?}{\geq} 624$ $700 \geq 624$ True
C: $(9.75, \infty)$	10	$-16\cdot10^2 + 220\cdot10 \overset{?}{\geq} 624$ $600 \geq 624$ False

The projectile will be at least 624 feet above ground between 4 sec and 9.75 sec (inclusive).

97. $2t^2 - 5t - 12 < 0$
Step 1: Find the values of x that satisfy
$$2t^2 - 5t - 12 = 0.$$
$$2t^2 - 5t - 12 = 0 \Rightarrow (2t + 3)(t - 4) = 0$$

$2t + 3 = 0 \Rightarrow t = -\frac{3}{2} = -1.5$ or
$t - 4 = 0 \Rightarrow t = 4$

Step 2: The two numbers divide a number line into three regions.

Step 3: Choose a test value to see if it satisfies the inequality, $2t^2 - 5t - 12 < 0$.

Interval	Test Value	Is $2t^2 - 5t - 12 < 0$ True or False?
A: $(-\infty, -1.5)$	-2	$2(-2)^2 - 5(-2) - 12 \overset{?}{<} 0$ $6 < 0$ False
B: $(-1.5, 4)$	0	$2\cdot0^2 - 5\cdot0 - 12 \overset{?}{<} 0$ $-12 < 0$ True
C: $(4, \infty)$	5	$2\cdot5^2 - 5\cdot5 - 12 \overset{?}{<} 0$ $13 < 0$ False

The velocity will be negative between -1.5 sec and 4 sec.

Copyright © 2013 Pearson Education, Inc.

99. Answers will vary.

The student's answer is not correct. The expression $x + 2$ is a variable expression, not a constant. You lose a critical value if you multiply through by $x + 2$. The correct solution can be found using the methods described in this section.

$$\frac{2x-3}{x+2} \leq 0$$

Since one side of the inequality is already 0, we start with Step 2.

Step 2: Determine the values that will cause either the numerator or denominator to equal 0.

$$2x - 3 = 0 \Rightarrow x = \tfrac{3}{2} \quad \text{or} \quad x + 2 = 0 \Rightarrow x = -2$$

The values -2 and $\tfrac{3}{2}$ divide the number line into three intervals. Use an open circle on -2 because it makes the denominator equal 0.

Step 3: Choose a test value to see if it satisfies the inequality, $\dfrac{2x-3}{x+2} \leq 0$.

Interval	Test Value	Is $\frac{2x-3}{x+2} \leq 0$ True or False?
A: $(-\infty, -2)$	-3	$\frac{2(-3)-3}{-3+2} \overset{?}{\leq} 0$ $7 \leq 0$ False
B: $\left(-2, \frac{3}{2}\right)$	0	$\frac{2 \cdot 0 - 3}{0+2} \overset{?}{\leq} 0$ $-\frac{3}{2} \leq 0$ True
C: $\left(\frac{3}{2}, \infty\right)$	2	$\frac{2 \cdot 2 - 3}{2+2} \overset{?}{\leq} 0$ $\frac{1}{4} \leq 0$ False

Interval B satisfies the inequality. The endpoint -2 is not included because it makes the denominator 0. Solution set: $\left(-2, \frac{3}{2}\right]$

Section 1.8 Absolute Value Equations and Inequalities

1. $|x| = 7$

The solution set includes any value of x whose absolute value is 7; thus $x = 7$ or $x = -7$ are both solutions. The correct graph is F.

3. $|x| > -7$

The solution set is all real numbers, since the absolute value of any real number is always greater than -7. The correct graph is D, which shows the entire number line.

5. $|x| < 7$

The solution set includes any value of x whose absolute value is less than 7; thus x must be between -7 and 7, not including -7 or 7. The correct graph is G.

7. $|x| \leq 7$

The solution set includes any value of x whose absolute value is less than or equal to 7; thus x must be between -7 and 7, including -7 and 7. The correct graph is C.

9. $|3x - 1| = 2$

$3x - 1 = 2 \Rightarrow 3x = 3 \Rightarrow x = 1$ or

$3x - 1 = -2 \Rightarrow 3x = -1 \Rightarrow x = -\frac{1}{3}$

Solution set: $\left\{-\frac{1}{3}, 1\right\}$

11. $|5 - 3x| = 3$

$5 - 3x = 3 \Rightarrow 2 = 3x \Rightarrow \frac{2}{3} = x$ or

$5 - 3x = -3 \Rightarrow 8 = 3x \Rightarrow \frac{8}{3} = x$

Solution set: $\left\{\frac{2}{3}, \frac{8}{3}\right\}$

13. $\left|\dfrac{x-4}{2}\right| = 5$

$\dfrac{x-4}{2} = 5 \Rightarrow x - 4 = 10 \Rightarrow x = 14$ or

$\dfrac{x-4}{2} = -5 \Rightarrow x - 4 = -10 \Rightarrow x = -6$

Solution set: $\{-6, 14\}$

15. $\left|\dfrac{5}{x-3}\right| = 10$

$\dfrac{5}{x-3} = 10 \Rightarrow 5 = 10(x - 3) \Rightarrow 5 = 10x - 30 \Rightarrow$

$35 = 10x \Rightarrow x = \frac{35}{10} = \frac{7}{2}$ or

$\dfrac{5}{x-3} = -10 \Rightarrow 5 = -10(x - 3) \Rightarrow$

$5 = -10x + 30 \Rightarrow -25 = -10x \Rightarrow x = \frac{-25}{-10} = \frac{5}{2}$

Solution set: $\left\{\frac{5}{2}, \frac{7}{2}\right\}$

17. $\left|\dfrac{6x+1}{x-1}\right| = 3$

$\dfrac{6x+1}{x-1} = 3 \Rightarrow 6x+1 = 3(x-1) \Rightarrow$

$6x+1 = 3x-3 \Rightarrow 3x = -4 \Rightarrow x = -\frac{4}{3}$ or

$\dfrac{6x+1}{x-1} = -3 \Rightarrow 6x+1 = -3(x-1) \Rightarrow$

$6x+1 = -3x+3 \Rightarrow 9x = 2 \Rightarrow x = \frac{2}{9}$

Solution set: $\left\{-\frac{4}{3}, \frac{2}{9}\right\}$

19. $|2x-3| = |5x+4|$

$2x-3 = 5x+4 \Rightarrow -7 = 3x \Rightarrow -\frac{7}{3} = x$ or

$2x-3 = -(5x+4) \Rightarrow 2x-3 = -5x-4 \Rightarrow$

$7x = -1 \Rightarrow x = \frac{-1}{7} = -\frac{1}{7}$

Solution set: $\left\{-\frac{7}{3}, -\frac{1}{7}\right\}$

21. $|4-3x| = |2-3x|$

$4-3x = 2-3x \Rightarrow 4 = 2$ False or

$4-3x = -(2-3x) \Rightarrow 4-3x = -2+3x \Rightarrow$

$6 = 6x \Rightarrow 1 = x$

Solution set: $\{1\}$

23. $|5x-2| = |2-5x|$

$5x-2 = 2-5x \Rightarrow 10x = 4 \Rightarrow x = \frac{4}{10} = \frac{2}{5}$ or

$5x-2 = -(2-5x) \Rightarrow 5x-2 = -2+5x \Rightarrow$

$0 = 0$ True

Solution set: $(-\infty, \infty)$

25. Answers will vary.
If x is positive, then $-5x$ will be negative.
Since the outcome of an absolute value can
never be negative, a positive value of x is not
possible.

27. $|2x+5| < 3$

$-3 < 2x+5 < 3$

$-8 < 2x < -2$

$-4 < x < -1$

Solution set: $(-4, -1)$

29. $|2x+5| \geq 3$

$2x+5 \leq -3 \Rightarrow 2x \leq -8 \Rightarrow x \leq -4$ or

$2x+5 \geq 3 \Rightarrow 2x \geq -2 \Rightarrow x \geq -1$

Solution set: $(-\infty, -4] \cup [-1, \infty)$

31. $\left|\frac{1}{2} - x\right| < 2$

$-2 < \frac{1}{2} - x < 2$

$2(-2) < 2\left(\frac{1}{2} - x\right) < 2(2)$

$-4 < 1 - 2x < 4$

$-5 < -2x < 3$

$\frac{5}{2} > x > -\frac{3}{2}$

Solution set: $\left(-\frac{3}{2}, \frac{5}{2}\right)$

33. $4|x-3| > 12 \Rightarrow |x-3| > 3$

$x-3 < -3 \Rightarrow x < 0$ or $x-3 > 3 \Rightarrow x > 6$

Solution set: $(-\infty, 0) \cup (6, \infty)$

35. $|5-3x| > 7$

$5-3x < -7 \Rightarrow -3x < -12 \Rightarrow x > 4$ or

$5-3x > 7 \Rightarrow -3x > 2 \Rightarrow x < -\frac{2}{3}$

Solution set: $\left(-\infty, -\frac{2}{3}\right) \cup (4, \infty)$

37. $|5-3x| \leq 7$

$-7 \leq 5-3x \leq 7$

$-12 \leq -3x \leq 2$

$4 \geq x \geq -\frac{2}{3}$

$-\frac{2}{3} \leq x \leq 4$

Solution set: $\left[-\frac{2}{3}, 4\right]$

39. $\left|\frac{2}{3}x + \frac{1}{2}\right| \leq \frac{1}{6}$

$-\frac{1}{6} \leq \frac{2}{3}x + \frac{1}{2} \leq \frac{1}{6}$

$6\left(-\frac{1}{6}\right) \leq 6\left(\frac{2}{3}x + \frac{1}{2}\right) \leq 6\left(\frac{1}{6}\right)$

$-1 \leq 4x+3 \leq 1$

$-4 \leq 4x \leq -2$

$-1 \leq x \leq -\frac{1}{2}$

Solution set: $\left[-1, -\frac{1}{2}\right]$

41. $|0.01x+1| < 0.01$

$-0.01 < 0.01x+1 < 0.01$

$-1 < x+100 < 1$

$-101 < x < -99$

Solution set: $(-101, -99)$

43. $|4x+3| - 2 = -1 \Rightarrow |4x+3| = 1$

$4x+3 = 1 \Rightarrow 4x = -2 \Rightarrow x = \frac{-2}{4} = -\frac{1}{2}$ or

$4x+3 = -1 \Rightarrow 4x = -4 \Rightarrow x = -1$

Solution set: $\left\{-\frac{1}{2}, -1\right\}$

Copyright © 2013 Pearson Education, Inc.

45. $|6-2x|+1=3 \Rightarrow |6-2x|=2$

$6-2x=2 \Rightarrow -2x=-4 \Rightarrow x=2$ or
$6-2x=-2 \Rightarrow -2x=-8 \Rightarrow x=4$
Solution set: $\{2,4\}$

47. $|3x+1|-1<2 \Rightarrow |3x+1|<3$

$-3<3x+1<3$
$-4<3x<2$
$-\frac{4}{3}<x<\frac{2}{3}$
Solution set: $\left(-\frac{4}{3},\frac{2}{3}\right)$

49. $\left|5x+\frac{1}{2}\right|-2<5 \Rightarrow \left|5x+\frac{1}{2}\right|<7$

$-7<5x+\frac{1}{2}<7$
$2(-7)<2\left(5x+\frac{1}{2}\right)<2(7)$
$-14<10x+1<14$
$-15<10x<13$
$-\frac{15}{10}<x<\frac{13}{10} \Rightarrow -\frac{3}{2}<x<\frac{13}{10}$
Solution set: $\left(-\frac{3}{2},\frac{13}{10}\right)$

51. $|10-4x|+1\ge5 \Rightarrow |10-4x|\ge4$

$10-4x\le-4 \Rightarrow -4x\le-14 \Rightarrow x\ge\frac{-14}{-4} \Rightarrow x\ge\frac{7}{2}$
or
$10-4x\ge4 \Rightarrow -4x\ge-6 \Rightarrow x\le\frac{-6}{-4} \Rightarrow x\le\frac{3}{2}$
Solution set: $\left(-\infty,\frac{3}{2}\right]\cup\left[\frac{7}{2},\infty\right)$

53. $|3x-7|+1<-2 \Rightarrow |3x-7|<-3$

An absolute value cannot be negative.
Solution set: $\varnothing$

55. Since the absolute value of a number is always nonnegative, the inequality $|10-4x|\ge-4$ is always true. The solution set is $(-\infty,\infty)$.

57. There is no number whose absolute value is less than any negative number. The solution set of $|6-3x|<-11$ is $\varnothing$.

59. The absolute value of a number will be 0 if that number is 0. Therefore $|8x+5|=0$ is equivalent to $8x+5=0$, which has solution set $\left\{-\frac{5}{8}\right\}$.

61. Any number less than zero will be negative. There is no number whose absolute value is a negative number. The solution set of $|4.3x+9.8|<0$ is $\varnothing$.

63. Since the absolute value of a number is always nonnegative, $|2x+1|<0$ is never true, so $|2x+1|\le0$ is only true when $|2x+1|=0$.

$|2x+1|=0 \Rightarrow 2x+1=0 \Rightarrow 2x=-1 \Rightarrow x=-\frac{1}{2}$
Solution set: $\left\{-\frac{1}{2}\right\}$

65. $|3x+2|>0$ will be false only when $3x+2=0$, which occurs when $x=-\frac{2}{3}$. So the solution set for $|3x+2|>0$ is $\left(-\infty,-\frac{2}{3}\right)\cup\left(-\frac{2}{3},\infty\right)$.

67. 6 and the opposite of 6, namely -6.

69. $x^2-x=-6 \Rightarrow x^2-x+6=0$

The quadratic formula, $x=\dfrac{-b\pm\sqrt{b^2-4ac}}{2a}$, can be evaluated with $a=1, b=-1,$ and $c=6$.

$x=\dfrac{-b\pm\sqrt{b^2-4ac}}{2a}$
$=\dfrac{-(-1)\pm\sqrt{(-1)^2-4(1)(6)}}{2(1)}=\dfrac{1\pm\sqrt{1-24}}{2}$
$=\dfrac{1\pm\sqrt{-23}}{2}=\dfrac{1\pm i\sqrt{23}}{2}=\dfrac{1}{2}\pm\dfrac{\sqrt{23}}{2}i$
Solution set: $\left\{\frac{1}{2}\pm\frac{\sqrt{23}}{2}i\right\}$

71. $|3x^2+x|=14 \Rightarrow 3x^2+x=14$ or $3x^2+x=-14$

$3x^2+x=14$
$3x^2+x-14=0$
$(3x+7)(x-2)=0$
$3x+7=0 \Rightarrow x=-\frac{7}{3}$
$x-2=0 \Rightarrow x=2$
$3x^2+x=-14$
$3x^2+x+14=0$
We must use the quadratic formula with $a=3,$ $b=1,$ and $c=14$.
$x=\dfrac{-b\pm\sqrt{b^2-4ac}}{2a}$
$x=\dfrac{-1\pm\sqrt{1^2-4\cdot3\cdot14}}{2\cdot3}=\dfrac{-1\pm\sqrt{-167}}{6}$
$=\dfrac{-1\pm i\sqrt{167}}{6}=-\dfrac{1}{6}\pm\dfrac{i\sqrt{167}}{6}$
Solution set: $\left\{-\frac{7}{3}, 2, -\frac{1}{6}\pm\frac{i\sqrt{167}}{6}\right\}$

Copyright © 2013 Pearson Education, Inc.

73. $|4x^2 - 23x - 6| = 0$

Because 0 and the opposite of 0 represent the same value, only one equation needs to be solved.

$4x^2 - 23x - 6 = 0 \Rightarrow (4x+1)(x-6) = 0$

$4x + 1 = 0$ or $x - 6 = 0$

$x = -\frac{1}{4}$ or $x = 6$

Solution set: $\left\{ -\frac{1}{4}, 6 \right\}$

75. $|x^2 + 1| - |2x| = 0$

$|x^2 + 1| - |2x| = 0 \Rightarrow |x^2 + 1| = |2x|$

$x^2 + 1 = 2x \Rightarrow x^2 - 2x + 1 = 0$

$(x-1)^2 = 0 \Rightarrow x - 1 = 0 \Rightarrow x = 1$ or

$x^2 + 1 = -2x \Rightarrow x^2 + 2x + 1 = 0 \Rightarrow (x+1)^2 = 0$

$x + 1 = 0 \Rightarrow x = -1$

Solution set: $\{-1, 1\}$

77. Any number less than zero will be negative. There is no number whose absolute value is a negative number. The solution set of $|x^4 + 2x^2 + 1| < 0$ is $\varnothing$.

79. $\left| \dfrac{x-4}{3x+1} \right| \geq 0$

This inequality will be true, except where $\frac{x-4}{3x+1}$ is undefined. This occurs when $3x + 1 = 0$, or $x = -\frac{1}{3}$.

Solution set: $\left(-\infty, -\frac{1}{3} \right) \cup \left(-\frac{1}{3}, \infty \right)$

81. $|p - q| = 2$, which is equivalent to $|q - p| = 2$, indicates that the distance between p and q is 2 units.

83. "m is no more than 2 units from 7" means that m is 2 units or less from 7. Thus the distance between m and 7 is less than or equal to 2, or $|m - 7| \leq 2$.

85. "p is within 0.0001 unit of 9" means that p is less than 0.0001 unit from 9. Thus the distance between p and 9 is less than 0.0001, or $|p - 9| < 0.0001$.

87. "r is no less than 1 unit from 29" means that r is 1 unit or more from 29. Thus the distance between r and 29 is greater than or equal to 1, or $|r - 29| \geq 1$.

89. Since we want y to be within 0.002 unit of 6, we have $|y - 6| < 0.002$ or $|5x + 1 - 6| < 0.002$.

$|5x - 5| < 0.002$

$-0.002 < 5x - 5 < 0.002$

$4.998 < 5x < 5.002$

$0.9996 < x < 1.0004$

Values of x in the interval $(0.9996, 1.0004)$ would satisfy the condition.

91. $|x - 8.2| \leq 1.5$

$-1.5 \leq x - 8.2 \leq 1.5$

$6.7 \leq x \leq 9.7$

The range of weights, in pounds, is $[6.7, 9.7]$.

93. 780 is 50 more than 730 and 680 is 50 less than 730, so all of the temperatures in the acceptable range are within 50° of 730°. That is $|F - 730| \leq 50$.

95. $|R_L - 26.75| \leq 1.42$

$-1.42 \leq R_L - 26.75 \leq 1.42$

$25.33 \leq R_L \leq 28.17$

$|R_E - 38.75| \leq 2.17$

$-2.17 \leq R_E - 38.75 \leq 2.17$

$36.58 \leq R_E \leq 40.92$

Chapter 1   Review Exercises

1. $2x + 8 = 3x + 2 \Rightarrow 8 = x + 2 \Rightarrow 6 = x$

Solution set: $\{6\}$

3. $5x - 2(x+4) = 3(2x+1)$

$5x - 2x - 8 = 6x + 3 \Rightarrow 3x - 8 = 6x + 3 \Rightarrow$

$-8 = 3x + 3 \Rightarrow -11 = 3x \Rightarrow -\dfrac{11}{3} = x$

Solution set: $\left\{ -\frac{11}{3} \right\}$

5. $A = \dfrac{24f}{B(p+1)}$ for f (approximate annual interest rate)

$B(p+1)A = B(p+1)\left(\dfrac{24f}{B(p+1)} \right)$

$AB(p+1) = 24f$

$\dfrac{AB(p+1)}{24} = f$

$f = \dfrac{AB(p+1)}{24}$

Copyright © 2013 Pearson Education, Inc.

7. A and B cannot be equations used to find the number of pennies in a jar. The number of pennies must be a whole number.

A. $5x+3=11 \Rightarrow 5x=8 \Rightarrow x=\frac{8}{5}$

B. $12x+6=-4 \Rightarrow 12x=-10 \Rightarrow$
$x=-\frac{10}{12}=-\frac{5}{6}$

C. $100x=50(x+3)$
$100x=50x+150 \Rightarrow 50x=150 \Rightarrow x=3$

D. $6(x+4)=x+24$
$6x+24=x+24$
$5x+24=24 \Rightarrow 5x=0 \Rightarrow x=0$

9. Let x = the original length of the square (in inches). Since the perimeter of a square is 4 times the length of one side, we have
$4(x-4)=\frac{1}{2}(4x)+10$. Solve this equation for x to determine the length of each side of the original square.
$4x-16=2x+10$
$2x-16=10 \Rightarrow 2x=26 \Rightarrow x=13$
The original square is 13 in. on each side.

11. Let x = the amount of 100% alcohol solution (in liters).

Strength	Liters of Solution	Liters of Pure Alcohol
100%	x	$1x=x$
10%	12	$0.10 \cdot 12 = 1.2$
30%	$x+12$	$0.30(x+12)$

The number of liters of pure alcohol in the 100% solution plus the number of liters of pure alcohol in the 10% solution must equal the number of liters of pure alcohol in the 30% solution.
$x+1.2=0.30(x+12)$
$x+1.2=0.30x+3.6$
$0.7x+1.2=3.6$
$0.7x=2.4$
$x=\frac{2.4}{.7}=\frac{24}{7}=3\frac{3}{7} \text{ L}$

$3\frac{3}{7}$ L of the 100% solution should be added.

13. Let x = average speed upriver.
Then $x+5$ = average speed on return trip.

	r	t	d
Upriver	x	1.2	$1.2x$
Downriver	$x+5$	0.9	$0.9(x+5)$

Since the distance upriver and downriver are the same, we solve the following.
$1.2x=0.9(x+5)$
$1.2x=0.9x+4.5 \Rightarrow 0.3x=4.5 \Rightarrow x=15$
The average speed of the boat upriver is 15 mph.

15. (a) In one year, the maximum amount of lead ingested would be

$$0.05 \; \frac{\text{mg}}{\text{liter}} \cdot 2 \; \frac{\text{liters}}{\text{day}} \cdot 365.25 \; \frac{\text{days}}{\text{year}}$$

$$= 36.525 \; \frac{\text{mg}}{\text{year}}.$$

The maximum amount A of lead (in milligrams) ingested in x years would be $A=36.525x$.

(b) If $x=72$, then
$A=36.525(72)=2629.8$ mg. The EPA maximum lead intake from water over a lifetime is 2629.8 mg.

17. (a) Using 1956 for $x=0$, then for 1990, $x=34$.
$y=0.1132x+0.4609$
$y=0.1132 \cdot 34+0.4609 \approx 4.31$
According to the model, the minimum wage in 1990 was $4.31. This is $0.51 more than the actual value of $3.80.

(b) Let $y=\$5.85$ and then solve for x.
$y=0.1132x+0.4609$
$5.85=0.1132x+0.4609$
$5.3891=0.1132x$
$47.6 \approx x$
The model predicts the minimum wage to be $5.85 about 47.6 years after 1956, which is mid-2003. This is close to the minimum wage changing to $5.85 in 2007.

19. $(6-i)+(7-2i)=(6+7)+[-1+(-2)]i$
$=13+(-3)i=13-3i$

21. $15i-(3+2i)-11=(-3-11)+(15-2)i$
$=-14+13i$

Copyright © 2013 Pearson Education, Inc.

23. $(5-i)(3+4i) = 5(3)+5(4i)-i(3)-i(4i)$
$= 15+20i-3i-4i^2$
$= 15+17i-4(-1)$
$= 15+17i+4 = 19+17i$

25. $(5-11i)(5+11i) = 5^2-(11i)^2$ Product of the sum and difference of two terms
$= 25-121i^2$
$= 25-121(-1)$
$= 25+121 = 146$

27. $-5i(3-i)^2 = -5i\left[3^2-2(3)(i)+i^2\right]$
$= -5i\left[9-6i+(-1)\right]$
$= -5i(8-6i) = -40i+30i^2$
$= -40i+30(-1) = -40i+(-30)$
$= -30-40i$

29. $\dfrac{-12-i}{-2-5i} = \dfrac{(-12-i)(-2+5i)}{(-2-5i)(-2+5i)}$
$= \dfrac{24-60i+2i-5i^2}{(-2)^2-(5i)^2} = \dfrac{24-58i-5(-1)}{4-25i^2}$
$= \dfrac{24-58i+5}{4-25(-1)} = \dfrac{29-58i}{4+25} = \dfrac{29-58i}{29}$
$= \dfrac{29}{29}-\dfrac{58}{29}i = 1-2i$

31. $i^{11} = i^8 \cdot i^3 = 1 \cdot (-i) = -i$

33. $i^{1001} = i^{1000} \cdot i = \left(i^4\right)^{250} \cdot i = 1^{250} \cdot i = i$

35. $i^{-27} = i^{-28} \cdot i = \left(i^4\right)^{-7} \cdot i = 1^{-7} \cdot i = i$

37. $(x+7)^2 = 5 \Rightarrow x+7 = \pm\sqrt{5} \Rightarrow x = -7\pm\sqrt{5}$
Solution set: $\{-7\pm\sqrt{5}\}$

39. $2x^2+x-15 = 0$
$(x+3)(2x-5) = 0$
$x+3 = 0 \Rightarrow x = -3$ or $2x-5 = 0 \Rightarrow x = \frac{5}{2}$
Solution set: $\left\{-3, \frac{5}{2}\right\}$

41. $-2x^2+11x = -21 \Rightarrow -2x^2+11x+21 = 0$
$2x^2-11x-21 = 0 \Rightarrow (2x+3)(x-7) = 0$
$2x+3 = 0 \Rightarrow x = -\frac{3}{2}$ or $x-7 = 0 \Rightarrow x = 7$
Solution set: $\left\{-\frac{3}{2}, 7\right\}$

43. $(2x+1)(x-4) = x \Rightarrow 2x^2-8x+x-4 = x \Rightarrow$
$2x^2-7x-4 = x \Rightarrow 2x^2-8x-4 = 0 \Rightarrow$
$x^2-4x-2 = 0$
Solve by completing the square.
$x^2-4x-2 = 0$
$x^2-4x+4 = 2+4$
Note: $\left[\frac{1}{2}\cdot(-4)\right]^2 = (-2)^2 = 4$
$(x-2)^2 = 6 \Rightarrow x-2 = \pm\sqrt{6} \Rightarrow x = 2\pm\sqrt{6}$
Solve by the quadratic formula.
Let $a = 1$, $b = -4$, and $c = -2$.
$x = \dfrac{-b\pm\sqrt{b^2-4ac}}{2a}$
$= \dfrac{-(-4)\pm\sqrt{(-4)^2-4(1)(-2)}}{2(1)} = \dfrac{4\pm\sqrt{16+8}}{2}$
$= \dfrac{4\pm\sqrt{24}}{2} = \dfrac{4\pm2\sqrt{6}}{2} = 2\pm\sqrt{6}$
Solution set: $\{2\pm\sqrt{6}\}$

45. $x^2-\sqrt{5}x-1 = 0$
Using the quadratic formula would be the most direct approach.
$a = 1, b = -\sqrt{5}$, and $c = -1$.
$x = \dfrac{-b\pm\sqrt{b^2-4ac}}{2a}$
$= \dfrac{-\left(-\sqrt{5}\right)\pm\sqrt{(-\sqrt{5})^2-4\cdot1\cdot(-1)}}{2\cdot1}$
$= \dfrac{\sqrt{5}\pm\sqrt{5+4}}{2} = \dfrac{\sqrt{5}\pm\sqrt{9}}{2} = \dfrac{\sqrt{5}\pm3}{2}$
Solution set: $\left\{\frac{\sqrt{5}\pm3}{2}\right\}$

47. D; $(7x+4)^2 = 11$
This equation has two real, distinct solutions since the positive number 11 has a positive square root and a negative square root.

49. A; $(3x-4)^2 = -9$
This equation has two imaginary solutions since the negative number –9 has two imaginary square roots.

51. $-6x^2+2x = -3 \Rightarrow -6x^2+2x+3 = 0$
$a = -6, b = 2$, and $c = 3$
$b^2-4ac = 2^2-4(-6)(3) = 4+72 = 76$
The equation has two distinct irrational solutions since the discriminant is positive but not a perfect square.

Copyright © 2013 Pearson Education, Inc.

53. $-8x^2 + 10x = 7 \Rightarrow 0 = 8x^2 - 10x + 7$

$a = 8, b = -10,$ and $c = 7$

$b^2 - 4ac = (-10)^2 - 4(8)(7)$

$\qquad\qquad = 100 - 224 = -124$

The equation has two distinct nonreal complex solutions since the discriminant is negative.

55. $\quad x(9x + 6) = -1 \Rightarrow 9x^2 + 6x = -1 \Rightarrow$

$9x^2 + 6x + 1 = 0$

$a = 9, b = 6,$ and $c = 1$

$b^2 - 4ac = 6^2 - 4(9)(1) = 36 - 36 = 0$

The equation has one rational solution (a double solution) since the discriminant is equal to zero.

57. The projectile will be 750 ft above the ground whenever $220t - 16t^2 = 750$.

Solve this equation for t.

$220t - 16t^2 = 750 \Rightarrow 0 = 16t^2 - 220t + 750 \Rightarrow$

$0 = 8t^2 - 110t + 375 \Rightarrow 0 = (4t - 25)(2t - 15)$

$4t - 25 = 0 \qquad$ or $\qquad 2t - 15 = 0$

$t = \frac{25}{4} = 6.25 \quad$ or $\qquad t = \frac{15}{2} = 7.5$

The projectile will be 750 ft high at 6.25 sec and at 7.5 sec.

59. Let x = width of border.

Apply the formula $A = LW$ to both the outside and inside rectangles.

Inside area = Outside area − Border area

$(12 - 2x)(10 - 2x) = 12 \cdot 10 - 21$

$120 - 24x - 20x + 4x^2 = 120 - 21$

$120 - 44x + 4x^2 = 99$

$4x^2 - 44x + 120 = 99$

$4x^2 - 44x + 21 = 0$

$(2x - 21)(2x - 1) = 0$

$2x = 21 \Rightarrow x = \frac{21}{2} = 10\frac{1}{2} \quad$ or

$2x - 1 = 0 \Rightarrow x = \frac{1}{2}$

The border width cannot be $10\frac{1}{2}$ since this exceeds the width of the outside rectangle, so reject this solution. The width of the border is $\frac{1}{2}$ ft.

61. In 2005, $x = 15$.

$y = 1.717x^2 + 0.8179x + 167.3$

$y = 1.717 \cdot 15^2 + 0.8179 \cdot 15 + 167.3 \approx 565.9$

In 2005, the U.S. government spent approximately \$565.9 billion on medical care.

63. $4x^4 + 3x^2 - 1 = 0$

Let $u = x^2$; then $u^2 = x^4$.

With this substitution, the equation becomes

$4u^2 + 3u - 1 = 0$.

Solve this equation by factoring.

$(u + 1)(4u - 1) = 0$

$u + 1 \Rightarrow u = -1$ or $4u - 1 = 0 \Rightarrow u = \frac{1}{4}$

To find x, replace u with x^2.

$x^2 = -1 \Rightarrow x = \pm\sqrt{-1} \Rightarrow x = \pm i$ or

$x^2 = \frac{1}{4} \Rightarrow x = \pm\sqrt{\frac{1}{4}} \Rightarrow x = \pm\frac{1}{2}$

Solution set: $\left\{\pm i, \pm\frac{1}{2}\right\}$

65. $\dfrac{2}{x} - \dfrac{4}{3x} = 8 + \dfrac{3}{x}$

$3x\left(\dfrac{2}{x} - \dfrac{4}{3x}\right) = 3x\left(8 + \dfrac{3}{x}\right)$

$6 - 4 = 24x + 9$

$2 = 24x + 9$

$-7 = 24x$

$-\dfrac{7}{24} = x$

Solution set: $\left\{-\frac{7}{24}\right\}$

67. $\dfrac{10}{4x - 4} = \dfrac{1}{1 - x} \Rightarrow \dfrac{10}{4(x - 1)} = \dfrac{1}{1 - x} \Rightarrow$

$\dfrac{10}{4(x - 1)} = \dfrac{(-1) \cdot 1}{(-1)(1 - x)} \Rightarrow \dfrac{10}{4(x - 1)} = \dfrac{-1}{x - 1}$

Multiply each term in the equation by the least common denominator, $4(x - 1)$, assuming $x \neq 1$.

$4(x - 1)\left[\dfrac{10}{4(x - 1)}\right] = 4(x - 1)\left(\dfrac{-1}{x - 1}\right)$

$10 = -4 \Rightarrow 14 = 0$

This is a false statement, the solution set is $\varnothing$.

Alternate solution:

$\dfrac{10}{4x - 4} = \dfrac{1}{1 - x} \quad$ or $\quad \dfrac{10}{4(x - 1)} = \dfrac{1}{1 - x}$

Multiply each term in the equation by the least common denominator, $4(x - 1)(1 - x)$, assuming $x \neq 1$.

(continued on next page)

Copyright © 2013 Pearson Education, Inc.

(*continued*)

$$4(x-1)(1-x)\left[\frac{10}{4(x-1)}\right]$$

$$= 4(x-1)(1-x)\left(\frac{1}{1-x}\right)$$

$$10(1-x) = 4(x-1)$$
$$10-10x = 4x-4$$
$$10 = 14x-4$$
$$14 = 14x$$
$$1 = x$$

Because of the restriction $x \neq 1$, the solution set is $\varnothing$.

69. $\dfrac{x}{x+2} + \dfrac{1}{x} + 3 = \dfrac{2}{x^2+2x} \Rightarrow$

$$\dfrac{x}{x+2} + \dfrac{1}{x} + 3 = \dfrac{2}{x(x+2)}$$

Multiply each term in the equation by the least common denominator, $x(x+2)$, assuming $x \neq 0, -2$.

$$x(x+2)\left[\frac{x}{x+2}+\frac{1}{x}+3\right] = x(x+2)\left(\frac{2}{x(x+2)}\right)$$

$$x^2 + (x+2) + 3x(x+2) = 2$$
$$x^2 + x + 2 + 3x^2 + 6x = 2$$
$$4x^2 + 7x + 2 = 2$$
$$4x^2 + 7x = 0 \Rightarrow x(4x+7) = 0$$

$$x = 0 \quad \text{or} \quad 4x+7 = 0 \Rightarrow x = -\tfrac{7}{4}$$

Because of the restriction $x \neq 0$, the only valid solution is $-\tfrac{7}{4}$. The solution set is $\left\{-\tfrac{7}{4}\right\}$.

71. $(2x+3)^{2/3} + (2x+3)^{1/3} - 6 = 0$

Let $u = (2x+3)^{1/3}$. Then

$$u^2 = [(2x+3)^{1/3}]^2 = (2x+3)^{2/3}.$$

With this substitution, the equation becomes $u^2 + u - 6 = 0$. Solve by factoring.

$$(u+3)(u-2) = 0$$

$$u+3 = 0 \Rightarrow u = -3 \quad \text{or} \quad u-2 = 0 \Rightarrow u = 2$$

To find x, replace u with $(2x+3)^{1/3}$.

$$(2x+3)^{1/3} = -3 \Rightarrow \left[(2x+3)^{1/3}\right]^3 = (-3)^3 \Rightarrow$$

$$2x+3 = -27 \Rightarrow 2x = -30 \Rightarrow x = -15$$

or

$$(2x+3)^{1/3} = 2 \Rightarrow \left[(2x+3)^{1/3}\right]^3 = 2^3 \Rightarrow$$

$$2x+3 = 8 \Rightarrow 2x = 5 \Rightarrow x = \tfrac{5}{2}$$

Check $x = -15$.

$$(2x+3)^{2/3} + (2x+3)^{1/3} = 6$$

$$[2(-15)+3]^{2/3} + [2(-15)+3]^{1/3} = 6$$

$$(-30+3)^{2/3} + (-30+3)^{1/3} = 6$$

$$(-27)^{2/3} + (-27)^{1/3} = 6$$

$$\left[(-27)^{1/3}\right]^2 + (-3) = 6$$

$$(-3)^2 - 3 = 6$$

$$9 - 3 = 6 \Rightarrow 6 = 6$$

This is a true statement. -15 is a solution.

Check $x = \tfrac{5}{2}$.

$$(2x+3)^{2/3} + (2x+3)^{1/3} = 6$$

$$\left[2\left(\tfrac{5}{2}\right)+3\right]^{2/3} + \left[2\left(\tfrac{5}{2}\right)+3\right]^{1/3} = 6$$

$$(5+3)^{2/3} + (5+3)^{1/3} = 6$$

$$8^{2/3} + 8^{1/3} = 6$$

$$\left[8^{1/3}\right]^2 + 2 = 6$$

$$(2)^2 + 2 = 6$$

$$4 + 2 = 6 \Rightarrow 6 = 6$$

This is a true statement. $\tfrac{5}{2}$ is a solution.

Solution set: $\left\{-15, \tfrac{5}{2}\right\}$

73. $\sqrt{4x-2} = \sqrt{3x+1}$

$$\left(\sqrt{4x-2}\right)^2 = \left(\sqrt{3x+1}\right)^2$$

$$4x-2 = 3x+1$$
$$x-2 = 1$$
$$x = 3$$

Check $x = 3$.

$$\sqrt{4x-2} = \sqrt{3x+1}$$

$$\sqrt{4(3)-2} = \sqrt{3(3)+1}$$

$$\sqrt{12-2} = \sqrt{9+1}$$

$$\sqrt{10} = \sqrt{10}$$

This is a true statement.
Solution set: $\{3\}$

75. $\sqrt{x+2} - x = 2 \Rightarrow \sqrt{x+2} = 2 + x$

$$\left(\sqrt{x+2}\right)^2 = (2+x)^2 \Rightarrow x+2 = 4 + 4x + x^2$$

$$0 = x^2 + 3x + 2 \Rightarrow 0 = (x+2)(x+1)$$

$$x+2 = 0 \Rightarrow x = -2 \quad \text{or} \quad x+1 = 0 \Rightarrow x = -1$$

Check $x = -2$.

$$\sqrt{x+2} = 2 + x$$

$$\sqrt{-2+2} = 2 + (-2)$$

$$\sqrt{0} = 0 \Rightarrow 0 = 0$$

This is a true statement. -2 is a solution.

(*continued on next page*)

Copyright © 2013 Pearson Education, Inc.

(continued)

Check $x = -1$.

$$\sqrt{x+2} = 2+x$$
$$\sqrt{-1+2} = 2+(-1)?$$
$$\sqrt{1} = 1 \Rightarrow 1 = 1$$

This is a true statement. -1 is a solution.

Solution set: $\{-2, -1\}$

77. $\sqrt{x+3} - \sqrt{3x+10} = 1$

$$\sqrt{x+3} = 1 + \sqrt{3x+10}$$
$$\left(\sqrt{x+3}\right)^2 = \left(1 + \sqrt{3x+10}\right)^2$$
$$x+3 = 1 + 2\sqrt{3x+10} + (3x+10)$$
$$x+3 = 3x+11 + 2\sqrt{3x+10}$$
$$-2x-8 = 2\sqrt{3x+10}$$
$$x+4 = -\sqrt{3x+10}$$
$$(x+4)^2 = \left(-\sqrt{3x+10}\right)^2$$
$$x^2 + 8x + 16 = 3x+10$$
$$x^2 + 5x + 6 = 0 \Rightarrow (x+2)(x+3) = 0$$

$x+3 = 0 \Rightarrow x = -3$ or $x+2 = 0 \Rightarrow x = -2$

Check $x = -3$.

$$\sqrt{x+3} - \sqrt{3x+10} = 1$$
$$\sqrt{-3+3} - \sqrt{3(-3)+10} = 1$$
$$\sqrt{0} - \sqrt{-9+10} = 1$$
$$0 - \sqrt{1} = 1$$
$$0 - 1 = 1 \Rightarrow -1 = 1$$

This is a false statement. -3 is not a solution.

Check $x = -2$.

$$\sqrt{x+3} - \sqrt{3x+10} = 1$$
$$\sqrt{-2+3} - \sqrt{3(-2)+10} = 1$$
$$\sqrt{1} - \sqrt{-6+10} = 1$$
$$1 - \sqrt{4} = 1$$
$$1 - 2 = 1 \Rightarrow -1 = 1$$

This is a false statement. -2 is not a solution. Since neither of the proposed solutions satisfies the original equation, the equation has no solution.

Solution set: $\varnothing$

79. $\sqrt{x^2 + 3x} - 2 = 0$

$$\sqrt{x^2 + 3x} = 2 \Rightarrow \left(\sqrt{x^2 + 3x}\right)^2 = 2^2$$
$$x^2 + 3x = 4 \Rightarrow x^2 + 3x - 4 = 0$$
$$(x-1)(x+4) = 0$$
$$x-1 = 0 \Rightarrow x = 1 \quad \text{or} \quad x+4 = 0 \Rightarrow x = -4$$

Check $x = -4$.

$$\sqrt{x^2 + 3x} - 2 = 0$$
$$\sqrt{(-4)^2 + 3(-4)} - 2 = 0$$
$$\sqrt{16 + (-12)} - 2 = 0$$
$$\sqrt{4} - 2 = 0$$
$$2 - 2 = 0$$
$$0 = 0$$

This is a true statement. -4 is a solution.

Check $x = 1$.

$$\sqrt{x^2 + 3x} - 2 = 0$$
$$\sqrt{1^2 + 3(1)} - 2 = 0$$
$$\sqrt{1+3} - 2 = 0$$
$$\sqrt{4} - 2 = 0$$
$$2 - 2 = 0$$
$$0 = 0$$

This is a true statement. 1 is a solution.

Solution set: $\{-4, 1\}$

81. $\sqrt[3]{6x+2} - \sqrt[3]{4x} = 0$

$$\sqrt[3]{6x+2} = \sqrt[3]{4x}$$
$$\left(\sqrt[3]{6x+2}\right)^3 = \left(\sqrt[3]{4x}\right)^3$$
$$6x+2 = 4x \Rightarrow 2 = -2x \Rightarrow -1 = x$$

Check $x = -1$.

$$\sqrt[3]{6x+2} - \sqrt[3]{4x} = 0$$
$$\sqrt[3]{6(-1)+2} - \sqrt[3]{4(-1)} = 0$$
$$\sqrt[3]{-6+2} - \sqrt[3]{-4} = 0$$
$$\sqrt[3]{-4} - \left(-\sqrt[3]{4}\right) = 0$$
$$-\sqrt[3]{4} + \sqrt[3]{4} = 0 \Rightarrow 0 = 0$$

This is a true statement.

Solution set: $\{-1\}$

83. $-9x + 3 < 4x + 10$

$$-13x < 7$$
$$x > -\tfrac{7}{13}$$

Solution set: $\left(-\tfrac{7}{13}, \infty\right)$

85. $-5x - 4 \geq 3(2x - 5)$

$$-5x - 4 \geq 6x - 15$$
$$-11x - 4 \geq -15$$
$$-11x \geq -11$$
$$x \leq 1$$

Solution set: $(-\infty, 1]$

87. $5 \leq 2x - 3 \leq 7$

$$8 \leq 2x \leq 10$$
$$4 \leq x \leq 5$$

Solution set: $[4, 5]$

Copyright © 2013 Pearson Education, Inc.

89. $x^2 + 3x - 4 \le 0$

Step 1: Find the values of x that satisfy
$x^2 + 3x - 4 = 0$.

$$x^2 + 3x - 4 = 0$$
$$(x+4)(x-1) = 0$$

$x + 4 = 0 \Rightarrow x = -4$ or $x - 1 = 0 \Rightarrow x = 1$

Step 2: The two numbers divide a number line into three regions.

Interval A   Interval B   Interval C
$(-\infty, -4)$ $(-4, 1)$ $(1, \infty)$

-4 0 1

Step 3: Choose a test value to see if it satisfies the inequality, $x^2 + 3x - 4 \le 0$.

Interval	Test Value	Is $x^2 + 3x - 4 \le 0$ True or False?
A: $(-\infty, -4)$	-5	$(-5)^2 + 3(-5) - 4 \overset{?}{\le} 0$ $6 \le 0$ False
B: $(-4, 1)$	0	$0^2 + 3(0) - 4 \overset{?}{\le} 0$ $-4 \le 0$ True
C: $(1, \infty)$	2	$2^2 + 3(2) - 4 \overset{?}{\le} 0$ $6 \le 0$ False

Solution set: $[-4, 1]$

91. $6x^2 - 11x < 10$

Step 1: Find the values of x that satisfy
$6x^2 - 11x = 10$.

$$6x^2 - 11x = 10$$
$$6x^2 - 11x - 10 = 0$$
$$(3x+2)(2x-5) = 0$$

$3x + 2 = 0 \Rightarrow x = -\frac{2}{3}$ or $2x - 5 = 0 \Rightarrow x = \frac{5}{2}$

Step 2: The two numbers divide a number line into three regions.

Interval A   Interval B   Interval C
$(-\infty, -\frac{2}{3})$ $(-\frac{2}{3}, \frac{5}{2})$ $(\frac{5}{2}, \infty)$

$-\frac{2}{3}$ 0 1 2 $\frac{5}{2}$

Step 3: Choose a test value to see if it satisfies the inequality, $6x^2 - 11x < 10$

Interval	Test Value	Is $6x^2 - 11x < 10$ True or False?
A: $(-\infty, -\frac{2}{3})$	-1	$6(-1)^2 - 11(-1) \overset{?}{<} 10$ $17 < 10$ False
B: $(-\frac{2}{3}, \frac{5}{2})$	0	$6 \cdot 0^2 - 11 \cdot 0 - 10 \overset{?}{<} 10$ $-10 < 10$ True
C: $(\frac{5}{2}, \infty)$	3	$6 \cdot 3^2 - 11 \cdot 3 - 10 \overset{?}{<} 10$ $11 < 10$ False

Solution set: $\left(-\frac{2}{3}, \frac{5}{2}\right)$

93. $x^3 - 16x \le 0$

Step 1: Solve $x^3 - 16x = 0$.

$$x^3 - 16x = 0 \Rightarrow x(x^2 - 16) = 0 \Rightarrow$$
$$x(x+4)(x-4) = 0$$

Set each factor to zero and solve.
$x = 0$ or $x + 4 = 0 \Rightarrow x = -4$ or
$x - 4 = 0 \Rightarrow x = 4$

Step 2: The values -4, 0, and 4 divide the number line into four intervals.

Interval A   Interval B   Interval C   Interval D
$(-\infty, -4)$ $(-4, 0)$ $(0, 4)$ $(4, \infty)$

-4 0 4

Step 3: Choose a test value to see if it satisfies the inequality, $x^3 - 16x \le 0$.

Interval	Test Value	Is $x^3 - 16x \le 0$ True or False?
A: $(-\infty, -4)$	-5	$(-5)^3 - 16(-5) \overset{?}{\le} 0$ $-45 \le 0$ True
B: $(-4, 0)$	-1	$(-1)^3 - 16(-1) \overset{?}{\le} 0$ $15 \le 0$ False
C: $(0, 4)$	1	$1^3 - 16 \cdot 1 \overset{?}{\le} 0$ $-15 \le 0$ True
D: $(4, \infty)$	5	$5^3 - 16 \cdot 5 \overset{?}{\le} 0$ $45 \le 0$ False

Solution set: $(-\infty, -4] \cup [0, 4]$

Copyright © 2013 Pearson Education, Inc.

95. $\dfrac{3x+6}{x-5} > 0$

Since one side of the inequality is already 0, we start with Step 2.

Step 2: Determine the values that will cause either the numerator or denominator to equal 0.

$3x+6 = 0 \Rightarrow x = -2$ or $x - 5 = 0 \Rightarrow x = 5$

The values −2 and 5 to divide the number line into three regions.

| Interval A
(−∞,−2) | Interval B
(−2,5) | Interval C
(5,∞) |

Step 3: Choose a test value to see if it satisfies the inequality, $\dfrac{3x+6}{x-5} > 0$.

Interval	Test Value	Is $\frac{3x+6}{x-5} > 0$ True or False?
A: $(-\infty,-2)$	−3	$\frac{3(-3)+6}{-3-5} \overset{?}{>} 0$ $\frac{3}{8} > 0$ True
B: $(-2,5)$	0	$\frac{3(0)+6}{0-5} \overset{?}{>} 0$ $-\frac{6}{5} > 0$ False
C: $(5,\infty)$	6	$\frac{3(6)+6}{6-5} \overset{?}{>} 0$ $24 > 0$ True

Solution set: $(-\infty,-2) \cup (5,\infty)$

97. $\dfrac{3x-2}{x} - 4 > 0$

Step 1: Rewrite the inequality so that 0 is on one side and there is a single fraction on the other side.

$\dfrac{3x-2}{x} - 4 > 0 \Rightarrow \dfrac{3x-2}{x} - \dfrac{4x}{x} > 0 \Rightarrow$

$\dfrac{3x-2-4x}{x} > 0 \Rightarrow \dfrac{-x-2}{x} > 0$

Step 2: Determine the values that will cause either the numerator or denominator to equal 0.

$-x-2 = 0 \Rightarrow x = -2$ or $x = 0$

The values −2 and 0 divide the number line into three regions.

| Interval A
(−∞,−2) | Interval B
(−2, 0) | Interval C
(0,∞) |

Step 3: Choose a test value to see if it satisfies the inequality, $\dfrac{3x-2}{x} - 4 > 0$.

Interval	Test Value	Is $\frac{3x-2}{x} - 4 > 0$ True or False?
A: $(-\infty,-2)$	−3	$\frac{3(-3)-2}{-3} - 4 \overset{?}{>} 0$ $-\frac{1}{3} > 0$ False
B: $(-2,0)$	−1	$\frac{3(-1)-2}{-1} - 4 \overset{?}{>} 0$ $1 > 0$ True
C: $(0,\infty)$	1	$\frac{3\cdot1-2}{1} - 4 \overset{?}{>} 0$ $-3 > 0$ False

Solution set: (−2, 0)

99. $\dfrac{3}{x-1} \le \dfrac{5}{x+3}$

Step 1: Rewrite the inequality so that 0 is on one side and there is a single fraction on the other side.

$$\dfrac{3}{x-1} - \dfrac{5}{x+3} \le 0$$

$$\dfrac{3(x+3)}{(x-1)(x+3)} - \dfrac{5(x-1)}{(x+3)(x-1)} \le 0$$

$$\dfrac{3(x+3) - 5(x-1)}{(x-1)(x+3)} \le 0$$

$$\dfrac{3x+9-5x+5}{(x-1)(x+3)} \le 0$$

$$\dfrac{-2x+14}{(x-1)(x+3)} \le 0$$

Step 2: Determine the values that will cause either the numerator or denominator to equal 0.

$-2x+14 = 0 \Rightarrow x = 7$ or $x-1 = 0 \Rightarrow x = 1$ or $x+3 = 0 \Rightarrow x = -3$

The values −3, 1 and 7 divide the number line into four regions. Use an open circle on −3 and 1 because they make the denominator equal 0.

| Interval A
(−∞,−3) | Interval B
(−3, 1) | Interval C
(1, 7) | Interval D
(7,∞) |

Step 3: Choose a test value to see if it satisfies the inequality, $\dfrac{3}{x-1} \le \dfrac{5}{x+3}$.

Interval	Test Value	Is $\frac{3}{x-1} \le \frac{5}{x+3}$ True or False?
A: $(-\infty,-3)$	−4	$\frac{3}{-4-1} \overset{?}{\le} \frac{5}{-4+3}$ $-\frac{3}{5} \le -5$ False

(continued on next page)

Copyright © 2013 Pearson Education, Inc.

(*continued*)

Interval	Test Value	Is $\frac{3}{x-1} \le \frac{5}{x+3}$ True or False?
B: $(-3,1)$	0	$\frac{3}{0-1} \overset{?}{\le} \frac{5}{0+3}$ $-3 \le \frac{5}{3}$ True
C: $(1,7)$	2	$\frac{3}{2-1} \overset{?}{\le} \frac{5}{2+3}$ $3 \le 1$ False
D: $(7,\infty)$	8	$\frac{3}{8-1} \overset{?}{\le} \frac{5}{8+3}$ $\frac{3}{7} \overset{?}{\le} \frac{5}{11}$ $\frac{33}{77} \le \frac{35}{77}$ True

Intervals B and D satisfy the inequality. The endpoints -3 and 1 are not included because they make the denominator 0.

Solution set: $(-3,1) \cup [7,\infty)$

101. **(a)** Let x = the ozone concentration after the Purafil air filter is used.

$x = 140 - 0.43(140) = 79.8$

The ozone concentration after the Purafil air filter is used is 79.8 ppb.

(b) Let x = the maximum initial concentration of ozone.

$x - 0.43x \le 50 \Rightarrow 0.57x \le 50$

$x \le 87.7$ (approximately)

The filter will reduce ozone concentrations that don't exceed 87.7 ppb.

103. $s = 320 - 16t^2$

(a) When $s = 0$, the projectile will be at ground level.

$0 = 320t - 16t^2 \Rightarrow 16t^2 - 320t = 0 \Rightarrow$
$t^2 - 20t = 0 \Rightarrow t(t - 20) = 0 \Rightarrow$
$t = 0$ or $t = 20$

The projectile will return to the ground after 20 sec.

(b) Solve $s > 576$ for t.

$320t - 16t^2 > 576$
$0 > 16t^2 - 320t + 576$
$0 > t^2 - 20t + 36$

Step 1: Find the values of t that satisfy $t^2 - 20t + 36 = 0$.

$t^2 - 20t + 36 = 0 \Rightarrow (t - 2)(t - 18) = 0 \Rightarrow$
$t - 2 = 0 \Rightarrow t = 2$ or $t - 18 = 0 \Rightarrow t = 18$

Step 2: The two numbers divide a number line into three regions.

| | Interval A (0,2) | Interval B (2,18) | Interval C (18,∞) |

Step 3: Choose a test value to see if it satisfies the inequality,

$320t - 16t^2 > 576$.

Interval	Test Value	Is $320t - 16t^2 > 576$ True or False?
A: $(0,2)$	1	$320(1) - 16(1)^2 \overset{?}{>} 576$ $304 > 576$ False
B: $(2,18)$	3	$320(3) - 16(3)^2 \overset{?}{>} 576$ $816 > 576$ True
C: $(18,\infty)$	20	$320(20) - 16(20)^2 \overset{?}{>} 576$ $0 > 576$ False

The projectile will be more than 576 ft above the ground between 2 and 18 sec.

105. Answers will vary. 3 cannot be in the solution set because when 3 is substituted into $\frac{14x+9}{x-3}$, division by zero occurs.

107. "at least 30" means that the number is 30 or greater; $c \ge 30$.

109. "more than 14,000"; $x > 14,000$

111. $|x + 4| = 7$

$x + 4 = 7 \Rightarrow x = 3$ or $x + 4 = -7 \Rightarrow x = -11$

Solution set: $\{-11, 3\}$

113. $\left|\dfrac{7}{2-3x}\right| - 9 = 0 \Rightarrow \left|\dfrac{7}{2-3x}\right| = 9$

$\dfrac{7}{2-3x} = 9 \Rightarrow 7 = 9(2 - 3x) \Rightarrow 7 = 18 - 27x \Rightarrow$
$-11 = -27x \Rightarrow \dfrac{-11}{-27} = x \Rightarrow x = \dfrac{11}{27}$ or

$\dfrac{7}{2-3x} = -9 \Rightarrow 7 = -9(2 - 3x) \Rightarrow$
$7 = -18 + 27x \Rightarrow 25 = 27x \Rightarrow \dfrac{25}{27} = x \Rightarrow x = \dfrac{25}{27}$

Solution set: $\left\{\dfrac{11}{27}, \dfrac{25}{27}\right\}$

Copyright © 2013 Pearson Education, Inc.

115. $|5x-1|=|2x+3|$

$5x-1=2x+3 \Rightarrow 3x-1=3 \Rightarrow 3x=4 \Rightarrow x=\frac{4}{3}$
or
$5x-1=-(2x+3) \Rightarrow 5x-1=-2x-3 \Rightarrow$
$7x-1=-3 \Rightarrow 7x=-2 \Rightarrow x=-\frac{2}{7}$

Solution set: $\left\{-\frac{2}{7},\frac{4}{3}\right\}$

117. $|2x+9|\le 3$

$-3\le 2x+9\le 3$
$-12\le 2x\le -6$
$-6\le x\le -3$

Solution set: $[-6,-3]$

119. $|7x-3|>4$

$7x-3<-4 \Rightarrow 7x<-1 \Rightarrow x<-\frac{1}{7}$ or
$7x-3>4 \Rightarrow 7x>7 \Rightarrow x>1$

Solution set: $\left(-\infty,-\frac{1}{7}\right)\cup(1,\infty)$

121. $|3x+7|-5=0 \Rightarrow |3x+7|=5$

$3x+7=5 \Rightarrow 3x=-2 \Rightarrow x=-\frac{2}{3}$ or
$3x+7=-5 \Rightarrow 3x=-12 \Rightarrow x=-4$

Solution set: $\left\{-4,-\frac{2}{3}\right\}$

123. Since the absolute value of a number is always nonnegative, the inequality $|4x-12|\ge -3$ is always true. The solution set is $(-\infty,\infty)$.

125. Since the absolute value of a number is always nonnegative, $|x^2+4x|<0$ is never true, so $|x^2+4x|\le 0$ is only true when $|x^2+4x|=0$.

$|x^2+4x|=0 \Rightarrow x^2+4x=0 \Rightarrow x(x+4)=0$
$\qquad x=0 \quad$ or $\quad x+4=0$
$\qquad x-0 \quad$ or $\quad x=-4$

Solution set: $\{-4,0\}$

127. "k is 12 units from 6 on the number line" means that the distance between k and 6 is 12 units, or $|k-6|=12$ or $|6-k|=12$.

129. "t is no less than 0.01 unit from 5" means that t is 0.01 unit or more from 5. Thus, the distance between t and 5 is greater than or equal to 0.01, or $|t-5|\ge 0.01$ or $|5-t|\ge 0.01$

Chapter 1   Test

1. $3(x-4)-5(x+2)=2-(x+24)$
$\qquad 3x-12-5x-10=2-x-24$
$\qquad\qquad -2x-22=-x-22$
$\qquad\qquad\qquad -22=x-22 \Rightarrow 0=x$

Solution set: $\{0\}$

2. $\frac{2}{3}x+\frac{1}{2}(x-4)=x-4$

$6\left[\frac{2}{3}x+\frac{1}{2}(x-4)\right]=6(x-4)$
$\qquad 4x+3(x-4)=6x-24$
$\qquad 4x+3x-12=6x-24$
$\qquad\qquad 7x-12=6x-24$
$\qquad\qquad\quad x-12=-24$
$\qquad\qquad\qquad\quad x=-12$

Solution set: $\{-12\}$

3. $\qquad 6x^2-11x-7=0$
$\qquad (2x+1)(3x-7)=0$

$2x+1=0 \Rightarrow x=-\frac{1}{2}$ or $3x-7=0 \Rightarrow x=\frac{7}{3}$

Solution set: $\left\{-\frac{1}{2},\frac{7}{3}\right\}$

4. $(3x+1)^2=8$
$\qquad 3x+1=\pm\sqrt{8}=\pm 2\sqrt{2}$
$\qquad 3x=-1\pm 2\sqrt{2} \Rightarrow x=\frac{-1\pm 2\sqrt{2}}{3}$

Solution set: $\left\{\frac{-1\pm 2\sqrt{2}}{3}\right\}$

5. $3x^2+2x=-2$

Solve by completing the square.
$\qquad 3x^2+2x=-2$
$\qquad 3x^2+2x+2=0$
$\qquad x^2+\frac{2}{3}x+\frac{2}{3}=0 \Rightarrow x^2+\frac{2}{3}x+\frac{1}{9}=-\frac{2}{3}+\frac{1}{9}$

$\qquad$ Note: $\left[\frac{1}{2}\cdot\frac{2}{3}\right]^2=\left(\frac{1}{3}\right)^2=\frac{1}{9}$

$\left(x+\frac{1}{3}\right)^2=-\frac{5}{9} \Rightarrow x+\frac{1}{3}=\pm\sqrt{-\frac{5}{9}} \Rightarrow$
$\qquad x+\frac{1}{3}=\pm\frac{\sqrt{5}}{3}i \Rightarrow x=-\frac{1}{3}\pm\frac{\sqrt{5}}{3}i$

Solve by the quadratic formula.
Let $a=3$, $b=2$, and $c=2$.

$x=\dfrac{-b\pm\sqrt{b^2-4ac}}{2a}$

$\quad =\dfrac{-2\pm\sqrt{2^2-4(3)(2)}}{2(3)}=\dfrac{-2\pm\sqrt{4-24}}{6}$

(continued on next page)

Copyright © 2013 Pearson Education, Inc.

(*continued*)

$$x = \frac{-2 \pm \sqrt{-20}}{6} = \frac{-2 \pm 2i\sqrt{5}}{6}$$

$$= -\frac{2}{6} \pm \frac{2\sqrt{5}}{6}i = -\frac{1}{3} \pm \frac{\sqrt{5}}{3}i$$

Solution set: $\left\{ -\frac{1}{3} \pm \frac{\sqrt{5}}{3}i \right\}$

6.
$$\frac{12}{x^2 - 9} = \frac{2}{x-3} - \frac{3}{x+3}$$

$$\frac{12}{(x+3)(x-3)} + \frac{3}{x+3} = \frac{2}{x-3}$$

Multiply each term in the equation by the least common denominator, $(x+3)(x-3)$ assuming $x \neq -3, 3$.

$$(x+3)(x-3)\left[\frac{12}{(x+3)(x-3)} + \frac{3}{x+3} \right]$$
$$= (x+3)(x-3)\left(\frac{2}{x-3} \right)$$
$$12 + 3(x-3) = 2(x+3)$$
$$12 + 3x - 9 = 2x + 6$$
$$3x + 3 = 2x + 6$$
$$x + 3 = 6 \Rightarrow x = 3$$

The only possible solution is 3. However, the variable is restricted to real numbers except -3 and 3. Therefore, the solution set is $\varnothing$.

7. $\frac{4x}{x-2} + \frac{3}{x} = \frac{-6}{x^2 - 2x}$ or $\frac{4x}{x-2} + \frac{3}{x} = \frac{-6}{x(x-2)}$

Multiply each term in the equation by the least common denominator, $x(x-2)$, assuming $x \neq 0, 2$.

$$x(x-2)\left[\frac{4x}{x-2} + \frac{3}{x} \right] = x(x-2)\left(\frac{-6}{x(x-2)} \right)$$
$$4x^2 + 3(x-2) = -6 \Rightarrow 4x^2 + 3x - 6 = -6$$
$$4x^2 + 3x = 0 \Rightarrow x(4x+3) = 0$$

$$x = 0 \quad \text{or} \quad 4x + 3 = 0 \Rightarrow x = -\frac{3}{4}$$

Because of the restriction $x \neq 0$, the only valid solution is $-\frac{3}{4}$. The solution set is $\left\{ -\frac{3}{4} \right\}$.

8. $\sqrt{3x+4} + 5 = 2x + 1 \Rightarrow \sqrt{3x+4} = 2x - 4$
$$\left(\sqrt{3x+4} \right)^2 = (2x-4)^2$$
$$3x + 4 = 4x^2 - 16x + 16$$
$$0 = 4x^2 - 19x + 12$$
$$0 = (4x-3)(x-4)$$

$$4x - 3 = 0 \Rightarrow x = \frac{3}{4} \quad \text{or} \quad x - 4 = 0 \Rightarrow x = 4$$

Check $x = \frac{3}{4}$.

$$\sqrt{3x+4} + 4 = 2x$$
$$\sqrt{3\left(\frac{3}{4}\right)+4} + 4 = 2\left(\frac{3}{4}\right)$$
$$\sqrt{\frac{9}{4}+4} + 4 = \frac{3}{2} \Rightarrow \sqrt{\frac{25}{4}} + 4 = \frac{3}{2}$$
$$\frac{5}{2} + 4 = \frac{3}{2} \Rightarrow \frac{13}{2} = \frac{3}{2}$$

This is a false statement. $\frac{3}{4}$ is a not solution.

Check $x = 4$.

$$\sqrt{3x+4} + 4 = 2x$$
$$\sqrt{3(4)+4} + 4 = 2(4)$$
$$\sqrt{12+4} + 4 = 8$$
$$\sqrt{16} + 4 = 8$$
$$4 + 4 = 8 \Rightarrow 8 = 8$$

This is a true statement. 4 is a solution.
Solution set: {4}

9. $\sqrt{-2x+3} + \sqrt{x+3} = 3$
$$\sqrt{-2x+3} = 3 - \sqrt{x+3}$$
$$\left(\sqrt{-2x+3} \right)^2 = \left(3 - \sqrt{x+3} \right)^2$$
$$-2x + 3 = 9 - 6\sqrt{x+3} + (x+3)$$
$$-2x + 3 = 12 + x - 6\sqrt{x+3}$$
$$-3x - 9 = -6\sqrt{x+3}$$
$$x + 3 = 2\sqrt{x+3}$$
$$(x+3)^2 = \left(2\sqrt{x+3} \right)^2$$
$$x^2 + 6x + 9 = 4(x+3)$$
$$x^2 + 6x + 9 = 4x + 12$$
$$x^2 + 2x - 3 = 0 \Rightarrow (x+3)(x-1) = 0$$
$$x + 3 = 0 \Rightarrow x = -3 \quad \text{or} \quad x - 1 = 0 \Rightarrow x = 1$$

Check $x = -3$.

$$\sqrt{-2x+3} + \sqrt{x+3} = 3$$
$$\sqrt{-2(-3)+3} + \sqrt{-3+3} = 3$$
$$\sqrt{6+3} + \sqrt{0} = 3$$
$$\sqrt{9} + 0 = 3$$
$$3 + 0 = 3 \Rightarrow 3 = 3$$

This is a true statement. -3 is a solution.

Check $x = 1$.

$$\sqrt{-2x+3} + \sqrt{x+3} = 3$$
$$\sqrt{-2(1)+3} + \sqrt{1+3} = 3$$
$$\sqrt{-2+3} + \sqrt{4} = 3$$
$$\sqrt{1} + 2 = 3$$
$$1 + 2 = 3 \Rightarrow 3 = 3$$

This is a true statement. 1 is a solution.
Solution set: $\{-3, 1\}$

Copyright © 2013 Pearson Education, Inc.

10. $\sqrt[3]{3x-8} = \sqrt[3]{9x+4}$

$\left(\sqrt[3]{3x-8}\right)^3 = \left(\sqrt[3]{9x+4}\right)^3$

$3x-8 = 9x+4 \Rightarrow -8 = 6x+4 \Rightarrow$

$-12 = 6x \Rightarrow -2 = x$

Check $x = -2$.

$\sqrt[3]{3x-8} = \sqrt[3]{9x+4}$

$\sqrt[3]{3(-2)-8} = \sqrt[3]{9(-2)+4}$

$\sqrt[3]{-6-8} = \sqrt[3]{-18+4}$

$\sqrt[3]{-14} = \sqrt[3]{-14} \Rightarrow -\sqrt[3]{14} = -\sqrt[3]{14}$

This is a true statement.

Solution set: $\{-2\}$

11. $x^4 - 17x^2 + 16 = 0$

Let $u = x^2$; then $u^2 = x^4$.

With this substitution, the equation becomes

$u^2 - 17u + 16 = 0$.

Solve this equation by factoring.

$(u-1)(u-16) = 0$

$u-1 = 0 \Rightarrow u = 1$ or $u-16 = 0 \Rightarrow u = 16$

To find x, replace u with x^2.

$x^2 = 1 \Rightarrow x = \pm\sqrt{1} \Rightarrow x = \pm 1$ or

$x^2 = 16 \Rightarrow x = \pm\sqrt{16} \Rightarrow x = \pm 4$

Solution set: $\{\pm 1, \pm 4\}$

12. $(x+3)^{2/3} + (x+3)^{1/3} - 6 = 0$

Let $u = (x+3)^{1/3}$. Then

$u^2 = \left[(x+3)^{1/3}\right]^2 = (x+3)^{2/3}$.

$u^2 + u - 6 = 0 \Rightarrow (u+3)(u-2) = 0$

$u+3 = 0 \Rightarrow u = -3$ or $u-2 = 0 \Rightarrow u = 2$

To find x, replace u with $(x+3)^{1/3}$.

$(x+3)^{1/3} = -3 \Rightarrow \left[(x+3)^{1/3}\right]^3 = (-3)^3 \Rightarrow$

$x+3 = -27 \Rightarrow x = -30$ or

$(x+3)^{1/3} = 2 \Rightarrow \left[(x+3)^{1/3}\right]^3 = 2^3 \Rightarrow$

$x+3 = 8 \Rightarrow x = 5$

Check $x = -30$.

$(x+3)^{2/3} + (x+3)^{1/3} - 6 = 0$

$(-30+3)^{2/3} + (-30+3)^{1/3} - 6 = 0$

$(-27)^{2/3} + (-27)^{1/3} - 6 = 0$

$\left[(-27)^{1/3}\right]^2 + (-3) - 6 = 0$

$(-3)^2 - 3 - 6 = 0$

$9 - 3 - 6 = 0 \Rightarrow 0 = 0$

This is a true statement. -30 is a solution.

Check $x = 5$.

$(x+3)^{2/3} + (x+3)^{1/3} - 6 = 0$

$(5+3)^{2/3} + (5+3)^{1/3} - 6 = 0$

$8^{2/3} + 8^{1/3} - 6 = 0$

$\left[8^{1/3}\right]^2 + 2 - 6 = 0$

$2^2 + 2 - 6 = 0$

$4 + 2 - 6 = 0 \Rightarrow 0 = 0$

This is a true statement. 5 is a solution.

Solution set: $\{-30, 5\}$

13. $|4x+3| = 7$

$4x+3 = 7 \Rightarrow 4x = 4 \Rightarrow x = 1$ or

$4x+3 = -7 \Rightarrow 4x = -10 \Rightarrow x = -\frac{10}{4} = -\frac{5}{2}$

Solution set: $\left\{-\frac{5}{2}, 1\right\}$

14. $|2x+1| = |5-x|$

$2x+1 = 5-x \Rightarrow 3x+1 = 5 \Rightarrow 3x = 4 \Rightarrow x = \frac{4}{3}$

or

$2x+1 = -(5-x) \Rightarrow 2x+1 = -5+x \Rightarrow x = -6$

Solution set: $\left\{-6, \frac{4}{3}\right\}$

15. $\qquad S = 2HW + 2LW + 2LH$

$S - 2LH = 2HW + 2LW$

$S - 2LH = W(2H + 2L)$

$\dfrac{S - 2LH}{2H + 2L} = W$

$W = \dfrac{S - 2LH}{2H + 2L}$

16. (a) $(9-3i) - (4+5i) = (9-4) + (-3-5)i$

$= 5 - 8i$

(b) $(4+3i)(-5+3i) = -20 + 12i - 15i + 9i^2$

$= -20 - 3i + 9(-1)$

$= -20 - 3i - 9 = -29 - 3i$

Copyright © 2013 Pearson Education, Inc.

(c) $(8+3i)^2 = 8^2 + 2(8)(3i) + (3i)^2$
$= 64 + 48i + 9i^2$
$= 64 + 48i + 9(-1)$
$= 64 + 48i - 9 = 55 + 48i$

(d) $\dfrac{3+19i}{1+3i} = \dfrac{(3+19i)(1-3i)}{(1+3i)(1-3i)}$
$= \dfrac{3 - 9i + 19i - 57i^2}{1 - (3i)^2}$
$= \dfrac{3 + 10i - 57(-1)}{1 - 9i^2} = \dfrac{3 + 10i + 57}{1 - 9(-1)}$
$= \dfrac{60 + 10i}{1+9} = \dfrac{60 + 10i}{10} = 6 + i$

17. (a) $i^{42} = i^{40} \cdot i^2 = (i^4)^{10} \cdot (-1) = 1^{10} \cdot (-1) = -1$

(b) $i^{-31} = i^{-32} \cdot i = (i^4)^{-8} \cdot i = 1^{-8} \cdot i = i$

(c) $\frac{1}{i^{19}} = i^{-19} = i^{-20} \cdot i = (i^4)^{-5} \cdot i = 1^{-5} \cdot i = i$

18. (a) Minimum:
$1120 \dfrac{\text{gal}}{\text{min}} \cdot 60 \dfrac{\text{min}}{\text{hr}} \cdot 12 \dfrac{\text{hr}}{\text{day}} = 806,400 \dfrac{\text{gal}}{\text{day}}$
The equation that will calculate the minimum amount of water pumped after x days would be $A = 806,400x$.

(b) $A = 806,400x$ when $x = 30$ would be $A = 806,400(30) = 24,192,000$ gal.

(c) Since there would be $806,400 \dfrac{\text{gal}}{\text{day}}$ minimum and each pool requires 20,000 gal, there would be a minimum of $\dfrac{806,400}{20,000} = 40.32$ pools that could be filled each day. The equation that will calculate the minimum number of pools that could be filled after x days would be $P = 40.32x$. Approximately 40 pools could be filled each day.

(d) Solve $P = 40.32x$ where $P = 1000$.
$1000 = 40.32x \Rightarrow x = \frac{1000}{40.32} \approx 24.8$ days.

19. Let $w =$ width of rectangle. Then
$2w - 20 =$ length of rectangle.
Use the formula for the perimeter of a rectangle.
$P = 2l + 2w$
$620 = 2(2w - 20) + 2w$
$620 = 4w - 40 + 2w$
$620 = 6w - 40 \Rightarrow 660 = 6w \Rightarrow 110 = w$
The width is 110 m and the length is
$2(110) - 20 = 220 - 20 = 200$ m.

20. Let $x =$ amount of cashews (in pounds). Then $35 - x =$ amount of walnuts (in pounds).

	Cost per Pound	Amount of Nuts	Total Cost
Cashews	7.00	x	$7.00x$
Walnuts	5.50	$35 - x$	$5.50(35 - x)$
Mixture	6.50	35	$35 \cdot 6.50$

Solve the following equation.
$7.00x + 5.50(35 - x) = 35 \cdot 6.50$
$7x + 192.5 - 5.5x = 227.5$
$1.5x + 192.5 = 227.5$
$1.5x = 35$
$x = \frac{35}{1.5} = \frac{350}{15} = \frac{70}{3} = 23\frac{1}{3}$

The fruit and nut stand owner should mix $23\frac{1}{3}$ lbs of cashews with $35 - 23\frac{1}{3} = 11\frac{2}{3}$ lbs of walnuts.

21. Let $x =$ time (in hours) the mother spent driving to meet plane.
Since Mary Lynn has been in the plane for 15 minutes, and 15 minutes is $\frac{1}{4}$ hr, she has been traveling by plane for $x + \frac{1}{4}$ hr.

	d	r	t
Mary Lynn by plane	420		$x + \frac{1}{4}$
Mother by car	20	40	x

The time driven by Mary Lynn's mother can be found by $20 = 40x \Rightarrow x = \frac{1}{2}$ hr. Mary Lynn, therefore, flew for $\frac{1}{2} + \frac{1}{4} = \frac{2}{4} + \frac{1}{4} = \frac{3}{4}$ hr. The rate of Mary Lynn's plane can be found by $r = \dfrac{d}{t} = \dfrac{420}{\frac{3}{4}} = 420 \cdot \frac{4}{3} = 560$ km per hour.

Copyright © 2013 Pearson Education, Inc.

22. $h = -16t^2 + 96t$

 (a) Let $h = 80$ and solve for t.

$$80 = -16t^2 + 96t \Rightarrow 16t^2 - 96t + 80 = 0$$
$$t^2 - 6t + 5 = 0$$
$$(t-1)(t-5) = 0$$
$$t - 1 = 0 \Rightarrow t = 1 \quad \text{or} \quad t - 5 = 0 \Rightarrow t = 5$$

The projectile will reach a height of 80 ft at 1 sec and 5 sec.

 (b) Let $h = 0$ and solve for t.

$$0 = -16t^2 + 96t$$
$$0 = -16t(t - 6)$$
$$t = 0 \quad \text{or} \quad t - 6 = 0 \Rightarrow t = 6$$

The projectile will return to the ground at 6 sec.

23. The table shows each equation evaluated at the years 1995 ($x = 0$), 2005 ($x = 10$), and 2009 ($x = 14$). Evaluating each equation for these x values shows that equation B best models the data.

$$y = -1.983x^2 + 38.91x + 547.8$$
$$y = -1.983 \cdot 0^2 + 38.91 \cdot 0 + 547.8 = 547.8$$
$$y = -1.983 \cdot 10^2 + 38.91 \cdot 10 + 547.8 = 738.6$$
$$y = -1.983 \cdot 14^2 + 38.91 \cdot 14 + 547.8 \approx 703.9$$

24. $-2(x-1) - 12 < 2(x+1)$

$$-2x + 2 - 12 < 2x + 2$$
$$-2x - 10 < 2x + 2$$
$$-4x - 10 < 2$$
$$-4x < 12$$
$$x > -3$$

Solution set: $(-3, \infty)$

25. $-3 \le \dfrac{1}{2}x + 2 \le 3$

$$2(-3) \le 2\left(\dfrac{1}{2}x + 2\right) \le 2(3)$$
$$-6 \le x + 4 \le 6$$
$$-10 \le x \le 2$$

Solution set: $[-10, 2]$

26. $2x^2 - x \ge 3$

Step 1: Find the values of x that satisfy $2x^2 - x = 3$.

$$2x^2 - x = 3$$
$$2x^2 - x - 3 = 0$$
$$(x+1)(2x-3) = 0$$
$$x + 1 = 0 \Rightarrow x = -1 \quad \text{or} \quad 2x - 3 = 0 \Rightarrow x = \tfrac{3}{2}$$

Step 2: The two numbers divide a number line into three regions.

Step 3: Choose a test value to see if it satisfies the inequality, $2x^2 - x \ge 3$

Interval	Test Value	Is $2x^2 - x \ge 3$ True or False?
A: $(-\infty, -1)$	-2	$2(-2)^2 - (-2) \overset{?}{\ge} 3$ $10 \ge 3$ True
B: $\left(-1, \tfrac{3}{2}\right)$	0	$2 \cdot 0^2 - 0 \overset{?}{\ge} 3$ $0 \ge 3$ False
C: $\left(\tfrac{3}{2}, \infty\right)$	2	$2 \cdot 2^2 - 2 \overset{?}{\ge} 3$ $6 \ge 3$ True

Solution set: $(-\infty, -1] \cup \left[\tfrac{3}{2}, \infty\right)$

27. $\dfrac{x+1}{x-3} < 5$

Step 1: Rewrite the inequality so that 0 is on one side and there is a single fraction on the other side.

$$\dfrac{x+1}{x-3} < 5 \Rightarrow \dfrac{x+1}{x-3} - 5 < 0$$
$$\dfrac{x+1}{x-3} - \dfrac{5(x-3)}{x-3} < 0 \Rightarrow \dfrac{x+1 - 5(x-3)}{x-3} < 0$$
$$\dfrac{x+1 - 5x + 15}{x-3} < 0 \Rightarrow \dfrac{-4x + 16}{x-3} < 0$$

Step 2: Determine the values that will cause either the numerator or denominator to equal 0.

$$-4x + 16 = 0 \Rightarrow x = 4 \quad \text{or} \quad x - 3 = 0 \Rightarrow x = 3$$

The values 3 and 4 divide the number line into three regions.

(*continued on next page*)

Copyright © 2013 Pearson Education, Inc.

(*continued*)

Step 3: Choose a test value to see if it satisfies the inequality, $\frac{x+1}{x-3} < 5$.

Interval	Test Value	Is $\frac{x+1}{x-3} < 5$ True or False?
A: $(-\infty, 3)$	0	$\frac{0+1}{0-3} \overset{?}{<} 5$ $-\frac{1}{3} < 5$ True
B: $(3, 4)$	3.5	$\frac{3.5+1}{3.5-3} \overset{?}{<} 5$ $9 < 5$ False
C: $(4, \infty)$	5	$\frac{5+1}{5-3} \overset{?}{<} 5$ $3 < 5$ True

Solution set: $(-\infty, 3) \cup (4, \infty)$

28. $|2x-5| < 9$
$-9 < 2x-5 < 9$
$-4 < 2x < 14$
$-2 < x < 7$
Solution set: $(-2, 7)$

29. $|2x+1| - 11 \geq 0 \Rightarrow |2x+1| \geq 11$
$2x+1 \leq -11$ or $2x+1 \geq 11$
$2x \leq -12$ $2x \geq 10$
$x \leq -6$ or $x \geq 5$
Solution set: $(-\infty, -6] \cup [5, \infty)$

30. $|3x+7| \leq 0 \Rightarrow 3x+7 \leq 0 \Rightarrow x \leq -\frac{7}{3}$

However, if $x < -\frac{7}{3}$, the expression inside the absolute value bars is negative, so x cannot be less than $-\frac{7}{3}$. The solution set of $|3x+7| \leq 0$ is $\left\{-\frac{7}{3}\right\}$.

Copyright © 2013 Pearson Education, Inc.

Chapter 2

GRAPHS AND FUNCTIONS

Section 2.1 Rectangular Coordinates and Graphs

1. False. $(-1, 3)$ lies in Quadrant II.

3. True. The origin has coordinates $(0,0)$. So, the distance from $(0,0)$ to (a,b) is
$$d = \sqrt{(a-0)^2 + (b-0)^2} = \sqrt{a^2 + b^2}$$

5. True. When $x = 0$, $y = 2(0) + 4 = 4$, so the y-intercept is 4. When $y = 0$, $0 = 2x + 4 \Rightarrow x = -2$, so the x-intercept is -2.

7. Any three of the following:
$(2,-5),(-1,7),(3,-9),(5,-17),(6,-21)$

9. Any three of the following: (1997, 36), (1999, 35), (2001, 29), (2003, 22), (2005, 23), (2007, 20)

11. $P(-5, -7), Q(-13, 1)$

   (a) $d(P, Q) = \sqrt{[-13-(-5)]^2 + [1-(-7)]^2}$
   $= \sqrt{(-8)^2 + 8^2} = \sqrt{128} = 8\sqrt{2}$

   (b) The midpoint M of the segment joining points P and Q has coordinates
   $$\left(\frac{-5+(-13)}{2}, \frac{-7+1}{2}\right) = \left(\frac{-18}{2}, \frac{-6}{2}\right)$$
   $$= (-9, -3).$$

13. $P(8, 2), Q(3, 5)$

   (a) $d(P, Q) = \sqrt{(3-8)^2 + (5-2)^2}$
   $= \sqrt{(-5)^2 + 3^2}$
   $= \sqrt{25+9} = \sqrt{34}$

   (b) The midpoint M of the segment joining points P and Q has coordinates
   $$\left(\frac{8+3}{2}, \frac{2+5}{2}\right) = \left(\frac{11}{2}, \frac{7}{2}\right).$$

15. $P(-6, -5), Q(6, 10)$

   (a) $d(P, Q) = \sqrt{[6-(-6)]^2 + [10-(-5)]^2}$
   $= \sqrt{12^2 + 15^2} = \sqrt{144 + 225}$
   $= \sqrt{369} = 3\sqrt{41}$

   (b) The midpoint M of the segment joining points P and Q has coordinates
   $$\left(\frac{-6+6}{2}, \frac{-5+10}{2}\right) = \left(\frac{0}{2}, \frac{5}{2}\right) = \left(0, \frac{5}{2}\right).$$

17. $P\left(3\sqrt{2}, 4\sqrt{5}\right), \ Q\left(\sqrt{2}, -\sqrt{5}\right)$

   (a) $d(P, Q)$
   $= \sqrt{\left(\sqrt{2} - 3\sqrt{2}\right)^2 + \left(-\sqrt{5} - 4\sqrt{5}\right)^2}$
   $= \sqrt{\left(-2\sqrt{2}\right)^2 + \left(-5\sqrt{5}\right)^2}$
   $= \sqrt{8 + 125} = \sqrt{133}$

   (b) The midpoint M of the segment joining points P and Q has coordinates
   $$\left(\frac{3\sqrt{2} + \sqrt{2}}{2}, \frac{4\sqrt{5} + (-\sqrt{5})}{2}\right)$$
   $$= \left(\frac{4\sqrt{2}}{2}, \frac{3\sqrt{5}}{2}\right) = \left(2\sqrt{2}, \frac{3\sqrt{5}}{2}\right).$$

19. Label the points $A(-6, -4)$, $B(0, -2)$, and $C(-10, 8)$. Use the distance formula to find the length of each side of the triangle.
   $$d(A, B) = \sqrt{[0-(-6)]^2 + [-2-(-4)]^2}$$
   $$= \sqrt{6^2 + 2^2} = \sqrt{36+4} = \sqrt{40}$$
   $$d(B, C) = \sqrt{(-10-0)^2 + [8-(-2)]^2}$$
   $$= \sqrt{(-10)^2 + 10^2} = \sqrt{100+100}$$
   $$= \sqrt{200}$$
   $$d(A, C) = \sqrt{[-10-(-6)]^2 + [8-(-4)]^2}$$
   $$= \sqrt{(-4)^2 + 12^2} = \sqrt{16+144} = \sqrt{160}$$

   Since $\left(\sqrt{40}\right)^2 + \left(\sqrt{160}\right)^2 = \left(\sqrt{200}\right)^2$, triangle ABC is a right triangle.

21. Label the points $A(-4, 1)$, $B(1, 4)$, and $C(-6, -1)$.
   $$d(A, B) = \sqrt{[1-(-4)]^2 + (4-1)^2}$$
   $$= \sqrt{5^2 + 3^2} = \sqrt{25+9} = \sqrt{34}$$
   $$d(B, C) = \sqrt{(-6-1)^2 + (-1-4)^2}$$
   $$= \sqrt{(-7)^2 + (-5)^2} = \sqrt{49+25} = \sqrt{74}$$

(continued on next page)

Copyright © 2013 Pearson Education, Inc.

(continued)

$$d(A, C) = \sqrt{[-6-(-4)]^2 + (-1-1)^2}$$

$$= \sqrt{(-2)^2 + (-2)^2} = \sqrt{4+4} = \sqrt{8}$$

Since $(\sqrt{8})^2 + (\sqrt{34})^2 \neq (\sqrt{74})^2$ because $8 + 34 = 42 \neq 74$, triangle ABC is not a right triangle.

23. Label the points $A(-4, 3)$, $B(2, 5)$, and $C(-1, -6)$.

$$d(A, B) = \sqrt{[2-(-4)]^2 + (5-3)^2}$$

$$= \sqrt{6^2 + 2^2} = \sqrt{36+4} = \sqrt{40}$$

$$d(B, C) = \sqrt{(-1-2)^2 + (-6-5)^2}$$

$$= \sqrt{(-3)^2 + (-11)^2}$$

$$= \sqrt{9+121} = \sqrt{130}$$

$$d(A, C) = \sqrt{[-1-(-4)]^2 + (-6-3)^2}$$

$$= \sqrt{3^2 + (-9)^2} = \sqrt{9+81} = \sqrt{90}$$

Since $\left(\sqrt{40}\right)^2 + \left(\sqrt{90}\right)^2 = \left(\sqrt{130}\right)^2$, triangle ABC is a right triangle.

25. Label the given points $A(0, -7)$, $B(-3, 5)$, and $C(2, -15)$. Find the distance between each pair of points.

$$d(A, B) = \sqrt{(-3-0)^2 + [5-(-7)]^2}$$

$$= \sqrt{(-3)^2 + 12^2} = \sqrt{9+144}$$

$$= \sqrt{153} = 3\sqrt{17}$$

$$d(B, C) = \sqrt{[2-(-3)]^2 + (-15-5)^2}$$

$$= \sqrt{5^2 + (-20)^2} = \sqrt{25+400}$$

$$= \sqrt{425} = 5\sqrt{17}$$

$$d(A, C) = \sqrt{(2-0)^2 + [-15-(-7)]^2}$$

$$= \sqrt{2^2 + (-8)^2} = \sqrt{68} = 2\sqrt{17}$$

Since $d(A, B) + d(A, C) = d(B, C)$ or $3\sqrt{17} + 2\sqrt{17} = 5\sqrt{17}$, the points are collinear.

27. Label the points $A(0, 9)$, $B(-3, -7)$, and $C(2, 19)$.

$$d(A, B) = \sqrt{(-3-0)^2 + (-7-9)^2}$$

$$= \sqrt{(-3)^2 + (-16)^2} = \sqrt{9+256}$$

$$= \sqrt{265} \approx 16.279$$

$$d(B, C) = \sqrt{[2-(-3)]^2 + [19-(-7)]^2}$$

$$= \sqrt{5^2 + 26^2} = \sqrt{25+676}$$

$$= \sqrt{701} \approx 26.476$$

$$d(A, C) = \sqrt{(2-0)^2 + (19-9)^2}$$

$$= \sqrt{2^2 + 10^2} = \sqrt{4+100}$$

$$= \sqrt{104} \approx 10.198$$

Since $d(A, B) + d(A, C) \neq d(B, C)$

or $\quad \sqrt{265} + \sqrt{104} \neq \sqrt{701}$

$$16.279 + 10.198 \neq 26.476,$$

$$26.477 \neq 26.476,$$

the three given points are not collinear. (Note, however, that these points are very close to lying on a straight line and may appear to lie on a straight line when graphed.)

29. Label the points $A(-7, 4)$, $B(6, -2)$, and $C(-1, 1)$.

$$d(A, B) = \sqrt{[6-(-7)]^2 + (-2-4)^2}$$

$$= \sqrt{13^2 + (-6)^2} = \sqrt{169+36}$$

$$= \sqrt{205} \approx 14.3178$$

$$d(B, C) = \sqrt{(-1-6)^2 + [1-(-2)]^2}$$

$$= \sqrt{(-7)^2 + 3^2} = \sqrt{49+9}$$

$$= \sqrt{58} \approx 7.6158$$

$$d(A, C) = \sqrt{[-1-(-7)]^2 + (1-4)^2}$$

$$= \sqrt{6^2 + (-3)^2} = \sqrt{36+9}$$

$$= \sqrt{45} \approx 6.7082$$

Since $d(B, C) + d(A, C) \neq d(A, B)$ or

$$\sqrt{58} + \sqrt{45} \neq \sqrt{205}$$

$$7.6158 + 6.7082 \neq 14.3178$$

$$14.3240 \neq 14.3178,$$

the three given points are not collinear. (Note, however, that these points are very close to lying on a straight line and may appear to lie on a straight line when graphed.)

31. Midpoint $(5, 8)$, endpoint $(13, 10)$

$$\frac{13+x}{2} = 5 \quad \text{and} \quad \frac{10+y}{2} = 8$$

$$13 + x = 10 \quad \text{and} \quad 10 + y = 16$$

$$x = -3 \quad \text{and} \quad y = 6.$$

The other endpoint has coordinates $(-3, 6)$.

Copyright © 2013 Pearson Education, Inc.

33. Midpoint (12, 6), endpoint (19, 16)

$$\frac{19+x}{2}=12 \quad\text{and}\quad \frac{16+y}{2}=6$$

$$19+x=24 \quad\text{and}\quad 16+y=12$$

$$x=5 \quad\text{and}\quad y=-4.$$

The other endpoint has coordinates (5, –4).

35. Midpoint (a, b), endpoint (p, q)

$$\frac{p+x}{2}=a \qquad\text{and}\qquad \frac{q+y}{2}=b$$

$$p+x=2a \qquad\text{and}\quad q+y=2b$$

$$x=2a-p \quad\text{and}\qquad y=2b-q$$

The other endpoint has coordinates $(2a-p,\,2b-q)$.

37. The endpoints of the segment are (1990, 21.3) and (20086, 29.4).

$$M=\left(\frac{1990+2008}{2},\frac{21.3+29.4}{2}\right)$$

$$=\left(1999,\,25.35\right)$$

The estimate is 25.35%. This is close to the actual figure of 25.2%.

39. The points to use would be (2004, 19,307) and (2008, 22,025). Their midpoint is

$$\left(\frac{2004+2008}{2},\,\frac{19,307+22,025}{2}\right)$$

$$=(2006,\,20666).$$

In 2006, the poverty level cutoff was approximately $20,666.

41. The midpoint M has coordinates

$$\left(\frac{x_1+x_2}{2},\frac{y_1+y_2}{2}\right).$$

$$d(P,M)$$

$$=\sqrt{\left(\frac{x_1+x_2}{2}-x_1\right)^2+\left(\frac{y_1+y_2}{2}-y_1\right)^2}$$

$$=\sqrt{\left(\frac{x_1+x_2}{2}-\frac{2x_1}{2}\right)^2+\left(\frac{y_1+y_2}{2}-\frac{2y_1}{2}\right)^2}$$

$$=\sqrt{\left(\frac{x_2-x_1}{2}\right)^2+\left(\frac{y_2-y_1}{2}\right)^2}$$

$$=\sqrt{\frac{\left(x_2-x_1\right)^2}{4}+\frac{\left(y_2-y_1\right)^2}{4}}$$

$$=\sqrt{\frac{\left(x_2-x_1\right)^2+\left(y_2-y_1\right)^2}{4}}$$

$$=\tfrac{1}{2}\sqrt{\left(x_2-x_1\right)^2+\left(y_2-y_1\right)^2}$$

$$d(M,Q)$$

$$=\sqrt{\left(x_2-\frac{x_1+x_2}{2}\right)^2+\left(y_2-\frac{y_1+y_2}{2}\right)^2}$$

$$=\sqrt{\left(\frac{2x_2}{2}-\frac{x_1+x_2}{2}\right)^2+\left(\frac{2y_2}{2}-\frac{y_1+y_2}{2}\right)^2}$$

$$=\sqrt{\left(\frac{x_2-x_1}{2}\right)^2+\left(\frac{y_2-y_1}{2}\right)^2}$$

$$=\sqrt{\frac{\left(x_2-x_1\right)^2}{4}+\frac{\left(y_2-y_1\right)^2}{4}}$$

$$=\sqrt{\frac{\left(x_2-x_1\right)^2+\left(y_2-y_1\right)^2}{4}}$$

$$=\tfrac{1}{2}\sqrt{\left(x_2-x_1\right)^2+\left(y_2-y_1\right)^2}$$

$$d(P,Q)=\sqrt{\left(x_2-x_1\right)^2+\left(y_2-y_1\right)^2}$$

Since $\tfrac{1}{2}\sqrt{\left(x_2-x_1\right)^2+\left(y_2-y_1\right)^2}$

$$+\tfrac{1}{2}\sqrt{\left(x_2-x_1\right)^2+\left(y_2-y_1\right)^2}$$

$$=\sqrt{\left(x_2-x_1\right)^2+\left(y_2-y_1\right)^2},$$

this shows $d(P,M)+d(M,Q)=d(P,Q)$ and $d(P,M)=d(M,Q)$.

In exercises 43–53, other ordered pairs are possible.

43. (a)

x	y
0	–2
4	0
2	–1

y-intercept:
$x=0\Rightarrow$
$y=\tfrac{1}{2}(0)-2=-2$

x-intercept:
$y=0\Rightarrow$
$0=\tfrac{1}{2}x-2\Rightarrow$
$2=\tfrac{1}{2}x\Rightarrow4=x$

additional point

(b)

$y=\tfrac{1}{2}x-2$

Copyright © 2013 Pearson Education, Inc.

45. (a)

x	y	
0	$\frac{5}{3}$	y-intercept: $x = 0 \Rightarrow$ $2(0) + 3y = 5 \Rightarrow$ $3y = 5 \Rightarrow y = \frac{5}{3}$
$\frac{5}{2}$	0	x-intercept: $y = 0 \Rightarrow$ $2x + 3(0) = 5 \Rightarrow$ $2x = 5 \Rightarrow x = \frac{5}{2}$
4	−1	additional point

(b)

47. (a)

x	y	
0	0	x- and y-intercept: $0 = 0^2$
1	1	additional point
−2	4	additional point

(b)

49. (a)

x	y	
3	0	x-intercept: $y = 0 \Rightarrow$ $0 = \sqrt{x - 3} \Rightarrow$ $0 = x - 3 \Rightarrow 3 = x$
4	1	additional point
7	2	additional point

no y-intercept:
$$x = 0 \Rightarrow y = \sqrt{0 - 3} \Rightarrow y = \sqrt{-3}$$

(b)

51. (a)

x	y	
0	2	y-intercept: $x = 0 \Rightarrow$ $y = |0 - 2| \Rightarrow$ $y = |-2| \Rightarrow y = 2$
2	0	x-intercept: $y = 0 \Rightarrow$ $0 = |x - 2| \Rightarrow$ $0 = x - 2 \Rightarrow 2 = x$
−2	4	additional point
4	2	additional point

(b)

53. (a)

x	y	
0	0	x- and y-intercept: $0 = 0^3$
−1	−1	additional point
2	8	additional point

(b)

55. Points on the x-axis have y-coordinates equal to 0. The point on the x-axis will have the same x-coordinate as point (4, 3). Therefore, the line will intersect the x-axis at (4, 0).

Copyright © 2013 Pearson Education, Inc.

57. Since (a, b) is in the second quadrant, a is negative and b is positive. Therefore, $(a, -b)$ will have a negative x–coordinate and a negative y-coordinate and will lie in quadrant III. $(-a, b)$ will have a positive x-coordinate and a positive y-coordinate and will lie in quadrant I. Also, $(-a, -b)$ will have a positive x-coordinate and a negative y-coordinate and will lie in quadrant IV. Finally, (b, a) will have a positive x-coordinate and a negative y-coordinate and will lie in quadrant IV.

59. To determine which points form sides of the quadrilateral (as opposed to diagonals), plot the points.

Use the distance formula to find the length of each side.

$$d(A, B) = \sqrt{(5-1)^2 + (2-1)^2}$$
$$= \sqrt{4^2 + 1^2} = \sqrt{16+1} = \sqrt{17}$$

$$d(B, C) = \sqrt{(3-5)^2 + (4-2)^2}$$
$$= \sqrt{(-2)^2 + 2^2} = \sqrt{4+4} = \sqrt{8}$$

$$d(C, D) = \sqrt{(-1-3)^2 + (3-4)^2}$$
$$= \sqrt{(-4)^2 + (-1)^2} = \sqrt{16+1} = \sqrt{17}$$

$$d(D, A) = \sqrt{[1-(-1)]^2 + (1-3)^2}$$
$$= \sqrt{2^2 + (-2)^2} = \sqrt{4+4} = \sqrt{8}$$

Since $d(A, B) = d(C, D)$ and $d(B, C) = d(D, A)$, the points are the vertices of a parallelogram. Since $d(A, B) \neq d(B, C)$, the points are not the vertices of a rhombus.

Section 2.2   Circles

1. (a) Center $(0, 0)$, radius 6

$$\sqrt{(x-0)^2 + (y-0)^2} = 6$$
$$(x-0)^2 + (y-0)^2 = 6^2 \Rightarrow x^2 + y^2 = 36$$

(b)

$$x^2 + y^2 = 36$$

3. (a) Center $(2, 0)$, radius 6

$$\sqrt{(x-2)^2 + (y-0)^2} = 6$$
$$(x-2)^2 + (y-0)^2 = 6^2$$
$$(x-2)^2 + y^2 = 36$$

(b)

$$(x-2)^2 + y^2 = 36$$

5. (a) Center $(0, 4)$, radius 4

$$\sqrt{(x-0)^2 + (y-4)^2} = 4$$
$$x^2 + (y-4)^2 = 16$$

(b)

$$x^2 + (y-4)^2 = 16$$

7. (a) Center $(-2, 5)$, radius 4

$$\sqrt{[x-(-2)]^2 + (y-5)^2} = 4$$
$$[x-(-2)]^2 + (y-5)^2 = 4^2$$
$$(x+2)^2 + (y-5)^2 = 16$$

Copyright © 2013 Pearson Education, Inc.

(b)

$(x + 2)^2 + (y - 5)^2 = 16$

9. **(a)** Center $(5, -4)$, radius 7

$$\sqrt{(x-5)^2 + \left[y-(-4)\right]^2} = 7$$
$$(x-5)^2 + [y-(-4)]^2 = 7^2$$
$$(x-5)^2 + (y+4)^2 = 49$$

(b)

$(x - 5)^2 + (y + 4)^2 = 49$

11. **(a)** Center $\left(\sqrt{2}, \sqrt{2}\right)$, radius $\sqrt{2}$

$$\sqrt{\left(x-\sqrt{2}\right)^2 + \left(y-\sqrt{2}\right)^2} = \sqrt{2}$$
$$\left(x-\sqrt{2}\right)^2 + \left(y-\sqrt{2}\right)^2 = 2$$

(b)

$(x - \sqrt{2})^2 + (y - \sqrt{2})^2 = 2$

13. **(a)** The center of the circle is located at the midpoint of the diameter determined by the points $(1, 1)$ and $(5, 1)$. Using the midpoint formula, we have

$$C = \left(\frac{1+5}{2}, \frac{1+1}{2}\right) = (3,1).$$

The radius is one-half the length of the diameter: $r = \dfrac{1}{2}\sqrt{(5-1)^2 + (1-1)^2} = 2$

The equation of the circle is

$$(x-3)^2 + (y-1)^2 = 4$$

(b) Expand $(x-3)^2 + (y-1)^2 = 4$ to find the equation of the circle in general form:

$$(x-3)^2 + (y-1)^2 = 4$$
$$x^2 - 6x + 9 + y^2 - 2y + 1 = 4$$
$$x^2 + y^2 - 6x - 2y + 6 = 0$$

15. **(a)** The center of the circle is located at the midpoint of the diameter determined by the points $(-2, 4)$ and $(-2, 0)$. Using the midpoint formula, we have

$$C = \left(\frac{-2+(-2)}{2}, \frac{4+0}{2}\right) = (-2,2).$$

The radius is one-half the length of the diameter:

$$r = \frac{1}{2}\sqrt{\left[-2-(-2)\right]^2 + (4-0)^2} = 2$$

The equation of the circle is

$$(x+2)^2 + (y-2)^2 = 4$$

(b) Expand $(x+2)^2 + (y-2)^2 = 4$ to find the equation of the circle in general form:

$$(x+2)^2 + (y-2)^2 = 4$$
$$x^2 + 4x + 4 + y^2 - 4y + 4 = 4$$
$$x^2 + y^2 + 4x - 4y + 4 = 0$$

17. Since the center $(-3, 5)$ is in quadrant II, choice B is the correct graph.

19. $x^2 + y^2 + 6x + 8y + 9 = 0$

Complete the square on x and y separately.

$$\left(x^2 + 6x\right) + \left(y^2 + 8y\right) = -9$$
$$\left(x^2 + 6x + 9\right) + \left(y^2 + 8y + 16\right) = -9 + 9 + 16$$
$$(x+3)^2 + (y+4)^2 = 16$$

Yes, it is a circle. The circle has its center at $(-3, -4)$ and radius 4.

21. $x^2 + y^2 - 4x + 12y = -4$

Complete the square on x and y separately.

$$\left(x^2 - 4x\right) + \left(y^2 + 12y\right) = -4$$
$$\left(x^2 - 4x + 4\right) + \left(y^2 + 12y + 36\right) = -4 + 4 + 36$$
$$(x-2)^2 + (y+6)^2 = 36$$

Yes, it is a circle. The circle has its center at $(2, -6)$ and radius 6.

Copyright © 2013 Pearson Education, Inc.

23. $4x^2 + 4y^2 + 4x - 16y - 19 = 0$

Complete the square on x and y separately.

$$4\left(x^2 + x\right) + 4\left(y^2 - 4y\right) = 19$$

$$4\left(x^2 + x + \tfrac{1}{4}\right) + 4\left(y^2 - 4y + 4\right) =$$
$$19 + 4\left(\tfrac{1}{4}\right) + 4(4)$$

$$4\left(x + \tfrac{1}{2}\right)^2 + 4\left(y - 2\right)^2 = 36$$

$$\left(x + \tfrac{1}{2}\right)^2 + \left(y - 2\right)^2 = 9$$

Yes, it is a circle with center $\left(-\tfrac{1}{2}, 2\right)$ and radius 3.

25. $x^2 + y^2 + 2x - 6y + 14 = 0$

Complete the square on x and y separately.

$$\left(x^2 + 2x\right) + \left(y^2 - 6y\right) = -14$$

$$\left(x^2 + 2x + 1\right) + \left(y^2 - 6y + 9\right) = -14 + 1 + 9$$

$$\left(x + 1\right)^2 + \left(y - 3\right)^2 = -4$$

The graph is nonexistent.

27. $x^2 + y^2 - 6x - 6y + 18 = 0$

Complete the square on x and y separately.

$$\left(x^2 - 6x\right) + \left(y^2 - 6y\right) = -18$$

$$\left(x^2 - 6x + 9\right) + \left(y^2 - 6y + 9\right) = -18 + 9 + 9$$

$$\left(x - 3\right)^2 + \left(y - 3\right)^2 = 0$$

The graph is the point (3, 3).

29. $9x^2 + 9y^2 - 6x + 6y - 23 = 0$

Complete the square on x and y separately.

$$\left(9x^2 - 6x\right) + \left(9y^2 + 6y\right) = 23$$

$$9\left(x^2 - \tfrac{2}{3}x\right) + 9\left(y^2 + \tfrac{2}{3}y\right) = 23$$

$$\left(x^2 - \tfrac{2}{3}x + \tfrac{1}{9}\right) + \left(y^2 + \tfrac{2}{3}y + \tfrac{1}{9}\right) = \tfrac{23}{9} + \tfrac{1}{9} + \tfrac{1}{9}$$

$$\left(x - \tfrac{1}{3}\right)^2 + \left(y + \tfrac{1}{3}\right)^2 = \tfrac{25}{9} = \left(\tfrac{5}{3}\right)^2$$

Yes, it is a circle with center $\left(\tfrac{1}{3}, -\tfrac{1}{3}\right)$ and radius $\tfrac{5}{3}$.

31. The midpoint M has coordinates

$$\left(\frac{-1 + 5}{2}, \frac{3 + (-9)}{2}\right) = \left(\frac{4}{2}, \frac{-6}{2}\right) = (2, -3).$$

33. Use points $C(2, -3)$ and $Q(5, -9)$.

$$d(C, Q) = \sqrt{(5 - 2)^2 + \left[-9 - (-3)\right]^2}$$
$$= \sqrt{3^2 + (-6)^2} = \sqrt{9 + 36} = \sqrt{45} = 3\sqrt{5}$$

The radius is $3\sqrt{5}$.

35. The center-radius form for this circle is

$$(x - 2)^2 + (y + 3)^2 = (3\sqrt{5})^2 \Rightarrow$$
$$(x - 2)^2 + (y + 3)^2 = 45.$$

37. Label the endpoints of the diameter $P(-1, 2)$ and $Q(11, 7)$. The midpoint M of the segment joining P and Q has coordinates

$$\left(\frac{-1 + 11}{2}, \frac{2 + 7}{2}\right) = \left(5, \frac{9}{2}\right).$$

The center is $C\left(5, \tfrac{9}{2}\right)$. To find the radius, we can use points $C\left(5, \tfrac{9}{2}\right)$ and $P(-1, 2)$.

$$d(C, P) = \sqrt{\left[5 - (-1)\right]^2 + \left(\tfrac{9}{2} - 2\right)^2}$$
$$= \sqrt{6^2 + \left(\tfrac{5}{2}\right)^2} = \sqrt{\tfrac{169}{4}} = \tfrac{13}{2}$$

We could also use points $C\left(5, \tfrac{9}{2}\right)$ and $Q(11, 7)$.

$$d(C, Q) = \sqrt{(5 - 11)^2 + \left(\tfrac{9}{2} - 7\right)^2}$$
$$= \sqrt{(-6)^2 + \left(-\tfrac{5}{2}\right)^2} = \sqrt{\tfrac{169}{4}} = \tfrac{13}{2}$$

Using the points P and Q to find the length of the diameter, we have

$$d(P, Q) = \sqrt{(-1 - 11)^2 + (2 - 7)^2}$$
$$= \sqrt{(-12)^2 + (-5)^2} = \sqrt{169} = 13$$

$$\frac{1}{2}d(P, Q) = \frac{1}{2}(13) = \frac{13}{2}$$

The center-radius form of the equation of the circle is

$$\left(x - 5\right)^2 + \left(y - \tfrac{9}{2}\right)^2 = \left(\tfrac{13}{2}\right)^2$$
$$\left(x - 5\right)^2 + \left(y - \tfrac{9}{2}\right)^2 = \tfrac{169}{4}$$

39. Label the endpoints of the diameter $P(1, 4)$ and $Q(5, 1)$. The midpoint M of the segment joining P and Q has coordinates

$$\left(\frac{1 + 5}{2}, \frac{4 + 1}{2}\right) = \left(3, \tfrac{5}{2}\right).$$

The center is $C\left(3, \tfrac{5}{2}\right)$.

The length of the diameter PQ is

$$\sqrt{(1 - 5)^2 + (4 - 1)^2} = \sqrt{(-4)^2 + 3^2} = \sqrt{25} = 5.$$

The length of the radius is $\tfrac{1}{2}(5) = \tfrac{5}{2}$.

The center-radius form of the equation of the circle is

$$\left(x - 3\right)^2 + \left(y - \tfrac{5}{2}\right)^2 = \left(\tfrac{5}{2}\right)^2 = \tfrac{25}{4}$$

Copyright © 2013 Pearson Education, Inc.

41. The equations of the three circles are
$(x-7)^2 + (y-4)^2 = 25$,

$(x+9)^2 + (y+4)^2 = 169$, and

$(x+3)^2 + (y-9)^2 = 100$. From the graph of the three circles, it appears that the epicenter is located at (3, 1).

Check algebraically:

$(x-7)^2 + (y-4)^2 = 25$
$(3-7)^2 + (1-4)^2 = 25$
$4^2 + 3^2 = 25 \Rightarrow 25 = 25$

$(x+9)^2 + (y+4)^2 = 169$
$(3+9)^2 + (1+4)^2 = 169$
$12^2 + 5^2 = 169 \Rightarrow 169 = 169$

$(x+3)^2 + (y-9)^2 = 100$
$(3+3)^2 + (1-9)^2 = 100$
$6^2 + (-8)^2 = 100 \Rightarrow 100 = 100$

(3, 1) satisfies all three equations, so the epicenter is at (3, 1).

43. From the graph of the three circles, it appears that the epicenter is located at (−2, −2).

Check algebraically:

$(x-2)^2 + (y-1)^2 = 25$
$(-2-2)^2 + (-2-1)^2 = 25$
$(-4)^2 + (-3)^2 = 25$
$25 = 25$

$(x+2)^2 + (y-2)^2 = 16$
$(-2+2)^2 + (-2-2)^2 = 16$
$0^2 + (-4)^2 = 16$
$16 = 16$

$(x-1)^2 + (y+2)^2 = 9$
$(-2-1)^2 + (-2+2)^2 = 9$
$(-3)^2 + 0^2 = 9$
$9 = 9$

(−2, −2) satisfies all three equations, so the epicenter is at (−2, −2).

45. The radius of this circle is the distance from the center $C(3, 2)$ to the x-axis. This distance is 2, so $r = 2$.

$(x-3)^2 + (y-2)^2 = 2^2 \Rightarrow$
$(x-3)^2 + (y-2)^2 = 4$

47. Label the points $P(x, y)$ and $Q(1, 3)$.

If $d(P, Q) = 4$, $\sqrt{(1-x)^2 + (3-y)^2} = 4 \Rightarrow$
$(1-x)^2 + (3-y)^2 = 16$.

If $x = y$, then we can either substitute x for y or y for x. Substituting x for y we solve the following:

$(1-x)^2 + (3-x)^2 = 16$
$1 - 2x + x^2 + 9 - 6x + x^2 = 16$
$2x^2 - 8x + 10 = 16$
$2x^2 - 8x - 6 = 0$
$x^2 - 4x - 3 = 0$

To solve this equation, we can use the quadratic formula with $a = 1$, $b = -4$, and $c = -3$.

$x = \dfrac{-(-4) \pm \sqrt{(-4)^2 - 4(1)(-3)}}{2(1)}$

$= \dfrac{4 \pm \sqrt{16+12}}{2} = \dfrac{4 \pm \sqrt{28}}{2}$

$= \dfrac{4 \pm 2\sqrt{7}}{2} = 2 \pm \sqrt{7}$

Since $x = y$, the points are
$\left(2+\sqrt{7}, 2+\sqrt{7}\right)$ and $\left(2-\sqrt{7}, 2-\sqrt{7}\right)$.

49. Let $P(x, y)$ be a point whose distance from $A(1, 0)$ is $\sqrt{10}$ and whose distance from $B(5, 4)$ is $\sqrt{10}$. $d(P, A) = \sqrt{10}$, so

$\sqrt{(1-x)^2 + (0-y)^2} = \sqrt{10} \Rightarrow$
$(1-x)^2 + y^2 = 10$. $d(P, B) = \sqrt{10}$, so

$\sqrt{(5-x)^2 + (4-y)^2} = \sqrt{10} \Rightarrow$
$(5-x)^2 + (4-y)^2 = 10$.

(continued on next page)

Copyright © 2013 Pearson Education, Inc.

(*continued*)

Thus,

$$(1-x)^2 + y^2 = (5-x)^2 + (4-y)^2$$
$$1 - 2x + x^2 + y^2 =$$
$$25 - 10x + x^2 + 16 - 8y + y^2$$
$$1 - 2x = 41 - 10x - 8y$$
$$8y = 40 - 8x$$
$$y = 5 - x$$

Substitute $5 - x$ for y in the equation $(1-x)^2 + y^2 = 10$ and solve for x.

$$(1-x)^2 + (5-x)^2 = 10 \Rightarrow$$
$$1 - 2x + x^2 + 25 - 10x + x^2 = 10$$
$$2x^2 - 12x + 26 = 10 \Rightarrow 2x^2 - 12x + 16 = 0$$
$$x^2 - 6x + 8 = 0 \Rightarrow (x-2)(x-4) = 0 \Rightarrow$$
$$x - 2 = 0 \quad \text{or} \quad x - 4 = 0$$
$$x = 2 \quad \text{or} \quad x = 4$$

To find the corresponding values of y use the equation $y = 5 - x$. If $x = 2$, then $y = 5 - 2 = 3$. If $x = 4$, then $y = 5 - 4 = 1$. The points satisfying the conditions are $(2, 3)$ and $(4, 1)$.

51. Label the points $A(3, y)$ and $B(-2, 9)$. If $d(A, B) = 12$, then

$$\sqrt{(-2-3)^2 + (9-y)^2} = 12$$
$$\sqrt{(-5)^2 + (9-y)^2} = 12$$
$$(-5)^2 + (9-y)^2 = 12^2$$
$$25 + 81 - 18y + y^2 = 144$$
$$y^2 - 18y - 38 = 0$$

Solve this equation by using the quadratic formula with $a = 1$, $b = -18$, and $c = -38$:

$$y = \frac{-(-18) \pm \sqrt{(-18)^2 - 4(1)(-38)}}{2(1)}$$
$$= \frac{18 \pm \sqrt{324 + 152}}{2(1)} = \frac{18 \pm \sqrt{476}}{2}$$
$$= \frac{18 \pm \sqrt{4(119)}}{2} = \frac{18 \pm 2\sqrt{119}}{2} = 9 \pm \sqrt{119}$$

The values of y are $9 + \sqrt{119}$ and $9 - \sqrt{119}$.

53. Let $P(x, y)$ be the point on the circle whose distance from the origin is the shortest. Complete the square on x and y separately to write the equation in center-radius form:

$$x^2 - 16x + y^2 - 14y + 88 = 0$$
$$x^2 - 16x + 64 + y^2 - 14y + 49 =$$
$$-88 + 64 + 49$$
$$(x-8)^2 + (y-7)^2 = 25$$

So, the center is $(8, 7)$ and the radius is 5.

$d(C, O) = \sqrt{8^2 + 7^2} = \sqrt{113}$. Since the length of the radius is 5, $d(P, O) = \sqrt{113} - 5$.

55. Using compasses, draw circles centered at Wickenburg, Kingman, Phoenix, and Las Vegas with scaled radii of 50, 75, 105, and 180 miles respectively. The four circles should intersect at the location of Nothing.

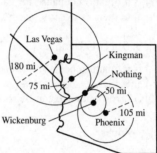

Section 2.3   Functions

1. The relation is a function because for each different x-value there is exactly one y-value. This correspondence can be shown as follows.

$\{5, 3, 4, 7\}$ x-values

$\{1, 2, 9, 8\}$ y-values

3. Two ordered pairs, namely $(2, 4)$ and $(2, 6)$, have the same x-value paired with different y-values, so the relation is not a function.

5. The relation is a function because for each different x-value there is exactly one y-value. This correspondence can be shown as follows.

$\{-3, 4, -2\}$ x-values

$\{1, 7\}$ y-values

Copyright © 2013 Pearson Education, Inc.

7. The relation is a function because for each different x-value there is exactly one y-value. This correspondence can be shown as follows.

$\{3, 7, 10\}$ x-values

$\{-4\}$ y-values

9. Two sets of ordered pairs, namely $(1, 1)$ and $(1, -1)$ as well as $(2, 4)$ and $(2, -4)$ have the same x-value paired with different y-values, so the relation is not a function.
domain: $\{0, 1, 2\}$; range: $\{-4, -1, 0, 1, 4\}$

$\{2, 3, 5\}$ x-values

$\{5, 7, 9, 11\}$ y-values
domain: $\{2, 3, 5\}$; range: $\{5, 7, 9, 11\}$

11. The relation is a function because for each different x-value there is exactly one y-value.
domain: $\{2, 3, 5, 11, 17\}$; range: $\{1, 7, 20\}$

13. The relation is a function because for each different x-value there is exactly one y-value. This correspondence can be shown as follows.

$\{0, -1, -2\}$ x-values

$\{0, \ 1, \ 2\}$ y-values
Domain: $\{0, -1, -2\}$; range: $\{0, 1, 2\}$

15. The relation is a function because for each different year, there is exactly one number for visitors.
domain: $\{2005, 2006, 2007, 2008\}$
range: $\{63.5, 60.4, 62.3, 61.2\}$

17. This graph represents a function. If you pass a vertical line through the graph, one x-value corresponds to only one y-value.
domain: $(-\infty, \infty)$; range: $(-\infty, \infty)$

19. This graph does not represent a function. If you pass a vertical line through the graph, there are places where one value of x corresponds to two values of y.
domain: $[3, \infty)$; range: $(-\infty, \infty)$

21. This graph represents a function. If you pass a vertical line through the graph, one x-value corresponds to only one y-value.
domain: $(-\infty, \infty)$; range: $(-\infty, \infty)$

23. $y = x^2$ represents a function since y is always found by squaring x. Thus, each value of x corresponds to just one value of y. x can be any real number. Since the square of any real number is not negative, the range would be zero or greater.

domain: $(-\infty, \infty)$; range: $[0, \infty)$

25. The ordered pairs $(1, 1)$ and $(1, -1)$ both satisfy $x = y^6$. This equation does not represent a function. Because x is equal to the sixth power of y, the values of x are nonnegative. Any real number can be raised to the sixth power, so the range of the relation is all real numbers.

domain: $[0, \infty)$ range: $(-\infty, \infty)$

27. $y = 2x - 5$ represents a function since y is found by multiplying x by 2 and subtracting 5. Each value of x corresponds to just one value of y. x can be any real number, so the domain is all real numbers. Since y is twice x, less 5, y also may be any real number, and so the range is also all real numbers.

domain: $(-\infty, \infty)$; range: $(-\infty, \infty)$

Copyright © 2013 Pearson Education, Inc.

29. By definition, y is a function of x if every value of x leads to exactly one value of y. Substituting a particular value of x, say 1, into $x + y < 3$ corresponds to many values of y. The ordered pairs (0, 2) (1, 1) (1, 0) (1, −1) and so on, all satisfy the inequality. Note that the points on the graphed line do not satisfy the inequality and only indicate the boundary of the solution set. This does not represent a function. Any number can be used for x or for y, so the domain and range of this relation are both all real numbers.

domain: $(-\infty, \infty)$; range: $(-\infty, \infty)$

31. For any choice of x in the domain of $y = \sqrt{x}$, there is exactly one corresponding value of y, so this equation defines a function. Since the quantity under the square root cannot be negative, we have $x \geq 0$. Because the radical is nonnegative, the range is also zero or greater.

domain: $[0, \infty)$; range: $[0, \infty)$

33. Since $xy = 2$ can be rewritten as $y = \frac{2}{x}$, we can see that y can be found by dividing x into 2. This process produces one value of y for each value of x in the domain, so this equation is a function. The domain includes all real numbers except those that make the denominator equal to zero, namely $x = 0$. Values of y can be negative or positive, but never zero. Therefore, the range will be all real numbers except zero.

domain: $(-\infty, 0) \cup (0, \infty)$;
range: $(-\infty, 0) \cup (0, \infty)$

35. For any choice of x in the domain of $y = \sqrt{4x + 1}$ there is exactly one corresponding value of y, so this equation defines a function. Since the quantity under the square root cannot be negative, we have $4x + 1 \geq 0 \Rightarrow 4x \geq -1 \Rightarrow x \geq -\frac{1}{4}$. Because the radical is nonnegative, the range is also zero or greater.

domain: $\left[-\frac{1}{4}, \infty\right)$; range: $[0, \infty)$

37. Given any value in the domain of $y = \frac{2}{x-3}$, we find y by subtracting 3, then dividing into 2. This process produces one value of y for each value of x in the domain, so this equation is a function. The domain includes all real numbers except those that make the denominator equal to zero, namely $x = 3$. Values of y can be negative or positive, but never zero. Therefore, the range will be all real numbers except zero.

domain: $(-\infty, 3) \cup (3, \infty)$;
range: $(-\infty, 0) \cup (0, \infty)$

Copyright © 2013 Pearson Education, Inc.

39. B. The notation $f(3)$ means the value of the dependent variable when the independent variable is 3.

41. $f(x) = -3x + 4$
$f(0) = -3 \cdot 0 + 4 = 0 + 4 = 4$

43. $g(x) = -x^2 + 4x + 1$
$g(-2) = -(-2)^2 + 4(-2) + 1$
$= -4 + (-8) + 1 = -11$

45. $f(x) = -3x + 4$
$f\left(\frac{1}{3}\right) = -3\left(\frac{1}{3}\right) + 4 = -1 + 4 = 3$

47. $g(x) = -x^2 + 4x + 1$
$g\left(\frac{1}{2}\right) = -\left(\frac{1}{2}\right)^2 + 4\left(\frac{1}{2}\right) + 1$
$= -\frac{1}{4} + 2 + 1 = \frac{11}{4}$

49. $f(x) = -3x + 4$
$f(p) = -3p + 4$

51. $f(x) = -3x + 4$
$f(-x) = -3(-x) + 4 = 3x + 4$

53. $f(x) = -3x + 4$
$f(x + 2) = -3(x + 2) + 4$
$= -3x - 6 + 4 = -3x - 2$

55. $f(x) = -3x + 4$
$f(2m - 3) = -3(2m - 3) + 4$
$= -6m + 9 + 4 = -6m + 13$

57. (a) $f(2) = 2$ (b) $f(-1) = 3$

59. (a) $f(2) = 15$ (b) $f(-1) = 10$

61. (a) $f(2) = 3$ (b) $f(-1) = -3$

63. (a) $x + 3y = 12$
$3y = -x + 12$
$y = \dfrac{-x + 12}{3}$
$y = -\frac{1}{3}x + 4 \Rightarrow f(x) = -\frac{1}{3}x + 4$

(b) $f(3) = -\frac{1}{3}(3) + 4 = -1 + 4 = 3$

65. (a) $y + 2x^2 = 3 - x$
$y = -2x^2 - x + 3$
$f(x) = -2x^2 - x + 3$

(b) $f(3) = -2(3)^2 - 3 + 3$
$= -2 \cdot 9 - 3 + 3 = -18$

67. (a) $4x - 3y = 8$
$4x = 3y + 8$
$4x - 8 = 3y$
$\dfrac{4x - 8}{3} = y$
$y = \frac{4}{3}x - \frac{8}{3} \Rightarrow f(x) = \frac{4}{3}x - \frac{8}{3}$

(b) $f(3) = \frac{4}{3}(3) - \frac{8}{3} = \frac{12}{3} - \frac{8}{3} = \frac{4}{3}$

69. $f(3) = 4$

71. $f(3)$ is the y-component of the coordinate, which is -4.

73. (a) $f(-2) = 0$ (b) $f(0) = 4$
(c) $f(1) = 2$ (d) $f(4) = 4$

75. (a) $f(-2) = -3$ (b) $f(0) = -2$
(c) $f(1) = 0$ (d) $f(4) = 2$

77. (a) $[4, \infty)$ (b) $(-\infty, -1]$
(c) $[-1, 4]$

79. (a) $(-\infty, 4]$ (b) $[4, \infty)$
(c) none

81. (a) none (b) $(-\infty, -2]; [3, \infty)$
(c) $(-2, 3)$

83. (a) Yes, it is the graph of a function.

(b) $[0, 24]$

(c) When $t = 8$, $y = 1200$ from the graph. At 8 A.M., approximately 1200 megawatts is being used.

(d) The most electricity was used at 17 hr or 5 P.M. The least electricity was used at 4 A.M.

(e) $f(12) = 2000$; At 12 noon, electricity use is 2000 megawatts.

(f) increasing from 4 A.M. to 5 P.M.; decreasing from midnight to 4 A.M. and from 5 P.M. to midnight

85. (a) At $t = 12$ and $t = 20$, $y = 55$ from the graph. Therefore, after about 12 noon until about 8 P.M. the temperature was over 55°.

(b) At $t = 6$ and $t = 22$, $y = 40$ from the graph. Therefore, until about 6 A.M. and after 10 P.M. the temperature was below 40°.

Copyright © 2013 Pearson Education, Inc.

(c) The temperature at noon in Bratenahl, Ohio was 55°. Since the temperature in Greenville is 7° higher, we are looking for the time at which Bratenahl, Ohio was at or above 48°. This occurred at approximately 10 A.M and 8:30 P.M.

(d) The temperature is just below 40° from midnight to 6 A.M., when it begins to rise until it reaches a maximum of just below 65° at 4 P.M. It then begins to fall util it reaches just under 40° again at midnight.

Section 2.4   Linear Functions

1. B; $f(x) = 3x + 6$ is a linear function with y-intercept 6.

3. C; $f(x) = -8$ is a constant function.

5. A; $f(x) = 5x$ is a linear function whose graph passes through the origin, $(0, 0)$.
$f(0) = 2(0) = 0$.

7. $f(x) = x - 4$; Use the intercepts.
$f(0) = 0 - 4 = -4$: y-intercept
$0 = x - 4 \Rightarrow x = 4$: x-intercept
Graph the line through $(0, -4)$ and $(4, 0)$.

The domain and range are both $(-\infty, \infty)$.

9. $f(x) = \frac{1}{2}x - 6$; Use the intercepts.
$f(0) = \frac{1}{2}(0) - 6 = -6$: y-intercept
$0 = \frac{1}{2}x - 6 \Rightarrow 6 = \frac{1}{2}x \Rightarrow x = 12$: x-intercept
Graph the line through $(0, -6)$ and $(12, 0)$.

The domain and range are both $(-\infty, \infty)$.

11. $f(x) = 3x$
The x-intercept and the y-intercept are both zero. This gives us only one point, $(0, 0)$. If $x = 1$, $y = 3(1) = 3$. Another point is $(1, 3)$. Graph the line through $(0, 0)$ and $(1, 3)$.

The domain and range are both $(-\infty, \infty)$.

13. $f(x) = -4$ is a constant function.
The graph of $f(x) = -4$ is a horizontal line with a y-intercept of -4.

domain: $(-\infty, \infty)$; range: $\{-4\}$

15. $-4x + 3y = 12$
Use the intercepts.
$-4(0) + 3y = 12 \Rightarrow 3y = 12 \Rightarrow$
$y = 4$: y-intercept
$-4x + 3(0) = 12 \Rightarrow -4x = 12 \Rightarrow$
$x = -3$: x-intercept
Graph the line through $(0, 4)$ and $(-3, 0)$.

The domain and range are both $(-\infty, \infty)$.

Copyright © 2013 Pearson Education, Inc.

17. $3y - 4x = 0$; Use the intercepts.

$3y - 4(0) = 0 \Rightarrow 3y = 0 \Rightarrow y = 0$: y-intercept

$3(0) - 4x = 0 \Rightarrow -4x = 0 \Rightarrow x = 0$: x-intercept

The graph has just one intercept. Choose an additional value, say 3, for x.

$3y - 4(3) = 0 \Rightarrow 3y - 12 = 0$

$\quad 3y = 12 \Rightarrow y = 4$

Graph the line through $(0, 0)$ and $(3, 4)$:

The domain and range are both $(-\infty, \infty)$.

19. $x = 3$ is a vertical line, intersecting the x-axis at $(3, 0)$.

domain: $\{3\}$; range: $(-\infty, \infty)$

21. $2x + 4 = 0 \Rightarrow 2x = -4 \Rightarrow x = -2$ is a vertical line intersecting the x-axis at $(-2, 0)$.

domain: $\{-2\}$; range: $(-\infty, \infty)$

23. $-x + 5 = 0 \Rightarrow x = 5$ is a vertical line intersecting the x-axis at $(5, 0)$.

domain: $\{5\}$; range: $(-\infty, \infty)$

25. $y = 5$ is a horizontal line with y-intercept 5. Choice A resembles this.

27. $x = 5$ is a vertical line with x-intercept 5. Choice D resembles this.

29. $y = 3x + 4$; Use $Y_1 = 3X + 4$.

31. $3x + 4y = 6$; Solve for y.

$3x + 4y = 6$

$4y = -3x + 6$

$y = -\frac{3}{4}x + \frac{3}{2}$

Use $Y_1 = (-3/4)X + (3/2)$

or $Y_1 = -3/4X + 3/2$.

33. The rise is 2.5 feet while the run is 10 feet so the slope is $\frac{2.5}{10} = .25 = 25\% = \frac{1}{4}$. So A = 0.25, C = $\frac{2.5}{10}$, D = 25%, and E = $\frac{1}{4}$ are all expressions of the slope.

35. Through $(2, -1)$ and $(-3, -3)$

Let $x_1 = 2$, $y_1 = -1$, $x_2 = -3$, and $y_2 = -3$.

Then rise $= \Delta y = -3 - (-1) = -2$ and run $= \Delta x = -3 - 2 = -5$.

The slope is $m = \dfrac{\text{rise}}{\text{run}} = \dfrac{\Delta y}{\Delta x} = \dfrac{-2}{-5} = \dfrac{2}{5}$.

37. Through $(5, 8)$ and $(3, 12)$

Let $x_1 = 5$, $y_1 = 8$, $x_2 = 3$, and $y_2 = 12$.

Then rise $= \Delta y = 12 - 8 = 4$ and run $= \Delta x = 3 - 5 = -2$.

The slope is $m = \dfrac{\text{rise}}{\text{run}} = \dfrac{\Delta y}{\Delta x} = \dfrac{4}{-2} = -2$.

39. Through $(5, 9)$ and $(-2, 9)$

$m = \dfrac{\Delta y}{\Delta x} = \dfrac{y_2 - y_1}{x_2 - x_1} = \dfrac{9 - 9}{-2 - 5} = \dfrac{0}{-7} = 0$

Copyright © 2013 Pearson Education, Inc.

41. Horizontal, through (5, 1)
The slope of every horizontal line is zero, so
$m = 0$.

43. Vertical, through (4, –7)
The slope of every vertical line is undefined;
m is undefined.

45. **(a)** $y = 3x + 5$

Find two ordered pairs that are solutions
to the equation.
If $x = 0$, then $y = 3(0) + 5 \Rightarrow y = 5$.
If $x = -1$, then
$y = 3(-1) + 5 \Rightarrow y = -3 + 5 \Rightarrow y = 2$.
Thus two ordered pairs are (0, 5) and
(–1, 2)
$m = \dfrac{\text{rise}}{\text{run}} = \dfrac{y_2 - y_1}{x_2 - x_1} = \dfrac{2 - 5}{-1 - 0} = \dfrac{-3}{-1} = 3$.

(b)

47. $2y = -3x$

Find two ordered pairs that are solutions to the
equation. If $x = 0$, then $2y = 0 \Rightarrow y = 0$.

If $y = -3$, then $2(-3) = -3x \Rightarrow -6 = -3x \Rightarrow$

$x = 2$. Thus two ordered pairs are $(0, 0)$ and

$(2, -3)$.

$m = \dfrac{\text{rise}}{\text{run}} = \dfrac{y_2 - y_1}{x_2 - x_1} = \dfrac{-3 - 0}{2 - 0} = -\dfrac{3}{2}$.

(b)

49. $5x - 2y = 10$

Find two ordered pairs that are solutions to the
equation. If $x = 0$, then $5(0) - 2y = 10 \Rightarrow$

$\Rightarrow y = -5$. If $y = 0$, then $5x - 2(0) = 10 \Rightarrow$

$5x = 10 \Rightarrow x = 2$.

Thus two ordered pairs are $(0, -5)$ and $(2, 0)$.

$m = \dfrac{\text{rise}}{\text{run}} = \dfrac{y_2 - y_1}{x_2 - x_1} = \dfrac{0 - (-5)}{2 - 0} = \dfrac{5}{2}$.

(b)

51. Through (–1, 3), $m = \dfrac{3}{2}$

First locate the point (–1, 3). Since the slope is
$\frac{3}{2}$, a change of 2 units horizontally (2 units to
the right) produces a change of 3 units
vertically (3 units up). This gives a second
point, (1, 6), which can be used to complete
the graph.

53. Through (3, –4), $m = -\dfrac{1}{3}$. First locate the point

(3, –4). Since the slope is $-\frac{1}{3}$, a change of 3

units horizontally (3 units to the right)
produces a change of –1 unit vertically (1 unit
down). This gives a second point, (6, –5),
which can be used to complete the graph.

Copyright © 2013 Pearson Education, Inc.

55. Through $\left(-\frac{1}{2}, 4\right)$, $m = 0$.

The graph is the horizontal line through $\left(-\frac{1}{2}, 4\right)$.

57. Through $\left(-\frac{5}{2}, 3\right)$, undefined slope. The slope is undefined, so the line is vertical, intersecting the x-axis at $\left(-\frac{5}{2}, 0\right)$.

59. $m = \frac{1}{3}$ matches graph D because the line rises gradually as x increases.

61. $m = 0$ matches graph A because horizontal lines have slopes of 0.

63. $m = 3$ matches graph E because the line rises rapidly as x increases.

65. The average rate of change is $m = \dfrac{\Delta y}{\Delta x}$

$\dfrac{20 - 4}{0 - 4} = \dfrac{-16}{4} = -\$4$ (thousand) per year. The value of the machine is decreasing \$4000 each year during these years.

67. The average rate of change is $m = \dfrac{\Delta y}{\Delta x}$

$\dfrac{3 - 3}{4 - 0} = \dfrac{0}{4} = 0\%$ per year. The percent of pay raise is not changing - it is 3% each year.

69. For a constant function, the average rate of change is zero.

71. (a) Answers will vary.

(b) $m = \dfrac{13{,}977 - 2773}{2007 - 1950} = \dfrac{11{,}204}{57} \approx 196.6$

This means that the average rate of change in the number of radio stations per year is an increase of about 196.6 stations.

73. (a) $m = \dfrac{y_2 - y_1}{x_2 - x_1} = \dfrac{52.5 - 338.3}{2009 - 1999}$

$= \dfrac{-285.8}{10} = -28.58$

The average rate of change was -28.58 thousand mobile homes per year.

(b) The negative slope means that the number of mobile homes decreased by an average of 28.58 thousand each year from 1999 to 2009.

75. $m = \dfrac{y_2 - y_1}{x_2 - x_1} = \dfrac{3118 - 5212}{2008 - 1980} = \dfrac{-2094}{28}$

$= -74.8$ thousand per year

The number of high school dropouts decreased by an average of 74.8 thousand per year.

77. The first two points are $A(0, -6)$ and $B(1, -3)$.

$m = \dfrac{-3 - (-6)}{1 - 0} = \dfrac{3}{1} = 3$

79. If we use any two points on a line to find its slope, we find that the slope is <u>the same</u> in all cases.

81. The second and fourth points are $B(1, -3)$ and $D(3, 3)$.

$d(B, D) = \sqrt{(3 - 1)^2 + [3 - (-3)]^2}$

$= \sqrt{2^2 + 6^2} = \sqrt{4 + 36}$

$= \sqrt{40} = 2\sqrt{10}$

83. $\sqrt{10} + 2\sqrt{10} = 3\sqrt{10}$; The sum is $3\sqrt{10}$, which is equal to the answer in Exercise 82.

85. The midpoint of the segment joining $A(0, -6)$ and $G(6, 12)$ has coordinates $\left(\frac{0+6}{2}, \frac{-6+12}{2}\right) = \left(\frac{6}{2}, \frac{6}{2}\right) = (3, 3)$. The midpoint is $M(3, 3)$, which is the same as the middle entry in the table.

87. (a) $C(x) = 10x + 500$

(b) $R(x) = 35x$

(c) $P(x) = R(x) - C(x)$

$= 35x - (10x + 500)$

$= 35x - 10x - 500 = 25x - 500$

Copyright © 2013 Pearson Education, Inc.

(d) $C(x) = R(x)$
$$10x + 500 = 35x$$
$$500 = 25x$$
$$20 = x$$
20 units; do not produce

89. (a) $C(x) = 400x + 1650$

(b) $R(x) = 305x$

(c) $P(x) = R(x) - C(x)$
$$= 305x - (400x + 1650)$$
$$= 305x - 400x - 1650$$
$$= -95x - 1650$$

(d) $C(x) = R(x)$
$$400x + 1650 = 305x$$
$$95x + 1650 = 0$$
$$95x = -1650$$
$$x \approx -17.37 \text{ units}$$

This result indicates a negative "break-even point," but the number of units produced must be a positive number. A calculator graph of the lines
$Y_1 = 400X + 1650$ and $Y_2 = 305X$ on the same screen or solving the inequality $305x < 400x + 1650$ will show that $R(x) < C(x)$ for all positive values of x (in fact whenever x is greater than -17.4). Do not produce the product since it is impossible to make a profit.

91. $C(x) = R(x) \Rightarrow 200x + 1000 = 240x \Rightarrow$
$1000 = 40x \Rightarrow 25 = x$
The break-even point is 25 units.
$C(25) = 200(25) + 1000 = \$6000$ which is the same as $R(25) = 240(25) = \$6000$

Chapter 2 Quiz
(Sections 2.1–2.4)

1. $d(A, B) = \sqrt{(x_2 - x_1)^2 + (y_2 - y_1)^2}$
$$= \sqrt{(-8 - (-4))^2 + (-3 - 2)^2}$$
$$= \sqrt{(-4)^2 + (-5)^2} = \sqrt{16 + 25} = \sqrt{41}$$

3.

$$y = -x^2 + 4$$

5. $x^2 + y^2 - 4x + 8y + 3 = 0$
Complete the square on x and y separately.
$$(x^2 - 4x + 4) + (y^2 + 8y + 16) = -3 + 4 + 16 \Rightarrow$$
$$(x - 2)^2 + (y + 4)^2 = 17$$
The radius is $\sqrt{17}$ and the midpoint of the circle is $(2, -4)$.

7.   Domain: $(-\infty, \infty)$; range: $[0, \infty)$

9. (a) $m = \dfrac{11 - 5}{5 - 1} = \dfrac{6}{4} = \dfrac{3}{2}$

(b) $m = \dfrac{4 - 4}{-1 - (-7)} = \dfrac{0}{6} = 0$

(c) $m = \dfrac{-4 - 12}{6 - 6} = \dfrac{-16}{0} \Rightarrow$ the slope is undefined.

Section 2.5 Equations of Lines and Linear Models

1. $y = \frac{1}{4}x + 2$ is graphed in D.
The slope is $\frac{1}{4}$ and the y-intercept is 2.

3. $y - (-1) = \frac{3}{2}(x - 1)$ is graphed in C. The slope is $\frac{3}{2}$ and a point on the graph is $(1, -1)$.

5.   Through $(1, 3)$, $m = -2$.
Write the equation in point-slope form.
$$y - y_1 = m(x - x_1) \Rightarrow y - 3 = -2(x - 1)$$
Then, change to standard form.
$$y - 3 = -2x + 2 \Rightarrow 2x + y = 5$$

Copyright © 2013 Pearson Education, Inc.

7. Through $(-5, 4)$, $m = -\frac{3}{2}$

Write the equation in point-slope form.

$$y - 4 = -\frac{3}{2}\left[x - (-5)\right]$$

Change to standard form.

$$2(y - 4) = -3(x + 5)$$
$$2y - 8 = -3x - 15$$
$$3x + 2y = -7$$

9. Through $(-8, 4)$, undefined slope
Since undefined slope indicates a vertical line, the equation will have the form $x = a$. The equation of the line is $x = -8$.

11. Through $(5, -8)$, $m = 0$
This is a horizontal line through $(5, -8)$, so the equation is $y = -8$.

13. Through $(-1, 3)$ and $(3, 4)$
First find m.

$$m = \frac{4 - 3}{3 - (-1)} = \frac{1}{4}$$

Use either point and the point-slope form.

$$y - 4 = \frac{1}{4}(x - 3)$$
$$4y - 16 = x - 3$$
$$-x + 4y = 13$$
$$x - 4y = -13$$

15. x-intercept 3, y-intercept -2
The line passes through $(3, 0)$ and $(0, -2)$. Use these points to find m.

$$m = \frac{-2 - 0}{0 - 3} = \frac{2}{3}$$

Using slope-intercept form we have

$$y = \frac{2}{3}x - 2.$$

17. Vertical, through $(-6, 4)$
The equation of a vertical line has an equation of the form $x = a$. Since the line passes through $(-6, 4)$, the equation is $x = -6$. (Since the slope of a vertical line is undefined, this equation cannot be written in slope-intercept form.)

19. Horizontal, through $(-7, 4)$
The equation of a horizontal line has an equation of the form $y = b$. Since the line passes through $(-7, 4)$, the equation is $y = 4$.

21. $m = 5$, $b = 15$
Using slope-intercept form, we have
$y = 5x + 15$.

23. Through $(-2, 5)$, slope $= -4$

$$y - 5 = -4\left(x - (-2)\right)$$
$$y - 5 = -4(x + 2)$$
$$y - 5 = -4x - 8$$
$$y = -4x - 3$$

25. slope 0, y-intercept $\frac{3}{2}$

These represent $m = 0$ and $b = \frac{3}{2}$. Using slope-intercept form, we have

$$y = (0)x + \frac{3}{2} \Rightarrow y = \frac{3}{2}.$$

27. The line $x + 2 = 0$ has x-intercept $\underline{-2}$. It <u>does not</u> have a y-intercept. The slope of this line is <u>undefined</u>.

The line $4y = 2$ has y-intercept $\frac{1}{2}$. It <u>does not</u> have an x-intercept. The slope of this line is $\underline{0}$.

29. $y = 3x - 1$
This equation is in the slope-intercept form, $y = mx + b$.
slope: 3;
y-intercept: -1

31. $4x - y = 7$
Solve for y to write the equation in slope-intercept form.

$$-y = -4x + 7 \Rightarrow y = 4x - 7$$

slope: 4; y-intercept: -7

33. $4y = -3x \Rightarrow y = -\frac{3}{4}x$ or $y = -\frac{3}{4}x + 0$

slope: $-\frac{3}{4}$; y-intercept 0

Copyright © 2013 Pearson Education, Inc.

35. $x + 2y = -4$

Solve the equation for y to write the equation in slope-intercept form.

$2y = -x - 4 \Rightarrow y = -\frac{1}{2}x - 2$

slope: $-\frac{1}{2}$; y-intercept: -2

37. $y - \frac{3}{2}x - 1 = 0$

Solve the equation for y to write the equation in slope-intercept form.

$y - \frac{3}{2}x - 1 = 0 \Rightarrow y = \frac{3}{2}x + 1$

slope: $\frac{3}{2}$; y-intercept: 1

39. (a) The line falls 2 units each time the x value increases by 1 unit. Therefore the slope is -2. The graph intersects the y-axis at the point (0, 1) and intersects the x-axis at $\left(\frac{1}{2}, 0\right)$, so the y-intercept is 1 and the x-intercept is $\frac{1}{2}$.

(b) The equation defining f is $y = -2x + 1$.

41. (a) The line falls 1 unit each time the x value increases by 3 units. Therefore the slope is $-\frac{1}{3}$. The graph intersects the y-axis at the point (0, 2), so the y-intercept is 2. The graph passes through (3, 1) and will fall 1 unit when the x value increases by 3, so the x-intercept is 6.

(b) The equation defining f is $y = -\frac{1}{3}x + 2$.

43. (a) The line falls 200 units each time the x value increases by 1 unit. Therefore the slope is -200. The graph intersects the y-axis at the point (0, 300) and intersects the x-axis at $\left(\frac{3}{2}, 0\right)$, so the y-intercept is 300 and the x-intercept is $\frac{3}{2}$.

(b) The equation defining f is $y = -200x + 300$.

45. (a) through $(-1, 4)$, parallel to $x + 3y = 5$
Find the slope of the line $x + 3y = 5$ by writing this equation in slope-intercept form.

$x + 3y = 5 \Rightarrow 3y = -x + 5 \Rightarrow$
$y = -\frac{1}{3}x + \frac{5}{3}$

The slope is $-\frac{1}{3}$.

Since the lines are parallel, $-\frac{1}{3}$ is also the slope of the line whose equation is to be found. Substitute $m = -\frac{1}{3}$, $x_1 = -1$, and $y_1 = 4$ into the point-slope form.

$y - y_1 = m(x - x_1)$
$y - 4 = -\frac{1}{3}\left[x - (-1)\right]$
$y - 4 = -\frac{1}{3}(x + 1)$
$3y - 12 = -x - 1 \Rightarrow x + 3y = 11$

(b) Solve for y.
$3y = -x + 11 \Rightarrow y = -\frac{1}{3}x + \frac{11}{3}$

47. (a) through (1, 6), perpendicular to $3x + 5y = 1$
Find the slope of the line $3x + 5y = 1$ by writing this equation in slope-intercept form.

$3x + 5y = 1 \Rightarrow 5y = -3x + 1 \Rightarrow$
$y = -\frac{3}{5}x + \frac{1}{5}$

This line has a slope of $-\frac{3}{5}$. The slope of any line perpendicular to this line is $\frac{5}{3}$, since $-\frac{3}{5}\left(\frac{5}{3}\right) = -1$. Substitute $m = \frac{5}{3}$, $x_1 = 1$, and $y_1 = 6$ into the point-slope form.

$y - 6 = \frac{5}{3}(x - 1)$
$3(y - 6) = 5(x - 1)$
$3y - 18 = 5x - 5$
$-13 = 5x - 3y$ or $5x - 3y = -13$

(b) Solve for y.
$3y = 5x + 13 \Rightarrow y = \frac{5}{3}x + \frac{13}{3}$

Copyright © 2013 Pearson Education, Inc.

49. (a) through (4, 1), parallel to $y = -5$
Since $y = -5$ is a horizontal line, any line parallel to this line will be horizontal and have an equation of the form $y = b$. Since the line passes through (4, 1), the equation is $y = 1$.

(b) The slope-intercept form is $y = 1$.

51. (a) through (−5, 6), perpendicular to $x = -2$.
Since $x = -2$ is a vertical line, any line perpendicular to this line will be horizontal and have an equation of the form $y = b$. Since the line passes through (−5, 6), the equation is $y = 6$.

(b) The slope-intercept form is $y = 6$.

53. (a) Find the slope of the line $3y + 2x = 6$.
$3y + 2x = 6 \Rightarrow 3y = -2x + 6 \Rightarrow$
$y = -\frac{2}{3}x + 2$

Thus, $m = -\frac{2}{3}$. A line parallel to
$3y + 2x = 6$ also has slope $-\frac{2}{3}$.
Solve for k using the slope formula.
$$\frac{2 - (-1)}{k - 4} = -\frac{2}{3}$$
$$\frac{3}{k - 4} = -\frac{2}{3}$$
$$3(k - 4)\left(\frac{3}{k - 4}\right) = 3(k - 4)\left(-\frac{2}{3}\right)$$
$$9 = -2(k - 4)$$
$$9 = -2k + 8$$
$$2k = -1 \Rightarrow k = -\frac{1}{2}$$

(b) Find the slope of the line $2y - 5x = 1$.
$2y - 5x = 1 \Rightarrow 2y = 5x + 1 \Rightarrow$
$y = \frac{5}{2}x + \frac{1}{2}$

Thus, $m = \frac{5}{2}$. A line perpendicular to $2y$
$- 5x = 1$ will have slope $-\frac{2}{5}$, since
$\frac{5}{2}\left(-\frac{2}{5}\right) = -1$.
Solve this equation for k.
$$\frac{3}{k - 4} = -\frac{2}{5}$$
$$5(k - 4)\left(\frac{3}{k - 4}\right) = 5(k - 4)\left(-\frac{2}{5}\right)$$
$$15 = -2(k - 4)$$
$$15 = -2k + 8$$
$$2k = -7 \Rightarrow k = -\frac{7}{2}$$

55. (0, 5492), (4, 7050)
$$m = \frac{7050 - 5492}{4 - 0} = \frac{1558}{4} = 389.5$$
The y-intercept is (0, 5492), so the equation is $y = 389.5x + 5492$.
The year 2010 is represented by $x = 5$, so
$y = 389.5(5) + 5492 \approx 7440.$
This is $\$7605 - 7440 = \$165$ less than the actual figure.

57. (a) (0, 12881), (12, 22449)
$$m = \frac{22{,}449 - 12{,}881}{12 - 0} = \frac{9568}{12} \approx 797.3$$
From the point (0, 12881), the value of b is 12,881. Therefore we have
$f(x) \approx 797.3x + 12{,}881.$

$f(x) \approx 797.3x + 12{,}881$

The average tuition increase is about $\$875$ per year for the period, because this is the slope of the line.

(b) 2007 corresponds to $x = 11$.
$f(11) \approx 797.3(11) + 12{,}881 \approx \$21{,}651$
This is a fairly good approximation.

(c) From the calculator,
$f(x) \approx 802.3x + 12{,}432$

59. (a) The ordered pairs are (0, 32) and (100, 212).

The slope is $m = \dfrac{212 - 32}{100 - 0} = \dfrac{180}{100} = \dfrac{9}{5}$.

Use $(x_1, y_1) = (0, 32)$ and $m = \frac{9}{5}$ in the point-slope form.
$$y - y_1 = m(x - x_1)$$
$$y - 32 = \frac{9}{5}(x - 0)$$
$$y - 32 = \frac{9}{5}x$$
$$y = \frac{9}{5}x + 32 \Rightarrow F = \frac{9}{5}C + 32$$

Copyright © 2013 Pearson Education, Inc.

(b) $F = \frac{9}{5}C + 32$

$5F = 9(C + 32)$

$5F = 9C + 160 \Rightarrow 9C = 5F - 160 \Rightarrow$

$9C = 5(F - 32) \Rightarrow C = \frac{5}{9}(F - 32)$

(c) $F = C \Rightarrow F = \frac{5}{9}(F - 32) \Rightarrow$

$9F = 5(F - 32) \Rightarrow 9F = 5F - 160 \Rightarrow$

$4F = -160 \Rightarrow F = -40$

$F = C$ when F is $-40°$.

61. (a) Since we are wanting to find C as a function of I, use the points (9937, 8285) and (12026, 10089), where the first component represents the independent variable, I. First find the slope of the line.

$$m = \frac{8285 - 10089}{9937 - 12026} = \frac{1804}{2089} \approx 0.8636$$

Now use either point, say (9937, 8285), and the point-slope form to find the equation.

$y - 8285 = 0.8636(x - 9937)$

$y - 8285 \approx 0.8636x - 8584.6$

$\quad\quad y \approx 0.8636x - 296.6$

or $C = 0.8636I - 296.6$

(b) Since the slope is 0.8636, the marginal propensity to consume is 0.8636.

63. Write the equation as an equivalent equation with 0 on one side: $7x - 2x + 4 - 5 = 3x + 1 \Rightarrow$ $7x - 2x + 4 - 5 - 3x - 1 = 0$. Now graph $Y = 7X - 2X + 4 - 5 - 3X - 1$ to find the x-intercept:

$Y = 7X - 2X + 4 - 5 - 3X - 1$

Solution set: {1}

65. Write the equation as an equivalent equation with 0 on one side:

$4x - 3(4 - 2x) = 2(x - 3) + 6x + 2 \Rightarrow$

$4x - 3(4 - 2x) - 2(x - 3) - 6x - 2 = 0$.

Now graph

$Y = 4X - 3(4 - 2X) - 2(X - 3) - 6X - 2$ to find the x-intercept:

$Y = 4X - 3(4 - 2X) - 2(X - 3) - 6X - 2$

Solution set: {4}

67. (a) $-2(x - 5) = -x - 2$

$-2x + 10 = -x - 2$

$\quad\quad 10 = x - 2$

$\quad\quad 12 = x$

Solution set: {12}

(b) Answers will vary. The largest value of x that is displayed in the standard viewing window is 10. As long as 12 is either a minimum or a maximum, or between the minimum and maximum, then the solution will be seen.

69. $d(O, P) = \sqrt{(x_1 - 0)^2 + (m_1 x_1 - 0)^2}$

$\quad\quad = \sqrt{x_1^2 + m_1^2 x_1^2}$

71. $d(P, Q) = \sqrt{(x_2 - x_1)^2 + (m_2 x_2 - m_1 x_1)^2}$

73. $-2m_1 m_2 x_1 x_2 - 2x_1 x_2 = 0$

$-2x_1 x_2 (m_1 m_2 + 1) = 0$

75. If two nonvertical lines are perpendicular, then the product of the slopes of these lines is -1.

77. $A(-1, 4), B(-2, -1), C(1, 14)$

For A and B, $m = \dfrac{-1 - 4}{-2 - (-1)} = \dfrac{-5}{-1} = 5$

For B and C, $m = \dfrac{14 - (-1)}{1 - (-2)} = \dfrac{15}{3} = 5$

For A and C, $m = \dfrac{14 - 4}{1 - (-1)} = \dfrac{10}{2} = 5$

Since all three slopes are the same, the points are collinear.

79. $A(-1, -3), B(-5, 12), C(1, -11)$

For A and B, $m = \dfrac{12 - (-3)}{-5 - (-1)} = -\dfrac{15}{4}$

For B and C, $m = \dfrac{-11 - 12}{1 - (-5)} = -\dfrac{23}{6}$

For A and C, $m = \dfrac{-11 - (-3)}{1 - (-1)} = -\dfrac{8}{2} = -4$

Since all three slopes are not the same, the points are not collinear.

Copyright © 2013 Pearson Education, Inc.

Summary Exercises on Graphs, Circles, Functions, and Equations

1. $P(3, 5)$, $Q(2, -3)$

(a) $d(P, Q) = \sqrt{(2-3)^2 + (-3-5)^2}$
$= \sqrt{(-1)^2 + (-8)^2}$
$= \sqrt{1 + 64} = \sqrt{65}$

(b) The midpoint M of the segment joining points P and Q has coordinates
$$\left(\frac{3+2}{2}, \frac{5+(-3)}{2}\right) = \left(\frac{5}{2}, \frac{2}{2}\right) = \left(\frac{5}{2}, 1\right).$$

(c) First find m: $m = \dfrac{-3-5}{2-3} = \dfrac{-8}{-1} = 8$
Use either point and the point-slope form.
$y - 5 = 8(x - 3)$
Change to slope-intercept form.
$y - 5 = 8x - 24 \Rightarrow y = 8x - 19$

3. $P(-2, 2)$, $Q(3, 2)$

(a) $d(P, Q) = \sqrt{[3-(-2)]^2 + (2-2)^2}$
$= \sqrt{5^2 + 0^2} = \sqrt{25 + 0} = \sqrt{25} = 5$

(b) The midpoint M of the segment joining points P and Q has coordinates
$$\left(\frac{-2+3}{2}, \frac{2+2}{2}\right) = \left(\frac{1}{2}, \frac{4}{2}\right) = \left(\frac{1}{2}, 2\right).$$

(c) First find m: $m = \dfrac{2-2}{3-(-2)} = \dfrac{0}{5} = 0$
All lines that have a slope of 0 are horizontal lines. The equation of a horizontal line has an equation of the form $y = b$. Since the line passes through $(3, 2)$, the equation is $y = 2$.

5. $P(5, -1)$, $Q(5, 1)$

(a) $d(P, Q) = \sqrt{(5-5)^2 + [1-(-1)]^2}$
$= \sqrt{0^2 + 2^2} = \sqrt{0 + 4} = \sqrt{4} = 2$

(b) The midpoint M of the segment joining points P and Q has coordinates
$$\left(\frac{5+5}{2}, \frac{-1+1}{2}\right) = \left(\frac{10}{2}, \frac{0}{2}\right) = (5, 0).$$

(c) First find m.
$$m = \frac{1-(-1)}{5-5} = \frac{2}{0} = \text{undefined}$$
All lines that have an undefined slope are vertical lines. The equation of a vertical line has an equation of the form $x = a$. Since the line passes through $(5, 1)$, the equation is $x = 5$. (Since the slope of a vertical line is undefined, this equation cannot be written in slope-intercept form.)

7. $P(2\sqrt{3}, 3\sqrt{5})$, $Q(6\sqrt{3}, 3\sqrt{5})$

(a) $d(P, Q) = \sqrt{(6\sqrt{3} - 2\sqrt{3})^2 + (3\sqrt{5} - 3\sqrt{5})^2}$
$= \sqrt{(4\sqrt{3})^2 + 0^2} = \sqrt{48} = 4\sqrt{3}$

(b) The midpoint M of the segment joining points P and Q has coordinates
$$\left(\frac{2\sqrt{3} + 6\sqrt{3}}{2}, \frac{3\sqrt{5} + 3\sqrt{5}}{2}\right)$$
$$= \left(\frac{8\sqrt{3}}{2}, \frac{6\sqrt{5}}{2}\right) = (4\sqrt{3}, 3\sqrt{5}).$$

(c) First find m: $m = \dfrac{3\sqrt{5} - 3\sqrt{5}}{6\sqrt{3} - 2\sqrt{3}} = \dfrac{0}{4\sqrt{3}} = 0$
All lines that have a slope of 0 are horizontal lines. The equation of a horizontal line has an equation of the form $y = b$. Since the line passes through $(2\sqrt{3}, 3\sqrt{5})$, the equation is $y = 3\sqrt{5}$.

9. Through $(-2, 1)$ and $(4, -1)$

First find m: $m = \dfrac{-1-1}{4-(-2)} = \dfrac{-2}{6} = -\dfrac{1}{3}$
Use either point and the point-slope form.
$y - (-1) = -\frac{1}{3}(x - 4)$
Change to slope-intercept form.
$3(y + 1) = -(x - 4) \Rightarrow 3y + 3 = -x + 4 \Rightarrow$
$3y = -x + 1 \Rightarrow y = -\frac{1}{3}x + \frac{1}{3}$

Copyright © 2013 Pearson Education, Inc.

11. the circle with center $(2, -1)$ and radius 3

$$(x-2)^2 + [y-(-1)]^2 = 3^2$$
$$(x-2)^2 + (y+1)^2 = 9$$

13. the line through $(3, -5)$ with slope $-\frac{5}{6}$

Write the equation in point-slope form.

$$y - (-5) = -\frac{5}{6}(x-3)$$

Change to standard form.

$$6(y+5) = -5(x-3) \Rightarrow 6y + 30 = -5x + 15$$
$$6y = -5x - 15 \Rightarrow y = -\frac{5}{6}x - \frac{15}{6}$$
$$y = -\frac{5}{6}x - \frac{5}{2}$$

15. a line through $(-3, 2)$ and parallel to the line $2x + 3y = 6$

First, find the slope of the line $2x + 3y = 6$ by writing this equation in slope-intercept form.

$$2x + 3y = 6 \Rightarrow 3y = -2x + 6 \Rightarrow y = -\frac{2}{3}x + 2$$

The slope is $-\frac{2}{3}$. Since the lines are parallel, $-\frac{2}{3}$ is also the slope of the line whose equation is to be found. Substitute $m = -\frac{2}{3}$, $x_1 = -3$, and $y_1 = 2$ into the point-slope form.

$$y - y_1 = m(x - x_1) \Rightarrow y - 2 = -\frac{2}{3}[x - (-3)] \Rightarrow$$
$$3(y-2) = -2(x+3) \Rightarrow 3y - 6 = -2x - 6 \Rightarrow$$
$$3y = -2x \Rightarrow y = -\frac{2}{3}x$$

17. $x^2 - 4x + y^2 + 2y = 4$

Complete the square on x and y separately.

$$(x^2 - 4x) + (y^2 + 2y) = 4$$
$$(x^2 - 4x + 4) + (y^2 + 2y + 1) = 4 + 4 + 1$$
$$(x-2)^2 + (y+1)^2 = 9$$

Yes, it is a circle. The circle has its center at $(2, -1)$ and radius 3.

19. $x^2 - 12x + y^2 + 20 = 0$

Complete the square on x and y separately.

$$(x^2 - 12x) + y^2 = -20$$
$$(x^2 - 12x + 36) + y^2 = -20 + 36$$
$$(x-6)^2 + y^2 = 16$$

Yes, it is a circle. The circle has its center at $(6, 0)$ and radius 4.

21. $x^2 - 2x + y^2 + 10 = 0$

Complete the square on x and y separately.

$$(x^2 - 2x) + y^2 = -10$$
$$(x^2 - 2x + 1) + y^2 = -10 + 1$$
$$(x-1)^2 + y^2 = -9$$

No, it is not a circle.

23. The equation of the circle is $(x-4)^2 + (y-5)^2 = 4^2$. Let $y = 2$ and solve for x: $(x-4)^2 + (2-5)^2 = 4^2 \Rightarrow$

$$(x-4)^2 + (-3)^2 = 4^2 \Rightarrow (x-4)^2 = 7 \Rightarrow$$
$$x - 4 = \pm\sqrt{7} \Rightarrow x = 4 \pm \sqrt{7}$$

The points of intersection are $\left(4 + \sqrt{7}, 2\right)$ and $\left(4 - \sqrt{7}, 2\right)$.

25. (a) The equation can be rewritten as

$-4y = -x - 6 \Rightarrow y = \frac{1}{4}x + \frac{6}{4} \Rightarrow y = \frac{1}{4}x + \frac{3}{2}$.

x can be any real number, so the domain is all real numbers and the range is also all real numbers.

domain: $(-\infty, \infty)$; range: $(-\infty, \infty)$

(b) Each value of x corresponds to just one value of y. $x - 4y = -6$ represents a function.

$$y = \frac{1}{4}x + \frac{3}{2} \Rightarrow f(x) = \frac{1}{4}x + \frac{3}{2}$$
$$f(-2) = \frac{1}{4}(-2) + \frac{3}{2} = -\frac{1}{2} + \frac{3}{2} = \frac{2}{2} = 1$$

Copyright © 2013 Pearson Education, Inc.

27. (a) $(x+2)^2 + y^2 = 25$ is a circle centered at $(-2, 0)$ with a radius of 5. The domain will start 5 units to the left of -2 and end 5 units to the right of -2. The domain will be $[-2 - 5, 2 + 5] = [-7, 3]$. The range will start 5 units below 0 and end 5 units above 0. The range will be $[0 - 5, 0 + 5] = [-5, 5]$.

(b) Since $(-2, 5)$ and $(-2, -5)$ both satisfy the relation, $(x+2)^2 + y^2 = 25$ does not represent a function.

Section 2.6   Graphs of Basic Functions

1. The equation $y = x^2$ matches graph E. The domain is $(-\infty, \infty)$.

3. The equation $y = x^3$ matches graph A. The range is $(-\infty, \infty)$.

5. Graph F is the graph of the identity function. Its equation is $y = x$.

7. The equation $y = \sqrt[3]{x}$ matches graph H. No, there is no interval over which the function is decreasing.

9. The graph in B is discontinuous at many points. Assuming the graph continues, the range would be $\{\ldots, -3, -2, -1, 0, 1, 2, 3, \ldots\}$.

11. The function is continuous over the entire domain of real numbers $(-\infty, \infty)$.

13. The function is continuous over the interval $[0, \infty)$.

15. The function has a point of discontinuity at $(3, 1)$. It is continuous over the interval $(-\infty, 3)$ and the interval $(3, \infty)$.

17. $f(x) = \begin{cases} 2x & \text{if } x \le -1 \\ x-1 & \text{if } x > -1 \end{cases}$

(a) $f(-5) = 2(-5) = -10$

(b) $f(-1) = 2(-1) = -2$

(c) $f(0) = 0 - 1 = -1$

(d) $f(3) = 3 - 1 = 2$

19. $f(x) = \begin{cases} 2+x & \text{if } x < -4 \\ -x & \text{if } -4 \le x \le 2 \\ 3x & \text{if } x > 2 \end{cases}$

(a) $f(-5) = 2 + (-5) = -3$

(b) $f(-1) = -(-1) = 1$

(c) $f(0) = -0 = 0$

(d) $f(3) = 3 \cdot 3 = 9$

21. $f(x) = \begin{cases} x-1 & \text{if } x \le 3 \\ 2 & \text{if } x > 3 \end{cases}$

Draw the graph of $y = x - 1$ to the left of $x = 3$, including the endpoint at $x = 3$. Draw the graph of $y = 2$ to the right of $x = 3$, and note that the endpoint at $x = 3$ coincides with the endpoint of the other ray.

23. $f(x) = \begin{cases} 4-x & \text{if } x < 2 \\ 1+2x & \text{if } x \ge 2 \end{cases}$

Draw the graph of $y = 4 - x$ to the left of $x = 2$, but do not include the endpoint. Draw the graph of $y = 1 + 2x$ to the right of $x = 2$, including the endpoint.

25. $f(x) = \begin{cases} -3 & \text{if } x \le 1 \\ -1 & \text{if } x > 1 \end{cases}$

Graph the line $y = -3$ to the left of $x = 1$, including the endpoint. Draw $y = -1$ to the right of $x = 1$, but do not include the endpoint.

Copyright © 2013 Pearson Education, Inc.

27. $f(x) = \begin{cases} 2+x & \text{if } x < -4 \\ -x & \text{if } -4 \le x \le 5 \\ 3x & \text{if } x > 5 \end{cases}$

Draw the graph of $y = 2 + x$ to the left of –4, but do not include the endpoint at $x = 4$. Draw the graph of $y = -x$ between –4 and 5, including both endpoints. Draw the graph of $y = 3x$ to the right of 5, but do not include the endpoint at $x = 5$.

29. $f(x) = \begin{cases} -\frac{1}{2}x^2+2 & \text{if } x \le 2 \\ \frac{1}{2}x & \text{if } x > 2 \end{cases}$

Graph the curve $y = -\frac{1}{2}x^2 + 2$ to the left of $x = 2$, including the endpoint at (2, 0). Graph the line $y = \frac{1}{2}x$ to the right of $x = 2$, but do not include the endpoint at (2, 1). Notice that the endpoints of the pieces do not coincide.

$f(x) = \begin{cases} -\frac{1}{2}x^2 + 2 & \text{if } x \le 2 \\ \frac{1}{2}x & \text{if } x > 2 \end{cases}$

31. $f(x) = \begin{cases} 2x & \text{if } -5 \le x < -1 \\ -2 & \text{if } -1 \le x < 0 \\ x^2 - 2 & \text{if } 0 \le x \le 2 \end{cases}$

Graph the line $y = 2x$ between $x = -5$ and $x = -1$, including the left endpoint at (–5, –10), but not including the right endpoint at (–1, –2). Graph the line $y = -2$ between $x = -1$ and $x = 0$, including the left endpoint at (–1, –2) and not including the right endpoint at (0, –2). Note that (–1, –2) coincides with the first two sections, so it is included. Graph the curve $y = x^2 - 2$ from $x = 0$ to $x = 2$, including the endpoints at (0, –2) and (2, 2).

Note that (0, –2) coincides with the second two sections, so it is included. The graph ends at $x = -5$ and $x = 2$.

$f(x) = \begin{cases} 2x & \text{if } -5 \le x < -1 \\ -2 & \text{if } -1 \le x < 0 \\ x^2 - 2 & \text{if } 0 \le x \le 2 \end{cases}$

33. $f(x) = \begin{cases} x^3 + 3 & \text{if } -2 \le x \le 0 \\ x+3 & \text{if } 0 < x < 1 \\ 4+x-x^2 & \text{if } 1 \le x \le 3 \end{cases}$

Graph the curve $y = x^3 + 3$ between $x = -2$ and $x = 0$, including the endpoints at (–2, –5) and (0, 3). Graph the line $y = x + 3$ between $x = 0$ and $x = 1$, but do not include the endpoints at (0, 3) and (1, 4). Graph the curve $y = 4 + x - x^2$ from $x = 1$ to $x = 3$, including the endpoints at (1, 4) and (3, –2). The graph ends at $x = -2$ and $x = 3$.

$f(x) = \begin{cases} x^3 + 3 & \text{if } -2 \le x \le 0 \\ x+3 & \text{if } 0 < x < 1 \\ 4+x-x^2 & \text{if } 1 \le x \le 3 \end{cases}$

35. The solid circle on the graph shows that the endpoint (0, –1) is part of the graph, while the open circle shows that the endpoint (0, 1) is not part of the graph. The graph is made up of parts of two horizontal lines. The function which fits this graph is

$f(x) = \begin{cases} -1 & \text{if } x \le 0 \\ 1 & \text{if } x > 0. \end{cases}$

domain: $(-\infty, \infty)$; range: $\{-1, 1\}$

Copyright © 2013 Pearson Education, Inc.

37. The graph is made up of parts of two horizontal lines. The solid circle shows that the endpoint (0, 2) of the one on the left belongs to the graph, while the open circle shows that the endpoint (0, –1) of the one on the right does not belong to the graph. The function that fits this graph is

$$f(x) = \begin{cases} 2 \text{ if } x \le 0 \\ -1 \text{ if } x > 1. \end{cases}$$

domain: $(-\infty, 0] \cup (1, \infty)$; range: $\{-1, 2\}$

39. For $x \le 0$, that piece of the graph goes through the points (–1, –1) and (0, 0). The slope is 1, so the equation of this piece is $y = x$. For $x > 0$, that piece of the graph is a horizontal line passing through (2, 2), so its equation is $y = 2$. We can write the function as

$$f(x) = \begin{cases} x \text{ if } x \le 0 \\ 2 \text{ if } x > 0 \end{cases}$$

domain: $(-\infty, \infty)$ range: $(-\infty, 0] \cup \{2\}$

41. For $x < 1$, that piece of the graph is a curve passes through (–8, –2), (–1, –1) and (1, 1), so the equation of this piece is $y = \sqrt[3]{x}$. The right piece of the graph passes through (1, 2) and (2, 3). $m = \dfrac{2-3}{1-2} = 1$, and the equation of the line is $y - 2 = x - 1 \Rightarrow y = x + 1$. We can write the function as $f(x) = \begin{cases} \sqrt[3]{x} \text{ if } x < 1 \\ x + 1 \text{ if } x \ge 1 \end{cases}$

domain: $(-\infty, \infty)$ range: $(-\infty, 1) \cup [2, \infty)$

43. $f(x) = [\![-x]\!]$

Plot points.

x	$-x$	$f(x) = [\![-x]\!]$
–2	2	2
–1.5	1.5	1
–1	1	1
–0.5	0.5	0
0	0	0
0.5	–0.5	–1
1	–1	–1
1.5	–1.5	–2
2	–2	–2

More generally, to get $y = 0$, we need $0 \le -x < 1 \Rightarrow 0 \ge x > -1 \Rightarrow -1 < x \le 0$. To get $y = 1$, we need $1 \le -x < 2 \Rightarrow -1 \ge x > -2 \Rightarrow -2 < x \le -1$. Follow this pattern to graph the step function.

domain: $(-\infty, \infty)$; range: $\{...,-2,-1,0,1,2,...\}$

45. $f(x) = [\![2x]\!]$

To get $y = 0$, we need $0 \le 2x < 1 \Rightarrow 0 \le x < \frac{1}{2}$.

To get $y = 1$, we need $1 \le 2x < 2 \Rightarrow \frac{1}{2} \le x < 1$.

To get $y = 2$, we need $2 \le 2x < 3 \Rightarrow 1 \le x < \frac{3}{2}$.

Follow this pattern to graph the step function.

domain: $(-\infty, \infty)$; range: $\{...,-2,-1,0,1,2,...\}$

47. The cost of mailing a letter that weighs more than 1 ounce and less than 2 ounces is the same as the cost of a 2-ounce letter, and the cost of mailing a letter that weighs more than 2 ounces and less than 3 ounces is the same as the cost of a 3-ounce letter, etc.

Copyright © 2013 Pearson Education, Inc.

49.

51. (a) For $0 \le x \le 4$, $m = \dfrac{39.2 - 42.8}{4 - 0} = -0.9$,

so $y = -0.9x + 42.8$. For $4 < x \le 8$,

$m = \dfrac{32.7 - 39.2}{8 - 4} = -1.625$, so the

equation is $y - 32.7 = -1.625(x - 8) \Rightarrow$

$y = -1.625x + 45.7$

(b) $f(x) = \begin{cases} -0.9x + 42.8 \text{ if } 0 \le x \le 4 \\ -1.625x + 45.7 \text{ if } 4 < x \le 8 \end{cases}$

53. (a) The initial amount is 50,000 gallons. The final amount is 30,000 gallons.

(b) The amount of water in the pool remained constant during the first and fourth days.

(c) $f(2) \approx 45,000$; $f(4) = 40,000$

(d) The slope of the segment between $(1, 50000)$ and $(3, 40000)$ is -5000, so the water was being drained at 5000 gallons per day.

55. (a) Since there is no charge for additional length, we use the greatest integer function. The cost is based on multiples of two feet, so $f(x) = 0.8 \left[\!\left[\frac{x}{2} \right]\!\right]$ if $6 \le x \le 18$.

(b) $f(8.5) = 0.8 \left[\!\left[\frac{8.5}{2} \right]\!\right] = 0.8(4) = \$3.20$

$f(15.2) = 0.8 \left[\!\left[\frac{15.2}{2} \right]\!\right] = 0.8(7) = \$5.60$

Section 2.7   Graphing Techniques

1. (a) B; $y = (x - 7)^2$ is a shift of $y = x^2$, 7 units to the right.

(b) D; $y = x^2 - 7$ is a shift of $y = x^2$, 7 units downward.

(c) E; $y = 7x^2$ is a vertical stretch of $y = x^2$, by a factor of 7.

(d) A; $y = (x + 7)^2$ is a shift of $y = x^2$, 7 units to the left.

(e) C; $y = x^2 + 7$ is a shift of $y = x^2$, 7 units upward.

3. (a) B; $y = x^2 + 2$ is a shift of $y = x^2$, 2 units upward.

(b) A; $y = x^2 - 2$ is a shift of $y = x^2$, 2 units downward.

(c) G; $y = (x + 2)^2$ is a shift of $y = x^2$, 2 units to the left.

(d) C; $y = (x - 2)^2$ is a shift of $y = x^2$, 2 units to the right.

(e) F; $y = 2x^2$ is a vertical stretch of $y = x^2$, by a factor of 2.

(f) D; $y = -x^2$ is a reflection of $y = x^2$, across the x-axis.

(g) H; $y = (x - 2)^2 + 1$ is a shift of $y = x^2$, 2 units to the right and 1 unit upward.

(h) E; $y = (x + 2)^2 + 1$ is a shift of $y = x^2$, 2 units to the left and 1 unit upward.

(i) I; $y = (x + 2)^2 - 1$ is a shift of $y = x^2$, 2 units to the left and 1 unit down.

5. (a) F; $y = |x - 2|$ is a shift of $y = |x|$ 2 units to the right.

(b) C; $y = |x| - 2$ is a shift of $y = |x|$ 2 units downward.

(c) H; $y = |x| + 2$ is a shift of $y = |x|$ 2 units upward.

(d) D; $y = 2|x|$ is a vertical stretch of $y = |x|$ by a factor of 2.

(e) G; $y = -|x|$ is a reflection of $y = |x|$ across the x-axis.

(f) A; $y = |-x|$ is a reflection of $y = |x|$ across the y-axis.

(g) E; $y = -2|x|$ is a reflection of $y = 2|x|$ across the x-axis. $y = 2|x|$ is a vertical stretch of $y = |x|$ by a factor of 2.

(h) I; $y = |x - 2| + 2$ is a shift of $y = |x|$ 2 units to the right and 2 units upward.

Copyright © 2013 Pearson Education, Inc.

(i) B; $y = |x + 2| - 2$ is a shift of $y = |x|$ 2
units to the left and 2 units downward.

7. $y = 3|x|$

x	$y = |x|$	$y = 3|x|$
-2	2	6
-1	1	3
0	0	0
1	1	3
2	2	6

9. $y = \frac{2}{3}|x|$

x	$y = |x|$	$y = \frac{2}{3}|x|$
-3	3	2
-2	2	$\frac{4}{3}$
-1	1	$\frac{2}{3}$
0	0	0
1	1	$\frac{2}{3}$
2	2	$\frac{4}{3}$
3	3	2

11. $y = 2x^2$

x	$y = x^2$	$y = 2x^2$
-2	4	8
-1	1	2

x	$y = x^2$	$y = 2x^2$
0	0	0
1	1	2
2	4	8

13. $y = \frac{1}{2}x^2$

x	$y = x^2$	$y = \frac{1}{2}x^2$
-2	4	2
-1	1	$\frac{1}{2}$
0	0	0
1	1	$\frac{1}{2}$
2	4	2

15. $y = -\frac{1}{2}x^2$

x	$y = x^2$	$y = -\frac{1}{2}x^2$
-3	9	$-\frac{9}{2}$
-2	4	-2
-1	1	$-\frac{1}{2}$
0	0	0

x	$y = x^2$	$y = -\frac{1}{2}x^2$
1	1	$-\frac{1}{2}$
2	4	-2
3	9	$-\frac{9}{2}$

(continued on next page)

Copyright © 2013 Pearson Education, Inc.

(*continued*)

17. $y = -3|x|$

x	$y = |x|$	$y = -3|x|$
−2	2	−6
−1	1	−3
0	0	0
1	1	−3
2	2	−6

19. $y = \left|-\frac{1}{2}x\right|$

x	$y = |x|$	$y = \left|-\frac{1}{2}x\right|$ $= \left|-\frac{1}{2}\right||x| = \frac{1}{2}|x|$
−4	4	2
−3	3	$\frac{3}{2}$
−2	2	1
−1	1	$\frac{1}{2}$
0	0	0
1	1	$\frac{1}{2}$
2	2	1
3	3	$\frac{3}{2}$
4	4	2

21. $y = \sqrt{4x}$

x	$y = \sqrt{x}$	$y = \sqrt{4x} = 2\sqrt{x}$
0	0	0
1	1	2
2	$\sqrt{2}$	$2\sqrt{2}$
3	$\sqrt{3}$	$2\sqrt{3}$
4	2	4

23. $y = -\sqrt{-x}$

x	$y = \sqrt{-x}$	$y = -\sqrt{-x}$
−4	2	−2
−3	$\sqrt{3}$	$-\sqrt{3}$
−2	$\sqrt{2}$	$-\sqrt{2}$
−1	1	−1
0	0	0

25. (a) $y = f(x+4)$ is a horizontal translation of f, 4 units to the left. The point that corresponds to (8, 12) on this translated function would be $(8-4, 12) = (4, 12)$.

Copyright © 2013 Pearson Education, Inc.

(b) $y = f(x) + 4$ is a vertical translation of f, 4 units up. The point that corresponds to $(8, 12)$ on this translated function would be $(8, 12 + 4) = (8, 16)$.

27. (a) $y = f(4x)$ is a horizontal shrinking of f, by a factor of 4. The point that corresponds to $(8, 12)$ on this translated function is $\left(8 \cdot \frac{1}{4}, 12\right) = (2, 12)$.

(b) $y = f\left(\frac{1}{4}x\right)$ is a horizontal stretching of f, by a factor of 4. The point that corresponds to $(8, 12)$ on this translated function is $(8 \cdot 4, 12) = (32, 12)$.

29. (a) The point that is symmetric to $(5, -3)$ with respect to the x-axis is $(5, 3)$.

(b) The point that is symmetric to $(5, -3)$ with respect to the y-axis is $(-5, -3)$.

(c) The point that is symmetric to $(5, -3)$ with respect to the origin is $(-5, 3)$.

31. (a) The point that is symmetric to $(-4, -2)$ with respect to the x-axis is $(-4, 2)$.

(b) The point that is symmetric to $(-4, -2)$ with respect to the y-axis is $(4, -2)$.

(c) The point that is symmetric to $(-4, -2)$ with respect to the origin is $(4, 2)$.

33. The graph of $y = |x - 2|$ is symmetric with respect to the line $x = 2$.

35. $y = x^2 + 5$

Replace x with $-x$ to obtain
$y = (-x)^2 + 5 = x^2 + 5$. The result is the same as the original equation, so the graph is symmetric with respect to the y-axis. Since y is a function of x, the graph cannot be symmetric with respect to the x-axis. Replace x with $-x$ and y with $-y$ to obtain
$-y = (-x)^2 + 2 \Rightarrow -y = x^2 + 2 \Rightarrow y = -x^2 - 2$.
The result is not the same as the original equation, so the graph is not symmetric with respect to the origin. Therefore, the graph is symmetric with respect to the y-axis only.

37. $x^2 + y^2 = 12$

Replace x with $-x$ to obtain
$(-x)^2 + y^2 = 12 \Rightarrow x^2 + y^2 = 12$.
The result is the same as the original equation, so the graph is symmetric with respect to the y-axis. Replace y with $-y$ to obtain
$x^2 + (-y)^2 = 12 \Rightarrow x^2 + y^2 = 12$
The result is the same as the original equation, so the graph is symmetric with respect to the x-axis. Since the graph is symmetric with respect to the x-axis and y-axis, it is also symmetric with respect to the origin.

39. $y = -4x^3 + x$

Replace x with $-x$ to obtain
$y = -4(-x)^3 + (-x) \Rightarrow y = -4(-x^3) - x \Rightarrow y = 4x^3 - x.$
The result is not the same as the original equation, so the graph is not symmetric with respect to the y-axis. Replace y with $-y$ to obtain $-y = -4x^3 + x \Rightarrow y = 4x^3 - x.$
The result is not the same as the original equation, so the graph is not symmetric with respect to the x-axis. Replace x with $-x$ and y with $-y$ to obtain
$-y = -4(-x)^3 + (-x) \Rightarrow -y = -4(-x^3) - x \Rightarrow$
$-y = 4x^3 - x \Rightarrow y = -4x^3 + x.$
The result is the same as the original equation, so the graph is symmetric with respect to the origin. Therefore, the graph is symmetric with respect to the origin only.

Copyright © 2013 Pearson Education, Inc.

41. $y = x^2 - x + 8$

Replace x with $-x$ to obtain

$y = (-x)^2 - (-x) + 8 \Rightarrow y = x^2 + x + 8.$

The result is not the same as the original equation, so the graph is not symmetric with respect to the y-axis. Since y is a function of x, the graph cannot be symmetric with respect to the x-axis. Replace x with $-x$ and y with $-y$ to obtain $-y = (-x)^2 - (-x) + 8 \Rightarrow$

$-y = x^2 + x + 8 \Rightarrow y = -x^2 - x - 8.$

The result is not the same as the original equation, so the graph is not symmetric with respect to the origin. Therefore, the graph has none of the listed symmetries.

43. $f(x) = -x^3 + 2x$

$f(-x) = -(-x)^3 + 2(-x)$
$\quad = x^3 - 2x = -(-x^3 + 2x) = -f(x)$

The function is odd.

45. $f(x) = 0.5x^4 - 2x^2 + 6$

$f(-x) = 0.5(-x)^4 - 2(-x)^2 + 6$
$\quad = 0.5x^4 - 2x^2 + 6 = f(x)$

The function is even.

47. $f(x) = x^3 - x + 9$

$f(x) = (-x)^3 - (-x) + 9$
$\quad = -x^3 + x + 9 = -(x^3 - x - 9) \neq -f(x)$

The function is neither.

49. $y = x^2 - 1$

This graph may be obtained by translating the graph of $y = x^2$, 1 unit downward.

51. $y = x^2 + 2$

This graph may be obtained by translating the graph of $y = x^2$, 2 units upward.

53. $y = (x - 4)^2$

This graph may be obtained by translating the graph of $y = x^2$, 4 units to the right.

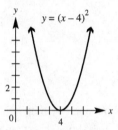

55. $y = (x + 2)^2$

This graph may be obtained by translating the graph of $y = x^2$, 2 units to the left.

57. $y = |x| - 1$

The graph is obtained by translating the graph of $y = |x|$, 1 unit downward.

Copyright © 2013 Pearson Education, Inc.

59. $y = -(x+1)^3$

This graph may be obtained by translating the graph of $y = x^3$, 1 unit to the left. It is then reflected across the x-axis.

61. $y = 2x^2 - 1$

This graph may be obtained by translating the graph of $y = x^2$, 1 unit down. It is then stretched vertically by a factor of 2.

63. $f(x) = 2(x-2)^2 - 4$

This graph may be obtained by translating the graph of $y = x^2$, 2 units to the right and 4 units down. It is then stretched vertically by a factor of 2.

65. $f(x) = \sqrt{x+2}$

This graph may be obtained by translating the graph of $y = \sqrt{x}$ two units to the left.

67. $f(x) = -\sqrt{x}$

This graph may be obtained by reflecting the graph of $y = \sqrt{x}$ across the x-axis.

69. $f(x) = 2\sqrt{x} + 1$

This graph may be obtained by stretching the graph of $y = \sqrt{x}$ vertically by a factor of two and then translating the resulting graph one unit up.

71. $y = \frac{1}{2}x^3 - 4$

This graph may be obtained by stretching the graph of $y = x^3$ vertically by a factor of $\frac{1}{2}$, then shifting the resulting graph down four units.

Copyright © 2013 Pearson Education, Inc.

73. $y = (x+3)^3$

This graph may be obtained by shifting the graph of $y = x^3$ three units left.

75. $y = \frac{2}{3}(x-2)^2$

This graph may be obtained by translating the graph of $y = x^2$ two units to the right, then stretching the resulting graph vertically by a factor of $\frac{2}{3}$.

77. (a) $y = g(-x)$

The graph of $g(x)$ is reflected across the y-axis.

(b) $y = g(x-2)$

The graph of $g(x)$ is translated to the right 2 units.

(c) $y = -g(x) + 2$

The graph of $g(x)$ is reflected across the x-axis and translated 2 units up.

79. It is the graph of $f(x) = |x|$ translated 1 unit to the left, reflected across the x-axis, and translated 3 units up. The equation is $y = -|x+1| + 3$.

81. It is the graph of $f(x) = \sqrt{x}$ translated one unit right and then three units down. The equation is $y = \sqrt{x-1} - 3$.

83. It is the graph of $g(x) = \sqrt{x}$ translated 4 units to the left, stretched vertically by a factor of 2, and translated 4 units down. The equation is $y = 2\sqrt{x+4} - 4$.

85. Since $f(3) = 6$, the point $(3, 6)$ is on the graph. Since the graph is symmetric with respect to the origin, the point $(-3, -6)$ is on the graph. Therefore, $f(-3) = -6$.

87. Since $f(3) = 6$, the point $(3, 6)$ is on the graph. Since the graph is symmetric with respect to the line $x = 6$ and since the point $(3, 6)$ is 3 units to the left of the line $x = 6$, the image point of $(3, 6)$, 3 units to the right of the line $x = 6$, is $(9, 6)$. Therefore, $f(9) = 6$.

89. An odd function is a function whose graph is symmetric with respect to the origin. Since $(3, 6)$ is on the graph, $(-3, -6)$ must also be on the graph. Therefore, $f(-3) = -6$.

Copyright © 2013 Pearson Education, Inc.

91. $f(x) = 2x + 5$: Translate the graph of $f(x)$ up 2 units to obtain the graph of $t(x) = (2x + 5) + 2 = 2x + 7$.

Now translate the graph of $t(x) = 2x + 7$ left 3 units to obtain the graph of $g(x) = 2(x + 3) + 7 = 2x + 6 + 7 = 2x + 13$.

(Note that if the original graph is first translated to the left 3 units and then up 2 units, the final result will be the same.)

93. **(a)** Since $f(-x) = f(x)$, the graph is symmetric with respect to the y-axis.

(b) Since $f(-x) = -f(x)$, the graph is symmetric with respect to the origin.

Chapter 2 Quiz
(Sections 2.5–2.7)

1. **(a)** First, find the slope: $m = \dfrac{9 - 5}{-1 - (-3)} = 2$

Choose either point, say, $(-3, 5)$, to find the equation of the line:
$y - 5 = 2(x - (-3)) \Rightarrow y = 2(x + 3) + 5 \Rightarrow$
$y = 2x + 11$.

(b) To find the x-intercept, let $y = 0$ and solve for x: $0 = 2x + 11 \Rightarrow x = -\frac{11}{2}$. The x-intercept is $-\frac{11}{2}$.

3. **(a)** $x = -8$ **(b)** $y = 5$

5. $f(x) = 0.40[\![x]\!] + 0.75$
$f(5.5) = 0.40[\![5.5]\!] + 0.75$
$= 0.40(5) + 0.75 = 2.75$
A 5.5-minute call costs \$2.75.

7. $f(x) = -x^3 + 1$

Reflect the graph of $f(x) = x^3$ across the x-axis, and then translate the resulting graph one unit up.

9. This is the graph of $f(x) = \sqrt{x}$, translated four units to the left, reflected across the x-axis, and then translated two units down. The equation is $y = -\sqrt{x + 4} - 2$.

Section 2.8 Function Operations and Composition

In Exercises 1–7, $f(x) = x^2 + 3$ and $g(x) = -2x + 6$.

1. $(f + g)(3) = f(3) + g(3)$
$= \left[(3)^2 + 3\right] + \left[-2(3) + 6\right]$
$= 12 + 0 = 12$

3. $(f - g)(-1) = f(-1) - g(-1)$
$= [(-1)^2 + 3] - [-2(-1) + 6]$
$= 4 - 8 = -4$

5. $(fg)(4) = f(4) \cdot g(4)$
$= [4^2 + 3] \cdot [-2(4) + 6]$
$= 19 \cdot (-2) = -38$

7. $\left(\dfrac{f}{g}\right)(-1) = \dfrac{f(-1)}{g(-1)} = \dfrac{(-1)^2 + 3}{-2(-1) + 6} = \dfrac{4}{8} = \dfrac{1}{2}$

9. $f(x) = 3x + 4, g(x) = 2x - 5$

i) $(f + g)(x) = f(x) + g(x)$
$= (3x + 4) + (2x - 5) = 5x - 1$

ii) $(f - g)(x) = f(x) - g(x)$
$= (3x + 4) - (2x - 5) = x + 9$

iii) $(fg)(x) = f(x) \cdot g(x) = (3x + 4)(2x - 5)$
$= 6x^2 - 15x + 8x - 20$
$= 6x^2 - 7x - 20$

Copyright © 2013 Pearson Education, Inc.

iv) $\left(\dfrac{f}{g}\right)(x) = \dfrac{f(x)}{g(x)} = \dfrac{3x+4}{2x-5}$

The domains of both f and g are the set of all real numbers, so the domains of $f+g$, $f-g$, and fg are all $(-\infty,\infty)$. The domain of $\frac{f}{g}$ is the set of all real numbers for which $g(x)\neq 0$. This is the set of all real numbers except $\frac{5}{2}$, which is written in interval notation as $\left(-\infty,\frac{5}{2}\right)\cup\left(\frac{5}{2},\infty\right)$.

11. $f(x) = 2x^2 - 3x,\ g(x) = x^2 - x + 3$

i) $(f+g)(x) = f(x)+g(x)$
$= (2x^2-3x)+(x^2-x+3)$
$= 3x^2-4x+3$

ii) $(f-g)(x) = f(x)-g(x)$
$= (2x^2-3x)-(x^2-x+3)$
$= 2x^2-3x-x^2+x-3$
$= x^2-2x-3$

iii) $(fg)(x) = f(x)\cdot g(x)$
$= (2x^2-3x)(x^2-x+3)$
$= 2x^4-2x^3+6x^2-3x^3+3x^2-9x$
$= 2x^4-5x^3+9x^2-9x$

iv) $\left(\dfrac{f}{g}\right)(x) = \dfrac{f(x)}{g(x)} = \dfrac{2x^2-3x}{x^2-x+3}$

The domains of both f and g are the set of all real numbers, so the domains of $f+g$, $f-g$, and fg are all $(-\infty,\infty)$. The domain of $\frac{f}{g}$ is the set of all real numbers for which $g(x)\neq 0$. If $x^2-x+3=0$, then by the quadratic formula $x=\frac{1\pm i\sqrt{11}}{2}$. The equation has no real solutions. There are no real numbers which make the denominator zero. Thus, the domain of $\frac{f}{g}$ is also $(-\infty,\infty)$.

13. $f(x) = \sqrt{4x-1},\ g(x) = \dfrac{1}{x}$

i) $(f+g)(x) = f(x)+g(x) = \sqrt{4x-1}+\dfrac{1}{x}$

ii) $(f-g)(x) = f(x)-g(x) = \sqrt{4x-1}-\dfrac{1}{x}$

iii) $(fg)(x) = f(x)\cdot g(x)$
$= \sqrt{4x-1}\left(\dfrac{1}{x}\right) = \dfrac{\sqrt{4x-1}}{x}$

iv) $\left(\dfrac{f}{g}\right)(x) = \dfrac{f(x)}{g(x)} = \dfrac{\sqrt{4x-1}}{\frac{1}{x}} = x\sqrt{4x-1}$

Since $4x-1\geq 0 \Rightarrow 4x\geq 1 \Rightarrow x\geq\frac{1}{4}$, the domain of f is $\left[\frac{1}{4},\infty\right)$. The domain of g is $(-\infty,0)\cup(0,\infty)$. Considering the intersection of the domains of f and g, the domains of $f+g$, $f-g$, and fg are all $\left[\frac{1}{4},\infty\right)$. Since $\frac{1}{x}\neq 0$ for any value of x, the domain of $\frac{f}{g}$ is also $\left[\frac{1}{4},\infty\right)$.

15. Numerical answers may vary.
$M(2004)\approx 260$ and $F(2004)\approx 400$, thus
$T(2004) = M(2004)+F(2004)$
$= 260+400 = 660$ (thousand).

17. Looking at the graphs of the functions, the slopes of the line segments for the period 2000–2004 are much steeper than the slopes of the corresponding line segments for the period 2004–2008. Thus, the number of associate's degrees increased more rapidly during the period 2000–2004.

19. Numerical answers may vary.
$(T-S)(2000) = T(2000)-S(2000)$
$= 19-13 = 6$
It represents the dollars in billions spent for general science in 2000.

21. In space and other technologies spending was almost static in the years 1995–2000.

23. (a) $(f+g)(2) = f(2)+g(2)$
$= 4+(-2) = 2$

(b) $(f-g)(1) = f(1)-g(1) = 1-(-3) = 4$

(c) $(fg)(0) = f(0)\cdot g(0) = 0(-4) = 0$

(d) $\left(\dfrac{f}{g}\right)(1) = \dfrac{f(1)}{g(1)} = \dfrac{1}{-3} = -\dfrac{1}{3}$

25. (a) $(f+g)(-1) = f(-1)+g(-1) = 0+3 = 3$

(b) $(f-g)(-2) = f(-2)-g(-2)$
$= -1-4 = -5$

(c) $(fg)(0) = f(0)\cdot g(0) = 1\cdot 2 = 2$

Copyright © 2013 Pearson Education, Inc.

(d) $\left(\dfrac{f}{g}\right)(2) = \dfrac{f(2)}{g(2)} = \dfrac{3}{0} = $ undefined

27. (a) $(f+g)(2) = f(2)+g(2) = 7+(-2) = 5$

(b) $(f-g)(4) = f(4)-g(4) = 10-5 = 5$

(c) $(fg)(-2) = f(-2)\cdot g(-2) = 0\cdot 6 = 0$

(d) $\left(\dfrac{f}{g}\right)(0) = \dfrac{f(0)}{g(0)} = \dfrac{5}{0} = $ undefined

29.

x	$f(x)$	$g(x)$	$(f+g)(x)$	$(f-g)(x)$	$(fg)(x)$	$\left(\dfrac{f}{g}\right)(x)$
-2	0	6	$0+6=6$	$0-6=-6$	$0\cdot 6=0$	$\frac{0}{6}=0$
0	5	0	$5+0=5$	$5-0=5$	$5\cdot 0=0$	$\frac{5}{0}=$ undefined
2	7	-2	$7+(-2)=5$	$7-(-2)=9$	$7(-2)=-14$	$\frac{7}{-2}=-3.5$
4	10	5	$10+5=15$	$10-5=5$	$10\cdot 5=50$	$\frac{10}{5}=2$

31. Answers will vary.

The difference quotient, $\dfrac{f(x+h)-f(x)}{h}$, represents the slope of the secant line which passes through points $(x, f(x))$ and $(x+h, f(x+h))$. The formula is derived by applying the rule that slope represents a change in y to a change in x.

33. $f(x) = 2-x$

(a) $f(x+h) = 2-(x+h) = 2-x-h$

(b) $f(x+h)-f(x) = (2-x-h)-(2-x)$
$= 2-x-h-2+x = -h$

(c) $\dfrac{f(x+h)-f(x)}{h} = \dfrac{-h}{h} = -1$

35. $f(x) = 6x+2$

(a) $f(x+h) = 6(x+h)+2 = 6x+6h+2$

(b) $f(x+h)-f(x)$
$= (6x+6h+2)-(6x+2)$
$= 6x+6h+2-6x-2 = 6h$

(c) $\dfrac{f(x+h)-f(x)}{h} = \dfrac{6h}{h} = 6$

37. $f(x) = -2x+5$

(a) $f(x+h) = -2(x+h)+5$
$= -2x-2h+5$

(b) $f(x+h)-f(x)$
$= (-2x-2h+5)-(-2x+5)$
$= -2x-2h+5+2x-5$
$= -2h$

(c) $\dfrac{f(x+h)-f(x)}{h} = \dfrac{-2h}{h} = -2$

39. $f(x) = \dfrac{1}{x}$

(a) $f(x+h) = \dfrac{1}{x+h}$

(b) $f(x+h)-f(x)$
$= \dfrac{1}{x+h}-\dfrac{1}{x} = \dfrac{x-(x+h)}{x(x+h)} = \dfrac{-h}{x(x+h)}$

(c) $\dfrac{f(x+h)-f(x)}{h} = \dfrac{\frac{-h}{x(x+h)}}{h} = \dfrac{-h}{hx(x+h)}$
$= -\dfrac{1}{x(x+h)}$

41. $f(x) = x^2$

(a) $f(x+h) = (x+h)^2 = x^2+2xh+h^2$

Copyright © 2013 Pearson Education, Inc.

(b) $f(x+h)-f(x) = x^2+2xh+h^2-x^2$
$$= 2xh+h^2$$

(c) $\dfrac{f(x+h)-f(x)}{h} = \dfrac{2xh+h^2}{h}$
$$= \dfrac{h(2x+h)}{h}$$
$$= 2x+h$$

43. $f(x) = 1-x^2$

   (a) $f(x+h) = 1-(x+h)^2$
$$= 1-(x^2+2xh+h^2)$$
$$= 1-x^2-2xh-h^2$$

   (b) $f(x+h)-f(x)$
$$= (1-x^2-2xh-h^2)-(1-x^2)$$
$$= 1-x^2-2xh-h^2-1+x^2$$
$$= -2xh-h^2$$

   (c) $\dfrac{f(x+h)-f(x)}{h} = \dfrac{-2xh-h^2}{h}$
$$= \dfrac{h(-2x-h)}{h}$$
$$= -2x-h$$

45. $f(x) = x^2+3x+1$

   (a) $f(x+h) = (x+h)^2+3(x+h)+1$
$$= x^2+2xh+h^2+3x+3h+1$$

   (b) $f(x+h)-f(x)$
$$= \left(x^2+2xh+h^2+3x+3h+1\right)$$
$$-\left(x^2+3x+1\right)$$
$$= x^2+2xh+h^2+3x+3h+1-x^2-3x-1$$
$$= 2xh+h^2+3h$$

   (c) $\dfrac{f(x+h)-f(x)}{h} = \dfrac{2xh+h^2+3h}{h}$
$$= \dfrac{h(2x+h+3)}{h}$$
$$= 2x+h+3$$

47. Since $g(x) = -x+3,\ g(4) = -4+3 = -1$.
Therefore, $(f \circ g)(4) = f\left[g(4)\right] = f(-1)$
$$= 2(-1)-3 = -2-3 = -5.$$

49. Since $g(x) = -x+3,\ g(-2) = -(-2)+3 = 5$.
Therefore, $(f \circ g)(-2) = f\left[g(-2)\right] = f(5)$
$$= 2(5)-3 = 10-3 = 7.$$

51. Since $f(x) = 2x-3$,
$$f(0) = 2(0)-3 = 0-3 = -3.$$
Therefore, $(g \circ f)(0) = g\left[f(0)\right]$
$$= g(-3) = -(-3)+3 = 3+3 = 6.$$

53. Since $f(x) = 2x-3$,
$$f(2) = 2(2)-3 = 4-3 = 1.$$
Therefore, $(f \circ f)(2) = f\left[f(2)\right]$
$$= f(1) = 2(1)-3 = 2-3 = -1.$$

55. $(f \circ g)(2) = f[g(2)] = f(3) = 1$

57. $(g \circ f)(3) = g[f(3)] = g(1) = 9$

59. $(f \circ f)(4) = f[f(4)] = f(3) = 1$

61. $(f \circ g)(1) = f[g(1)] = f(9)$
However, $f(9)$ cannot be determined from the table given.

63. **(a)** $(f \circ g)(x) = f(g(x)) = f(5x+7)$
$$= -6(5x+7)+9$$
$$= -30x-42+9 = -30x-33$$
The domain and range of both f and g are $(-\infty,\infty)$, so the domain of $f \circ g$ is $(-\infty,\infty)$.

   (b) $(g \circ f)(x) = g(f(x)) = g(-6x+9)$
$$= 5(-6x+9)+7$$
$$= -30x+45+7 = -30x+52$$
The domain of $g \circ f$ is $(-\infty,\infty)$.

65. **(a)** $(f \circ g)(x) = f(g(x)) = f(x+3) = \sqrt{x+3}$
The domain and range of g are $(-\infty,\infty)$, however, the domain and range of f are $[0,\infty)$. So, $x+3 \ge 0 \Rightarrow x \ge -3$.
Therefore, the domain of $f \circ g$ is $[-3,\infty)$.

   (b) $(g \circ f)(x) = g(f(x)) = g\left(\sqrt{x}\right) = \sqrt{x}+3$
The domain and range of g are $(-\infty,\infty)$, however, the domain and range of f are $[0,\infty)$. Therefore, the domain of $g \circ f$ is $[0,\infty)$.

Copyright © 2013 Pearson Education, Inc.

67. (a) $(f \circ g)(x) = f(g(x)) = f(x^2 + 3x - 1)$
$= (x^2 + 3x - 1)^3$
The domain and range of f and g are $(-\infty, \infty)$, so the domain of $f \circ g$ is $(-\infty, \infty)$.

(b) $(g \circ f)(x) = g(f(x)) = g(x^3)$
$= \left(x^3\right)^2 + 3\left(x^3\right) - 1$
$= x^6 + 3x^3 - 1$
The domain and range of f and g are $(-\infty, \infty)$, so the domain of $g \circ f$ is $(-\infty, \infty)$.

69. (a) $(f \circ g)(x) = f(g(x)) = f(3x) = \sqrt{3x - 1}$
The domain and range of g are $(-\infty, \infty)$, however, the domain of f is $[1, \infty)$, while the range of f is $[0, \infty)$. So,
$3x - 1 \ge 0 \Rightarrow x \ge \frac{1}{3}$. Therefore, the
domain of $f \circ g$ is $\left[\frac{1}{3}, \infty\right)$.

(b) $(g \circ f)(x) = g(f(x)) = g\left(\sqrt{x - 1}\right)$
$= 3\sqrt{x - 1}$
The domain and range of g are $(-\infty, \infty)$, however, the range of f is $[0, \infty)$. So
$x - 1 \ge 0 \Rightarrow x \ge 1$. Therefore, the domain
of $g \circ f$ is $[1, \infty)$.

71. (a) $(f \circ g)(x) = f(g(x)) = f(x + 1) = \frac{2}{x+1}$
The domain and range of g are $(-\infty, \infty)$, however, the domain of f is
$(-\infty, 0) \cup (0, \infty)$. So, $x + 1 \ne 0 \Rightarrow x \ne -1$.
Therefore, the domain of $f \circ g$ is
$(-\infty, -1) \cup (-1, \infty)$.

(b) $(g \circ f)(x) = g(f(x)) = g\left(\frac{2}{x}\right) = \frac{2}{x} + 1$
The domain and range of f is
$(-\infty, 0) \cup (0, \infty)$, however, the domain
and range of g are $(-\infty, \infty)$. So $x \ne 0$.
Therefore, the domain of $g \circ f$ is
$(-\infty, 0) \cup (0, \infty)$.

73. (a) $(f \circ g)(x) = f(g(x)) = f\left(-\frac{1}{x}\right) = \sqrt{-\frac{1}{x} + 2}$
The domain and range of g
are $(-\infty, 0) \cup (0, \infty)$, however, the domain
of f is $[-2, \infty)$.

So, $-\frac{1}{x} + 2 \ge 0 \Rightarrow x < 0$ or $x \ge \frac{1}{2}$ (using
test intervals). Therefore, the domain of
$f \circ g$ is $(-\infty, 0) \cup \left[\frac{1}{2}, \infty\right)$.

(b) $(g \circ f)(x) = g(f(x)) = g\left(\sqrt{x + 2}\right) = -\frac{1}{\sqrt{x+2}}$
The domain of f is $[-2, \infty)$ and its range is
$[0, \infty)$. The domain and range of g
are $(-\infty, 0) \cup (0, \infty)$. So
$x + 2 > 0 \Rightarrow x > -2$. Therefore, the domain
of $g \circ f$ is $(-2, \infty)$.

75. (a) $(f \circ g)(x) = f(g(x)) = f\left(\frac{1}{x+5}\right) = \sqrt{\frac{1}{x+5}}$
The domain of g is $(-\infty, -5) \cup (-5, \infty)$,
and the range of g is $(-\infty, 0) \cup (0, \infty)$.
The domain of f is $[0, \infty)$. Therefore, the
domain of $f \circ g$ is $(-5, \infty)$.

(b) $(g \circ f)(x) = g(f(x)) = g\left(\sqrt{x}\right) = \frac{1}{\sqrt{x} + 5}$
The domain and range of f is $[0, \infty)$. The
domain of g is $(-\infty, -5) \cup (-5, \infty)$.
Therefore, the domain of $g \circ f$ is $[0, \infty)$.

77. (a) $(f \circ g)(x) = f(g(x)) = f\left(\frac{1}{x}\right) = \frac{1}{1/x - 2} = \frac{x}{1 - 2x}$
The domain and range of g are
$(-\infty, 0) \cup (0, \infty)$. The domain of f is
$(-\infty, -2) \cup (-2, \infty)$, and the range of f is
$(-\infty, 0) \cup (0, \infty)$. So, $\frac{x}{1 - 2x} < 0 \Rightarrow x < 0$
or $0 < x < \frac{1}{2}$ or $x > \frac{1}{2}$ (using test
intervals). Thus, $x \ne 0$ and
$x \ne \frac{1}{2}$. Therefore, the domain of $f \circ g$ is
$(-\infty, 0) \cup \left(0, \frac{1}{2}\right) \cup \left(\frac{1}{2}, \infty\right)$.

(b) $(g \circ f)(x) = g(f(x)) = g\left(\frac{1}{x-2}\right) = \frac{1}{1/(x-2)}$
$= x - 2$
The domain and range of g are
$(-\infty, 0) \cup (0, \infty)$. The domain of f is
$(-\infty, 2) \cup (2, \infty)$, and the range of f is
$(-\infty, 0) \cup (0, \infty)$. Therefore, the domain of
$g \circ f$ is $(-\infty, 2) \cup (2, \infty)$.

Copyright © 2013 Pearson Education, Inc.

79. $g[f(2)] = g(1) = 2$ and $g[f(3)] = g(2) = 5$
Since $g[f(1)] = 7$ and $f(1) = 3$, $g(3) = 7$.

x	$f(x)$	$g(x)$	$g[f(x)]$
1	3	2	7
2	1	5	**2**
3	2	7	**5**

81. Answers will vary. In general, composition of functions is not commutative.
$$(f \circ g)(x) = f(2x-3) = 3(2x-3) - 2$$
$$= 6x - 9 - 2 = 6x - 11$$
$$(g \circ f)(x) = g(3x-2) = 2(3x-2) - 3$$
$$= 6x - 4 - 3 = 6x - 7$$
Thus, $(f \circ g)(x)$ is not equivalent to $(g \circ f)(x)$.

83. $(f \circ g)(x) = f[g(x)] = 4[\frac{1}{4}(x-2)] + 2$
$$= (4 \cdot \frac{1}{4})(x-2) + 2$$
$$= (x-2) + 2 = x - 2 + 2 = x$$
$(g \circ f)(x) = g[f(x)] = \frac{1}{4}[(4x+2) - 2]$
$$= \frac{1}{4}(4x + 2 - 2) = \frac{1}{4}(4x) = x$$

85. $(f \circ g)(x) = f[g(x)] = \sqrt[3]{5(\frac{1}{5}x^3 - \frac{4}{5}) + 4}$
$$= \sqrt[3]{x^3 - 4 + 4} = \sqrt[3]{x^3} = x$$
$(g \circ f)(x) = g[f(x)] = \frac{1}{5}(\sqrt[3]{5x+4})^3 - \frac{4}{5}$
$$= \frac{1}{5}(5x+4) - \frac{4}{5} = \frac{5x}{5} + \frac{4}{5} - \frac{4}{5}$$
$$= \frac{5x}{5} = x$$

In Exercises 87–91, we give only one of many possible ways.

87. $h(x) = (6x-2)^2$
Let $g(x) = 6x - 2$ and $f(x) = x^2$.
$(f \circ g)(x) = f(6x-2) = (6x-2)^2 = h(x)$

89. $h(x) = \sqrt{x^2 - 1}$
Let $g(x) = x^2 - 1$ and $f(x) = \sqrt{x}$.
$(f \circ g)(x) = f(x^2 - 1) = \sqrt{x^2 - 1} = h(x)$.

91. $h(x) = \sqrt{6x} + 12$
Let $g(x) = 6x$ and $f(x) = \sqrt{x} + 12$.
$(f \circ g)(x) = f(6x) = \sqrt{6x} + 12 = h(x)$

93. $f(x) = 12x$, $g(x) = 5280x$
$(f \circ g)(x) = f[g(x)] = f(5280x)$
$$= 12(5280x) = 63,360x$$
The function $f \circ g$ computes the number of inches in x miles.

95. $A(x) = \frac{\sqrt{3}}{4}x^2$

(a) $A(2x) = \frac{\sqrt{3}}{4}(2x)^2 = \frac{\sqrt{3}}{4}(4x^2) = \sqrt{3}x^2$

(b) $A(16) = A(2 \cdot 8) = \sqrt{3}(8)^2$
$$= 64\sqrt{3} \text{ square units}$$

97. (a) $r(t) = 4t$ and $A(r) = \pi r^2$
$(A \circ r)(t) = A[r(t)]$
$$= A(4t) = \pi(4t)^2 = 16\pi t^2$$

(b) $(A \circ r)(t)$ defines the area of the leak in terms of the time t, in minutes.

(c) $A(3) = 16\pi(3)^2 = 144\pi \text{ ft}^2$

99. Let $x =$ the number of people less than 100 people that attend.

(a) x people fewer than 100 attend, so $100 - x$ people do attend $N(x) = 100 - x$

(b) The cost per person starts at \$20 and increases by \$5 for each of the x people that do not attend. The total increase is \$5x, and the cost per person increases to \$20 + \$5x. Thus, $G(x) = 20 + 5x$.

(c) $C(x) = N(x) \cdot G(x) = (100 - x)(20 + 5x)$

(d) If 80 people attend, $x = 100 - 80 = 20$.
$C(20) = (100 - 20)[20 + 5(20)]$
$$= (80)(20 + 100)$$
$$= (80)(120) = \$9600$$

Copyright © 2013 Pearson Education, Inc.

Chapter 2   Review Exercises

1. $P(3, -1), Q(-4, 5)$

$$d(P, Q) = \sqrt{(-4-3)^2 + [5-(-1)]^2}$$
$$= \sqrt{(-7)^2 + 6^2} = \sqrt{49+36} = \sqrt{85}$$

Midpoint:

$$\left(\frac{3+(-4)}{2}, \frac{-1+5}{2}\right) = \left(\frac{-1}{2}, \frac{4}{2}\right) = \left(-\frac{1}{2}, 2\right)$$

2. $M(-8, 2), N(3, -7)$

$$d(M, N) = \sqrt{[3-(-8)]^2 + (-7-2)^2}$$
$$= \sqrt{11^2 + (-9)^2} = \sqrt{121+81} = \sqrt{202}$$

Midpoint: $\left(\frac{-8+3}{2}, \frac{2+(-7)}{2}\right) = \left(-\frac{5}{2}, -\frac{5}{2}\right)$

3. $A(-6, 3), B(-6, 8)$

$$d(A, B) = \sqrt{[-6-(-6)]^2 + (8-3)^2}$$
$$= \sqrt{0+5^2} = \sqrt{25} = 5$$

Midpoint:

$$\left(\frac{-6+(-6)}{2}, \frac{3+8}{2}\right) = \left(\frac{-12}{2}, \frac{11}{2}\right) = \left(-6, \frac{11}{2}\right)$$

5. Let B have coordinates (x, y). Using the midpoint formula, we have

$$(8, 2) = \left(\frac{-6+x}{2}, \frac{10+y}{2}\right) \Rightarrow$$

$$\begin{array}{c|c} \dfrac{-6+x}{2} = 8 & \dfrac{10+y}{2} = 2 \\ -6+x = 16 & 10+y = 4 \\ x = 22 & y = -6 \end{array}$$

The coordinates of B are $(22, -6)$.

7. Center $(-2, 3)$, radius 15

$$(x-h)^2 + (y-k)^2 = r^2$$
$$[x-(-2)]^2 + (y-3)^2 = 15^2$$
$$(x+2)^2 + (y-3)^2 = 225$$

9. Center $(-8, 1)$, passing through $(0, 16)$
The radius is the distance from the center to any point on the circle. The distance between $(-8, 1)$ and $(0, 16)$ is

$$r = \sqrt{(-8-0)^2 + (1-16)^2} = \sqrt{(-8)^2 + (-15)^2}$$
$$= \sqrt{64+225} = \sqrt{289} = 17.$$

The equation of the circle is

$$[x-(-8)]^2 + (y-1)^2 = 17^2$$
$$(x+8)^2 + (y-1)^2 = 289$$

11. The center of the circle is $(0, 0)$. Use the distance formula to find the radius:

$$r^2 = (3-0)^2 + (5-0)^2 = 9+25 = 34$$

The equation is $x^2 + y^2 = 34$.

13. The center of the circle is $(0, 3)$. Use the distance formula to find the radius:

$$r^2 = (-2-0)^2 + (6-3)^2 = 4+9 = 13$$

The equation is $x^2 + (y-3)^2 = 13$.

15. $x^2 - 4x + y^2 + 6y + 12 = 0$

Complete the square on x and y to put the equation in center-radius form.

$$\left(x^2 - 4x\right) + \left(y^2 + 6y\right) = -12$$
$$\left(x^2 - 4x + 4\right) + \left(y^2 + 6y + 9\right) = -12 + 4 + 9$$
$$(x-2)^2 + (y+3)^2 = 1$$

The circle has center $(2, -3)$ and radius 1.

17.
$$2x^2 + 14x + 2y^2 + 6y + 2 = 0$$
$$x^2 + 7x + y^2 + 3y + 1 = 0$$
$$\left(x^2 + 7x\right) + \left(y^2 + 3y\right) = -1$$
$$\left(x^2 + 7x + \tfrac{49}{4}\right) + \left(y^2 + 3y + \tfrac{9}{4}\right) = -1 + \tfrac{49}{4} + \tfrac{9}{4}$$
$$\left(x + \tfrac{7}{2}\right)^2 + \left(y + \tfrac{3}{2}\right)^2 = -\tfrac{4}{4} + \tfrac{49}{4} + \tfrac{9}{4}$$
$$\left(x + \tfrac{7}{2}\right)^2 + \left(y + \tfrac{3}{2}\right)^2 = \tfrac{54}{4}$$

The circle has center $\left(-\tfrac{7}{2}, -\tfrac{3}{2}\right)$ and radius

$$\sqrt{\tfrac{54}{4}} = \frac{\sqrt{54}}{\sqrt{4}} = \frac{\sqrt{9 \cdot 6}}{\sqrt{4}} = \frac{3\sqrt{6}}{2}.$$

19. Find all possible values of x so that the distance between $(x, -9)$ and $(3, -5)$ is 6.

$$\sqrt{(3-x)^2 + (-5+9)^2} = 6$$
$$\sqrt{9 - 6x + x^2 + 16} = 6$$
$$\sqrt{x^2 - 6x + 25} = 6$$
$$x^2 - 6x + 25 = 36$$
$$x^2 - 6x - 11 = 0$$

Apply the quadratic formula where $a = 1$, $b = -6$, and $c = -11$.

$$x = \frac{6 \pm \sqrt{36 - 4(1)(-11)}}{2} = \frac{6 \pm \sqrt{36+44}}{2}$$
$$= \frac{6 \pm \sqrt{80}}{2} = \frac{6 \pm 4\sqrt{5}}{2} = \frac{2(3 \pm 2\sqrt{5})}{2}$$
$$x = 3 + 2\sqrt{5} \text{ or } x = 3 - 2\sqrt{5}$$

21. This is not the graph of a function because a vertical line can intersect it in two points. domain: $[-6, 6]$; range: $[-6, 6]$

Copyright © 2013 Pearson Education, Inc.

23. This is not the graph of a function because a vertical line can intersect it in two points.
domain: $(-\infty,\infty)$; range: $(-\infty,-1]\cup[1,\infty)$

25. This is not the graph of a function because a vertical line can intersect it in two points.
domain: $[0,\infty)$; range: $(-\infty,\infty)$

27. $y=6-x^2$
Each value of x corresponds to exactly one value of y, so this equation defines a function.

29. The equation $y=\pm\sqrt{x-2}$ does not define y as a function of x. For some values of x, there will be more than one value of y. For example, ordered pairs (3, 1) and (3, −1) satisfy the relation.

31. $f(x)=\dfrac{8+x}{8-x}$
x can be any real number except 8, since this will give a denominator of zero. Thus, the domain is $(-\infty,8)\cup(8,\infty)$.

33. (a) As x is getting larger on the interval $[2,\infty)$, the value of y is increasing.

(b) As x is getting larger on the interval $(-\infty,-2]$, the value of y is decreasing.

(c) $f(x)$ is constant on $[-2, 2]$.

35. $f(x)=-2x^2+3x-6$
$$f(3)=-2\cdot3^2+3\cdot3-6$$
$$=-2\cdot9+3\cdot3-6$$
$$=-18+9-6=-15$$

37. $f(x)=-2x^2+3x-6\Rightarrow f(k)=-2k^2+3k-6$

39. $2x-5y=5\Rightarrow-5y=-2x+5\Rightarrow y=\frac{2}{5}x-1$
The graph is the line with slope $\frac{2}{5}$ and y-intercept −1. It may also be graphed using intercepts. To do this, locate the x-intercept: x-intercept: $y=0$
$$2x-5(0)=5\Rightarrow2x=5\Rightarrow x=\tfrac{5}{2}$$

41. $2x+5y=20\Rightarrow5y=-2x+20\Rightarrow y=-\frac{2}{5}x+4$
The graph is the line with slope of $-\frac{2}{5}$ and y-intercept 4. It may also be graphed using intercepts. To do this, locate the x-intercept: x-intercept: $y=0$
$$2x+5(0)=20\Rightarrow2x=20\Rightarrow x=10$$

43. $f(x)=x$
The graph is the line with slope 1 and y-intercept 0, which means that it passes through the origin. Use another point such as (1, 1) to complete the graph.

45. $x=-5$
The graph is the vertical line through (−5, 0).
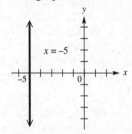

47. The equation of the line that lies along the x-axis is $y=0$.

49. Line through (0, 5), $m=-\frac{2}{3}$
Note that $m=-\frac{2}{3}=\frac{-2}{3}$.
Begin by locating the point (0, 5). Since the slope is $\frac{-2}{3}$, a change of 3 units horizontally (3 units to the right) produces a change of −2 units vertically (2 units down). This gives a second point, (3, 3), which can be used to complete the graph.

(continued on next page)

Copyright © 2013 Pearson Education, Inc.

(*continued*)

51. through $(2, -2)$ and $(3, -4)$

$$m = \frac{y_2 - y_1}{x_2 - x_1} = \frac{-4 - (-2)}{3 - 2} = \frac{-2}{1} = -2$$

53. through $(0, -7)$ and $(3, -7)$

$$m = \frac{-7 - (-7)}{3 - 0} = \frac{0}{3} = 0$$

55. $11x + 2y = 3$

Solve for y to put the equation in slope-intercept form.

$$2y = -11x + 3 \Rightarrow y = -\frac{11}{2}x + \frac{3}{2}$$

Thus, the slope is $-\frac{11}{2}$.

57. $x - 2 = 0 \Rightarrow x = 2$

The graph is a vertical line, through $(2, 0)$. The slope is undefined.

59. Initially, the car is at home. After traveling for 30 mph for 1 hr, the car is 30 mi away from home. During the second hour the car travels 20 mph until it is 50 mi away. During the third hour the car travels toward home at 30 mph until it is 20 mi away. During the fourth hour the car travels away from home at 40 mph until it is 60 mi away from home. During the last hour, the car travels 60 mi at 60 mph until it arrived home.

61. (a) We need to first find the slope of a line that passes between points $(0, 30.7)$ and $(8, 67.2)$

$$m = \frac{y_2 - y_1}{x_2 - x_1} = \frac{67.2 - 30.7}{8 - 0} = \frac{36.5}{8} \approx 4.56$$

Now use the point-intercept form with $b = 30.7$ and $m = 4.56$.
$y = 4.56x + 30.7$
The slope, 4.56, indicates that the number of e-filing taxpayers increased by 4.56% each year from 2001 to 2009.

(b) For 2007, we evaluate the function for $x = 6$. $y = 4.56(6) + 30.7 \approx 58.1$
58.1% of the tax returns are predicted to have been filed electronically.

63. (a) through $(3, -5)$ with slope -2
Use the point-slope form.
$$y - y_1 = m(x - x_1)$$
$$y - (-5) = -2(x - 3)$$
$$y + 5 = -2(x - 3)$$
$$y + 5 = -2x + 6$$
$$y = -2x + 1$$

(b) Standard form: $y = -2x + 1 \Rightarrow 2x + y = 1$

65. (a) through $(2, -1)$ parallel to $3x - y = 1$
Find the slope of $3x - y = 1$.
$$3x - y = 1 \Rightarrow -y = -3x + 1 \Rightarrow y = 3x - 1$$
The slope of this line is 3. Since parallel lines have the same slope, 3 is also the slope of the line whose equation is to be found. Now use the point-slope form with $(x_1, y_1) = (2, -1)$ and $m = 3$.
$$y - y_1 = m(x - x_1)$$
$$y - (-1) = 3(x - 2)$$
$$y + 1 = 3x - 6 \Rightarrow y = 3x - 7$$

(b) Standard form:
$$y = 3x - 7 \Rightarrow -3x + y = -7 \Rightarrow 3x - y = 7$$

67. (a) through $(2, -10)$, perpendicular to a line with an undefined slope
A line with an undefined slope is a vertical line. Any line perpendicular to a vertical line is a horizontal line, with an equation of the form $y = b$. Since the line passes through $(2, -10)$, the equation of the line is $y = -10$.

(b) Standard form: $y = -10$

69. (a) through $(-7, 4)$, perpendicular to $y = 8$
The line $y = 8$ is a horizontal line, so any line perpendicular to it will be a vertical line. Since x has the same value at all points on the line, the equation is $x = -7$. It is not possible to write this in slope-intercept form.

(b) Standard form: $x = -7$

71. $f(x) = |x| - 3$

The graph is the same as that of $y = |x|$, except that it is translated 3 units downward.

Copyright © 2013 Pearson Education, Inc.

73. $f(x) = -(x+1)^2 + 3$

The graph of $f(x) = -(x+1)^2 + 3$ is a

translation of the graph of $y = x^2$ to the left 1 unit, reflected over the x-axis and translated up 3 units.

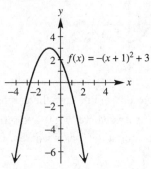

75. $f(x) = [\![x-3]\!]$

To get $y = 0$, we need $0 \le x - 3 < 1 \Rightarrow$ $3 \le x < 4$. To get $y = 1$, we need $1 \le x - 3 < 2 \Rightarrow 4 \le x < 5$. Follow this pattern to graph the step function.

77. $f(x) = \begin{cases} -4x + 2 & \text{if } x \le 1 \\ 3x - 5 & \text{if } x > 1 \end{cases}$

Draw the graph of $y = -4x + 2$ to the left of $x = 1$, including the endpoint at $x = 1$. Draw the graph of $y = 3x - 5$ to the right of $x = 1$, but do not include the endpoint at $x = 1$. Observe that the endpoints of the two pieces coincide.

79. Since x represents an integer, $[\![x]\!] = x$.

Therefore, $[\![x]\!] + x = x + x = 2x$.

81. The graph of an even function is symmetric with respect to the y-axis. This statement is true.

83. If (a, b) is on the graph of an even function, so is $(a, -b)$. The statement is false. For example, $f(x) = x^2$ is even, and $(2, 4)$ is on the graph but $(2, -4)$ is not.

85. The constant function $f(x) = 0$ is both even and odd. Since $f(-x) = 0 = f(x)$, the function is even. Also since $f(-x) = 0 = -0 = -f(x)$, the function is odd. This statement is true.

87. $x + y^2 = 10$

Replace x with $-x$ to obtain $(-x) + y^2 = 10$. The result is not the same as the original equation, so the graph is not symmetric with respect to the y-axis. Replace y with $-y$ to obtain $x + (-y)^2 = 10 \Rightarrow x + y^2 = 10$. The result is the same as the original equation, so the graph is symmetric with respect to the x-axis. Replace x with $-x$ and y with $-y$ to obtain $(-x) + (-y)^2 = 10 \Rightarrow (-x) + y^2 = 10$. The result is not the same as the original equation, so the graph is not symmetric with respect to the origin. The graph is symmetric with respect to the x-axis only.

89. $x^2 = y^3$

Replace x with $-x$ to obtain $(-x)^2 = y^3 \Rightarrow x^2 = y^3$. The result is the same as the original equation, so the graph is symmetric with respect to the y-axis. Replace y with $-y$ to obtain $x^2 = (-y)^3 \Rightarrow x^2 = -y^3$. The result is not the same as the original equation, so the graph is not symmetric with respect to the x-axis. Replace x with $-x$ and y with $-y$ to obtain $(-x)^2 = (-y)^3 \Rightarrow x^2 = -y^3$. The result is not the same as the original equation, so the graph is not symmetric with respect to the origin. Therefore, the graph is symmetric with respect to the y-axis only.

Copyright © 2013 Pearson Education, Inc.

91. $6x + y = 4$

Replace x with $-x$ to obtain $6(-x) + y = 4 \Rightarrow$ $-6x + y = 4$. The result is not the same as the original equation, so the graph is not symmetric with respect to the y-axis. Replace y with $-y$ to obtain $6x + (-y) = 4 \Rightarrow 6x - y = 4$. The result is not the same as the original equation, so the graph is not symmetric with respect to the x-axis. Replace x with $-x$ and y with $-y$ to obtain $6(-x) + (-y) = 4 \Rightarrow -6x - y = 4$. This equation is not equivalent to the original one, so the graph is not symmetric with respect to the origin. Therefore, the graph has none of the listed symmetries.

93. $y = 1$

This is the graph of a horizontal line through $(0, 1)$. It is symmetric with respect to the y-axis, but not symmetric with respect to the x-axis and the origin.

95. To obtain the graph of $g(x) = -|x|$, reflect the graph of $f(x) = |x|$ across the x-axis.

97. To obtain the graph of $k(x) = 2|x - 4|$, translate the graph of $f(x) = |x|$ to the right 4 units and stretch vertically by a factor of 2.

99. If the graph of $f(x) = 3x - 4$ is reflected about the y-axis, we obtain a graph whose equation is $y = f(-x) = 3(-x) - 4 = -3x - 4$.

101. (a) To graph $y = f(x) + 3$, translate the graph of $y = f(x)$, 3 units up.

(b) To graph $y = f(x - 2)$, translate the graph of $y = f(x)$, 2 units to the right.

(c) To graph $y = f(x + 3) - 2$, translate the graph of $y = f(x)$, 3 units to the left and 2 units down.

(d) To graph $y = |f(x)|$, keep the graph of $y = f(x)$ as it is where $y \geq 0$ and reflect the graph about the x-axis where $y < 0$.

For Exercises 103–109, $f(x) = 3x^2 - 4$ and $g(x) = x^2 - 3x - 4$.

103. $(fg)(x) = f(x) \cdot g(x)$
$= (3x^2 - 4)(x^2 - 3x - 4)$
$= 3x^4 - 9x^3 - 12x^2 - 4x^2 + 12x + 16$
$= 3x^4 - 9x^3 - 16x^2 + 12x + 16$

105. $(f + g)(-4) = f(-4) + g(-4)$
$= [3(-4)^2 - 4] + [(-4)^2 - 3(-4) - 4]$
$= [3(16) - 4] + [16 - 3(-4) - 4]$
$= [48 - 4] + [16 + 12 - 4]$
$= 44 + 24 = 68$

107. $\left(\dfrac{f}{g}\right)(3) = \dfrac{f(3)}{g(3)} = \dfrac{3 \cdot 3^2 - 4}{3^2 - 3 \cdot 3 - 4} = \dfrac{3 \cdot 9 - 4}{9 - 3 \cdot 3 - 4}$
$= \dfrac{27 - 4}{9 - 9 - 4} = \dfrac{23}{-4} = -\dfrac{23}{4}$

109. The domain of $(fg)(x)$ is the intersection of the domain of $f(x)$ and the domain of $g(x)$. Both have domain $(-\infty, \infty)$, so the domain of $(fg)(x)$ is $(-\infty, \infty)$.

Copyright © 2013 Pearson Education, Inc.

111. $f(x) = \dfrac{1}{x}, g(x) = x^2 + 1$

Since
$$(f \circ g)(x) = f[g(x)] \text{ and } (f \circ g)(x) = \dfrac{1}{x^2 + 1},$$
choices (C) and (D) are not equivalent to $(f \circ g)(x)$.

113. $f(x) = x^2 - 5x + 3$

$f(x + h) = (x + h)^2 - 5(x + h) + 3$
$\qquad = x^2 + 2xh + h^2 - 5x - 5h + 3$

$f(x + h) - f(x)$
$\quad = (x^2 + 2xh + h^2 - 5x - 5h + 3) - (x^2 - 5x + 3)$
$\quad = x^2 + 2xh + h^2 - 5x - 5h + 3 - x^2 + 5x - 3$
$\quad = 2xh + h^2 - 5h$

$\dfrac{f(x+h) - f(x)}{h} = \dfrac{2xh + h^2 - 5h}{h}$
$\qquad\qquad\qquad = \dfrac{h(2x + h - 5)}{h} = 2x + h - 5$

For Exercises 115–119,
$f(x) = \sqrt{x - 2}$ and $g(x) = x^2$.

115. $(g \circ f)(x) = g[f(x)] = g\left(\sqrt{x-2}\right)$
$\qquad\qquad = \left(\sqrt{x-2}\right)^2 = x - 2$

117. Since $f(x) = \sqrt{x-2}$, $f(3) = \sqrt{3 - 2} = \sqrt{1} = 1$.
Therefore, $(g \circ f)(3) = g\left[f(3)\right] = g(1)$
$\qquad = 1^2 = 1.$

119. To find the domain of $f \circ g$, we must consider the domain of g as well as the composed function, $f \circ g$. Since
$$(f \circ g)(x) = f\left[g(x)\right] = \sqrt{x^2 - 2} \text{ we need to}$$
determine when $x^2 - 2 \ge 0$.
Step 1: Find the values of x that satisfy
$x^2 - 2 = 0$
$x^2 = 2 \Rightarrow x = \pm\sqrt{2}$
Step 2: The two numbers divide a number line into three regions.

Step 3 Choose a test value to see if it satisfies the inequality, $x^2 - 2 \ge 0$.

Interval	Test Value	Is $x^2 - 2 \ge 0$ true or false?
$\left(-\infty, -\sqrt{2}\right)$	-2	$(-2)^2 - 2 \ge 0$? $2 \ge 0$ True
$\left(-\sqrt{2}, \sqrt{2}\right)$	0	$0^2 - 2 \ge 0$? $-2 \ge 0$ False
$\left(\sqrt{2}, \infty\right)$	2	$2^2 - 2 \ge 0$? $2 \ge 0$ True

The domain of $f \circ g$ is
$$\left(-\infty, -\sqrt{2}\right] \cup \left[\sqrt{2}, \infty\right).$$

121. $(f - g)(3) = f(3) - g(3) = 9 - 9 = 0$

123. $\left(\dfrac{f}{g}\right)(0) = \dfrac{f(0)}{g(0)} = \dfrac{5}{0} = $ undefined

125. $(g \circ f)(-2) = g[f(-2)] = g(1) = 2$

127. $(f \circ g)(2) = f[g(2)] = f(2) = 1$

129. Let x = number of yards.
$f(x) = 36x$, where $f(x)$ is the number of inches.
$g(x) = 1760x$, where $g(x)$ is the number of yards. Then
$(g \circ f)(x) = g\left[f(x)\right] = 1760(36x) = 63,360x.$
There are $63,360x$ inches in x miles

131. If $V(r) = \frac{4}{3}\pi r^3$ and if the radius is increased by 3 inches, then the amount of volume gained is given by
$V_g(r) = V(r + 3) - V(r) = \frac{4}{3}\pi(r + 3)^3 - \frac{4}{3}\pi r^3.$

Chapter 2   Test

1. (a) The domain of $f(x) = \sqrt{x} + 3$ occurs when $x \ge 0$. In interval notation, this correlates to the interval in D, $[0, \infty)$.

(b) The range of $f(x) = \sqrt{x} - 3$ is all real numbers greater than or equal to 0. In interval notation, this correlates to the interval in D, $[0, \infty)$.

(c) The domain of $f(x) = x^2 - 3$ is all real numbers. In interval notation, this correlates to the interval in C, $(-\infty, \infty)$.

Copyright © 2013 Pearson Education, Inc.

(d) The range of $f(x) = x^2 + 3$ is all real numbers greater than or equal to 3. In interval notation, this correlates to the interval in B, $[3, \infty)$.

(e) The domain of $f(x) = \sqrt[3]{x-3}$ is all real numbers. In interval notation, this correlates to the interval in C, $(-\infty, \infty)$.

(f) The range of $f(x) = \sqrt[3]{x} + 3$ is all real numbers. In interval notation, this correlates to the interval in C, $(-\infty, \infty)$.

(g) The domain of $f(x) = |x| - 3$ is all real numbers. In interval notation, this correlates to the interval in C, $(-\infty, \infty)$.

(h) The range of $f(x) = |x + 3|$ is all real numbers greater than or equal to 0. In interval notation, this correlates to the interval in D, $[0, \infty)$.

(i) The domain of $x = y^2$ is $x \geq 0$ since when you square any value of y, the outcome will be nonnegative. In interval notation, this correlates to the interval in D, $[0, \infty)$.

(j) The range of $x = y^2$ is all real numbers. In interval notation, this correlates to the interval in C, $(-\infty, \infty)$.

2. Consider the points $(-2, 1)$ and $(3, 4)$.
$$m = \frac{4-1}{3-(-2)} = \frac{3}{5}$$

3. We label the points $A(-2, 1)$ and $B(3, 4)$.
$$d(A, B) = \sqrt{[3-(-2)]^2 + (4-1)^2}$$
$$= \sqrt{5^2 + 3^2} = \sqrt{25+9} = \sqrt{34}$$

4. The midpoint has coordinates
$$\left(\frac{-2+3}{2}, \frac{1+4}{2}\right) = \left(\frac{1}{2}, \frac{5}{2}\right).$$

5. Use the point-slope form with $(x_1, y_1) = (-2, 1)$ and $m = \frac{3}{5}$.
$$y - y_1 = m(x - x_1)$$
$$y - 1 = \frac{3}{5}[x - (-2)]$$
$$y - 1 = \frac{3}{5}(x + 2) \Rightarrow 5(y-1) = 3(x+2) \Rightarrow$$
$$5y - 5 = 3x + 6 \Rightarrow 5y = 3x + 11 \Rightarrow$$
$$-3x + 5y = 11 \Rightarrow 3x - 5y = -11$$

6. Solve $3x - 5y = -11$ for y.
$$3x - 5y = -11$$
$$-5y = -3x - 11$$
$$y = \frac{3}{5}x + \frac{11}{5}$$
Therefore, the linear function is
$$f(x) = \frac{3}{5}x + \frac{11}{5}.$$

7. (a) The center is at (0, 0) and the radius is 2, so the equation of the circle is
$$x^2 + y^2 = 4.$$

(b) The center is at (1, 4) and the radius is 1, so the equation of the circle is
$$(x-1)^2 + (y-4)^2 = 1$$

8. $x^2 + y^2 + 4x - 10y + 13 = 0$
Complete the square on x and y to write the equation in standard form:
$$x^2 + y^2 + 4x - 10y + 13 = 0$$
$$\left(x^2 + 4x + \quad\right) + \left(y^2 - 10y + \quad\right) = -13$$
$$\left(x^2 + 4x + 4\right) + \left(y^2 - 10y + 25\right) = -13 + 4 + 25$$
$$(x+2)^2 + (y-5)^2 = 16$$
The circle has center $(-2, 5)$ and radius 4.

$x^2 + y^2 + 4x - 10y + 13 = 0$

9. (a) This is not the graph of a function because some vertical lines intersect it in more than one point. The domain of the relation is [0, 4]. The range is [−4, 4].

Copyright © 2013 Pearson Education, Inc.

(b) This is the graph of a function because no vertical line intersects the graph in more than one point. The domain of the function is $(-\infty, -1) \cup (-1, \infty)$. The range is $(-\infty, 0) \cup (0, \infty)$. As x is getting larger on the intervals $(-\infty, -1)$ and $(-1, \infty)$, the value of y is decreasing, so the function is decreasing on these intervals. (The function is never increasing or constant.)

10. Point A has coordinates $(5, -3)$.
    (a) The equation of a vertical line through A is $x = 5$.

    (b) The equation of a horizontal line through A is $y = -3$.

11. The slope of the graph of $y = -3x + 2$ is -3.
    (a) A line parallel to the graph of $y = -3x + 2$ has a slope of -3. Use the point-slope form with $(x_1, y_1) = (2, 3)$ and $m = -3$.
    $$y - y_1 = m(x - x_1)$$
    $$y - 3 = -3(x - 2)$$
    $$y - 3 = -3x + 6 \Rightarrow y = -3x + 9$$

    (b) A line perpendicular to the graph of $y = -3x + 2$ has a slope of $\frac{1}{3}$ since $-3\left(\frac{1}{3}\right) = -1$.
    $$y - 3 = \frac{1}{3}(x - 2)$$
    $$3(y - 3) = x - 2 \Rightarrow 3y - 9 = x - 2 \Rightarrow$$
    $$3y = x + 7 \Rightarrow y = \frac{1}{3}x + \frac{7}{3}$$

12. (a) $(-\infty, -3)$

    (b) $(4, \infty)$

    (c) $[-3, 4]$

    (d) $(-\infty, -3); [-3, 4]; (4, \infty)$

    (e) $(-\infty, \infty)$

    (f) $(-\infty, 2)$

13. To graph $y = |x - 2| - 1$, we translate the graph of $y = |x|$, 2 units to the right and 1 unit down.

$$y = |x - 2| - 1$$

14. $f(x) = [\![x + 1]\!]$
    To get $y = 0$, we need $0 \le x + 1 < 1 \Rightarrow$ $-1 \le x < 0$. To get $y = 1$, we need $1 \le x + 1 < 2 \Rightarrow 0 \le x < 1$. Follow this pattern to graph the step function.

$$f(x) = [\![x + 1]\!]$$

15. $f(x) = \begin{cases} 3 & \text{if } x < -2 \\ 2 - \frac{1}{2}x & \text{if } x \ge -2 \end{cases}$
    For values of x with $x < -2$, we graph the horizontal line $y = 3$. For values of x with $x \ge -2$, we graph the line with a slope of $-\frac{1}{2}$ and a y-intercept of 2. Two points on this line are $(-2, 3)$ and $(0, 2)$.

$$f(x) = \begin{cases} 3 & \text{if } x < -2 \\ 2 - \frac{1}{2}x & \text{if } x \ge -2 \end{cases}$$

16. (a) Shift $f(x)$, 2 units vertically upward.

$$y = f(x) + 2$$

Copyright © 2013 Pearson Education, Inc.

(b) Shift $f(x)$, 2 units horizontally to the left.

(c) Reflect $f(x)$, across the x-axis.

(d) Reflect $f(x)$, across the y-axis.

(e) Stretch $f(x)$, vertically by a factor of 2.

17. Answers will vary. Starting with $y = \sqrt{x}$, we shift it to the left 2 units and stretch it vertically by a factor of 2. The graph is then reflected over the x-axis and then shifted down 3 units.

18. $3x^2 - y^2 = 3$

(a) Replace y with $-y$ to obtain
$$3x^2 - (-y)^2 = 3 \Rightarrow 3x^2 - y^2 = 3.$$
The result is the same as the original equation, so the graph is symmetric with respect to the x-axis.

(b) Replace x with $-x$ to obtain
$$3(-x)^2 - y^2 = 3 \Rightarrow 3x^2 - y^2 = 3.$$
The result is the same as the original equation, so the graph is symmetric with respect to the y-axis.

(c) Since the graph is symmetric with respect to the x-axis and with respect to the y-axis, it must also be symmetric with respect to the origin.

19. $f(x) = 2x^2 - 3x + 2,\ g(x) = -2x + 1$

(a) $(f - g)(x) = f(x) - g(x)$
$$= \left(2x^2 - 3x + 2\right) - \left(-2x + 1\right)$$
$$= 2x^2 - 3x + 2 + 2x - 1$$
$$= 2x^2 - x + 1$$

(b) $\left(\dfrac{f}{g}\right)(x) = \dfrac{f(x)}{g(x)} = \dfrac{2x^2 - 3x + 2}{-2x + 1}$

(c) We must determine which values solve the equation $-2x + 1 = 0$.
$$-2x + 1 = 0 \Rightarrow -2x = -1 \Rightarrow x = \tfrac{1}{2}$$
Thus, $\tfrac{1}{2}$ is excluded from the domain, and the domain is $\left(-\infty, \tfrac{1}{2}\right) \cup \left(\tfrac{1}{2}, \infty\right)$.

(d) $f(x) = 2x^2 - 3x + 2$
$$f(x + h) = 2(x + h)^2 - 3(x + h) + 2$$
$$= 2\left(x^2 + 2xh + h^2\right) - 3x - 3h + 2$$
$$= 2x^2 + 4xh + 2h^2 - 3x - 3h + 2$$
$$f(x + h) - f(x)$$
$$= (2x^2 + 4xh + 2h^2 - 3x - 3h + 2)$$
$$\qquad - (2x^2 - 3x + 2)$$
$$= 2x^2 + 4xh + 2h^2 - 3x$$
$$\qquad -3h + 2 - 2x^2 + 3x - 2$$
$$= 4xh + 2h^2 - 3h$$
$$\frac{f(x + h) - f(x)}{h} = \frac{4xh + 2h^2 - 3h}{h}$$
$$= \frac{h(4x + 2h - 3)}{h}$$
$$= 4x + 2h - 3$$

20. (a) $(f + g)(1) = f(1) + g(1)$
$$= (2 \cdot 1^2 - 3 \cdot 1 + 2) + (-2 \cdot 1 + 1)$$
$$= (2 \cdot 1 - 3 \cdot 1 + 2) + (-2 \cdot 1 + 1)$$
$$= (2 - 3 + 2) + (-2 + 1)$$
$$= 1 + (-1) = 0$$

Copyright © 2013 Pearson Education, Inc.

(b) $(fg)(2) = f(2) \cdot g(2)$
$$= (2 \cdot 2^2 - 3 \cdot 2 + 2) \cdot (-2 \cdot 2 + 1)$$
$$= (2 \cdot 4 - 3 \cdot 2 + 2) \cdot (-2 \cdot 2 + 1)$$
$$= (8 - 6 + 2) \cdot (-4 + 1)$$
$$= 4(-3) = -12$$

(c) $g(x) = -2x + 1 \Rightarrow g(0) = -2(0) + 1$
$= 0 + 1 = 1$. Therefore,
$$(f \circ g)(0) = f\left[g(0)\right]$$
$$= f(1) = 2 \cdot 1^2 - 3 \cdot 1 + 2$$
$$= 2 \cdot 1 - 3 \cdot 1 + 2$$
$$= 2 - 3 + 2 = 1$$

21. $(f \circ g) = f\left(g(x)\right) = f(2x - 7)$
$$= \sqrt{(2x - 7) + 1} = \sqrt{2x - 6}$$

The domain and range of g are $(-\infty, \infty)$, while the domain of f is $[0, \infty)$. We need to find the values of x which fit the domain of f: $2x - 6 \geq 0 \Rightarrow x \geq 3$. So, the domain of $f \circ g$ is $[3, \infty)$.

22. $(g \circ f) = g\left(f(x)\right) = g\left(\sqrt{x + 1}\right)$
$$= 2\sqrt{x + 1} - 7$$

The domain and range of g are $(-\infty, \infty)$, while the domain of f is $[0, \infty)$. We need to find the values of x which fit the domain of f: $x + 1 \geq 0 \Rightarrow x \geq -1$. So, the domain of $g \circ f$ is $[-1, \infty)$.

23. $f(x) = 0.75[\![x]\!] + 1.5$
$f(7.5) = 0.75[\![7.5]\!] + 1.5$
$$= 0.75(7) + 1.5 = \$6.75$$

24. (a) $C(x) = 3300 + 4.50x$

(b) $R(x) = 10.50x$

(c) $P(x) = R(x) - C(x)$
$$= 10.50x - (3300 + 4.50x)$$
$$= 6.00x - 3300$$

(d) $P(x) > 0$
$6.00x - 3300 > 0$
$6.00x > 3300$
$x > 550$

She must produce and sell 551 items before she earns a profit.

Copyright © 2013 Pearson Education, Inc.

Chapter 3

POLYNOMIAL AND RATIONAL FUNCTIONS

Section 3.1   Quadratic Functions and Models

1. $f(x) = (x+3)^2 - 4$

    (a) domain: $(-\infty, \infty)$; range: $[-4, \infty)$

    (b) vertex: $(h, k) = (-3, -4)$

    (c) axis: $x = -3$

    (d) To find the y-intercept, let $x = 0$.
    $$y = (0+3)^2 - 4 = 3^2 - 4 = 9 - 4 = 5$$
    y-intercept: 5

    (e) To find the x-intercepts, let $f(x) = 0$.
    $$0 = (x+3)^2 - 4$$
    $$(x+3)^2 = 4$$
    $$x + 3 = \pm\sqrt{4} = \pm 2$$
    $$x = -3 \pm 2$$
    $$x = -3 - 2 = -5 \text{ or } x = -3 + 2 = -1$$
    x-intercepts: -5 and -1

3. $f(x) = -2(x+3)^2 + 2$

    (a) domain: $(-\infty, \infty)$; range: $(-\infty, 2]$

    (b) vertex: $(h, k) = (-3, 2)$

    (c) axis: $x = -3$

    (d) To find the y-intercept, let $x = 0$.
    $$y = -2(0+3)^2 + 2 = -2 \cdot 3^2 + 2$$
    $$= -2 \cdot 9 + 2 = -18 + 2 = -16$$
    y-intercept: -16

    (e) To find the x-intercepts, let $f(x) = 0$.
    $$0 = -2(x+3)^2 + 2$$
    $$(x+3)^2 = 1$$
    $$x + 3 = \pm\sqrt{1} = \pm 1$$
    $$x = -3 \pm 1$$
    $$x = -3 - 1 = -4 \text{ or } x = -3 + 1 = -2$$
    x-intercepts: -4 and -2

5. $f(x) = (x-4)^2 - 3$

    Since $a > 0$, the parabola opens upward. The vertex is at $(4, -3)$. The correct graph, therefore, is B.

7. $f(x) = (x+4)^2 - 3$

    Since $a > 0$, the parabola opens upward. The vertex is at $(-4, -3)$. The correct graph, therefore, is D.

9. For parts (a), (b), and (c), see the following graph.

    (d) If the absolute value of the coefficient is greater than 1, it causes the graph to be stretched vertically, so it is narrower. If the absolute value of the coefficient is between 0 and 1, it causes the graph to shrink vertically, so it is wider.

11. For parts (a), (b), and (c), see the following graph.

    (d) The graph of $y = (x-h)^2$ is translated h units to the right if h is positive and $|h|$ units to the left if h is negative.

    Copyright © 2013 Pearson Education, Inc.

13. $f(x) = (x-2)^2$

$f(x) = (x-2)^2$

This equation is of the form $y = (x-h)^2$, with $h = 2$. The graph opens upward and has the same shape as that of $y = x^2$. It is a horizontal translation of the graph of $y = x^2$, 2 units to the right. Two points on the graph are (1, 1) and (3, 1).

(a) The vertex is (2, 0).

(b) The axis is the vertical line $x = 2$.

(c) The domain is $(-\infty, \infty)$.

(d) Since the smallest value of y is 0 and the graph opens upward, the range is $[0, \infty)$.

(e) The function is increasing on $[2, \infty)$.

(f) The function is decreasing on $(-\infty, 2]$.

15. $f(x) = (x+3)^2 - 4 = [x-(-3)]^2 + (-4)$

$f(x) = (x+3)^2 - 4$

This equation is of the form $y = (x-h)^2 + k$, with $h = -3$ and $k = -4$. The graph opens upward and has the same shape as $y = x^2$. It is a translation of $y = x^2$, 3 units to the left and 4 units down. Two points on the graph are $(-4, -3)$ and $(-2, -3)$.

(a) The vertex is $(-3, -4)$.

(b) The axis is the vertical line $x = -3$.

(c) The domain is $(-\infty, \infty)$.

(d) Since the smallest value of y is -4 and the graph opens upward, the range is $[-4, \infty)$.

(e) The function is increasing on $[-3, \infty)$.

(f) The function is decreasing on $(-\infty, -3]$.

17. $f(x) = -\frac{1}{2}(x+1)^2 - 3 = -\frac{1}{2}[x-(-1)]^2 + (-3)$

$f(x) = -\frac{1}{2}(x+1)^2 - 3$

This equation is of the form $y = a(x-h)^2 + k$, with $h = -1$, $k = -3$, $a = -\frac{1}{2}$. The graph opens downward and is wider than $y = x^2$. It is a translation of the graph $y = -\frac{1}{2}x^2$, 1 unit to the left and 3 units down. Additional points on the graph are $\left(-2, -3\frac{1}{2}\right)$ and $\left(0, -3\frac{1}{2}\right)$.

(a) The vertex is $(-1, -3)$.

(b) The axis is the vertical line $x = -1$.

(c) The domain is $(-\infty, \infty)$.

(d) Since the largest value of y is -3 and the graph opens downward, the range is $(-\infty, -3]$.

(e) The function is increasing on $(-\infty, -1]$.

(f) The function is decreasing on $[-1, \infty)$.

19. $f(x) = x^2 - 2x + 3$

Rewrite by completing the square on x.

$f(x) = x^2 - 2x + 3 = (x^2 - 2x + 1 - 1) + 3$

Note: $\left[\frac{1}{2}(-2)\right]^2 = (-1)^2 = 1$

$= (x^2 - 2x + 1) - 1 + 3 = (x-1)^2 + 2$

$f(x) = x^2 - 2x + 3$

This equation is of the form $y = (x-h)^2 + k$, with $h = 1$ and $k = 2$. The graph opens upward and has the same shape as $y = x^2$.

(continued on next page)

Copyright © 2013 Pearson Education, Inc.

(continued)

It is a translation of the graph $y = x^2$, 1 unit to the right and 2 units up. Two points on the graph are (0, 3) and (2, 3).

(a) The vertex is (1, 2).

(b) The axis is the vertical line $x = 1$.

(c) The domain is $(-\infty, \infty)$.

(d) Since the smallest value of y is 2 and the graph opens upward, the range is $[2, \infty)$.

(e) The function is increasing on $[1, \infty)$.

(f) The function is decreasing on $(-\infty, 1]$.

21. $f(x) = x^2 - 10x + 21$

Rewrite by completing the square on x.

$$f(x) = x^2 - 10x + 21 = (x^2 - 10x + 25 - 25) + 21$$

Note: $\left[\frac{1}{2}(-10)\right]^2 = (-5)^2 = 25$

$$= (x^2 - 10x + 25) - 25 + 21 = (x - 5)^2 - 4$$
$$= (x - 5)^2 + (-4)$$

This equation is of the form $y = (x - h)^2 + k$, with $h = 5$ and $k = -4$. The graph opens upward and has the same shape as $y = x^2$. It is a translation of the graph $y = x^2$, 5 units to the right and 4 units down. Two points on the graph are (4, -3) and (6, -3).

(a) The vertex is (5, -4).

(b) The axis is the vertical line $x = 5$.

(c) The domain is $(-\infty, \infty)$.

(d) Since the smallest value of y is -4 and the graph opens upward, the range is $[-4, \infty)$.

(e) The function is increasing on $[5, \infty)$.

(f) The function is decreasing on $(-\infty, 5]$.

23. $f(x) = -2x^2 - 12x - 16$

Rewrite by completing the square on x.

$$f(x) = -2x^2 - 12x - 16 = -2(x^2 + 6x) - 16$$
$$= -2(x^2 + 6x + 9 - 9) - 16$$

Note: $\left[\frac{1}{2}(6)\right]^2 = 3^2 = 9$

$$= -2(x^2 + 6x + 9) + 18 - 16$$
$$= -2(x + 3)^2 + 2 = -2[x - (-3)]^2 + 2$$

$f(x) = -2x^2 - 12x - 16$

This equation is of the form $y = a(x - h)^2 + k$, with $h = -3$, $k = 2$, and $a = -2$. The graph opens downward and is narrower than $y = x^2$.

It is a translation of the graph $y = -2x^2$, 3 units to the left and 2 units up. Additional points on the graph are (-4, 0) and (-2, 0).

(a) The vertex is (-3, 2).

(b) The axis is the vertical line $x = -3$.

(c) The domain is $(-\infty, \infty)$.

(d) Since the largest value of y is 2 and the graph opens downward, the range is $(-\infty, 2]$.

(e) The function is increasing on $(-\infty, -3]$.

(f) The function is decreasing on $[-3, \infty)$.

25. $f(x) = -\frac{1}{2}x^2 - 3x - \frac{1}{2}$

Rewrite by completing the square on x.

$$f(x) = -\frac{1}{2}x^2 - 3x - \frac{1}{2}$$
$$= -\frac{1}{2}(x^2 + 6x + 1 + 8) - \frac{1}{2}(-8)$$

Note: $\left[\frac{1}{2}(6)\right]^2 = 3^2 = 9$

$$= -\frac{1}{2}(x^2 + 6x + 9) + 4 = -(x + 3)^2 + 4$$
$$= -\frac{1}{2}[x - (-3)]^2 + 4$$

This equation is of the form $y = a(x - h)^2 + k$, with $h = -3$, $k = 4$, and $a = -1$. The graph opens downward and is wider than $y = x^2$. It is a translation of the graph $y = -x^2$, 3 units to the left and 4 units up. Two points on the graph are (-1, 2) and (1, -4).

(continued on next page)

Copyright © 2013 Pearson Education, Inc.

(continued)

$$f(x) = \tfrac{1}{2}x^2 - 3x - \tfrac{1}{2}$$

(a) The vertex is $(-3, 4)$.

(b) The axis is the vertical line $x = -3$.

(c) The domain is $(-\infty, \infty)$.

(d) Since the largest value of y is 4 and the graph opens downward, the range is $(-\infty, 4]$.

(e) The function is increasing on $(-\infty, -3]$.

(f) The function is decreasing on $[-3, \infty)$.

27. The minimum value of $f(x)$ is $f(-3) = 3$.

29. There are no real solutions to the equation $f(x) = 1$ since the value of $f(x)$ is never less than 3.

31. $a < 0,\ b^2 - 4ac = 0$

    The correct choice is E. $a < 0$ indicates that the parabola will open downward, while $b^2 - 4ac = 0$ indicates that the graph will have exactly one x-intercept.

33. $a < 0,\ b^2 - 4ac < 0$

    The correct choice is D. $a < 0$ indicates that the parabola will open downward, while $b^2 - 4ac < 0$ indicates that the graph will have no x-intercepts.

35. $a > 0,\ b^2 - 4ac > 0$

    The correct choice is C. $a > 0$ indicates that the parabola will open upward, while $b^2 - 4ac > 0$ indicates that the graph will have two x-intercepts.

37. The vertex of the parabola in the figure is $(2, -1)$ and the y-intercept is 0. The equation takes the form $f(x) = a(x - 2)^2 - 1$. When $x = 0, f(x) = 0$, so $0 = a(0 - 2)^2 - 1 \Rightarrow$
    $0 = 4a - 1 \Rightarrow a = \tfrac{1}{4}$. The equation is
    $$f(x) = \tfrac{1}{4}(x - 2)^2 - 1.$$

This function may also be written as
$$f(x) = \tfrac{1}{4}(x - 2)^2 - 1$$
$$= \tfrac{1}{4}\left(x^2 - 4x + 4\right) - 1 = \tfrac{1}{4}x^2 - x.$$

39. The vertex of the parabola in the figure is $(1, 4)$ and the y-intercept is 2. The equation takes the form $f(x) = a(x - 1)^2 + 4$. When $x = 0, f(x) = 2$, so $2 = a(0 - 1)^2 + 4 \Rightarrow$
    $2 = a + 4 \Rightarrow a = -2$. The equation is
    $f(x) = -2(x - 1)^2 + 4$. This function may also be written as $f(x) = -2(x^2 - 2x + 1) + 4$
    $$= -2x^2 + 4x - 2 + 4 = -2x^2 + 4x + 2.$$

41. Linear; the points lie in a pattern that suggests a positive slope.

43. Quadratic; the points lie in a pattern suggesting a parabola opening upward, so $a > 0$.

45. Quadratic; the points lie in a pattern that suggests a parabola opening downward, so $a < 0$.

47. Let $x =$ one number. Then $20 - x =$ the other number. Now find the maximum of $x(20 - x)$ by finding the vertex of the function:
    $$f(x) = x(20 - x) = 20x - x^2 = -(x^2 - 20x)$$
    $$= -(x^2 - 20x + 100) + 100 \quad \text{complete the square}$$
    $$= -(x - 10)^2 + 100$$
    The maximum occurs at the vertex. So the two numbers are 10 and $20 - 10 = 10$. Confirm graphically:

49. To find the number of units to be sold to minimize her costs, find the vertex of the function:
    $$C(x) = x^2 - 40x + 610$$
    $$= (x^2 - 40x + 400) + 610 - 400 \quad \text{complete the square}$$
    $$= (x - 20)^2 + 210$$
    The vertex occurs at $x = 20$, so she should sell 20 units to minimize her costs. The minimum cost is the value of the function at the vertex, $\$210$. Confirm graphically.

(continued on next page)

Copyright © 2013 Pearson Education, Inc.

(*continued*)

51. To find the amount of rainfall that will maximize the number of mosquitos, find the vertex of the function:

$$M(x) = 10x - x^2 = -(x^2 - 10x)$$
$$= -(x^2 - 10x + 25) + 25 \quad \begin{array}{l}\text{complete}\\\text{the square}\end{array}$$
$$= -(x - 5)^2 + 25$$

The vertex occurs at $x = 5$, which means that the maximum number of mosquitoes occurs when there are 5 inches of rain. The maximum number of mosquitoes is the value of the function at the vertex, 25 million. Confirm graphically.

53. (a) Since $v_0 = 200$, and $s_0 = 50$, and

$$s(t) = -16t^2 + v_0t + s_0 \text{ we have}$$

$$s(t) \text{ or } f(t) = -16t^2 + 200t + 50.$$

(b) Algebraic Solution:
Find the coordinates of the vertex of the parabola. Using the vertex formula with $a = -16$ and $b = 200$.

$$x = -\frac{b}{2a} = -\frac{200}{2(-16)} = 6.25 \text{ and}$$

$$y = -16(6.25)^2 + 200(6.25) + 50 = 675.$$

The vertex is (6.25, 675). Since $a < 0$, this is the maximum point.
Graphing Calculator Solution:

Thus, the number of seconds to reach maximum height is 6.25 seconds. The maximum height is 675 ft.

(c) Algebraic Solution:
To find the time interval in which the rocket will be more than 300 ft above ground level, solve the inequality

$$-16t^2 + 200t + 50 > 300.$$

$$-16t^2 + 200t + 50 > 300 \Rightarrow$$
$$-16t^2 + 200t - 250 > 0 \Rightarrow$$
$$-8t^2 + 100t - 125 > 0$$

Solve the corresponding equation

$$-8t^2 + 100t - 125 = 0.$$

Use the quadratic formula with $a = -8$, $b = 100$, and $c = -125$.

$$t = \frac{-100 \pm \sqrt{100^2 - 4(-8)(-125)}}{2(-8)}$$

$$= \frac{-100 \pm \sqrt{10,000 - 4000}}{-16}$$

$$= \frac{-100 \pm \sqrt{6000}}{-16}$$

$$t = \frac{-100 + \sqrt{6000}}{-16} \approx 1.41 \text{ or}$$

$$t = \frac{-100 - \sqrt{6000}}{-16} \approx 11.09$$

The values 1.4 and 11.1 divide the number line into three intervals: $(-\infty, 1.41)$, $(1.40, 11.09)$, and $(11.09, \infty)$. Use a test point in each interval to determine where the inequality is satisfied.

Interval	Test Value	Is $-16t^2 + 200t + 50 > 300$ True or False?
$(-\infty, 1.41)$	0	$-16 \cdot 0^2 + 200 \cdot 0 + 50 \overset{?}{>} 300$ $50 > 300$ False
$(1.4, 11.09)$	2	$-16 \cdot 2^2 + 200 \cdot 2 + 50 \overset{?}{>} 300$ $386 > 300$ True
$(11.09, \infty)$	12	$-16 \cdot 12^2 + 200 \cdot 12 + 50 \overset{?}{>} 300$ $146 > 300$ False

(*continued on next page*)

Copyright © 2013 Pearson Education, Inc.

(*continued*)

Graphing Calculator Solution:

The rocket will be more than 300 ft above the ground between 1.41 sec and 11.09 sec.

(d) Algebraic Solution:
To find the number of seconds for the toy rocket to hit the ground, let $f(t) = 0$ and solve for t.

$$-16t^2 + 200t + 50 = 0$$

Use the quadratic formula with $a = -16$, $b = 200$, and $c = 50$.

$$t = \frac{-200 \pm \sqrt{200^2 - 4(-16)(50)}}{2(-16)}$$

$$= \frac{-200 \pm \sqrt{40,000 + 3200}}{-32}$$

$$= \frac{-200 \pm \sqrt{43,200}}{-32}$$

$$t = \frac{-200 + \sqrt{43,200}}{-32} \approx -0.25 \text{ or}$$

$$t = \frac{-200 - \sqrt{43,200}}{-32} \approx 12.75$$

We reject the negative solution.

Graphing Calculator Solution:

It will take approximately 12.75 seconds for the toy rocket to hit the ground.

55. (a) The length of the other side would be $640 - 2x$.

(b) In order for the two lengths to be positive, $0 < x < 320$.

(c) $\mathcal{A}(x) = x(640 - 2x) = 640x - 2x^2$
$$= -2x^2 + 640x$$

(d) Algebraic Solution:
Solve the inequality
$30,000 < -2x^2 + 640x < 40,000$. Treat this as two inequalities,

$$-2x^2 + 640x > 30,000 \text{ and}$$

$$-2x^2 + 640x < 40,000.$$

For $-2x^2 + 640x > 30,000$, we solve the corresponding equation

$$-2x^2 + 640x = 30,000.$$

This equation is equivalent to

$$-2x^2 + 640x - 30,000 = 0 \text{ or}$$

$$x^2 - 320x + 15,000 = 0.$$

Use the quadratic formula with $a = 1$, $b = -320$, and $c = 15,000$.

$$x = \frac{-(-320) \pm \sqrt{(-320)^2 - 4(1)(15,000)}}{2(1)}$$

$$= \frac{320 \pm \sqrt{102,400 - 60,000}}{2}$$

$$= \frac{320 \pm \sqrt{42,400}}{2}$$

$$x = \frac{320 - \sqrt{42,400}}{2} \approx 57.04 \text{ or}$$

$$x = \frac{320 + \sqrt{42,400}}{2} \approx 262.96$$

The values 57.04 and 262.96 divide the number line into three intervals:
$(-\infty, 57.04)$, $(57.04, 262.96)$, and
$(262.96, \infty)$. Use a test point in each interval to determine where the inequality is satisfied.

Interval	Test Value	Is $-2x^2 + 640x > 30,000$ True or False?
$(-\infty, 57.04)$	0	$-2 \cdot 0^2 + 640 \cdot 0 \overset{?}{>} 30,000$ $\qquad\qquad 0 > 30,000$ False
$(57.04, 262.96)$	60	$-2 \cdot 60^2 + 640 \cdot 60 \overset{?}{>} 30,000$ $\qquad\qquad 31,200 > 30,000$ True
$(262.96, \infty)$	300	$-2 \cdot 300^2 + 640 \cdot 300 \overset{?}{>} 30,000$ $\qquad\qquad 12,000 > 30,000$ False

Thus, the first inequality is satisfied when the measure of x is in the interval $(57.04, 262.96)$.

(*continued on next page*)

Copyright © 2013 Pearson Education, Inc.

(*continued*)

For $-2x^2 + 640x < 40,000,$ we solve the corresponding equation

$-2x^2 + 640x = 40,000.$

This equation is equivalent to

$-2x^2 + 640x - 40,000 = 0$ or

$x^2 - 320x + 20,000 = 0$

Use the quadratic formula with $a = 1,$ $b = -320,$ and $c = 20,000.$

$$x = \frac{-(-320) \pm \sqrt{(-320)^2 - 4(1)(20,000)}}{2(1)}$$

$$= \frac{320 \pm \sqrt{102,400 - 80,000}}{2}$$

$$= \frac{320 \pm \sqrt{22,400}}{2}$$

$$x = \frac{320 - \sqrt{22,400}}{2} \approx 85.17 \text{ or}$$

$$x = \frac{320 + \sqrt{22,400}}{2} \approx 234.83$$

The values 85.17 and 234.83 divide the number line into three intervals:

$(-\infty, 85.17),$ $(85.17, 234.83),$ and

$(234.83, \infty).$ Use a test point in each interval to determine where the inequality is satisfied.

Interval	Test Value	Is $-2x^2 + 640x < 40,000$ True or False?
$(-\infty, 85.17)$	0	$-2 \cdot 0^2 + 640 \cdot 0 \overset{?}{<} 40,000$ $0 < 40,000$ True
$(85.17, 234.83)$	90	$-2 \cdot 90^2 + 640 \cdot 90 \overset{?}{<} 40,000$ $41,400 < 40,000$ False
$(234.83, \infty)$	300	$-2 \cdot 300^2 + 640 \cdot 300 \overset{?}{<} 40,000$ $12,000 < 40,000$ True

Thus, the second inequality is satisfied when the measure of x is in the interval $(-\infty, 85.17)$ or $(234.83, \infty).$ We must now seek the intersection of the intervals $(57.04, 262.96),$ and $(-\infty, 85.17)$ or $(234.83, \infty).$ If we use the real number line as an aid, we can see that the solution would be between 57.04 ft and 85.17 ft or 234.83 ft and 262.96 ft.

57.04 85.17   234.83 262.96

Graphing Calculator Solution:

We can see from the graphs that the quadratic function lies between the lines when x is between 57.04 ft and 85.17 ft or 234.83 ft and 262.96 ft.

(e) Algebraic Solution:
Find the coordinates of the vertex of the parabola, $A(x) = -2x^2 + 640x.$ Using the vertex formula with $a = -2$ and $b = 640.$ We have

$x = -\frac{b}{2a} = -\frac{640}{2(-2)} = 160$ and

$y = -2(160)^2 + 640(160) = 51,200.$

Since $a < 0,$ this is the maximum point.

Graphing Calculator Solution:

Thus, the length of the two parallel sides would be 160 ft and the third side would be $640 - 2(160) = 620 - 320 = 320$ ft.

The maximum area would be 51,200 ft^2.

Copyright © 2013 Pearson Education, Inc.

57. (a) The length of the original piece of cardboard would be $2x$.

(b) The length of the rectangular box would be $2x - 4$ and the width would be $x - 4$, where $x > 4$.

(c) $V(x) = (2x - 4)(x - 4)(2)$
$$= (2x^2 - 12x + 16)(2)$$
$$= 4x^2 - 24x + 32$$

(d) Algebraic Solution:

Solve the equation $4x^2 - 24x + 32 = 320$.
$$4x^2 - 24x + 32 = 320$$
$$4x^2 - 24x - 288 = 0$$
$$x^2 - 6x - 72 = 0$$
$$(x + 6)(x - 12) = 0$$
$$x + 6 = 0 \Rightarrow x = -6 \text{ or}$$
$$x - 12 = 0 \Rightarrow x = 12$$
We discard the negative solution.

Graphing Calculator Solution:

If $x = 12$, then the dimensions of the bottom of the box will be $12 - 4 = 8$ in. by $2(12) - 4 = 24 - 4 = 20$ in.

(e) Algebraic Solution:

Solve the inequality
$400 < 4x^2 - 24x + 32 < 500$. Treat this as two inequalities, $4x^2 - 24x + 32 > 400$ and $4x^2 - 24x + 32 < 500$.

For $4x^2 - 24x + 32 > 400$, we solve the corresponding equation
$4x^2 - 24x + 32 = 400$.
This equation is equivalent to
$4x^2 - 24x - 368 = 0$ or $x^2 - 6x - 92 = 0$
Use the quadratic formula with $a = 1$, $b = -6$, and $c = -92$.

$$x = \frac{-(-6) \pm \sqrt{(-6)^2 - 4(1)(-92)}}{2(1)}$$
$$= \frac{6 \pm \sqrt{36 + 368}}{2} = \frac{6 \pm \sqrt{404}}{2}$$
$$= \frac{6 \pm 2\sqrt{101}}{2} = 3 \pm \sqrt{101}$$
$$x = 3 - \sqrt{101} \approx -7.0 \text{ or}$$
$$x = 3 + \sqrt{101} \approx 13.0$$

Since $x > 4$, we need only check the intervals: $(4, 13.0)$ and $(13.0, \infty)$. Use a test point in each interval to determine where the inequality is satisfied.

Interval	Test Value	Is $4x^2 - 24x + 32 > 400$ True or False?
$(4, 13.0)$	5	$4 \cdot 5^2 - 24 \cdot 5 + 32 \overset{?}{>} 400$ $12 > 400$ False
$(13.0, \infty)$	14	$4 \cdot 14^2 - 24 \cdot 14 + 32 \overset{?}{>} 400$ $480 > 400$ True

Thus, the first inequality is satisfied when the length, x, is in the interval $(13.0, \infty)$.

For $4x^2 - 24x + 32 < 500$, we solve the corresponding equation
$4x^2 - 24x + 32 = 500$. This equation is equivalent to $4x^2 - 24x - 468 = 0$ or $x^2 - 6x - 117 = 0$. Use the quadratic formula with $a = 1$, $b = -6$, and $c = -117$.

$$x = \frac{-(-6) \pm \sqrt{(-6)^2 - 4(1)(-117)}}{2(1)}$$
$$= \frac{6 \pm \sqrt{36 + 468}}{2} = \frac{6 \pm \sqrt{504}}{2}$$
$$= \frac{6 \pm 2\sqrt{126}}{2} = 3 \pm \sqrt{126}$$
$$x = 3 - \sqrt{126} \approx -8.2 \text{ or}$$
$$x = 3 + \sqrt{126} \approx 14.2$$

Since $x > 4$, we need only check the intervals: $(4, 14.2)$ and $(14.2, \infty)$. Use a test point in each interval to determine where the inequality is satisfied.

(continued on next page)

Copyright © 2013 Pearson Education, Inc.

(continued)

Interval	Test Value	Is $4x^2 - 24x + 32 < 500$ True or False?
$(4, 14.2)$	5	$4 \cdot 5^2 - 24 \cdot 5 + 32 \overset{?}{<} 500$ $12 < 500$ True
$(14.2, \infty)$	15	$4 \cdot 15^2 - 24 \cdot 15 + 32 \overset{?}{<} 500$ $572 < 500$ False

Thus, the second inequality is satisfied when the length, x, is in the interval $(4, 14.2)$. We must now seek the intersection of the intervals $(13.0, \infty)$ and $(4, 14.2)$. This intersection is $(13.0, 14.2)$.

Graphing Calculator Solution:

We can see from the graphs that the quadratic function lies between the lines when x is between 13.0 in and 14.2 in. Thus, the volume will be between 400 in.^{3} and 500 in.^{3}, when the original width is between 13.0 in. and 14.2 in.

59. $y = \dfrac{-16x^2}{0.434v^2} + 1.15x + 8$

(a) Let $y = 10$ and $x = 15$.

$$10 = \frac{-16(15)^2}{0.434v^2} + 1.15(15) + 8$$

$$10 = \frac{-16(225)}{0.434v^2} + 1.15(15) + 8$$

$$10 = \frac{-3600}{0.434v^2} + 17.25 + 8$$

$$10 = \frac{-3600}{0.434v^2} + 25.25$$

$$\frac{3600}{0.434v^2} = 15.25$$

$$3600 = 6.6185v^2 \Rightarrow \frac{3600}{6.6185} = v^2$$

$$\pm\sqrt{\frac{3600}{6.6185}} = v \Rightarrow v \approx \pm 23.32$$

Since v represents a velocity, only the positive square root is meaningful. The basketball should have an initial velocity of 23.32 ft per sec.

(b) $y = \dfrac{-16x^2}{0.434(23.32)^2} + 1.15x + 8$

Algebraic Solution:

Since the parabola opens downward, the vertex is the maximum point. Use the vertex formula to find the x-coordinate of the vertex.

$$x = -\frac{b}{2a} = -\frac{1.15}{2\left(\dfrac{-16}{0.434(23.32)^2}\right)} \approx 8.482$$

To find the y-coordinate of the vertex, evaluate the quadratic function when $x = 8.482$.

$$y = \frac{-16(8.482)^2}{0.434(23.32)^2} + 1.15(8.482) + 8$$

$$\approx 12.88$$

Graphing Calculator Solution:

The y-coordinate of the vertex represents the maximum height of the basketball, which is approximately 12.88 ft.

61. $f(x) = 0.0773x^2 + 0.2115x + 32.64$

The year 2012 implies $x = 16$

$f(16) = 0.0773(16)^2 - 0.2115(16) + 32.64$

≈ 49.0

The model predicts that approximately 49.0% of births will be to unmarried women for 2012.

Copyright © 2013 Pearson Education, Inc.

63. $f(x) = -4.979x^2 + 71.73x + 97.29$

We seek the x-coordinate of the vertex.

Algebraic Solution:

Since the parabola opens downward, the vertex is the maximum point. Use the vertex formula to find the x-coordinate of the vertex.

$$x = -\frac{b}{2a} = -\frac{71.73}{2(-4.979)} \approx 7.2$$

Graphing Calculator Solution:

Based on the model, spending on shoes and clothing reached a maximum about seven years after 2000, during the year 2007.

65. (a) Plot the 18 points given.

(b) A quadratic function would model the data better because the data increases at a different rate each year.

(c) Use the quadratic regression function on the graphing calculator to find the equation:

The equation is

$f(x) = -1328x^2 + 71,882x + 181,195.$

(d) Plotting the points together with $f(x)$, we see that f models the data almost exactly.

(e) $x = 19$ corresponds to the year 2009, and $x = 20$ corresponds to the year 2010.

Algebraic Solution:

$$f(19) \approx -1328(19)^2 + 71,882(19) + 181,195$$
$$\approx 1,067,545$$

$$f(20) \approx -1328(20)^2 + 71,882(20) + 181,195$$
$$\approx 1,087,635$$

Graphing Calculator Solution:

In the year 2009, approximately 1,067,439 people will have been diagnosed with AIDS since 1990. In the year 2010, approximately 1,087,518 people will have been diagnosed with AIDS since 1990.

(f) The number of new cases in the year 2010 will be approximately

$$f(20) - f(19) \approx 1,087,635 - 1,067,545$$
$$\approx 20,090$$

67. (a) Plot the 9 points given. Note that the year 2008 is represented by $x = 78$.

(b) Let the point (40, 4.7), be the vertex.

Then, $f(x) = a(x - 40)^2 + 4.7.$

Next let the point (20, 6.9) lie on the graph of the function and solve for a.

$$f(20) = a(20 - 40)^2 + 4.7 = 6.9$$
$$a(-20)^2 + 4.7 = 6.9$$
$$400a + 4.7 = 6.9$$
$$400a = 2.2 \Rightarrow a = \frac{2.2}{400} = 0.0055$$

Thus, $f(x) = 0.0055(x - 40)^2 + 4.7.$

Copyright © 2013 Pearson Education, Inc.

(c) Plotting the points together with $f(x)$, we see that there is a relatively good fit.

(d) The quadratic regression curve is
$$g(x) = 0.0045x^2 - 0.3357x + 11.62$$

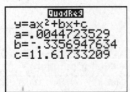

(e) The year 2015 corresponds to $x = 85$.

Algebraic Solution:

$$f(85) = 0.0055(85 - 40)^2 + 4.7 \approx 15.8$$
$$g(85) = 0.0045(85)^2 - 0.3357(85) + 11.62$$
$$\approx 15.6$$

Graphing Calculator Solution:

Using f, about 15.8% of the U.S. population will be foreign-born in 2015. Using g, about 15.4% of the U.S. population will be foreign-born in 2015. The unrounded regression equation gives 15.4%.

69. $y = x^2 - 10x + c$

An x-intercept occurs where $y = 0$, or $0 = x^2 - 10x + c$. There will be exactly one x-intercept if this equation has exactly one solution, or if the discriminant is zero.
$$b^2 - 4ac = 0 \Rightarrow (-10)^2 - 4(1)c = 0 \Rightarrow$$
$$100 - 4c = 0 \Rightarrow 100 = 4c \Rightarrow c = 25$$

71. x-intercepts 2 and 5, and y-intercept 5
Since we have x-intercepts 2 and 5, f has linear factors of $x - 2$ and $x - 5$.
$$f(x) = a(x - 2)(x - 5) = a(x^2 - 7x + 10)$$
Since the y-intercept is 5, $f(0) = 5$.

$$f(x) = a(0^2 - 7(0) + 10) = 5 \Rightarrow$$
$$10a = 5 \Rightarrow a = \tfrac{1}{2}$$
The required quadratic function is
$$f(x) = \tfrac{1}{2}(x^2 - 7x + 10) = \tfrac{1}{2}x^2 - \tfrac{7}{2}x + 5.$$

73. Use the distance formula,
$$d(P, R) = \sqrt{(x_1 - x_2)^2 + (y_1 - y_2)^2}, \text{ to find the}$$
distance between the points
$P(x, 2x)$ and $R(1, 7)$, where P is any point on the line $y = 2x$.
$$d(P, R) = \sqrt{(x - 1)^2 + (2x - 7)^2}$$
$$= \sqrt{(x^2 - 2x + 1) + (4x^2 - 28x + 49)}$$
$$= \sqrt{5x^2 - 30x + 50}$$
Consider the equation $y = 5x^2 - 30x + 50$.
This is the equation of a parabola opening upward, so the expression $5x^2 - 30x + 50$ has a minimum value, which is the y-value of the vertex. Complete the square to find the vertex.
$$y = 5x^2 - 30x + 50 = 5(x^2 - 6x) + 50$$
$$= 5(x^2 - 6x + 9 - 9) + 50$$
$$= 5(x^2 - 6x + 9) - 45 + 50 = 5(x - 3)^2 + 5$$
The vertex is (3, 5), so the minimum value of $5x^2 - 30x + 50$ is 5 when $x = 3$. Thus, the minimum value of $\sqrt{5x^2 - 30x + 50}$ is $\sqrt{5}$ when $x = 3$. The point on the line $y = 2x$ for which $x = 3$ is (3, 6). Thus, the closest point on the line $y = 2x$ to the point (1, 7) is the point (3, 6).

75. Graph the function $f(x) = x^2 + 2x - 8$.
Complete the square to find the vertex, $(-1, -9)$.
$$y = x^2 - 2x - 8 = (x^2 - 2x) - 8$$
$$= (x^2 - 2x + 1 - 1) - 8$$
$$= (x^2 - 2x + 1) - 1 - 8 = (x - 1)^2 - 9$$
Use a table of values to find points on the graph.

x	y
-5	7
-4	0
-3	-5
-2	-8
-1	-9
0	-8
1	-5
2	0
3	7

$f(x) = x^2 + 2x - 8$

Copyright © 2013 Pearson Education, Inc.

77. The solution set of the inequality

$x^2 + 2x - 8 < 0$ consists of all x-values for which the graph of f lies below the x-axis. By examining the graph of the function

$f(x) = x^2 + 2x - 8$ from Exercise 75, we see that the graph lies below the x-axis when x is between the -4 and 2. Thus, the solution set is the open interval $(-4, 2)$.

79. $f(x) = x^2 - x - 6$

Complete the square to find the vertex..

$$y = x^2 - x - 6 = (x^2 - x) - 6$$
$$= \left(x^2 - x + \tfrac{1}{4} - \tfrac{1}{4}\right) - 6$$
$$= \left(x^2 - x + \tfrac{1}{4}\right) - \tfrac{1}{4} - 6 = \left(x - \tfrac{1}{2}\right)^2 - \tfrac{25}{4}$$

The vertex is $\left(\tfrac{1}{2}, -\tfrac{25}{4}\right)$. Now use a table of values to find points on the graph.

x	y
-4	14
-3	6
-2	0
-1	-4
0	-6
1	-6
2	-4
3	0
4	6

$$f(x) = x^2 - x - 6$$

The zeros of the function are at $x = -2$ and $x = 3$. The value of the function is less than 0 between these values. Thus, the solution set of $x^2 - x - 6 < 0$ is $(-2, 3)$.

81. $2x^2 - 9x \geq 18 \Rightarrow f(x) = 2x^2 - 9x - 18$

Complete the square to find the vertex.

$$y = 2x^2 - 9x - 18$$
$$y = 2\left(x^2 - \tfrac{9}{2}x\right) - 9 = 2\left(x^2 - \tfrac{9}{2}x + \tfrac{81}{16} - \tfrac{81}{16}\right) - 9$$
$$= 2\left(x^2 - \tfrac{9}{2}x + \tfrac{81}{16}\right) - 2\left(\tfrac{81}{16}\right) - 9$$
$$= 2\left(x - \tfrac{9}{4}\right)^2 - \tfrac{81}{8} - 9 = 2\left(x - \tfrac{9}{4}\right)^2 - \tfrac{153}{8}$$

The vertex is $\left(\tfrac{9}{4}, -\tfrac{153}{8}\right)$. Now use a table of values to find points on the graph.

x	y
-2	8
-1.5	0
-1	-7
0	-18
1	-25
2	-28
3	-27
4	-22
5	-13
6	0
7	17

$$f(x) = 2x^2 - 9x - 18$$

The zeros of the function are at $x = -1.5$ and $x = 6$. The value of the function is less than 0 between these values. Thus, the solution set of $2x^2 - 9x \geq 18$ or $2x^2 - 9x - 18 \geq 0$ is $(-\infty, -1.5] \cup [6, \infty)$.

83. $f(x) = -x^2 + 4x + 1$

Complete the square to find the vertex.

$$y = -x^2 + 4x + 1$$
$$y = -\left(x^2 - 4x\right) + 1 = -\left(x^2 - 4x + 4 - 4\right) + 1$$
$$= -\left(x^2 - 4x + 4\right) + 4 + 1$$
$$= -(x - 2)^2 + 5$$

The vertex is $(2, 5)$ and the graph opens downward. Now use a table of values to find points on the graph.

x	y
-2	11
-1	-4
0	1
1	4
2	5
3	4
4	1
5	-4
6	-11

$$f(x) = -x^2 + 4x + 1$$

Use the quadratic formula to determine the zeros of the function.

$$x = \frac{-b \pm \sqrt{b^2 - 4ac}}{2a} = \frac{-4 \pm \sqrt{4^2 - 4(-1)(1)}}{2(-1)}$$
$$= \frac{-4 \pm \sqrt{20}}{-2} = \frac{-4 \pm 2\sqrt{5}}{-2} = 2 \pm \sqrt{5}$$

The value of the function is greater than or equal to 0 between these values. Thus, the solution set of $-x^2 + 4x + 1 \geq 0$ is $\left[2 - \sqrt{5}, 2 + \sqrt{5}\right]$.

Copyright © 2013 Pearson Education, Inc.

Section 3.2   Synthetic Division

1. Since $x + 1$ is in the form $x + k$, $k = -1$.

$$-1)\overline{\begin{array}{cccc} 1 & 3 & 11 & 9 \end{array}}$$
$$\begin{array}{cccc} & -1 & -2 & -9 \end{array}$$
$$\overline{\begin{array}{cccc} 1 & 2 & 9 & 0 \end{array}}$$

$$\frac{x^3 + 3x^2 + 11x + 9}{x + 1} = x^2 + 2x + 9$$

3. Express $x + 1$ in the form $x - k$ by writing it as $x - (-1)$. Thus $k = -1$.

$$-1)\overline{\begin{array}{ccccc} 5 & 5 & 2 & -1 & -3 \end{array}}$$
$$\begin{array}{ccccc} & -5 & 0 & -2 & 3 \end{array}$$
$$\overline{\begin{array}{ccccc} 5 & 0 & 2 & -3 & 0 \end{array}}$$

$$\frac{5x^4 + 5x^3 + 2x^2 - x - 3}{x + 1} = 5x^3 + 2x - 3$$

5. Express $x + 4$ in the form $x - k$ by writing it as $x - (-4)$. Thus $k = -4$.

$$-4)\overline{\begin{array}{ccccc} 1 & 4 & 2 & 9 & 4 \end{array}}$$
$$\begin{array}{ccccc} & -4 & 0 & -8 & -4 \end{array}$$
$$\overline{\begin{array}{ccccc} 1 & 0 & 2 & 1 & 0 \end{array}}$$

$$\frac{x^4 + 4x^3 + 2x^2 + 9x + 4}{x + 4} = x^3 + 2x + 1$$

7. Express $x + 2$ in the form $x - k$ by writing it as $x - (-2)$. Thus $k = -2$.

$$-2)\overline{\begin{array}{cccccc} 1 & 3 & 2 & 2 & 3 & 1 \end{array}}$$
$$\begin{array}{cccccc} & -2 & -2 & 0 & -4 & 2 \end{array}$$
$$\overline{\begin{array}{cccccc} 1 & 1 & 0 & 2 & -1 & 3 \end{array}}$$

$$\frac{x^5 + 3x^4 + 2x^3 + 2x^2 + 3x + 1}{x + 2}$$
$$= x^4 + x^3 + 2x - 1 + \frac{3}{x + 2}$$

9. Since $x - 2$ is in the form $x - k$, $k = 2$.

$$2)\overline{\begin{array}{cccc} -9 & 8 & -7 & 2 \end{array}}$$
$$\begin{array}{cccc} & -18 & -20 & -54 \end{array}$$
$$\overline{\begin{array}{cccc} -9 & -10 & -27 & -52 \end{array}}$$

$$\frac{-9x^3 + 8x^2 - 7x + 2}{x - 2} = -9x^2 - 10x - 27 + \frac{-52}{x - 2}$$

11. Since $x - \frac{1}{3}$ is in the form $x - k$, $k = \frac{1}{3}$.

$$\frac{1}{3})\overline{\begin{array}{cccc} \frac{1}{3} & -\frac{2}{9} & \frac{2}{27} & -\frac{1}{81} \end{array}}$$
$$\begin{array}{cccc} & \frac{1}{9} & -\frac{1}{27} & \frac{1}{81} \end{array}$$
$$\overline{\begin{array}{cccc} \frac{1}{3} & -\frac{1}{9} & \frac{1}{27} & 0 \end{array}}$$

$$\frac{\frac{1}{3}x^3 - \frac{2}{9}x^2 + \frac{2}{27}x - \frac{1}{81}}{x - \frac{1}{3}} = \frac{1}{3}x^2 - \frac{1}{9}x + \frac{1}{27}$$

13. Since $x - 2$ is in the form $x - k$, $k = 2$. The constant term is missing, so include a 0.

$$2)\overline{\begin{array}{ccccc} 1 & -3 & -4 & 12 & 0 \end{array}}$$
$$\begin{array}{ccccc} & 2 & -2 & -12 & 0 \end{array}$$
$$\overline{\begin{array}{ccccc} 1 & -1 & -6 & 0 & 0 \end{array}}$$

$$\frac{x^4 - 3x^3 - 4x^2 + 12x}{x - 2} = x^3 - x^2 - 6x$$

15. Since $x - 1$ is in the form $x - k$, $k = 1$. The x^2- and x-terms are missing, so include 0's as their coefficients.

$$1)\overline{\begin{array}{cccc} 1 & 0 & 0 & -1 \end{array}}$$
$$\begin{array}{cccc} & 1 & 1 & 1 \end{array}$$
$$\overline{\begin{array}{cccc} 1 & 1 & 1 & 0 \end{array}}$$

$$\frac{x^3 - 1}{x - 1} = x^2 + x + 1$$

17. Express $x + 1$ in the form $x - k$ by writing it as $x - (-1)$. Thus $k = -1$. Since the x^4, x^3, x^2 and x-terms are missing, include 0's as their coefficients.

$$-1)\overline{\begin{array}{ccccc} 1 & 0 & 0 & 0 & 1 \end{array}}$$
$$\begin{array}{ccccc} & -1 & 1 & -1 & 1 & -1 \end{array}$$
$$\overline{\begin{array}{ccccc} 1 & -1 & 1 & -1 & 1 & 0 \end{array}}$$

$$\frac{x^5 + 1}{x + 1} = x^4 - x^3 + x^2 - x + 1$$

19. $f(x) = 2x^3 + x^2 + x - 8$; $k = -1$
Use synthetic division to write the polynomial in the form $f(x) = (x - k)q(x) + r$.

$$-1)\overline{\begin{array}{cccc} 2 & 1 & 1 & -8 \end{array}}$$
$$\begin{array}{cccc} & -2 & 1 & -2 \end{array}$$
$$\overline{\begin{array}{cccc} 2 & -1 & 2 & -10 \end{array}}$$

$$f(x) = [x - (-1)](2x^2 - x + 2) - 10$$
$$= (x + 1)(2x^2 - x + 2) - 10$$

Copyright © 2013 Pearson Education, Inc.

21. $f(x) = x^3 + 4x^2 + 5x + 2;\ k = -2$

$$-2\overline{)1\ \ 4\ \ 5\ \ \ 2}$$
$$\underline{\ \ -2\ -4\ -2\ }$$
$$1\ \ 2\ \ 1\ \ \ 0$$

$f(x) = [x - (-2)](x^2 + 2x + 1) + 0$
$\quad = (x + 2)(x^2 + 2x + 1) + 0$

23. $f(x) = 4x^4 - 3x^3 - 20x^2 - x;\ k = 3$

$$3\overline{)4\ -3\ -20\ -1\ \ \ 0}$$
$$\underline{\ \ \ 12\ \ \ 27\ \ 21\ 60\ }$$
$$4\ \ \ 9\ \ \ \ 7\ \ 20\ 60$$

$f(x) = (x - 3)(4x^3 + 9x^2 + 7x + 20) + 60$

25. $f(x) = 3x^4 + 4x^3 - 10x^2 + 15;\ k = -1$

$$-1\overline{)3\ \ \ 4\ -10\ \ \ 0\ \ \ 15}$$
$$\underline{\ \ \ -3\ -1\ \ 11\ -11\ }$$
$$3\ \ \ 1\ -11\ \ 11\ \ \ 4$$

$f(x) = [x - (-1)](3x^3 + x^2 - 11x + 11) + 4$
$\quad = (x + 1)(3x^3 + x^2 - 11x + 11) + 4$

27. $f(x) = x^2 + 5x + 6;\ k = -2$

$$-2\overline{)1\ \ 5\ \ 6}$$
$$\underline{\ \ -2\ -6\ }$$
$$1\ \ 3\ \ 0$$

$f(-2) = 0$

29. $f(x) = 2x^2 - 3x - 3;\ k = 2$

$$2\overline{)2\ -3\ -3}$$
$$\underline{\ \ \ \ 4\ \ \ 2\ }$$
$$2\ \ \ 1\ -1$$

$f(2) = -1$

31. $f(x) = x^3 - 4x^2 + 2x + 1;\ k = -1$

$$-1\overline{)1\ -4\ \ 2\ \ \ 1}$$
$$\underline{\ \ \ -1\ \ 5\ -7\ }$$
$$1\ -5\ \ 7\ -6$$

$f(-1) = -6$

33. $f(x) = 2x^5 - 10x^3 - 19x^2 - 50;\ k = 3$

$$3\overline{)2\ \ \ 0\ -10\ -19\ \ \ 0\ -50}$$
$$\underline{\ \ \ \ 6\ \ 18\ \ 24\ \ 15\ \ 45\ }$$
$$2\ \ \ 6\ \ \ \ 8\ \ \ \ 5\ \ 15\ -5$$

$f(3) = -5$

35. $f(x) = 6x^4 + x^3 - 8x^2 + 5x + 6;\ k = \frac{1}{2}$

$$\tfrac{1}{2}\overline{)6\ \ \ 1\ -8\ \ 5\ \ 6}$$
$$\underline{\ \ \ \ 3\ \ \ 2\ -3\ \ 1\ }$$
$$6\ \ \ 4\ -6\ \ 2\ \ 7$$

$f\left(\tfrac{1}{2}\right) = 7$

37. $f(x) = x^2 - 5x + 1;\ k = 2 + i$

$$2 + i\overline{)1\ -5\ \qquad\ 1}$$
$$\underline{\qquad\ 2 + i\ -7 - i\ }$$
$$1\ -3 + i\ -6 - i$$

$f(2 + i) = -6 - i$

39. $f(x) = x^2 + 4;\ k = 2i$

$$2i\overline{)1\ \ \ 0\qquad\ \ 4}$$
$$\underline{\qquad\ 2i\ -4\ }$$
$$1\ \ 0 - 2i\ \ \ 0$$

$f(2i) = 0$

41. To determine if $k = 2$ is a zero of
$f(x) = x^2 + 2x - 8$, divide synthetically.

$$2\overline{)1\ \ 2\ -8}$$
$$\underline{\ \ \ \ 2\ \ \ 8\ }$$
$$1\ \ 4\ \ \ 0$$

Yes, 2 is a zero of $f(x)$ because $f(2) = 0$.

43. To determine if $k = 2$ is a zero of
$f(x) = x^3 - 3x^2 + 4x - 4$, divide synthetically.

$$2\overline{)1\ -3\ \ \ 4\ -4}$$
$$\underline{\ \ \ \ \ 2\ -2\ \ \ 4\ }$$
$$1\ -1\ \ \ 2\ \ \ 0$$

Yes, 2 is a zero of $f(x)$ because $f(2) = 0$.

45. To determine if $k = 1$ is a zero of
$f(x) = 2x^3 - 6x^2 - 9x + 4$, divide
synthetically.

$$1\overline{)2\ -6\ \ -9\ \ \ 4}$$
$$\underline{\ \ \ \ \ 2\ \ -4\ -13\ }$$
$$2\ -4\ -13\ -9$$

No, 1 is not a zero of $f(x)$ because
$f(1) = -9$.

Copyright © 2013 Pearson Education, Inc.

47. To determine if $k = 0$ is a zero of
$f(x) = x^3 + 7x^2 + 10x$, divide synthetically.

$$
\begin{array}{r|rrrr}
0 & 1 & 7 & 10 & 0 \\
  &   & 0 & 0 & 0 \\
\hline
  & 1 & 7 & 10 & 0
\end{array}
$$

Yes, 0 is a zero of $f(x)$ because $f(0) = 0$.

49. To determine if $k = -\frac{3}{2}$ is a zero of
$f(x) = 4x^4 + x^2 + 17x + 3$, divide
synthetically. The x^3 term is missing, so insert
a 0.

$$
\begin{array}{r|rrrrr}
-\frac{3}{2} & 4 & 0 & 1 & 17 & 3 \\
             &   & -6 & 9 & -15 & -3 \\
\hline
             & 4 & -6 & 10 & 2 & 0
\end{array}
$$

Yes, $-\frac{3}{2}$ is a zero of $f(x)$ because
$f\left(-\frac{3}{2}\right) = 0$.

51. To determine if $k = \frac{2}{5}$ is a zero of
$f(x) = 5x^4 + 2x^3 - x + 3$, divide synthetically.

$$
\begin{array}{r|rrrrr}
\frac{2}{5} & 5 & 2 & 0 & -1 & 3 \\
            &   & 2 & \frac{8}{5} & \frac{16}{25} & -\frac{18}{125} \\
\hline
            & 5 & 4 & \frac{8}{5} & -\frac{9}{25} & \frac{357}{125}
\end{array}
$$

No, $\frac{2}{5}$ is not a zero of $f(x)$ because
$f\left(\frac{2}{5}\right) = \frac{357}{125}$.

53. To determine if $k = 1 - i$ is a zero of
$f(x) = x^2 - 2x + 2$, divide synthetically.

$$
\begin{array}{r|rrr}
1-i & 1 & -2 & 2 \\
    &   & 1-i & -2 \\
\hline
    & 1 & -1-i & 0
\end{array}
$$

Yes, $1 - i$ is a zero of $f(x)$ because
$f(1 - i) = 0$.

55. To determine if $k = 2 + i$ is a zero of
$f(x) = x^2 + 3x + 4$, divide synthetically.

$$
\begin{array}{r|rrr}
2+i & 1 & 3 & 4 \\
    &   & 2+i & 9+7i \\
\hline
    & 1 & 5+i & 13+7i
\end{array}
$$

No, $2 + i$ is not a zero of $f(x)$ because
$f(2 + i) = 13 + 7i$.

57. To determine if $k = 1 + i$ is a zero of
$f(x) = x^3 + 3x^2 - x + 1$, divide synthetically.

$$
\begin{array}{r|rrrr}
1+i & 1 & 3 & -1 & 1 \\
    &   & 1+i & 3+5i & -3+7i \\
\hline
    & 1 & 4+i & 2+5i & -2+7i
\end{array}
$$

No, $1 + i$ is not a zero of $f(x)$ because
$f(1 + i) = -2 + 7i$.

59.
$$
\begin{array}{r|rrrr}
-2 & 1 & -2 & -1 & 2 \\
   &   & -2 & 8 & -14 \\
\hline
   & 1 & -4 & 7 & -12
\end{array}
$$

$f(-2) = -12$
The coordinates of the corresponding point are
$(-2, -12)$.

61.
$$
\begin{array}{r|rrrr}
-\frac{1}{2} & 1 & -2 & -1 & 2 \\
             &   & -\frac{1}{2} & \frac{5}{4} & -\frac{1}{8} \\
\hline
             & 1 & -\frac{5}{2} & \frac{1}{4} & \frac{15}{8}
\end{array}
$$

$f\left(-\frac{1}{2}\right) = \frac{15}{8}$
The coordinates of the corresponding point are
$\left(-\frac{1}{2}, \frac{15}{8}\right)$.

63.
$$
\begin{array}{r|rrrr}
1 & 1 & -2 & -1 & 2 \\
  &   & 1 & -1 & -2 \\
\hline
  & 1 & -1 & -2 & 0
\end{array}
$$

$f(1) = 0$
The coordinates of the corresponding point are
$(1, 0)$.

65.
$$
\begin{array}{r|rrrr}
2 & 1 & -2 & -1 & 2 \\
  &   & 2 & 0 & -2 \\
\hline
  & 1 & 0 & -1 & 0
\end{array}
$$

$f(2) = 0$
The coordinates of the corresponding point are
$(2, 0)$.

67.

$f(x) = x^3 - 2x^2 - x + 2$

Copyright © 2013 Pearson Education, Inc.

Section 3.3 Zeros of Polynomial Functions

1. Since $x - 1$ is a factor of
   $f(x) = x^6 - x^4 + 2x^2 - 2,$ we are assured that
   $f(1) = 0.$ This statement is justified by the
   factor theorem; therefore, it is true.

3. For the function $f(x) = (x + 2)^4(x - 3),$ 2 is a
   zero of multiplicity 4. To find the zero, set the
   factor equal to 0: $x + 2 = 0 \Rightarrow x = -2$
   2 is not a zero of the function; therefore, the
   statement is false. (It would be true to say that
   -2 is a zero of multiplicity 4.)

5. $x^3 - 5x^2 + 3x + 1; \; x - 1$

   Let $f(x) = x^3 - 5x^2 + 3x + 1.$ By the factor
   theorem, $x - 1$ will be a factor of $f(x)$ if and
   only if $f(1) = 0.$ Use synthetic division and the
   remainder theorem.

   ```
   1)1  -5   3   1
        1  -4  -1
      1  -4  -1   0
   ```

   Since $f(1) = 0,$ $x - 1$ is a factor of $f(x).$

7. $2x^4 + 5x^3 - 8x^2 + 3x + 13; \; x + 1$

   Let $f(x) = 2x^4 + 5x^3 - 8x^2 + 3x + 13.$ By the
   factor theorem, $x + 1$ will be a factor of $f(x)$
   if and only if $f(-1) = 0.$ Use synthetic
   division and the remainder theorem.

   ```
   -1)2   5   -8    3    13
         -2   -3   11   -14
       2   3  -11   14   -1
   ```

   Since the remainder is $-1,$ $f(-1) = -1,$ so
   $x + 1$ is not a factor of $f(x).$

9. $-x^3 + 3x - 2; \; x + 2$

   Let $f(x) = -x^3 + 3x - 2.$ By the factor
   theorem, $x + 2$ will be a factor of $f(x)$ if and
   only if $f(-2) = 0.$ Use synthetic division and
   the remainder theorem.

   ```
   -2)-1  0   3  -2
          2  -4   2
      -1  2  -1   0
   ```

   Since $f(-2) = 0,$ $x + 2$ is a factor of $f(x).$

11. $4x^2 + 2x + 54; \; x - 4$

    Let $f(x) = 4x^2 + 2x + 54.$ By the factor
    theorem, $x - 4$ will be a factor of $f(x)$ if and
    only if $f(4) = 0.$ Use synthetic division and
    the remainder theorem.

    ```
    4)4   2   54
         16   72
      4  18  126
    ```

    Since the remainder is 126, $f(4) = 126,$ so
    $x - 4$ is not a factor of $f(x).$

13. $x^3 + 2x^2 + 3; \; x - 1$

    Let $f(x) = x^3 + 2x^2 + 1.$ By the factor
    theorem, $x - 1$ will be a factor of $f(x)$ if and
    only if $f(1) = 0.$ Use synthetic division and
    the remainder theorem.

    ```
    1)1  2  0  3
         1  3  3
      1  3  3  6
    ```

    Since $f(1) = 6,$ $x - 1$ is not a factor of $f(x).$

15. $2x^4 + 5x^3 - 2x^2 + 5x + 6; \; x + 3$

    Let $f(x) = 2x^4 + 5x^3 - 2x^2 + 5x + 6.$ By the
    factor theorem, $x + 3$ will be a factor of $f(x)$ if
    and only if $f(-3) = 0.$ Use synthetic division
    and the remainder theorem.

    ```
    -3)2   5   -2   5   6
          -6    3  -3  -6
       2  -1    1   2   0
    ```

    Since $f(-3) = 0,$ $x + 3$ is a factor of $f(x).$

17. $f(x) = 2x^3 - 3x^2 - 17x + 30; \; k = 2$

    Since 2 is a zero of $f(x),$ $x - 2$ is a factor.
    Divide $f(x)$ by $x - 2.$

    ```
    2)2  -3  -17   30
         4    2  -30
      2   1  -15    0
    ```

    Thus,
    $$f(x) = (x - 2)(2x^2 + x - 15)$$
    $$= (x - 2)(2x - 5)(x + 3).$$

Copyright © 2013 Pearson Education, Inc.

19. $f(x) = 6x^3 + 13x^2 - 14x + 3;\ k = -3$
Since -3 is a zero of $f(x)$, $x + 3$ is a factor.
Divide $f(x)$ by $x + 3$.

$$-3 \overline{)\ 6\quad 13\quad -14\quad 3}$$
$$\quad\ -18\quad 15\quad -3$$
$$\overline{\ 6\ -5\quad\ 1\quad\ 0}$$

Thus,
$$f(x) = (x + 3)(6x^2 - 5x + 1)$$
$$= (x + 3)(3x - 1)(2x - 1)$$

21. $f(x) = 6x^3 + 25x^2 + 3x - 4;\ k = -4$
Since -4 is a zero of $f(x)$, $x + 4$ is a factor.
Divide $f(x)$ by $x + 4$.

$$-4 \overline{)\ 6\quad 25\quad\ 3\ -4}$$
$$\quad\ -24\ -4\quad\ 4$$
$$\overline{\ 6\quad\ 1\ -1\quad\ 0}$$

Thus,
$$f(x) = (x + 4)(6x^2 + x - 1)$$
$$= (x + 4)(3x - 1)(2x + 1)$$

23. $f(x) = x^3 + (7 - 3i)x^2 + (12 - 21i)x - 36i$
$k = 3i$
Since $3i$ is a zero of $f(x)$, $x - 3i$ is a factor.
Divide $f(x)$ by $x - 3i$.

$$3i \overline{)\ 1\quad 7 - 3i\quad 12 - 21i\quad -36i}$$
$$\qquad\quad 3i\qquad\ 21i\qquad 36i$$
$$\overline{\ 1\quad 7\qquad\ 12\qquad\quad\ 0}$$

Thus,
$$f(x) = (x - 3i)(x^2 + 7x + 12)$$
$$= (x - 3i)(x + 4)(x + 3)$$

25. $f(x) = 2x^3 + (3 - 2i)x^2 + (-8 - 5i)x + (3 + 3i)$
$k = 1 + i$
Since $1 + i$ is a zero of $f(x)$, $x - (1 + i)$ is a
factor. Divide $f(x)$ by $x - (1 + i)$.

$$1+i \overline{)\ 2\quad 3 - 2i\quad -8 - 5i\quad 3 + 3i}$$
$$\qquad 2 + 2i\quad 5 + 5i\ -3 - 3i$$
$$\overline{\ 2\quad 5\qquad\ -3\qquad\quad 0}$$

Thus,
$$f(x) = [x - (1 + i)](2x^2 + 5x - 3)$$
$$= [x - (1 + i)](2x - 1)(x + 3)$$

27. $f(x) = x^4 + 2x^3 - 7x^2 - 20x - 12;\ k = -2$
(multiplicity 2)
Since -2 is a zero of $f(x)$, $x + 2$ is a factor.
Divide $f(x)$ by $x + 2$.

$$-2 \overline{)\ 1\quad 2\ -7\ -20\ -12}$$
$$\quad\ -2\quad\ 0\quad 14\quad 12$$
$$\overline{\ 1\quad 0\ -7\ -6\qquad 0}$$

Thus, $f(x) = x^4 + 2x^3 - 7x^2 - 20x - 12$
$$= (x + 2)(x^3 - 7x - 6)$$

Since -2 has multiplicity 2, divide the quotient
polynomial by $x + 2$.

$$-2 \overline{)\ 1\quad\ 0\ -7\ -6}$$
$$\quad\ -2\quad\ 4\quad 6$$
$$\overline{\ 1\ -2\ -3\quad\ 0}$$

Thus,
$$f(x) = (x + 2)(x^3 - 7x - 6)$$
$$= (x + 2)^2(x^2 - 2x - 3)$$
$$= (x + 2)^2(x + 1)(x - 3)$$

29. $f(x) = x^3 - x^2 - 4x - 6;\ 3$
Since 3 is a zero, first divide $f(x)$ by $x - 3$.

$$3 \overline{)\ 1\ -1\ -4\ -6}$$
$$\quad\ 3\quad\ 6\quad 6$$
$$\overline{\ 1\quad 2\quad 2\quad 0}$$

This gives $f(x) = (x - 3)(x^2 + 2x + 2)$. Since
$x^2 + 2x + 2$ cannot be factored, use the
quadratic formula with $a = 1$, $b = 2$, and $c = 2$
to find the remaining two zeros.

$$x = \frac{-2 \pm \sqrt{4 - 4(1)(2)}}{2(1)} = \frac{-2 \pm \sqrt{4 - 8}}{2}$$
$$= \frac{-2 \pm \sqrt{-4}}{2} = \frac{-2 \pm 2i}{2} = -1 \pm i$$

The remaining zeros are $-1 \pm i$.

31. $f(x) = x^3 - 7x^2 + 17x - 15;\ 2 - i$
Since $2 - i$ is also a zero, first divide $f(x)$ by
$x - (2 - i)$.

$$2-i \overline{)\ 1\ -7\qquad 17\qquad\ -15}$$
$$\qquad 2 - i\ -11 + 3i\quad 15$$
$$\overline{\ 1\ -5 - i\quad 6 + 3i\qquad 0}$$

By the conjugate zeros theorem, $2 + i$ is also a
zero, so divide the quotient polynomial from the
first synthetic division by $x - (2 + i)$.

$$2+i \overline{)\ 1\ -5 - i\quad 6 + 3i}$$
$$\qquad 2 + i\ -6 - 3i$$
$$\overline{\ 1\ -3\qquad\quad 0}$$

This gives
$$f(x) = [x - (2 - i)][x - (2 + i)](x - 3).$$
The remaining zeros are $2 + i$ and 3.

Copyright © 2013 Pearson Education, Inc.

33. $f(x) = x^4 + 5x^2 + 4; -i$

Since $-i$ is a zero, first divide $f(x)$ by $x + i$.

$$-i \overline{)1 \quad 0 \quad 5 \quad 0 \quad 4}$$
$$\underline{\quad -i \quad -1 - 4i \quad -4}$$
$$1 \quad -i \quad 4 \quad -4i \quad 0$$

By the conjugate zeros theorem, i is also a zero, so divide the quotient polynomial from the first synthetic division by $x - i$.

$$i \overline{)1 \quad -i \quad 4 \quad -4i}$$
$$\underline{\quad i \quad 0 \quad 4i}$$
$$1 \quad 0 \quad 4 \quad 0$$

The remaining zeros will be zeros of the new quotient polynomial, $x^2 + 4$. Find the remaining zeros by using the square root property.

$$x^2 + 4 = 0 \Rightarrow x^2 = -4 \Rightarrow x = \pm\sqrt{-4} \Rightarrow x = \pm 2i$$

The other zeros are i and $\pm 2i$.

35. (a) $f(x) = x^3 - 2x^2 - 13x - 10$

p must be a factor of $a_0 = -10$ and q must be a factor of $a_3 = 1$. Thus, p can be $\pm 1, \pm 2, \pm 5, \pm 10$ and q can be ± 1. The possible rational zeros, $\frac{p}{q}$, are $\pm 1, \pm 2, \pm 5, \pm 10$.

(b) The remainder theorem shows that -1 is a zero.

$$-1 \overline{)1 \quad -2 \quad -13 \quad -10}$$
$$\underline{\quad -1 \quad 3 \quad 10}$$
$$1 \quad -3 \quad -10 \quad 0$$

The new quotient polynomial will be $x^2 - 3x - 10$.

$$x^2 - 3x - 10 = 0$$
$$(x + 2)(x - 5) = 0$$
$$x + 2 = 0 \Rightarrow x = -2 \text{ or } x - 5 = 0 \Rightarrow x = 5$$

The rational zeros are $-1, -2,$ and 5.

(c) Since the three zeros are $-1, -2,$ and 5, the factors are $x + 1, x + 2,$ and $x - 5$.
$$f(x) = (x+1)(x+2)(x-5)$$

37. (a) $f(x) = x^3 + 6x^2 - x - 30$

p must be a factor of $a_0 = -30$ and q must be a factor of $a_3 = 1$. Thus, p can be $\pm 1, \pm 2, \pm 3, \pm 5, \pm 6, \pm 10, \pm 15, \pm 30$ and q can be ± 1. The possible zeros, $\frac{p}{q}$, are $\pm 1, \pm 2, \pm 3, \pm 5, \pm 6, \pm 10, \pm 15, \pm 30$.

(b) The remainder theorem shows that -5 is a zero.

$$-5 \overline{)1 \quad 6 \quad -1 \quad -30}$$
$$\underline{\quad -5 \quad -5 \quad 30}$$
$$1 \quad 1 \quad -6 \quad 0$$

The new quotient polynomial will be $x^2 + x - 6$.

$$x^2 + x - 6 = 0$$
$$(x + 3)(x - 2) = 0$$
$$x + 3 = 0 \Rightarrow x = -3 \text{ or } x - 2 = 0 \Rightarrow x = 2$$

The rational zeros are $-5, -3,$ and 2.

(c) Since the three zeros are $-5, -3,$ and 2, the factors are $x + 5, x + 3,$ and $x - 2$.
$$f(x) = (x+5)(x+3)(x-2)$$

39. (a) $f(x) = 6x^3 + 17x^2 - 31x - 12$

p must be a factor of $a_0 = -12$ and q must be a factor of $a_3 = 6$. Thus, p can be $\pm 1, \pm 2, \pm 3, \pm 4, \pm 6, \pm 12$ and q can be $\pm 1, \pm 2, \pm 3, \pm 6$. The possible zeros, $\frac{p}{q}$, are $\pm 1, \pm 2, \pm 3, \pm 4, \pm 6, \pm 12, \pm\frac{1}{2}, \pm\frac{3}{2}, \pm\frac{1}{3}, \pm\frac{2}{3}, \pm\frac{4}{3}, \pm\frac{1}{6}$.

(b) The remainder theorem shows that -4 is a zero.

$$-4 \overline{)6 \quad 17 \quad -31 \quad -12}$$
$$\underline{\quad -24 \quad 28 \quad 12}$$
$$6 \quad -7 \quad -3 \quad 0$$

The new quotient polynomial is $6x^2 - 7x - 3$.

$$6x^2 - 7x - 3 = 0 \Rightarrow (3x+1)(2x-3) = 0$$
$$3x + 1 = 0 \Rightarrow x = -\frac{1}{3} \text{ or}$$
$$2x - 3 = 0 \Rightarrow x = \frac{3}{2}$$

The rational zeros are $-4, -\frac{1}{3},$ and $\frac{3}{2}$.

(c) Since the three zeros are $-4, -\frac{1}{3},$ and $\frac{3}{2}$, the factors are $x + 4, 3x + 1,$ and $2x - 3$. $f(x) = (x+4)(3x+1)(2x-3)$

Copyright © 2013 Pearson Education, Inc.

41. (a) $f(x) = 24x^3 + 40x^2 - 2x - 12$

p must be a factor of $a_0 = -12$ and q must be a factor of $a_3 = 24$. Thus, p can be ± 1, $\pm 2, \pm 3, \pm 4, \pm 6, \pm 12$ and q can be $\pm 1, \pm 2, \pm 3, \pm 4, \pm 6, \pm 8, \pm 12, \pm 24$. The possible zeros, $\frac{p}{q}$, are $\pm 1, \pm 2, \pm 3, \pm 4, \pm 6, \pm 12,$

$$\pm \tfrac{1}{2}, \pm \tfrac{3}{2}, \pm \tfrac{1}{3}, \pm \tfrac{2}{3}, \pm \tfrac{4}{3}, \pm \tfrac{1}{4},$$

$$\pm \tfrac{3}{4}, \pm \tfrac{1}{6}, \pm \tfrac{1}{8}, \pm \tfrac{3}{8}, \pm \tfrac{1}{12}, \pm \tfrac{1}{24}.$$

(b) The remainder theorem shows that $-\frac{3}{2}$ is a zero.

$$
\begin{array}{r|rrrr}
-\frac{3}{2} & 24 & 40 & -2 & -12 \\
 & & -36 & -6 & 12 \\
\hline
 & 24 & 4 & -8 & 0
\end{array}
$$

The new quotient polynomial is $24x^2 + 4x - 8$.

$24x^2 + 4x - 8 = 0 \Rightarrow 4(6x^2 + x - 2) = 0$
$\Rightarrow 4(3x + 2)(2x - 1) = 0$

$3x + 2 = 0 \Rightarrow x = -\frac{2}{3}$ or
$2x - 1 = 0 \Rightarrow x = \frac{1}{2}$

The rational zeros are $-\frac{3}{2}, -\frac{2}{3}$, and $\frac{1}{2}$.

(c) Since the three rational zeros are $-\frac{3}{2}, -\frac{2}{3}$, and $\frac{1}{2}$, the factors are $3x + 2$, $2x + 3$, and $2x - 1$.
$f(x) = 2(2x + 3)(3x + 2)(2x - 1)$

Note: Since $-\frac{3}{2}$ is a zero,

$$
\begin{aligned}
f(x) &= 24x^3 + 40x^2 - 2x - 12 \\
&= \left[x - \left(-\tfrac{3}{2}\right)\right]\left(24x^2 + 4x - 8\right) \\
&= \left(x + \tfrac{3}{2}\right)\left[4(6x^2 + x - 2)\right] \\
&= 4\left(x + \tfrac{3}{2}\right)(3x + 2)(2x - 1) \\
&= 2\left[2\left(x + \tfrac{3}{2}\right)\right](3x + 2)(2x - 1) \\
&= 2(2x + 3)(3x + 2)(2x - 1)
\end{aligned}
$$

43. $f(x) = (x - 2)^3 \left(x^2 - 7\right)$

To find the zeros, let $f(x) = 0$. Set each factor equal to zero and solve for x.

$(x - 2)^3 \left(x^2 - 7\right) = 0$

$(x - 2)^3 = 0$ or $x^2 - 7 = 0$
$x - 2 = 0$ $\quad\quad x^2 = 7$
$x = 2$ $\quad\quad\quad x = \pm\sqrt{7}$

The zeros are 2 (multiplicity 3) and $\pm\sqrt{7}$.

45. $f(x) = 3x(x - 2)(x + 3)(x^2 - 1)$

To find the zeros, let $f(x) = 0$.
Set each factor equal to zero and solve for x.
$3x = 0 \Rightarrow x = 0$
$x - 2 = 0 \Rightarrow x = 2$
$x + 3 = 0 \Rightarrow x = -3$
$x^2 - 1 = 0 \Rightarrow x^2 = 1 \Rightarrow x = \pm 1$
The zeros are 0, 2, –3, 1, and –1.

47. $f(x) = (x^2 + x - 2)^5(x - 1 + \sqrt{3})^2$

To find the zeros, let $f(x) = 0$.
Set each factor equal to zero and solve for x.
$(x^2 + x - 2)^5 = 0 \Rightarrow x^2 + x - 2 = 0 \Rightarrow$
$(x + 2)(x - 1) = 0 \Rightarrow$
$x = -2$, multiplicity 5 or $x = 1$, multiplicity 5

$(x - 1 + \sqrt{3})^2 = 0 \Rightarrow x - 1 + \sqrt{3} = 0 \Rightarrow$
$x = 1 - \sqrt{3}$, multiplicity 2

The zeros are –2 (multiplicity 5), 1 (multiplicity 5) and $1 - \sqrt{3}$ (multiplicity 2).

49. Zeros of –3, 1, and 4, $f(2) = 30$

These three zeros give $x - (-3) = x + 3$, $x - 1$, and $x - 4$ as factors of $f(x)$. Since $f(x)$ is to have degree 3, these are the only possible factors by the number of zeros theorem. Therefore, $f(x)$ has the form
$f(x) = a(x + 3)(x - 1)(x - 4)$ for some real number a. To find a, use the fact that $f(2) = 30$.

$f(2) = a(2 + 3)(2 - 1)(2 - 4) = 30 \Rightarrow$
$a(5)(1)(-2) = 30 \Rightarrow -10a = 30 \Rightarrow a = -3$

Thus,
$$
\begin{aligned}
f(x) &= -3(x + 3)(x - 1)(x - 4) \\
&= -3(x^2 + 2x - 3)(x - 4) \\
&= -3(x^3 - 2x^2 - 11x + 12) \\
&= -3x^3 + 6x^2 + 33x - 36
\end{aligned}
$$

51. Zeros of –2, 1, and 0; $f(-1) = -1$

These three zeros give $x - (-2) = x + 2$, $x - 1$, and $x - 0 = x$ as factors of $f(x)$. Since $f(x)$ is to have degree 3, these are the only possible factors by the number of zeros theorem. Therefore, $f(x)$ has the form
$f(x) = a(x + 2)(x - 1)x$ for some real number a. To find a, use the fact that $f(-1) = -1$.
$f(-1) = a(-1 + 2)(-1 - 1)(-1) = -1 \Rightarrow$
$a(1)(-2)(-1) = -1 \Rightarrow 2a = -1 \Rightarrow a = -\frac{1}{2}$

Thus,
$$
\begin{aligned}
f(x) &= -\tfrac{1}{2}(x + 2)(x - 1)x \\
&= -\tfrac{1}{2}(x^2 + x - 2)x = -\tfrac{1}{2}x^3 - \tfrac{1}{2}x^2 + x
\end{aligned}
$$

Copyright © 2013 Pearson Education, Inc.

53. Zero of –3 having multiplicity 3; $f(3) = 36$

These three zeros give $x - (-3) = x + 3$, $x - (-3) = x + 3$, and $x - (-3) = x + 3$ as factors of $f(x)$. Since $f(x)$ is to have degree 3, these are the only possible factors by the number of zeros theorem. Therefore, $f(x)$ has the form

$$f(x) = a(x+3)(x+3)(x+3) = a(x+3)^3 \text{ for some real number } a. \text{ To find } a, \text{ use the fact that} f(3) = 36.$$

$$f(3) = a(3+3)^3 = 36 \Rightarrow a(6)^3 = 36 \Rightarrow 216a = 36 \Rightarrow a = \tfrac{1}{6}$$

Thus,

$$f(x) = \tfrac{1}{6}(x+3)^3 = \tfrac{1}{6}(x+3)^2(x+3) = \tfrac{1}{6}(x^2 + 6x + 9)(x+3)$$
$$= \tfrac{1}{6}(x^3 + 9x^2 + 27x + 27) = \tfrac{1}{6}x^3 + \tfrac{3}{2}x^2 + \tfrac{9}{2}x + \tfrac{9}{2}$$

55. Zero of 0 and zero of 1 having multiplicity 2; $f(2) = 10$

These three zeros give x, $x - 1$, and $x - 1$ as factors of $f(x)$. Since $f(x)$ is to have degree 3, these are the only possible factors by the number of zeros theorem. Therefore, $f(x)$ has the form

$$f(x) = ax(x-1)(x-1) = ax(x-1)^2 \text{ for some real number } a. \text{ To find } a, \text{ use the fact that } f(2) = 10.$$

$$f(2) = a(2)(2-1)^2 = 10 \Rightarrow 2a = 10 \Rightarrow a = 5$$

Thus, $f(x) = 5x(x-1)^2 = 5x(x^2 - 2x + 1) = 5x^3 - 10x^2 + 5x.$

In Exercises 57–73, we must find a polynomial of least degree with real coefficients having the given zeros. For each of these exercises, other answers are possible.

57. $5 + i$ and $5 - i$

$$f(x) = \left[x - (5+i)\right]\left[x - (5-i)\right] = (x - 5 - i)(x - 5 + i) = \left[(x-5) - i\right]\left[(x-5) + i\right] = (x-5)^2 - i^2$$
$$= (x^2 - 10x + 25) - i^2 = x^2 - 10x + 25 + 1 = x^2 - 10x + 26$$

59. $0, i,$ and $1 + i$

By the conjugate zeros theorem, $-i$ and $1 - i$ must also be zeros.

$$f(x) = x(x-i)(x+i)\left[x - (1+i)\right]\left[x - (1-i)\right] = x(x^2 - i^2)(x - 1 - i)(x - 1 + i) = x(x^2 + 1)(x - 1 - i)(x - 1 + i)$$
$$= x(x^2 + 1)\left[(x-1) - i\right]\left[(x-1) + i\right] = x(x^2 + 1)\left[(x-1)^2 - i^2\right] = x(x^2 + 1)\left[(x^2 - 2x + 1) - i^2\right]$$
$$= x(x^2 + 1)(x^2 - 2x + 1 + 1) = (x^3 + x)(x^2 - 2x + 2) = x^5 - 2x^4 + 3x^3 - 2x^2 + 2x$$

61. $1 + \sqrt{2}, 1 - \sqrt{2},$ and 1

$$f(x) = \left[x - (1 + \sqrt{2})\right]\left[x - (1 - \sqrt{2})\right](x - 1) = (x - 1 - \sqrt{2})(x - 1 + \sqrt{2})(x - 1)$$
$$= \left[(x-1) - \sqrt{2}\right]\left[(x-1) + \sqrt{2}\right](x-1) = \left[(x-1)^2 - (\sqrt{2})^2\right](x-1) = (x^2 - 2x + 1 - 2)(x - 1)$$
$$= (x^2 - 2x - 1)(x - 1) = x^3 - 3x^2 + x + 1$$

63. $2 - i, 3,$ and -1

By the conjugate zeros theorem, $2 + i$ must also be a zero.

$$f(x) = \left[x - (2+i)\right]\left[x - (2-i)\right](x - 3)(x + 1) = \left[(x - 2 - i)(x - 2 + i)\right]\left[(x - 3)(x + 1)\right]$$
$$= \left[(x-2) - i\right]\left[(x-2) + i\right](x^2 - 2x - 3) = \left[(x-2)^2 - i^2\right](x^2 - 2x - 3)$$
$$= \left[(x^2 - 4x + 4) - i^2\right](x^2 - 2x - 3) = (x^2 - 4x + 4 + 1)(x^2 - 2x - 3)$$
$$= (x^2 - 4x + 5)(x^2 - 2x - 3) = x^4 - 6x^3 + 10x^2 + 2x - 15$$

Copyright © 2013 Pearson Education, Inc.

65. 2 and $3 + i$

By the conjugate zeros theorem, $3 - i$ must also be a zero.

$$f(x) = (x-2)\big[x-(3+i)\big]\big[x-(3-i)\big] = (x-2)\big[(x-3)-i\big]\big[(x-3)+i\big] = (x-2)\Big[(x-3)^2-i^2\Big]$$

$$= (x-2)\big(x^2-6x+9+1\big) = (x-2)\big(x^2-6x+10\big) = x^3-8x^2+22x-20$$

67. $1-\sqrt{2}$, $1+\sqrt{2}$, and $1-i$

By the conjugate zeros theorem, $1+i$ must also be a zero.

$$f(x) = \Big[x-\big(1-\sqrt{2}\big)\Big]\Big[x-\big(1+\sqrt{2}\big)\Big]\cdot\Big[x-(1-i)\Big]\Big[x-(1+i)\Big]$$

$$= \big(x-1+\sqrt{2}\big)\big(x-1-\sqrt{2}\big)\cdot(x-1+i)(x-1-i) = \Big[(x-1)+\sqrt{2}\Big]\Big[(x-1)-\sqrt{2}\Big]\cdot\Big[(x-1)+i\Big]\Big[(x-1)-i\Big]$$

$$= \Big[(x-1)^2-\big(\sqrt{2}\big)^2\Big]\Big[(x-1)^2-i^2\Big] = \big(x^2-2x+1-2\big)\Big[\big(x^2-2x+1\big)-i^2\Big] = \big(x^2-2x-1\big)\big(x^2-2x+1+1\big)$$

$$= \big(x^2-2x-1\big)\big(x^2-2x+2\big) = x^4-4x^3+5x^2-2x-2$$

69. $2-i$ and $6-3i$

By the conjugate zeros theorem, $2+i$ and $6+3i$ must also be zeros.

$$f(x) = \big[x-(2-i)\big]\big[x-(2+i)\big]\cdot\big[x-(6-3i)\big]\big[x-(6+3i)\big] = \big[(x-2+i)(x-2-i)\big]\cdot\big[(x-6+3i)(x-6-3i)\big]$$

$$= \big[(x-2)+i\big]\big[(x-2)-i\big]\cdot\big[(x-6)+3i\big]\big[(x-6)-3i\big] = \Big[(x-2)^2-i^2\Big]\Big[(x-6)^2-(3i)^2\Big]$$

$$= \Big[\big(x^2-4x+4\big)-i^2\Big]\cdot\Big[\big(x^2-12x+36\big)-9i^2\Big] = \big(x^2-4x+4+1\big)\big(x^2-12x+36+9\big)$$

$$= \big(x^2-4x+5\big)\big(x^2-12x+45\big) = x^4-16x^3+98x^2-240x+225$$

71. 4, $1-2i$, and $3+4i$

By the conjugate zeros theorem, $1+2i$ and $3-4i$ must also be zeros.

$$f(x) = (x-4)\big[x-(1-2i)\big]\big[x-(1+2i)\big]\cdot\big[x-(3+4i)\big]\big[x-(3-4i)\big]$$

$$= (x-4)(x-1+2i)(x-1-2i)\cdot(x-3-4i)(x-3+4i)$$

$$= (x-4)\big[(x-1)+2i\big]\big[(x-1)-2i\big]\cdot\big[(x-3)-4i\big]\big[(x-3)+4i\big]$$

$$= (x-4)\Big[(x-1)^2-(2i)^2\Big]\Big[(x-3)^2-(4i)^2\Big] = (x-4)\big(x^2-2x+1-4i^2\big)\cdot\big(x^2-6x+9-16i^2\big)$$

$$= (x-4)\big(x^2-2x+1+4\big)\big(x^2-6x+9+16\big) = (x-4)\big(x^2-2x+5\big)\big(x^2-6x+25\big)$$

$$= (x-4)\big(x^4-8x^3+42x^2-80x+125\big) = x^5-12x^4+74x^3-248x^2+445x-500$$

73. $1+2i$, 2 (multiplicity 2).

By the conjugate zeros theorem, $1-2i$ must also be a zero.

$$f(x) = (x-2)^2\big[x-(1+2i)\big]\big[x-(1-2i)\big]$$

$$= \big(x^2-4x+4\big)\big[(x-1)-2i\big]\big[(x-1)+2i\big]$$

$$= \big(x^2-4x+4\big)\Big[(x-1)^2-(2i)^2\Big]$$

$$= \big(x^2-4x+4\big)\big(x^2-2x+1-4i^2\big)$$

$$= \big(x^2-4x+4\big)\big(x^2-2x+1+4\big)$$

$$= \big(x^2-4x+4\big)\big(x^2-2x+5\big)$$

$$= x^4-6x^3+17x^2-28x+20$$

75. $f(x) = 2x^3-4x^2+2x+7$

$f(x) = 2x^3 \underset{1}{-} 4x^2 \underset{2}{+} 2x+7$ has 2 variations in sign. f has either 2 or $2-2=0$ positive real zeros.

$f(-x) = -2x^3-4x^2 \underset{1}{-2x+7}$ has 1 variation in sign. f has 1 negative real zero. Since the degree of the function is 3, the remaining zeros are nonreal complex. Thus, the different possibilities are

Positive	Negative	Nonreal Complex
2	1	0
0	1	2

Copyright © 2013 Pearson Education, Inc.

77. $f(x) = 4x^3 - x^2 + 2x - 7$

$f(x) = \underbrace{4x^3}_{1} \underbrace{- x^2}_{2} \underbrace{+ 2x}_{3} - 7$ has 3 variations in

sign. f has either 3 or 3 − 2 = 1 positive real zeros.

$f(-x) = -4x^3 - x^2 - 2x - 7$ has no variations in sign, so. f has no negative real zeros. Since the degree of the function is 3, the remaining zeros are nonreal complex. Thus, the different possibilities are

Positive	Negative	Nonreal Complex
3	0	0
1	0	2

79. $f(x) = 5x^4 + 3x^2 + 2x - 9$

$f(x) = 5x^4 + 3x^2 \underbrace{+ 2x}_{1} - 9$ has 1 variation in

sign. f has 1 positive real zero.

$f(-x) = 5x^4 \underbrace{+ 3x^2}_{1} - 2x - 9$ has 1 variation in

sign. f has 1 negative real zero. Since the degree of the function is 4, the remaining zeros are nonreal complex. Thus, the different possibilities are

Positive	Negative	Nonreal Complex
1	1	2

81. $f(x) = -8x^4 + 3x^3 - 6x^2 + 5x - 7$

$f(x) = \underbrace{-8x^4}_{} \underbrace{+ 3x^3}_{1} \underbrace{- 6x^2}_{2} \underbrace{+ 5x}_{3} \underbrace{- 7}_{4}$ has 4

variation in sign. f has 4 or 4 − 2 = 2 or 4 − 4 = 0 positive real zeros.

$f(-x) = -8x^4 - 3x^3 - 6x^2 - 5x - 7$ has no variations in sign, so. f has no negative real zeros. Since the degree of the function is 4, the remaining zeros are nonreal complex. Thus, the different possibilities are

Positive	Negative	Nonreal Complex
4	0	0
2	0	2
0	0	4

83. $f(x) = x^5 + 3x^4 - x^3 + 2x + 3$

$f(x) = x^5 + 3x^4 \underbrace{- x^3}_{1} \underbrace{+ 2x}_{2} + 3$ has 2 variations

in sign. f has 2 or 2 − 2 = 0 positive real zeros.

$f(-x) = \underbrace{-x^5}_{1} \underbrace{+ 3x^4}_{} \underbrace{+ x^3}_{2} \underbrace{- 2x}_{3} + 3$ has 3

variations in sign. f has 3 or 3 − 2 = 1 negative real zeros. Since the degree of the function is 5, the remaining zeros are nonreal complex. Thus, the different possibilities are

Positive	Negative	Nonreal Complex
2	3	0
2	1	2
0	3	2
0	1	4

85. $f(x) = 7x^5 + 6x^4 + 2x^3 + 9x^2 + x + 5$

$f(x) = 7x^5 + 6x^4 + 2x^3 + 9x^2 + x + 5$ has no variations in sign, so f has no positive real zeros.

$f(-x) = \underbrace{-7x^5}_{1} \underbrace{+ 6x^4}_{2} \underbrace{- 2x^3}_{3} \underbrace{+ 9x^2}_{4} \underbrace{- x}_{5} + 5$ has

five variations in sign, so f has 5 or 5 − 2 = 3 or 5 − 4 = 1 negative real zeros. Since the degree of the function is 5, the remaining zeros are nonreal complex. Thus, the different possibilities are

Positive	Negative	Nonreal Complex
0	5	0
0	3	2
0	1	4

87. $f(x) = 2x^5 - 7x^3 + 6x + 8$

$f(x) = \underbrace{2x^5}_{1} \underbrace{- 7x^3}_{2} + 6x + 8$ has two variations

in sign, so f has 2 or 2 − 2 = 0 positive real zeros.

$f(-x) = \underbrace{-2x^5}_{1} \underbrace{+ 7x^3}_{2} \underbrace{- 6x}_{3} + 8$ has three

variations in sign, so f has 3 or 3 − 2 = 1 negative real zeros. Since the degree of the function is 5, the remaining zeros are nonreal complex. Thus, the different possibilities are

Positive	Negative	Nonreal Complex
2	3	0
2	1	2
0	3	2
0	1	4

Copyright © 2013 Pearson Education, Inc.

89. $f(x) = 5x^6 - 6x^5 + 7x^3 - 4x^2 + x + 2$

$f(x) = \underbrace{5x^6 - 6x^5}_{1}\underbrace{+ 7x^3}_{2}\underbrace{- 4x^2}_{3}\underbrace{+ x}_{4} + 2$ has four

variations in sign, so f has 4 or $4 - 2 = 2$ or $4 - 4 = 0$ positive real zeros.

$f(-x) = 5x^6 \underbrace{+ 6x^5}_{1} \underbrace{- 7x^3 - 4x^2 - x}_{2} + 2$ has two

variations in sign, so f has 2 or $2 - 2 = 0$ negative real zeros. Since the degree of the function is 6, the remaining zeros are nonreal complex.

Positive	Negative	Nonreal Complex
4	2	0
4	0	2
2	2	2
2	0	4
0	2	4
0	0	6

91. $f(x) = x^4 + 2x^3 - 3x^2 + 24x - 180$

p must be a factor of $a_0 = -180$ and q must be a factor of $a_4 = 1$. Thus, p can be ±1, ±2, ±3, ±4, ±5, ±6, ±9, ±10, ±12, ±15, ±18, ±20, ±30, ±36, ±45, ±60, ±90, and ±180. q can be ±1. The possible rational zeros, $\frac{p}{q}$, are ±1, ±2, ±3, ±4, ±5, ±6, ±9, ±10, ±12, ±15, ±18, ±20, ±30, ±36, ±45, ±60, ±90, and ±180.

Using the remainder theorem and synthetic division, we find that one zero is $x = -5$.

$$-5\overline{)\begin{array}{ccccc} 1 & 2 & -3 & 24 & -180 \\ & -5 & 15 & -60 & 180 \\ \hline 1 & -3 & 12 & -36 & 0 \end{array}}$$

Setting the quotient $x^3 - 3x^2 + 12x - 36$ equal to zero and factoring by grouping, we have:
$0 = x^3 - 3x^2 + 12x - 36$
$= x^2(x - 3) + 12(x - 3)$
$= (x^2 + 12)(x - 3) \Rightarrow$
$x^2 + 12 = 0 \Rightarrow x^2 = -12 \Rightarrow x = \sqrt{-12} \Rightarrow$
$x = \pm 2i\sqrt{3}$ or $x - 3 = 0 \Rightarrow x = 3$

Thus, the zeros of $f(x)$ are $\left\{-5, 3, \pm 2i\sqrt{3}\right\}$.

93. $f(x) = x^4 + x^3 - 9x^2 + 11x - 4$

p must be a factor of $a_0 = -4$ and q must be a factor of $a_4 = 1$. Thus, p can be ±1, ±2, and ±4. q can be ±1.

The possible rational zeros, $\frac{p}{q}$, are ±1, ±2, and ±4. Using the remainder theorem and synthetic division, we find that one zero is $x = -4$.

$$-4\overline{)\begin{array}{ccccc} 1 & 1 & -9 & 11 & -4 \\ & -4 & 12 & -12 & 4 \\ \hline 1 & -3 & 3 & -1 & 0 \end{array}}$$

Setting the quotient $x^3 - 3x^2 + 3x - 1$ equal to zero and factoring, we have:
$x^3 - 3x^2 + 3x - 1 = 0$
$(x - 1)^3 = 0 \Rightarrow x = 1$ (with multiplicity 3)

Thus, the zeros of $f(x)$ are $\{-4, 1, 1, 1\}$.

95. $f(x) = 2x^5 + 11x^4 + 16x^3 + 15x^2 + 36x$

$a_0 = 0$, so 0 is a zero of the function. Factoring, we have
$2x^5 + 11x^4 + 16x^3 + 15x^2 + 36x$
$= x(2x^4 + 11x^3 + 16x^2 + 15x + 36)$

To find a zero of the quotient, $2x^4 + 11x^3 + 16x^2 + 15x + 36$, p must be a factor of $a_0 = 36$ and q must be a factor of $a_4 = 2$. Thus, p can be ±1, ±2, ±3, ±4, ±6, ±9, ±12, ±18, or ±36. q can be ±1 or ±2. The possible rational zeros, $\frac{p}{q}$, are

$\pm\frac{1}{2}, \pm\frac{3}{2}, \pm\frac{9}{2}, \pm 1, \pm 2, \pm 3, \pm 4, \pm 6, \pm 8, \pm 9, \pm 12$ ±18, or ±36. Using the remainder theorem and synthetic division, we find that one zero is $x = -3$.

$$-3\overline{)\begin{array}{ccccc} 2 & 11 & 16 & 15 & 36 \\ & -6 & -15 & -3 & -36 \\ \hline 2 & 5 & 1 & 12 & 0 \end{array}}$$

Now find a zero of the quotient $2x^3 + 5x^2 + x + 12$. p must be a factor of $a_0 = 12$ and q must be a factor of $a_3 = 2$. Thus, p can be ±1, ±2, ±3, ±4, ±6, or ±12. q can be ±1 or ±2. The possible rational zeros, $\frac{p}{q}$, are $\pm\frac{1}{2}, \pm\frac{3}{2}, \pm 1, \pm 2, \pm 3, \pm 4, \pm 6,$ or ±12.

Using the remainder theorem and synthetic division, we find that one zero is $x = -3$

$$-3\overline{)\begin{array}{cccc} 2 & 5 & 1 & 12 \\ & -6 & 3 & -12 \\ \hline 2 & -1 & 4 & 0 \end{array}}$$

(continued on next page)

Copyright © 2013 Pearson Education, Inc.

(continued)

Setting the quotient $2x^2 - x + 4$ equal to zero and using the quadratic equation with $a = 2$, $b = -1$, and $c = 4$ to solve this, we have:

$$x = \frac{-(-1) \pm \sqrt{(-1)^2 - 4(2)(4)}}{2(2)}$$

$$= \frac{1 \pm \sqrt{1 - 32}}{4} = \frac{1 \pm i\sqrt{31}}{4}$$

Thus, the zeros of $f(x)$ are

$$\left\{ -3, -3, 0, \frac{1 \pm i\sqrt{31}}{4} \right\}.$$

97. $f(x) = x^5 - 6x^4 + 14x^3 - 20x^2 + 24x - 16$

p must be a factor of $a_0 = -16$ and q must be a factor of $a_3 = 2$. Thus, p can be ± 1, ± 2, ± 4, ± 8, or ± 16. q can be ± 1. The possible rational zeros, $\frac{p}{q}$, are ± 1, ± 2, ± 4, ± 8, or ± 16. Using the remainder theorem and synthetic division, we find that one zero is $x = 2$

$$\begin{array}{r|rrrrrr} 2 & 1 & -6 & 14 & -20 & 24 & -16 \\ & & 2 & -8 & 12 & -16 & 16 \\ \hline & 1 & -4 & 6 & -8 & 8 & 0 \end{array}$$

To find a zero of the quotient, $x^4 - 4x^3 + 6x^2 - 8x + 8$, p must be a factor of $a_0 = 8$ and q must be a factor of $a_4 = 1$. Thus, p can be ± 1, ± 2, ± 4, or ± 8. q can be ± 1. The possible rational zeros, $\frac{p}{q}$, are ± 1, ± 2, ± 4, or ± 8. Using the remainder theorem and synthetic division, we find that one zero is $x = 2$.

$$\begin{array}{r|rrrrr} 2 & 1 & -4 & 6 & -8 & 8 \\ & & 2 & -4 & 4 & -8 \\ \hline & 1 & -2 & 2 & -4 & 0 \end{array}$$

Now find a zero of the quotient $x^3 - 2x^2 + 2x - 4$. p must be a factor of $a_0 = -4$ and q must be a factor of $a_3 = 1$. Thus, p can be ± 1, ± 2, or ± 4. q can be ± 1. The possible rational zeros, $\frac{p}{q}$, are ± 1, ± 2, or ± 4.

Using the remainder theorem and synthetic division, we find that one zero is $x = 2$.

$$\begin{array}{r|rrrr} 2 & 1 & -2 & 2 & -4 \\ & & 2 & 0 & 4 \\ \hline & 1 & 0 & 2 & 0 \end{array}$$

Setting the quotient $x^2 + 2$ equal to zero, we have: $x^2 + 2 = 0 \Rightarrow x^2 = -2 \Rightarrow x = \pm i\sqrt{2}$

Thus, the zeros of $f(x)$ are $\left\{ 2, 2, 2, \pm i\sqrt{2} \right\}$.

99. $f(x) = 2x^4 - x^3 + 7x^2 - 4x - 4$

p must be a factor of $a_0 = -4$ and q must be a factor of $a_4 = 2$. Thus, p can be ± 1, ± 2, and ± 4. q can be ± 1 or ± 2. The possible rational zeros, $\frac{p}{q}$, are $\pm \frac{1}{2}$, ± 1, ± 2, and ± 4. Using the remainder theorem and synthetic division, we find that one zero is $x = 1$.

$$\begin{array}{r|rrrrr} 1 & 2 & -1 & 7 & -4 & -4 \\ & & 2 & 1 & 8 & 4 \\ \hline & 2 & 1 & 8 & 4 & 0 \end{array}$$

Setting the quotient $2x^3 + x^2 + 8x + 4$ equal to zero and factoring by grouping, we have:

$$2x^3 + x^2 + 8x + 4 = 0$$
$$x^2(2x + 1) + 4(2x + 1) = 0$$
$$(x^2 + 4)(2x + 1) = 0 \Rightarrow$$
$$x^2 + 4 = 0 \Rightarrow x^2 = -4 \Rightarrow x = \pm 2i \text{ or}$$
$$2x + 1 = 0 \Rightarrow x = -\tfrac{1}{2}$$

Thus, the zeros of $f(x)$ are $\left\{ -\frac{1}{2}, 1, \pm 2i \right\}$.

101. $f(x) = 5x^3 - 9x^2 + 28x + 6$

p must be a factor of $a_0 = 6$ and q must be a factor of $a_3 = 5$. Thus, p can be ± 1, ± 2, ± 3, or ± 6. q can be ± 1 or ± 5. The possible rational zeros, $\frac{p}{q}$, are $\pm 1, \pm 2, \pm 3, \pm 6, \pm \frac{1}{5}, \pm \frac{2}{5}, \pm \frac{3}{5}$, or $\pm \frac{6}{5}$. Using the remainder theorem and synthetic division, we find that one zero is $x = -\frac{1}{5}$

$$\begin{array}{r|rrrr} -\frac{1}{5} & 5 & -9 & 28 & 6 \\ & & -1 & 2 & -6 \\ \hline & 5 & -10 & 30 & 0 \end{array}$$

Setting the quotient $5x^2 - 10x + 30 = 5(x^2 - 2x + 6)$ equal to zero and using the quadratic equation with $a = 1$, $b = -2$, and $c = 6$ to solve this, we have:

$$x = \frac{-(-2) \pm \sqrt{(-2)^2 - 4(1)(6)}}{2(1)} = \frac{2 \pm \sqrt{4 - 24}}{2}$$

$$= \frac{2 \pm \sqrt{-20}}{2} = \frac{2 \pm 2i\sqrt{5}}{2} = 1 \pm i\sqrt{5}$$

Thus, the zeros of $f(x)$ are $\left\{ -\frac{1}{5}, 1 \pm i\sqrt{5} \right\}$.

Copyright © 2013 Pearson Education, Inc.

103. $f(x) = x^4 + 29x^2 + 100$

Letting $u = x^2$, we have

$x^4 + 29x^2 + 100 = u^2 + 29u + 100$. Set the function equal to zero, and solve for u using the quadratic formula with $a = 1$, $b = 29$ and $c = 100$.

$$u = \frac{-29 \pm \sqrt{29^2 - 4(1)(100)}}{2(1)}$$

$$= \frac{-29 \pm \sqrt{441}}{2} = \frac{-29 \pm 21}{2} = -25 \text{ or } u = -4$$

$u = -25 \Rightarrow x^2 = -25 \Rightarrow x = \pm 5i$

$u = -4 \Rightarrow x^2 = -4 \Rightarrow x = \pm 2i$

Thus, the zeros of $f(x)$ are $\{\pm 2i, \pm 5i\}$.

105. $f(x) = x^4 + 2x^2 + 1$

Setting $f(x)$ equal to 0, then factoring to solve for x, we have

$x^4 + 2x^2 + 1 = 0$

$(x^2 + 1)^2 = 0$

$x^2 + 1 = 0 \Rightarrow x^2 = -1 \Rightarrow$

$x = \pm i$ (with multiplicity 2)

Thus, the zeros of $f(x)$ are $\{\pm i, \pm i\}$.

107. $f(x) = x^4 - 6x^3 + 7x^2$

Setting $f(x)$ equal to zero and solving for x, we have $x^4 - 6x^3 + 7x^2 = 0 \Rightarrow x^2(x^2 - 6x + 7) = 0 \Rightarrow$

$x^2 = 0 \Rightarrow$

$x = 0$ (with multiplicity 2) or

$x^2 - 6x + 7 = 0 \Rightarrow$

$$x = \frac{-(-6) \pm \sqrt{(-6)^2 - 4(1)(7)}}{2(1)}$$

$$= \frac{6 \pm \sqrt{36 - 28}}{2} = \frac{6 \pm \sqrt{8}}{2} = 3 \pm \sqrt{2}$$

Thus, the zeros of $f(x)$ are $\{0, 0, 3 \pm \sqrt{2}\}$.

109. $f(x) = x^4 - 8x^3 + 29x^2 - 66x + 72$

p must be a factor of $a_0 = 72$ and q must be a factor of $a_4 = 1$. Thus, p can be $\pm 1, \pm 2, \pm 3,$ $\pm 4, \pm 6, \pm 8, \pm 9, \pm 12, \pm 18, \pm 24, \pm 36,$ or ± 72.

q can be ± 1. The possible rational zeros, $\frac{p}{q}$, are $\pm 1, \pm 2, \pm 3, \pm 4, \pm 6, \pm 8, \pm 9, \pm 12, \pm 18, \pm 24,$ $\pm 36,$ or ± 72.

Using the remainder theorem and synthetic division, we find that one zero is $x = 3$.

$$\begin{array}{r|rrrrr} 3 & 1 & -8 & 29 & -66 & 72 \\ & & 3 & -15 & 42 & -72 \\ \hline & 1 & -5 & 14 & -24 & 0 \end{array}$$

Now find a zero of the quotient

$x^3 - 5x^2 + 14x - 24$. p must be a factor of $a_0 = -24$ and q must be a factor of $a_3 = 1$.

Thus, p can be $\pm 1, \pm 2, \pm 3, \pm 4, \pm 6, \pm 8, \pm 12,$ or ± 24, while q can be ± 1.

Using the remainder theorem and synthetic division, we find that one zero is $x = 3$.

$$\begin{array}{r|rrrr} 3 & 1 & -5 & 14 & -24 \\ & & 3 & -6 & 24 \\ \hline & 1 & -2 & 8 & 0 \end{array}$$

Setting the quotient $x^2 - 2x + 8$ equal to zero and solving for x using the quadratic formula with $a = 1$, $b = -2$, and $c = 8$, we have

$$x = \frac{-(-2) \pm \sqrt{(-2)^2 - 4(1)(8)}}{2(1)}$$

$$= \frac{2 \pm \sqrt{4 - 32}}{2} = \frac{2 \pm \sqrt{-28}}{2} = 1 \pm i\sqrt{7}$$

Thus, the zeros of $f(x)$ are $\{3, 3, 1 \pm i\sqrt{7}\}$.

111. $f(x) = x^6 - 9x^4 - 16x^2 + 144$

Let $u = x^2$. Then $x^6 - 9x^4 - 16x^2 + 144 \Rightarrow$ $u^3 - 9u^2 - 16u + 144$. Set $f(u)$ equal to 0, then factor by grouping to solve for u:

$u^3 - 9u^2 - 16u + 144 = 0$

$u^2(u - 9) - 16(u - 9) = 0$

$(u^2 - 16)(u - 9) = 0 \Rightarrow$

$u^2 - 16 = 0 \Rightarrow u^2 = 16 \Rightarrow u = \pm 4$ or

$u - 9 = 0 \Rightarrow u = 9$

$u = 4 \Rightarrow x^2 = 4 \Rightarrow x = \pm 2$

$u = -4 \Rightarrow x^2 = -4 \Rightarrow x = \pm 2i$

$u = 9 \Rightarrow x^2 = 9 \Rightarrow x = \pm 3$

Thus, the zeros of $f(x)$ are $\{\pm 2, \pm 3, \pm 2i\}$.

For exercises 113–115, let $c = a + bi$ and $d = m + ni$.

113. $\overline{c + d} = \overline{(a + bi) + (m + ni)}$

$= \overline{(a + m) + (b + n)i} = (a + m) - (b + n)i$

$= a + m - bi - ni = (a - bi) + (m - ni)$

$= \overline{c} + \overline{d}$

115. If a is a real number, then a is of the form $a + 0 \cdot i$. $\overline{a} = \overline{a + 0 \cdot i} = a - 0i = a$. Therefore, the statement is true.

Copyright © 2013 Pearson Education, Inc.

117. $x = \sqrt[3]{\frac{n}{2} + \sqrt{\left(\frac{n}{2}\right)^2 + \left(\frac{m}{3}\right)^3}} - \sqrt[3]{\frac{-n}{2} + \sqrt{\left(\frac{n}{2}\right)^2 + \left(\frac{m}{3}\right)^3}}$ can be used to solve a cubic equation $x^3 + mx = n$.

For $x^3 + 9x = 26$, $m = 9$ and $n = 26$.
Substitute

$x = \sqrt[3]{\frac{26}{2} + \sqrt{\left(\frac{26}{2}\right)^2 + \left(\frac{9}{3}\right)^3}} - \sqrt[3]{-\frac{26}{2} + \sqrt{\left(\frac{26}{2}\right)^2 + \left(\frac{9}{3}\right)^3}} = \sqrt[3]{13 + \sqrt{13^2 + 3^3}} - \sqrt[3]{-13 + \sqrt{13^2 + 3^3}}$

$= \sqrt[3]{13 + \sqrt{169 + 27}} - \sqrt[3]{-13 + \sqrt{169 + 27}} = \sqrt[3]{13 + 14} - \sqrt[3]{-13 + 14} = \sqrt[3]{27} - \sqrt[3]{1} = 3 - 1 = 2$

Section 3.4 Polynomial Functions: Graphs, Applications, and Models

1. $y = x^3 - 3x^2 - 6x + 8$

The range of an odd-degree polynomial is $(-\infty, \infty)$. The y-intercept of the graph is 8. The graph fitting these criteria is A.

3. Since graph C crosses the x-axis at one point, the function has one real zero.

5. A polynomial of degree 3 can have at most 2 turning points. Graphs B and D have more than 2 turning points, so they cannot be graphs of cubic polynomial functions.

7. Since graph B touches the x-axis at –5, the function has 2 real zeros of –5. The two other real zeros are where the graph crosses the x-axis, at 0 and 3.
$f(x) = x^4 + 7x^3 - 5x^2 - 75x = x(x+5)^2(x-3)$

9. $f(x) = 2x^4$ is in the form $f(x) = ax^n$.
$|a| = 2 > 1$, so the graph is narrower than $f(x) = x^4$. It includes the points $(-2, 32)$, $(-1, 2)$, $(0, 0)$, $(1, 2)$, and $(2, 32)$. Connect these points with a smooth curve.

(a) The function is increasing on $[0, \infty)$.

(b) The function is decreasing on $(-\infty, 0]$

11. $f(x) = -\frac{2}{3}x^5$ is in the form $f(x) = ax^n$.
$|a| = \frac{2}{3} < 1$, so the graph is broader than that of $f(x) = x^5$. Since $a = -\frac{2}{3}$ is a negative, the graph is the reflection of $f(x) = \frac{2}{3}x^5$ about the x-axis. It includes the points $\left(-2, \frac{64}{3}\right)$, $\left(-1, \frac{2}{3}\right)$, $(0, 0)$, $\left(1, -\frac{2}{3}\right)$, and $\left(2, -\frac{64}{3}\right)$.
Connect these points with a smooth curve.

(a) The function never increases.

(b) The function is decreasing on $(-\infty, \infty)$.

13. $f(x) = \frac{1}{2}x^3 + 1$ is in the form $f(x) = ax^n + k$, with $|a| = \frac{1}{2} < 0$ and $k = 1$. The graph of $f(x) = \frac{1}{2}x^3 + 1$ looks like $y = x^3$ but is broader and is translated 1 unit up. The graph includes the points $(-2, -3)$, $\left(-1, \frac{1}{2}\right)$, $(0, 1)$, $\left(1, \frac{3}{2}\right)$, and $(2, 5)$.

(a) The function is increasing on $(-\infty, \infty)$.

(b) The function never decreases.

Copyright © 2013 Pearson Education, Inc.

15. $f(x) = -(x+1)^3 + 1 = -\left[x - (-1)\right]^3 + 1$

The graph can be obtained by reflecting the graph of $f(x) = x^3$ about the x-axis and then translating it 1 unit to the left and 1 unit up.

(a) The function never increases.

(b) The function is decreasing on $(-\infty, \infty)$.

17. $f(x) = (x-1)^4 + 2$ This graph has the same shape as $y = x^4$, but is translated 1 unit to the right and 2 units up.

(a) The function is increasing on $[1, \infty)$.

(b) The function is decreasing on $(-\infty, 1]$

19. $f(x) = \frac{1}{2}(x-2)^2 + 4$ is in the form

$f(x) = a(x-h)^n + k$, with $|a| = \frac{1}{2} < 1$.

The graph is broader than that of $f(x) = x^2$. Since $h = 2$, the graph has been translated 2 units to the right. Also, since $k = 4$, the graph has been translated 4 units up.

$f(x) = \frac{1}{2}(x-2)^2 + 4$

(a) The function is increasing on $[2, \infty)$.

(b) The function is decreasing on $(-\infty, 2]$

21. **23.**

25. **27.**

29. $f(x) = x^3 + 5x^2 + 2x - 8$

Step 1: p must be a factor of $a_0 = -8$ and q must be a factor of $a_3 = 1$. Thus, p can be ± 1, ± 2, ± 4, ± 8 and q can be ± 1. The possible rational zeros, $\frac{p}{q}$, are ± 1, ± 2, ± 4, and ± 8. The remainder theorem shows that -4 is a zero.

$$-4)\overline{\begin{array}{rrrr} 1 & 5 & 2 & -8 \\ & -4 & -4 & 8 \\ \hline 1 & 1 & -2 & 0 \end{array}}$$

The new quotient polynomial is $x^2 + x - 2$.

$x^2 + x - 2 = 0$
$(x+2)(x-1) = 0$
$x + 2 = 0 \Rightarrow x = -2$ or $x - 1 = 0 \Rightarrow x = 1$

The rational zeros are -4, -2, and 1. Since the three zeros are -4, -2, and 1, the factors are $x + 4$, $x + 2$, and $x - 1$ and thus $f(x) = (x+4)(x+2)(x-1)$.

Step 2: $f(0) = -8$, so plot $(0, -8)$.

Step 3: The x-intercepts divide the x-axis into four intervals:

Interval	Test Point	Value of $f(x)$	Sign of f (x)	Graph Above or Below x-Axis
$(-\infty, -4)$	-5	-18	Negative	Below
$(-4, -2)$	-3	4	Positive	Above
$(-2, 1)$	0	-8	Negative	Below
$(1, \infty)$	2	24	Positive	Above

Plot the x-intercepts, y-intercept and test points (the y-intercept is one of the test points) with a smooth curve to get the graph.

$f(x) = x^3 + 5x^2 + 2x - 8$

Copyright © 2013 Pearson Education, Inc.

31. $f(x) = 2x(x-3)(x+2)$

Step 1: Set each factor equal to 0 and solve the resulting equations to find the zeros of the function.

$2x = 0 \Rightarrow x = 0$ or $x - 3 = 0 \Rightarrow x = 3$ or $x + 2 = 0 \Rightarrow x = -2$

The three zeros, -2, 0, and 3, divide the x-axis into four regions. Test a point in each region to find the sign of $f(x)$ in that region.

Step 2: $f(0) = 0$, so plot $(0,0)$.

Step 3: The x-intercepts divide the x-axis into four intervals.

Interval	Test Point	Value of $f(x)$	Sign of $f(x)$	Graph Above or Below x-Axis
$(-\infty, -2)$	-3	-36	Negative	Below
$(-2, 0)$	-1	8	Positive	Above
$(0, 3)$	1	-12	Negative	Below
$(3, \infty)$	4	48	Positive	Above

Plot the x-intercepts, y-intercept (which is also an x-intercept in this exercise), and test points with a smooth curve to get the graph.

$f(x) = 2x(x-3)(x+2)$

33. $f(x) = x^2(x-2)(x+3)^2$

Step 1: Set each factor equal to 0 and solve the resulting equations to find the zeros of the function.

$x^2 = 0 \Rightarrow x = 0$ or $x - 2 = 0 \Rightarrow x = 2$ or $(x+3)^2 = 0 \Rightarrow x + 3 = 0 \Rightarrow x = -3$

The zeros are -3, 0, and 2; divide the x-axis into four regions. Test a point in each region to find the sign of $f(x)$ in that region.

Step 2: $f(0) = 0$, so plot $(0,0)$.

Step 3: The x-intercepts divide the x-axis into four intervals.

Interval	Test Point	Value of $f(x)$	Sign of $f(x)$	Graph Above or Below x-Axis
$(-\infty, -3)$	-4	-96	Negative	Below
$(-3, 0)$	-1	-12	Negative	Below
$(0, 2)$	1	-16	Negative	Below
$(2, \infty)$	3	324	Positive	Above

Plot the x-intercepts, y-intercept (which is also an x-intercept in this exercise), and test points with a smooth curve to get the graph.

$f(x) = x^2(x-2)(x+3)^2$

35. $f(x) = (3x-1)(x+2)^2$

Step 1: Set each factor equal to 0 and solve the resulting equations to find the zeros of the function.

$3x - 1 = 0 \Rightarrow x = \frac{1}{3}$ or $x + 2 = 0 \Rightarrow x = -2$

The zeros are -2 and $\frac{1}{3}$, which divide the x-axis into three regions. Test a point in each region to find the sign of $f(x)$ in that region.

Step 2: $f(0) = -4$, so plot $(0,-4)$.

Step 3: The x-intercepts divide the x-axis into three intervals.

Interval	Test Point	Value of $f(x)$	Sign of $f(x)$	Graph Above or Below x-Axis
$(-\infty, -2)$	-3	-10	Negative	Below
$\left(-2, \frac{1}{3}\right)$	0	-4	Negative	Below
$\left(\frac{1}{3}, \infty\right)$	1	18	Positive	Above

Plot the x-intercepts, y-intercept, and test points (the y-intercept is one of the test points) with a smooth curve to get the graph.

(*continued on next page*)

Copyright © 2013 Pearson Education, Inc.

(*continued*)

$$f(x) = (3x - 1)(x + 2)^2$$

37. $f(x) = x^3 + 5x^2 - x - 5$

Step 1: Find the zeros of the function by factoring by grouping.

$$f(x) = x^3 + 5x^2 - x - 5 = x^2(x+5) - 1(x+5)$$
$$= (x+5)(x^2-1) = (x+5)(x+1)(x-1)$$

$x + 5 = 0 \Rightarrow x = -5$ or $x + 1 = 0 \Rightarrow x = -1$ or $x - 1 = 0 \Rightarrow x = 1$

The zeros are –5, –1, and 1, which divide the *x*-axis into four regions. Test a point in each region to find the sign of $f(x)$ in that region.

Step 2: $f(0) = -5$, so plot $(0, -5)$.

Step 3: The *x*-intercepts divide the *x*-axis into four intervals.

Interval	Test Point	Value of f (x)	Sign of f (x)	Graph Above or Below *x*-Axis
$(-\infty, -5)$	–6	–35	Negative	Below
$(-5, -1)$	–2	9	Positive	Above
$(-1, 1)$	0	–5	Negative	Below
$(1, \infty)$	2	21	Positive	Above

Plot the *x*-intercepts, *y*-intercept, and test points (the *y*-intercept is one of the test points) with a smooth curve to get the graph.

$$f(x) = x^3 + 5x^2 - x - 5$$

39. $f(x) = x^3 - x^2 - 2x$

Step 1: Find the zeros of the function by factoring out the common factor, *x*, and then factoring the resulting quadratic factor. Set each factor equal to 0 and solve the resulting equations.

$$f(x) = x^3 - x^2 - 2x$$
$$= x(x^2 - x - 2) = x(x+1)(x-2)$$

$x = 0$ or $x + 1 = 0$ or $x - 2 = 0$
$x = -1$ $x = 2$

The zeros are –1, 0, and 2, which divide the *x*-axis into four regions. Test a point in each region to find the sign of $f(x)$ in that region.

Step 2: $f(0) = 0$, so plot $(0, 0)$.

Step 3: The *x*-intercepts divide the *x*-axis into four intervals.

Interval	Test Point	Value of $f(x)$	Sign of $f(x)$	Graph Above or Below *x*-Axis
$(-\infty, -1)$	-2	-8	Negative	Below
$(-1, 0)$	$-\frac{1}{2}$	$\frac{5}{8}$	Positive	Above
$(0, 2)$	1	-2	Negative	Below
$(2, \infty)$	3	12	Positive	Above

Plot the *x*-intercepts, *y*-intercept (which is also an *x*-intercept in this exercise), and test points with a smooth curve to get the graph.

$$f(x) = x^3 - x^2 - 2x$$

41. $f(x) = 2x^3\left(x^2 - 4\right)(x-1)$

Step 1: Find the zeros of the function by factoring the difference of two squares in the polynomial. Set each factor equal to 0 and solve the resulting equations.

$$f(x) = 2x^3\left(x^2 - 4\right)(x-1)$$
$$= 2x^3(x+2)(x-2)(x-1)$$

$2x^3 = 0 \Rightarrow x = 0$ or $x + 2 = 0 \Rightarrow x = -2$ or
$x - 2 = 0 \Rightarrow x = 2$ or $x - 1 = 0 \Rightarrow x = 1$

The zeros are –2, 0, 1, and 2, which divide the *x*-axis into five regions. Test a point in each region to find the sign of $f(x)$ in that region.

Step 2: $f(0) = 0$, so plot $(0, 0)$.

Step 3: The *x*-intercepts divide the *x*-axis into four intervals.

(*continued on next page*)

Copyright © 2013 Pearson Education, Inc.

(*continued*)

Interval	Test Point	Value of $f(x)$	Sign of $f(x)$	Graph Above or Below x-Axis
$(-\infty, -2)$	-3	1080	Positive	Above
$(-2, 0)$	-1	-12	Negative	Below
$(0, 1)$	$\frac{1}{2}$	$\frac{15}{32} \approx 0.5$	Positive	Above
$(1, 2)$	$\frac{3}{2}$	$-\frac{189}{32} \approx -5.9$	Negative	Below
$(2, \infty)$	3	540	Positive	Above

Plot the x-intercepts, y-intercept (which is also an x-intercept in this exercise), and test points with a smooth curve to get the graph.

$f(x) = 2x^3(x^2 - 4)(x - 1)$

43. $f(x) = 2x^3 - 5x^2 - x + 6$

Step 1: Find the zeros of the function.
p must be a factor of $a_0 = 6$ and q must be a factor of $a_3 = 2$. Thus, p can be ±1, ±2, ±3, or ±6 and q can be ±1 or ±2. The possible rational zeros, $\frac{p}{q}$, are $\pm\frac{1}{2}$, ±1, $\pm\frac{3}{2}$, ±2, ±3, or ±6. The remainder theorem shows that 2 is a zero.

$$2\overline{)\begin{array}{rrrr} 2 & -5 & -1 & 6 \\ & 4 & -2 & -6 \\ \hline 2 & -1 & -3 & 0 \end{array}}$$

The new quotient polynomial is $2x^2 - x - 3$.
$$2x^2 - x - 3 = 0$$
$$(2x - 3)(x + 1) = 0$$
$$2x - 3 = 0 \Rightarrow x = \frac{3}{2} \quad \text{or} \quad x + 1 = 0 \Rightarrow x = -1$$

The rational zeros are -1, $\frac{3}{2}$, and 2. Since the three zeros are -1, $\frac{3}{2}$, and 2, the factors are $x + 1$, $x - \frac{3}{2}$, and $x - 2$ and thus

$$f(x) = (x + 1)\left(x - \frac{3}{2}\right)(x - 2) \text{ or}$$
$$f(x) = (x + 1)(2x - 3)(x - 2).$$

Step 2: $f(0) = 6$, so plot $(0, 6)$.
Step 3: The x-intercepts divide the x-axis into four intervals.

Interval	Test Point	Value of $f(x)$	Sign of $f(x)$	Graph Above or Below x-Axis
$(-\infty, -1)$	-2	-28	Negative	Below
$\left(-1, \frac{3}{2}\right)$	1	2	Positive	Above
$\left(\frac{3}{2}, 2\right)$	$\frac{7}{4}$	$-\frac{11}{32}$	Negative	Below
$(2, \infty)$	3	12	Positive	Above

Plot the x-intercepts, y-intercept, and test points with a smooth curve to get the graph.

$f(x) = 2x^3 - 5x^2 - x + 6$

45. $f(x) = 3x^4 - 7x^3 - 6x^2 + 12x + 8$

Step 1: Find the zeros of the function.
p must be a factor of $a_0 = 8$ and q must be a factor of $a_4 = 3$. Thus, p can be ±1, ±2, ±4, or ±8, and q can be ±1 or ±3. The possible rational zeros, $\frac{p}{q}$, are $\pm\frac{1}{3}, \pm\frac{2}{3}, \pm1, \pm\frac{4}{3}, \pm2,$ $\pm\frac{8}{3}, \pm4,$ or ±8. The remainder theorem shows that 2 is a zero.

$$2\overline{)\begin{array}{rrrrr} 3 & -7 & -6 & 12 & 8 \\ & 6 & -2 & -16 & -8 \\ \hline 3 & -1 & -8 & -4 & 0 \end{array}}$$

The new quotient polynomial is $3x^3 - x^2 - 8x - 4$. Find the zeros of this function. p must be a factor of $a_0 = -4$ and q must be a factor of $a_3 = 3$. Thus, p can be ±1, ±2, or ±4, and q can be ±1 or ±3. The possible rational zeros, $\frac{p}{q}$, are $\pm\frac{1}{3}, \pm\frac{2}{3}, \pm1, \pm\frac{4}{3}, \pm2,$ or ±4. The remainder theorem shows that -1 is a zero.

$$-1\overline{)\begin{array}{rrrr} 3 & -1 & -8 & -4 \\ & -3 & 4 & 4 \\ \hline 3 & -4 & -4 & 0 \end{array}}$$

(continued on next page)

Copyright © 2013 Pearson Education, Inc.

(continued)

The new quotient polynomial is $3x^2 - 4x - 4$.

$$3x^2 - 4x - 4 = 0$$
$$(3x + 2)(x - 2) = 0$$
$$3x + 2 = 0 \Rightarrow x = -\tfrac{2}{3} \text{ or } x - 2 = 0 \Rightarrow x = 2$$

The rational zeros are -1, $-\tfrac{2}{3}$, 2, and 2. Since the four zeros are -1, $-\tfrac{2}{3}$, 2, and 2, the factors are $x + 1$, $x + \tfrac{2}{3}$, $x - 2$, and $x - 2$, and thus $f(x) = (x+1)\left(x + \tfrac{2}{3}\right)(x-2)^2$ or

$$f(x) = (x+1)(3x+2)(x-2)^2.$$

Step 2: $f(0)$ so plot $(0, 8)$.

Step 3: The x-intercepts divide the x-axis into four intervals.

Interval	Test Point	Value of $f(x)$	Sign of $f(x)$	Graph Above or Below x-Axis
$(-\infty, -1)$	-2	64	Positive	Above
$\left(-1, -\tfrac{2}{3}\right)$	$-\tfrac{5}{6}$	$-\tfrac{289}{432}$	Negative	Below
$\left(-\tfrac{2}{3}, 2\right)$	1	10	Positive	Above
$(2, \infty)$	3	44	Positive	Above

Plot the x-intercepts, y-intercept, and test points with a smooth curve to get the graph.

$f(x) = 3x^4 - 7x^3 - 6x^2 + 12x + 8$

47. $f(x) = 2x^2 - 7x + 4$; 2 and 3

Use synthetic division to find $f(2)$ and $f(3)$.

```
2)2 -7  4      3)2 -7  4
     4 -6           6 -3
  2 -3 -2        2 -1  1
```

Since $f(2)$ is negative and $f(3)$ is positive, there is a zero between 2 and 3.

49. $f(x) = 2x^3 - 5x^2 - 5x + 7$; 0 and 1

Since $f(0) = 7$ can easily be determined, use synthetic division only to find $f(1)$.

```
1)2 -5 -5  7
    2 -3 -8
  2 -3 -8 -1
```

Since $f(0) = 7$ is positive and $f(1) = -1$ is negative, there is a zero between 0 and 1.

51. $f(x) = 2x^4 - 4x^2 + 4x - 8$; 1 and 2

Use synthetic division to find $f(1)$ and $f(2)$.

```
1)2  0 -4  4 -8      2)2  0 -4  4 -8
    2  2 -2  2           4  8  8 24
  2  2 -2  2 -6        2  4  4 12 16
```

Since $f(1) = -6$ is negative and $f(2) = 16$ is positive, there is a zero between 1 and 2.

53. $f(x) = x^4 + x^3 - 6x^2 - 20x - 16$; 3.2 and 3.3

Use synthetic division to find $f(3.2)$ and $f(3.3)$.

```
3.2)1  1    -6     -20      -16
       3.2  13.44  23.808   12.1856
    1  4.2  7.44   3.808    -3.8144

3.3)1  1    -6     -20      -16
       3.3  14.19  27.027   23.1891
    1  4.3  8.19   7.027    7.1891
```

Since $f(3.2) = -3.8144$ is negative and $f(3.3) = 7.1891$ is positive, there is a zero between 3.2 and 3.3.

55. $f(x) = x^4 - 4x^3 - 20x^2 + 32x + 12$; -1 and 0

Since $f(0) = 12$ can easily be determined, use synthetic division only to find $f(-1)$.

```
-1)1 -4 -20  32  12
     -1   5  15 -47
   1 -5 -15  47 -35
```

Since $f(-1) = -35$ is negative and $f(0) = 12$ is positive, there is a zero between -1 and 0.

57. $f(x) = x^4 - x^3 + 3x^2 - 8x + 8$; no real zero greater than 2.

Since $f(x)$ has real coefficients and the leading coefficient, 1, is positive, use the boundedness theorem. Divide $f(x)$ synthetically by $x - 2$. Since $2 > 0$ and all numbers in the last row are nonnegative, $f(x)$ has no zero greater than 2.

```
2)1 -1  3 -8  8
    2  2 10  4
  1  1  5  2 12
```

Copyright © 2013 Pearson Education, Inc.

59. $f(x) = x^4 + x^3 - x^2 + 3$; no real zero less than
-2.

Since $f(x)$ has real coefficients, use the
boundedness theorem. Divide $f(x)$
synthetically by $x + 2 = x - (-2)$. Since $-2 < 0$
and the numbers in the last row alternate in
sign, $f(x)$ has no zero less than -2.

$$\begin{array}{r|rrrrr} -2) & 1 & 1 & -1 & 0 & 3 \\ & & -2 & 2 & -2 & 4 \\ \hline & 1 & -1 & 1 & -2 & 7 \end{array}$$

61. $f(x) = 3x^4 + 2x^3 - 4x^2 + x - 1$; no real zero
greater than 1
Since $f(x)$ has real coefficients and the
leading coefficient, 3, is positive, use the
boundedness theorem. Divide $f(x)$
synthetically by $x - 1$. Since $1 > 0$ and all
numbers in the last row are nonnegative,
$f(x)$ has no zero greater than 1.

$$\begin{array}{r|rrrrr} 1) & 3 & 2 & -4 & 1 & -1 \\ & & 3 & 5 & 1 & 2 \\ \hline & 3 & 5 & 1 & 2 & 1 \end{array}$$

63. $f(x) = x^5 - 3x^3 + x + 2$; no real zero greater
than 2.
Since $f(x)$ has real coefficients and the
leading coefficient, 1, is positive, use the
boundedness theorem. Divide $f(x)$
synthetically by $x - 2$. Since $2 > 0$ and all
numbers in the last row are nonnegative,
$f(x)$ has no zero greater than 2.

$$\begin{array}{r|rrrrrr} 2) & 1 & 0 & -3 & 0 & 1 & 2 \\ & & 2 & 4 & 2 & 4 & 10 \\ \hline & 1 & 2 & 1 & 2 & 5 & 12 \end{array}$$

65. The graph shows that the zeros are -6, 2, and
5. The polynomial function has the form
$f(x) = a(x + 6)(x - 2)(x - 5)$. Since $(0, 30)$ is
on the graph, $f(0) = 30$.
$f(0) = a(0 + 6)(0 - 2)(0 - 5) \Rightarrow 30 = 60a \Rightarrow$
$\frac{1}{2} = a$
A cubic polynomial that has the graph shown
is $f(x) = \frac{1}{2}(x + 6)(x - 2)(x - 5)$ or
$f(x) = \frac{1}{2}x^3 - \frac{1}{2}x^2 - 16x + 30.$

67. The graph shows that the zeros are -1 and 1.
There is a turning point at both points. Since
the graph crosses and is tangent to the x-axis
at both $x = -1$ and $x = 1$, these are zeros with
odd multiplicity. This has to be 3 in order to
find a polynomial of least degree. The
polynomial function has the form
$f(x) = a(x - 1)^3(x + 1)^3$. Since $(0, -1)$ is on
the graph, $f(0) = -1$.
$f(0) = -1 = a(0 - 1)^3(0 + 1)^3 \Rightarrow a = 1$
So the function is
$$\begin{aligned} f(x) &= (x - 1)^3 (x + 1)^3 \\ &= (x^3 - 3x^2 + 3x - 1)(x^3 + 3x^2 + 3x + 1) \\ &= x^6 - 3x^4 + 3x^2 - 1 \end{aligned}$$

69. The graph shows that the zeros are -3 and 3.
The graph is tangent to the x-axis at both
$x = -3$ and $x = 3$, so these are zeros with even
multiplicity. This has to be 2 in order to find a
polynomial of least degree. The polynomial
function has the form
$f(x) = a(x - 3)^2(x + 3)^2$. Since $(0, 81)$ is on
the graph, $f(0) = 81$.
$f(0) = 81 = a(0 - 3)^2(0 + 3)^2 \Rightarrow 81 = 81a \Rightarrow$
$a = 1$.
So the function is
$$\begin{aligned} f(x) &= (x - 3)^2 (x + 3)^2 \\ &= (x^2 - 6x + 9)(x^2 + 6x + 9) \\ &= x^4 - 18x^2 + 81 \end{aligned}$$

71. $f(1.25) \approx -14.21875$
In order to have the calculator give the third
screen, enter your function and the window,
then press the GRAPH button. Press the
TRACE button, then type in 1.25 then hit the
ENTER key.

Copyright © 2013 Pearson Education, Inc.

73. $f(1.25) \approx 29.046875$

75. $f(x) = 2x^2 - 7x + 4$; 2 and 3

The real zero between 2 and 3 is approximately 2.7807764.

77. $f(x) = 2x^4 - 4x^2 + 4x - 8$; 1 and 2

The real zero between 1 and 2 is approximately 1.543689.

79. $f(x) = x^3 + 3x^2 - 2x - 6$

The highest degree term is x^3, so the graph will have end behavior similar to the graph of $f(x) = x^3$, which is downward at the left and upward at the right. There is at least one real zero because the polynomial is of odd degree.

There are at most three real zeros because the polynomial is third-degree. A graphing calculator can be used to approximate each zero.

The graphs show that the zeros are approximately –3.0, –1.4 and 1.4.

81. $f(x) = -2x^4 - x^2 + x + 5$

The highest degree term is $-2x^4$ so the graph will have the same end behavior as the graph of $f(x) = -x^4$, which is downward at both the left and the right. Since $f(0) = 5 > 0$, the end behavior and the intermediate value theorem tell us that there must be at least one zero on each side of the y-axis, that is, at least one negative and one positive zero. A graphing calculator can be used to approximate each zero.

The graphs show that the only zeros are approximately –1.1 and 1.2

Copyright © 2013 Pearson Education, Inc.

83. $f(x) = x^3 + 4x^2 - 8x - 8; \; [-3.8, -3]$

The turning point is $(-3.44, 26.15)$.

85. $f(x) = 2x^3 - 5x^2 - x + 1; [-1, 0]$

The turning point is $(-0.09, 1.05)$.

87. $f(x) = x^4 - 7x^3 + 13x^2 + 6x - 28; [-1, 0]$

Plot1 Plot2 Plot3
\Y1◻X^4-7X3+13X2
+6X-28
\Y2=
\Y3=
\Y4=
\Y5=
\Y6=

WINDOW
Xmin=-4
Xmax=4
Xscl=1
Ymin=-40
Ymax=10
Yscl=5
Xres=1

Minimum
X=-.1979331 Y=-28.6??4?

The turning point is $(-0.20, -28.62)$.

89. Answers will vary.

91. $f(x) = x^3 - 3x^2 - 6x + 8 = (x-4)(x-1)(x+2)$

Step 1: Set each factor equal to 0 and solve the resulting equations to find the zeros of the function.

$x - 4 = 0 \Rightarrow x = 4$ or $x - 1 = 0 \Rightarrow x = 1$ or $x + 2 = 0 \Rightarrow x = -2$

The zeros are -2, 1, and 4, which divide the x-axis into four regions. Test a point in each region to find the sign of $f(x)$ in that region.

Step 2: $f(0) = 8$, so plot $(0, 8)$.

Step 3: The x-intercepts divide the x-axis into four intervals.

Interval	Test Point	Value of $f(x)$	Sign of $f(x)$	Graph Above or Below x-Axis
$(-\infty, -2)$	-3	-28	Negative	Below
$(-2, 1)$	0	8	Positive	Above
$(1, 4)$	2	-8	Negative	Below
$(4, \infty)$	5	28	Positive	Above

Plot the x-intercepts, y-intercept, and test points (the y-intercept is one of the test points) with a smooth curve to get the graph.

$f(x) = x^3 - 3x^2 - 6x + 8$
$\qquad = (x-4)(x-1)(x+2)$

(a) $f(x) = 0$ has the solution set $\{-2, 1, 4\}$.

(b) $f(x) < 0$ has the solution set
$(-\infty, -2) \cup (1, 4)$.

(c) $f(x) > 0$ has the solution set
$(-2, 1) \cup (4, \infty)$.

93. $f(x) = 2x^4 - 9x^3 - 5x^2 + 57x - 45$
$\qquad = (x-3)^2(2x+5)(x-1)$

Step 1: Set each factor equal to 0 and solve the resulting equations to find the zeros of the function.

$(x-3)^2 = 0 \Rightarrow x = 3$ or $2x + 5 = 0 \Rightarrow x = -\frac{5}{2}$
or $x - 1 = 0 \Rightarrow x = 1$

The zeros are $-\frac{5}{2} = -2.5$, 1, and 3, which divide the x-axis into four regions. Test a point in each region to find the sign of $f(x)$ in that region.

(*continued on next page*)

Copyright © 2013 Pearson Education, Inc.

(*continued*)

Step 2: $f(0) = -45$, so plot $(0, -45)$.

Step 3: The *x*-intercepts divide the *x*-axis into four intervals.

Interval	Test Point	Value of $f(x)$	Sign of $f(x)$	Graph Above or Below *x*-Axis
$(-\infty, -2.5)$	-3	144	Positive	Above
$(-2.5, 1)$	0	-45	Negative	Below
$(1, 3)$	2	9	Positive	Above
$(3, \infty)$	4	39	Positive	Above

Plot the *x*-intercepts, *y*-intercept, and test points (the *y*-intercept is one of the test points) with a smooth curve to get the graph.

$$f(x) = 2x^4 - 9x^3 - 5x^2 + 57x - 45$$
$$= (x - 3)^2(2x + 5)(x - 1)$$

(a) $f(x) = 0$ has the solution set $\{-2.5, 1, 3$ (multiplicity 2)$\}$.

(b) $f(x) < 0$ has the solution set $(-2.5, 1)$.

(c) $f(x) > 0$ has the solution set
$(-\infty, -2.5) \cup (1, 3) \cup (3, \infty)$.

95. $f(x) = -x^4 - 4x^3 + 3x^2 + 18x$
$$= x(2 - x)(x + 3)^2$$

Step 1: Set each factor equal to 0 and solve the resulting equations to find the zeros of the function.
$x = 0$ or $2 - x = 0 \Rightarrow 2 = x$ or
$(x + 3)^2 = 0 \Rightarrow x = -3$

The zeros are 0, 2, and -3, which divide the *x*-axis into four regions. Test a point in each region to find the sign of $f(x)$ in that region.

Step 2: $f(0) = 0$, so plot $(0, 0)$.

Step 3: The *x*-intercepts divide the *x*-axis into four intervals.

Interval	Test Point	Value of $f(x)$	Sign of $f(x)$	Graph Above or Below *x*-Axis
$(-\infty, -3)$	-4	-24	Negative	Below
$(-3, 0)$	-1	-12	Negative	Below
$(0, 2)$	1	16	Positive	Above
$(2, \infty)$	3	-108	Negative	Below

Plot the *x*-intercepts, *y*-intercept (which is also an *x*-intercept in this exercise), and test points with a smooth curve to get the graph.

$$f(x) = -x^4 - 4x^3 + 3x^2 + 18x$$
$$= x(2 - x)(x + 3)^2$$

(a) $f(x) = 0$ has the solution set
$\{-3$ (multiplicity 2)$, 0, 2\}$.

(b) $f(x) \geq 0$ has the solution set
$\{-3\} \cup [0, 2]$.

(c) $f(x) \leq 0$ has the solution set
$(-\infty, 0] \cup [2, \infty)$.

97. (a) In order for the length, width and height of the box to be positive quantities, $0 < x < 6$.

(b) The width of the rectangular base of the box would be $12 - 2x$. The length would be $18 - 2x$ and the height would be x. The volume would be the product of these three measures; therefore,
$$V(x) = x(18 - 2x)(12 - 2x)$$
$$= 4x^3 - 60x^2 + 216x$$

Copyright © 2013 Pearson Education, Inc.

(c) When x is approximately 2.35 in. the maximum volume will be approximately 228.16 in.^{3}.

(d) When x is between 0.42 in. (approximately) and 5 in., the volume will be greater than 80 in.^{3}.

99. (a) length of the leg = $x - 1$. The domain is $x > 1$ or $(1, \infty)$.

(b) By the Pythagorean theorem,
$$a^2 + b^2 = c^2 \Rightarrow a^2 + (x-1)^2 = x^2 \Rightarrow$$
$$a^2 = x^2 - (x-1)^2 \Rightarrow a = \sqrt{x^2 - (x-1)^2}$$
Thus, the length of the other leg is $\sqrt{x^2 - (x-1)^2}$.

(c) $A = \frac{1}{2}bh \Rightarrow 84 = \frac{1}{2}(x-1)\sqrt{x^2 - (x-1)^2}$

Multiply by 2: $168 = (x-1)\sqrt{x^2 - (x-1)^2}$

Square both sides.
$$28,224 = (x-1)^2\left[x^2 - (x-1)^2\right]$$
$$28,224 = (x^2 - 2x + 1) \cdot \left[x^2 - (x^2 - 2x + 1)\right]$$
$$28,224 = (x^2 - 2x + 1)(2x - 1)$$
$$28,224 = 2x^3 - 5x^2 + 4x - 1$$
$$2x^3 - 5x^2 + 4x - 28,225 = 0$$

(d) Solving this cubic equation graphically, we obtain $x = 25$. If $x = 25$, $x - 1 = 24$, and $\sqrt{x^2 - (x-1)^2} = \sqrt{625 - 576} = 7$.

The hypotenuse is 25 in.; the legs are 24 in. and 7 in.

101. Use the following volume formulas:
$$V_{\text{cylinder}} = \pi r^2 h \text{ and}$$
$$V_{\text{hemisphere}} = \frac{1}{2}V_{\text{sphere}} = \frac{1}{2}\left(\frac{4}{3}\pi r^3\right) = \frac{2}{3}\pi r^3$$
$$\pi r^2 h + 2\left(\frac{2}{3}\pi r^3\right) = \text{Total volume of tank } (V)$$

Let $V = 144\pi$, $h = 12$, and $r = x$.
$$\pi x^2 (12) + \frac{4}{3}\pi x^3 = 144\pi$$
$$\frac{4}{3}x^3 + 12x^2 = 144$$
$$\frac{4}{3}x^3 + 12x^2 - 144 = 0$$
$$4x^3 + 36x^2 - 432 = 0$$
$$x^3 + 9x^2 - 108 = 0$$

The end behavior of the graph of $f(x) = x^3 + 9x^2 - 108$, together with the negative y-intercept, tells us that this cubic polynomial must have one positive zero. (We are not interested in negative zeros because x represents the radius.)

Algebraic Solution:

Use synthetic division or a graphing calculator to locate this zero.

```
3) 1  9   0  -108
      3  36   108
   ─────────────────
   1 12  36    0
```

Note that the bottom line of the synthetic division above implies the polynomial function $g(x) = x^2 + 12x + 36 = (x + 6)^2$, which has the zero −6 (multiplicity 2).

(continued on next page)

Copyright © 2013 Pearson Education, Inc.

(*continued*)

Graphing Calculator Solution:

A radius of 3 feet would cause the volume of the tank to be 144π ft^3.

103. $f(x) = \dfrac{\pi}{3}x^3 - 5\pi x^2 + \dfrac{500\pi d}{3}$

Use a graphing calculator for the three parts of this exercise.

(a) When $d = 0.8$ we have

$$f(x) = \dfrac{\pi}{3}x^3 - 5\pi x^2 + \dfrac{500\pi(0.8)}{3}.$$

The approximate depth is 7.13 cm. The ball floats partly above the surface.

(b) When $d = 2.7$ we have

$$f(x) = \dfrac{\pi}{3}x^3 - 5\pi x^2 + \dfrac{500\pi(2.7)}{3}.$$

There is no *x*-intercept in this window. The sphere is more dense than water and sinks below the surface.

(c) When $d = 1$ we have

$$f(x) = \dfrac{\pi}{3}x^3 - 5\pi x^2 + \dfrac{500\pi}{3}.$$

By tracing on the curve we see that the approximate depth is 10 cm. The balloon is submerged with its top even with the surface.

105. (a)

(b) The best-fitting linear function would be
$y = 33.93x + 113.4$.

(c) The best-fitting cubic function would be
$y = -0.0032x^3 + 0.4245x^2 + 16.64x + 323.1$.

Copyright © 2013 Pearson Education, Inc.

(d) linear: approximately 1572 ft;
cubic: approximately 1569 ft

(e) The cubic function appears slightly better because only one data point is not on the curve.

107. The function in C,
$$h(x) = 3.5\sqrt{x - 2002} + 21.4$$ provides the best model.

A.

B.

C.

Summary Exercises on Polynomial Functions, Zeros, and Graphs

1. $f(x) = 6x^3 - 41x^2 + 26x + 24$

(a) $f(x) = \underbrace{6x^3 - 41x^2}_{1\quad 2} + 26x + 24$ has two

variations in sign, so f has 2 or $2 - 2 = 0$ positive real zeros.
$$f(-x) = -6x^3 - 41x^2 \underbrace{-26x + 24}_{1}$$ has one

variation in sign, so f has 1 negative real zero. Since the degree of the function is 3, the remaining zeros are nonreal complex.

Positive	Negative	Nonreal Complex
2	1	0
0	1	2

(b) p must be a factor of $a_0 = 24$ and q must be a factor of $a_3 = 6$. Thus, p can be ± 1, $\pm 2, \pm 3, \pm 4, \pm 6, \pm 8, \pm 12,$ or ± 24, and q can be $\pm 1, \pm 2, \pm 3,$ or ± 6. The possible rational zeros, $\frac{p}{q}$, are $\pm 1, \pm 2, \pm 3, \pm 4, \pm 6, \pm 8, \pm 12,$ $\pm 24, \pm\frac{1}{2}, \pm\frac{3}{2}, \pm\frac{1}{3}, \pm\frac{2}{3}, \pm\frac{4}{3}, \pm\frac{8}{3}, \pm\frac{1}{6}.$

(c) The remainder theorem shows that 6 is a zero.

$$
\begin{array}{r|rrrr}
6 & 6 & -41 & 26 & 24 \\
  &   & 36 & -30 & -24 \\
\hline
  & 6 & -5 & -4 & 0
\end{array}
$$

Setting the new quotient polynomial $6x^2 - 5x - 4$ equal to 0 and factoring, we have

$$6x^2 - 5x - 4 = 0$$
$$(2x + 1)(3x - 4) = 0$$

$$
\begin{array}{c|c}
2x + 1 = 0 & 3x - 4 = 0 \\
x = -\frac{1}{2} & x = \frac{4}{3}
\end{array}
$$

Thus, the rational zeros of $f(x)$ are $\left\{6, -\frac{1}{2}, \frac{4}{3}\right\}.$

(d) Since we identified three rational zeros, there are no nonreal complex zeros.

Copyright © 2013 Pearson Education, Inc.

3. $f(x) = 3x^4 - 5x^3 + 14x^2 - 20x + 8$

(a) $f(x) = \underbrace{3x^4}_{} \underbrace{- 5x^3}_{1} \underbrace{+ 14x^2}_{2} \underbrace{- 20x}_{3} \underbrace{+ 8}_{4}$ has

four variations in sign, so f has 4 or $4 - 2 = 2$ or $4 - 4 = 0$ positive real zeros.
$f(-x) = 3x^4 + 5x^3 + 14x^2 + 20x + 8$ has

no variations in sign, so f has no negative real zeros. Since the degree of the function is 4, the remaining zeros are nonreal complex.

Positive	Negative	Nonreal Complex
4	0	0
2	0	2
0	0	4

(b) p must be a factor of $a_0 = 8$ and q must be a factor of $a_3 = 3$. Thus, p can be ±1, ±2, ±4, or ±8, and q can be ±1 or ±3. The possible rational zeros, $\frac{p}{q}$, are ±1, ±2, ±4, ±8, $\pm\frac{1}{3}$, $\pm\frac{2}{3}$, $\pm\frac{4}{3}$, $\pm\frac{8}{3}$.

(c) The remainder theorem shows that 1 is a zero.

$$1)\overline{\begin{array}{ccccc} 3 & -5 & 14 & -20 & 8 \\ & 3 & -2 & 12 & -8 \\ \hline 3 & -2 & 12 & -8 & 0 \end{array}}$$

Setting the new quotient polynomial $3x^3 - 2x^2 + 12x - 8$ equal to 0 and factoring by grouping, we have
$$3x^3 - 2x^2 + 12x - 8 = 0$$
$$x^2(3x - 2) + 4(3x - 2) = 0$$
$$(3x - 2)(x^2 + 4) = 0$$

$$\begin{array}{c|c} 3x - 2 = 0 & x^2 + 4 = 0 \\ x = \frac{2}{3} & x^2 = -4 \\ & x = \pm 2i \end{array}$$

Thus, the rational zeros of $f(x)$ are 1 and $\frac{2}{3}$.

(d) The nonreal complex zeros are $\pm2i$.

5. $f(x) = 6x^4 - 5x^3 - 11x^2 + 10x - 2$

(a) $f(x) = \underbrace{6x^4}_{} \underbrace{- 5x^3}_{1} \underbrace{- 11x^2}_{} \underbrace{+ 10x}_{2} \underbrace{- 2}_{3}$ has

three variations in sign, so f has 3 or $3 - 2 = 1$ positive real zeros.

$f(-x) = 6x^4 + 5x^3 + 11x^2 + 10x - 2$ has
$$\underbrace{}_{1}$$

one variation in sign, so f has 1 negative real zero. Since the degree of the function is 4, the remaining zeros are nonreal complex.

Positive	Negative	Nonreal Complex
3	1	0
1	1	2

(b) p must be a factor of $a_0 = -2$ and q must be a factor of $a_4 = 6$. Thus, p can be ±1 or ±2, and q can be ±1, ±2, ±3, or ±6. The possible rational zeros, $\frac{p}{q}$, are ±1, ±2, $\pm\frac{1}{2}$, $\pm\frac{1}{3}$, $\pm\frac{2}{3}$, $\pm\frac{1}{6}$.

(c) The remainder theorem shows that $\frac{1}{3}$ is a zero.

$$\tfrac{1}{3})\overline{\begin{array}{ccccc} 6 & -5 & -11 & 10 & -2 \\ & 2 & -1 & -4 & 2 \\ \hline 6 & -3 & -12 & 6 & 0 \end{array}}$$

Setting the new quotient polynomial $6x^3 - 3x^2 - 12x + 6$ equal to 0 and factoring by grouping, we have
$$6x^3 - 3x^2 - 12x + 6 = 0$$
$$3x^2(2x - 1) - 6(2x - 1) = 0$$
$$(3x^2 - 6)(2x - 1) = 0$$

$$\begin{array}{c|c} 3x^2 - 6 = 0 & 2x - 1 = 0 \\ 3x^2 = 6 & x = \frac{1}{2} \\ x^2 = 2 & \\ x = \pm\sqrt{2} & \end{array}$$

Thus, the rational zeros of $f(x)$ are $\frac{1}{3}$ and $\frac{1}{2}$.

(d) The real complex zeros are $\pm\sqrt{2}$.

7. $f(x) = x^5 - 6x^4 + 16x^3 - 24x^2 + 16x$

(a) $f(x) = \underbrace{x^5}_{} \underbrace{- 6x^4}_{1} \underbrace{+ 16x^3}_{2} \underbrace{- 24x^2}_{3} + 16x$ has
$$\underbrace{}_{4}$$

four variations in sign, so f has 4 or $4 - 2 = 2$ or $4 - 4 = 0$ positive real zeros.
$f(-x) = -x^5 - 6x^4 - 16x^3 - 24x^2 - 16x$

has no variations in sign, so f has no negative real zeros. Since the degree of the function is 5, the remaining zeros are nonreal complex.

(continued on next page)

Copyright © 2013 Pearson Education, Inc.

(continued)

However, nonreal complex roots occur in pairs, so the fifth zero is not positive, negative, or nonreal complex.

Positive	Negative	Nonreal Complex
4	0	0
2	0	2
0	0	4

If we factor out x, we have
$$f(x) = x^5 - 6x^4 + 16x^3 - 24x^2 + 16x$$
$$= x\left(x^4 - 6x^3 + 16x^2 - 24x + 16\right)$$
If we set $f(x) = 0$, then $x = 0$. Thus 0 is the remaining zero.

(b) Since one factor is x, we must find the zeros of the other factor, $x^4 - 6x^3 + 16x^2 - 24x + 16$. p must be a factor of $a_0 = 16$ and q must be a factor of $a_4 = 1$. Thus, p can be ±1, ±2, ±4, ±8, or ±16, and q can be ±1. The possible rational zeros $\frac{p}{q}$, of this factor, are ±1, ±2, ±4, ±8, or ±16. The possible rational zeros of the original function $f(x)$ are 0, ±1, ±2, ±4, ±8, or ±16.

(c) The remainder theorem shows that 2 is a zero.

```
2)1  -6   16  -24   16
      2   -8   16  -16
   1  -4    8   -8    0
```
The new quotient polynomial is $x^3 - 4x^2 + 8x - 8$. For this polynomial, p must be a factor of $a_0 = -8$ and q must be a factor of $a_3 = 1$. Thus, p can be ±1, ±2, ±4, or ±8, and q can be ±1 The possible rational zeros $\frac{p}{q}$, are ±1, ±2, ±4, or ±8. The remainder theorem shows that 2 is a zero.

```
2)1  -4   8  -8
      2  -4   8
   1  -2   4   0
```
Now use the quadratic formula to find the zeros of the quotient polynomial $x^2 - 2x + 4$.

$$x = \frac{-(-2) \pm \sqrt{(-2)^2 - 4(1)(4)}}{2(1)}$$
$$= \frac{2 \pm \sqrt{-12}}{2} = \frac{2 \pm 2i\sqrt{3}}{2} = 1 \pm i\sqrt{3}$$
Thus, the rational zeros are 0 and 2 (multiplicity 2).

(d) The nonreal complex zeros are $1 \pm i\sqrt{3}$.

9. $f(x) = 8x^4 + 8x^3 - x - 1$

(a) $f(x) = 8x^4 \underbrace{+8x^3 - x}_{1} - 1$ has one variation in sign, so f has 1 positive real zero.
$f(-x) = \underbrace{8x^4}_{} \underbrace{-8x^3}_{1} \underbrace{+8x}_{2} \underbrace{-1}_{3}$ has three variations in sign, so f has 3 or $3 - 1 = 0$ negative real zeros. Since the degree of the function is 4, the remaining zeros are nonreal complex.

Positive	Negative	Nonreal Complex
1	3	0
1	1	2

(b) p must be a factor of $a_0 = -1$ and q must be a factor of $a_4 = 8$. Thus, p can be ±1, and q can be ±1, ±2, ±4, or ±8. The possible rational zeros, $\frac{p}{q}$, are ±1, $\pm\frac{1}{2}$, $\pm\frac{1}{4}$, $\pm\frac{1}{8}$.

(c) Since we can factor $f(x)$, set $f(x) = 0$, then solve to find the rational zeros.
$$8x^4 + 8x^3 - x - 1 = 0$$
$$8x^3(x+1) - (x+1) = 0$$
$$(8x^3 - 1)(x+1) = 0$$
$$(4x^2 + 2x + 1)(2x-1)(x+1) = 0$$
This gives $2x - 1 = 0 \Rightarrow x = \frac{1}{2}$ and $x + 1 = 0 \Rightarrow x = -1$. Thus, the rational zeros are $\frac{1}{2}$ and -1.

Copyright © 2013 Pearson Education, Inc.

(d) To find the zeros of the factor $4x^2 + 2x + 1$, use the quadratic formula.

$$x = \frac{-2 \pm \sqrt{2^2 - 4(4)(1)}}{2(4)} = \frac{-2 \pm \sqrt{-12}}{8}$$

$$= \frac{-2 \pm 2i\sqrt{3}}{8} = -\frac{1}{4} \pm \frac{\sqrt{3}}{4}i$$

The real complex zeros are $-\frac{1}{4} \pm \frac{\sqrt{3}}{4}i$.

11. $f(x) = x^4 + 3x^3 - 3x^2 - 11x - 6$

(a) $f(x) = x^4 \underbrace{+3x^3 - 3x^2 - 11x - 6}_{1}$ has 1

variation in sign. f has 1 positive real zero.

$f(-x) = \underbrace{x^4 - 3x^3}_{1} \underbrace{-3x^2 + 11x}_{2} \underbrace{-6}_{3}$ has 3

variations in sign. f has 3 or $3 - 2 = 1$ negative real zeros. Thus, the different possibilities are

Positive	Negative	Nonreal Complex
1	3	0
1	1	2

(b) p must be a factor of $a_0 = -6$ and q must be a factor of $a_4 = 1$. Thus, p can be ± 1, $\pm 2, \pm 3, \pm 6$ and q can be ± 1. The possible zeros, $\frac{p}{q}$, are $\pm 1, \pm 2, \pm 3, \pm 6$.

(c) The remainder theorem shows that 2 is zero.

$$2)\overline{1 \quad 3 \quad -3 \quad -11 \quad -6}$$
$$ \underline{2 \quad 10 \quad 14 \quad 6}$$
$$ 1 \quad 5 \quad 7 \quad 3 \quad 0$$

The new quotient polynomial is $x^3 + 5x^2 + 7x + 3$. Since f has 1 positive real zero, try negative values. The remainder theorem shows that -3 is zero.

$$-3)\overline{1 \quad 5 \quad 7 \quad 3}$$
$$ \underline{-3 \quad -6 \quad -3}$$
$$ 1 \quad 2 \quad 1 \quad 0$$

The new quotient polynomial is $x^2 + 2x + 1$. Factor this polynomial and set equal to zero to find the remaining zeros.

$$x^2 + 2x + 1 = 0 \Rightarrow (x+1)^2 = 0 \Rightarrow$$
$$x + 1 = 0 \Rightarrow x = -1$$

The rational zeros are $-3, -1$ (multiplicity 2), and 2.

(d) All zeros have been found, and they are all rational.

(e) All zeros have been found, and they are all real.

(f) The x-intercepts occur when $x = -3, -1$, and 2.

(g) $f(0) = -6$

(h)
$$4)\overline{1 \quad 3 \quad -3 \quad -11 \quad -6}$$
$$ \underline{4 \quad 28 \quad 100 \quad 356}$$
$$ 1 \quad 7 \quad 25 \quad 89 \quad 350$$

The corresponding point on the graph is $(4, 350)$.

(i)

(j) The x-intercepts divide the x-axis into four intervals:

Interval	Test Point	Value of $f(x)$	Sign of $f(x)$	Graph Above or Below x-Axis
$(-\infty, -3)$	-4	54	Positive	Above
$(-3, -1)$	-2	-4	Negative	Below
$(-1, 2)$	0	-6	Negative	Below
$(2, \infty)$	3	96	Positive	Above

Plot the x-intercepts, y-intercept and test points with a smooth curve to get the graph.

$$f(x) = x^4 + 3x^3 - 3x^2 - 11x - 6$$

13. $f(x) = 2x^5 - 10x^4 + x^3 - 5x^2 - x + 5$

(a) $f(x) = \underbrace{2x^5 - 10x^4}_{1} \underbrace{+ x^3 - 5x^2}_{2} \underbrace{- x}_{3} \underbrace{+ 5}_{4}$ has

4 variations in sign. f has 4 or $4 - 2 = 2$ or $2 - 2 = 0$ positive real zeros.

$f(-x) = -2x^5 - 10x^4 - x^3 \underbrace{- 5x^2 + x}_{1} + 5$

has 1 variation in sign. f has 1 negative real zero.

(continued on next page)

Copyright © 2013 Pearson Education, Inc.

(*continued*)

Thus, the different possibilities are

Positive	Negative	Nonreal Complex
4	1	0
2	1	2
0	1	4

(b) p must be a factor of $a_0 = 5$ and q must be a factor of $a_5 = 2$. Thus, p can be ± 1, ± 5 and q can be ± 1, ± 2. The possible zeros, $\frac{p}{q}$, are ± 1, ± 5, $\pm \frac{1}{2}$, $\pm \frac{5}{2}$.

(c) The remainder theorem shows that 5 is zero.

$$5)\overline{2 \quad -10 \quad 1 \quad -5 \quad -1 \quad 5}$$
$$\;\; 10 \quad 0 \quad 5 \quad 0 \quad -5$$
$$\overline{2 \quad\;\; 0 \quad 1 \quad 0 \quad -1 \quad 0}$$

The new quotient polynomial will be $2x^4 + x^2 - 1$. At this point, if we try to determine which of the possible rational zero will result in a remainder of zero, we will find that none of them work. We have found the only rational zero of our polynomial function. The rational zero is 5.

(d) Since the new quotient polynomial is $2x^4 + x^2 - 1$, we can factor this polynomial and set equal to zero to find the remaining zeros.

$$2x^4 + x^2 - 1 = 0 \Rightarrow \left(x^2 + 1\right)\left(2x^2 - 1\right) = 0$$

Since $x^2 + 1 = 0$ does not yield any real solutions, we will examine $2x^2 - 1 = 0$ first.

$$2x^2 - 1 = 0 \Rightarrow 2x^2 = 1 \Rightarrow x^2 = \tfrac{1}{2} \Rightarrow$$
$$x = \pm\sqrt{\tfrac{1}{2}} = \pm\tfrac{\sqrt{2}}{2}$$

The other real zeros are $-\frac{\sqrt{2}}{2}$ and $\frac{\sqrt{2}}{2}$.

(e) Examining $x^2 + 1 = 0$, we have $x^2 = -1 \Rightarrow x = \pm\sqrt{-1} = \pm i$. Thus, the complex zeros are $-i$ and i.

(f) The x-intercepts occur when $x = -\frac{\sqrt{2}}{2} \approx -0.71$, $\frac{\sqrt{2}}{2} \approx 0.71$, and 5.

(g) $f(0) = 5$

(h)
$$4)\overline{2 \quad -10 \quad\;\; 1 \quad\;\; -5 \quad\;\; -1 \quad\;\; 5}$$
$$\;\; 8 \quad -8 \quad -28 \quad -132 \quad -532$$
$$\overline{2 \quad -2 \quad -7 \quad -33 \quad -133 \quad -527}$$

The corresponding point on the graph is $(4, -527)$.

(i)

(j) The x-intercepts divide the x-axis into four intervals:

Interval	Test Point	Value of $f(x)$	Sign of $f(x)$	Graph Above or Below x-Axis
$\left(-\infty, -\frac{\sqrt{2}}{2}\right)$	-1	-12	Negative	Below
$\left(-\frac{\sqrt{2}}{2}, \frac{\sqrt{2}}{2}\right)$	0	5	Positive	Above
$\left(\frac{\sqrt{2}}{2}, 5\right)$	1	-8	Negative	Below
$(5, \infty)$	6	2627	Positive	Above

Plot the x-intercepts, y-intercept and test points (the y-intercept is also a test point) with a smooth curve to get the graph.

$$f(x) = 2x^5 - 10x^4 + x^3 - 5x^2 - x + 5$$

15. $f(x) = -2x^4 - x^3 + x + 2$

(a) $f(x) = -2x^4 \underbrace{-x^3 + x}_{1} + 2$ has 1 variation in sign. f has 1 positive real zero.

$f(x) = \underbrace{-2x^4 + x^3}_{1} \underbrace{- x}_{2} \underbrace{+ 2}_{3}$ has 3 variations in sign. f has 3 or $3 - 2 = 1$ negative real zeros. Thus, the different possibilities are

Positive	Negative	Nonreal Complex
1	3	0
1	1	2

Copyright © 2013 Pearson Education, Inc.

(b) p must be a factor of $a_0 = 2$ and q must be a factor of $a_4 = -2$. Thus, p can be ±1, ±2 and q can be ±1, ±2. The possible zeros, $\frac{p}{q}$, are ±1, ±2, $\pm\frac{1}{2}$.

(c) The remainder theorem shows that -1 is zero.

$$-1)\overline{\begin{array}{rrrrr} -2 & -1 & 0 & 1 & 2 \\ & 2 & -1 & 1 & -2 \\ \hline -2 & 1 & -1 & 2 & 0 \end{array}}$$

The new quotient polynomial will be $-2x^3 + x^2 - x + 2$. Since the signs are alternating in the bottom row of the synthetic division, we know that -1 is a lower bound. The remainder theorem shows that 1 is zero.

$$1)\overline{\begin{array}{rrrr} -2 & 1 & -1 & 2 \\ & -2 & -1 & -2 \\ \hline -2 & -1 & -2 & 0 \end{array}}$$

The new quotient polynomial is $-2x^2 - x - 2 = -(2x^2 + x + 2)$. At this point, if we try to determine which of the possible rational zeros will result in a remainder of zero, we will find that none of them work. We have found the only rational zeros of our polynomial function. The rational zeros are -1 and 1.

(d) If we examine the discriminant of $2x^2 + x + 2$, $1^2 - 4(2)(2) = 1 - 16 = -15$, we see that it is negative, which implies there are no other real zeros.

(e) To find the remaining complex zeros, we must solve $2x^2 + x + 2 = 0$. To solve this equation, we use the quadratic formula with $a = 2$, $b = 1$, and $c = 2$.

$$x = \frac{-1 \pm \sqrt{1^2 - 4(2)(2)}}{2(2)} = \frac{-1 \pm \sqrt{1-16}}{4}$$
$$= \frac{-1 \pm \sqrt{-15}}{4} = \frac{-1 \pm i\sqrt{15}}{4}$$

Thus, the other complex zeros are $-\frac{1}{4} + \frac{\sqrt{15}}{4}i$ and $-\frac{1}{4} - \frac{\sqrt{15}}{4}i$

(f) The x-intercepts occur when $x = -1$ and $x = 1$.

(g) $f(0) = 2$

(h)
$$4)\overline{\begin{array}{rrrrr} -2 & -1 & 0 & 1 & 2 \\ & -8 & -36 & -144 & -572 \\ \hline -2 & -9 & -36 & -143 & -570 \end{array}}$$

The corresponding point on the graph is $(4, -570)$.

(i)

(j) The x-intercepts divide the x-axis into three intervals:

Interval	Test Point	Value of $f(x)$	Sign of $f(x)$	Graph Above or Below x-Axis
$(-\infty, -1)$	-2	-24	Negative	Below
$(-1, 1)$	0	2	Positive	Above
$(1, \infty)$	2	-36	Negative	Below

Plot the x-intercepts, y-intercept and test points (the y-intercept is also a test point) with a smooth curve to get the graph.

$f(x) = -2x^4 - x^3 + x + 2$

17. $f(x) = 3x^4 - 14x^2 - 5$

(a) $f(x) = \underset{1}{3x^4 - 14x^2} - 5$ has 1 variation in sign. f has 1 positive real zero.

$f(-x) = \underset{1}{3x^4 - 14x^2} - 5$ has 1 variation in sign. f has 1 negative real zeros. Thus, the different possibilities are

Positive	Negative	Nonreal Complex
1	1	2

(b) p must be a factor of $a_0 = -5$ and q must be a factor of $a_4 = 3$. Thus, p can be ±1, ±5 and q can be $\pm1, \pm3$. The possible zeros, $\frac{p}{q}$, are ±1, ±5, $\pm\frac{1}{3}$, $\pm\frac{5}{3}$.

(c) If we try to determine which of the possible rational zeros will result in a remainder of zero, we will find that none of them work. There are no rational zeros.

Copyright © 2013 Pearson Education, Inc.

(d) We can factor the polynomial,

$3x^4 - 14x^2 - 5,$ and set equal to zero to the remaining zeros.

$$3x^4 - 14x^2 - 5 = 0$$
$$(3x^2 + 1)(x^2 - 5) = 0$$

Since $3x^2 + 1 = 0$ does not yield any real solutions, we will examine $x^2 - 5 = 0$ first.

$$x^2 - 5 = 0 \Rightarrow x^2 = 5 \Rightarrow x = \pm\sqrt{5}$$

The real zeros are $-\sqrt{5}$ and $\sqrt{5}$.

(e) Examining $3x^2 + 1 = 0,$ we have

$$3x^2 = -1 \Rightarrow x^2 = -\tfrac{1}{3} \Rightarrow x = \pm\sqrt{-\tfrac{1}{3}} = \pm i\tfrac{\sqrt{3}}{3}.$$

Thus, the complex zeros are $-\tfrac{\sqrt{3}}{3}i$ and $\tfrac{\sqrt{3}}{3}i.$

(f) The x-intercepts occur when

$$x = -\sqrt{5} \approx -2.24 \text{ and } \sqrt{5} \approx 2.24.$$

(g) $f(0) = -5$

(h)
$$4\overline{)\,3 \quad 0 \quad -14 \quad 0 \quad -5\,}$$
$$\underline{\quad\; 12 \quad 48 \quad 136 \quad 544}$$
$$\; 3 \quad 12 \quad 34 \quad 136 \quad 539$$

The corresponding point on the graph is $(4, 539).$

(i)

(j) The x-intercepts divide the x-axis into three intervals:

Interval	Test Point	Value of $f(x)$	Sign of $f(x)$	Graph Above or Below x-Axis
$(-\infty, -\sqrt{5})$	-3	112	Positive	Above
$(-\sqrt{5}, \sqrt{5})$	0	-5	Negative	Below
$(\sqrt{5}, \infty)$	3	112	Positive	Above

Plot the x-intercepts, y-intercept and test points (the y-intercept is also a test point) with a smooth curve to get the graph.

$f(x) = 3x^4 - 14x^2 - 5$

19. $f(x) = -3x^4 + 22x^3 - 55x^2 + 52x - 12$

(a) $f(x) = \underbrace{-3x^4}_{} + \underbrace{22x^3}_{1} - \underbrace{55x^2}_{2} + \underbrace{52x}_{3} - 12_{\;4}$

has 4 variations in sign. f has 4 or $4 - 2 = 2$ or $2 - 2 = 0$ positive real zeros.

$$f(-x) = -3x^4 - 22x^3 - 55x^2 - 52x - 12$$

has 0 variations in sign. f has 0 negative real zeros. Thus, the different possibilities are

Positive	Negative	Nonreal Complex
4	0	0
2	0	2
0	0	4

(b) p must be a factor of $a_0 = -12$ and q must be a factor of $a_4 = -3.$ Thus, p can be $\pm1, \pm2, \pm3, \pm4, \pm6, \pm12$ and q can be $\pm1, \pm3.$ The possible zeros, $\tfrac{p}{q},$ are

$$\pm1, \;\pm2, \;\pm3, \;\pm4, \;\pm6, \;\pm12, \;\pm\tfrac{1}{3},$$
$$\pm\tfrac{2}{3}, \;\pm\tfrac{4}{3}.$$

(c) The remainder theorem shows that 2 is a zero.

$$2\overline{)\,-3 \quad 22 \quad -55 \quad 52 \quad -12\,}$$
$$\underline{\quad\;\; -6 \quad 32 \quad -46 \quad 12}$$
$$\; -3 \quad 16 \quad -23 \quad 6 \quad 0$$

The new quotient polynomial will be $-3x^3 + 16x^2 - 23x + 6.$ The remainder theorem shows that 3 is zero.

$$3\overline{)\,-3 \quad 16 \quad -23 \quad 6\,}$$
$$\underline{\quad\;\; -9 \quad 21 \quad -6}$$
$$\; -3 \quad 7 \quad -2 \quad 0$$

The new quotient polynomial is

$$-3x^2 + 7x - 2 = -(3x^2 - 7x + 2).$$ Factor

$3x^2 - 7x + 2$ and set equal to zero to find the remaining zeros.

(continued on next page)

Copyright © 2013 Pearson Education, Inc.

(*continued*)

$$3x^2 - 7x + 2 = 0 \Rightarrow (3x - 1)(x - 2) = 0$$

$$3x - 1 = 0 \Rightarrow x = \tfrac{1}{3} \text{ or } x - 2 = 0 \Rightarrow x = 2$$

The rational zeros are

$\tfrac{1}{3}$, 2 (multiplicity 2), and 3.

(d) All zeros have been found, and they are all rational.

(e) All zeros have been found, and they are all real.

(f) The *x*-intercepts occur when $x = \tfrac{1}{3}$, 2, and 3.

(g) $f(0) = -12$

(h)
$$
\begin{array}{r}
4)\overline{\begin{array}{rrrrr} -3 & 22 & -55 & 52 & -12 \end{array}} \\
\begin{array}{rrrrr} & -12 & 40 & -60 & -32 \end{array} \\
\hline
\begin{array}{rrrrr} -3 & 10 & -15 & -8 & -44 \end{array}
\end{array}
$$

The corresponding point on the graph is $(4, -44)$.

(i)

(j) The *x*-intercepts divide the *x*-axis into four intervals:

Interval	Test Point	Value of $f(x)$	Sign of $f(x)$	Graph Above or Below *x*-Axis
$\left(-\infty, \tfrac{1}{3}\right)$	0	-12	Negative	Below
$\left(\tfrac{1}{3}, 2\right)$	1	4	Positive	Above
$(2, 3)$	$\tfrac{5}{2}$	$\tfrac{13}{16}$	Positive	Above
$(3, \infty)$	4	-44	Negative	Below

Plot the *x*-intercepts, *y*-intercept and test points (the *y*-intercept is also a test point) with a smooth curve to get the graph.

$f(x) = -3x^4 + 22x^3 - 55x^2 + 52x - 12$

Section 3.5 Rational Functions: Graphs, Applications, and Models

1. $f(x) = \dfrac{1}{x}$

Domain: $(-\infty, 0) \cup (0, \infty)$

Range: $(-\infty, 0) \cup (0, \infty)$

3. $f(x) = \dfrac{1}{x}$

Increasing: nowhere

Decreasing: $(-\infty, 0) \cup (0, \infty)$

Constant: nowhere

5. $y = \dfrac{1}{x - 3} + 2$

Vertical asymptote: $x = 3$

Horizontal asymptote: $y = 2$

7. Since $f(-x) = \dfrac{1}{(-x)^2} = \dfrac{1}{x^2} = f(x)$, the function is an even function. It exhibits symmetry with respect to *y*-axis.

9. Graphs A, B, and C have a domain of $(-\infty, 3) \cup (3, \infty)$.

11. Graph A has a range of $(-\infty, 0) \cup (0, \infty)$.

13. Graph A has a single solution to the equation $f(x) = 3$.

15. Graphs A, C, and D have the *x*-axis as a horizontal asymptote.

17. $f(x) = \dfrac{2}{x}$

To obtain the graph of $f(x) = \tfrac{2}{x}$, stretch the graph of $f(x) = \tfrac{1}{x}$, vertically by a factor of 2. Just as with the graph of $f(x) = \tfrac{1}{x}$, $y = 0$ is the horizontal asymptote and $x = 0$ is the vertical asymptote.

$f(x) = \dfrac{2}{x}$

(a) Domain: $(-\infty, 0) \cup (0, \infty)$

Copyright © 2013 Pearson Education, Inc.

(b) Range: $(-\infty,0)\cup(0,\infty)$

(c) The function is never increasing.

(d) The function decreases over $(-\infty,0)$ and $(0,\infty)$.

19. $f(x)=\dfrac{1}{x+2}$

To obtain the graph of $f(x)=\frac{1}{x+2}$ shift the graph of $f(x)=\frac{1}{x}$, to the left 2 units. Just as with $f(x)=\frac{1}{x}$, $y=0$ is the horizontal asymptote, but this graph has $x=-2$ as its vertical asymptote (this affects the domain).

$f(x)=\frac{1}{x+2}$

(a) Domain: $(-\infty,-2)\cup(-2,\infty)$

(b) Range: $(-\infty,0)\cup(0,\infty)$

(c) The function is never increasing.

(d) The function decreases over $(-\infty,-2)$ and $(-2,\infty)$.

21. $f(x)=\dfrac{1}{x}+1$

To obtain the graph of $f(x)=\frac{1}{x}+1$, translate the graph of $f(x)=\frac{1}{x}$, 1 unit up. Just as with $f(x)=\frac{1}{x}$, $x=0$ is the vertical asymptote, but this graph has $y=1$ as its horizontal asymptote (this affects the range).

$f(x)=\frac{1}{x}+1$

a. Domain: $(-\infty,0)\cup(0,\infty)$

b. Range: $(-\infty,1)\cup(1,\infty)$

(c) The function is never increasing.

(d) The function decreases over $(-\infty,0)$ and $(0,\infty)$.

23. $f(x)=-\dfrac{2}{x^2}$

To obtain the graph of $f(x)=-\frac{2}{x^2}$, stretch the graph of $f(x)=\frac{1}{x^2}$, by a factor of 2, and then reflect the graph will be reflected across the x-axis.

$f(x)=-\frac{2}{x^2}$

(a) Domain: $(-\infty,0)\cup(0,\infty)$

(b) Range: $(-\infty,0)$

(c) The function increases over $(0,\infty)$.

(d) The function decreases over $(-\infty,0)$.

25. $f(x)=\dfrac{1}{(x-3)^2}$

To obtain the graph of $f(x)=\frac{1}{(x-3)^2}$, shift the graph of $f(x)=\frac{1}{x^2}$, 3 units to the right. Just as with $f(x)=\frac{1}{x^2}$, $y=0$ is the horizontal asymptote, but this graph has $x=3$ as its vertical asymptote.

$f(x)=\frac{1}{(x-3)^2}$

(a) Domain: $(-\infty,3)\cup(3,\infty)$

(b) Range: $(0,\infty)$

(c) The function increases over $(-\infty,3)$.

(d) The function decreases over $(3,\infty)$.

Copyright © 2013 Pearson Education, Inc.

27. $f(x) = \dfrac{-1}{(x+2)^2} - 3$

To obtain the graph of $f(x) = \dfrac{-1}{(x+2)^2} - 3$, shift

the graph of $f(x) = \dfrac{1}{x^2}$, 2 units to the left,

reflect the graph across the x-axis, and then shfit the graph 3 units down. Unlike the graph

of $f(x) = \dfrac{1}{x^2}$, the vertical asymptote of

$f(x) = \dfrac{-1}{(x+2)^2} - 3$ will be $x = -2$ and the

horizontal asymptote is $y = -3$.

$f(x) = \dfrac{-1}{(x+2)^2} - 3$

(a) Domain: $(-\infty, -2) \cup (-2, \infty)$

(b) Range: $(-\infty, -3)$

(c) The function increases over $(-2, \infty)$.

(d) The function decreases over $(-\infty, -2)$.

29. D. The graph of $f(x) = \frac{x+7}{x+1}$ has the vertical asymptote at $x = -1$.

31. G. The graph of $f(x) = \frac{1}{x+4}$ has the x-axis as its horizontal asymptote, and the y-axis is not its vertical asymptote. The line $x = -4$ is its vertical asymptote.

33. E. The graph of $f(x) = \frac{x^2-16}{x+4}$ has a "hole" in its graph located at $x = -4$ since

$f(x) = \frac{x^2-16}{x+4} = \frac{(x+4)(x-4)}{x+4} = x - 4, x \neq -4.$

35. F. The graph of $f(x) = \frac{x^2+3x+4}{x-5}$ has an

oblique asymptote since $f(x) = x + 8 + \frac{44}{x-5}$.

37. $f(x) = \dfrac{3}{x-5}$. To find the vertical asymptote,

set the denominator equal to zero.

$x - 5 = 0 \Rightarrow x = 5$

The equation of the vertical asymptote is $x = 5$. To find the horizontal asymptote, divide each term by the largest power of x in the expression.

$f(x) = \dfrac{\frac{3}{x}}{\frac{x}{x} - \frac{5}{x}} = \dfrac{\frac{3}{x}}{1 - \frac{5}{x}}$

As $|x| \to \infty$, $\frac{1}{x}$ approaches 0, thus $f(x)$

approaches $\frac{0}{1-0} = \frac{0}{1} = 0$. The line $y = 0$ (that

is, the x-axis) is the horizontal asymptote.

39. $f(x) = \dfrac{4 - 3x}{2x+1}$

To find the vertical asymptote, set the denominator equal to zero.

$2x + 1 = 0 \Rightarrow x = -\frac{1}{2}$

The equation of the vertical asymptote is

$x = -\frac{1}{2}$. To find the horizontal asymptote,

divide each term by the largest power of x in the expression.

$f(x) = \dfrac{\frac{4}{x} - \frac{3x}{x}}{\frac{2x}{x} + \frac{1}{x}} = \dfrac{\frac{4}{x} - 3}{2 + \frac{1}{x}}$

As $|x| \to \infty$, $\frac{1}{x}$ approaches 0, so $f(x)$

approaches $\frac{0-3}{2+0} = -\frac{3}{2}$. The equation of the

horizontal asymptote is $y = -\frac{3}{2}$.

41. $f(x) = \dfrac{x^2 - 1}{x+3}$

The vertical asymptote is $x = -3$, found by solving $x + 3 = 0$. Since the numerator is of degree exactly one more than the denominator, there is no horizontal asymptote, but there may be an oblique asymptote. To find it, divide the numerator by the denominator.

$$-3 \overline{)\begin{array}{ccc} 1 & 0 & -1 \\ & -3 & 9 \\ \hline 1 & -3 & 8 \end{array}}$$

Thus, $f(x) = \frac{x^2-1}{x+3} = x - 3 + \frac{8}{x+3}$.

For very large values of $|x|$, $\frac{8}{x+3}$ is close to 0,

and the graph approaches the line $y = x - 3$.

Copyright © 2013 Pearson Education, Inc.

43. $f(x) = \dfrac{(x-3)(x+1)}{(x+2)(2x-5)}$

To find the vertical asymptotes, set the denominator equal to zero and solve.

$x+2=0 \Rightarrow x=-2$ and $2x-5=0 \Rightarrow x=\frac{5}{2}$

Thus, the vertical asymptotes are $x=-2$ and $x=\frac{5}{2}$.

To determine the horizontal asymptote, first multiply the factors in the numerator and denominator to get $f(x) = \frac{x^2-2x-3}{2x^2-x-10}$. Divide the numerator and denominator by x^2.

$$f(x) = \dfrac{\frac{x^2}{x^2} - \frac{2x}{x^2} - \frac{3}{x^2}}{\frac{2x^2}{x^2} - \frac{x}{x^2} - \frac{10}{x^2}} = \dfrac{1 - \frac{2}{x} - \frac{3}{x^2}}{2 - \frac{1}{x} - \frac{10}{x^2}}$$

As $\left|x\right| \to \infty$, $\frac{1}{x}$ and $\frac{1}{x^2}$ approach 0, so $f(x)$ approaches $\frac{1-0-0}{2-0-0} = \frac{1}{2}$. Thus, the equation of the horizontal asymptote is $y = \frac{1}{2}$.

45. $f(x) = \dfrac{x^2+1}{x^2+9}$

To find the vertical asymptotes, set the denominator equal to zero and solve.

$x^2+9=0 \Rightarrow x^2=-9 \Rightarrow x=\pm\sqrt{-9}=\pm 3i$

Thus, there are no vertical asymptotes. To determine the horizontal asymptote, divide the numerator and denominator by x^2.

$$f(x) = \dfrac{\frac{x^2}{x^2} + \frac{1}{x^2}}{\frac{x^2}{x^2} + \frac{9}{x^2}} = \dfrac{1 + \frac{1}{x^2}}{1 + \frac{9}{x^2}}$$

As $\left|x\right| \to \infty$, $\frac{1}{x^2}$ approaches 0, so $f(x)$ approaches $\frac{1+0}{1+0} = \frac{1}{1} = 1$. Thus, the equation of the horizontal asymptote is $y = 1$.

47. (a) Translating $y = \frac{1}{x}$ three units to the right gives $y = \frac{1}{x-3}$. Translating $y = \frac{1}{x-3}$ two units up yields

$y = \frac{1}{x-3} + 2 = \frac{1}{x-3} + \frac{2x-6}{x-3} = \frac{2x-5}{x-3}$. So $f(x) = \frac{2x-5}{x-3}$.

(b) $f(x) = \frac{2x-5}{x-3}$ has a zero when $2x - 5 = 0$ or $x = \frac{5}{2}$.

(c) $f(x) = \frac{2x-5}{x-3}$ has a horizontal asymptote at $y = 2$ and a vertical asymptote at $x = 3$.

49. (a) $f(x) = x+1+\frac{x^2-x}{x^4+1}$ has an oblique asymptote of $y = x + 1$.

(b) In order to determine when the function intersects the oblique asymptote, we must determine the values of that that make $\frac{x^2-x}{x^4+1} = 0$. Since

$\frac{x^2-x}{x^4+1} = \frac{x(x-1)}{x^4+1}$, and $x(x-1)=0$ when $x = 0$ or $x = 1$, the function crosses its asymptote at $x = 0$ and $x = 1$.

(c) For large values of x, $\frac{x^2-x}{x^4+1} > 0$. Thus as $x \to \infty$, the function approaches its asymptote from above.

51. Function A, because the denominator can never be equal to 0.

53. From the graph, the vertical asymptote is $x = 2$, the horizontal asymptote is $y = 4$, and there is no oblique asymptote. The function is defined for all real numbers x such that $x \neq 2$, therefore the domain is $(-\infty, 2) \cup (2, \infty)$.

55. From the graph, the vertical asymptotes are $x = -2$ and $x = 2$, the horizontal asymptote is $y = -4$, and there is no oblique asymptote. The function is defined for all real numbers x such that $x \neq \pm 2$, therefore the domain is $(-\infty, -2) \cup (-2, 2) \cup (2, \infty)$.

57. From the graph, there is no vertical asymptote, the horizontal asymptote is $y = 0$, and there is no oblique asymptote. The function is defined for all x; therefore, the domain is $(-\infty, \infty)$.

59. From the graph, the vertical asymptote is $x = -1$, there is no horizontal asymptote, and the oblique asymptote passes through the points $(0, -1)$ and $(1, 0)$. Thus, the equation of the oblique asymptote is $y = x - 1$. The function is defined for all $x \neq -1$, therefore the domain is $(-\infty, -1) \cup (-1, \infty)$.

Copyright © 2013 Pearson Education, Inc.

For exercises 61–99, follow the 7 steps outlined on page 346 to sketch the graph. The solution for exercise 61 shows all of the steps. The solutions for exercises 63–99 are abbreviated.

61. $f(x) = \dfrac{x+1}{x-4}$

Step 1: Find the vertical asymptote by setting the denominator equal to 0: $x - 4 = 0 \Rightarrow x = 4$

Step 2: Find the horizontal asymptote. Since the numerator and denominator have the same degree, the horizontal asymptote has equation $y = \dfrac{a_n}{b_n} = \dfrac{1}{1} = 1$. There is no oblique asymptote.

Step 3: Find the y-intercept: $f(0) = \dfrac{0+1}{0-4} = -\dfrac{1}{4}$

Step 4: Find the x-intercept:
$\dfrac{x+1}{x-4} = 0 \Rightarrow x+1 = 0 \Rightarrow x = -1$

Step 5: Determine whether the graph will intersect the horizontal asymptote:

$f(x) = \dfrac{x+1}{x-4} = 1 \Rightarrow x+1 = x-4 \Rightarrow 0 = -5 \Rightarrow$ the graph does not intersect its horizontal asymptote.

Step 6: Plot a point in each of the intervals $(-\infty, -1), (-1, 4),$ and $(4, \infty)$.

Interval	Test Point	Value of $f(x)$	Sign of $f(x)$	Graph Above or Below x-Axis
$(-\infty, -1)$	−5	$\frac{4}{9}$	Positive	Above
$(-1, 4)$	1	$-\frac{2}{3}$	Negative	Below
$(4, \infty)$	5	6	Positive	Above

Step 7: Plot the intercepts and test points to sketch the graph.

$f(x) = \dfrac{x+1}{x-4}$

63. $f(x) = \dfrac{x+2}{x-3}$

vertical asymptote: $x = 3$
horizontal asymptote: $y = 1$

y-intercept: $-\dfrac{2}{3}$; x-intercept: -2

$f(x)$ does not intersect the horizontal asymptote

Interval	Test Point	Value of $f(x)$	Sign of $f(x)$	Graph Above or Below x-Axis
$(-\infty, -2)$	−5	$\frac{3}{8}$	Positive	Above
$(-2, 3)$	1	$-\frac{3}{2}$	Negative	Below
$(3, \infty)$	5	$\frac{7}{2}$	Positive	Above

$f(x) = \dfrac{x+2}{x-3}$

65. $f(x) = \dfrac{4-2x}{8-x}$

vertical asymptote: $x = 8$
horizontal asymptote: $y = 2$

y-intercept: $\dfrac{1}{2}$; x-intercept: 2

$f(x)$ does not intersect the horizontal asymptote

Interval	Test Point	Value of $f(x)$	Sign of $f(x)$	Graph Above or Below x-Axis
$(-\infty, 2)$	−5	$\frac{14}{13}$	Positive	Above
$(2, 8)$	4	−1	Negative	Below
$(8, \infty)$	10	8	Positive	Above

$f(x) = \dfrac{4-2x}{8-x}$

Copyright © 2013 Pearson Education, Inc.

67. $f(x) = \frac{3x}{(x+1)(x-2)}$

vertical asymptotes: $x = -1$, $x = 2$
horizontal asymptote: $y = 0$
y-intercept: 0; x-intercept: 0
$f(x)$ intersects the horizontal asymptote

Interval	Test Point	Value of $f(x)$	Sign of $f(x)$	Graph Above or Below x-Axis
$(-\infty, -1)$	-5	$-\frac{15}{28}$	Negative	Below
$(-1, 0)$	$-\frac{1}{2}$	$\frac{6}{5}$	Positive	Above
$(0, 2)$	1	$-\frac{3}{2}$	Negative	Below
$(2, \infty)$	5	$\frac{5}{6}$	Positive	Above

$f(x) = \dfrac{3x}{x^2 - x - 2}$

69. $f(x) = \dfrac{5x}{x^2 - 1}$

vertical asymptotes: $x = -1$, $x = 1$
horizontal asymptote: $y = 0$
y-intercept: 0; x-intercept: 0
$f(x)$ intersects the horizontal asymptote

Interval	Test Point	Value of $f(x)$	Sign of $f(x)$	Graph Above or Below x-Axis
$(-\infty, -1)$	-2	$-\frac{10}{3}$	Negative	Below
$(-1, 0)$	$-\frac{1}{2}$	$\frac{10}{3}$	Positive	Above
$(0, 1)$	$\frac{1}{2}$	$-\frac{10}{3}$	Negative	Below
$(1, \infty)$	2	$\frac{10}{3}$	Positive	Above

$f(x) = \dfrac{5x}{x^2 - 1}$

71. $f(x) = \dfrac{(x+6)(x-2)}{(x+3)(x-4)}$

vertical asymptotes: $x = -3$, $x = 4$
horizontal asymptote: $y = 1$
y-intercept: 1; x-intercepts: -6, 2
$f(x)$ intersects the horizontal asymptote

Interval	Test Point	Value of $f(x)$	Sign of $f(x)$	Graph Above or Below x-Axis
$(-\infty, -6)$	-8	$\frac{1}{3}$	Positive	Above
$(-6, -3)$	-4	$-\frac{3}{2}$	Negative	Below
$(-3, 2)$	1	$\frac{7}{12}$	Positive	Above
$(2, 4)$	3	$-\frac{3}{2}$	Negative	Below
$(4, \infty)$	6	$\frac{8}{3}$	Positive	Above

$f(x) = \dfrac{(x+6)(x-2)}{(x+3)(x-4)}$

73. $f(x) = \dfrac{3x^2 + 3x - 6}{x^2 - x - 12} = \dfrac{3(x+2)(x-1)}{(x-4)(x+3)}$

vertical asymptotes: $x = -3$, $x = 4$
horizontal asymptote: $y = 3$
y-intercept: $\frac{1}{2}$; x-intercepts: -2, 1
$f(x)$ intersects the horizontal asymptote

(continued on next page)

Copyright © 2013 Pearson Education, Inc.

(*continued*)

Interval	Test Point	Value of $f(x)$	Sign of $f(x)$	Graph Above or Below x-Axis
$(-\infty, -3)$	-4	$\frac{15}{4}$	Positive	Above
$(-3, -2)$	$-\frac{5}{2}$	$-\frac{21}{13}$	Negative	Below
$(-2, 1)$	-1	$\frac{3}{5}$	Positive	Above
$(1, 4)$	2	$-\frac{6}{5}$	Negative	Below
$(4, \infty)$	6	$\frac{20}{3}$	Positive	Above

$$f(x) = \frac{3x^2 + 3x - 6}{x^2 - x - 12}$$

75. $f(x) = \dfrac{9x^2 - 1}{x^2 - 4} = \dfrac{(3x+1)(3x-1)}{(x+2)(x-2)}$

vertical asymptotes: $x = -2$, $x = 2$
horizontal asymptote: $y = 9$

y-intercept: $\frac{1}{4}$; x-intercepts: $-\frac{1}{3}, \frac{1}{3}$

$f(x)$ does not intersect the horizontal asymptote

Interval	Test Point	Value of $f(x)$	Sign of $f(x)$	Graph Above or Below x-Axis
$(-\infty, -2)$	-3	16	Positive	Above
$\left(-2, -\frac{1}{3}\right)$	-1	$-\frac{8}{3}$	Negative	Below
$\left(-\frac{1}{3}, \frac{1}{3}\right)$	$\frac{1}{4}$	$\frac{1}{9}$	Positive	Above
$\left(\frac{1}{3}, 2\right)$	1	$-\frac{8}{3}$	Negative	Below
$(2, \infty)$	3	16	Positive	Above

$$f(x) = \frac{9x^2 - 1}{x^2 - 4}$$

77. $f(x) = \dfrac{(x-3)(x+1)}{(x-1)^2}$

vertical asymptote: $x = 1$
horizontal asymptote: $y = 1$
y-intercept: -3; x-intercepts: -1, 3
$f(x)$ intersects the horizontal asymptote

Interval	Test Point	Value of $f(x)$	Sign of $f(x)$	Graph Above or Below x-Axis
$(-\infty, -1)$	-2	$\frac{5}{9}$	Positive	Above
$(-1, 1)$	0	-3	Negative	Below
$(1, 3)$	2	-3	Negative	Below
$(3, \infty)$	4	$\frac{5}{9}$	Positive	Above

$$f(x) = \frac{(x-3)(x+1)}{(x-1)^2}$$

79. $f(x) = \dfrac{x}{x^2 - 9}$

vertical asymptotes: $x = -3$, $x = 3$
horizontal asymptote: $y = 0$
y-intercept: 0; x-intercept: 0
$f(x)$ intersects the horizontal asymptote

Interval	Test Point	Value of $f(x)$	Sign of $f(x)$	Graph Above or Below x-Axis
$(-\infty, -3)$	-4	$-\frac{4}{7}$	Negative	Below
$(-3, 0)$	-1	$\frac{1}{8}$	Positive	Above

(continued on next page)

Copyright © 2013 Pearson Education, Inc.

(*continued*)

Interval	Test Point	Value of $f(x)$	Sign of $f(x)$	Graph Above or Below x-Axis
$(0, 3)$	1	$-\frac{1}{8}$	Negative	Below
$(3, \infty)$	4	$\frac{4}{7}$	Positive	Above

$$f(x) = \frac{x}{x^2 - 9}$$

81. $f(x) = \dfrac{1}{x^2 + 1}$

vertical asymptote: none
horizontal asymptote: $y = 0$
y-intercept: 1; x-intercept: none
$f(x)$ does not intersect the horizontal asymptote

$$f(x) = \frac{1}{x^2 + 1}$$

83. $f(x) = \dfrac{(x+4)^2}{(x-1)(x+5)}$

vertical asymptotes: $x = -5$, $x = 1$
horizontal asymptote: $y = 1$
y-intercept: $-\frac{16}{5}$; x-intercept: -4

$f(x)$ intersects the horizontal asymptote

Interval	Test Point	Value of $f(x)$	Sign of $f(x)$	Graph Above or Below x-Axis
$(-\infty, -5)$	-6	$\frac{4}{7}$	Positive	Above
$(-5, -4)$	$-\frac{9}{2}$	$-\frac{1}{11}$	Negative	Below
$(-4, 1)$	-2	$-\frac{4}{9}$	Negative	Below
$(1, \infty)$	2	$\frac{36}{7}$	Positive	Above

$$f(x) = \frac{(x+4)^2}{(x-1)(x+5)}$$

85. $f(x) = \dfrac{20 + 6x - 2x^2}{8 + 6x - 2x^2} = \dfrac{(x-5)(x+2)}{(x-4)(x+1)}$

vertical asymptotes: $x = -1$, $x = 4$
horizontal asymptote: $y = 1$

y-intercept: $\frac{5}{2}$; x-intercepts: $-2, 5$

$f(x)$ does not intersect the horizontal asymptote

Interval	Test Point	Value of $f(x)$	Sign of $f(x)$	Graph Above or Below x-Axis
$(-\infty, -2)$	-4	$\frac{3}{4}$	Positive	Above
$(-2, -1)$	$-\frac{3}{2}$	$-\frac{13}{11}$	Negative	Below
$(-1, 4)$	2	2	Positive	Above
$(4, 5)$	$\frac{9}{2}$	$-\frac{13}{11}$	Negative	Below
$(5, \infty)$	6	$\frac{4}{7}$	Positive	Above

$$f(x) = \frac{20 + 6x - 2x^2}{8 + 6x - 2x^2}$$

87. $f(x) = \dfrac{x^2 + 1}{x + 3}$

vertical asymptote: $x = -3$
oblique asymptote: $y = x - 3$
y-intercept: $\frac{1}{3}$; x-intercepts: none

$f(x)$ does not intersect the oblique asymptote

(*continued on next page*)

Copyright © 2013 Pearson Education, Inc.

(*continued*)

Interval	Test Point	Value of $f(x)$	Sign of $f(x)$	Graph Above or Below x-Axis
$(-\infty, -3)$	-4	-17	Negative	Below
$(-3, \infty)$	4	$\frac{17}{7}$	Positive	Above

$$f(x) = \frac{x^2 + 1}{x + 3}$$

89. $f(x) = \dfrac{x^2 + 2x}{2x - 1} = \dfrac{x(x+2)}{2x-1}$

vertical asymptote: $x = \frac{1}{2}$

oblique asymptote: $y = \frac{x}{2} + \frac{5}{4}$

y-intercept: 0; x-intercepts: $-2, 0$

$f(x)$ does not intersect the oblique asymptote

Interval	Test Point	Value of $f(x)$	Sign of $f(x)$	Graph Above or Below x-Axis
$(-\infty, -2)$	-3	$-\frac{3}{7}$	Negative	Below
$(-2, 0)$	-1	$\frac{1}{3}$	Positive	Above
$\left(0, \frac{1}{2}\right)$	$\frac{1}{4}$	$-\frac{9}{8}$	Negative	Below
$\left(\frac{1}{2}, \infty\right)$	2	$\frac{8}{3}$	Positive	Above

$$f(x) = \frac{x^2 + 2x}{2x - 1}$$

91. $f(x) = \dfrac{x^2 - 9}{x + 3} = \dfrac{(x+3)(x-3)}{x+3}$

The function degenerates into the line
$f(x) = x - 3, x \neq -3$

$$f(x) = \frac{x^2 - 9}{x + 3}$$

93. $f(x) = \dfrac{2x^2 - 5x - 2}{x - 2}$

vertical asymptote: $x = 2$
oblique asymptote: $y = 2x - 1$
y-intercept: 1; x-intercepts: $\approx -.4, \approx 2.9$
$f(x)$ does not intersect the oblique asymptote

Interval	Test Point	Value of $f(x)$	Sign of $f(x)$	Graph Above or Below x-Axis
$(-\infty, -.4)$	-2	-4	Negative	Below
$(-.4, 2)$	1	5	Positive	Above
$(2, 2.9)$	2.5	-4	Negative	Below
$(2.9, \infty)$	4	5	Positive	Above

$$f(x) = \frac{2x^2 - 5x - 2}{x - 2}$$

95. $f(x) = \dfrac{x^2 - 1}{x^2 - 4x + 3} = \dfrac{(x+1)(x-1)}{(x-3)(x-1)}$

vertical asymptote: $x = 3$
horizontal asymptote: $y = 1$

y-intercept: $-\frac{1}{3}$; x-intercept: -1

$f(x)$ is not defined for $x = 1$, so there is a hole in the graph.
$f(x)$ does not intersect the horizontal asymptote

(*continued on next page*)

Copyright © 2013 Pearson Education, Inc.

(continued)

Interval	Test Point	Value of $f(x)$	Sign of $f(x)$	Graph Above or Below x-Axis
$(-\infty, -1)$	-4	$\frac{3}{7}$	Positive	Above
$(-1, 1)$	$\frac{1}{2}$	$-\frac{3}{5}$	Negative	Below
$(1, 3)$	2	-3	Negative	Below
$(3, \infty)$	4	5	Positive	Above

$$f(x) = \frac{x^2 - 1}{x^2 - 4x + 3}$$

97. $f(x) = \dfrac{(x^2 - 9)(2 + x)}{(x^2 - 4)(3 + x)} = \dfrac{(x+3)(x-3)(x+2)}{(x+2)(x-2)(x+3)}$

vertical asymptote: $x = 2$

horizontal asymptote: $y = 1$

y-intercept: $\frac{3}{2}$; x-intercept: 3

$f(x)$ is not defined for $x = -3$ and $x = -2$, so there are two holes in the graph.

$f(x)$ does not intersect the horizontal asymptote

Interval	Test Point	Value of $f(x)$	Sign of $f(x)$	Graph Above or Below x-Axis
$(-\infty, -3)$	-4	$\frac{7}{6}$	Positive	Above
$(-3, -2)$	$-\frac{5}{2}$	$\frac{11}{9}$	Positive	Above
$(-2, 2)$	1	2	Positive	Above
$(2, 3)$	$\frac{5}{2}$	-1	Negative	Below
$(3, \infty)$	4	$\frac{1}{2}$	Positive	Above

$$f(x) = \frac{(x^2 - 9)(2 + x)}{(x^2 - 4)(3 + x)}$$

99. $f(x) = \dfrac{x^4 - 20x^2 + 64}{x^4 - 10x^2 + 9}$

$\qquad = \dfrac{(x+4)(x-4)(x+2)(x-2)}{(x+3)(x-3)(x+1)(x-3)}$

vertical asymptotes: $x = -3$, $x = -1$, $x = 1$, $x = 3$

horizontal asymptote: $y = 1$

y-intercept: $\frac{64}{9}$; x-intercepts: $-4, -2, 2, 4$

$f(x)$ intersects the horizontal asymptote

Interval	Test Point	Value of $f(x)$	Sign of $f(x)$	Graph Above or Below x-Axis
$(-\infty, -4)$	-5	0.49	Positive	Above
$(-4, -3)$	-3.5	-0.85	Negative	Below
$(-3, -2)$	-2.5	1.52	Positive	Above
$(-2, -1)$	-1.5	-2.85	Negative	Below
$(-1, 1)$	$.5$	9	Positive	Above
$(1, 2)$	1.5	-2.85	Negative	Below
$(2, 3)$	2.5	1.52	Positive	Above
$(3, 4)$	3.5	-0.85	Negative	Below
$(4, \infty)$	5	0.49	Positive	Above

$$f(x) = \frac{x^4 - 20x^2 + 64}{x^4 - 10x^2 + 9}$$

101. The graph has a vertical asymptote, $x = 2$, so the factor $x - 2$ is in the denominator of the function. The horizontal asymptote is $y = 1$, so the numerator and denominator have the same degree. There is a "hole" in the graph at $x = -2$, so the factor $x + 2$ is in the denominator and numerator also. The x-intercept is 3, so that when $f(x) = 0$, $x = 3$. This condition exists if $x - 3$ is a factor of the numerator. Putting these conditions together, we have a possible function

$$f(x) = \frac{(x-3)(x+2)}{(x-2)(x+2)} \text{ or } f(x) = \frac{x^2 - x - 6}{x^2 - 4}.$$

Copyright © 2013 Pearson Education, Inc.

103. The graph has vertical asymptotes at $x = 4$ and $x = 0$, so $x - 4$ and x are factors in the denominator of the function. The only x-intercept is 2, so that when $f(x) = 0$, $x = 2$. This condition exists if $x - 2$ is a factor of the numerator. The graph has a horizontal asymptote $y = 0$, so the degree of the denominator is larger than the degree of the numerator. Putting these conditions together, we have a possible function

$$f(x) = \frac{x-2}{x(x-4)} \text{ or } f(x) = \frac{x-2}{x^2-4x}.$$

105. The graph has a vertical asymptote $x = 1$ so $x - 1$ is a factor in the denominator of the function. The x-intercepts are 0 and 2, so that when $f(x) = 0$, $x = 0$ or $x = 2$. Such conditions exist if x and $x - 2$ are factors of the numerator. The horizontal asymptote is $y = -1$, so the numerator and denominator have the same degree. Since the numerator will have degree 2, we must make the denominator also have degree 2. Also, from the horizontal asymptote, we have $y = -1 = \frac{a_n}{b_n}$. Putting these conditions together, we have a possible function

$$f(x) = \frac{-x(x-2)}{(x-1)^2} \text{ or } f(x) = \frac{-x^2+2x}{x^2-2x+1}.$$

107. Several answers are possible. One answer is

$$f(x) = \frac{(x-3)(x+1)}{(x-1)^2}.$$

109. $f(x) = \dfrac{x+1}{x-4}$

From the last screen we can see that $f(1.25) = -0.\overline{81}$. If you wanted the fractional equivalent, you can go back to your homescreen and obtain Y_1 from the VARS menu. Thus $f(1.25) = -\frac{9}{11}$.

111. $f(x) = \dfrac{x^2+2x}{2x-1}$

From the last screen we can see that $f(1.25) = 2.708\overline{3}$. If you wanted the fractional equivalent, you can go back to your homescreen and obtain Y_1 from the VARS menu. Thus $f(1.25) = \frac{65}{24}$.

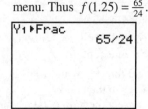

113. (a) $T(r) = \dfrac{2r-k}{2r^2-2kr}; k = 25 \Rightarrow$

$$T(r) = \frac{2r-25}{2r^2-2(25)r} = \frac{2r-25}{2r^2-50r}$$

Graph $Y_1 = \frac{2x-25}{2x^2-50x}$ and $Y_2 = 0.5$ (since 30 sec $= 0.5$ min) on the same screen.

(continued on next page)

Copyright © 2013 Pearson Education, Inc.

(*continued*)

Using the "intersect" option in the CALC menu, we find that the graphs intersect at $x \approx 26$, which represents $r \approx 26$ in the given function. Therefore, there must be an average admittance rate of 26 vehicles per minute.

(b) $\frac{26}{5.3} \approx 4.9$ or 5 parking attendants must be on duty to keep the wait less than 30 seconds.

115. (a) Graph

$$y = d(x) = \frac{8710x^2 - 69,400x + 470,000}{1.08x^2 - 324x + 82,200}$$

and $y = 300$ on the same calculator screen.

The graphs intersect when $x \approx 52.1$ miles per hour.

(b)

x	$d(x)$	x	$d(x)$
20	**34**	50	**273**
25	**56**	55	**340**
30	**85**	60	**415**
35	**121**	65	**499**
40	**164**	70	**591**
45	**215**		

(c) By comparing values in the table generated in part b, it appears that when the speed is doubled, the breaking distance is more than doubled.

(d) If the breaking distance doubled whenever the speed doubled, then there would be a linear relationship between the speed and the stopping distance.

117. $R(x) = \dfrac{80x - 8000}{x - 110}$

(a) $R(55) = \dfrac{80(55) - 8000}{55 - 110}$
$\approx \$65.5$ tens of millions

(b) $R(60) = \dfrac{80(60) - 8000}{60 - 110}$
$= \$64$ tens of millions

(c) $R(70) = \dfrac{80(70) - 8000}{70 - 110}$
$= \$60$ tens of millions

(d) $R(90) = \dfrac{80(90) - 8000}{90 - 110}$
$= \$40$ tens of millions

(e) $R(100) = \dfrac{80(100) - 8000}{100 - 110} = \$0$

(f)

119. Since the degree of the numerator equals the degree of the denominator in

$f(x) = \dfrac{x^4 - 3x^3 - 21x^2 + 43x + 60}{x^4 - 6x^3 + x^2 + 24x - 20}$, the graph has a

horizontal asymptote at $y = \frac{1}{1} = 1$.

121. (a) Use synthetic division where

$g(x) = x^4 - 6x^3 + x^2 + 24x - 20$

and $k = 1$.

$$\begin{array}{r|rrrrr} 1 & 1 & -6 & 1 & 24 & -20 \\ & & 1 & -5 & -4 & 20 \\ \hline & 1 & -5 & -4 & 20 & 0 \end{array}$$

Now use synthetic division where

$h(x) = x^3 - 5x^2 - 4x + 20$ and $k = 2$.

$$\begin{array}{r|rrrr} 2 & 1 & -5 & -4 & 20 \\ & & 2 & -6 & -20 \\ \hline & 1 & -3 & -10 & 0 \end{array}$$

(*continued on next page*)

Copyright © 2013 Pearson Education, Inc.

(*continued*)

The resulting polynomial quotient is $x^2 - 3x - 10$, which factors to $(x+2)(x-5)$. Thus, the complete factorization of the denominator would be $(x-1)(x-2)(x+2)(x-5)$.

(b) $f(x) = \dfrac{(x+4)(x+1)(x-3)(x-5)}{(x-1)(x-2)(x+2)(x-5)}$

123. Although $x - 5$ is a factor of the numerator, it will not yield an x-intercept because it is also a factor of the denominator. The other three factors of the numerator, namely $(x + 4)$, $(x + 1)$, and $(x - 3)$ will yield x-intercepts of -4, -1, and 3 because any of these will yield $f(x) = 0$.

125. Although $x - 5$ is a factor of the denominator, it will not yield a vertical asymptote because it is also a factor of the numerator. The other three factors of the denominator (namely $x - 1$, $x - 2$, and $x + 2$) will yield vertical asymptotes of $x = 1$, $x = 2$, and $x = -2$ because any of these will yield a denominator of zero.

127. The vertical asymptotes are $x = 1$, $x = 2$, and $x = -2$ and the x-intercepts occur at -4, -1, and 3. We should also consider the "hole" in the graph which occurs at $x = 5$ This divides the real number line into eight intervals.

Interval	Test Point	Value of $f(x)$	Sign of $f(x)$	Graph Above or Below x-Axis
$(-\infty, -4)$	-5	$\frac{16}{63}$	Positive	Above
$(-4, -2)$	-3	$-\frac{3}{5}$	Negative	Below
$(-2, -1)$	$-\frac{3}{2}$	$\frac{9}{7}$	Positive	Above
$(-1, 1)$	0	-3	Negative	Below
$(1, 2)$	$\frac{3}{2}$	$\frac{165}{7}$	Positive	Above
$(2, 3)$	$\frac{5}{2}$	$-\frac{91}{27}$	Negative	Below
$(3, 5)$	4	$\frac{10}{9}$	Positive	Above
$(5, \infty)$	6	$\frac{21}{16}$	Positive	Above

$$f(x) = \dfrac{x^4 - 3x^3 - 21x^2 + 43x + 60}{x^4 - 6x^3 + x^2 + 24x - 20}$$

Chapter 3 Quiz
(Sections 3.1–3.5)

1. (a) $f(x) = -2(x+3)^2 - 1$

This equation is of the form $y = a(x-h)^2 + k$, with $h = -3$, $k = -1$, and $a = -2$. The graph opens downward and is narrower than $y = x^2$. It is a horizontal translation of the graph of $y = -2x^2$, 3 units to the left and 1 unit down. The vertex is $(-3, -1)$. The axis is $x = -3$. The domain is $(-\infty, \infty)$. Since the largest value of y is -1 and the graph opens downward, the range is $(-\infty, -1]$. The function is increasing on $(-\infty, -3]$ and decreasing on $[-3, \infty)$.

$f(x) = -2(x + 3)^2 - 1$

(b) $f(x) = 2x^2 - 8x + 3$

Rewrite by completing the square on x

$f(x) = 2x^2 - 8x + 3$

$\qquad = 2(x^2 - 4x + 4) + (3 - 8)$

Note: $2\left[\frac{1}{2}(-4)\right]^2 = 2(-2)^2 = 8$

$\qquad = 2(x - 2)^2 - 5$

This equation is of the form $y = a(x-h)^2 + k$, with $h = 2$, $k = -5$, and $a = 2$. The graph opens upward and is narrower than $y = x^2$.

(*continued on next page*)

Copyright © 2013 Pearson Education, Inc.

(*continued*)

It is a horizontal translation of the graph of $y = 2x^2$, 2 units to the right and 5 units down. The vertex is (2, –5). The axis is $x = 2$. The domain is $(-\infty, \infty)$. Since the smallest value of y is –5 and the graph opens upward, the range is $[-5, \infty)$. The function is increasing on $[2, \infty)$ and decreasing on $(-\infty, 2]$.

$f(x) = 2x^2 - 8x + 3$

3. $f(x) = 2x^4 + x^3 - 3x + 4;\ k = 2$

Use synthetic division to write the polynomial in the form $f(x) = (x - k)q(x) + r$. Since the x^2-term is missing, include 0 as its coefficient.

$$\begin{array}{r|rrrrr} 2 & 2 & 1 & 0 & -3 & 4 \\ & & 4 & 10 & 20 & 34 \\ \hline & 2 & 5 & 10 & 17 & 38 \end{array}$$

$f(x) = (x - 2)(2x^3 + 5x^2 + 10x + 17) + 38$

The remainder is 38, so $k = 2$ is not a zero of the function. $k(2) = 38$.

5. Since $3 - i$ is a factor, $3 + i$ is also a factor. So
$f(x)$
$= \left[(x - (3 - i)]\left[(x - (3 + i)\right]\left[(x - (-2)\right](x - 3)$
$= (x - 3 + i)(x - 3 - i)(x + 2)(x - 3)$
$= (x^2 - 6x + 1)(x^2 - x - 6)$
$= x^4 - 7x^3 + 10x^2 + 26x - 60$

7. $f(x) = 2x^4 - 9x^3 - 5x^2 + 57x - 45$

Step 1: p must be a factor of $a_0 = -45$ and q must be a factor of $a_4 = 2$. Thus, p can be ± 1, ± 3, ± 5, ± 9, ± 15, or ± 45, and q can be ± 1 or ± 2. The possible rational zeros, $\frac{p}{q}$, are

$\pm\frac{1}{2}, \pm 1,\ \pm\frac{3}{2}, \pm\frac{5}{2}, \pm 3, \pm\frac{9}{2}, \pm 5, \pm\frac{15}{2}, \pm 9,\ \pm 15,$

$\pm\frac{45}{2}$, or ± 45. The remainder theorem shows that 1 is a zero.

$$\begin{array}{r|rrrrr} 1 & 2 & -9 & -5 & 57 & -45 \\ & & 2 & -7 & -12 & 45 \\ \hline & 2 & -7 & -12 & 45 & 0 \end{array}$$

The new quotient polynomial is
$2x^3 - 7x^2 - 12x + 45$.

p must be a factor of $a_0 = -45$ and q must be a factor of $a_3 = 2$. Thus, p can be ± 1, ± 3, ± 5, ± 9, ± 15, or ± 45, and q can be ± 1 or ± 2. The possible rational zeros, $\frac{p}{q}$, are $\pm\frac{1}{2}, \pm 1,$

$\pm\frac{3}{2}, \pm\frac{5}{2}, \pm 3, \pm\frac{9}{2}, \pm 5, \pm\frac{15}{2},\ \pm 9,\ \pm 15, \pm\frac{45}{2}$, or ± 45. The remainder theorem shows that 3 is a zero.

$$\begin{array}{r|rrrr} 3 & 2 & -7 & -12 & 45 \\ & & 6 & -3 & -45 \\ \hline & 2 & -1 & -15 & 0 \end{array}$$

The new quotient polynomial is
$2x^2 - x - 15 = 0$.
$(2x + 5)(x - 3) = 0$

$2x + 5 = 0 \Rightarrow x = -\frac{5}{2}\quad$ or $\quad x - 3 = 0 \Rightarrow x = 3$

The rational zeros are $-\frac{5}{2}$, 1, and 3. Note that 3 is a zero of multiplicity 2, so the graph is tangent to the x-axis at $x = 3$. The factors are $2x + 5$, $x - 1$, and $(x - 3)^2$ and thus

$f(x) = (2x + 5)(x - 1)(x - 3)^2$.

Step 2: $f(0) = -45$, so plot $(0, -45)$.

Step 3: The x-intercepts divide the x-axis into four intervals:

Interval	Test Point	Value of $f(x)$	Sign of $f(x)$	Graph Above or Below x-Axis
$\left(-\infty, -\frac{5}{2}\right)$	–3	144	Positive	Above
$\left(-\frac{5}{2}, 1\right)$	–1	–96	Negative	Below
$(1, 3)$	2	9	Positive	Above
$(3, \infty)$	4	39	Positive	Above

Plot the x-intercepts, y-intercept and test points (the y-intercept is one of the test points) with a smooth curve to get the graph.

$f(x) = 2x^4 - 9x^3 - 5x^2 + 57x - 45$
$\quad = (x - 3)^2(2x + 5)(x - 1)$

Copyright © 2013 Pearson Education, Inc.

9. $f(x) = \dfrac{3x+1}{x^2+7x+10} = \dfrac{3x+1}{(x+5)(x+2)}$

Step 1: The graph has vertical asymptotes where the denominator equals zero, that is, at $x = -5$ and $x = -2$.

Step 2: Since the degree of the numerator is less than the degree of the denominator, the graph has a horizontal asymptote at $y = 0$ (the x-axis).

Step 3: The y-intercept is

$$f(0) = \dfrac{3(0)+1}{0^2 + 7(0) + 10} = \dfrac{1}{10}$$

Step 4: Find any x-intercepts by solving $f(x) = 0$:

$$\dfrac{3x+1}{x^2+7x+10} = \dfrac{3x+1}{(x+2)(x+5)} = 0 \Rightarrow$$

$$3x + 1 = 0 \Rightarrow x = -\tfrac{1}{3}$$

Step 5: The graph will intersect the horizontal asymptote, $y = 0$, when $f(x) = 0$. From step 4, that is at $x = -\tfrac{1}{3}$.

Step 6: Plot a point in each of the intervals determined by the x-intercept and the vertical asymptotes. There are four intervals,

$(-\infty, -5), (-5, -2), \left(-2, -\tfrac{1}{3}\right),$ and $\left(-\tfrac{1}{3}, \infty\right)$.

Interval	Test Point	Value of $f(x)$	Sign of $f(x)$	Graph Above or Below x-Axis
$(-\infty, -5)$	-6	-4.25	Negative	Below
$(-5, -2)$	-4	5.5	Positive	Above
$\left(-2, -\tfrac{1}{3}\right)$	-1	-0.5	Negative	Below
$\left(-\tfrac{1}{3}, \infty\right)$	3	0.25	Positive	Above

Use the asymptotes, intercepts, and these points to sketch the graph.

$f(x) = \dfrac{3x+1}{x^2+7x+10}$

Summary Exercises on Solving Equations and Inequalities

1. $\dfrac{5x+8}{-2} = 2x - 10$

$5x + 8 = -4x + 20$

$9x = 12$

$x = \dfrac{12}{9} = \dfrac{4}{3}$

Solution set: $\left\{\tfrac{4}{3}\right\}$

3. $(x-5)^{-4} - 13(x-5)^{-2} = -36$

Let $u = (x-5)^{-2}$. Then we have

$$u^2 - 13u = -36$$

$$u^2 - 13u + 36 = 0$$

$$(u-4)(u-9) = 0 \Rightarrow u = 4, 9$$

Now substitute.

If $u = 4$, then

$$4 = (x-5)^{-2} \Rightarrow \dfrac{1}{4} = (x-5)^2 \Rightarrow$$

$$\pm\sqrt{\dfrac{1}{4}} = x - 5 \Rightarrow \pm\dfrac{1}{2} = x - 5 \Rightarrow$$

$$x = 5 \pm \dfrac{1}{2} = \dfrac{11}{2} \text{ or } \dfrac{9}{2}$$

If $u = 9$, then

$$9 = (x-5)^{-2} \Rightarrow \dfrac{1}{9} = (x-5)^2 \Rightarrow \pm\sqrt{\dfrac{1}{9}} = x - 5$$

$$\pm\dfrac{1}{3} = x - 5 \Rightarrow x = 5 \pm \dfrac{1}{3} = \dfrac{16}{3} \text{ or } \dfrac{14}{3}$$

Solution set: $\left\{\tfrac{9}{2}, \tfrac{11}{2}, \tfrac{14}{3}, \tfrac{16}{3}\right\}$

5. $\sqrt{2x-5} - \sqrt{x-3} = 1$

$$\sqrt{2x-5} = 1 + \sqrt{x-3}$$

$$\left(\sqrt{2x-5}\right)^2 = \left(1 + \sqrt{x-3}\right)^2$$

$$2x - 5 = 1 + 2\sqrt{x-3} + x - 3$$

$$x - 3 = 2\sqrt{x-3}$$

$$(x-3)^2 = \left(2\sqrt{x-3}\right)^2$$

$$x^2 - 6x + 9 = 4x - 12$$

$$x^2 - 10x + 21 = 0$$

$$(x-3)(x-7) = 0 \Rightarrow x = 3, \; x = 7$$

Be sure to check each proposed solution since extraneous solutions may have been introduced when both sides of the equation were squared.

Solution set: $\{3, 7\}$

Copyright © 2013 Pearson Education, Inc.

7. $x^{2/3} + \dfrac{1}{2} = \dfrac{3}{4} \Rightarrow x^{2/3} = \dfrac{1}{4} \Rightarrow$

$\left(x^{2/3}\right)^{3/2} = \pm\left(\dfrac{1}{4}\right)^{3/2} \Rightarrow x = \left(\pm\dfrac{1}{2}\right)^3 = \pm\dfrac{1}{8}$

Be sure to check each proposed solution since extraneous solutions may have been introduced when both sides of the equation were raised to the power.

Solution set: $\left\{\pm\frac{1}{8}\right\}$

9. (a) $f(x) > 0$ for $\left\{x \mid x \in (-1, 1) \cup (1, \infty)\right\}$

(b) $f(x) \le 0$ for $\left\{x \mid x \in (-\infty, -1] \cup \{1\}\right\}$

11. (a) $f(x) = 0$ for $x = 0$

(b) $f(x) > 0$ for $\left\{x \mid x \in (-3, 0) \cup (3, \infty)\right\}$

For exercise 13, review the characteristics of the graph of a quadratic function outlined on page 289. For exercises 15–21, follow the three steps outlined on page 323 or the seven steps outlined on page 346 (as appropriate) to sketch the graph. The solutions shown here may be abbreviated.

13. $25x^2 - 20x + 4 > 0$

The expression is an inequality. We will graph the associated function

$f(x) = 25x^2 - 20x + 4$, which is a parabola.

$h = -\dfrac{b}{2a} = -\dfrac{-20}{2(25)} = \dfrac{2}{5}$ and

$k = f(h) = f\left(\dfrac{2}{5}\right) = 25\left(\dfrac{2}{5}\right)^2 - 20\left(\dfrac{2}{5}\right) + 4 = 0.$

Thus, the vertex is $\left(\dfrac{2}{5}, 0\right)$. The parabola opens upward and is narrower than the graph of $y = x^2$. Since the vertex is the x-intercept and the parabola opens upward, all values of the function except the value for $x = \dfrac{2}{5}$ are greater than 0. The graph is given for additional explanation. Since

$f(0) = 25 \cdot 0^2 - 20 \cdot 0 + 4 = 4,$ the

y-intercept is $(0, 4)$. Another point on the graph is $\left(\dfrac{4}{5}, 4\right)$.

$f(x) = 25x^2 - 20x + 4$

15. $x^4 - 2x^3 - 3x^2 + 4x + 4 = 0$

The expression is an equation.

Step 1: Find the zeros of the function $f(x) = x^4 - 2x^3 - 3x^2 + 4x + 4$. p must be a factor of $a_0 = 4$ and q must be a factor of $a_4 = 1$. Thus, p can be ± 1, ± 2, or ± 4, and q can be ± 1. The possible rational zeros, $\dfrac{p}{q}$, are ± 1, ± 2, or ± 4. The remainder theorem shows that 2 is a zero.

$$\begin{array}{r|rrrrr} 2) & 1 & -2 & -3 & 4 & 4 \\ & & 2 & 0 & -6 & -4 \\ \hline & 1 & 0 & -3 & -2 & 0 \end{array}$$

The new quotient polynomial is $x^3 - 3x - 2$. We must find the zeros of this function. p must be a factor of $a_0 = -2$ and q must be a factor of $a_3 = 1$.

Thus, p can be ± 1 or ± 2, and q can be ± 1. The possible rational zeros, $\dfrac{p}{q}$, are ± 1 or ± 2. The remainder theorem shows that -1 is a zero.

$$\begin{array}{r|rrrr} -1) & 1 & 0 & -3 & -2 \\ & & -1 & 1 & 2 \\ \hline & 1 & -1 & -2 & 0 \end{array}$$

The new quotient polynomial is $x^2 - x - 2$.

$x^2 - x - 2 = 0 \Rightarrow (x+1)(x-2) = 0 \Rightarrow$
$x = -1, 2$

The rational zeros are -1 (multiplicity 2) and 2 (multiplicity 2). The solution set of the equation is $\{-1, 2\}$. We now complete the graphing process.

Step 2: $f(0) = 4$, so the y-intercept is $(0, 4)$.

(continued on next page)

Copyright © 2013 Pearson Education, Inc.

(*continued*)

Step 3: The *x*-intercepts divide the *x*-axis into three intervals.

Interval	Test Point	Value of $f(x)$	Sign of $f(x)$	Graph Above or Below *x*-Axis
$(-\infty, -1)$	-2	16	Positive	Above
$(-1, 2)$	1	4	Positive	Above
$(2, \infty)$	3	16	Positive	Above

Plot the *x*-intercepts, *y*-intercept, and test points with a smooth curve to get the graph.

$f(x) = x^4 - 2x^3 - 3x^2 + 4x + 4$

17. $-x^4 - x^3 + 12x^2 = 0$

The expression is an equation.
Step 1: Find the zeros of the function
$f(x) = -x^4 - x^3 + 12x^2 = 0$.

$$-x^4 - x^3 + 12x^2 = 0$$
$$-x^2 \left(x^2 + x - 12 \right) = 0$$
$$x^2 (x+4)(x-3) = 0 \Rightarrow x = 0, -4, 3$$

The solution set of the equation is $\{0, -4, 3\}$.
We now complete the graphing process.
Step 2: $f(0) = 0$, so the *y*-intercept is $(0, 0)$.
Step 3: The *x*-intercepts divide the *x*-axis into four intervals.

Interval	Test Point	Value of $f(x)$	Sign of $f(x)$	Graph Above or Below *x*-Axis
$(-\infty, -4)$	-5	-200	Negative	Below
$(-4, 0)$	-1	12	Positive	Above
$(0, 3)$	1	10	Positive	Above
$(3, \infty)$	4	-128	Negative	Below

Plot the *x*-intercepts, *y*-intercept, and test points with a smooth curve to get the graph.

$f(x) = -x^4 - x^3 + 12x^2$

19. $\dfrac{2x^2 - 13x + 15}{x^2 - 3x} \geq 0$

The expression is an inequality. We will graph

$$f(x) = \frac{2x^2 - 13x + 15}{x^2 - 3x}$$

Step 1: Find the vertical asymptotes by setting the denominator equal to 0.

$x^2 - 3x = 0 \Rightarrow x(x-3) = 0 \Rightarrow x = 0, \ x = 3$

The vertical asymptotes are $x = 0$ (the *y*-axis) and $x = 3$.
Step 2: Find the horizontal asymptote. Since the numerator and denominator have the same degree, the horizontal asymptote has equation

$y = \frac{a_n}{b_n} = \frac{2}{1} = 2$. There is no oblique

asymptote.
Step 3: There is no the *y*-intercept since the *y*-axis is one of the vertical asymptotes.
Step 4: Find the *x*-intercept:

$$\frac{2x^2 - 13x + 15}{x^2 - 3x} = 0 \Rightarrow 2x^2 - 13x + 15 = 0 \Rightarrow$$
$$(2x-3)(x-5) = 0 \Rightarrow x = \tfrac{3}{2}, \ 5$$

The *x*-intercepts are $\left(\tfrac{3}{2}, 0 \right)$ and $(5, 0)$.

Step 5: Determine whether the graph will intersect the horizontal asymptote:

$$\frac{2x^2 - 13x + 15}{x^2 - 3x} = 2 \Rightarrow$$
$$2x^2 - 13x + 15 = 2x^2 - 6x \Rightarrow 15 = 7x \Rightarrow$$
$$x = \tfrac{15}{7}$$

$f(x)$ intersects the horizontal asymptote
Step 6: Plot a point in each of the intervals
$(-\infty, 0), \left(0, \tfrac{3}{2}\right), \left(\tfrac{3}{2}, 3\right), (3, 5)$, and $(5, \infty)$.

(*continued on next page*)

Copyright © 2013 Pearson Education, Inc.

(continued)

Interval	Test Point	Value of $f(x)$	Sign of $f(x)$	Graph Above or Below x-Axis
$(-\infty, 0)$	-1	$\frac{15}{2}$	Positive	Above
$\left(0, \frac{3}{2}\right)$	1	-2	Negative	Below
$\left(\frac{3}{2}, 3\right)$	2	$\frac{3}{2}$	Positive	Above
$(3, 5)$	4	$-\frac{5}{4}$	Negative	Below
$(5, \infty)$	6	$\frac{1}{2}$	Positive	Above

$$f(x) = \frac{2x^2 - 13x + 15}{x^2 - 3x}$$

By examining the graph, we find that the solution set for the original inequality is

$$\left(-\infty, 0\right) \cup \left[\frac{3}{2}, 3\right) \cup \left[5, \infty\right).$$

21. $\dfrac{x-1}{(x-3)^2} = 0$

The expression is an equation. We will graph

$$f(x) = \frac{x-1}{(x-3)^2}$$

Step 1: Find the vertical asymptotes by setting the denominator equal to 0.

$$(x-3)^2 = 0 \Rightarrow x - 3 = 0 \Rightarrow x = 3$$

The vertical asymptote is $x = 3$.

Step 2: Since the numerator has lesser degree than the denominator, the x-axis is the horizontal asymptote. There is no oblique asymptote.

Step 3: Find the y-intercept:

$$f(0) = \frac{0-1}{(0-3)^2} = -\frac{1}{9}$$

Step 4: Find the x-intercept:

$$\frac{x-1}{(x-3)^2} = 0 \Rightarrow x - 1 = 0 \Rightarrow x = 1$$

The x-intercept is $(1, 0)$.

Step 5: The graph will intersect the horizontal asymptote.

Step 6: Plot a point in each of the intervals $(-\infty, 1)$, $(1, 3)$, and $(3, \infty)$.

Interval	Test Point	Value of $f(x)$	Sign of $f(x)$	Graph Above or Below x-Axis
$(-\infty, 1)$	-2	$-\frac{3}{25}$	Negative	Below
$(1, 3)$	2	1	Positive	Above
$(3, \infty)$	4	3	Positive	Above

$$f(x) = \frac{x-1}{(x-3)^2}$$

By examining the graph, we find that the solution set for the original equation is $\{1\}$.

Section 3.6 Variation

1. $C = 2\pi r$, where C is the circumference of a circle of radius r

The circumference of a circle varies directly as (or is proportional to) its radius.

3. $r = \frac{d}{t}$, where r is the speed when traveling d miles in t hours

The average speed varies directly as (or is proportional to) the distance traveled and inversely as the time.

5. $s = kx^3$, where s is the strength of a muscle of length x

The strength of a muscle varies directly as (or is proportional to) the cube of its length.

7. y varies directly as x, $y = kx$, is a straight-line model. It matches graph C.

9. y varies directly as the second power of x, $y = kx^2$, matches graph A.

11. *Step 1*: $y = kx$

Step 2: Substitute $x = 4$ and $y = 20$ to find k.

$$20 = k(4) \Rightarrow k = 5$$

Step 3: $y = 5x$

Step 4: Now find y when $x = -6$.

$$y = 5(-6) = -30$$

Copyright © 2013 Pearson Education, Inc.

13. *Step 1*: $m = kxy$

Step 2: Substitute $m = 10$, $x = 2$, and $y = 14$ to find k.

$10 = k(2)(14) \Rightarrow 10 = 28k \Rightarrow k = \frac{5}{14}$

Step 3: $m = \frac{5}{14}xy$

Step 4: Now find m when $x = 21$ and $y = 8$.

$m = \frac{5}{14}(21)(8) = 60$

15. *Step 1*: $y = \dfrac{k}{x}$

Step 2: Substitute $x = 3$ and $y = 10$ to find k.

$10 = \frac{k}{3} \Rightarrow k = 30$

Step 3: $y = \dfrac{30}{x}$

Step 4: Now find y when $x = 20$.

$y = \frac{30}{20} = \frac{3}{2}$

17. *Step 1*: $r = \dfrac{km^2}{s}$

Step 2: Substitute $r = 12$, $m = 6$, and $s = 4$ to find k.

$12 = \frac{k \cdot 6^2}{4} \Rightarrow 12 = \frac{36k}{4} \Rightarrow 12 = 9k \Rightarrow k = \frac{4}{3}$

Step 3: $r = \dfrac{4}{3} \cdot \dfrac{m^2}{s} = \dfrac{4m^2}{3s}$

Step 4: Now find r when $m = 6$ and $s = 20$.

$r = \frac{4(6^2)}{3(20)} = \frac{4(36)}{60} = \frac{144}{60} = \frac{12}{5}$

19. *Step 1*: $a = \dfrac{kmn^2}{y^3}$

Step 2: Substitute $a = 9$, $m = 4$, $n = 9$, and $y = 3$ to find k.

$9 = \dfrac{k(4)(9^2)}{3^3} \Rightarrow 9 = \dfrac{324k}{27} \Rightarrow 9 = 12k \Rightarrow$

$k = \frac{3}{4}$

Step 3: $a = \dfrac{3}{4} \cdot \dfrac{mn^2}{y^3} = \dfrac{3mn^2}{4y^3}$

Step 4: Now find a when $m = 6$, $n = 2$, and $y = 5$.

$a = \frac{3(6)(2^2)}{4(5^3)} = \frac{3(6)(4)}{4(125)} = \frac{18}{125}$

21. *Step 1*: Let C be the circumference of the circle (in inches), r is the radius of the circle (in inches). $C = kr$

Step 2: Substitute $C = 43.96$ and $r = 7$ to find k.

$43.96 = k(7) \Rightarrow k = 6.28$

Step 3: $C = 6.28r$

Step 4: Now find C when $r = 11$.

$C = 6.28(11) = 69.08$

A radius of 11 in. yields a circumference of 69.08 in.

23. *Step 1*: Let R be the resistance (in ohms); t is the temperature (in degrees Kelvin, K). $R = kt$

Step 2: Substitute $R = 646$ and $t = 190$ to find k: $646 = k(190) \Rightarrow k = \frac{646}{190} = \frac{17}{5}$

Step 3: $R = \frac{17}{5}t$

Step 4: Now find R when $t = 250$.

$R = \frac{17}{5}(250) = 850$

A resistance of 850 ohms occurs at 250K.

25. *Step 1*: Let d be the distance a person can see (in km), h is the height above Earth (in meters). $d = k\sqrt{h}$

Step 2: Substitute $d = 18$ and $h = 144$ to find k.

$18 = k\sqrt{144} \Rightarrow 18 = 12k \Rightarrow k = \frac{18}{12} = \frac{3}{2}$

Step 3: $d = \frac{3}{2}\sqrt{h}$

Step 4: Now find d when $h = 64$.

$d = \frac{3}{2}\sqrt{64} = \frac{3}{2}(8) = 12$

A person can see 12 km from a point 64 m above earth.

27. *Step 1*: Let d be the distance the spring stretches (in inches), f is the force applied (in lbs). $d = kf$

Step 2: Substitute $d = 8$ and $f = 15$ to find k.

$8 = k(15) \Rightarrow k = \frac{8}{15}$

Step 3: $d = \frac{8}{15}f$

Step 4: Now find d when $f = 30$.

$d = \frac{8}{15}(30) = 16$

The spring will stretch 16 in.

29. *Step 1*: Let v be the speed of the pulley (in rpm). D is the diameter of the pulley (in inches). $v = \dfrac{k}{D}$

Step 2: Substitute $v = 150$ and $D = 3$ to find k.

$150 = \dfrac{k}{3} \Rightarrow k = 450$

Step 3: $v = \dfrac{450}{d}$

Step 4: Now find v when $D = 5$: $v = \frac{450}{5} = 90$

The 5-in. diameter pulley will have a speed of 90 revolutions per minute.

Copyright © 2013 Pearson Education, Inc.

31. *Step 1*: Let R be the resistance (in ohms), D is the diameter (in inches). $R = \dfrac{k}{D^2}$

Step 2: Substitute $D = 0.01$ and $R = 0.4$ to find k.

$0.4 = \dfrac{k}{0.01^2} \Rightarrow 0.4 = \dfrac{k}{0.0001} \Rightarrow k = 0.00004$

Step 3: $R = \dfrac{0.00004}{D^2}$

Step 4: Now find R when $D = 0.03$

$R = \dfrac{0.00004}{0.03^2} = \dfrac{0.00004}{0.0009} \approx 0.0444$

The resistance is approximately 0.0444 ohm.

33. *Step 1*: Let i be the interest (in dollars). p is the principal (in dollars). t is the time (in years). $i = kpt$

Step 2: Substitute $i = 70$, $p = 1000$, and $t = 2$ to find k.

$70 = k(1000)(2) \Rightarrow 70 = 2000k$

$k = \dfrac{70}{2000} = .035 \ (3.5\%)$

Step 3: $i = 0.035pt$

Step 4: Now find i when $p = 5000$ and $t = 5$.

$i = 0.035(5000)(5) = 875$

The amount of interest is \$875.

35. *Step 1*: Let F be the force of the wind (in lbs), A is the area (in ft^2), v is the velocity of the wind (in mph). $F = kAv^2$

Step 2: Substitute $F = 50$, $A = \frac{1}{2}$, and $v = 40$ to find k.

$50 = k\left(\frac{1}{2}\right)\left(40^2\right) = k\left(\frac{1}{2}\right)(1600)$

$50 = 800k \Rightarrow k = \dfrac{50}{800} = \dfrac{1}{16}$

Step 3: $F = \frac{1}{16}Av^2$

Step 4: Now find F when $v = 80$, and $A = 2$.

$F = \frac{1}{16}(2)\left(80^2\right) = \frac{1}{16}(2)(6400) = 800$

The force would be 800 pounds.

37. *Step 1*: Let L be the load (in metric tons), D is the diameter (in meters), h is the height (in meters). $L = \dfrac{kD^4}{h^2}$

Step 2: Substitute $h = 9$, $D = 1$, and $L = 8$ to find k.

$8 = \dfrac{k\left(1^4\right)}{9^2} \Rightarrow k = 648$

Step 3: $L = \dfrac{648D^4}{h^2}$

Step 4: Now find L when $D = \frac{2}{3}$ and $h = 12$.

$L = \dfrac{648\left(\frac{2}{3}\right)^4}{12^2} = \dfrac{648\left(\frac{16}{81}\right)}{144} = \dfrac{128}{144} = \dfrac{8}{9}$

A column 12 m high and $\frac{2}{3}$ m in diameter will support $\frac{8}{9}$ metric ton.

39. *Step 1*: Let p be the period of pendulum (in sec), l is the length of pendulum (in cm), a is the acceleration due to gravity $\left(\text{in } \frac{\text{cm}}{\text{sec}^2}\right)$.

$p = \dfrac{k\sqrt{l}}{\sqrt{a}}$

Step 2: Substitute $p = 6\pi$, $l = 289$, and $a = 980$ to find k.

$6\pi = \dfrac{k\sqrt{289}}{\sqrt{980}} \Rightarrow 6\pi = \dfrac{17k}{14\sqrt{5}} \Rightarrow$

$84\sqrt{5}\pi = 17k \Rightarrow k = \dfrac{84\sqrt{5}\pi}{17}$

Step 3: $p = \dfrac{84\sqrt{5}\pi}{17} \cdot \dfrac{\sqrt{l}}{\sqrt{a}} = \dfrac{84\sqrt{5}\pi\sqrt{l}}{17\sqrt{a}}$

Step 4: Now find p when $l = 121$ and $a = 980$.

$p = \dfrac{84\pi\sqrt{5}\sqrt{121}}{17\sqrt{980}} = \dfrac{84\pi\sqrt{5}(11)}{17(14\sqrt{5})} = \dfrac{924\pi\sqrt{5}}{238\sqrt{5}} = \dfrac{66\pi}{17}$

The period is $\frac{66\pi}{17}$ sec.

41. *Step 1*: Let B be the BMI, w is the weight (in lbs), h is the height (in inches). $B = \dfrac{kw}{h^2}$

Step 2: Substitute $w = 177$, $B = 24$, and $h = 72$ (6 feet) to find k.

$24 = \dfrac{k(177)}{72^2} \Rightarrow 124,416 = 177k \Rightarrow$

$k = \dfrac{124,416}{177} = \dfrac{41,472}{59}$

Step 3: $B = \dfrac{41,472}{59} \cdot \dfrac{w}{h^2} = \dfrac{41,472w}{59h^2}$

Step 4: Now find B when $w = 130$ and $h = 66$.

$B = \dfrac{41,472(130)}{59\left(66^2\right)} = \dfrac{5,391,360}{59(4356)} = \dfrac{5,391,360}{257,004} \approx 20.98$

The BMI would be approximately 21.

Copyright © 2013 Pearson Education, Inc.

43. *Step 1*: Let R be the radiation, t is the temperature (in degrees kelvin, K). $R = kt^4$:

Step 2: Substitute $R = 213.73$ and $t = 293$ to find k.

$$213.73 = k\left(293^4\right) \Rightarrow k = \frac{213.73}{293^4} \approx 2.9 \times 10^{-8}$$

Step 3: $R = (2.9 \times 10^{-8})t^4$

Step 4: Now find R when $t = 335$.

$$R = \left(2.9 \times 10^{-8}\right) \cdot 335^4 \approx 365.24$$

The radiation of heat would be 365.24.

45. *Step 1*: Let p be the person's pelidisi, w is the person's weight (in g), h is the person's sitting height (in cm). $p = \dfrac{k\sqrt[3]{w}}{h}$

Step 2: Substitute $w = 48{,}820$, $h = 78.7$, and $p = 100$ to find k.

$$100 = \frac{k\sqrt[3]{48{,}820}}{78.7} \Rightarrow 7870 = \sqrt[3]{48{,}820}k \Rightarrow$$

$$k = \frac{7870}{\sqrt[3]{48{,}820}} \approx 215.33$$

Step 3: $p = \dfrac{215.33\sqrt[3]{w}}{h}$

Step 4: Now find p when $w = 54{,}430$ and $h = 88.9$

$$p = \frac{215.33\sqrt[3]{54{,}430}}{88.9} \approx 92$$

This person's pelidisi is 92. The individual is undernourished since his pelidisi is below 100.

47. For $k > 0$, if y varies directly as x, when x increases, y <u>increases</u>, and when x decreases, y <u>decreases</u>.

49. $y = \dfrac{k}{x}$

If x is doubled, then $y_1 = \dfrac{k}{2x} = \dfrac{1}{2} \cdot \dfrac{k}{x} = \dfrac{1}{2}y$.

Thus, y is half as large as it was before.

51. $y = kx$

If x is replaced by $\frac{1}{3}x$, then we have

$y_1 = k\left(\frac{1}{3}x\right) = \frac{1}{3} \cdot kx = \frac{1}{3}y$. Thus, y is one-third as large as it was before.

53. $p = \dfrac{kr^3}{t^2}$

If r is replaced by $\frac{1}{2}r$ and t is replaced by $2t$, then we have the following.

$$p_1 = \frac{k\left(\frac{1}{2}r\right)^3}{\left(2t\right)^2} = \frac{k\left(\frac{1}{8}r^3\right)}{4t^2} = \frac{kr^3}{32t^2} = \frac{1}{32} \cdot \frac{kr^3}{t^2}$$

Thus, so p is $\frac{1}{32}$ as large as it was before.

Chapter 3   Review Exercises

1. $f(x) = 3(x+4)^2 - 5$

Since

$$f(x) = 3(x+4)^2 - 5 = 3\left[x - (-4)\right]^2 + (-5),$$

the function has the form $f(x) = a(x-h)^2 + k$ with $a = 3$, $h = -4$, and $k = -5$. The graph is a parabola that opens upward with vertex $(h, k) = (-4, -5)$. The axis is $x = h \Rightarrow x = -4$. To find the x-intercepts, let $f(x) = 0$.

$$3(x+4)^2 - 5 = 0 \Rightarrow 3(x+4)^2 = 5 \Rightarrow$$

$$(x+4)^2 = \frac{5}{3} \Rightarrow x + 4 = \pm\sqrt{\frac{5}{3}} \Rightarrow$$

$$x = -4 \pm \sqrt{\frac{5}{3}} = -4 \pm \frac{\sqrt{15}}{3} = \frac{-12 \pm \sqrt{15}}{3}$$

Thus, the x-intercepts are $\frac{-12 \pm \sqrt{15}}{3}$ or approximately -5.3 and -2.7. To find the y-intercept, let $x = 0$.

$$f(0) = 3(0+4)^2 - 5 = 3(16) - 5 = 48 - 5 = 43$$

The y-intercept is 43.

The domain is $(-\infty, \infty)$. Since the lowest point on the graph is $(-4, -5)$, the range is $[-5, \infty)$. The function is increasing on $[-4, \infty)$ and decreasing on $(-\infty, -4]$.

Copyright © 2013 Pearson Education, Inc.

3. $f(x) = -3x^2 - 12x - 1$

Complete the square so the function has the form $f(x) = a(x-h)^2 + k$.

$$f(x) = -3x^2 - 12x - 1 = -3(x^2 + 4x) - 1$$
$$= -3(x^2 + 4x + 4 - 4) - 1$$
$$f(x) = -3(x+2)^2 + 12 - 1$$
$$= -3[x - (-2)]^2 + 11$$

This parabola is in now in standard form with $a = -3$, $h = -2$, and $k = 11$. It opens downward with vertex $(h, k) = (-2, 11)$. The axis is $x = h \Rightarrow x = -2$. To find the x-intercepts, let $f(x) = 0$.

$$-3(x+2)^2 + 11 = 0 \Rightarrow -3(x+2)^2 = -11$$
$$(x+2)^2 = \tfrac{11}{3} \Rightarrow x + 2 = \pm\sqrt{\tfrac{11}{3}}$$
$$x = -2 \pm \sqrt{\tfrac{11}{3}} = -2 \pm \tfrac{\sqrt{33}}{3}$$
$$= \tfrac{-6 \pm \sqrt{33}}{3}$$

Thus, the x-intercepts are $\tfrac{-6 \pm \sqrt{33}}{3}$ or approximately -3.9 and -0.1.
To find the y-intercept, let $x = 0$.

$$f(0) = -3(0^2) - 12(0) - 1 = -1$$

The y-intercept is -1.

The domain is $(-\infty, \infty)$. Since the highest point on the graph is $(-2, 11)$, the range is $(-\infty, 11]$. The function is increasing on $(-\infty, -2]$ and decreasing on $[-2, \infty)$.

5. $f(x) = a(x-h)^2 + k;\ a > 0$

The graph is a parabola that opens upward. The coordinates of the lowest point of the graph are represented by the vertex, (h, k).

7. For the graph to have one or more x-intercepts, $f(x) = 0$ must have real number solutions.

$$a(x-h)^2 + k = 0 \Rightarrow a(x-h)^2 = -k$$
$$(x-h)^2 = \tfrac{-k}{a} \Rightarrow x - h = \pm\sqrt{\tfrac{-k}{a}}$$
$$x = h \pm \sqrt{\tfrac{-k}{a}}$$

These solutions are real only if $\tfrac{-k}{a} \geq 0$. Since $a > 0$, this condition is equivalent to $-k \geq 0$ or $k \leq 0$. The x-intercepts for these conditions are given by $h \pm \sqrt{\tfrac{-k}{a}}$.

9. Let x = the width of the rectangular region. $180 - 2x$ = the length of the region.

Since $A = LW$, we have
$A(x) = (180 - 2x)x = -2x^2 + 180x$. Since the graph of $A(x)$ is a parabola that opens downward, the maximum area is the y-coordinate of the vertex.

$$A(x) = -2x^2 + 180x = -2(x^2 - 90x)$$
$$= -2(x^2 - 90x + 2025) + 4050$$
$$= -2(x - 45)^2 + 4050$$

The x-coordinate of the vertex, $x = 45$, is the width that gives the maximum area. The length will be $L = 180 - 2(45) = 90$. Thus, the dimensions of the region are 90 m $\times$ 45 m.

11. Because $V(x)$ has different equations on two different intervals, it is a piecewise-defined function.

$$V(x) = \begin{cases} 2x^2 - 32x + 150 & \text{if } 1 \leq x < 8 \\ 31x - 226 & \text{if } 8 \leq x \leq 12 \end{cases}$$

Note: Since $2 \cdot 8^2 - 32 \cdot 8 + 150 = 22$ and $31 \cdot 8 - 226 = 22$, the point $(8, 22)$ can be found using either rule, and the graph will have no breaks.

(a) In January, $x = 1$.
$$V(x) = 2x^2 - 32x + 150$$
$$V(1) = 2(1)^2 - 32(1) + 150 = 120$$

(b) In May, $x = 5$.
$$V(x) = 2x^2 - 32x + 150$$
$$V(5) = 2(5)^2 - 32(5) + 150 = 40$$

Copyright © 2013 Pearson Education, Inc.

(c)   In August, $x = 8$.
$$V(x) = 31x - 226$$
$$V(8) = 31(8) - 226 = 22$$

(d)   In October, $x = 10$.
$$V(x) = 31x - 226$$
$$V(10) = 31(10) - 226 = 84$$

(e)   In December, $x = 12$.
$$V(x) = 31x - 226$$
$$V(12) = 31(12) - 226 = 146$$

(f)

The graph shows a minimum point at $(8, 22)$. Thus, in August $(x = 8)$ the fewest volunteers are available.

13. $f(x) = -2.64x^2 + 5.47x + 3.54$

The discriminant is $b^2 - 4ac$ in the standard quadratic equation $y = ax^2 + bx + c$. Here we have $a = -2.64$, $b = 5.47$, and $c = 3.54$. Thus,
$b^2 - 4ac = (5.47)^2 - 4(-2.64)(3.54) = 67.3033$.
Because the discriminant is 67.3033, a positive number, there are two x-intercepts.

15.   (a) $f(x) > 0$ on the open interval $(-0.52, 2.59)$.

   (b) $f(x) < 0$ on $(-\infty, -0.52) \cup (2.59, \infty)$.

17. $\dfrac{x^3 + x^2 - 11x - 10}{x - 3}$

Since $x - 3$ is in the form $x - k$, $k = 3$.

$$\begin{array}{r} 3)\overline{1 \quad 1 \quad -11 \quad -10} \\ \underline{3 \quad 12 \quad 3} \\ 1 \quad 4 \quad 1 \quad -7 \end{array}$$

$$\frac{x^3 + x^2 - 11x - 10}{x - 3} = x^2 + 4x + 1 + \frac{-7}{x - 3}$$

19. $\dfrac{2x^3 - x + 6}{x + 4}$

Express $x + 4$ in the form $x - k$ by writing it as $x - (-4)$. Thus we have $k = -4$. Since the x^2-term is missing, include a 0 as its coefficient.

$$\begin{array}{r} -4)\overline{2 \quad 0 \quad -1 \quad 6} \\ \underline{-8 \quad 32 \quad -124} \\ 2 \quad -8 \quad 31 \quad -118 \end{array}$$

$$\frac{2x^3 - x + 6}{x + 4} = 2x^2 - 8x + 31 + \frac{-118}{x + 4}$$

21. $f(x) = 5x^3 - 3x^2 + 2x - 6$; $k = 2$
Use synthetic division to write the polynomial in the form $f(x) = (x - k)q(x) + r$.

$$\begin{array}{r} 2)\overline{5 \quad -3 \quad 2 \quad -6} \\ \underline{10 \quad 14 \quad 32} \\ 5 \quad 7 \quad 16 \quad 26 \end{array}$$

$$f(x) = (x - 2)(5x^2 + 7x + 16) + 26$$

23. $f(x) = -x^3 + 5x^2 - 7x + 1$; find $f(2)$.

$$\begin{array}{r} 2)\overline{-1 \quad 5 \quad -7 \quad 1} \\ \underline{-2 \quad 6 \quad -2} \\ -1 \quad 3 \quad -1 \quad -1 \end{array}$$

The synthetic division shows that $f(2) = -1$.

25. $f(x) = 5x^4 - 12x^2 + 2x - 8$; find $f(2)$.

$$\begin{array}{r} 2)\overline{5 \quad 0 \quad -12 \quad 2 \quad -8} \\ \underline{10 \quad 20 \quad 16 \quad 36} \\ 5 \quad 10 \quad 8 \quad 18 \quad 28 \end{array}$$

The synthetic division shows that $f(2) = 28$.

27. $f(x) = x^3 + 2x^2 + 3x + 2$; $k = -1$

$$\begin{array}{r} -1)\overline{1 \quad 2 \quad 3 \quad 2} \\ \underline{-1 \quad -1 \quad -2} \\ 1 \quad 1 \quad 2 \quad 0 \end{array}$$

Since $f(-3) = 0$, we have that -3 is a zero of the function.

29.   By the conjugate zeros theorem, $7 - 2i$ is also a zero.

In Exercises 31 through 33, other answers are possible.

31.   Zeros: $-1, 4, 7$
$$\begin{aligned} f(x) &= \left[x - (-1)\right](x - 4)(x - 7) \\ &= (x + 1)(x - 4)(x - 7) \\ &= (x + 1)\left[(x - 4)(x - 7)\right] \\ &= (x + 1)\left(x^2 - 11x + 28\right) \\ &= x^3 - 10x^2 + 17x + 28 \end{aligned}$$

Copyright © 2013 Pearson Education, Inc.

33. Zeros: $\sqrt{3}, -\sqrt{3}, 2, 3$.

$$\begin{aligned} f(x) &= \left(x - \sqrt{3}\right)\left(x + \sqrt{3}\right)(x-2)(x-3) \\ &= \left[\left(x - \sqrt{3}\right)\left(x + \sqrt{3}\right)\right]\left[(x-2)(x-3)\right] \\ &= \left(x^2 - 3\right)\left(x^2 - 5x + 6\right) \\ &= x^4 - 5x^3 + 3x^2 + 15x - 18 \end{aligned}$$

35. Zeros: $2, 4, -i$

By the conjugate zeros theorem, if $-i$ is a zero of a polynomial function, then i is also a zero of the function.

$$\begin{aligned} f(x) &= (x-2)(x-4)(x-i)(x+i) \\ &= \left[(x-2)(x-4)\right]\left[(x-i)(x+i)\right] \\ &= (x^2 - 6x + 8)\left(x^2 + 1\right) \\ &= x^4 - 6x^3 + 9x^2 - 6x + 8 \end{aligned}$$

37. $f(x) = 2x^3 - 9x^2 - 6x + 5$

p must be a factor of $a_0 = 5$ and q must be a factor of $a_3 = 2$. Thus, p can be $\pm 1, \pm 5,$ and q can be $\pm 1, \pm 2$. The possible zeros, $\frac{p}{q}$, are $\pm 1, \pm 5, \pm\frac{1}{2}, \pm\frac{5}{2}$. The remainder theorem shows that $\frac{1}{2}$ is a zero.

$$\frac{1}{2}\overline{)\begin{array}{rrrr} 2 & -9 & -6 & 5 \\ & 1 & -4 & -5 \end{array}}$$
$$\begin{array}{rrrr} 2 & -8 & -10 & 0 \end{array}$$

The new quotient polynomial is therefore $2x^2 - 8x - 10$.

$$2x^2 - 8x - 10 = 0$$
$$x^2 - 4x - 5 = 0$$
$$(x+1)(x-5) = 0$$
$$x + 1 = 0 \Rightarrow x = -1 \quad \text{or} \quad x - 5 = 0 \Rightarrow x = 5$$

The rational zeros are $\frac{1}{2}, -1,$ and 5.

39. $f(x) = 3x^3 - 8x^2 + x + 2$

(a) To show there is a zero between -1 and 0, we need to find $f(-1)$ and $f(0)$. Find $f(-1)$ by synthetic division and $f(0)$ by evaluating the function.

$$-1\overline{)\begin{array}{rrrr} 3 & -8 & 1 & 2 \\ & -3 & 11 & -12 \end{array}}$$
$$\begin{array}{rrrr} 3 & -11 & 12 & -10 \end{array}$$

$$f(0) = 3\left(0^3\right) - 8\left(0^2\right) + (0) + 2 = 2$$

Since $f(-1) = -10 < 0$ and $f(0) = 2 > 0,$ there must be a zero between -1 and 0.

(b) To show there is a zero between 2 and 3, we need to find $f(2)$ and $f(3)$. Find these by synthetic division.

$$2\overline{)\begin{array}{rrrr} 3 & -8 & 1 & 2 \\ & 6 & -4 & -6 \end{array}}$$
$$\begin{array}{rrrr} 3 & -2 & -3 & -4 \end{array}$$

$$3\overline{)\begin{array}{rrrr} 3 & -8 & 1 & 2 \\ & 9 & 3 & 12 \end{array}}$$
$$\begin{array}{rrrr} 3 & 1 & 4 & 14 \end{array}$$

Since $f(2) = -4 < 0$ and $f(3) = 14 > 0,$ there must be a zero between 2 and 3.

(c) Using a graphing calculator, we find that the zero between 2 and 3 is about 2.414.

41. $f(x) = 6x^4 + 13x^3 - 11x^2 - 3x + 5$

(a) To show that there is no zero greater than 1, we must show that 1 is an upper bound.

$$1\overline{)\begin{array}{rrrrr} 6 & 13 & -11 & -3 & 5 \\ & 6 & 19 & 8 & 5 \end{array}}$$
$$\begin{array}{rrrrr} 6 & 19 & 8 & 5 & 10 \end{array}$$

Since the 1 is positive and the bottom row has all positive numbers, there is no zero greater than 1.

(b) To show that there is no zero less than -3, we must show that -3 is a lower bound.

$$-3\overline{)\begin{array}{rrrrr} 6 & 13 & -11 & -3 & 5 \\ & -18 & 15 & -12 & 45 \end{array}}$$
$$\begin{array}{rrrrr} 6 & -5 & 4 & -15 & 50 \end{array}$$

Since the -3 is negative and the bottom row alternates in sign, there is no zero less than -3.

43. To determine if $x + 1$ is a factor of $f(x) = x^3 + 2x^2 + 3x + 2,$ we can find $f(-1)$ by synthetic division

$$-1\overline{)\begin{array}{rrrr} 1 & 2 & 3 & 2 \\ & -1 & -1 & -2 \end{array}}$$
$$\begin{array}{rrrr} 1 & 1 & 2 & 0 \end{array}$$

Since $f(-1) = 0,$ $x + 1$ is a factor of $f(x)$.

Copyright © 2013 Pearson Education, Inc.

45. Since $f(x) = a[x - (-2)](x-1)(x-4)$

$= a(x+2)(x-1)(x-4)$, we can solve for a.

$16 = a(2+2)(2-1)(2-4)$

$16 = -8a \Rightarrow a = -2$

The polynomial function is

$f(x) = -2(x+2)(x-1)(x-4)$. It can be expanded as follows.

$f(x) = -2(x+2)(x-1)(x-4)$

$= -2[(x+2)(x-1)](x-4)$

$= -2(x^2 + x - 2)(x-4)$

$= -2(x^3 - 3x^2 - 6x + 8)$

$= -2x^3 + 6x^2 + 12x - 16$

47. $f(x) = 2x^4 - x^3 + 7x^2 - 4x - 4$; 1 and $-2i$ are zeros.

Use synthetic division where $k = 1$.

$$\begin{array}{r|rrrr} 1 & 2 & -1 & 7 & -4 & -4 \\ & & 2 & 1 & 8 & 4 \\ \hline & 2 & 1 & 8 & 4 & 0 \end{array}$$

Synthetically divide $-2i$ into the quotient polynomial.

$$\begin{array}{r|rrrr} -2i & 2 & 1 & 8 & 4 \\ & & -4i & -8-2i & -4 \\ \hline & 2 & 1-4i & -2i & 0 \end{array}$$

Since $-2i$ is a zero, $2i$ is also a zero. Synthetically divide $2i$ into the quotient polynomial.

$$\begin{array}{r|rrr} 2i & 2 & 1-4i & -2i \\ & & 4i & 2i \\ \hline & 2 & 1 & 0 \end{array}$$

Since the quotient polynomial represented is $2x + 1$, we can set it equal to zero to find the remaining zero.

$2x + 1 = 0 \Rightarrow x = -\frac{1}{2}$

Thus, all the zeros are 1, $-\frac{1}{2}$ and $\pm 2i$.

49. To determine the value of k such that when the polynomial $x^3 - 3x^2 + kx - 4$ is divided by $x - 2$, the remainder is 5, we perform synthetic division.

$$\begin{array}{r|rrrr} 2 & 1 & -3 & k & -4 \\ & & 2 & -2 & 2k-4 \\ \hline & 1 & -1 & k-2 & 2k-8 \end{array}$$

To ensure that 5 is the remainder set $2k - 8$ equal to 5 and solve.

$2k - 8 = 5 \Rightarrow 2k = 13 \Rightarrow k = \frac{13}{2}$

Thus, the value of s is $\frac{13}{2}$.

51. Any polynomial that can be factored as $a(x-b)^3$, where a and b are real numbers, will be a cubic polynomial function having exactly one real zero. One example is $f(x) = 2(x-1)^3$.

53. The polynomial has a leading term of $10x^7$.

(a) The domain is $(-\infty, \infty)$.

(b) The range is $(-\infty, \infty)$.

(c) As $x \to \infty$, $f(x) \to \infty$ and as $x \to -\infty$, $f(x) \to -\infty$.

(d) There are at most 7 zeros.

(e) There are at most 6 turning points.

55. $f(x) = (x-2)^2(x+3)$

Step 1: Set each factor equal to 0 and solve the resulting equations to find the zeros of the function.

$(x-2)^2 = 0 \Rightarrow x = 2$ or $x+3 = 0 \Rightarrow x = -3$

The zeros are -3 and 2, which divide the x-axis into three regions. Test a point in each region to find the sign of $f(x)$ in that region.

Step 2: $f(0) = (0-2)^2(0+3) = 4 \cdot 3 = 12$, so plot $(0, 12)$.

(*continued on next page*)

Copyright © 2013 Pearson Education, Inc.

(continued)

Step 3: The *x*-intercepts divide the *x*-axis into three intervals.

Interval	Test Point	Value of $f(x)$	Sign of $f(x)$	Graph Above or Below *x*-Axis
$(-\infty, -3)$	-4	-36	Negative	Below
$(-3, 2)$	1	4	Positive	Above
$(2, \infty)$	3	6	Positive	Above

Plot the *x*-intercepts, *y*-intercept, and test points with a smooth curve to get the graph.

$$f(x) = (x-2)^2(x+3)$$

57. $f(x) = 2x^3 + x^2 - x$

Step 1: Factor the polynomial function and set each factor equal to 0 and solve the resulting equations to find the zeros of the function.

$$2x^3 + x^2 - x = x(2x^2 + x - 1)$$
$$= x(x+1)(2x-1)$$

$x = 0$ or $x + 1 = 0 \Rightarrow x = -1$ or
$2x - 1 = 0 \Rightarrow x = \frac{1}{2}$

The zeros are -1, 0, and $\frac{1}{2}$, which divide the *x*-axis into four regions. Test a point in each region to find the sign of $f(x)$ in that region.

Step 2: $f(0) = 0$, so plot $(0, 0)$.

Step 3: The *x*-intercepts divide the *x*-axis into four intervals.

Interval	Test Point	Value of $f(x)$	Sign of $f(x)$	Graph Above or Below *x*-Axis
$(-\infty, -1)$	-2	-10	Negative	Below
$(-1, 0)$	$-\frac{1}{2}$	$\frac{1}{2}$	Positive	Above
$\left(0, \frac{1}{2}\right)$	$\frac{1}{4}$	$-\frac{5}{32}$	Negative	Below
$\left(\frac{1}{2}, \infty\right)$	1	2	Positive	Above

Plot the *x*-intercepts, *y*-intercept (the *y*-intercept is also an *x*-intercept), and test points with a smooth curve to get the graph.

$f(x) = 2x^3 + x^2 - x$

59. $f(x) = x^4 + x^3 - 3x^2 - 4x - 4$

Step 1: The first step is to find the zeros of the polynomial function. *p* must be a factor of $a_0 = -4$ and *q* must be a factor of $a_4 = 1$. Thus, *p* can be $\pm1, \pm2$, or ±4, and *q* can be ±1. The possible zeros, $\frac{p}{q}$, are $\pm1, \pm2$, or ±4. The remainder theorem shows that 2 is a zero.

$$
\begin{array}{r|rrrr}
2 & 1 & 1 & -3 & -4 & -4 \\
  &   & 2 & 6 & 6 & 4 \\
\hline
  & 1 & 3 & 3 & 2 & 0
\end{array}
$$

The new quotient polynomial therefore is $x^3 + 3x^2 + 3x + 2$.

p must be a factor of $a_0 = 2$ and *q* must be a factor of $a_3 = 1$. Thus, *p* can be ±1, or ±2, and *q* can be ±1. The possible zeros, $\frac{p}{q}$, are ±1, or ±2. The remainder theorem shows that -2 is a zero.

$$
\begin{array}{r|rrrr}
-2 & 1 & 3 & 3 & 2 \\
   &   & -2 & -2 & -2 \\
\hline
   & 1 & 1 & 1 & 0
\end{array}
$$

The new quotient polynomial therefore is $x^2 + x + 1$. Setting this equal to 0 and using the quadratic formula with $a = 1$, $b = 1$, and $c = 1$, we have

$$x = \frac{-1 \pm \sqrt{1^2 - 4(1)(1)}}{2(1)} = \frac{-1 \pm \sqrt{-4}}{2}$$
$$= -\frac{1}{2} \pm 2i$$

The rational zeros are -2 and 2, which divide the *x*-axis into three regions.

Step 2: $f(0) = -4$, so plot $(0, -4)$.

(continued on next page)

Copyright © 2013 Pearson Education, Inc.

(continued)

Step 3: The *x*-intercepts divide the *x*-axis into three intervals.

Interval	Test Point	Value of $f(x)$	Sign of $f(x)$	Graph Above or Below *x*-Axis
$(-\infty, -2)$	-3	35	Positive	Above
$(-2, 2)$	-1	-3	Negative	Below
$(2, \infty)$	3	65	Positive	Above

Plot the *x*-intercepts, *y*-intercept, and test points with a smooth curve to get the graph.

$$f(x) = x^4 + x^3 - 3x^2 - 4x - 4$$

61. $f(x) = (x-2)^2(x-5)$ is a cubic polynomial with positive *y*-values for $x > 5$, so it matches graph C.

63. $f(x) = (x-2)^2(x-5)^2$ is a quartic polynomial that is always greater than or equal to 0, so it matches graph E.

65. $f(x) = -(x-2)(x-5)$ is a quadratic polynomial that opens down, so it matches graph B.

67. Using the calculator, we find that the real zeros are 7.6533119, 1, and −0.6533119. For the two approximations of the zeros, a higher accuracy can be found by going to the home screen and displaying the stored *x*-value.

69. (a)

(b) $f(x) = -0.0109x^2 + 0.8693x + 11.85$

(c) $f(x) = -0.00087x^3 + 0.0456x^2 - 0.2191x + 17.83$

(d)

(e) Both functions approximate the data well. The quadratic function is probably better for prediction, because it is unlikely that the percent of out-of-pocket spending would decrease after 2025 (as the cubic function shows) unless changes were made in Medicare law.

Copyright © 2013 Pearson Education, Inc.

71. $V_{box} = LWH$

If we let $L = x + 11$, $W = 3x$, and $H = x$, we have the volume of the rectangular box is $V(x) = (x + 11)(3x)x$. Since the volume is 720 in.^{3}, we need to solve the equation

$$720 = (x + 11)(3x)x \Rightarrow 720 = 3x^3 + 33x^2$$

$$0 = 3x^3 + 33x^2 - 720 \Rightarrow x^3 + 11x^2 - 240 = 0$$

Algebraic Solution:

p must be a factor of $a_0 = -240$ and q must be a factor of $a_3 = 1$. Including the $\pm$, there are 40 factors of p and 2 factors of q. There are 80 possible zeros, $\frac{p}{q}$. Because this list is so long, we will simply show that 4 is a zero by the remainder theorem.

$$
\begin{array}{r|rrr}
4) & 1 & 11 & 0 & -240 \\
   &   & 4 & 60 & 240 \\
\hline
   & 1 & 15 & 60 & 0
\end{array}
$$

The new quotient polynomial is $x^2 + 15x + 60$. If we examine the discriminant, $b^2 - 4ac$, we will see that the quotient polynomial cannot be factored further over the reals

$$b^2 - 4ac = 15^2 - 4(1)(60)$$
$$= 225 - 240 = -15 < 0$$

Since the discriminant is negative, 4 is our only real solution to the equation $x^3 + 11x^2 - 240 = 0$.

Graphing Calculator Solution:

Graph $Y_1 = x^3 + 11x^2 - 240$ on a graphing calculator.

The graph will show that 4 is the only real zero. Since the only real solution is 4, the dimensions of the rectangular box are 4 in. by $3(4) = 12$ in. by $4 + 11 = 15$ in. or 12 in. × 4 in. × 15 in.

73. $f(x) = \dfrac{4}{x - 1}$

Step 1: The graph has a vertical asymptote where $x - 1 = 0 \Rightarrow x = 1$

Step 2: Since the degree of the numerator (which is considered degree zero) is less than the degree of the denominator, the graph has a horizontal asymptote at $y = 0$ (the x-axis).

Step 3: The y-intercept is $f(0) = \frac{4}{0-1} = \frac{4}{-1} = -4$.

Step 4: Any x-intercepts are found by solving

$$f(x) = 0: \quad \frac{4}{x - 1} = 0 \Rightarrow 4 = 0$$

This is a false statement, so there is no x-intercept.

Step 5: The graph will intersect the horizontal asymptote when $\frac{4}{x-1} = 0 \Rightarrow 4 = 0$

This is a false statement, so the graph does not intersect the horizontal asymptote.

Step 6: Since the vertical asymptote is $x = 1$ and there is no x-intercept, we must determine values in two intervals.

Interval	Test Point	Value of $f(x)$	Sign of $f(x)$	Graph Above or Below x-Axis
$(-\infty, 1)$	-2	$-\frac{4}{3}$	Negative	Below
$(1, \infty)$	2	4	Positive	Above

Step 7: Use the asymptotes, intercepts, and these points to sketch the graph.

75. $f(x) = \dfrac{6x}{x^2 + x - 2} = \dfrac{6x}{(x + 2)(x - 1)}$

Step 1: The graph has a vertical asymptote where $(x + 2)(x - 1) = 0$, that is, when $x = -2$ and $x = 1$.

Step 2: Since the degree of the numerator is less than the degree of the denominator, the graph has a horizontal asymptote at $y = 0$ (the x-axis).

Step 3: The y-intercept is

$$f(0) = \frac{6(0)}{0^2 + 0 - 2} = \frac{0}{0 + 0 - 2} = \frac{0}{-2} = 0.$$

Step 4: Any x-intercepts are found by solving $f(x) = 0$.

$$\frac{6x}{x^2 + x - 2} = 0 \Rightarrow 6x = 0 \Rightarrow x = 0$$

The only x-intercept is 0.

(continued on next page)

Copyright © 2013 Pearson Education, Inc.

(continued)

Step 5: The graph will intersect the horizontal asymptote when $\dfrac{6x}{x^2+x-2}=0$.

$\frac{6x}{x^2+x-2}=0 \Rightarrow 6x=0 \Rightarrow x=0$

Step 6: Since the vertical asymptotes are $x=-2$ and $x=1$ and the x-intercept occurs at 0, we must determine values in four intervals.

Interval	Test Point	Value of $f(x)$	Sign of $f(x)$	Graph Above or Below x-Axis
$(-\infty,-2)$	-3	$-\frac{9}{2}$	Negative	Below
$(-2,0)$	-1	3	Positive	Above
$(0,1)$	$\frac{1}{2}$	$-\frac{12}{5}$	Negative	Below
$(1,\infty)$	2	3	Positive	Above

Step 7: Use the asymptotes, intercepts, and these points to sketch the graph.

$$f(x)=\frac{6x}{x^2+x-2}$$

77. $f(x)=\dfrac{x^2+4}{x+2}$

Step 1: The graph has a vertical asymptote where $x+2=0 \Rightarrow x=-2$.

Step 2: Since the degree of the numerator is one more than the degree of the denominator, the graph has an oblique asymptote. Divide x^2+4 by $x+2$.

$$-2\overline{)\begin{array}{ccc} 1 & 0 & 4 \\ & -2 & 4 \\ \hline 1 & -2 & 8 \end{array}}$$

$f(x)=\dfrac{x^2+4}{x+2}=x-2+\dfrac{8}{x+3}$

The oblique asymptote is the line $y=x-2$.

Step 3: The y-intercept is

$f(0)=\frac{0^2+4}{0+2}=\frac{0+4}{2}=\frac{4}{2}=2.$

Step 4: Any x-intercepts are found by solving $f(x)=0$.

$\frac{x^2+4}{x+2}=0 \Rightarrow x^2+4=0 \Rightarrow x=\pm 2i$

There are no x-intercepts.

Step 5: The graph will intersect the oblique asymptote when $\frac{x^2+4}{x+2}=x-2$.

$\dfrac{x^2+4}{x+2}=x-2 \Rightarrow x^2+4=(x+2)(x-2)$

$x^2+4=x^2-4 \Rightarrow 4=-4$

This is a false statement, so the graph does not intersect the oblique asymptote.

Step 6: Since the vertical asymptote is $x=-2$ and there are no x-intercepts, we must determine values in two intervals.

Interval	Test Point	Value of $f(x)$	Sign of $f(x)$	Graph Above or Below x-Axis
$(-\infty,-2)$	-3	-13	Negative	Below
$(-2,\infty)$	3	$\frac{13}{5}$	Positive	Above

Step 7: Use the asymptotes, y-intercept, and these points to sketch the graph.

$$f(x)=\frac{x^2+4}{x+2}$$

79. $f(x)=\dfrac{-2}{x^2+1}$

Step 1: The graph has a vertical asymptote where $x^2+1=0$. Since we have $x^2+1=0 \Rightarrow x=\pm i$, there are no vertical asymptotes.

Step 2: Since the degree of the numerator (which is considered degree zero) is less than the degree of the denominator, the graph has a horizontal asymptote at $y=0$ (the x-axis).

Step 3: The y-intercept is

$f(0)=\frac{-2}{0^2+1}=\frac{-2}{0+1}=\frac{-2}{1}=-2.$

(continued on next page)

Copyright © 2013 Pearson Education, Inc.

(*continued*)

Step 4: Any x-intercepts are found by solving $f(x) = 0$.

$$\frac{-2}{x^2+1} = 0 \Rightarrow -2 = 0$$

This is a false statement, so there is no x-intercept.

Step 5: The graph will intersect the horizontal asymptote when $\frac{-2}{x^2+1} = 0$.

$$\frac{-2}{x^2+1} = 0 \Rightarrow -2 = 0$$

This is a false statement, so the graph does not intersect the horizontal asymptote.

Step 6: Since there are no vertical asymptotes and no x-intercept, we can see from the y-intercept that the function is always negative and, therefore, below the x-axis. Because of the limited information we have, we should explore symmetry and calculate a few more points on the graph.

x	y
-3	$-\frac{1}{5}$
$\frac{1}{2}$	$-\frac{8}{5}$
1	-1

Since $f(-x) = \frac{-2}{(-x)^2+1} = \frac{-2}{x^2+1} = f(x)$, the graph is symmetric about the y-axis.

Step 7: Use the asymptote, y-intercept, these points, and symmetry to sketch the graph.

$f(x) = \frac{-2}{x^2+1}$

81. (a)

$f(x) = \frac{(x-2)(x-4)}{(x-3)^2}$

(b) The graph has a vertical asymptote $x = 3$, so $x - 3$ is in the denominator of the function. The x-intercepts are 2 and 4, so that when $f(x) = 0$, $x = 2$ or $x = 4$. This would exist if $x - 2$ and $x - 4$ were factors of the numerator. The horizontal asymptote is $y = 1$, so the numerator and denominator have the same degree. Since the numerator will have degree 2, we must make the denominator also have degree 2. Putting these conditions together, we get a possible function $f(x) = \frac{(x-2)(x-4)}{(x-3)^2}$.

83. The graph has a vertical asymptote $x = 1$, so $x - 1$ is in the denominator of the function. The x-intercept is 2, so that when $f(x) = 0$, $x = 2$. This would exist if $x - 2$ was a factor of the numerator. The horizontal asymptote is $y = -3$, so the numerator and denominator have the same degree. They both have degree 1. Also, from the horizontal asymptote, we have $y = -3 = \frac{a_n}{b_n} = \frac{-3}{1}$. Putting these conditions together, we have a possible function $f(x) = \frac{-3(x-2)}{x-1}$ or $f(x) = \frac{-3x+6}{x-1}$.

85. $C(x) = \dfrac{6.7x}{100-x}$

(a)

(b) To find the cost to remove 95% of the pollutant, find $C(95)$.

$$C(95) = \frac{6.7(95)}{100-95} = 127.3$$

This can also be found using the graphing calculator.

It would cost $127.3 thousand to remove 95% of the pollutant.

Copyright © 2013 Pearson Education, Inc.

87. *Step 1:* $x = ky$

Step 2: Substitute $x = 20$ and $y = 14$ to find

k. $20 = k(14) \Rightarrow k = \frac{20}{14} = \frac{10}{7}$

Step 3: $x = \frac{10}{7}y$

Step 4: Now find y when $x = 50$.

$50 = \frac{10}{7}y \Rightarrow 350 = 10y \Rightarrow y = 35$

89. *Step 1:* $t = \dfrac{k}{s}$

Step 2: Substitute $t = 3$ & $s = 5$ to find k.

$3 = \frac{k}{5} \Rightarrow k = 15$

Step 3: $t = \dfrac{15}{s}$

Step 4: Now find s when $t = 20$.

$20 = \frac{15}{s} \Rightarrow 20s = 15 \Rightarrow s = \frac{15}{20} = \frac{3}{4}$

91. *Step 1:* $f = kg^2h$

Step 2: Substitute $f = 50$, $g = 5$, and $h = 4$ to find k.

$50 = k(5^2)(4) \Rightarrow 50 = 100k \Rightarrow k = \frac{50}{100} = \frac{1}{2}$

Step 3: $f = \frac{1}{2}g^2h$

Step 4: Now find f when $g = 3$ and $h = 6$.

$f = \frac{1}{2}(3^2)(6) = 27$

93. *Step 1:* Let f be the force to keep car from skidding (in lbs), r is the radius of curve (in ft), w is the weight of car (in lbs), s is the speed of the car (in mph). $f = \dfrac{kws^2}{r}$

Step 2: $f = 3000$, $w = 2000$, $r = 500$, and $s = 30$ to find k.

$3000 = \dfrac{k(2000)(30^2)}{500} \Rightarrow 3000 = 3600k \Rightarrow$

$k = \frac{3000}{3600} = \frac{5}{6}$

Step 3: $f = \dfrac{5}{6} \cdot \dfrac{ws^2}{r} = \dfrac{5ws^2}{6r}$

Step 4: Now find f when $w = 2000$, $r = 800$, and $s = 60$

$f = \dfrac{5(2000)(60^2)}{6(800)} = 7500$

A force of 7500 lb is needed.

Chapter 3   Test

1. $f(x) = -2x^2 + 6x - 3$

Complete the square so the function has the form $f(x) = a(x - h)^2 + k$.

$f(x) = -2x^2 + 6x - 3 = -2(x^2 - 3x) - 3$

$\qquad = -2\left(x^2 - 3x + \frac{9}{4} - \frac{9}{4}\right) - 3$

Note: $\left[\frac{1}{2}(-3)\right]^2 = \frac{9}{4}$

$\qquad = -2\left(x - \frac{3}{2}\right)^2 + \frac{9}{2} - 3 = -2\left(x - \frac{3}{2}\right)^2 + \frac{3}{2}$

This parabola is in now in standard form with $a = -2$, $h = \frac{3}{2}$, and $k = \frac{3}{2}$. It opens downward with vertex $(h, k) = \left(\frac{3}{2}, \frac{3}{2}\right)$. The axis is $x = h \Rightarrow x = \frac{3}{2}$. To find the x-intercepts, let $f(x) = 0$.

$-2\left(x - \frac{3}{2}\right)^2 + \frac{3}{2} = 0 \Rightarrow -2\left(x - \frac{3}{2}\right)^2 = -\frac{3}{2}$

$\left(x - \frac{3}{2}\right)^2 = \frac{3}{4} \Rightarrow x - \frac{3}{2} = \pm\sqrt{\frac{3}{4}}$

$x = \frac{3}{2} \pm \sqrt{\frac{3}{4}} = \frac{3}{2} \pm \frac{\sqrt{3}}{2} = \frac{3 \pm \sqrt{3}}{2}$

Thus the x-intercepts are $\frac{3 \pm \sqrt{3}}{2}$ or approximately 0.63 and 2.37.

Note: If you use the quadratic formula to solve $-2x^2 + 6x - 3 = 0$ where $a = -2$, $b = 6$, and $c = -3$, you will arrive at the equivalent answer for x-intercepts: $\frac{-3 \pm \sqrt{3}}{-2}$.

To find the y-intercept, let $x = 0$.

$f(0) = -2(0^2) + 6(0) - 3 = -3$

The y-intercept is -3. The domain is $(-\infty, \infty)$.

Since the highest point on the graph is $\left(\frac{3}{2}, \frac{3}{2}\right)$, the range is $\left(-\infty, \frac{3}{2}\right]$. The function is increasing on $\left(-\infty, \frac{3}{2}\right]$ and decreasing on $\left[\frac{3}{2}, \infty\right)$.

$f(x) = -2x^2 + 6x - 3$

Copyright © 2013 Pearson Education, Inc.

2. $s(t) = -16t^2 + 88t + 48$

   (a) The rocket reaches its maximum height at the vertex of the function.

$$h = -\frac{b}{2a} = -\frac{88}{2(-16)} = 2.75$$

   The rocket reaches its maximum height at 2.75 seconds.

   (b) The maximum height is given by $s(2.75)$.

$$s(2.75) = -16(2.75)^2 + 88(2.75) + 48$$
$$= 169$$

   The maximum height is 169 ft.

   (c) We must solve $-16t^2 + 88t + 48 > 100$.

$$-16t^2 + 88t + 48 = 100$$
$$-16t^2 + 88t - 52 = 0$$
$$t = \frac{-88 \pm \sqrt{88^2 - 4(-16)(-52)}}{2(-16)} \approx 0.7,\ 4.8$$

   Since the vertex lies between these two t-values, we know that the rocket will be more than 100 ft above ground level between 0.7 sec and 4.8 sec. We can verify this graphically.

   (d) The rocket returns to the ground when $s = 0$.

$$-16t^2 + 88t + 48 = 0$$
$$t = \frac{-88 \pm \sqrt{88^2 - 4(-16)(48)}}{2(-16)} = -0.5,\ 6$$

   The negative solution has no meaning, so we disregard it. The rocket returns to ground after 6 seconds.

3. $\dfrac{3x^3 + 4x^2 - 9x + 6}{x + 2}$

Express $x + 2$ in the form $x - k$ by writing it as $x - (-2)$. Thus $k = -2$.

$$\begin{array}{r} -2\overline{)3 \quad 4 \ -9 \quad 6} \\ -6 \quad 4 \quad 10 \\ \hline 3 \ -2 \ -5 \quad 16 \end{array}$$

Thus,

$$\frac{3x^3 + 4x^2 - 9x + 6}{x + 2} = 3x^2 - 2x - 5 + \frac{16}{x + 2}$$

4. $\dfrac{2x^3 - 11x^2 + 25}{x - 5}$

Since $x - 5$ is in the form $x - k$, $k = 5$.

$$\begin{array}{r} 5\overline{)2 \ -11 \quad 0 \quad 25} \\ 10 \ -5 \ -25 \\ \hline 2 \ -1 \ -5 \quad 0 \end{array}$$

$$\frac{2x^3 - 11x^2 + 28}{x - 5} = 2x^2 - x - 5$$

5. $f(x) = 2x^3 - 9x^2 + 4x + 8;\ k = 5$

$$\begin{array}{r} 5\overline{)2 \ -9 \ 4 \quad 8} \\ 10 \ 5 \ 45 \\ \hline 2 \quad 1\ 9 \ 53 \end{array}$$

Thus, $f(5) = 53$.

6. $6x^4 - 11x^3 - 35x^2 + 34x + 24;\ x - 3$

Let $f(x) = 6x^4 - 11x^3 - 35x^2 + 34x + 24$. By the factor theorem, $x - 3$ will be a factor of $f(x)$ only if $f(3) = 0$.

$$\begin{array}{r} 3\overline{)6 \ -11 \ -35 \quad 34 \quad 24} \\ 18 \quad 21 \ -42 \ -24 \\ \hline 6 \quad 7 \ -14 \ -8 \quad 0 \end{array}$$

Since $f(3) = 0$, $x - 3$ is a factor of $f(x)$. The other factor is $6x^3 + 7x^2 - 14x - 8$.

7. $f(x) = x^3 + 8x^2 + 25x + 26;\ -2$

Since -2 is a zero, first divide $f(x)$ by $x + 2$.

$$\begin{array}{r} -2\overline{)1 \ 8 \ 25 \quad 26} \\ -2 \ -12 \ -26 \\ \hline 1 \ 6 \quad 13 \quad 0 \end{array}$$

This gives $f(x) = (x + 2)(x^2 + 6x + 13)$. Since $x^2 + 6x + 13$ cannot be factored, use the quadratic formula with $a = 1$, $b = 6$, and $c = 13$ to find the remaining two zeros.

$$x = \frac{-6 \pm \sqrt{6^2 - 4(1)(13)}}{2(1)} = \frac{-6 \pm \sqrt{36 - 52}}{2}$$

$$= \frac{-6 \pm \sqrt{-16}}{2} = \frac{-6 \pm 4i}{2} = -3 \pm 2i$$

The zeros are -2, $-3 - 2i$, and $-3 + 2i$.

Copyright © 2013 Pearson Education, Inc.

8. Zeros of -1, 2 and i; $f(3) = 80$

By the conjugate zeros theorem, $-i$ is also a zero. The polynomial has the following form.

$$f(x) = a(x+1)(x-2)(x-i)(x+i)$$

Use the condition $f(3) = 80$ to find a.

$$80 = a(3+1)(3-2)(3-i)(3+i)$$
$$80 = a(4)(1)[(3-i)(3+i)]$$
$$80 = a[(4)(1)][3^2 - i^2]$$
$$80 = a(4)[9 - (-1)]$$
$$80 = a(4)(10) \Rightarrow 80 = 40a \Rightarrow a = 2$$

Thus we have the following.

$$f(x) = 2(x+1)(x-2)(x-i)(x+i)$$
$$= 2[(x+1)(x-2)][(x-i)(x+i)]$$
$$= 2(x^2 - x - 2)(x^2 + 1)$$
$$= 2(x^4 - x^3 - x^2 - x - 2)$$
$$= 2x^4 - 2x^3 - 2x^2 - 2x - 4$$

9. Since $f(x) = x^4 + 8x^2 + 12$

$= (x^2 + 6)(x^2 + 2)$, the zeros are $\pm i\sqrt{6}$ and $\pm i\sqrt{2}$. Moreover, $f(x) > 0$ for all x; thus, the graph never crosses or touches the x-axis, so $f(x)$ has no real zeros.

10. (a) $f(x) = x^3 - 5x^2 + 2x + 7$; find $f(1)$ and $f(2)$ synthetically.

$$\begin{array}{r|rrrr} 1) & 1 & -5 & 2 & 7 \\ & & 1 & -4 & -2 \\ \hline & 1 & -4 & -2 & 5 \end{array}$$

$$\begin{array}{r|rrrr} 2) & 1 & -5 & 2 & 7 \\ & & 2 & -6 & -8 \\ \hline & 1 & -3 & -4 & -1 \end{array}$$

By the intermediate value theorem, since $f(1) = 5 > 0$ and $f(2) = -1 < 0$, there must be at least one real zero between 1 and 2.

(b) $f(x) = \underset{1}{x^3 - \underset{2}{5x^2}} + 2x + 7$ has 2 variations in sign. f has 2 or $2 - 2 = 0$ positive real zeros.

$f(-x) = -x^3 - 5x^2 \underset{1}{-2x + 7}$ has 1 variation in sign. f has 1 negative real zero. Thus, the different possibilities are

Positive	Negative	Nonreal Complex
2	1	0
0	1	2

(c) Using the calculator, we find that the real zeros are 4.0937635, 1.8370381, and -0.9308016. For these approximations, a higher accuracy can be found by going to the home screen and displaying the stored x-value.

11. To obtain the graph of g, shift the graph of f 5 units to the left, stretch by a factor of 2, reflect across the x-axis, and shift 3 units up.

12. $f(x) = -x^7 + x - 4$

Since $f(x)$ is of odd degree and the sign of a_n is negative, the end behavior is ⌐↘. The correct graph is C.

13. $f(x) = x^3 - 5x^2 + 3x + 9$

Step 1: The first step is to find the zeros of the polynomial function. p must be a factor of $a_0 = 9$ and q must be a factor of $a_3 = 1$. Thus, p can be $\pm 1, \pm 3, \pm 9$ and q can be ± 1.

The possible zeros, $\frac{p}{q}$, are $\pm 1, \pm 3, \pm 9$. The remainder theorem shows that 3 is a zero.

$$\begin{array}{r|rrrr} 3) & 1 & -5 & 3 & 9 \\ & & 3 & -6 & -9 \\ \hline & 1 & -2 & -3 & 0 \end{array}$$

The new quotient polynomial is $x^2 - 2x - 3$.

(continued on next page)

Copyright © 2013 Pearson Education, Inc.

(*continued*)

$$x^2 - 2x - 3 = 0 \Rightarrow (x+1)(x-3) = 0$$

$$x+1 = 0 \Rightarrow x = -1 \quad \text{or} \quad x-3 = 0 \Rightarrow x = 3$$

The rational zeros are −1 and 3 (multiplicity 2) which divide the *x*-axis into three regions. Test a point in each region to find the sign of $f(x)$ in that region.

Step 2: $f(0) = 9$, so plot $(0,9)$.

Step 3: The *x*-intercepts divide the *x*-axis into three intervals.

Interval	Test Point	Value of $f(x)$	Sign of $f(x)$	Graph Above or Below *x*-Axis
$(-\infty, -1)$	−2	−25	Negative	Below
$(-1, 3)$	0	9	Positive	Above
$(3, \infty)$	4	5	Positive	Above

Plot the *x*-intercepts, *y*-intercept, and test points (the *y*-intercept is one of the test points) with a smooth curve to get the graph.

$$f(x) = x^3 - 5x^2 + 3x + 9$$

14. $f(x) = 2x^2 (x-2)^2$

Step 1: Set each factor equal to 0 and solve the resulting equations to find the zeros of the function.

$$2x^2 = 0 \Rightarrow x = 0 \quad \text{or} \quad (x-2)^2 = 0 \Rightarrow x = 2$$

The zeros are 0 (multiplicity 2) and 2 (multiplicity 2), which divide the *x*-axis into three regions. Test a point in each region to find the sign of $f(x)$ in that region.

Step 2: $f(0) = 0$, so plot $(0, 0)$.

Step 3: The *x*-intercepts divide the *x*-axis into three intervals.

Interval	Test Point	Value of $f(x)$	Sign of $f(x)$	Graph Above or Below *x*-Axis
$(-\infty, 0)$	−1	18	Positive	Above
$(0, 2)$	1	2	Positive	Above
$(2, \infty)$	3	18	Positive	Above

Plot the *x*-intercepts, *y*-intercept (the *y*-intercept is an *x*-intercept), and test points with a smooth curve to get the graph.

$$f(x) = 2x^2 (x-2)^2$$

15. $f(x) = -x^3 - 4x^2 + 11x + 30$

Step 1: The first step is to find the zeros of the polynomial function. *p* must be a factor of $a_0 = 30$ and *q* must be a factor of $a_3 = -1$. Thus, *p* can be $\pm 1,\ \pm 2,\ \pm 3,\ \pm 5,\ \pm 6,\ \pm 10,$ $\pm 15,\ \pm 30$ and *q* can be ± 1. The possible zeros, $\frac{p}{q}$, are $\pm 1,\ \pm 2,\ \pm 3,\ \pm 5,\ \pm 6,\ \pm 10,$ $\pm 15, \pm 30$.

The remainder theorem shows that 3 is a zero.

$$3 \overline{)\begin{array}{rrrr} -1 & -4 & 11 & 30 \\ & -3 & -21 & -30 \end{array}}$$
$$\begin{array}{rrrr} -1 & -7 & -10 & 0 \end{array}$$

The new quotient polynomial is $-x^2 - 7x - 10$.

$$-x^2 - 7x - 10 = 0 \Rightarrow x^2 + 7x + 10 = 0$$
$$(x+5)(x+2) = 0$$

$$x+5 = 0 \Rightarrow x = -5 \quad \text{or} \quad x+2 = 0 \Rightarrow x = -2$$

The rational zeros are −5, −2, and 3, which divide the *x*-axis into four regions. Test a point in each region to find the sign of $f(x)$ in that region.

Step 2: $f(0) = 30$, so plot $(0, 30)$.

Step 3: The *x*-intercepts divide the *x*-axis into four intervals.

Interval	Test Point	Value of $f(x)$	Sign of $f(x)$	Graph Above or Below *x*-Axis
$(-\infty, -5)$	−6	36	Positive	Above
$(-5, -2)$	−3	−12	Negative	Below
$(-2, 3)$	0	30	Positive	Above
$(3, \infty)$	4	−54	Negative	Below

(*continued on next page*)

Copyright © 2013 Pearson Education, Inc.

(*continued*)

Plot the *x*-intercepts, *y*-intercept, and test points (the *y*-intercept is one of the test points) with a smooth curve to get the graph.

$$f(x) = -x^3 - 4x^2 + 11x + 30$$

16. The zeros are –3 and 2. Since the graph of *f* touches the *x*-axis at 2, the zero 2 has multiplicity 2. Thus, $f(x) = a(x-2)^2 (x+3)$.

Also, since the point $(0, 24)$ is on the graph, we have $f(0) = 24$.

$$24 = a(0-2)^2 (0+3) \Rightarrow 24 = a(-2)^2 (3)$$
$$24 = a(4)(3) \Rightarrow 24 = 12a \Rightarrow a = 2$$

The polynomial function is the following.

$$\begin{aligned} f(x) &= 2(x-2)^2 (x+3) \\ &= 2(x^2 - 4x + 4)(x+3) \\ &= 2(x^3 - x^2 - 8x + 12) \\ &= 2x^3 - 2x^2 - 16x + 24 \end{aligned}$$

17. $f(t) = 1.06t^3 - 24.6t^2 + 180t$

(a) $f(2) = 1.06(2^3) - 24.6(2)^2 + 180(2)$
$= 270.08$

This can also be found using the graphing calculator.

(b) From the graph we see that the amount of change is increasing from $t = 0$ to $t = 5.9$ and from $t = 9.5$ to $t = 15$ and decreasing from $t = 5.9$ to $t = 9.5$.

18. $f(x) = \dfrac{3x-1}{x-2}$

Step 1: The graph has a vertical asymptote where $x - 2 = 0$, that is, when $x = 2$.

Step 2: Since the degree of the numerator equals the degree of the denominator, the graph has a horizontal asymptote at $y = \frac{3}{1} = 3$.

Step 3: The *y*-intercept is

$$f(0) = \frac{3(0)-1}{0-2} = \frac{0-1}{-2} = \frac{-1}{-2} = \frac{1}{2}.$$

Step 4: Any *x*-intercepts are found by solving $f(x) = 0$.

$$\frac{3x-1}{x-2} = 0 \Rightarrow 3x - 1 = 0 \Rightarrow 3x = 1 \Rightarrow x = \frac{1}{3}$$

The only *x*-intercept is $\frac{1}{3}$.

Step 5: The graph will intersect the horizontal asymptote when $\frac{3x-1}{x-2} = 3$.

$$\frac{3x-1}{x-2} = 3 \Rightarrow 3x - 1 = 3(x-2) \Rightarrow$$
$$3x - 1 = 3x - 6 \Rightarrow -1 = -6$$

This is a false statement, so the graph does not intersect the horizontal asymptote.

Step 6: Since the vertical asymptote is $x = 2$ and the *x*-intercept occurs at $\frac{1}{3}$, we must determine values in three intervals.

Interval	Test Point	Value of $f(x)$	Sign of $f(x)$	Graph Above or Below *x*-Axis
$\left(-\infty, \frac{1}{3}\right)$	0	$\frac{1}{2}$	Positive	Above
$\left(\frac{1}{3}, 2\right)$	1	–2	Negative	Below
$(2, \infty)$	3	8	Positive	Above

Step 7: Use the asymptotes, intercepts, and test points (the *y*-intercept is one of the test points) to sketch the graph.

Copyright © 2013 Pearson Education, Inc.

19. $f(x) = \frac{x^2-1}{x^2-9} = \frac{(x+1)(x-1)}{(x+3)(x-3)}$

Step 1: The graph has a vertical asymptote where $(x+3)(x-3) = 0$, that is, when $x = -3$ and $x = 3$.

Step 2: Since the degree of the numerator equals the degree of the denominator, the graph has a horizontal asymptote at $y = \frac{1}{1} = 1$.

Step 3: The y-intercept is

$f(0) = \frac{0^2-1}{0^2-9} = \frac{-1}{-9} = \frac{1}{9}$.

Step 4: Any x-intercepts are found by solving $f(x) = 0$.

$\frac{(x+1)(x-1)}{(x+3)(x-3)} = 0 \Rightarrow (x+1)(x-1) = 0$

$x = -1$ or $x = 1$

The x-intercepts are -1 and 1.

Step 5: The graph will intersect the horizontal asymptote when $\frac{x^2-1}{x^2-9} = 1$.

$\frac{x^2-1}{x^2-9} = 1 \Rightarrow x^2 - 1 = x^2 - 9 \Rightarrow -1 = -9$

This is a false statement, so the graph does not intersect the horizontal asymptote.

Step 6: Since the vertical asymptotes are $x = -3$ and $x = 3$, and the x-intercepts occur at -1 and 1, we must determine values in five intervals.

Interval	Test Point	Value of $f(x)$	Sign of $f(x)$	Graph Above or Below x-Axis
$(-\infty, -3)$	-4	$\frac{15}{7}$	Positive	Above
$(-3, -1)$	-2	$-\frac{3}{5}$	Negative	Below
$(-1, 1)$	0	$\frac{1}{9}$	Positive	Above
$(1, 3)$	2	$-\frac{3}{5}$	Negative	Below
$(3, \infty)$	4	$\frac{15}{7}$	Positive	Above

Step 7: Use the asymptotes, intercepts, and test points (the y-intercept is one of the test points) to sketch the graph. Note: This function is an even function, thus the graph is symmetric with respect to the y-axis.

$f(x) = \frac{x^2-1}{x^2-9}$

20. $f(x) = \frac{2x^2 + x - 6}{x-1}$

(a) Since the degree of the numerator is one more than the degree of the denominator, the graph has an oblique asymptote.

Divide $2x^2 + x - 6$ by $x - 1$.

$$1\overline{)\begin{array}{rrr} 2 & 1 & -6 \\ & 2 & 3 \\ \hline 2 & 3 & -3 \end{array}}$$

$f(x) = \frac{2x^2+x-6}{x-1} = 2x + 3 - \frac{3}{x-1}$

The oblique asymptote is the line $y = 2x + 3$.

(b) To find the x-intercepts, let $f(x) = 0$.

$\frac{2x^2+x-6}{x-1} = 0 \Rightarrow 2x^2 + x - 6 = 0$

$(x+2)(2x-3) = 0$

$x = -2$ or $x = \frac{3}{2}$

The x-intercepts are -2 and $\frac{3}{2}$.

(c) The y-intercept is

$f(0) = \frac{2(0^2)+0-6}{0-1} = \frac{2(0)+0-6}{-1}$

$= \frac{0+0-6}{-1} = \frac{-6}{-1} = 6$.

(d) To find the vertical asymptote, set the denominator equal to zero and solve for x.

$x - 1 = 0 \Rightarrow x = 1$

The equation of the vertical asymptote is $x = 1$.

(e) Use the information from (a)–(d) and a few additional points to graph the function.

Since the vertical asymptote is $x = 1$ and the x-intercepts occur at -2 and $\frac{3}{2}$, we must determine values in four intervals.

(continued on next page)

Copyright © 2013 Pearson Education, Inc.

(*continued*)

Interval	Test Point	Value of $f(x)$	Sign of $f(x)$	Graph Above or Below x-Axis
$(-\infty, -2)$	-3	$-\frac{9}{4}$	Negative	Below
$(-2, 1)$	0	6	Positive	Above
$\left(1, \frac{3}{2}\right)$	$\frac{5}{4}$	$-\frac{13}{2}$	Negative	Below
$\left(\frac{3}{2}, \infty\right)$	2	4	Positive	Above

$y = 2x + 3 \qquad x = 1$

$$f(x) = \frac{2x^2 + x - 6}{x - 1}$$

21. *Step 1*: $y = k\sqrt{x}$

    Step 2: Substitute $y = 12$ and $x = 4$ to find k.

    $12 = k\sqrt{4} \Rightarrow 12 = 2k \Rightarrow k = 6$

    Step 3: $y = 6\sqrt{x}$

    Step 4: Now find y when $x = 100$.

    $y = 6\sqrt{100} = 6 \cdot 10 = 60$

22. *Step 1*: Let w be the weight of the object (in kg), d is the distance from the center of Earth (in km). $w = \dfrac{k}{d^2}$

    Step 2: Substitute $w = 90$ and $d = 6400$ to find k.

    $90 = \dfrac{k}{6400^2} \Rightarrow k = 3{,}686{,}400{,}000$

    Step 3: $w = \dfrac{3{,}686{,}400{,}000}{d^2}$

    Step 4: Now find w when $d = 800 + 6400 = 7200$.

    $w = \frac{3{,}686{,}400{,}000}{7200^2} = \frac{3{,}686{,}400{,}000}{51{,}840{,}000} = \frac{640}{9} \approx 71.1$

    The man weighs $\frac{640}{9}$ kg or approximately 71.1 kg.

Copyright © 2013 Pearson Education, Inc.

Chapter 4

INVERSE, EXPONENTIAL, AND LOGARITHMIC FUNCTIONS

Section 4.1   Inverse Functions

1. Yes, it is one-to-one, because every number in the list of registered passenger cars is used only once.

3. This is a one-to-one function since every horizontal line intersects the graph in no more than one point.

5. This function is not one-to-one because there are infinitely many horizontal lines that intersect the graph in two points.

7. This function is one-to-one because every horizontal line will intersect the graph in exactly one point.

9. $y = 2x - 8$

    Using the definition of a one-to-one function, we have $f(a) = f(b) \Rightarrow 2a - 8 = 2b - 8 \Rightarrow$ $2a = 2b \Rightarrow a = b$. So the function is one-to-one.

11. $y = \sqrt{36 - x^2}$

    If $x = 6$, $y = \sqrt{36 - 6^2} = \sqrt{36 - 36} = \sqrt{0} = 0$.
    If $x = -6$,
    $$y = \sqrt{36 - (-6)^2} = \sqrt{36 - 36} = \sqrt{0} = 0.$$
    Since two different values of x lead to the same value of y, the function is not one-to-one.

13. $y = 2x^3 - 1$

    Looking at this function graphed on a TI-83, we can see that it appears that any horizontal line passed through the function will intersect the graph in at most one place.

Another way of showing that a function is one-to-one is to assume that you have two equal y-values $\big(f(a) = f(b)\big)$ and show that they must have come from the same x-value $(a = b)$.

$f(a) = f(b) \Rightarrow 2a^3 - 1 = 2b^3 - 1 \Rightarrow$
$2a^3 = 2b^3 \Rightarrow a^3 = b^3 \Rightarrow \sqrt[3]{a^3} = \sqrt[3]{b^3} \Rightarrow a = b$
So, the function is one-to-one.

15. $y = -\dfrac{1}{x + 2}$

    Looking at this function graphed on a TI-83, we can see that it appears that any horizontal line passed through the function will intersect the graph in at most one place.

We could also show that $f(a) = f(a)$ implies $a = b$.

$f(a) = f(b) \Rightarrow -\frac{1}{a+2} = -\frac{1}{b+2} \Rightarrow$
$b + 2 = a + 2 \Rightarrow b = a$
So, the function is one-to-one.

17. $y = 2(x + 1)^2 - 6$

    Looking at this function graphed on a TI-83, we can see that it appears that any horizontal line passed through the function will intersect the graph in two places, except a horizontal line through the vertex. For example,
    $f(0) = 2(0 + 1)^2 - 6 = -4$ and
    $f(-2) = 2(-2 + 1)^2 - 6 = -4$

So, the function is not one-to-one.

Copyright © 2013 Pearson Education, Inc.

19. $y = \sqrt[3]{x+1} - 3$

Looking at this function graphed on a TI-84, we can see that it appears that any horizontal line passed through the function will intersect the graph in at most one place.

We could also show that $f(a) = f(a)$ implies $a = b$.

$f(a) = f(b) \Rightarrow \sqrt[3]{a+1} - 3 = \sqrt[3]{b+1} - 3 \Rightarrow$
$\sqrt[3]{a+1} = \sqrt[3]{b+1} \Rightarrow a+1 = b+1 \Rightarrow a = b$

So, the function is one-to-one.

21. For a constant function defined over the set of real numbers, $f(x) = c$ for all values of c. Therefore, the function is not one-to-one.

23. In order for a function to have an inverse, it must be <u>one-to-one</u>.

25. The domain of f is equal to the <u>range</u> of f^{-1}, and the range of f is equal to the <u>domain</u> of f^{-1}.

27. It is false that if $f(x) = x^2$, then $f^{-1}(x) = \sqrt{x}$ because f is not a one-to-one function and, hence, does not have an inverse.

29. If a function f has an inverse and $f(-3) = 6$, then $f^{-1}(6) = \underline{-3}$.

31. The inverse operation of tying your shoelaces would be untying your shoelaces, since untying "undoes" tying.

33. The inverse operation of entering a room would be leaving a room, since leaving "undoes" entering.

35. The inverse operation of screwing in a light bulb would be unscrewing the light bulb.

37. For each point (x, y) for the first function, there is a point (y, x) for the second function, so $f(x)$ and $g(x)$ are inverses of each other.

39. The point $(3, 5)$ is on $f(x)$, but the point $(5, 3)$ is not on $g(x)$ (there is another example), so the functions are not inverses of each other.

41. $f(x) = 2x + 4, g(x) = \frac{1}{2}x - 2$

$(f \circ g)(x) = 2\left(\frac{1}{2}x - 2\right) + 4 = x - 4 + 4 = x$

$(g \circ f)(x) = \frac{1}{2}(2x + 4) - 2 = x + 2 - 2 = x$

Since $(f \circ g)(x) = x$ and $(g \circ f)(x) = x$, these functions are inverses.

43. $f(x) = -3x + 12, g(x) = -\frac{1}{3}x - 12$

$(f \circ g)(x) = -3\left(-\frac{1}{3}x - 12\right) + 12$
$= x + 36 + 12 = x + 48$

Since $(f \circ g)(x) \neq x$, the functions are not inverses. It is not necessary to check $(g \circ f)(x)$.

45. $f(x) = \frac{x+1}{x-2}, g(x) = \frac{2x+1}{x-1}$

$(f \circ g)(x) = \dfrac{\frac{2x+1}{x-1} + 1}{\frac{2x+1}{x-1} - 2} = \dfrac{\frac{2x+1+x-1}{x-1}}{\frac{2x+1-2(x-1)}{x-1}} = \dfrac{3x}{3} = x$

$(g \circ f)(x) = \dfrac{2\left(\frac{x+1}{x-2}\right) + 1}{\frac{x+1}{x-2} - 1} = \dfrac{\frac{2x+2}{x-2} + 1}{\frac{x+1-(x-2)}{x-2}}$

$= \dfrac{\frac{2x+2+x-2}{x-2}}{\frac{x+1-(x-2)}{x-2}} = \dfrac{3x}{3} = x$

Since $(f \circ g)(x) = x$ and $(g \circ f)(x) = x$, these functions are inverses.

47. $f(x) = \frac{2}{x+6}, g(x) = \frac{6x+2}{x}$

$(f \circ g)(x) = \dfrac{2}{\frac{6x+2}{x} + 6} = \dfrac{2}{\frac{6x+2+6x}{x}} = \dfrac{2}{1} \cdot \dfrac{x}{12x+2}$

$= \dfrac{2x}{12x+2} = \dfrac{x}{6x+1} \neq x$

Since $(f \circ g)(x) \neq x$, the functions are not inverses. It is not necessary to check $(g \circ f)(x)$.

49. $f(x) = x^2 + 3$, domain $[0, \infty)$;
$g(x) = \sqrt{x-3}$, domain $[3, \infty)$

$(f \circ g)(x) = f\left(\sqrt{x-3}\right) = \left(\sqrt{x-3}\right)^2 + 3 = x$

$(g \circ f)(x) = g\left(x^2 + 3\right) = \sqrt{x^2 + 3 - 3}$
$= \sqrt{x^2} = |x| = x$ for $[0, \infty)$

Since $(f \circ g)(x) = x$ and $(g \circ f)(x) = x$, these functions are inverses.

51. Since each y-value corresponds to only one x-value, this function is one-to one and has an inverse. The inverse is: $\{(6, -3), (1, 2), (8, 5)\}$.

Copyright © 2013 Pearson Education, Inc.

53. Since the y-value -3 corresponds to two different x-values, this function is not one-to-one.

55. These functions are inverses since their graphs are symmetric with respect to the line $y = x$.

57. These functions are not inverses since their graphs are not symmetric with respect to the line $y = x$.

59. $y = 3x - 4$

The function, $f(x) = 3x - 4$, is one-to-one.

(a) *Step 1*: Interchange x and y: $x = 3y - 4$

Step 2: Solve for y.

$x = 3y - 4 \Rightarrow x + 4 = 3y \Rightarrow \frac{x+4}{3} = y \Rightarrow$

$y = \frac{x+4}{3} = \frac{1}{3}x + \frac{4}{3}$

Step 3: Replace y with $f^{-1}(x)$.

$f^{-1}(x) = \frac{1}{3}x + \frac{4}{3}$

(b) The graph of the original function, $f(x) = 3x - 4$, is a line with slope 3 and y-intercept -4. Since $f^{-1}(x) = \frac{1}{3}x + \frac{4}{3}$, the graph of the inverse function is a line with slope $\frac{1}{3}$ and y-intercept $\frac{4}{3}$.

(c) For both $f(x)$ and $f^{-1}(x)$, the domain is $(-\infty, \infty)$ and the range is $(-\infty, \infty)$.

61. $f(x) = -4x + 3$

This function is one-to-one.

(a) *Step 1*: Replace $f(x)$ with y and interchange x and y.

$y = -4x + 3$

$x = -4y + 3$

Step 2: Solve for y.

$x = -4y + 3 \Rightarrow x - 3 = -4y \Rightarrow \frac{x-3}{-4} = y \Rightarrow$

$y = \frac{x-3}{-4} = -\frac{1}{4}x + \frac{3}{4}$

Step 3: Replace y with $f^{-1}(x)$.

$f^{-1}(x) = -\frac{1}{4}x + \frac{3}{4}$

(b) The graph of the original function, $f(x) = -4x + 3$, is a line with slope -4 and y-intercept 3. Since $f^{-1}(x) = -\frac{1}{4}x + \frac{3}{4}$, the graph of the inverse function is a line with slope $-\frac{1}{4}$ and y-intercept $\frac{3}{4}$.

(c) For both $f(x)$ and $f^{-1}(x)$, the domain is $(-\infty, \infty)$ and the range is $(-\infty, \infty)$.

63. $f(x) = x^3 + 1$

This function is one-to-one.

(a) *Step 1*: Replace $f(x)$ with y and interchange x and y.

$y = x^3 + 1 \Rightarrow x = y^3 + 1$

Step 2: Solve for y.

$x = y^3 + 1 \Rightarrow x - 1 = y^3 \Rightarrow$

$\sqrt[3]{x-1} = y \Rightarrow y = \sqrt[3]{x-1}$

Step 3: Replace y with $f^{-1}(x)$.

$f^{-1}(x) = \sqrt[3]{x-1}$

(b) Tables of ordered pairs will be helpful in drawing the graphs of these functions.

x	$f(x)$	x	$f^{-1}(x)$
-2	-7	-7	-2
-1	0	0	-1
0	1	1	0
1	2	2	1
2	9	9	2

(c) For both $f(x)$ and $f^{-1}(x)$, the domain is $(-\infty, \infty)$ and the range is $(-\infty, \infty)$.

Copyright © 2013 Pearson Education, Inc.

65. $y = x^2$

This is not a one-to-one function since two different x-values can correspond to the same y-value ($2^2 = 4$ and $(-2)^2 = 4$, for example), so this function is not one-to-one. Thus, the function has no inverse function.

67. $y = \dfrac{1}{x}$, $x \neq 0$

The function, $f(x) = \frac{1}{x}$, is one-to-one.

(a) *Step 1*: Interchange x and y.

$$y = \frac{1}{x} \Rightarrow x = \frac{1}{y}$$

Step 2: Solve for y.

$$x = \frac{1}{y} \Rightarrow xy = 1 \Rightarrow y = \frac{1}{x}$$

Step 3: Replace y with $f^{-1}(x)$.

$$f^{-1}(x) = \frac{1}{x}, \ x \neq 0 = f(x).$$

(b) Tables of ordered pairs will be helpful in drawing the graph of this function (in this case, $f(x) = f^{-1}(x)$).

x	$f(x) = f^{-1}(x)$
-2	$-\frac{1}{2}$
-1	-1
$-\frac{1}{2}$	-2
$\frac{1}{2}$	2
1	1
2	$\frac{1}{2}$

(c) For both $f(x)$ and $f^{-1}(x)$, the domain and range are both $(-\infty, 0) \cup (0, \infty)$.

69. $f(x) = \dfrac{1}{x-3}$, $x \neq 3$

This function is one-to-one.

(a) *Step 1*: Replace $f(x)$ with y and interchange x and y.

$$y = \frac{1}{x-3} \Rightarrow x = \frac{1}{y-3}$$

Step 2: Solve for y.

$$x = \frac{1}{y-3} \Rightarrow x(y-3) = 1 \Rightarrow xy - 3x = 1 \Rightarrow$$

$$xy = 1 + 3x \Rightarrow y = \frac{1+3x}{x}$$

Step 3: Replace y with $f^{-1}(x)$.

$$f^{-1}(x) = \frac{1+3x}{x}, \ x \neq 0$$

(b) To graph $f(x) = \frac{1}{x-3}$, we can determine that there are no x-intercepts. The y-intercept is $f(0) = \frac{1}{0-3} = -\frac{1}{3}$. There is a vertical aymptote when the denominator is zero, $x - 3 = 0$, which implies $x = 3$ is the vertical asymptote. Also, the horizontal asymptote is $y = 0$ since the degree of the numerator is less than the denominator. Examining the following intervals, we have test points which will be helpful in drawing the graph of f as well as f^{-1}.

Interval	Test Point	Value of $f(x)$	Sign of $f(x)$	Graph Above or Below x-Axis
$(-\infty, 3)$	2	-1	Negative	Below
$(3, \infty)$	4	1	Positive	Above

Plot the vertical asymptote, y-intercept, and test points with a smooth curve to get the graph of f.

To graph $f^{-1}(x) = \frac{1+3x}{x}$, we can determine that the x-intercept occurs when $1 + 3x = 0 \Rightarrow x = -\frac{1}{3}$. There is no y-intercept since 0 is not in the domain f^{-1}. There is a vertical aymptote when the denominator is zero, namely $x = 0$. Also, since the degree of the numerator is the same as the denominator, the horizontal asymptote is $y = \frac{3}{1} = 3$.

Using this information along with the points $(-1, 2)$ and $(1, 4)$, we can sketch f^{-1}.

(c) Domain of f = range of $f^{-1} = (-\infty, 3) \cup (3, \infty)$; Domain of f^{-1} = range of $f = (-\infty, 0) \cup (0, \infty)$

Copyright © 2013 Pearson Education, Inc.

71. $f(x) = \dfrac{x+1}{x-3},\ x \neq 3$

This function is one-to-one.

(a) *Step 1:* Replace $f(x)$ with y and interchange x and y.

$$y = \frac{x+1}{x-3} \Rightarrow x = \frac{y+1}{y-3}$$

Step 2: Solve for y.

$$x = \frac{y+1}{y-3} \Rightarrow x(y-3) = y+1 \Rightarrow$$
$$xy - 3x = y + 1 \Rightarrow xy - y = 3x + 1 \Rightarrow$$
$$y(x-1) = 3x + 1 \Rightarrow y = \frac{3x+1}{x-1}$$

Step 3: Replace y with $f^{-1}(x)$.

$$f^{-1}(x) = \frac{3x+1}{x-1},\ x \neq 1$$

(b) To graph $f(x) = \frac{x+1}{x-3}$, find the x-intercept:

$\frac{x+1}{x-3} = 0 \Rightarrow x + 1 = 0 \Rightarrow x = -1.$

The y-intercept is $f(0) = \frac{0+1}{0-3} = -\frac{1}{3}$.

There is a vertical asymptote when the denominator is zero, $x - 3 = 0 \Rightarrow x = 3$ is the vertical asymptote. Since the degree of the numerator equals the degree of the denominator, the horizontal asymptote is $y = \frac{1}{1} = 1$. Examining the following intervals, we have test points which will be helpful in drawing the graph of f as well as f^{-1}.

Interval	Test Point	Value of $f(x)$	Sign of $f(x)$	Graph Above or Below x-Axis
$(-\infty, -1)$	-3	$\frac{1}{3}$	Positive	Above
$(-1, 3)$	$\frac{1}{2}$	$-\frac{3}{5}$	Negative	Below
$(3, \infty)$	4	5	Positive	Above

Plot the vertical asymptote, y-intercept, and test points with a smooth curve to get the graph of f.

To graph $f^{-1}(x) = \frac{3x+1}{x-1}$ we can determine that the x-intercept occurs when $3x + 1 = 0 \Rightarrow x = -\frac{1}{3}$. The y-intercept is $f(0) = \frac{3(0)+1}{0-1} = -1$.

There is a vertical asymptote when the denominator is zero, $x - 1 = 0 \Rightarrow x = 1$. Also, since the degree of the numerator is the same as the denominator, the horizontal asymptote is $y = \frac{3}{1} = 3$. Examining the following intervals, we have test points which will be helpful in drawing the graph of f as well as f^{-1}.

Interval	Test Point	Value of $f(x)$	Sign of $f(x)$	Graph Above or Below x-Axis
$\left(-\infty, -\frac{1}{3}\right)$	-2	$\frac{5}{3}$	Positive	Above
$\left(-\frac{1}{3}, 1\right)$	$\frac{1}{2}$	-5	Negative	Below
$(1, \infty)$	2	7	Positive	Above

Plot the vertical asymptote, y-intercept, and test points with a smooth curve to get the graph of f^{-1}.

(c) Domain of f = range of $f^{-1} = (-\infty, 3) \cup (3, \infty);$

Domain of f^{-1} = range of $f = (-\infty, 1) \cup (1, \infty)$

73. $f(x) = \dfrac{2x+6}{x-3},\ x \neq 3$

This function is one-to-one.

(a) *Step 1:* Replace $f(x)$ with y and interchange x and y.

$$y = \frac{2x+6}{x-3} \Rightarrow x = \frac{2y+6}{y-3}$$

Step 2: Solve for y.

$$x = \frac{2y+6}{y-3} \Rightarrow x(y-3) = 2y+6 \Rightarrow$$
$$xy - 3x = 2y + 6 \Rightarrow xy - 2y = 3x + 6 \Rightarrow$$
$$y(x-2) = 3x + 6 \Rightarrow y = \frac{3x+6}{x-2}$$

(continued on next page)

Copyright © 2013 Pearson Education, Inc.

224 Chapter 4 Inverse, Exponential, and Logarithmic Functions

(continued)

Step 3: Replace y with $f^{-1}(x)$.

$$f^{-1}(x) = \frac{3x+6}{x-2},\ x \neq 2.$$

(b) To graph $f(x) = \frac{2x+6}{x-3}$, find the x-intercept:

$\frac{2x+6}{x-3} = 0 \Rightarrow 2x+6 = 0 \Rightarrow 2x = -6 \Rightarrow x = -3$

The y-intercept is $f(0) = \frac{2(0)+6}{0-3} = -2$.

There is a vertical aymptote when the denominator is zero, $x-3 = 0 \Rightarrow x = 3$ is the vertical asymptote. Since the degree of the numerator equals the degree of the denominator, the horizontal asymptote is $y = \frac{2}{1} = 2$. Examining the following intervals, we have test points which will be helpful in drawing the graph of f as well as f^{-1}.

Interval	Test Point	Value of $f(x)$	Sign of $f(x)$	Graph Above or Below x-Axis
$(-\infty,-3)$	-9	1	Positive	Above
$(-3, 3)$	-1	-1	Negative	Below
$(3, \infty)$	6	6	Positive	Above

Plot the vertical asymptote, y-intercept, and test points with a smooth curve to get the graph of f.

To graph $f^{-1}(x) = \frac{3x+6}{x-2}$, find the x-intercept:

$\frac{3x+6}{x-2} = 0 \Rightarrow 3x+6 = 0 \Rightarrow 3x = -6 \Rightarrow x = -2$

The y-intercept is $f(0) = \frac{3(0)+6}{0-2} = -3$.

There is a vertical aymptote when the denominator is zero, $x-2 = 0 \Rightarrow x = 2$ is the vertical asymptote. Since the degree of the numerator equals the degree of the denominator, the horizontal asymptote is $y = \frac{3}{1} = 3$. Examining the following intervals, we have test points which will be helpful in drawing the graph of f as well as f^{-1}.

Interval	Test Point	Value of $f(x)$	Sign of $f(x)$	Graph Above or Below x-Axis
$(-\infty,-2)$	-4	1	Positive	Above
$(-2, 2)$	-1	-1	Negative	Below
$(2, \infty)$	3	15	Positive	Above

Plot the vertical asymptote, y-intercept, and test points with a smooth curve to get the graph of f.

(c) Domain of f = range of $f^{-1} = (-\infty, 3) \cup (3, \infty)$;

Domain of f^{-1} = range of $f = (-\infty, 2) \cup (2, \infty)$

75. $f(x) = \sqrt{6+x}$

This function is one-to-one.

(a) *Step 1:* Replace $f(x)$ with y and interchange x and y.

$y = \sqrt{6+x} \Rightarrow x = \sqrt{6+y}$

Step 2: Solve for y. In this problem we must consider that the range of f will be the domain of f^{-1}.

$x = \sqrt{6+y}$

$x^2 = \left(\sqrt{6+y}\right)^2$, for $x \geq 0$

$x^2 = 6+y$, for $x \geq 0$

$x^2 - 6 = y$, for $x \geq 0$

Step 3: Replace y with $f^{-1}(x)$.

$f^{-1}(x) = x^2 - 6$, for $x \geq 0$

(b) Tables of ordered pairs will be helpful in drawing the graphs of these functions.

x	$f(x)$	x	$f^{-1}(x)$
-6	0	0	-6
-5	1	1	-5
-2	2	2	-2
3	3	3	3

(continued on next page)

Copyright © 2013 Pearson Education, Inc.

(continued)

(c) Domain of f = range of f^{-1} = $[-6, \infty)$;

Range of f = domain of f^{-1} = $[0, \infty)$

77. Draw the mirror image of the original graph across the line $y = x$.

79. Carefully draw the mirror image of the original graph across the line $y = x$.

81. Draw the mirror image of the original graph across the line $y = x$.

83. To find $f^{-1}(4)$, find the point with y-coordinate equal to 4. That point is (4, 4). The graph of f^{-1} contains (4, 4). Hence

$f^{-1}(4) = 4$.

85. To find $f^{-1}(0)$, find the point with y-coordinate equal to 0. That point is $(2, 0)$. The graph of f^{-1} contains $(0, 2)$. Hence $f^{-1}(0) = 2$.

87. To find $f^{-1}(-3)$, find the point with y-coordinate equal to -3. That point is $(-2, -3)$. The graph of f^{-1} contains $(-3, -2)$. Hence $f^{-1}(-3) = -2$.

89. $f^{-1}(1000)$ represents the number of dollars required to build 1000 cars.

91. If a line has slope a, the slope of its reflection in the line $y = x$ will be reciprocal of a, which is $\frac{1}{a}$.

93. $f(x) = 6x^3 + 11x^2 - 6$

The horizontal line test will show that this function is not one-to-one.

95. $f(x) = \dfrac{x-5}{x+3}$, $x \ne -3$

The horizontal line test will show that this function is one-to-one.

Find the equation of f^{-1}.

Step 1: Replace $f(x) = \frac{x-5}{x+3}$ with y and interchange x and y.

$y = \frac{x-5}{x+3} \Rightarrow x = \frac{y-5}{y+3}$

Step 2: Solve for y.

$x = \frac{y-5}{y+3} \Rightarrow x(y+3) = y - 5 \Rightarrow$

$xy + 3x = y - 5 \Rightarrow xy - y = -5 - 3x \Rightarrow$

$y(x-1) = -5 - 3x \Rightarrow y = \frac{-5-3x}{x-1}$

Step 3: Replace y with $f^{-1}(x)$.

$f^{-1}(x) = \frac{-5-3x}{x-1}$, $x \ne 1$

Copyright © 2013 Pearson Education, Inc.

97. Given $f(x) = 3x - 2$, find $f^{-1}(x)$.

Step 1: Replace $f(x) = 3x - 2$ with y and interchange x and y.

$y = 3x - 2 \Rightarrow x = 3y - 2$

Step 2: Solve for y.

$x = 3y - 2 \Rightarrow x + 2 = 3y \Rightarrow \frac{x+2}{3} = y$

Step 3: Replace y with $f^{-1}(x)$.

$f^{-1}(x) = \frac{x+2}{3}$

$37;\ f^{-1}(37) = \frac{37+2}{3}$ $= \frac{39}{3} = 13;\ \text{M}$	$25;\ f^{-1}(25) = \frac{25+2}{3}$ $= \frac{27}{3} = 9;\ \text{I}$
$19;\ f^{-1}(19) = \frac{19+2}{3}$ $= \frac{21}{3} = 7;\ \text{G}$	$61;\ f^{-1}(61) = \frac{61+2}{3}$ $= \frac{63}{3} = 21;\ \text{U}$
$13;\ f^{-1}(13) = \frac{13+2}{3}$ $= \frac{15}{3} = 5;\ \text{E}$	$34;\ f^{-1}(34) = \frac{34+2}{3}$ $= \frac{36}{3} = 12;\ \text{L}$
$22;\ f^{-1}(22) = \frac{22+2}{3}$ $= \frac{24}{3} = 8;\ \text{H}$	$1;\ f^{-1}(1) = \frac{1+2}{3}$ $= \frac{3}{3} = 1;\ \text{A}$
$55;\ f^{-1}(55) = \frac{55+2}{3}$ $= \frac{57}{3} = 19;\ \text{S}$	$1;\ f^{-1}(1) = \frac{1+2}{3}$ $= \frac{3}{3} = 1;\ \text{A}$
$52;\ f^{-1}(52) = \frac{52+2}{3}$ $= \frac{54}{3} = 18;\ \text{R}$	$52;\ f^{-1}(52) = \frac{52+2}{3}$ $= \frac{54}{3} = 18;\ \text{R}$
$25;\ f^{-1}(25) = \frac{25+2}{3}$ $= \frac{27}{3} = 9;\ \text{I}$	$64;\ f^{-1}(64) = \frac{64+2}{3}$ $= \frac{66}{3} = 22;\ \text{V}$
$13;\ f^{-1}(13) = \frac{13+2}{3}$ $= \frac{15}{3} = 5;\ \text{E}$	$10;\ f^{-1}(10) = \frac{10+2}{3}$ $= \frac{12}{3} = 4;\ \text{D}$

The message is MIGUEL HAS ARRIVED.

99. Given $f(x) = x^3 - 1$, we have the following.

$S = 19;$ $f(19) = 19^3 - 1$ $= 6859 - 1$ $= 6858$	$E = 5;$ $f(5) = 5^3$ $= 125 - 1 = 124$
$N = 14;$ $f(14) = 14^3 - 1$ $= 2744 - 1$ $= 2743$	$D = 4;$ $f(4) = 4^3 - 1$ $= 64 - 1 = 63$
$H = 8;$ $f(8) = 8^3 - 1$ $= 512 - 1 = 511$	$E = 5;$ $f(5) = 5^3 - 1$ $= 125 - 1 = 124$
$L = 12;$ $f(12) = 12^3 - 1$ $= 1728 - 1$ $= 1727$	$P = 16;$ $f(16) = 16^3 - 1$ $= 4096 - 1$ $= 4095$

Given $f(x) = x^3 - 1$, find $f^{-1}(x)$.

Step 1: Replace $f(x) = x^3 - 1$ with y and interchange x and y.

$y = x^3 - 1 \Rightarrow x = y^3 - 1$

Step 2: Solve for y.

$x = y^3 - 1 \Rightarrow x + 1 = y^3 \Rightarrow \sqrt[3]{x+1} = y$

Step 3: Replace y with $f^{-1}(x)$.

$f^{-1}(x) = \sqrt[3]{x+1}$

Section 4.2   Exponential Functions

For exercises 1–15, $f(x) = 3^x$ and $g(x) = \left(\frac{1}{4}\right)^x$.

1. $f(2) = 3^2 = 9$

3. $f(-2) = 3^{-2} = \frac{1}{3^2} = \frac{1}{9}$

5. $g(2) = \left(\frac{1}{4}\right)^2 = \frac{1}{16}$

7. $g(-2) = \left(\frac{1}{4}\right)^{-2} = 4^2 = 16$

9. $f\left(\frac{3}{2}\right) = \left(\sqrt{3}\right)^3 = \sqrt{27} = \sqrt{9 \cdot 3} = 3\sqrt{3}$

11. $g\left(\frac{3}{2}\right) = \left(\frac{1}{4}\right)^{3/2} = \frac{1}{\left(\sqrt{4}\right)^3} = \frac{1}{2^3} = \frac{1}{8}$

13. $f(2.34) = 3^{2.34} \approx 13.076$

15. $g(-1.68) = \left(\frac{1}{4}\right)^{-1.68} \approx 10.267$

17. The y-intercept of $f(x) = 3^x$ is 1, and the x-axis is a horizontal asymptote. Make a table of values.

x	$f(x)$
-2	$\frac{1}{9} \approx 0.1$
-1	$\frac{1}{3} \approx 0.3$
$-\frac{1}{2}$	≈ 0.6
0	1
$\frac{1}{2}$	≈ 1.7
1	3
2	9

(continued on next page)

Copyright © 2013 Pearson Education, Inc.

(continued)

Plot these points and draw a smooth curve through them. This is an increasing function. The domain is $(-\infty, \infty)$ and the range is $(0, \infty)$ and is one-to-one.

19. The y-intercept of $f(x) = \left(\frac{1}{3}\right)^x$ is 1, and the x-axis is a horizontal asymptote. Make a table of values.

x	$f(x)$
-2	9
-1	3
$-\frac{1}{2}$	≈ 1.7
0	1
$\frac{1}{2}$	≈ 0.6
1	$\frac{1}{3} \approx 0.3$
2	$\frac{1}{9} \approx 0.1$

Plot these points and draw a smooth curve through them. This is a decreasing function. The domain is $(-\infty, \infty)$ and the range is $(0, \infty)$ and is one-to-one. Note: Since

$$f(x) = \left(\frac{1}{3}\right)^x = \left(3^{-1}\right)^x = 3^{-x},$$ the graph of

$f(x) = \left(\frac{1}{3}\right)^x$ is the reflection of the graph of

$f(x) = 3^x$ (Exercise 17) about the y-axis.

21. The y-intercept of $f(x) = \left(\frac{3}{2}\right)^x$ is 1, and the x-axis is a horizontal asymptote. Make a table of values.

x	$f(x)$
-2	≈ 0.4
-1	≈ 0.7
$-\frac{1}{2}$	≈ 0.8
0	1
$\frac{1}{2}$	≈ 1.2
1	1.5
2	2.25

Plot these points and draw a smooth curve through them. This is an increasing function. The domain is $(-\infty, \infty)$ and the range is $(0, \infty)$ and is one-to-one.

23. The y-intercept of $f(x) = \left(\frac{1}{10}\right)^{-x}$ is 1, and the x-axis is a horizontal asymptote. Make a table of values.

x	$f(x)$
-2	0.01
-1	0.1
$-\frac{1}{2}$	≈ 0.3
0	1
$\frac{1}{2}$	≈ 3.2
1	10
2	100

Plot these points and draw a smooth curve through them. This is an increasing function. The domain is $(-\infty, \infty)$ and the range is $(0, \infty)$ and is one-to-one.

25. The y-intercept of $f(x) = 4^{-x}$ is 1, and the x-axis is a horizontal asymptote. Make a table of values.

x	$f(x)$
-2	16
-1	4
$-\frac{1}{2}$	2
0	1
$\frac{1}{2}$	0.5
1	0.25
2	0.0625

Plot these points and draw a smooth curve through them. This is a decreasing function. The domain is $(-\infty, \infty)$ and the range is $(0, \infty)$ and is one-to-one. Note: The graph of

$f(x) = 4^{-x}$ is the reflection of the graph of

$f(x) = 4^x$ (Exercise 18) about the y-axis.

27. The y intercept of $f(x) = 2^{|x|}$ is 1. Make a table of values.

x	$f(x)$
-2	4
-1	2
$-\frac{1}{2}$	≈ 1.4
0	1
$\frac{1}{2}$	≈ 1.4
1	2
2	4

(continued on next page)

Copyright © 2013 Pearson Education, Inc.

(continued)

Plot these points and draw a smooth curve through them. The domain is $(-\infty, \infty)$ and the range is $[1, \infty)$ and is not one-to-one. Note: For $x < 0$, $|x| = -x$, so the graph is the same as that of $f(x) = 2^{-x}$. For $x \geq 0$, we have $|x| = x$, so the graph is the same as that of $f(x) = 2^x$. Since $|-x| = |x|$, the graph is symmetric with respect to the y-axis.

For Exercises 29–39, refer to the following graph of $f(x) = 2^x$.

$f(x) = 2^x$

29. The graph of $f(x) = 2^x + 1$ is obtained by translating the graph of $f(x) = 2^x$ up one unit.

31. Since $f(x) = 2^{x+1} = 2^{x-(-1)}$, the graph is obtained by translating the graph of $f(x) = 2^x$ to the left one unit.

33. The graph of $f(x) = -2^{x+2}$ is obtained by translating the graph of $f(x) = 2^x$ to the left 2 units and then reflecting the graph across the x-axis.

35. The graph of $f(x) = 2^{-x}$ is obtained by reflecting the graph across the y-axis.

37. The graph of $f(x) = 2^{x-1} + 2$ is obtained by translating the graph of $f(x) = 2^x$ to the right one unit and up two units.

39. The graph of $f(x) = 2^{x+2} - 4$ is obtained by translating the graph of $f(x) = 2^x$ to the left two units and down four units.

Copyright © 2013 Pearson Education, Inc.

For Exercises 41–51, refer to the following graph of $f(x) = \left(\frac{1}{3}\right)^x$.

41. The graph of $f(x) = \left(\frac{1}{3}\right)^x - 2$ obtained by translating the graph of $f(x) = \left(\frac{1}{3}\right)^x$ down 2 units.

43. Since $f(x) = \left(\frac{1}{3}\right)^{x+2} = \left(\frac{1}{3}\right)^{x-(-2)}$, the graph is obtained by translating the graph of $f(x) = \left(\frac{1}{3}\right)^x$ 2 units to the left.

45. The graph of $f(x) = \left(\frac{1}{3}\right)^{-x+1}$ is obtained by translating the graph of $f(x) = \left(\frac{1}{3}\right)^x$ left one unit and then reflecting the resulting graph across the y-axis.

47. The graph of $f(x) = \left(\frac{1}{3}\right)^{-x}$ is obtained by reflecting the graph of $f(x) = \left(\frac{1}{3}\right)^x$ across the y-axis.

49. The graph of $f(x) = \left(\frac{1}{3}\right)^{x-2} + 2$ is obtained by translating the graph of $f(x) = \left(\frac{1}{3}\right)^x$ two units to the right and two units up.

51. The graph of $f(x) = \left(\frac{1}{3}\right)^{x+2} - 1$ is obtained by translating the graph of $f(x) = \left(\frac{1}{3}\right)^x$ two units to the left and one unit down.

53. The graph of $f(x) = a^{-x}$ is the same as $g(x) = \left(\frac{1}{a}\right)^x$.

55. Since the horizontal asymptote is $y = -1$, the graph has been shifted down one unit. So the general form of the equation is $f(x) = a^{x+b} - 1$. The base is either 2 or 3, so try $a = 2$. Then substitute the coordinates of a point in the equation and solve for b:
$1 = 2^{-2+b} - 1 \Rightarrow 2 = 2^{-2+b} \Rightarrow 2^1 = 2^{-2+b} \Rightarrow$
$1 = -2 + b \Rightarrow 3 = b$
So, the equation is $f(x) = 2^{x+3} - 1$. Verify that the coordinates of other two points given satisfy the equation.

(continued on next page)

Copyright © 2013 Pearson Education, Inc.

(*continued*)

Alternate solution: Working backward and shifting the graph up one unit and right three units to transform the given graph into the graph of $y = 2^x$, it goes through the points (3, 8), (1, 2), and (0, 1), which is the y-intercept. $8 = 2^3$, so $a = 2$, and the equation is $f(x) = 2^{x+3} - 1$. Verify by checking that the coordinates of the points satisfy the equation.

57. Since the horizontal asymptote is $y = 3$, the graph has been shifted up three units. The graph has also been reflected across the x-axis. So the general form of the equation is

$f(x) = -a^{x+b} + 3$. The base is either 2 or 3, so try $a = 2$. Then substitute the coordinates of a point in the equation and solve for b:

$-1 = -2^{0+b} + 3 \Rightarrow -4 = -2^b \Rightarrow 4 = 2^b \Rightarrow$
$2^2 = 2^b \Rightarrow 2 = b$

So, the equation is $f(x) = -2^{x+2} + 3$. Verify that the coordinates of other two points given satisfy the equation.
Alternate solution: Working backward and shifting the graph down three units and right two units to transform the given graph into the graph of $y = -\left(2^x\right)$, it goes through the points (0, −1), (1, −2), and (2, −4). The y-intercept is (0, −1). $-2 = -\left(2^1\right)$, so $a = 2$, and the equation is $f(x) = -2^{x+1} + 3$. Verify by checking that the coordinates of the points satisfy the equation.

59. Since the horizontal asymptote is $y = 1$, the graph has been shifted up one unit. The graph has also been reflected across the y-axis. So the general form of the equation is

$f(x) = a^{-x+b} + 1$. The base is either 2 or 3, so try $a = 3$. Then substitute the coordinates of a point in the equation and solve for b:

$4 = 3^{-(-1)+b} + 1 \Rightarrow 3 = 3^{1+b} \Rightarrow 3^1 = 3^{1+b} \Rightarrow$
$1 = 1 + b \Rightarrow 0 = b$

So, the equation is $f(x) = 3^{-x} + 1$. Verify by checking that the coordinates of the other two points satisfy the equation.
Alternate solution: Working backward and shifting the graph down one unit to transform the given graph into the graph of $y = 3^{-x}$, it goes through the points (−1, 3), (0, 1), and $\left(1, \frac{1}{3}\right)$. The y-intercept is (0, 1). $3 = 3^{-(-1)}$, so $a = 3$, and the equation is $f(x) = 3^{-x} + 1$. Verify by checking that the coordinates of the points satisfy the equation.

61. $4^x = 2 \Rightarrow \left(2^2\right)^x = 2^1 \Rightarrow 2^{2x} = 2^1 \Rightarrow$
$2x = 1 \Rightarrow x = \frac{1}{2}$

Solution set: $\left\{\frac{1}{2}\right\}$

Copyright © 2013 Pearson Education, Inc.

63. $\left(\dfrac{5}{2}\right)^x = \dfrac{4}{25} \Rightarrow \left(\dfrac{2}{5}\right)^{-x} = \left(\dfrac{2}{5}\right)^2 \Rightarrow -x = 2 \Rightarrow$
$x = -2$
Solution set: $\{-2\}$

65. $2^{3-2x} = 8 \Rightarrow 2^{3-2x} = 2^3 \Rightarrow 3 - 2x = 3 \Rightarrow$
$-2x = 0 \Rightarrow x = 0$
Solution set: $\{0\}$

67. $e^{4x-1} = \left(e^2\right)^x \Rightarrow e^{4x-1} = e^{2x} \Rightarrow 4x - 1 = 2x \Rightarrow$
$-1 = -2x \Rightarrow \dfrac{1}{2} = x$
Solution set: $\left\{\dfrac{1}{2}\right\}$

69. $27^{4x} = 9^{x+1} \Rightarrow \left(3^3\right)^{4x} = \left(3^2\right)^{x+1} \Rightarrow$
$3^{3(4x)} = 3^{2(x+1)} \Rightarrow 3^{12x} = 3^{2x+2} \Rightarrow$
$12x = 2x + 2 \Rightarrow 10x = 2 \Rightarrow x = \dfrac{1}{5}$
Solution set: $\left\{\dfrac{1}{5}\right\}$

71. $4^{x-2} = 2^{3x+3} \Rightarrow \left(2^2\right)^{x-2} = 2^{3x+3} \Rightarrow$
$2^{2(x-2)} = 2^{3x+3} \Rightarrow 2^{2x-4} = 2^{3x+3} \Rightarrow$
$2x - 4 = 3x + 3 \Rightarrow -4 = x + 3 \Rightarrow -7 = x$
Solution set: $\{-7\}$

73. $\left(\dfrac{1}{e}\right)^{-x} = \left(\dfrac{1}{e^2}\right)^{x+1} \Rightarrow \left(e^{-1}\right)^{-x} = \left(e^{-2}\right)^{x+1} \Rightarrow$
$e^x = e^{-2(x+1)} \Rightarrow e^x = e^{-2x-2} \Rightarrow$
$x = -2x - 2 \Rightarrow 3x = -2 \Rightarrow x = -\dfrac{2}{3}$
Solution set: $\left\{-\dfrac{2}{3}\right\}$

75. $\left(\sqrt{2}\right)^{x+4} = 4^x \Rightarrow \left(2^{1/2}\right)^{x+4} = \left(2^2\right)^x \Rightarrow$
$2^{(1/2)(x+4)} = 2^{2x} \Rightarrow 2^{(1/2)x+2} = 2^{2x} \Rightarrow$
$\dfrac{1}{2}x + 2 = 2x \Rightarrow 2 = \dfrac{3}{2}x \Rightarrow \dfrac{2}{3} \cdot 2 = x \Rightarrow x = \dfrac{4}{3}$
Solution set: $\left\{\dfrac{4}{3}\right\}$

77. $\dfrac{1}{27} = x^{-3} \Rightarrow 3^{-3} = x^{-3} \Rightarrow x = 3$
Alternate solution:
$\dfrac{1}{27} = x^{-3} \Rightarrow \dfrac{1}{27} = \dfrac{1}{x^3} \Rightarrow 27 = x^3 \Rightarrow$
$b = \sqrt[3]{27} = 3$
Solution set: $\{3\}$

79. $x^{2/3} = 4 \Rightarrow \left(x^{2/3}\right)^{3/2} = \pm 4^{3/2} \Rightarrow$
$x = \left(\pm\sqrt{4}\right)^3 \Rightarrow x = \left(\pm 2\right)^3 \Rightarrow x = \pm 8$

Recall from Chapter 1 that it is necessary to check all proposed solutions in the original equation when you raise both sides to a power.
Check $x = -8$.
$4 = x^{2/3}$
$4 \overset{?}{=} (-8)^{2/3}$
$4 = \left(\sqrt[3]{-8}\right)^2 \Rightarrow 4 = (-2)^2 \Rightarrow 4 = 4$
This is a true statement. -8 is a solution.
Check $r = 8$.
$4 = x^{2/3}$
$4 \overset{?}{=} 8^{2/3}$
$4 = \left(\sqrt[3]{8}\right)^2 \Rightarrow 4 = 2^2 \Rightarrow 4 = 4$
This is a true statement. 8 is a solution.
Solution set: $\{-8, 8\}$

81. $x^{5/2} = 32 \Rightarrow \left(x^{5/2}\right)^{2/5} = 32^{2/5} \Rightarrow x = 32^{2/5} \Rightarrow$
$x = \left(\sqrt[5]{32}\right)^2 \Rightarrow x = 2^2 \Rightarrow x = 4$
Recall from Chapter 1 that it is necessary to check all proposed solutions in the original equation when you raise both sides to a power.
Check $x = 4$.
$x^{5/2} = 32$
$4^{5/2} \overset{?}{=} 32$
$\left(\sqrt{4}\right)^5 = 32 \Rightarrow 2^5 = 32 \Rightarrow 32 = 32$
This is a true statement.
Solution set: $\{4\}$

83. $x^{-6} = \dfrac{1}{64} \Rightarrow \dfrac{1}{x^6} = \dfrac{1}{64} \Rightarrow x^6 = 64 \Rightarrow$
$x^6 = (\pm 2)^6 \Rightarrow x = \pm 2$
Solution set: $\{-2, 2\}$

85. $x^{5/3} = -243 \Rightarrow \left(x^{5/3}\right)^{3/5} = (-243)^{3/5} \Rightarrow$
$x = (-3)^3 = -27$
Recall from Chapter 1 that it is necessary to check all proposed solutions in the original equation when you raise both sides to a power.
Check $x = -27$.
$x^{5/3} = -243$
$(-27)^{5/3} \overset{?}{=} -243$
$(-3)^5 = -243$
This is a true statement. -27 is a solution.
Solution set: $\{-27\}$

Copyright © 2013 Pearson Education, Inc.

87. (a) Use the compound interest formula to find the future value, $A = P\left(1+\frac{r}{n}\right)^{tn}$, given $n = 2$, $P = 8906.54$, $r = 0.05$, and $t = 9$.

$$A = P\left(1+\frac{r}{m}\right)^{tm} = (8906.54)\left(1+\frac{0.05}{2}\right)^{9(2)}$$
$$= (8906.54)0(1+0.025)^{18}$$
$$\approx 13,891.16276$$

Rounding to the nearest cent, the future value is $13,891.16. The amount of interest would be $13,891.16 − $8906.54 = $4984.62.

(b) Use the continuous compounding interest formula to find the future value, $A = Pe^{rt}$, given $P = 8906.54$, $r = 0.05$, and $t = 9$.

$$A = Pe^{rt} = 8906.54e^{0.05(9)} = 8906.54e^{0.45}$$
$$\approx 8906.54(1.568312) \approx 13,968.23521$$

Rounding to the nearest cent, the future value is $13,968.24. The amount of interest would be $13,968.24 − $898806.54 = $5061.70.

89. Use the compound interest formula to find the present amount, $A = P\left(1+\frac{r}{n}\right)^{tn}$, given $n = 4$, $A = 25,000$, $r = 0.06$, and $t = \frac{11}{4}$.

$$A = P\left(1+\frac{r}{n}\right)^{tn}$$
$$25,000 = P\left(1+\frac{0.06}{4}\right)^{(11/4)(4)}$$
$$25,000 = P(1.015)^{11}$$
$$P = \frac{25,000}{(1.015)^{11}} \approx \$21,223.33083$$

Rounding to the nearest cent, the present value is $21,223.33.

91. Use the compound interest formula to find the present value, $A = P\left(1+\frac{r}{n}\right)^{tn}$, given $n = 4$, $A = 5000$, $r = 0.035$, and $t = 10$.

$$A = P\left(1+\frac{r}{n}\right)^{tn}$$
$$5,000 = P\left(1+\frac{0.035}{4}\right)^{10(4)} = P(1.00875)^{40}$$
$$P = \frac{5,000}{(1.00875)^{40}} \approx \$3528.808535$$

Rounding to the nearest cent, the present value is $3528.81.

93. Use the compound interest formula to find the interest rate, $A = P\left(1+\frac{r}{n}\right)^{tn}$, given $n = 4$, $A = 1500$, $P = 1200$, and $t = 5$.

$$A = P\left(1+\frac{r}{n}\right)^{tn}$$
$$1500 = 1200\left(1+\frac{r}{4}\right)^{5(4)}$$
$$1500 = 1200\left(1+\frac{r}{4}\right)^{20}$$
$$1.25 = \left(1+\frac{r}{4}\right)^{20}$$
$$(1.25)^{1/20} = 1+\frac{r}{4} \Rightarrow (1.25)^{1/20} - 1 = \frac{r}{4} \Rightarrow$$
$$4\left[(1.25)^{1/20} - 1\right] = r \Rightarrow r \approx 0.044878604$$

The interest rate, to the nearest tenth, is 4.5%.

95. For each bank we need to calculate $\left(1+\frac{r}{n}\right)^n$.

Since the base, $1+\frac{r}{n}$, is greater than 1, we need only compare the three values calculated to determine which bank will yield the least amount of interest. It is understood that the amount of time, t, and the principal, P, are the same for all three banks.

Bank A: Calculate $\left(1+\frac{r}{n}\right)^n$ where $n = 1$ and $r = 0.064$.

$$\left(1+\frac{0.064}{1}\right)^1 = (1+0.064)^1 = (1.064)^1 = 1.064$$

Bank B: Calculate $\left(1+\frac{r}{n}\right)^n$ where $n = 12$ and $r = 0.063$.

$$\left(1+\frac{0.063}{12}\right)^{12} = (1+0.00525)^{12} = (1.00525)^{12}$$
$$\approx 1.064851339$$

Bank C: Calculate $\left(1+\frac{r}{n}\right)^n$ where $n = 4$ and $r = 0.0635$.

$$\left(1+\frac{0.0635}{4}\right)^4 = (1+0.015875)^4 = (1.015875)^4$$
$$\approx 1.06502816$$

Bank A will charge you the least amount of interest, even though it has the highest stated rate.

97. (a)

(b) From the graph above, we can see that the data are not linear but exponentially decreasing.

Copyright © 2013 Pearson Education, Inc.

(c)

(d) $P(x) = 1013e^{-0.0001341x}$

$P(1500) = 1013e^{-0.0001341(1500)}$

$\approx 1013(0.817790) \approx 828$

$P(11,000) = 1013e^{-0.0001341(11,000)}$

$\approx 1013(0.228756) \approx 232$

When the altitude is 1500 m, the function P gives a pressure of 828 mb, which is less than the actual value of 846 mb. When the altitude is 11,000 m, the function P gives a pressure of 232 mb, which is more than the actual value of 227 mb.

99. (a) Evaluate $T = 50,000(1+0.06)^n$ where $n = 4$.

$T = 50,000(1+0.06)^4$

$= 50,000(1.06)^4 \approx 63,123.848$

Total population after 4 years is about 63,000.

(b) Evaluate $T = 30,000(1+0.12)^n$ where $n = 3$.

$T = 30,000(1+0.12)^3$

$= 30,000(1.12)^3 = 42,147.84$

There would be about 42,000 deer after 3 years.

(c) Evaluate $T = 45,000(1+0.08)^n$ where $n = 5$.

$T = 45,000(1+0.08)^5$

$= 45,000(1.08)^5 \approx 66,119.76346$

There would be about 66,000 deer after 5 years. Thus, we can expect about $66,000 - 45,000 = 21,000$ additional deer after 5 years.

101. $5e^{3x} = 75$

Since $5e^{3x} = 75 \Rightarrow e^{3x} = 15$ we could also do the following.

Solution set: {0.9}

103. $3x + 2 = 4^x$

We can see from the following screens that there are two solutions.

Solution set: {−0.5, 1.3}

105. Answers will vary.

107. If the graph of the exponential function $f(x) = a^x$ contains the point $(3, 8)$, we have $a^3 = 8$. This implies $a = \sqrt[3]{8} = 2$. Thus, the equation which satisfies the given condition is $f(x) = 2^x$.

109. If the graph of the exponential function $f(x) = a^x$ contains the point $(-3, 64)$, we have $a^{-3} = 64$. This implies $a^3 = \frac{1}{64} \Rightarrow a = \sqrt[3]{\frac{1}{64}} = \frac{1}{4}$. Thus, the equation which satisfies the given condition is $f(x) = \left(\frac{1}{4}\right)^x$.

Copyright © 2013 Pearson Education, Inc.

111. $f(t) = 3^{2t+3} = 3^{2t} \cdot 3^3 = 27 \cdot \left(3^2\right)^t = 27 \cdot 9^t$

113. $f(t) = \left(\frac{1}{3}\right)^{1-2t} = \left(\frac{1}{3}\right)^1 \left(\frac{1}{3}\right)^{-2t}$

$= \left(\frac{1}{3}\right)(3)^{2t} = \left(\frac{1}{3}\right)\left(3^2\right)^t = \left(\frac{1}{3}\right)9^t$

115. $e^1 \approx 1 + 1 + \frac{1^2}{2 \cdot 1} + \frac{1^3}{3 \cdot 2 \cdot 1} + \frac{1^4}{4 \cdot 3 \cdot 2 \cdot 1} + \frac{1^5}{5 \cdot 4 \cdot 3 \cdot 2 \cdot 1}$

$= 1 + 1 + \frac{1}{2} + \frac{1}{6} + \frac{1}{24} + \frac{1}{120} \approx 2.717$

```
e^(1)
        2.718281828
```

117. Yes; $f(x) = a^x$ is a one-to-one function. Therefore, an inverse function exists for f.

119. Since $f(x) = a^x$ has an inverse, we find it as follows: $y = a^x \Rightarrow x = a^y$

121. If $a = e$, the equation for $f^{-1}(x)$ will be given by $x = e^y$.

Section 4.3 Logarithmic Functions

1. **(a)** C; $\log_2 16 = 4$ because $2^4 = 16$.

(b) A; $\log_3 1 = 0$ because $3^0 = 1$.

(c) E; $\log_{10} 0.1 = -1$ because $10^{-1} = 0.1$.

(d) B; $\log_2 \sqrt{2} = \frac{1}{2}$ because $2^{1/2} = \sqrt{2}$.

(e) F; $\log_e \left(\frac{1}{e^2}\right) = -2$ because $e^{-2} = \frac{1}{e^2}$.

(f) D; $\log_{1/2} 8 = -3$ because $\left(\frac{1}{2}\right)^{-3} = 8$.

3. $3^4 = 81$ is equivalent to $\log_3 81 = 4$.

5. $\left(\frac{2}{3}\right)^{-3} = \frac{27}{8}$ is equivalent to $\log_{2/3} \frac{27}{8} = -3$.

7. $\log_6 36 = 2$ is equivalent to $6^2 = 36$.

9. $\log_{\sqrt{3}} 81 = 8$ is equivalent to $\left(\sqrt{3}\right)^8 = 81$.

11. Answers will vary.

13. $x = \log_5 \frac{1}{625} \Rightarrow 5^x = \frac{1}{625} \Rightarrow 5^x = \frac{1}{5^4} \Rightarrow$

$5^x = 5^{-4} \Rightarrow x = -4$
Solution set: $\{-4\}$

15. $\log_x \frac{1}{32} = 5 \Rightarrow x^5 = \frac{1}{32} \Rightarrow x^5 = \frac{1}{2^5} = \left(\frac{1}{2}\right)^5 \Rightarrow$

$x = \frac{1}{2}$
Solution set: $\left\{\frac{1}{2}\right\}$

17. $x = \log_8 \sqrt[4]{8} \Rightarrow 8^x = \sqrt[4]{8} \Rightarrow 8^x = 8^{1/4} \Rightarrow x = \frac{1}{4}$
Solution set: $\left\{\frac{1}{4}\right\}$

19. $x = 3^{\log_3 8}$
Writing as a logarithmic equation, we have
$\log_3 8 = \log_3 x \Rightarrow x = 8$
Using the Theorem of Inverses on page 440, we can directly state that $x = 8$.
Solution set: $\{8\}$

21. $x = 2^{\log_2 9}$
Writing as a logarithmic equation, we have
$\log_2 9 = \log_2 x \Rightarrow x = 9$
Using the Theorem of Inverses on page 440, we can directly state that $x = 9$.
Solution set: $\{9\}$

23. $\log_x 25 = -2 \Rightarrow x^{-2} = 25 \Rightarrow x^{-2} = 5^2 \Rightarrow$
$\left(x^{-2}\right)^{-1/2} = \left(5^2\right)^{-1/2} \Rightarrow x = 5^{-1}$

Do not include a $\pm$ since the base, x, cannot be negative.

$x = \frac{1}{5}$
Solution set: $\left\{\frac{1}{5}\right\}$

25. $\log_4 x = 3 \Rightarrow 4^3 = x \Rightarrow 64 = x$
Solution set: $\{64\}$

27. $x = \log_4 \sqrt[3]{16} \Rightarrow 4^x = \sqrt[3]{16} \Rightarrow 4^x = (16)^{1/3} \Rightarrow$

$4^x = \left(4^2\right)^{1/3} \Rightarrow 4^x = 4^{2/3} \Rightarrow x = \frac{2}{3}$
Solution set: $\left\{\frac{2}{3}\right\}$

29. $\log_9 x = \frac{5}{2} \Rightarrow 9^{5/2} = x \Rightarrow (3)^5 = 243 = x$

Note that we do not include $\sqrt{9} = -3$ because logarithms are not defined for negative numbers.
Solution set: $\{243\}$

Copyright © 2013 Pearson Education, Inc.

31. $\log_{1/2}(x+3) = -4 \Rightarrow x+3 = \left(\dfrac{1}{2}\right)^{-4} \Rightarrow$

$x+3 = 2^4 \Rightarrow x+3 = 16 \Rightarrow x = 13$
Solution set: $\{13\}$

33. $\log_{(x+3)} 6 = 1 \Rightarrow (x+3)^1 = 6 \Rightarrow x+3 = 6 \Rightarrow$
$x = 3$
Solution set: $\{3\}$

35. $3x - 15 = \log_x 1 \ (x > 0, \ x \ne 1)$

Note that $\log_x 1 = 0$ since $x^0 = 1$ for any
number x. Thus,
$3x - 15 = \log_x 1 \Rightarrow 3x - 15 = 0 \Rightarrow 3x = 15 \Rightarrow$
$x = 5$
Solution set: $\{5\}$

37. Answers will vary.

For Exercises 39–41, refer to the following graph of
$f(x) = \log_2 x$.

39. The graph of $f(x) = (\log_2 x) + 3$ is obtained
by translating the graph of $f(x) = \log_2 x$ up 3
units.

Domain: $(0, \infty)$; range: $(-\infty, \infty)$

41. To find the graph of $f(x) = \left| \log_2(x+3) \right|$,
translate the graph of $f(x) = \log_2 x$ to the left
3 units to obtain the graph of $\log_2(x+3)$. (See
exercise 40.) For the portion of the graph
where $f(x) \ge 0$, that is, where $x \ge -2$, use the
same graph as in exercise 40. For the portion
of the graph in exercise 40 where $f(x) < 0$,
$-3 < x < -2$, reflect the graph about the
x-axis. In this way, each negative value of
$f(x)$ on the graph in exercise 40 is replaced
by its opposite, which is positive. The graph
has a vertical asymptote at $x = -3$.

$f(x) = \left| \log_2(x+3) \right|$
Domain: $(-3, \infty)$; range: $[0, \infty)$

43. The graph of $f(x) = \log_{1/2}(x-2)$ is obtained
by translating the graph of $f(x) = \log_{1/2} x$ to
the right 2 units. The graph has a vertical
asymptote at $x = 2$.

Domain: $(2, \infty)$; range: $(-\infty, \infty)$

45. Because $f(x) = \log_2 x$ has a vertical
asymptote, which is the y-axis (the line
$x = 0$), x-intercept of 1, and is increasing, the
correct choice is the graph in E.

47. Because $f(x) = \log_2 \dfrac{1}{x} = \log_2 x^{-1} = -\log_2 x$,
it has a vertical asymptote, which is the y-axis
(the line $x = 0$), has an x-intercept 1, and is
the reflection of $f(x) = \log_2 x$ across the
x-axis, it is decreasing and the correct choice
is the graph in B.

Copyright © 2013 Pearson Education, Inc.

49. Because $f(x) = \log_2(x-1)$ represents the horizontal shift of $f(x) = \log_2 x$ to the right 1 unit, the function has a vertical asymptote which is the line $x = 1$, has an x-intercept when $x - 1 = 1 \Rightarrow x = 2$, and is increasing, the correct choice is the graph in F.

51. $f(x) = \log_5 x$

Since $f(x) = y = \log_5 x$, we can write the exponential form as $x = 5^y$ to find ordered pairs that satisfy the equation. It is easier to choose values for y and find the corresponding values of x. Make a table of values.

x	$y = \log_5 x$
$\frac{1}{25} = 0.04$	-2
$\frac{1}{5} = 0.2$	-1
1	0
5	1
25	2

The graph can also be found by reflecting the graph of $f(x) = 5^x$ about the line $y = x$. The graph has the y-axis as a vertical asymptote.

53. $f(x) = \log_{1/2}(1-x)$

Since $f(x) = y = \log_{1/2}(1-x)$, we can write the exponential form as

$1 - x = \left(\frac{1}{2}\right)^y \Rightarrow x = 1 - \left(\frac{1}{2}\right)^y$ to find ordered pairs that satisfy the equation. It is easier to choose values for y and find the corresponding values of x. Make a table of values.

x	$y = \log_{1/2}(1-x)$
-3	-2
-1	-1
0	0
$\frac{1}{2} = 0.5$	1
$\frac{3}{4} = 0.75$	2

The graph has the line $x = 1$ as a vertical asymptote.

55. $f(x) = \log_3(x-1)+2$

Since $f(x) = y = \log_3(x-1)+2$, we can write the exponential form as

$y - 2 = \log_3(x-1) \Rightarrow x - 1 = 3^{y-2} \Rightarrow$

$x = 3^{y-2}+1$ to find ordered pairs that satisfy the equation.

It is easier to choose values for y and find the corresponding values of x. Make a table of values.

$x = 3^{y-2}+1$	$y = \log_3(x-1)+2$
$\frac{82}{81} \approx 1.01$	-2
$\frac{28}{27} \approx 1.04$	-1
$\frac{10}{9} \approx 1.11$	0
$\frac{4}{3} \approx 1.33$	1
1	2
4	3
10	4

57. Since the vertical asymptote is $x = -1$, the graph has been shifted left one unit. So the general form of the equation is

$f(x) = \log_a(x+1)+k$. The base is either 2 or 3, so try $a = 2$. Then substitute the coordinates of a point in the equation and solve for k:

$-2 = \log_2(1+1) + k \Rightarrow -2 - k = \log_2 2 \Rightarrow$

$2^{-2-k} = 2 \Rightarrow 2^{-2-k} = 2^1 \Rightarrow$

$-2 - k = 1 \Rightarrow -3 = k$

So, the equation is $f(x) = \log_2(x+1)-3$.

Verify that the coordinates of other two points given satisfy the equation.

Alternate solution: Working backward and shifting the graph up three units and right one unit to transform the given graph into the graph of $y = \log_2 x$, it goes through the points $(1, 0)$, which is the x-intercept, $(2, 1)$, and $(8, 3)$. $3 = \log_2 8$, so $a = 2$, and the equation is $f(x) = \log_2(x+1)-3$. Verify by checking that the coordinates of the points shown on the graph satisfy the equation.

Copyright © 2013 Pearson Education, Inc.

59. The graph has been reflected across the y-axis, so the general form of the equation is
$f(x) = \log_a(-x - b) + k$. Since the vertical asymptote of the graph is $x = 3$, the graph has been shifted right three units, and $b = -3$. So the general form of the equation is
$f(x) = \log_a(-x + 3) + k$. The base is either 2 or 3, so try $a = 2$. Then substitute the coordinates of a point in the equation and solve for k:
$-1 = \log_2(-1 + 3) + k \Rightarrow -1 - k = \log_2 2 \Rightarrow$
$2^{-1-k} = 2 \Rightarrow 2^{-1-k} = 2^1 \Rightarrow$
$-1 - k = 1 \Rightarrow -2 = k$
So, the equation is $f(x) = \log_2(-x + 3) - 2$.
Verify that the coordinates of other two points given satisfy the equation.
Alternate solution: Working backward and shifting the graph up two units and left three units to transform the given graph into the graph of $y = \log_2 x$, it goes through the points $(-1, 0)$, which is the x-intercept, $(-2, 1)$, and $(-4, 2)$. $2 = \log_2[-(-4)]$, so $a = 2$, and the equation is $f(x) = \log_2(-x + 3) - 2$. Verify by checking that the coordinates of the points shown on the graph satisfy the equation.

61. The graph has been reflected across the x-axis, so the general form of the equation is
$f(x) = -\log_a(x - b) + k$. Since the vertical asymptote of the original graph is $x = 1$, the graph has been shifted right one unit and $b = 1$. So the general form of the equation is
$f(x) = -\log_a(x - 1) + k$. The base is either 2 or 3, so try $a = 3$. Then substitute the coordinates of a point in the equation and solve for k:
$-1 = -\log_3(4 - 1) + k \Rightarrow$
$-(-1 - k) = \log_3 3 \Rightarrow 3^{1+k} = 3 \Rightarrow$
$3^{1+k} = 3^1 \Rightarrow 1 + k = 1 \Rightarrow k = 0$

So, the equation is $f(x) = -\log_3(x - 1)$.
Verify that the coordinates of other two points given satisfy the equation.
Alternate solution: Working backward and shifting the graph left one unit, it goes through the points $(1, 0)$, which is the
x-intercept, $\left(\frac{1}{3}, 1\right)$, and $(3, -1)$. $-1 = -\log_3 3$, so $a = 3$, and the equation is
$f(x) = -\log_3(x - 1)$. Verify by checking that the coordinates of the points shown on the graph satisfy the equation.

63. $\log_2 \frac{6x}{y} = \log_2 6x - \log_2 y$
$= \log_2 6 + \log_2 x - \log_2 y$

65. $\log_5 \frac{5\sqrt{7}}{3} = \log_5 5\sqrt{7} - \log_5 3$
$= \log_5 5 + \log_5 \sqrt{7} - \log_5 3$
$= 1 + \log_5 7^{1/2} - \log_5 3$
$= 1 + \frac{1}{2}\log_5 7 - \log_5 3$

67. $\log_4(2x + 5y)$
Since this is a sum, none of the logarithm properties apply, so this expression cannot be simplified.

69. $\log_m \sqrt{\frac{5r^3}{z^5}} = \log_m \left(\frac{5r^3}{z^5}\right)^{1/2} = \frac{1}{2}\log_m \frac{5r^3}{z^5}$
$= \frac{1}{2}\left(\log_m 5r^3 - \log_m z^5\right)$
$= \frac{1}{2}\left(\log_m 5 + \log_m r^3 - \log_m z^5\right)$
$= \frac{1}{2}\left(\log_m 5 + 3\log_m r - 5\log_m z\right)$

71. $\log_2 \frac{ab}{cd} = \log_2(ab) - \log_2(cd)$
$= \log_2 a + \log_2 b - \left(\log_2 c + \log_2 d\right)$
$= \log_2 a + \log_2 b - \log_2 c - \log_2 d$

Copyright © 2013 Pearson Education, Inc.

73. $\log_3 \frac{\sqrt{x}\sqrt[3]{y}}{w^2\sqrt{z}} = \log_3\left(x^{1/2}y^{1/3}\right) - \log_3\left(w^2 z^{1/2}\right)$

$= \log_3 x^{1/2} + \log_3 y^{1/3}$
$\quad - \left(\log_3 w^2 + \log_3 z^{1/2}\right)$
$= \frac{1}{2}\log_3 x + \frac{1}{3}\log_3 y$
$\quad - \left(2\log_3 w + \frac{1}{2}\log_3 z\right)$
$= \frac{1}{2}\log_3 x + \frac{1}{3}\log_3 y$
$\quad - 2\log_3 w - \frac{1}{2}\log_3 z$

75. $\log_a x + \log_a y - \log_a m = \log_a xy - \log_a m$
$= \log_a \frac{xy}{m}$

77. $\log_a m - \log_a n - \log_a t$
$= \log_a m - \left(\log_a n + \log_a t\right)$
$= \log_a m - \log_a (nt) = \log_a \frac{m}{nt}$

79. $\frac{1}{3}\log_b x^4 y^5 - \frac{3}{4}\log_b x^2 y$

$= \log_b\left(x^4 y^5\right)^{1/3} - \log_b\left(x^2 y\right)^{3/4}$
$= \log_b x^{4/3} y^{5/3} - \log_b x^{3/2} y^{3/4}$
$= \log_b \frac{x^{4/3} y^{5/3}}{x^{3/2} y^{3/4}} = \log_b x^{4/3-3/2} y^{5/3-3/4}$
$= \log_b x^{-1/6} y^{11/12}$

81. $2\log_a (z+1) + \log_a (3z+2)$
$= \log_a (z+1)^2 + \log_a (3z+2)$
$= \log_a\left[(z+1)^2(3z+2)\right]$

83. $-\frac{2}{3}\log_5 5m^2 + \frac{1}{2}\log_5 25m^2$
$= \log_5\left(5m^2\right)^{-2/3} + \log_5\left(25m^2\right)^{1/2}$
$= \log_5\left[\left(5m^2\right)^{-2/3}\cdot\left(25m^2\right)^{1/2}\right]$
$= \log_5\left(5^{-2/3}m^{-4/3}\cdot 5m\right)$
$= \log_5\left(5^{-2/3}\cdot 5^1\cdot m^{-4/3}\cdot m^1\right)$
$= \log_5\left(5^{1/3}\cdot m^{-1/3}\right) = \log_5 \frac{5^{1/3}}{m^{1/3}} = \log_5 \sqrt[3]{\frac{5}{m}}$

85. $\log_{10} 6 = \log_{10}(2\cdot 3) = \log_{10}2 + \log_{10}3$
$= 0.3010 + 0.4771 = 0.7781$

87. $\log_{10}\frac{3}{2} = \log_{10}3 - \log_{10}2$
$= 0.4771 - 0.3010 = 0.1761$

89. $\log_{10}\frac{9}{4} = \log_{10}9 - \log_{10}4 = \log_{10}3^2 - \log_{10}2^2$
$= 2\log_{10}3 - 2\log_{10}2$
$= 2(0.4771) - 2(0.3010)$
$= 0.9542 - 0.6020 = 0.3522$

91. $\log_{10}\sqrt{30} = \log_{10}30^{1/2} = \frac{1}{2}\log_{10}30$
$= \frac{1}{2}\log_{10}(10\cdot 3)$
$= \frac{1}{2}\left(\log_{10}10 + \log_{10}3\right)$
$= \frac{1}{2}(1+0.4771) = \frac{1}{2}(1.4771)$
≈ 0.7386

93. (a) If the x-values are representing years, 3 months is $\frac{3}{12} = \frac{1}{4} = 0.25$ yr and 6 months is $\frac{6}{12} = \frac{1}{2} = 0.5$ yr.

(b) A logarithmic function will model the data best.

95. $f(x) = \log_a x$ and $f(3) = 2$

$2 = \log_a 3 \Rightarrow a^2 = 3 \Rightarrow \left(a^2\right)^{1/2} = 3^{1/2} \Rightarrow a = \sqrt{3}$

(There is no $\pm$ because a must be positive and not equal to 1.) We now have $f(x) = \log_{\sqrt{3}} x$.

(a) $f\left(\frac{1}{9}\right) = \log_{\sqrt{3}}\frac{1}{9} \Rightarrow y = \log_{\sqrt{3}}\frac{1}{9} \Rightarrow$
$\left(\sqrt{3}\right)^y = \frac{1}{9} \Rightarrow \left(3^{1/2}\right)^y = \frac{1}{3^2} \Rightarrow$
$3^{y/2} = 3^{-2} \Rightarrow \frac{y}{2} = -2 \Rightarrow y = -4$

(b) $f(27) = \log_{\sqrt{3}}27 \Rightarrow y = \log_{\sqrt{3}}27 \Rightarrow$
$\left(\sqrt{3}\right)^y = 27 \Rightarrow \left(3^{1/2}\right)^y = 3^3 \Rightarrow$
$3^{y/2} = 3^3 \Rightarrow \frac{y}{2} = 3 \Rightarrow y = 6$

(c) $f(9) = \log_{\sqrt{3}}9 \Rightarrow y = \log_{\sqrt{3}}9 \Rightarrow$
$\left(\sqrt{3}\right)^y = 9 \Rightarrow \left(3^{1/2}\right)^y = 3^2 \Rightarrow$
$3^{y/2} = 3^2 \Rightarrow \frac{y}{2} = 2 \Rightarrow y = 4$

(d) $f\left(\frac{\sqrt{3}}{3}\right) = \log_{\sqrt{3}}\left(\frac{\sqrt{3}}{3}\right) \Rightarrow y = \log_{\sqrt{3}}\left(\frac{\sqrt{3}}{3}\right) \Rightarrow$
$\left(\sqrt{3}\right)^y = \frac{\sqrt{3}}{3} \Rightarrow \left(3^{1/2}\right)^y = \frac{3^{1/2}}{3^1} \Rightarrow$
$\left(3^{1/2}\right)^y = 3^{1/2-1} = 3^{-1/2} \Rightarrow$
$3^{y/2} = 3^{-1/2} \Rightarrow \frac{y}{2} = -\frac{1}{2} \Rightarrow y = -1$

Copyright © 2013 Pearson Education, Inc.

97. In the formula $A = P\left(1+\frac{r}{n}\right)^{tn}$ we substitute $A = 2P$ since we want the present value to be doubled in the future. Thus, we need to solve for t in the equation $2P = P\left(1+\frac{r}{n}\right)^{tn}$.

$$2P = P\left(1+\frac{r}{n}\right)^{tn} \Rightarrow 2 = \left(1+\frac{r}{n}\right)^{tn}$$

$$2^{\frac{1}{tn}} = \left[\left(1+\frac{r}{n}\right)^{tn}\right]^{\frac{1}{tn}} \Rightarrow 2^{\frac{1}{tn}} = \left(1+\frac{r}{n}\right)$$

$$\log_2\left(1+\frac{r}{n}\right) = \frac{1}{tn} \Rightarrow tn = \frac{1}{\log_2\left(1+\frac{r}{n}\right)}$$

$$t = \frac{1}{n\log_2\left(1+\frac{r}{n}\right)} \Rightarrow t = \frac{1}{\log_2\left(1+\frac{r}{n}\right)^n}$$

99. $f(x) = x \log_{10} x$

101. $\log_{10} x = x - 2$

The x-coordinates of the intersection points will be the solutions to the given equation. There are two points of intersection, hence there are two possible solutions.

Solution set: $\{0.01, 2.38\}$

103. Prove that $\log_a \frac{x}{y} = \log_a x - \log_a y$.

Let $m = \log_a x$ and $n = \log_a y$. Changing to exponential form we have $a^m = x$ and $a^n = y$. Since $\frac{x}{y} = \frac{a^m}{a^n}$ we have $\frac{x}{y} = a^{m-n}$. Changing to logarithmic form, we have $\log_a \frac{x}{y} = m - n$. Substituting for m and n we have $\log_a \frac{x}{y} = \log_a x - \log_a y$.

Summary Exercises on Inverse, Exponential, and Logarithmic Functions

1. $f(x) = 3x - 4$, $g(x) = \frac{1}{3}x + \frac{4}{3}$

$(f \circ g)(x) = f\left[g(x)\right] = 3\left(\frac{1}{3}x + \frac{4}{3}\right) - 4$ Since
$$= x + 4 - 4 = x$$
$(g \circ f)(x) = g\left[f(x)\right] = \frac{1}{3}(3x - 4) + \frac{4}{3}$
$$= \frac{3x}{3} = x$$

$(f \circ g)(x) = x$ and $(g \circ f)(x) = x$, these functions are inverses.

3. $f(x) = 1 + \log_2 x$, $g(x) = 2^{x-1}$

$(f \circ g)(x) = f\left[g(x)\right] = f\left(2^{x-1}\right) = 1 + \log_2 2^{x-1}$
$$= 1 + (x-1)\log_2 2 = 1 + x - 1 = x$$
$(g \circ f)(x) = g\left[f(x)\right] = g\left(1 + \log_2 x\right)$
$$= 2^{(1+\log_2 x - 1)} = 2^{\log_2 x} = x$$

Since $(f \circ g)(x) = x$ and $(g \circ f)(x) = x$, these functions are inverses.

5. Since any horizontal line when passed through the graph of this function will touch the graph in at most one place, the function is one-to-one. A sketch of the graph of the inverse function is as follows.

7. Since a horizontal line when passed through the graph of this function will touch the graph in more than one place, the function is not one-to-one.

9. Because $f(x) = \log_3(x+2)$ has a vertical asymptote when $x + 2 = 0 \Rightarrow x = -2$, x-intercept when $x + 2 = 1 \Rightarrow x = -1$, and is increasing, the correct choice is the graph in B.

11. Because $f(x) = \log_2(5 - x)$ has a vertical asymptote when $5 - x = 0 \Rightarrow x = 5$, x-intercept when $5 - x = 1 \Rightarrow x = 4$. The correct choice is the graph in C.

13. The functions in Exercises 9 and 12 are inverses. The functions in Exercises 10 and 11 are inverses.

Copyright © 2013 Pearson Education, Inc.

15. $f(x) = 3x - 6$

This function is one-to-one.

Step 1: Replace $f(x)$ with y and interchange x and y. $y = 3x - 6 \Rightarrow x = 3y - 6$

Step 2: Solve for y.

$x = 3y - 6 \Rightarrow x + 6 = 3y \Rightarrow \frac{x+6}{3} = y$

Step 3: Replace y with $f^{-1}(x)$:

$f^{-1}(x) = \frac{x+6}{3} = \frac{1}{3}x + 2$

For both $f(x)$ and $f^{-1}(x)$, the domain is $(-\infty, \infty)$ and the range is $(-\infty, \infty)$.

17. $f(x) = 3x^2$

If $x = 1$, $f(1) = 3(1)^2 = 3(1) = 3$. Also if

$x = -1$, $f(-1) = 3(-1)^2 = 3(1) = 3$.

Since two different values of x lead to the same value of y, the function is not one-to-one.

19. $f(x) = \sqrt[3]{5 - x^4}$

If $x = 1$, $f(1) = \sqrt[3]{5 - 1^4} = \sqrt[3]{5 - 1} = \sqrt[3]{4}$. Also if

$x = -1$, $f(-1) = \sqrt[3]{5 - (-1)^4} = \sqrt[3]{5 - 1} = \sqrt[3]{4}$.

Since two different values of x lead to the same value of y, the function is not one-to-one.

21. $\left(\frac{1}{10}\right)^{-3} = 1000$ is equivalent to

$\log_{1/10} 1000 = -3$.

23. $\left(\sqrt{3}\right)^4 = 9$ is equivalent to $\log_{\sqrt{3}} 9 = 4$.

25. $2^x = 32$ is equivalent to $\log_2 32 = x$.

27. $3x = 7^{\log_7 6} \Rightarrow 3x = 6 \Rightarrow x = 2$

Solution set: $\{2\}$

28. $x = \log_{10} 0.001 \Rightarrow 10^x = 0.001 \Rightarrow$

$10^x = 10^{-3} \Rightarrow x = -3$

Solution set: $\{-3\}$

29. $x = \log_6 \frac{1}{216} \Rightarrow 6^x = \frac{1}{216} \Rightarrow 6^x = 6^{-3} \Rightarrow$

$x = -3$

Solution set: $\{-3\}$

31. $\log_{10} 0.01 = x$

$10^x = 0.01 \Rightarrow 10^x = 10^{-2} \Rightarrow x = -2$

Solution set: $\{-2\}$

33. $\log_x 1 = 0 \Rightarrow x^0 = 1$

This is a true statement for all real numbers greater than 0, excluding 1.

Solution set: $(0,1) \cup (1,\infty)$

35. $\log_x \sqrt[3]{5} = \frac{1}{3} \Rightarrow x^{1/3} = \sqrt[3]{5} \Rightarrow x^{1/3} = 5^{1/3} \Rightarrow$

$\left(x^{1/3}\right)^3 = \left(5^{1/3}\right)^3 \Rightarrow x = 5$

Recall from Chapter 1 that it is necessary to check all proposed solutions in the original equation when you raise both sides to a power. Check $x = 5$.

$\overline{\log_x \sqrt[3]{5} = \frac{1}{3}}$

$\log_5 \sqrt[3]{5} \overset{?}{=} \frac{1}{3}$

$\log_5 5^{1/3} = \frac{1}{3} \Rightarrow \frac{1}{3}\log_5 5 = \frac{1}{3} \Rightarrow \frac{1}{3} \cdot 1 = \frac{1}{3} \Rightarrow \frac{1}{3} = \frac{1}{3}$

This is a true statement.

Solution set: $\{5\}$

37. $\log_{10}\left(\log_2 2^{10}\right) = x \Rightarrow$

$\log_{10}\left(10\log_2 2\right) = x \Rightarrow \log_{10}\left(10 \cdot 1\right) = x \Rightarrow$

$\log_{10} 10 = x \Rightarrow x = 1$

Solution set: $\{1\}$

39. $2x - 1 = \log_6 6^x \Rightarrow 2x - 1 = x \Rightarrow$

$-1 = -x \Rightarrow 1 = x$

Solution set: $\{1\}$

41. $2^x = \log_2 16 \Rightarrow 2^{\left(2^x\right)} = 16 \Rightarrow 2^{\left(2^x\right)} = 2^4 \Rightarrow$

$2^x = 4 \Rightarrow 2^x = 2^2 \Rightarrow x = 2$

Solution set: $\{2\}$

43. $\left(\frac{1}{3}\right)^{x+1} = 9^x \Rightarrow \left(3^{-1}\right)^{x+1} = \left(3^2\right)^x \Rightarrow$

$3^{(-1)(x+1)} = 3^{2x} \Rightarrow 3^{-x-1} = 3^{2x} \Rightarrow$

$-x - 1 = 2x \Rightarrow x = -\frac{1}{3}$

Solution set: $\left\{-\frac{1}{3}\right\}$

Copyright © 2013 Pearson Education, Inc.

ion 4.4 Evaluating Logarithms and the Change-of-Base Theorem**   **241**

Section 4.4   Evaluating Logarithms and the Change-of-Base Theorem

1. For $f(x) = a^x$, where $a > 0$, the function is *increasing* over its entire domain.

3. $f(x) = 5^x$

This function is one-to-one.

Step 1: Replace $f(x)$ with y and interchange x and y. $y = 5^x \Rightarrow x = 5^y$

Step 2: Solve for y.

$x = 5^y \Rightarrow y = \log_5 x$

Step 3: Replace y with $f^{-1}(x)$.

$f^{-1}(x) = \log_5 x$

5. A base e logarithm is called a <u>natural</u> logarithm, while a base 10 logarithm is called a <u>common</u> logarithm.

7. $\log_2 0$ is undefined because there is no power of 2 that yields a result of 0. In other words, the equation $2^x = 0$ has no solution.

9. $\log 8 = 0.90308999$

11. $\log 10^{12} = 12$

13. $\log 0.1 = \log 10^{-1} = -1$

15. $\log 63 \approx 1.7993$

17. $\log 0.0022 \approx -2.6576$

19. $\log(387 \times 23) \approx 3.9494$

21. $\log\left(\dfrac{518}{342}\right) \approx 0.1803$

23. $\log 387 + \log 23 \approx 3.9494$

25. $\log 518 - \log 342 \approx 0.1803$

27. $\log(387 \times 23) = \log 387 + \log 23$ by the product property of logarithms.

29. Grapefruit, $\left[H_3O^+\right] = 6.3 \times 10^{-4}$

$pH = -\log\left[H_3O^+\right] = -\log\left(6.3 \times 10^{-4}\right)$
$= -\left(\log 6.3 + \log 10^{-4}\right) = -\left(0.7793 - 4\right)$
$= -0.7993 + 4 \approx 3.2$

```
-log(6.3*10^(-4)
)
          3.200659451
```

The answer is rounded to the nearest tenth because it is customary to round pH values to the nearest tenth. The pH of grapefruit is 3.2.

31. Crackers, $\left[H_3O^+\right] = 3.9 \times 10^{-9}$

$pH = -\log\left[H_3O^+\right] = -\log\left(3.9 \times 10^{-9}\right)$
$= -\left(\log 3.9 + \log 10^{-9}\right) = -\left(0.59106 - 9\right)$
$= -(-8.409) \approx 8.4$

```
-log(3.9*10^(-9)
)
          8.408935393
```

The answer is rounded to the nearest tenth because it is customary to round pH values to the nearest tenth. The pH of crackers is 8.4.

33. Soda pop, 2.7

$pH = -\log\left[H_3O^+\right]$
$2.7 = -\log\left[H_3O^+\right]$
$-2.7 = \log\left[H_3O^+\right] \Rightarrow \left[H_3O^+\right] = 10^{-2.7}$

```
10^(-2.7)
          .0019952623
```

$\left[H_3O^+\right] \approx 2.0 \times 10^{-3}$

35. Beer, 4.8

$pH = -\log\left[H_3O^+\right]$
$4.8 = -\log\left[H_3O^+\right]$
$-4.8 = \log\left[H_3O^+\right] \Rightarrow \left[H_3O^+\right] = 10^{-4.8}$

```
10^(-4.8)
     1.584893192E-5
```

$\left[H_3O^+\right] \approx 1.6 \times 10^{-5}$

Copyright © 2013 Pearson Education, Inc.

37. Wetland, 2.49×10^{-5}

$\begin{aligned} pH &= -\log\left[H_3O^+\right] = -\log\left(2.49 \times 10^{-5}\right) \\ &= -\left(\log 2.49 + \log 10^{-5}\right) \\ &= -\log 2.49 - (-5) = -\log 2.49 + 5 \end{aligned}$

$pH \approx 4.6$

Since the pH is between 4.0 and 6.0, it is a poor fen.

39. Wetland, 2.49×10^{-2}

$\begin{aligned} pH &= -\log\left[H_3O^+\right] = -\log\left(2.49 \times 10^{-2}\right) \\ &= -\left(\log 2.49 + \log 10^{-2}\right) \\ &= -\log 2.49 - (-2) = -\log 2.49 + 2 \end{aligned}$

$pH \approx 1.6$

Since the pH is 3.0 or less, it is a bog.

41. Wetland, 2.49×10^{-7}

$\begin{aligned} pH &= -\log\left[H_3O^+\right] = -\log\left(2.49 \times 10^{-7}\right) \\ &= -\left(\log 2.49 + \log 10^{-7}\right) \\ &= -\log 2.49 - (-7) = -\log 2.49 + 7 \end{aligned}$

$pH \approx 6.6$

Since the pH is greater than 6.0, it is a rich fen.

43. (a) $\log 398.4 \approx 2.60031933$

(b) $\log 39.84 \approx 1.60031933$

(c) $\log 3.984 \approx 0.6003193298$

(d) The whole number parts will vary, but the decimal parts will be the same.

45. $\ln e^{1.6} = 1.6$

47. $\ln\left(\dfrac{1}{e^2}\right) = \ln e^{-2} = -2$

49. $\ln 28 \approx 3.3322$

51. $\ln 0.00013 \approx -8.9480$

53. $\ln(27 \times 943) \approx 10.1449$

55. $\ln\left(\dfrac{98}{13}\right) \approx 2.0200$

57. $\ln 27 + \ln 943 \approx 10.1449$

59. $\ln 98 - \ln 13 \approx 2.0200$

61. $\ln(27 \times 943) = \ln 27 + \ln 943$ by the product property of logarithms.

63. $d = 10 \log \dfrac{I}{I_0}$, where d is the decibel rating.

(a) $d = 10 \log \dfrac{100 I_0}{I_0}$

$= 10 \log_{10} 100 = 10(2) = 20$

(b) $d = 10 \log \dfrac{1000 I_0}{I_0}$

$= 10 \log_{10} 1000 = 10(3) = 30$

(c) $d = 10 \log \dfrac{100,000 I_0}{I_0}$

$= 10 \log_{10} 100,000 = 10(5) = 50$

(d) $d = 10 \log \dfrac{1,000,000 I_0}{I_0}$

$= 10 \log_{10} 1,000,000 = 10(6) = 60$

(e) $I = 2 I_0$

$d = 10 \log \dfrac{2 I_0}{I_0} = 10 \log 2 \approx 3.0103$

The described rating is increased by about 3 decimals.

65. $r = \log_{10} \dfrac{I}{I_0}$, where r is the Richter scale rating of an earthquake.

(a) $r = \log_{10} \dfrac{1000 I_0}{I_0} = \log_{10} 1000 = 3$

(b) $r = \log_{10} \dfrac{1,000,000 I_0}{I_0} = \log_{10} 1,000,000 = 6$

(c) $r = \log_{10} \dfrac{100,000,000 I_0}{I_0}$

$= \log_{10} 100,000,000 = 8$

67. From exercise 65, the magnitude of an earthquake, measured on the Richter scale, is $r = \log_{10} \dfrac{I}{I_0}$, where I is the amplitude registered on a seismograph 100 km fro the epicenter of the earthquake, and I_0 is the amplitude of an earthquake of a certain small size. So, $8.8 = \log_{10} \dfrac{I}{I_0} \Rightarrow \dfrac{I}{I_0} = 10^{8.8} \Rightarrow$

$I = 10^{8.8} I_0 \approx 603,957,345 I_0$

69. The year 2012 is represented by 62. $f(62) = -228.1 + 78.19 \ln 62 \approx 94.6$ thousand

We must assume that the model continues to be logarithmic.

71. If $a = 0.36$, then

$S(n) = a \ln\left(1 + \dfrac{n}{a}\right) = 0.36 \ln\left(1 + \dfrac{n}{0.36}\right)$.

(a) $S(100) = 0.36 \ln\left(1 + \dfrac{100}{0.36}\right) \approx 2.0269 \approx 2$

(b) $S(200) = 0.36 \ln\left(1 + \dfrac{200}{0.36}\right) \approx 2.2758 \approx 2$

Copyright © 2013 Pearson Education, Inc.

(c) $S(150) = 0.36 \ln\left(1 + \frac{150}{0.36}\right) \approx 2.1725 \approx 2$

(d) $S(10) = 0.36 \ln\left(1 + \frac{10}{0.36}\right) \approx 1.2095 \approx 1$

73. The index of diversity H for 2 species is given by $H = -\left[P_1 \log_2 P_1 + P_2 \log_2 P_2\right]$. When $P_1 = \frac{50}{100} = 0.5$ and $P_2 = \frac{50}{100} = 0.5$ we have $H = -\left[0.5 \log_2 0.5 + 0.5 \log_2 0.5\right]$.

Since $\log_2 0.5 = \log_2 \frac{1}{2} = \log_2 2^{-1} = -1$, we have $H = -[0.5(-1) + (-1)] = -(-1) = 1$. Thus, the index of diversity is 1.

75. $T(k) = 1.03k \ln\frac{C}{C_0}$

Since $10 \leq k \leq 16$ and $\frac{C}{C_0} = 2$, the range for $T = 1.03k \ln\frac{C}{C_0}$ will be between $T(10)$ and $T(16)$. Since $T(10) = 1.03(10)\ln 2 \approx 7.1$ and $T(16) = 1.03(16)\ln 2 \approx 11.4$, the predicted increased global temperature due to the greenhouse effect from a doubling of the carbon dioxide in the atmosphere is between $7°F$ and $11°F$.

77. $t = \left(1.26 \times 10^9\right)\frac{\ln[1 + 8.33(0.103)]}{\ln 2} \approx 1.126 \times 10^9$

The rock sample is approximately 1.126 billion years old.

For Exercises 79–89, the solutions will be evaluated at the intermediate steps to four decimal places. However, the final answers are obtained without rounding the intermediate steps.

79. $\log_2 5 = \frac{\ln 5}{\ln 2} \approx \frac{1.6094}{0.6931} \approx 2.3219$ or

$\log_2 5 = \frac{\log 5}{\log 2} \approx \frac{0.6990}{0.3010} \approx 2.3219$

81. $\log_8 0.59 = \frac{\log 0.59}{\log 8} \approx \frac{-0.2291}{0.9031} \approx -0.2537$ or

$\log_8 0.59 = \frac{\ln 0.59}{\ln 8} \approx \frac{-0.5276}{2.0794} \approx -0.2537$

83. $\log_{1/2} 3 = \frac{\log 3}{\log\left(\frac{1}{2}\right)} \approx \frac{0.4771}{-0.3010} \approx -1.5850$ or

$\log_{1/2} 3 = \frac{\ln 3}{\ln\left(\frac{1}{2}\right)} \approx \frac{1.0986}{-0.6931} \approx -1.5850$

85. $\log_\pi e = \frac{\ln e}{\ln \pi} \approx \frac{1}{1.1447} \approx 0.8736$ or

$\log_\pi e = \frac{\log e}{\log \pi} \approx \frac{0.4343}{0.4971} \approx 0.8736$

87. Since $\sqrt{13} = 13^{1/2}$, we have

$\log_{\sqrt{13}} 12 = \frac{\ln 12}{\ln \sqrt{13}} = \frac{\ln 12}{\frac{1}{2}\ln 13} \approx \frac{2.4849}{1.2825} \approx 1.9376$

or

$\log_{\sqrt{13}} 12 = \frac{\log 12}{\log \sqrt{13}} = \frac{\log 12}{\frac{1}{2}\log 13} \approx \frac{1.0792}{0.5570}$
≈ 1.9376

89. $\log_{0.32} 5 = \frac{\log 5}{\log 0.32} \approx \frac{0.6990}{-0.4949} \approx -1.4125$

or $\log_{0.32} 5 = \frac{\ln 5}{\ln 0.32} \approx \frac{1.6094}{-1.1394} \approx -1.4125$

91. $\ln\left(b^4 \sqrt{a}\right) = \ln\left(b^4 a^{1/2}\right) = \ln b^4 + \ln a^{1/2}$
$= 4 \ln b + \frac{1}{2}\ln a = 4v + \frac{1}{2}u$

93. $\ln\sqrt{\frac{a^3}{b^5}} = \ln\left(\frac{a^3}{b^5}\right)^{1/2} = \ln\left(\frac{a^{3/2}}{b^{5/2}}\right)$
$= \ln a^{3/2} - \ln b^{5/2} = \frac{3}{2}\ln a - \frac{5}{2}\ln b$
$= \frac{3}{2}u - \frac{5}{2}v$

95. $g(x) = e^x$

(a) $g(\ln 4) = e^{\ln 4} = 4$

(b) $g\left[\ln\left(5^2\right)\right] = e^{\ln 5^2} = 5^2$ or 25

(c) $g\left[\ln\left(\frac{1}{e}\right)\right] = e^{\ln(1/e)} = \frac{1}{e}$

97. $f(x) = \ln x$

(a) $f\left(e^6\right) = \ln e^6 = 6$

(b) $f\left(e^{\ln 3}\right) = \ln e^{\ln 3} = \ln 3$

(c) $f\left(e^{2\ln 3}\right) = \ln e^{2\ln 3} = 2\ln 3$ or $\ln 9$.

99. $2\ln 3x = \ln(3x)^2 = \ln\left(3^2 x^2\right) = \ln 9x^2$

It is equivalent to D.

Copyright © 2013 Pearson Education, Inc.

101. $f(x) = \ln|x|$

The domain of f is all real numbers except 0: $(-\infty, 0) \cup (0, \infty)$ and the range is $(-\infty, \infty)$.

Since $f(-x) = \ln|-x| = \ln|x| = f(x)$, this is an even function and symmetric with respect to the y-axis.

103. $f(x) = \ln(e^2 x) = \ln e^2 + \ln x = 2 + \ln x$

$f(x) = \ln(e^2 x)$ is a vertical shift of the graph of $g(x) = \ln x$, 2 units up.

105. $f(x) = \ln\frac{x}{e^2} = \ln x - 2\ln e = \ln x - 2$

$f(x) = \ln\frac{x}{e^2}$ is a vertical shift of the graph of $g(x) = \ln x$, 2 units down.

Chapter 4 Quiz
(Sections 4.1–4.4)

1. *Step 1*: Replace $f(x)$ with y and interchange x and y: $y = \sqrt[3]{3x - 6} \Rightarrow x = \sqrt[3]{3y - 6}$

Step 2: Solve for y.

$x = \sqrt[3]{3y - 6} \Rightarrow x^3 = 3y - 6 \Rightarrow \frac{x^3 + 6}{3} = y$

Step 3: Replace y with $f^{-1}(x)$.

$f^{-1}(x) = \frac{x^3 + 6}{3}$

3.

Domain: $(-\infty, \infty)$; range: $(-\infty, 0)$

5. Use the compound interest formula to find the future value (the amount in the account).

$A = P\left(1 + \frac{r}{n}\right)^{tn}$, $P = 15{,}000$, $r = 0.035$, $t = 8$

(a) $n = 1$

$A = 15{,}000\left(1 + \frac{0.035}{1}\right)^{8\cdot1}$

$= (15{,}000)(1.035)^8 \approx \$19{,}752.14$

There will be about $\$19{,}752.14$ after one year.

(b) $n = 4$

$A = 15{,}000\left(1 + \frac{0.035}{4}\right)^{8\cdot4}$

$= (15{,}000)(1.00875)^{32} \approx \$19{,}822.79$

There will be about $\$19{,}822.79$ after one year.

(c) $n = 12$

$A = 15{,}000\left(1 + \frac{0.035}{12}\right)^{8\cdot12}$

$= (15{,}000)(1.0029)^{96} \approx \$19{,}838.86$

There will be about $\$19{,}838.86$ after one year.

(d) $n = 365$

$A = 15{,}000\left(1 + \frac{0.035}{365}\right)^{8\cdot365}$

$= (15{,}000)(1.0001)^{2920} \approx \$19{,}846.68$

There will be about $\$19{,}846.68$ after one year.

7. The expression $\log_6 25$ represents the exponent to which 6 must be raised in order to obtain 25.

9. $\log_3 \dfrac{\sqrt{x} \cdot y}{pq^4} = \log_3\left(\sqrt{x} \cdot y\right) - \log_3\left(pq^4\right)$

$= \log_3\left(x^{1/2}\right) + \log_3 y$

$\qquad\qquad - \left(\log_3 p + \log_3 q^4\right)$

$= \tfrac{1}{2}\log_3 x + \log_3 y$

$\qquad\qquad - \log_3 p - 4\log_3 q$

11. $\log_3 40 = \frac{\log 40}{\log 3} \approx \frac{1.6021}{.4771} \approx 3.3578$

We could also have used the natural logarithm. $\log_3 40 = \frac{\ln 40}{\ln 3} \approx \frac{3.689}{1.0986} \approx 3.3578$

Section 4.5 Exponential and Logarithmic Equations

1. Since x is the exponent to which 7 must be raised in order to obtain 19, the solution is $\log_7 19$ or $\dfrac{\log 19}{\log 7}$ or $\dfrac{\ln 19}{\ln 7}$.

3. Since x is the exponent to which $\frac{1}{2}$ must be raised in order to obtain 12, the solution is $\log_{1/2} 12$ or $\dfrac{\log 12}{\log\left(\frac{1}{2}\right)}$ or $\dfrac{\ln 12}{\ln\left(\frac{1}{2}\right)}$.

Copyright © 2013 Pearson Education, Inc.

5. $3^x = 7$

$\ln 3^x = \ln 7$

$x \ln 3 = \ln 7$

$x = \dfrac{\ln 7}{\ln 3} \approx 1.771$

Solution set: $\{1.771\}$

7. $\left(\dfrac{1}{2}\right)^x = 5 \Rightarrow \ln\left(\dfrac{1}{2}\right)^x = \ln 5 \Rightarrow x \ln \dfrac{1}{2} = \ln 5$

$x = \dfrac{\ln 5}{\ln \frac{1}{2}} \approx -2.322$

Solution set: $\{-2.322\}$

9. $0.8^x = 4 \Rightarrow \ln\left(0.8^x\right) = \ln 4 \Rightarrow x \ln 0.8 = \ln 4 \Rightarrow$

$x = \dfrac{\ln 4}{\ln 0.8} \approx -6.213$

Solution set: $\{-6.213\}$

11. $\qquad 4^{x-1} = 3^{2x} \Rightarrow \ln\left(4^{x-1}\right) = \ln\left(3^{2x}\right) \Rightarrow$

$(x-1)\ln 4 = 2x \ln 3$

$x \ln 4 - \ln 4 = 2x \ln 3$

$x \ln 4 - 2x \ln 3 = \ln 4 \Rightarrow x(\ln 4 - 2 \ln 3) = \ln 4 \Rightarrow$

$x = \dfrac{\ln 4}{\ln 4 - 2 \ln 3} \approx -1.710$

Solution set: $\{-1.710\}$

13. $\qquad 6^{x+1} = 4^{2x-1} \Rightarrow \ln\left(6^{x+1}\right) = \ln\left(4^{2x-1}\right)$

$(x+1)\ln 6 = (2x-1)\ln 4$

$x \ln 6 + \ln 6 = 2x \ln 4 - \ln 4$

$\ln 6 + \ln 4 = 2x \ln 4 - x \ln 6$

$\ln 6 + \ln 4 = x(2 \ln 4 - \ln 6)$

$x = \dfrac{\ln 6 + \ln 4}{2 \ln 4 - \ln 6} \approx 3.240$

Solution set: $\{3.240\}$

15. $e^{x^2} = 100 \Rightarrow \ln\left(e^{x^2}\right) = \ln 100$

$x^2 = \ln 100 \Rightarrow x = \pm\sqrt{\ln 100} = \pm 2.146$

Solution set: $\left\{\pm 2.146\right\}$

17. $e^{3x-7} \cdot e^{-2x} = 4e \Rightarrow e^{x-7} = 4e$

$\ln\left(e^{x-7}\right) = \ln\left(4e\right) \Rightarrow (x-7)\ln e = \ln 4 + \ln e$

$x - 7 = \ln 4 + 1 \Rightarrow x = \ln 4 + 8 \approx 9.386$

Solution set: $\{9.386\}$

19. $\left(\frac{1}{3}\right)^x = -3$ has no solution since $\frac{1}{3}$ raised to any power is positive.

Solution set: $\varnothing$

21. $0.05(1.15)^x = 5 \Rightarrow 1.15^x = \dfrac{5}{0.05} = 100 \Rightarrow$

$\log\left(1.15^x\right) = \log 100 \Rightarrow x \log 1.15 = \log 100 \Rightarrow$

$x = \dfrac{\log 100}{\log 1.15} = \dfrac{2}{\log 1.15} \approx 32.950$

Solution set: $\{32.950\}$

23. $3(2)^{x-2} + 1 = 100 \Rightarrow 3(2)^{x-2} = 99 \Rightarrow$

$2^{x-2} = 33 \Rightarrow \ln\left(2^{x-2}\right) = \ln 33 \Rightarrow$

$(x-2)\ln 2 = \ln 33 \Rightarrow x = \dfrac{\ln 33}{\ln 2} + 2 \approx 7.044$

Solution set: $\{7.044\}$

25. $2(1.05)^x + 3 = 10 \Rightarrow 2(1.05)^x = 7 \Rightarrow$

$1.05^x = 3.5 \Rightarrow \ln\left(1.05^x\right) = \ln 3.5 \Rightarrow$

$x \ln 1.05 = \ln 3.5 \Rightarrow x = \dfrac{\ln 3.5}{\ln 1.05} \approx 25.677$

Solution set: $\{25.677\}$

27. $\qquad 5(1.015)^{x-1980} = 8 \Rightarrow 1.015^{(x-1980)} = 1.6$

$\ln\left(1.015^{(x-1980)}\right) = \ln 1.6$

$(x-1980)\ln 1.015 = \ln 1.6$

$x = \dfrac{\ln 1.6}{\ln 1.015} + 1980$

≈ 2011.568

Solution set: $\{2011.568\}$

29. $e^{2x} - 6e^x + 8 = 0$

Let $u = e^x$.

$u^2 - 6u + 8 = 0$

$(u-4)(u-2) = 0$

$u - 4 = 0$	$u - 2 = 0$
$u = 4$	$u = 2$
$e^x = 4$	$e^x = 2$
$\ln e^x = \ln 4$	$\ln e^x = \ln 2$
$x \ln e = \ln 4$	$x \ln e = \ln 2$
$x = \ln 4$	$x = \ln 2$

Solution set: $\{\ln 2, \ln 4\}$

Copyright © 2013 Pearson Education, Inc.

31. $2e^{2x} + e^x = 6 \Rightarrow 2e^{2x} + e^x - 6 = 0$

Let $u = e^x$.

$2u^2 + u - 6 = 0$

$(2u - 3)(u + 2) = 0$

$2u - 3 = 0$	$u + 2 = 0$
$2u = 3$	$u = -2$
$u = \frac{3}{2}$	$e^x = -2$
$e^x = \frac{3}{2}$	Disregard this value
$\ln e^x = \ln \frac{3}{2}$	because e^x is always
$x \ln e = \ln \frac{3}{2}$	positive.
$x = \ln \frac{3}{2}$	

Solution set: $\left\{ \ln \frac{3}{2} \right\}$

33. $5^{2x} + 3(5^x) = 28 \Rightarrow 5^{2x} + 3(5^x) - 28 = 0$

Let $u = 5^x$.

$u^2 + 3u - 28 = 0$

$(u - 4)(u + 7) = 0$

$u - 4 = 0$	$u + 7 = 0$
$u = 4$	$u = -7$
$5^x = 4$	$5^x = -7$
$\log 5^x = \log 4$	Disregard this
$x \log 5 = \log 4$	value because 5^x
$x = \frac{\log 4}{\log 5} = \log_5 4$	is always positive.

Solution set: $\left\{ \log_5 4 \right\}$

35. $5 \ln x = 10 \Rightarrow \ln x = 2 \Rightarrow e^2 = x$

Solution set: $\left\{ e^2 \right\}$

37. $\ln(4x) = 1.5 \Rightarrow 4x = e^{1.5} \Rightarrow x = \frac{e^{1.5}}{4}$

Solution set: $\left\{ \frac{e^{1.5}}{4} \right\}$

39. $\log(2 - x) = 0.5 \Rightarrow 2 - x = 10^{0.5} \Rightarrow$

$2 - x = \sqrt{10} \Rightarrow x = 2 - \sqrt{10}$

Solution set: $\left\{ 2 - \sqrt{10} \right\}$

41. $\log_6 (2x + 4) = 2 \Rightarrow 2x + 4 = 6^2 \Rightarrow$

$2x + 4 = 36 \Rightarrow 2x = 32 \Rightarrow x = 16$

Solution set: $\{16\}$

43. $\log_4 \left(x^3 + 37 \right) = 3 \Rightarrow x^3 + 37 = 4^3 \Rightarrow$

$x^3 = 27 \Rightarrow x = 3$

Solution set: $\{3\}$

45. $\ln x + \ln x^2 = 3 \Rightarrow \ln \left(x \cdot x^2 \right) = 3 \Rightarrow \ln x^3 = 3 \Rightarrow$

$x^3 = e^3 \Rightarrow x = e$

Solution set: $\{e\}$

47. $\log_3 \left[(x + 5)(x - 3) \right] = 2$

$(x + 5)(x - 3) = 3^2$

$x^2 + 2x - 15 = 9$

$x^2 + 2x - 24 = 0$

$(x + 6)(x - 4) = 0$

$x + 6 = 0$	$x - 4 = 0$
$x = -6$	$x = 4$

Solution set: $\{-6, 4\}$

49. $\log_2 \left[(2x + 8)(x + 4) \right] = 5$

$(2x + 8)(x + 4) = 2^5$

$2x^2 + 16x + 32 = 32$

$2x^2 + 16x = 0$

$2x(x + 8) = 0$

$2x = 0$	$x + 8 = 0$
$x = 0$	$x = -8$

Solution set: $\{-8, 0\}$

51. $\log x + \log(x + 15) = 2$

$\log \left[x(x + 15) \right] = 2$

$x^2 + 15x = 10^2 = 100$

$x^2 + 15x - 100 = 0$

$(x - 5)(x + 20) = 0$

$x - 5 = 0$	$x + 20 = 0$
$x = 5$	$x = -20$

Since $\log(-20)$ is undefined, we reject this proposed solution.

Solution set: $\{5\}$

53. $\log(x + 25) = \log(x + 10) + \log 4$

$\log(x + 25) = \log \left[4(x + 10) \right]$

$\log(x + 25) = \log(4x + 40)$

$x + 25 = 4x + 40$

$-3x = 15 \Rightarrow x = -5$

Solution set: $\{-5\}$

55. $\log(x - 10) - \log(x - 6) = \log 2 \Rightarrow$

$\log \left(\frac{x - 10}{x - 6} \right) = \log 2 \Rightarrow \frac{x - 10}{x - 6} = 10^2 = 100 \Rightarrow$

$x - 10 = 100(x - 6) \Rightarrow x - 10 = 100x - 600 \Rightarrow$

$590 = 99x \Rightarrow \frac{590}{99} \approx 5.96 = x$

When the proposed solution is substituted for x in $\log(x - 10)$, a negative argument results. Since this is not allowed, we reject this proposed solution. Solution set: $\varnothing$

Copyright © 2013 Pearson Education, Inc.

57. $\ln(5-x) + \ln(-3-x) = \ln(1-8x)$

$\ln[(5-x)(-3-x)] = \ln(1-8x)$

$-15 - 2x + x^2 = 1 - 8x$

$x^2 + 6x - 16 = 0$

$(x+8)(x-2) = 0 \Rightarrow x = -8, 2$

If 2 is substituted for x in $\ln(-3-x)$, the argument becomes -5. Since this is not allowed, we reject this proposed solution. Solution set: $\{-8\}$

59. $\log_8(x+2) + \log_8(x+4) = \log_8 8$

$\log_8[(x+2)(x+4)] = \log_8 8$

$x^2 + 6x + 8 = 8$

$x^2 + 6x = 0$

$x(x+6) = 0 \Rightarrow x = 0, -6$

If -6 is substituted for x in $\log_8(x+2)$, the argument becomes -4. Since this is not allowed, we reject this proposed solution. Solution set: $\{0\}$

61. $\log_2(x^2 - 100) - \log_2(x+10) = 1$

$$\log_2\left(\frac{x^2 - 100}{x+10}\right) = 1$$

$$\frac{x^2 - 100}{x+10} = 2^1 = 2$$

$$x^2 - 100 = 2x + 20$$

$$x^2 - 2x - 120 = 0$$

$$(x+10)(x-12) = 0$$

$$x = -10, 12$$

If -10 is substituted for x in $\log_2(x+10)$, the argument becomes 0. Since this is not allowed, we reject this proposed solution. Solution set: $\{12\}$

63. $\log x + \log(x - 21) = \log 100 \Rightarrow$

$\log[x(x-21)] = 2 \Rightarrow \log(x^2 - 21x) = 2 \Rightarrow$

$x^2 - 21x = 10^2 \Rightarrow x^2 - 21x - 100 = 0 \Rightarrow$

$(x-25)(x+4) = 0 \Rightarrow x = 25$ or $x = -4$

Since the negative solution $(x = -4)$ is not in the domain of $\log x$, it must be discarded. Solution set: $\{25\}$

65. $\log(9x+5) = 3 + \log(x+2)$

$\log(9x+5) - \log(x+2) = 3$

$$\log\frac{9x+5}{x+2} = 3$$

$$\frac{9x+5}{x+2} = 10^3 = 1000$$

$$9x + 5 = 1000x + 2000$$

$$-991x = 1995$$

$$x = -\frac{1995}{991} \approx -2.013$$

If $-\frac{1995}{991}$ is substituted for x in $\log_2(x+2)$, the argument is negative. Since this is not allowed, we reject this proposed solution. Solution set: $\varnothing$

67. $\ln(4x-2) - \ln 4 = -\ln(x-2)$

$\ln\frac{4x-2}{4} = -\ln(x-2) \Rightarrow \frac{4x-2}{4} = \frac{1}{x-2}$

$(4x-2)(x-2) = 4 \Rightarrow 4x^2 - 10x + 4 = 4$

$4x^2 - 10x = 0 \Rightarrow 2x(2x-5) = 0 \Rightarrow$

$2x = 0 \Rightarrow x = 0$ or $2x - 5 = 0 \Rightarrow x = \frac{5}{2} = 2.5$

Since $x = 0$ is not in the domain of $\ln(x-2)$, it must be discarded. Solution set: $\{2.5\}$

69. $\log_5(x+2) + \log_5(x-2) = 1 \Rightarrow$

$\log_5[(x+2)(x-2)] = 1 \Rightarrow x^2 - 4 = 5^1$

$x^2 - 9 = 0 \Rightarrow (x-3)(x+3) = 0 \Rightarrow x = \pm 3$

-3 is not in the domain, so reject it. Solution set: $\{3\}$

71. $\log_2(2x-3) + \log_2(x+1) = 1$

$\log_2[(2x-3)(x+1)] = 1$

$2x^2 - x - 3 = 2^1$

$2x^2 - x - 5 = 0$

$$x = \frac{-(-1) \pm \sqrt{(-1)^2 - 4(2)(-5)}}{2(2)} = \frac{1 \pm \sqrt{41}}{4}$$

Since the negative solution $\left(x = \frac{1-\sqrt{41}}{4}\right)$ is not in the domain of $\log(x+1)$, it must be discarded.

Solution set: $\left\{\frac{1+\sqrt{41}}{4}\right\}$

73. $\ln e^x - 2\ln e = \ln e^4 \Rightarrow x - 2 = 4 \Rightarrow x = 6$

Solution set: $\{6\}$

75. $\log_2(\log_2 x) = 1 \Rightarrow \log_2 x = 2^1 \Rightarrow$

$\log_2 x = 2 \Rightarrow x = 2^2 \Rightarrow x = 4$

Solution set: $\{4\}$

Copyright © 2013 Pearson Education, Inc.

77. $\log x^2 = (\log x)^2 \Rightarrow 2\log x = (\log x)^2 \Rightarrow$
$(\log x)^2 - 2\log x = 0 \Rightarrow \log x(\log x - 2) = 0$

$\log_{10} x = 0 \quad$ or $\quad \log_{10} x - 2 = 0$
$\qquad x = 10^0 \qquad\qquad \log_{10} x = 2$
$\qquad x = 1 \qquad\qquad\quad x = 10^2 = 100$

Solution set: $\{1, 100\}$

79. Answers will vary. One should not immediately reject a negative answer when solving equations involving logarithms. One must examine what happens to an answer when substituted back into the original equation. An answer (whether negative, positive, or zero) must not allow a nonpositive value in the argument of the logarithm. If it does, regardless of its sign, it must be rejected.

81. $p = a + \dfrac{k}{\ln x}$, for x

$p - a = \dfrac{k}{\ln x} \Rightarrow (\ln x)(p - a) = k \Rightarrow$

$\ln x = \dfrac{k}{p - a} \Rightarrow x = e^{k/(p-a)}$

83. $T = T_0 + (T_1 - T_0)10^{-kt}$, for t

$T - T_0 = (T_1 - T_0)10^{-kt}$

$\dfrac{T - T_0}{T_1 - T_0} = 10^{-kt}$

$\log_{10}\left(\dfrac{T - T_0}{T_1 - T_0}\right) = -kt$

$\dfrac{\log\left(\frac{T-T_0}{T_1-T_0}\right)}{-k} = \dfrac{-kt}{-k}$

$t = -\dfrac{1}{k}\log\left(\dfrac{T - T_0}{T_1 - T_0}\right)$

85. $I = \dfrac{E}{R}\left(1 - e^{-Rt/2}\right)$, for t

$RI = R\left[\dfrac{E}{R}\left(1 - e^{-Rt/2}\right)\right]$

$RI = E\left(1 - e^{-Rt/2}\right)$

$\dfrac{RI}{E} = 1 - e^{-Rt/2}$

$\dfrac{RI}{E} - 1 = -e^{-Rt/2}$

$1 - \dfrac{RI}{E} = e^{-Rt/2}$

$\ln\left(1 - \dfrac{RI}{E}\right) = \ln e^{-Rt/2}$

$-\dfrac{Rt}{2} = \ln\left(1 - \dfrac{RI}{E}\right)$

$-\dfrac{2}{R}\left(-\dfrac{Rt}{2}\right) = -\dfrac{2}{R}\ln\left(1 - \dfrac{RI}{E}\right)$

$t = -\dfrac{2}{R}\ln\left(1 - \dfrac{RI}{E}\right)$

87. $y = A + B\left(1 - e^{-Cx}\right)$, for x

$y - A = B\left(1 - e^{-Cx}\right)$

$\dfrac{y - A}{B} = 1 - e^{-Cx}$

$\dfrac{y - A}{B} - 1 = \dfrac{y - A - B}{B} = -e^{-Cx}$

$\dfrac{A + B - y}{B} = e^{-Cx}$

$\ln\left(\dfrac{A + B - y}{B}\right) = -Cx$

$\dfrac{\ln\left(\frac{A+B-y}{B}\right)}{-C} = x$

89. $\log A = \log B - C\log x$, for A

$\log A = \log B - C\log x$

$\log A = \log\dfrac{B}{x^C} \Rightarrow A = \dfrac{B}{x^C}$

91. $A = P\left(1 + \dfrac{r}{n}\right)^{tn}$, for t

$\dfrac{A}{P} = \left(1 + \dfrac{r}{n}\right)^{tn}$

$\log\left(\dfrac{A}{P}\right) = nt\log\left(1 + \dfrac{r}{n}\right)$

$\dfrac{\log\left(\frac{A}{P}\right)}{n\log\left(1 + \frac{r}{n}\right)} = t$

93. $A = P\left(1 + \dfrac{r}{n}\right)^{tn}$

To solve for A, substitute $P = 10{,}000$, $r = 0.03$, $n = 4$, and $t = 5$.

$A = 10{,}000\left(1 + \dfrac{0.03}{4}\right)^{(4)(5)}$

$A = 10{,}000(1.0075)^{20} \approx 11{,}611.84$

There will be $11,611.84 in the account.

Copyright © 2013 Pearson Education, Inc.

95. $A = P\left(1+\dfrac{r}{n}\right)^{tn}$

To solve for t, substitute $A = 30{,}000$, $P = 27{,}000$, $r = 0.04$, and $n = 4$.

$$30{,}000 = 27{,}000\left(1+\dfrac{0.04}{4}\right)^{t(4)}$$

$$\dfrac{30{,}000}{27{,}000} = (1+0.01)^{4t} \Rightarrow \dfrac{10}{9} = 1.01^{4t} \Rightarrow$$

$$\ln\dfrac{10}{9} = \ln\left(1.01^{4t}\right) \Rightarrow \ln\dfrac{10}{9} = 4t\ln 1.01 \Rightarrow$$

$$t = \dfrac{\ln\dfrac{10}{9}}{4\ln 1.01} \approx 2.6$$

To the nearest tenth of a year, Tom will be ready to buy a car in 2.6 yr.

97. $A = P\left(1+\dfrac{r}{n}\right)^{tn}$

To solve for r, substitute $A = 2500$, $P = 2000$, $t = 4.5$, and $n = 2$.

$$2500 = 2000\left(1+\dfrac{r}{2}\right)^{(4.5)(2)} \Rightarrow$$

$$1.25 = \left(1+\dfrac{r}{2}\right)^{9} \Rightarrow \sqrt[9]{1.25} = 1+\dfrac{r}{2}$$

$$\sqrt[9]{1.25} - 1 = \dfrac{r}{2} \Rightarrow r = 2\left(\sqrt[9]{1.25}-1\right) \approx 0.0502$$

The interest rate is about 5.02%.

99. **(a)** $f(3000) = 86.3\ln 3000 - 680 \approx 10.9$

  At 3000 ft, about 10.9% of the moisture falls as snow.

  (b) $f(4000) = 86.3\ln 4000 - 680 \approx 35.8$

  At 4000 ft, about 35.8% of the moisture falls as snow.

  (c) $f(7000) = 86.3\ln 7000 - 680 \approx 84.1$

  At 7000 ft, about 84.1% of the moisture falls as snow.

101. Double the 2002 value is $2(9{,}119) = 18{,}238$.

$$f(x) = 9318(1.06)^{x}$$

$$18{,}238 = 9318(1.06)^{x} \Rightarrow \dfrac{18{,}238}{9318} = 1.06^{x} \Rightarrow$$

$$\ln\dfrac{18{,}238}{9318} = \ln 1.06^{x} \Rightarrow \ln\dfrac{18{,}238}{9318} = x\ln 1.06$$

$$x = \dfrac{\ln\dfrac{18{,}238}{9318}}{\ln 1.06} \approx 11.53$$

During 2013, the cost of a year's tuition, room and board, and fees at a public university will be double the cost in 2002.

103. $f(x) = \dfrac{67}{1+1.081e^{-x/24.71}}$

  (a) In 2008, $x = 58$.

  $$f(58) = \dfrac{67}{1+1.081e^{-58/24.71}} \approx 60.7$$

  In 2008, about 61% of U.S. women were in the civilian labor force.

  (b) $\dfrac{67}{1+1.081e^{-x/24.71}} = 55$

  $$67 = 55\left(1+1.081e^{-x/24.71}\right)$$

  $$67 = 55 + 59.455e^{-x/24.71}$$

  $$12 = 59.455e^{-x/24.71}$$

  $$\dfrac{12}{59.455} = e^{-x/24.71}$$

  $$\ln\left(\dfrac{12}{59.455}\right) = \ln\left(e^{-x/24.71}\right)$$

  $$\ln\left(\dfrac{12}{59.455}\right) = -\dfrac{x}{24.71}$$

  $$-24.71\ln\left(\dfrac{12}{59.455}\right) \approx 39.5 = x$$

  55% of U.S. women were in the civilian labor force during 1989.

105. $\ln(1-P) = -0.0034 - 0.0053T$

  (a) Change this equation to exponential form, then isolate P.

  $$1 - P = e^{-0.0034-0.0053T}$$

  $$P(T) = 1 - e^{-0.0034-0.0053T}$$

  (b)

  From the graph one can see that initially there is a rapid reduction of carbon dioxide emissions. However, after a while there is little benefit in raising taxes further.

Copyright © 2013 Pearson Education, Inc.

(c) $P(60) = 1 - e^{-0.00340 - 0.0053(60)}$
$$\approx 0.275 \text{ or } 27.5\%$$

The reduction in carbon emissions from a tax of \$60 per ton of carbon is 27.5%.

(d) We must determine T when $P = .5$.
$$0.5 = 1 - e^{-0.0034 - 0.0053T}$$
$$0.5 - 1 = -e^{-0.0034 - 0.0053T}$$
$$-0.5 = -e^{-0.0034 - 0.0053T}$$
$$0.5 = e^{-0.0034 - 0.0053T}$$
$$\ln 0.5 = -0.0034 - 0.0053T$$
$$\ln 0.5 + 0.0034 = -0.0053T$$
$$T = \frac{\ln 0.5 + 0.0034}{-0.0053} \approx 130.14$$

The value $T = \$130.14$ will give a 50% reduction in carbon emissions.

107. *Step 1*: Replace $f(x) = e^{x+1} - 4$ with y and interchange x and y.
$$y = e^{x+1} - 4 \Rightarrow x = e^{y+1} - 4$$
Step 2: Solve for y.
$$x = e^{y+1} - 4 \Rightarrow x + 4 = e^{y+1}$$
$$\ln(x+4) = y + 1 \Rightarrow \ln(x+4) - 1 = y$$
Step 3: Replace y with $f^{-1}(x)$.
$$f^{-1}(x) = \ln(x+4) - 1$$
Domain: $(-4, \infty)$; range: $(-\infty, \infty)$

109. $e^x + \ln x = 5$

The two graphs intersect at approximately (1.52, 5). The x-coordinate of this point is the solution of the equation.
Solution set: $\{1.52\}$

111. $2e^x + 1 = 3e^{-x}$

The two curves intersect at the point (0, 3). The x-coordinate of this point is the solution of the equation. Solution set: $\{0\}$

113. $\log x = x^2 - 8x + 14$

The two graphs intersect at approximately (2.45, 0.39) and (5.66, 0.75). The x-coordinate of these points represent the solution of the equation. Solution set: $\{2.45, 5.66\}$

115. In the last line, $2 < 1$, the sign should have flipped because the value of $\log \frac{1}{3} \approx -.4771$ is negative and whenever you divide (or multiply) both sides of an inequality by a negative value, the sign flips.

Section 4.6 Applications and Models of Exponential Growth and Decay

1. The equation $2e^{0.02x} = 6$ represents an initial amount of 2 and a final amount 6 (triple that of 2). The correct choice is B.

3. The equation $y = 2e^{0.02(3)}$ represents the amount after 3 yr. The correct choice is C.

Copyright © 2013 Pearson Education, Inc.

For exercises 5–9, use the function for exponential growth or decay, $y = y_0 e^{kt}$.

5. $20 = 60e^{3k} \Rightarrow \frac{1}{3} = e^{3k} \Rightarrow \ln\frac{1}{3} = 3k \Rightarrow$
$k = \frac{1}{3}\ln\frac{1}{3}$

7. Since the half-life is 100 days, after 100 days there are 5 mg remaining.
$5 = 10e^{100k} \Rightarrow \frac{1}{2} = e^{100k} \Rightarrow \ln\frac{1}{2} = 100k \Rightarrow$
$k = \frac{1}{100}\ln\frac{1}{2}$

9. $0.64 = 2.56e^{2k} \Rightarrow \frac{1}{4} = e^{2k} \Rightarrow \ln\frac{1}{4} = 2k \Rightarrow$
$k = \frac{1}{2}\ln\frac{1}{4}$

11. $A(t) = 500e^{-0.032t}$

    (a) $t = 4 \Rightarrow A(4) = 500e^{-0.032(4)} \approx 440$
    After 4 years, about 440 g remain.

    (b) $t = 8 \Rightarrow A(8) = 500e^{-0.032(8)} \approx 387$
    After 8 years, about 387 g remain.

    (c) $t = 20 \Rightarrow A(20) = 500e^{-0.032(20)} \approx 264$
    After 20 years, about 264 g remain.

    (d) Find t when $A(t) = 250$.
    $$250 = 500e^{-0.032t}$$
    $$0.5 = e^{-0.032t}$$
    $$\ln 0.5 = \ln e^{-0.032t} \Rightarrow \ln 0.5 = -0.032t$$
    $$t = \frac{\ln 0.5}{-0.032} \approx 21.66$$
    The half-life is about 21.66 yr.

13. $A(t) = A_0 e^{-0.00043t}$
Find t when $A(t) = \frac{1}{2}A_0$.
$$\frac{1}{2}A_0 = A_0 e^{-0.00043t}$$
$$\frac{1}{2} = e^{-0.00043t} \Rightarrow \ln\frac{1}{2} = \ln e^{-0.00043t}$$
$$\ln\frac{1}{2} = -0.00043t \Rightarrow t = \frac{\ln\frac{1}{2}}{-0.00043} \approx 1611.97$$
The half-life is about 1611.97 yr.

15. First find the given values to find y_0 and then
k: $12 = y_0 e^{k(0)} \Rightarrow 12 = y_0$
$$y = 12e^{kt} \Rightarrow 6 = 12e^{4k} \Rightarrow 0.5 = e^{4k} \Rightarrow$$
$$\ln 0.5 = 4k \Rightarrow \frac{\ln 0.5}{4} = k \Rightarrow k \approx -0.173$$
The exponential decay equation is
$y = 12e^{-0.173t}$

To find the amount present after 7 years, let
$t = 7$. $y = 12e^{-0.173(7)} \approx 3.57$
After 7 years, about 3.57 g of the substance will be present.

17. From Example 5, we have the amount of radiocarbon present after t years is given by
$y = y_0 e^{-0.0001216t}$, where y_0 is the initial amount present. Letting $y = \frac{1}{3}y_0$, solve for t.
$$\frac{1}{3}y_0 = y_0 e^{-0.0001216t} \Rightarrow \frac{1}{3} = e^{-0.0001216t} \Rightarrow$$
$$\ln\frac{1}{3} = \ln e^{-0.0001216t} \Rightarrow \ln\frac{1}{3} = -0.0001216t \Rightarrow$$
$$\frac{1}{-0.0001216}\ln\frac{1}{3} \approx 9034.64 = t$$
The Egyptian died about 9000 yr ago.

19. Since $y = y_0 e^{-0.0001216t}$, where y_0 is the initial amount present. If we let $y = 0.15y_0$, we can solve for t.
$$0.15y_0 = y_0 e^{-0.0001216t}$$
$$0.15 = e^{-0.0001216t} \Rightarrow \ln 0.15 = \ln e^{-0.0001216t} \Rightarrow$$
$$\ln 0.15 = -0.0001216t \Rightarrow$$
$$\frac{1}{-0.0001216}\ln 0.15 \approx 15,601.32 = t$$
The paintings are about 15,600 yr old.

21. (a) A point on the graph of $f(x) = A_0 a^{x-1970}$ appears to be (1970, 31,000)
    Since $f(1970) = A_0 a^{1970-1970} = 31,000$,
    we have $A_0 a^0 = 31,000 \Rightarrow A_0 = 31,000$.
    Thus, we have the function
    $f(x) = 31,000a^{x-1970}$. Since the point (2008, 11,500) is also on the graph of the function, we have
    $$f(2008) = 31,000a^{2008-1970}$$
    $$11,500 = 31,000a^{2008-1970}$$
    $$a^{38} \approx 0.3710 \Rightarrow$$
    $$a = \sqrt[38]{0.3710} \approx 0.97$$
    Thus, a possible function is
    $f(x) = 31,000(0.97)^{x-1970}$.

Copyright © 2013 Pearson Education, Inc.

(b) If you think of

$$f(x) = 31,000(0.97)^{x-1970}$$ as a function

in the form of $A = P\left(1 + \frac{r}{n}\right)^{tn}$ where

$n = 1$ (compounded annually), we have

$P(1+r)^t = 31,000(0.97)^{x-1970} \Rightarrow$ Thus,

$1 + r = 0.97 \Rightarrow r = -0.03$

the average annual percent decrease is about 3%.

23. From Example 6, we have the temperature, $f(t)$, of a body at time t after being introduced into an environment having constant temperature T_0 is $f(t) = T_0 + Ce^{-kt}$, where C and k are constants. From the given information, when $t = 0$, $T_0 = 0$, and the temperature of the water is $f(0) = 100$.

$100 = 0 + Ce^{-0k} \Rightarrow 100 = C$

Thus, we have $f(t) = 0 + 100e^{-kt} = 100e^{-kt}$.

Also, when $t = \frac{24}{60} = \frac{2}{5}$ hr, $f\left(\frac{2}{5}\right) = 50$. Using this information, we have

$$50 = 100e^{-(2/5)k} \Rightarrow \frac{1}{2} = e^{-(2/5)k}$$

$$\ln\frac{1}{2} = \ln e^{-(2/5)k} \Rightarrow \ln\frac{1}{2} = -\frac{2}{5}k$$

$$\ln 1 - \ln 2 = -\frac{2}{5}k \Rightarrow \ln 2 = \frac{2}{5}k$$

$$k = \frac{5}{2}\ln 2 \approx 1.733$$

Thus, the model is $f(t) = 100e^{-1.733t}$.

To find the temperature after $\frac{96}{60} = \frac{8}{5}$ hrs, we find $f\left(\frac{8}{5}\right)$. Since

$f\left(\frac{8}{5}\right) = 100e^{-1.733(8/5)} \approx 6.25$, the temperature after 96 minutes is about $6.25°C$. Note: We could have used the exact value of k to perform the calculations.

$$f\left(\frac{8}{5}\right) = 100e^{-\left(\frac{5}{2}\ln 2\right)\left(\frac{8}{5}\right)} = 100e^{-4\ln 2}$$

$$= 100e^{\ln\frac{1}{16}} = 100 \cdot \frac{1}{16} = 6.25$$

25. Given $P = 60,000$ and $t = 5$, substitute $r = 0.05$ and $n = 4$ into the compound interest formula, $A = P\left(1 + \frac{r}{n}\right)^{tn}$. We have

$$A = 60,000\left(1 + \frac{0.05}{4}\right)^{5(4)}$$

$$= 60,000(1.0125)^{20} \approx 76,922.23$$

The interest from this investment would be $76,922.23 - $60,000 = $16,922.23.

Given $P = 60,000$ and $t = 5$, substitute $r = 0.0475$ into the continuous compounding formula, $A = Pe^{rt}$.

$$A = 60,000e^{0.0475(5)} = 60,000e^{0.2375}$$

$$\approx 76,084.50$$

The interest from this investment would be $76,084.50 - $60,000 = $16,084.50.

(a) The investment that offers 5% compounded quarterly will earn more interest than the investment that offers 4.75% compounded continuously.

(b) Note that $16,922.23 - $16,084.50 = $837.73. The investment that offers 5% compounded quarterly will earn $837.73 more in interest.

27. $A = Pe^{rt}$, $A = 2P$,

$$2P = Pe^{0.025t}$$

$$2 = e^{0.025t}$$

$$\ln 2 = 0.025t$$

$$27.73 \approx t$$

The doubling time is about 27.73 yr if interest is compounded continuously.

29. $A = Pe^{rt}$

$$3P = Pe^{0.05t} \Rightarrow 3 = e^{0.05t} \Rightarrow \ln 3 = \ln e^{0.05t}$$

$$\ln 3 = 0.05t \Rightarrow t = \frac{\ln 3}{0.05} \approx 21.97$$

It will take about 21.97 years for the investment to triple.

31. **(a)** 1969 is represented by $t = 4$

$$M(4) = 434e^{-0.08(4)} \approx 315$$

There were 315 continuously serving members in 1969.

(b) 1973 is represented by $t = 8$

$$M(8) = 434e^{-0.08(8)} \approx 229$$

There were 229 continuously serving members in 1973.

(c) 1979 is represented by $t = 14$

$$M(14) = 434e^{-0.08(14)} \approx 142$$

There were 142 continuously serving members in 1973.

Copyright © 2013 Pearson Education, Inc.

33. (a) A point associated with the graph of
$f(x) = P_0 a^{x-2000}$ is (2000, 1). Since

$f(2000) = P_0 a^{2000-2000} = 1$, we have

$P_0 a^0 = 1 \Rightarrow P_0 = 1$. Thus, we have

$f(x) = a^{x-2000}$. Since the point
(2025, 1.4) is projected to be on the
graph, we have the following.

$f(2025) = a^{2025-2000} = 1.4$

$a^{25} = 1.4 \Rightarrow a = \sqrt[25]{1.4} \approx 1.01355$

(b) From part (a) we have

$f(x) = (1.01355)^{x-2000}$. To find the
population projected for 2020, we must
find $f(2020)$. Thus

$f(2020) = (1.01355)^{2020-2000}$

$\qquad = (1.01355)^{20} \approx 1.3$ billion is

the estimated population in 2020.

(c) We must solve $(1.01355)^{x-2000} = 1.5$ for
x.

$(1.01355)^{x-2000} = 1.5$

$\ln(1.01355)^{x-2000} = \ln 1.5$

$(x - 2000)\ln(1.01355) = \ln 1.5$

$x - 2000 = \dfrac{\ln 1.5}{\ln 1.01355}$

$2000 + \dfrac{\ln 1.5}{\ln 1.01355} \approx 2030.13 = x$

In 2030, it is projected that the population
will reach 1.5 billion.

35. $f(x) = 2572e^{0.0359x}$

(a) $t = 4$ represents 2008.

$f(4) = 2572e^{0.0359(4)} \approx 2969.17$

In 2008, out-of-pocket spending on health
care was about \$2969 per household.

(b) Find t when $f(t) = 2775$.

$2572e^{0.0359x} = 2775 \Rightarrow e^{0.0359x} = \dfrac{2775}{2572} \Rightarrow$

$0.0359x = \ln\left(\dfrac{2775}{2572}\right) \Rightarrow$

$x = \dfrac{1}{0.0359}\ln\left(\dfrac{2775}{2572}\right) \approx 2.1$

Out-of-pocket spending on health care
reached \$2775 per household during 2006
(2 years after 2004).

37. $H = 8790e^{0.0382t}$

(a) For 2000, $t = 10$.

$H = 8790e^{0.0382(10)} \approx \$12,879$

(b) For 2005, $t = 15$.

$H = 8790e^{0.0382(15)} \approx \$15,590$

(c) For 2008, $t = 18$.

$H = 8790e^{0.0382(18)} \approx \$17,483$

39. $f(t) = 15,000e^{-0.05t}$

(a) At the beginning of the epidemic, $t = 0$.

$f(0) = 15,000e^{-0.05(0)} = 15,000$

At the beginning of the epidemic, 15,000
people were susceptible.

(b) After 10 days, $t = 10$.

$f(0) = 15,000e^{-0.05(10)} \approx 9098$

After 10 days, approximately 9098 people
were susceptible.

(c) After 3 weeks, $t = 21$.

$f(21) = 15,000e^{-0.05(21)} \approx 5249$

After three weeks, approximately 5249
people were susceptible.

41. $f(t) = 500e^{0.1t}$

(a) $f(2) = 500e^{0.1(2)} \approx 611$

At two days, the bacteria count is
approximately 611 million.

(b) $f(4) = 500e^{0.1(4)} \approx 746$

At four days, the bacteria count is
approximately 746 million.

(c) $f(7) = 500e^{0.1(7)} \approx 1007$

At one week (seven days), the bacteria
count is approximately 1007 million.

43. $f(t) = 200(0.90)^t$

Find t when $f(t) = 50$.

$50 = 200(0.90)^t$

$0.25 = (0.90)^t$

$\ln 0.25 = \ln\left[(0.90)^t\right]$

$\ln 0.25 = t \ln 0.90$

$t = \dfrac{\ln 0.25}{\ln 0.90} \Rightarrow t = \dfrac{\ln 0.25}{\ln 0.90} \approx 13.2$

The dose will reach a level of 50 mg in about
13.2 hr.

Copyright © 2013 Pearson Education, Inc.

45. $A(t) = 100e^{0.026t}$

We want to find the year in which the CPI will be 200.

$200 = 100e^{0.026t} \Rightarrow 2 = e^{0.026t} \Rightarrow$

$\ln 2 = \ln e^{0.026t} \Rightarrow \ln 2 = 0.026t \Rightarrow$

$t = \dfrac{\ln 2}{0.026} \approx 26.7$

Costs will be 100% higher than in 1990 in 2016, 26 years after 1990.

47. $S(t) = 50,000e^{-0.1t}$

Find t when $S(t) = 25,000$.

$25,000 = 50,000e^{-0.1t} \Rightarrow 0.5 = e^{-0.1t} \Rightarrow$

$\ln 0.5 = \ln e^{-0.1t} \Rightarrow \ln 0.5 = -0.1t \Rightarrow$

$t = \dfrac{\ln 0.5}{-0.1} \approx 6.9$

It will take about 6.9 yr for sales to fall to half the initial sales.

49. Use the formula for continuous compounding with $r = 0.06$.

$A = Pe^{rt} \Rightarrow 2P = Pe^{0.06t} \Rightarrow 2 = e^{0.06t} \Rightarrow$

$\ln 2 = \ln e^{0.06t} \Rightarrow \ln 2 = 0.06t \Rightarrow$

$t = \dfrac{\ln 2}{0.06} \approx 11.6$

It will take about 11.6 yr before twice as much electricity is needed.

51. $f(x) = \dfrac{0.9}{1 + 271e^{-0.122x}}$

(a) $f(25) = \dfrac{0.9}{1 + 271e^{-0.122(25)}} \approx 0.065$

$f(65) = \dfrac{0.9}{1 + 271e^{-0.122(65)}} \approx 0.820$

Among people age 25, 6.5% have some CHD, while among people age 65, 82% have some CHD.

(b) $0.50 = \dfrac{0.9}{1 + 271e^{-0.122x}}$

$0.50\left(1 + 271e^{-0.122x}\right) = 0.9$

$0.5 + 135.5e^{-0.122x} = 0.9$

$135.5e^{-0.122x} = 0.4$

$e^{-0.122x} = \dfrac{0.4}{135.5}$

$\ln e^{-0.122x} = \ln \dfrac{0.4}{135.5}$

$-0.122x = \ln \dfrac{0.4}{135.5}$

$x = \dfrac{\ln \frac{0.4}{135.5}}{-0.122} \approx 47.75$

At about 48, the likelihood of coronary heart disease is 50%.

Summary Exercises on Functions: Domains and Defining Equations

1. $f(x) = 3x - 6$

Domain: $(-\infty, \infty)$

3. $f(x) = |x + 4|$

Domain: $(-\infty, \infty)$

5. $f(x) = \dfrac{-2}{x^2 + 7}$

The domain is the set of all real numbers such that $x^2 + 7 \neq 0 \Rightarrow$ there is no real solution.

Domain: $(-\infty, \infty)$

7. $f(x) = \dfrac{x^2 + 7}{x^2 - 9}$

The domain is the set of all real numbers such that $x^2 - 9 \neq 0 \Rightarrow x \neq -3$ or $x \neq 3$

Domain: $(-\infty, -3) \cup (-3, 3) \cup (3, \infty)$

9. $f(x) = \log_5\left(16 - x^2\right)$

The domain is the set of all real numbers such that $16 - x^2 > 0 \Rightarrow 16 > x^2 \Rightarrow 4 > x$ and $-4 < x$.

Domain: $(-4, 4)$

11. $f(x) = \sqrt{x^2 - 7x - 8}$

The domain is the set of all real numbers such that $x^2 - 7x - 8 \geq 0$. Solve the equation to find the test intervals: $x^2 - 7x - 8 = 0 \Rightarrow$
$(x - 8)(x + 1) = 0 \Rightarrow x = 8$ or $x = -1$

Interval	Test Point	Value of $x^2 - 7x - 8$	Sign of $x^2 - 7x - 8$
$(-\infty, -1)$	-2	10	Positive
$(-1, 8)$	0	-8	Negative
$(8, \infty)$	10	22	Positive

Domain: $(-\infty, -1] \cup [8, \infty)$

Copyright © 2013 Pearson Education, Inc.

13. $f(x) = \dfrac{1}{2x^2 - x + 7}$

The domain is the set of all real numbers such that $2x^2 - x + 7 \neq 0$

$x = \dfrac{-(-1)\pm\sqrt{(-1)^2 - 4(2)(7)}}{2(2)} = \dfrac{1\pm\sqrt{-55}}{4} \Rightarrow$ there are no real solutions. Domain: $(-\infty, \infty)$

15. $f(x) = \sqrt{x^3 - 1}$

The domain is the set of all real numbers such that $x^3 - 1 \geq 0 \Rightarrow x^3 \geq 1 \Rightarrow x \geq 1$
Domain: $[1, \infty)$

17. $f(x) = e^{x^2 + x + 4}$

The domain is the set of values such that $x^2 + x + 4$ is real. Domain: $(-\infty, \infty)$

19. $f(x) = \sqrt{\dfrac{-1}{x^3 - 1}}$

The domain is the set of all real numbers such that $\sqrt{\dfrac{-1}{x^3-1}}$ is defined or $\dfrac{-1}{x^3-1} \geq 0$. $\dfrac{-1}{x^3-1}$ is defined for $x^3 - 1 \neq 0 \Rightarrow x^3 \neq 1 \Rightarrow x \neq 1$

Interval	Test Point	Value of $\frac{-1}{x^3-1}$	Sign of $\frac{-1}{x^3-1}$
$(-\infty, 1)$	0	1	Positive
$(1, \infty)$	2	$-\frac{1}{7}$	Negative

Domain: $(-\infty, 1)$

21. $f(x) = \ln\left(x^2 + 1\right)$

The domain is the set of all real numbers such that $x^2 + 1 > 0$. Domain: $(-\infty, \infty)$

23. $f(x) = \log\left(\dfrac{x+2}{x-3}\right)^2$

Since $\left(\frac{x+2}{x-3}\right)^2 \geq 0$ for all real numbers, the domain of $f(x)$ is the set of all real numbers such that $\frac{x+2}{x-3} \neq 0$. $\frac{x+2}{x-3} = 0$ when $x = -2$, and $\frac{x+2}{x-3}$ is undefined when $x = 3$.
Domain: $(-\infty, -2) \cup (-2, 3) \cup (3, \infty)$

25. $f(x) = e^{|1/x|}$

The domain is the set of all real numbers such that $\frac{1}{x}$ is defined, or $x \neq 0$
Domain: $(-\infty, 0) \cup (0, \infty)$

27. $f(x) = x^{100} - x^{50} + x^2 + 5$; Domain: $(-\infty, \infty)$

29. $f(x) = \sqrt[4]{16 - x^4}$

The domain is the set of all real numbers such that $16 - x^4 \geq 0$. Solve the equation to find the test intervals: $16 - x^4 = 0 \Rightarrow$
$(2-x)(2+x)(4+x^2) = 0 \Rightarrow x = 2$ or $x = -2$

Interval	Test Point	Value of $16 - x^4$	Sign of $16 - x^4$
$(-\infty, -2)$	-3	-65	Negative
$(-2, 2)$	0	16	Positive
$(2, \infty)$	3	-65	Negative

Domain: $[-2, 2]$

31. $f(x) = \sqrt{\dfrac{x^2 - 2x - 63}{x^2 + x - 12}}$

The domain is the set of real numbers such that $\frac{x^2-2x-63}{x^2+x-12} \geq 0$. $\frac{x^2-2x-63}{x^2+x-12}$ is not defined for
$x^2 + x - 12 = 0 \Rightarrow (x+4)(x-3) = 0 \Rightarrow$
$x \neq -4$ or $x \neq 3$. Solve $\frac{x^2-2x-63}{x^2+x-12} = 0$ to find the test intervals: $\frac{x^2-2x-63}{x^2+x-12} = 0 \Rightarrow$
$x^2 - 2x - 63 = 0 \Rightarrow (x-9)(x+7) = 0 \Rightarrow x = 9$ or $x = -7$.

Interval	Test Point	Value of $\frac{x^2-2x-63}{x^2+x-12}$	Sign
$(-\infty, -7)$	-10	$\frac{19}{26}$	Positive
$(-7, -4)$	-5	$-\frac{7}{2}$	Negative
$(-4, 3)$	0	$\frac{21}{4}$	Positive
$(3, 9)$	5	$-\frac{8}{3}$	Negative
$(9, \infty)$	10	$\frac{17}{98}$	Positive

Domain: $(-\infty, -7] \cup (-4, 3) \cup [9, \infty)$

33. $f(x) = \left|\sqrt{5 - x}\right|$

The domain is the set of real numbers such that $5 - x \geq 0 \Rightarrow 5 \geq x$
Domain: $(-\infty, 5]$

Copyright © 2013 Pearson Education, Inc.

35. $f(x) = \log \left| \dfrac{1}{4-x} \right|$

The domain is the set of real numbers such

that $\left| \dfrac{1}{4-x} \right| > 0 \Rightarrow \dfrac{1}{4-x} > 0 \Rightarrow 4 - x > 0 \Rightarrow 4 > x$

or $-\dfrac{1}{4-x} < 0 \Rightarrow -4 + x < 0 \Rightarrow x < 4$

Domain: $(-\infty, 4) \cup (4, \infty)$

37. $f(x) = 6^{\sqrt{x^2 - 25}}$

The domain is the set of real numbers such

that $\sqrt{x^2 - 25}$ is a real number.

$x^2 - 25 \geq 0 \Rightarrow x^2 \geq 25 \Rightarrow x \geq 5 \text{ or } x \leq -5$

Domain: $(-\infty, -5] \cup [5, \infty)$

39. $f(x) = \ln \left(\dfrac{-3}{(x+2)(x-6)} \right)$

The domain is the set of real numbers such

that $\dfrac{-3}{(x+2)(x-6)} > 0$ and $(x+2)(x-6) \neq 0 \Rightarrow$

$x \neq -2 \text{ or } x \neq 6$.

Interval	Test Point	Value of $\dfrac{-3}{(x+2)(x-6)}$	Sign of $\dfrac{-3}{(x+2)(x-6)}$
$(-\infty, -2)$	-3	$-\dfrac{1}{3}$	Negative
$(-2, 6)$	0	$\dfrac{1}{4}$	Positive
$(6, \infty)$	7	$-\dfrac{1}{3}$	Negative

Domain: $(-2, 6)$

41. Choice A can be written as a function of x.

$3x + 2y = 6 \Rightarrow y = f(x) = -\dfrac{3}{2}x + 3$

43. Choice C can be written as a function of x.

$x^3 + y^3 = 5 \Rightarrow y = f(x) = \sqrt[3]{5 - x^3}$

45. Choice A can be written as a function of x.

$x = \dfrac{2 - y}{y + 3} \Rightarrow y = f(x) = \dfrac{2 - 3x}{x + 1}$

47. Choice D can be written as a function of x.

$2x = \dfrac{1}{y^3} \Rightarrow y = f(x) = \sqrt[3]{\dfrac{1}{2x}}$

49. Choice C can be written as a function of x.

$\dfrac{x}{4} - \dfrac{y}{9} = 0 \Rightarrow y = f(x) = \dfrac{9x}{4}$

Chapter 4   Review Exercises

1. This is not a one-to-one function since a horizontal line can intersect the graph in more than one point.

3. $y = 5x - 4$

Looking at this function graphed on a TI-83, we can see that it appears that any horizontal line passed through the function will intersect the graph in at most one place.

If we attempt to find the inverse function, we see that this function can be found.

Step 1: Interchange x and y. $x = 5y - 4$

Step 2: Solve for y.

$x = 5y - 4 \Rightarrow x + 4 = 5y \Rightarrow y = \dfrac{x+4}{5}$

Step 3: Replace y with $f^{-1}(x)$. $f^{-1}(x) = \dfrac{x+4}{5}$

Also, an acceptable way of showing that a function is one-to-one is to assume that you have two equal y-values $\left(f(x_1) = f(x_2) \right)$ and show that they must have come from the same x-value $\left(x_1 = x_2 \right)$.

$f(x_1) = f(x_2) \Rightarrow 5x_1 - 4 = 5x_2 - 4 \Rightarrow$

$\quad 5x_1 = 5x_2 \Rightarrow x_1 = x_2$

So, the function is one-to-one.

5. $y = (x+3)^2$

If $x = -2$, $y = (-2 + 3)^2 = 1^2 = 1$.

If $x = -4$, $y = (-4 + 3)^2 = (-1)^2 = 1$.

Since two different values of x lead to the same value of y, the function is not one-to-one.

7. $f(x) = x^3 - 3$

This function is one-to-one.

Step 1: Replace $f(x)$ with y and interchange

x and y. $y = x^3 - 3 \Rightarrow x = y^3 - 3$

Step 2: Solve for y.

$x = y^3 - 3 \Rightarrow x + 3 = y^3 \Rightarrow y = \sqrt[3]{x+3}$

Step 3: Replace y with $f^{-1}(x)$.

$f^{-1}(x) = \sqrt[3]{x+3}$

Copyright © 2013 Pearson Education, Inc.

9. $f^{-1}(\$50,000)$ represents the number of years after 2004 required for the investment to reach $50,000.

11. To have an inverse, a function must be a <u>one-to-one</u> function.

13. $y = \log_{.3} x$

The point (1, 0) is on the graph of every function of the form $y = \log_a x,$ so the correct choice must be either B or C. Since the base is $a = 0.3$ and $0 < 0.3 < 1,$ $y = \log_{.3} x$ is a decreasing function, and so the correct choice must be B.

15. $y = \ln x = \log_e x$

The point (1, 0) is on the graph of every function of the form $y = \log_a x,$ so the correct choice must be either B or C. Since the base is $a = e$ and $e > 1,$ $y = \ln x$ is an increasing function, and so the correct choice must be C.

17. $2^5 = 32$ is written in logarithmic form as $\log_2 32 = 5.$

19. $\left(\frac{3}{4}\right)^{-1} = \frac{4}{3}$ is written in logarithmic form as

$\log_{3/4} \frac{4}{3} = -1.$

21. $\log 1000 = 3$ is written in exponential form as $10^3 = 1000.$

23. $\ln \sqrt{e} = \frac{1}{2}$ is written in exponential form as $e^{1/2} = \sqrt{e}.$

25. Let $f(x) = a^x$ be the required function. Then

$f(-4) = \frac{1}{16} \Rightarrow a^{-4} = \frac{1}{16} \Rightarrow a^{-4} = 2^{-4} \Rightarrow a = 2.$
The base is 2.

27. $\log_3 \frac{mn}{5r} = \log_3 mn - \log_3 5r$

$= \log_3 m + \log_3 n - \left(\log_3 5 + \log_3 r\right)$

$= \log_3 m + \log_3 n - \log_3 5 - \log_3 r$

29. $\log 0.0411 \approx -1.3862$

31. $\ln 144,000 \approx 11.8776$

33. To find $\log_{2/3} \frac{5}{8},$ use the change-of-base theorem. We have

$\log_{2/3} \frac{5}{8} = \frac{\log \frac{5}{8}}{\log \frac{2}{3}} = \frac{\ln \frac{5}{8}}{\ln \frac{2}{3}} \approx 1.1592.$

35. $16^{x+4} = 8^{3x-2} \Rightarrow \left(2^4\right)^{x+4} = \left(2^3\right)^{3x-2} \Rightarrow$

$2^{4x+16} = 2^{9x-6} \Rightarrow 4x + 16 = 9x - 6 \Rightarrow$

$22 = 5x \Rightarrow \frac{22}{5} = x$

Solution set: $\left\{\frac{22}{5}\right\}$

37. $3^{2x-5} = 13 \Rightarrow \ln 3^{2x-5} = \ln 13 \Rightarrow$

$(2x - 5)\ln 3 = \ln 13 \Rightarrow 2x - 5 = \frac{\ln 13}{\ln 3} \Rightarrow$

$2x = 5 + \frac{\ln 13}{\ln 3} \Rightarrow x = \frac{1}{2}\left(5 + \frac{\ln 13}{\ln 3}\right) \approx 3.667$

or

$3^{2x-5} = 13 \Rightarrow \ln 3^{2x-5} = \ln 13 \Rightarrow$

$(2x - 5)\ln 3 = \ln 13 \Rightarrow 2x \ln 3 - 5 \ln 3 = \ln 13 \Rightarrow$

$x \ln 3^2 - \ln 3^5 = \ln 13 \Rightarrow$

$x \ln 9 - \ln 243 = \ln 13 \Rightarrow$

$x \ln 9 = \ln 13 + \ln 243 \Rightarrow x \ln 9 = \ln 3159 \Rightarrow$

$x = \frac{\ln 3159}{\ln 9} \approx 3.667$

Solution set: {3.667}

39. $6^{x+3} = 4^x \Rightarrow \ln 6^{x+3} = \ln 4^x \Rightarrow$

$(x + 3)\ln 6 = x \ln 4$

$x \ln 6 + 3 \ln 6 = x \ln 4$

$x \ln 6 - x \ln 4 = -3 \ln 6$

$x(\ln 6 - \ln 4) = -3 \ln 6 \Rightarrow x\left(\ln \frac{6}{4}\right) = -\ln 6^3 \Rightarrow$

$x\left(\ln \frac{3}{2}\right) = -\ln 216$

$x = \frac{-\ln 216}{\ln \frac{3}{2}} \approx -13.257$

Solution set: {−13.257}

41. $e^{2-x} = 12 \Rightarrow \ln e^{2-x} = \ln 12 \Rightarrow$

$2 - x = \ln 12 \Rightarrow -x = -2 + \ln 12 \Rightarrow$

$x = 2 - \ln 12 \approx -0.485$

Solution set: {−0.485}

43. $10e^{3x-7} = 5 \Rightarrow e^{3x-7} = \frac{1}{2} \Rightarrow$

$\ln e^{3x-7} = \ln \frac{1}{2} \Rightarrow 3x - 7 = \ln \frac{1}{2} \Rightarrow$

$3x = \ln \frac{1}{2} + 7 \Rightarrow x = \frac{1}{3}\left(\ln \frac{1}{2} + 7\right) \approx 2.102$

Solution set: {2.102}

45. $6^{x-3} = 3^{4x+1}$

$\ln 6^{x-3} = \ln 3^{4x+1}$

$(x - 3)\ln 6 = (4x + 1)\ln 3$

$x \ln 6 - 3 \ln 6 = 4x \ln 3 + \ln 3$

$x \ln 6 - \ln 6^3 = x \ln 3^4 + \ln 3$

Copyright © 2013 Pearson Education, Inc.

$$x \ln 6 - \ln 216 = x \ln 81 + \ln 3$$
$$x \ln 6 - x \ln 81 = \ln 3 + \ln 216$$
$$x(\ln 6 - \ln 81) = \ln 3 + \ln 216$$
$$x = \frac{\ln 3 + \ln 216}{\ln 6 - \ln 81} = \frac{\ln(3 \cdot 216)}{\ln \frac{6}{81}}$$
$$= \frac{\ln 648}{\ln \frac{2}{27}} \; x \approx -2.487$$

Solution set: $\{-2.487\}$

47. $e^{6x} \cdot e^{x} = e^{21} \Rightarrow e^{7x} = e^{21} \Rightarrow 7x = 21 \Rightarrow x = 3$
Solution set: $\{3\}$

49. $2e^{2x} - 5e^{x} - 3 = 0$

Let $u = e^{x}$.
$$2u^2 - 5u - 3 = 0$$
$$(u - 3)(2u + 1) = 0$$

$$
\begin{array}{c|c}
u - 3 = 0 & 2u + 1 = 0 \\
u = 3 & 2u = -1 \\
e^{x} = 3 & u = -\frac{1}{2} \\
\ln e^{x} = \ln 3 & \\
x = \ln 3 & e^{x} = -\frac{1}{2}
\end{array}
$$

Disregard the proposed solution $-\frac{1}{2}$ because e^{x} is always positive.
Solution set: $\{\ln 3\}$

51. $4(1.06)^{x} + 2 = 8 \Rightarrow 4(1.06)^{x} = 6 \Rightarrow$
$(1.06)^{x} = 1.5 \Rightarrow \ln(1.06^{x}) = \ln 1.5 \Rightarrow$
$$x \ln 1.06 = \ln 1.5 \Rightarrow x = \frac{\ln 1.5}{\ln 1.06} \approx 6.959$$

53. $3 \ln x = 13 \Rightarrow \ln x = \frac{13}{3} \Rightarrow x = e^{13/3}$
Solution set: $\left\{ e^{13/3} \right\}$

55. $\log(2x + 7) = 0.25 \Rightarrow 2x + 7 = 10^{0.25} = \sqrt[4]{10} \Rightarrow$
$x = \frac{\sqrt[4]{10} - 7}{2}$
Solution set: $\left\{ \frac{\sqrt[4]{10} - 7}{2} \right\}$

57. $\log_2 (x^3 + 5) = 5 \Rightarrow x^3 + 5 = 2^5 \Rightarrow x^3 = 27 \Rightarrow$
$x = 3$
Solution set: $\{3\}$

59. $\log_4 \left[(3x + 1)(x - 4) \right] = 2 \Rightarrow$
$(3x + 1)(x - 4) = 4^2 \Rightarrow 3x^2 - 11x - 4 = 16 \Rightarrow$
$3x^2 - 11x - 20 = 0 \Rightarrow (3x + 4)(x - 5) = 0 \Rightarrow$
$x = -\frac{4}{3}$ or $x = 5$
Solution set: $\left\{ -\frac{4}{3}, 5 \right\}$

61. $\log x + \log(13 - 3x) = 1 \Rightarrow$
$\log[x(13 - 3x)] = 1 \Rightarrow 13x - 3x^2 = 10^{1} \Rightarrow$
$13x - 3x^2 = 10 \Rightarrow 3x^2 - 13x + 10 = 0 \Rightarrow$
$(3x - 10)(x - 1) = 0 \Rightarrow x = \frac{10}{3}$ or $x = 1$
Solution set: $\left\{ 1, \frac{10}{3} \right\}$

63. $\ln(6x) - \ln(x + 1) = \ln 4$
$$\ln \frac{6x}{x + 1} = \ln 4 \Rightarrow \frac{6x}{x + 1} = 4$$
$$6x = 4(x + 1)$$
$$6x = 4x + 4$$
$$2x = 4 \Rightarrow x = 2$$
Solution set: $\{2\}$

65. $\ln \left[\ln \left(e^{-x} \right) \right] = \ln 3$
$\ln(-x) = \ln 3 \Rightarrow -x = 3 \Rightarrow x = -3$
Solution set: $\{-3\}$

67. $\frac{d}{10} = \log \left(\frac{I}{I_0} \right) \Rightarrow 10^{d/10} = \frac{I}{I_0} \Rightarrow$
$$I_0 \left(10^{d/10} \right) = I \Rightarrow I_0 = \frac{I}{10^{d/10}}$$

69.

Solution set: $\{1.315\}$

71. (a) $8.3 = \log_{10} \frac{I}{I_0} \Rightarrow \frac{I}{I_0} = 10^{8.3} \Rightarrow$
$I = 10^{8.3} I_0 \approx 199,526,231.5 I_0$
The magnitude was about $200,000,000 I_0$.

(b) $7.1 = \log_{10} \frac{I}{I_0} \Rightarrow \frac{I}{I_0} = 10^{7.1} \Rightarrow$
$I = 10^{7.1} I_0 \approx 12,589,254.12 I_0$
The magnitude was about $13,000,000 I_0$.

(c) $\frac{200,000,000 I_0}{13,000,000 I_0} = \frac{200}{13} \approx 15.38$

The 1906 earthquake had a magnitude more than 15 times greater than the 1989 earthquake. Note: If the more precise values found in parts (a) and (b) were used, the 1906 earthquake had a magnitude of almost 16 times greater than the 1989 earthquake.

Copyright © 2013 Pearson Education, Inc.

73. Substitute $A = 5760$, $P = 3500$, $t = 10$, $n = 1$ into the formula $A = P\left(1 + \frac{r}{n}\right)^{tn}$.

$$5760 = 3500\left(1 + \frac{r}{1}\right)^{10(1)} \Rightarrow \frac{288}{175} = (1 + r)^{10} \Rightarrow$$

$$\left(\frac{288}{175}\right)^{1/10} = 1 + r \Rightarrow \left(\frac{288}{175}\right)^{1/10} - 1 = r \Rightarrow$$

$$r \approx 0.051$$

The annual interest rate, to the nearest tenth, is 5.1%.

75. First, substitute $P = 10{,}000$, $r = 0.04$, $t = 12$, and $n = 1$ into the formula $A = P\left(1 + \frac{r}{n}\right)^{tn}$.

$$A = 10{,}000\left(1 + \frac{0.04}{1}\right)^{12(1)} = 10{,}000(1.04)^{12}$$

$$\approx 16{,}010.32$$

After the first 12 years, there will be about $16,010.32 in the account. To finish off the 21-year period, substitute $P = 16{,}010.32$, $r = 0.05$, $t = 9$, and $n = 2$ into the formula $A = P\left(1 + \frac{r}{n}\right)^{tn}$.

$$A = 16{,}010.32\left(1 + \frac{0.05}{2}\right)^{9(2)}$$

$$= 16{,}010.32(1.025)^{18} \approx 24{,}970.64$$

At the end of the 21-year period, about $60,606.76 would be in the account. Note: If it was possible to transfer the money accrued after the 12 years to the new account without rounding, the amount after the 21-year period would be $60,606.77. The difference is not significant.

77. To find t, substitute $a = 2$, $P = 1$, and $r = 0.04$ into $A = Pe^{rt}$ and solve.

$$2 = 1 \cdot e^{0.04t} \Rightarrow 2 = e^{0.04t} \Rightarrow \ln 2 = \ln e^{0.04t} \Rightarrow$$

$$\ln 2 = 0.04t \Rightarrow t = \frac{\ln 2}{0.04} \approx 17.3$$

It would take about 17.3 yr.

79. Double the 2010 total payroll is $2(146.6) = 293.2$. Using the function $f(x) = 113.2e^{0.0836x}$, we solve for x when $f(x) = 293.2$.

$$113.2e^{0.0836x} = 293.2 \Rightarrow e^{0.0836x} = \frac{293.2}{113.2} \Rightarrow$$

$$\ln e^{0.0836x} = \ln \frac{293.2}{113.2} \Rightarrow 0.0836x = \ln \frac{293.2}{113.2} \Rightarrow$$

$$x = \frac{\ln \frac{293.2}{113.2}}{0.0836} \approx 11.4$$

Since x represents the number of years since 2007, in 2018 the total payroll will be double that in 2010.

81. (a) We must determine how much money will be in the account after 35 years. The money is being compounded continuously, so we use $A = Pe^{rt}$.

$$A = 5000e^{0.04 \cdot 35} \approx \$20{,}076$$

After paying 25% of this for taxes, there is $20{,}076 \cdot 0.75 = \$15{,}207$ remaining.

(b) $A = 3750e^{0.03 \cdot 35} \approx \$10{,}716.19$

You will have about $10,716.19.

(c) You will have $20,076 − $10,716 = $4491 more money with the IRA.

(d) $A = 3750e^{0.04 \cdot 35} \approx \$15{,}207$

The two balances are the same.

Chapter 4   Test

1. (a) $f(x) = \sqrt[3]{2x - 7}$

Since it is a cube root, $2x - 7$ may be any real number.

Domain: $(-\infty, \infty)$

Since the cube root of any real number is also any real number.

Range: $(-\infty, \infty)$

(b) $f(x) = \sqrt[3]{2x - 7}$

The graph of f passes the horizontal line test, and thus is a one-to-one function.

Copyright © 2013 Pearson Education, Inc.

(c) *Step 1:* Replace $f(x)$ with y and interchange x and y.

$$y = \sqrt[3]{2x - 7} \Rightarrow x = \sqrt[3]{2y - 7}$$

Step 2: Solve for y.

$$x = \sqrt[3]{2y - 7} \Rightarrow x^3 = \left(\sqrt[3]{2y - 7}\right)^3 \Rightarrow$$

$$x^3 = 2y - 7 \Rightarrow x^3 + 7 = 2y \Rightarrow \frac{x^3 + 7}{2} = y$$

Step 3: Replace y with $f^{-1}(x)$.

$$f^{-1}(x) = \frac{x^3 + 7}{2}$$

(d) Since the domain and range of f are $(-\infty, \infty)$, the domain and range of f^{-1} are also $(-\infty, \infty)$.

(e)

These graphs are reflections of each other across the line $y = x$.

2. (a) $y = \log_{1/3} x$

The point $(1, 0)$ is on the graph of every function of the form $y = \log_a x$, so the correct choice must be either B or C. Since the base is $a = \frac{1}{3}$ and $0 < \frac{1}{3} < 1$, $y = \log_{1/3} x$ is a decreasing function, and so the correct choice must be B.

(b) $y = e^x$

The point $(0, 1)$ is on the graph since $e^0 = 1$, so the correct choice must be either A or D. Since the base is e and $e > 1$, $y = e^x$ is an increasing function, and so the correct choice must be A.

(c) $y = \ln x$ or $y = \log_e x$

The point $(1, 0)$ is on the graph of every function of the form $y = \log_a x$, so the correct choice must be B or C. Since the base is $a = e$ and $e > 1$, $y = \ln x$ is an increasing function, and the correct choice must be C.

(d) $y = \left(\frac{1}{3}\right)^x$

The point $(0, 1)$ is on the graph since $\left(\frac{1}{3}\right)^0 = 1$, so the correct choice must be either A or D. Since the base is $\frac{1}{3}$ and $0 < \frac{1}{3} < 1$, $y = \left(\frac{1}{3}\right)^x$ is a decreasing function, and so the correct choice must be D.

3. $\left(\frac{1}{8}\right)^{2x-3} = 16^{x+1} \Rightarrow \left(2^{-3}\right)^{2x-3} = \left(2^4\right)^{x+1} \Rightarrow$
$2^{-3(2x-3)} = 2^{4(x+1)} \Rightarrow 2^{-6x+9} = 2^{4x+4} \Rightarrow$
$-6x + 9 = 4x + 4 \Rightarrow -10x + 9 = 4 \Rightarrow$
$-10x = -5 \Rightarrow x = \frac{1}{2}$

Solution set: $\left\{\frac{1}{2}\right\}$

4. (a) $4^{3/2} = 8$ is written in logarithmic form as $\log_4 8 = \frac{3}{2}$.

(b) $\log_8 4 = \frac{2}{3}$ is written in exponential form as $8^{2/3} = 4$.

5. They are inverses of each other.

6. $\log_7 \dfrac{x^2 \sqrt[4]{y}}{z^3} = \log_7 x^2 + \log_7 \sqrt[4]{y} - \log_7 z^3$
$$= \log_7 x^2 + \log_7 y^{1/4} - \log_7 z^3$$
$$= 2\log_7 x + \frac{1}{4}\log_7 y - 3\log_7 z$$

7. $\log 2388 \approx 3.3780$

8. $\ln 2388 \approx 7.7782$

9. $\log_9 13 = \dfrac{\ln 13}{\ln 9} = \dfrac{\log 13}{\log 9} \approx 1.1674$

10. $x^{2/3} = 25 \Rightarrow \left(x^{2/3}\right)^{3/2} = \pm 25^{3/2} \Rightarrow$
$x = (\pm 5)^3 = \pm 125$

Solution set: $\{\pm 125\}$

11. $12^x = 1 \Rightarrow x \log 12 = \log 1 \Rightarrow x \log 12 = 0 \Rightarrow$
$x = 0$

Solution set: $\{0\}$

Copyright © 2013 Pearson Education, Inc.

12. $9^x = 4 \Rightarrow x \log 9 = \log 4 \Rightarrow x = \dfrac{\log 4}{\log 9} \approx 0.631$

Solution set: $\{0.631\}$

13. $16^{2x+1} = 8^{3x} \Rightarrow 2^{4(2x+1)} = 2^{3(3x)} \Rightarrow$

$4(2x+1) = 3(3x) = 8x + 4 = 9x \Rightarrow x = 4$

Solution set: $\{4\}$

14.
$$2^{x+1} = 3^{x-4}$$
$$(x+1)\log 2 = (x-4)\log 3$$
$$x\log 2 + \log 2 = x\log 3 - 4\log 3$$
$$x\log 2 - x\log 3 = -\log 2 - 4\log 3$$
$$x(\log 2 - \log 3) = -\log 2 - 4\log 3$$
$$x = \frac{-\log 2 - 4\log 3}{\log 2 - \log 3} \approx 12.548$$

Solution set: $\{12.548\}$

15.
$$e^{0.4x} = 4^{x-2}$$
$$0.4x = (x-2)\ln 4$$
$$0.4x = x\ln 4 - 2\ln 4$$
$$0.4x - x\ln 4 = -2\ln 4$$
$$x(0.4 - \ln 4) = -2\ln 4$$
$$x = \frac{-2\ln 4}{0.4 - \ln 4} \approx 2.811$$

Solution set: $\{2.811\}$

16. $2e^{2x} - 5e^x + 3 = 0$

Let $u = e^x$.

$$2u^2 - 5u + 3 = 0$$
$$(u-1)(2u-3) = 0$$

$$
\begin{array}{c|c}
u - 1 = 0 & 2u - 3 = 0 \\
u = 1 & u = \frac{3}{2} \\
e^x = 1 & e^x = \frac{3}{2} \\
\ln e^x = \ln 1 & \ln e^x = \ln \frac{3}{2} \Rightarrow x = \ln \frac{3}{2} \\
x = 0 &
\end{array}
$$

Solution set: $\left\{0, \ln \frac{3}{2}\right\}$ or $\{0, 0.405\}$

17. $\log_x \dfrac{9}{16} = 2 \Rightarrow x^2 = \dfrac{9}{16} \Rightarrow x = \pm\dfrac{3}{4}$

Since it is not possible for a logarithm to have a negative base, we reject the negative solution. Solution set: $\left\{\frac{3}{4}\right\}$

18. $\log_2 (x-4)(x-2) = 3$
$$(x-4)(x-2) = 2^3$$
$$x^2 - 6x + 8 = 8$$
$$x^2 - 6x = 0$$
$$x(x-6) = 0 \Rightarrow x = 0 \text{ or } x = 6$$

Solution set: $\{0, 6\}$

19. $\log_2 x + \log_2(x+2) = 3 \Rightarrow \log_2[x(x+2)] = 3$
$$x^2 + 2x = 2^3 \Rightarrow x^2 + 2x - 8 = 0$$
$$(x+4)(x-2) = 0 \Rightarrow x = -4 \text{ or } x = 2$$

Since the negative solution is not in the domain it must be discarded.
Solution set: $\{2\}$

20. $\ln x - 4\ln 3 = \ln\left(\frac{1}{5}x\right)$

$$\ln x - \ln 3^4 = \ln \frac{x}{5}$$

$$\ln x - \ln 81 = \ln \frac{x}{5} \Rightarrow \ln \frac{x}{81} = \ln \frac{x}{5}$$

$$\frac{x}{81} = \frac{x}{5} \Rightarrow \frac{5}{81} = 1 \Rightarrow \text{ there is no}$$

solution.
 Solution set: $\varnothing$

21. $\log_3(x+1) - \log_3(x-3) = 2$

$$\log_3 \frac{x+1}{x-3} = 2$$

$$\frac{x+1}{x-3} = 3^2 \Rightarrow \frac{x+1}{x-3} = 9$$

$$x+1 = 9(x-3)$$
$$x+1 = 9x - 27$$
$$-8x = -28 \Rightarrow x = \frac{28}{8} = \frac{7}{2}$$

Solution set: $\left\{\frac{7}{2}\right\}$

22. Answers will vary.

$\log_5 27$ is the exponent to which 5 must be raised in order to obtain 27. To approximate $\log_5 27$ on your calculator, use the change-of-base formula; $\log_5 27 = \frac{\log 27}{\log 5} = \frac{\ln 27}{\ln 5} \approx 2.048$.

23. $v(t) = 176(1 - e^{-0.18t})$

Find the time t at which $v(t) = 147$.

$$147 = 176(1 - e^{-0.18t}) \Rightarrow \frac{147}{176} = 1 - e^{-0.18t} \Rightarrow$$
$$-e^{-0.18t} = \frac{147}{176} - 1 \Rightarrow -e^{-0.18t} = -\frac{29}{176}$$
$$e^{-0.18t} = \frac{29}{176} \Rightarrow \ln e^{-0.18t} = \ln \frac{29}{176} \Rightarrow$$
$$-0.18t = \ln \frac{29}{176} \Rightarrow t = \frac{\ln \frac{29}{176}}{-0.18} \approx 10.02$$

It will take the skydiver about 10 sec to attain the speed of 147 ft per sec (100 mph).

Copyright © 2013 Pearson Education, Inc.

24. (a) Substitute $P = 5000$, $A = 18,000$, $r = 0.038$, and $n = 12$ into the formula

$$A = P\left(1 + \frac{r}{n}\right)^{tn}.$$

$$18,000 = 5000\left(1 + \frac{0.038}{12}\right)^{t(12)}$$

$$3.6 = \left(1 + \frac{0.038}{12}\right)^{12t}$$

$$\ln 3.6 = \ln\left(1 + \frac{0.038}{12}\right)^{12t}$$

$$\ln 3.6 = 12t \ln\left(1 + \frac{0.038}{12}\right)$$

$$t = \frac{\ln 3.6}{12 \ln\left(1 + \frac{0.038}{12}\right)} \approx 33.8$$

It will take about 33.8 years.

(b) Substitute $P = 5000$, $A = 18,000$, and $r = 0.038$, and into the formula $A = Pe^{rt}$.

$$18,000 = 5000e^{0.038t} \Rightarrow 3.6 = e^{0.038t} \Rightarrow$$

$$\ln 3.6 = \ln e^{0.038t} \Rightarrow \ln 3.6 = 0.038t \Rightarrow$$

$$t = \frac{\ln 3.6}{0.038} \approx 33.7$$

It will take about 33.7 years.

25. Substitute $A = 3P$ and $r = 0.038$ into the continuous compounding formula $A = Pe^{rt}$, then solve for t:

$$A = Pe^{rt}$$

$$3P = Pe^{0.038t} \Rightarrow 3 = e^{0.038t} \Rightarrow$$

$$\ln 3 = \ln e^{0.038t} \Rightarrow \ln 3 = 0.038t \Rightarrow$$

$$\frac{\ln 3}{0.038} = t \Rightarrow t \approx 28.9$$

It will take about 28.9 years for any amount of money to triple at 3.8% annual interest.

26. $A(t) = 600e^{-0.05t}$

(a) $A(12) = 600e^{-0.05(12)} = 600e^{-0.6} \approx 329.3$

The amount of radioactive material present after 12 days is about 329.3 g.

(b) Since

$$A(0) = 600e^{-0.05(0)} = 600e^{0} = 600 \text{ g is}$$

the amount initially present, we seek to find t when $A(t) = \frac{1}{2}(600) = 300$ g is present.

$$300 = 600e^{-0.05t} \Rightarrow 0.5 = e^{-0.05t} \Rightarrow$$

$$\ln 0.5 = \ln e^{-0.05t} \Rightarrow \ln 0.5 = -0.05t \Rightarrow$$

$$t = \frac{\ln 0.5}{-0.05} \approx 13.9$$

The half-life of the material is about 13.9 days.

Chapter 5

SYSTEMS AND MATRICES

Section 5.1   Systems of Linear Equations

1. The population of the Jacksonville metropolitan area was greater than that of the New Orleans metropolitan area from 2006–2009.

3. approximately $(2005.2, 1.26)$

5. t would represent time in years and y would represent the population in millions.

7. $4x + 3y = -13$ (1)

    $-x + y = 5$ (2)

Solve equation (2) for y.

$-x + y = 5$ (2)

   $y = x + 5$ (3)

Replace y with $x + 5$ in equation (1), and solve for x.

$$4x + 3y = -13 \Rightarrow 4x + 3(x+5) = -13 \Rightarrow$$
$$4x + 3x + 15 = -13 \Rightarrow 7x + 15 = -13 \Rightarrow$$
$$7x = -28 \Rightarrow x = -4$$

Replace x with -4 in equation (3) to obtain $y = -4 + 5 = 1$.

Check:

$4x + 3y = -13$ (1)	$-x + y = 5$ (2)
$4(-4) + 3(1) = -13$?	$-(-4) + 1 = 5$?
$-16 + 3 = -13$	$4 + 1 = 5$
$-13 = -13$ True	$5 = 5$ True

Solution set: $\{(-4, 1)\}$

9. $x - 5y = 8$ (1)

    $x = 6y$ (2)

Replace x with $6y$ in equation (1), and solve for y: $x - 5y = 8 \rightarrow 6y - 5y = 8 \Rightarrow y = 8$

Replace y with 8 in equation (2) to obtain $x = 6(8) = 48$.

Check: $x - 5y = 8$ (1)	$x = 6y$ (2)
$48 - 5(8) = 8$?	$48 = 6(8)$?
$48 - 40 = 8$	$48 = 48$ True
$8 = 8$ True	

Solution set: $\{(48, 8)\}$

11. $8x - 10y = -22$ (1)

    $3x + y = 6$ (2)

Solve equation (2) for y.

$3x + y = 6$ (2)

    $y = -3x + 6$ (3)

Replace y with $-3x + 6$ in equation (1), and solve for x.

$$8x - 10y = -22$$
$$8x - 10(-3x + 6) = -22$$
$$8x + 30x - 60 = -22$$
$$38x - 60 = -22$$
$$38x = 38$$
$$x = 1$$

Replace x with 1 in equation (3) to obtain $y = -3(1) + 6 = 3$.

Check:

$8x - 10y = -22$ (1)	$3x + y = 6$ (2)
$8(1) - 10(3) = -22$?	$3(1) + 3 = 6$?
$8 - 30 = -22$	$3 + 3 = 6$
$-22 = -22$ True	$6 = 6$ True

Solution set: $\{(1, 3)\}$

13. $7x - y = -10$ (1)

    $3y - x = 10$ (2)

Solve equation (1) for y.

$7x - y = -10$ (1)

    $y = 7x + 10$ (3)

Replace y with $7x + 10$ in equation (2), and solve for x.

$$3y - x = 10 \Rightarrow 3(7x + 10) - x = 10 \Rightarrow$$
$$21x + 30 - x = 10 \Rightarrow 20x + 30 = 10 \Rightarrow$$
$$20x = -20 \Rightarrow x = -1$$

Replace x with -1 in equation (3) to obtain $y - 7(-1) + 10 = 3$.

Check:

$7x - y = -10$ (1)	$3y - x = 10$ (2)
$7(-1) - 3 = -10$?	$3(3) - (-1) = 10$?
$-7 - 3 = -10$	$9 + 1 = 10$
$-10 = -10$ True	$10 = 10$ True

Solution set: $\{(-1, 3)\}$

15. $-2x = 6y + 18$ (1)
$-29 = 5y - 3x$ (2)
Solve equation (1) for x.
$-2x = 6y + 18$ (1)
$x = -3y - 9$ (3)
Replace x with $-3y - 9$ in equation (2), and solve for y.
$-29 = 5y - 3(-3y - 9)$
$-29 = 5y + 9y + 27 \Rightarrow -29 = 14y + 27 \Rightarrow$
$-56 = 14y \Rightarrow -4 = y$
Replace y with -4 in equation (3) to obtain
$x = -3(-4) - 9 = 12 - 9 = 3$.
Check: $-2x = 6y + 18$ (1)
$-2(3) = 6(-4) + 18$?
$-6 = -24 + 18 \Rightarrow -6 = -6$ True
$-29 = 5y - 3x$ (2)
$-29 = 5(-4) - 3(3)$?
$-29 = -20 - 9 \Rightarrow -29 = -29$ True
Solution set: $\{(3, -4)\}$

17. $3y = 5x + 6$ (1)
$x + y = 2$ (2)
Solve equation (2) for y. (You could solve equation (2) for x just as easily.)
$x + y = 2$ (2)
$y = 2 - x$ (3)
Replace y with $2 - x$ in equation (1), and solve for x.
$3y = 5x + 6 \Rightarrow 3(2 - x) = 5x + 6 \Rightarrow$
$6 - 3x = 5x + 6 \Rightarrow 6 = 8x + 6 \Rightarrow$
$0 = 8x \Rightarrow 0 = x$
Replace x with 0 in equation (3) to obtain $y = 2 - 0 = 2$.
Check: $3y = 5x + 6$ (1) | $x + y = 2$ (2)
$3(2) = 5(0) + 6$? | $0 + 2 = 2$?
$6 = 0 + 6$ | $2 = 2$ True
$6 = 6$ True
Solution set: $\{(0, 2)\}$

19. $3x - y = -4$ (1)
$x + 3y = 12$ (2)
Multiply equation (2) by -3 and add the result to equation (1).
$3x - y = -4$
$-3x - 9y = -36$
$-10y = -40 \Rightarrow y = 4$

Substitute 4 for y in equation (2) and solve for x: $x + 3(4) = 12 \Rightarrow x + 12 = 12 \Rightarrow x = 0$
Check:
$3x - y = -4$ (1) | $x + 3y = 12$ (2)
$3(0) - 4 = -4$? | $0 + 3(4) = 12$?
$0 - 4 = -4$ | $0 + 12 = 12$
$-4 = -4$ True | $12 = 12$ True
Solution set: $\{(0, 4)\}$

21. $2x - 3y = -7$ (1)
$5x + 4y = 17$ (2)
Multiply equation (1) by 4 and equation (2) by 3 and then add the resulting equations.
$8x - 12y = -28$
$15x + 12y = 51$
$23x = 23 \Rightarrow x = 1$
Substitute 1 for x in equation (2) and solve for y.
$5(1) + 4y = 17 \Rightarrow 5 + 4y = 17 \Rightarrow$
$4y = 12 \Rightarrow y = 3$
Check:
$2x - 3y = -7$ (1) | $5x + 4y = 17$ (2)
$2(1) - 3(3) = -7$? | $5(1) + 4(3) = 17$?
$2 - 9 = -7$ | $5 + 12 = 17$
$-7 = -7$ True | $17 = 17$ True
Solution set: $\{(1, 3)\}$

23. $5x + 7y = 6$ (1)
$10x - 3y = 46$ (2)
Multiply equation (1) by -2 and add to equation (2).
$-10x - 14y = -12$
$10x - 3y = 46$
$-17y = 34 \Rightarrow y = -2$
Substitute -2 for y in equation (2) and solve for x.
$10x - 3(-2) = 46 \Rightarrow 10x + 6 = 46$
$10x = 40 \Rightarrow x = 4$
Check: $5x + 7y = 6$ (1)
$5(4) + 7(-2) = 6$?
$20 - 14 = 6$
$6 = 6$ True
$10x - 3y = 46$ (2)
$10(4) - 3(-2) = 46$?
$40 + 6 = 46$
$46 = 46$ True
Solution set: $\{(4, -2)\}$

Copyright © 2013 Pearson Education, Inc.

25. $6x + 7y + 2 = 0$ (1)

$\quad\;\; 7x - 6y - 26 = 0$ (2)

Multiply equation (1) by 6 and equation (2) by 7 and then add the resulting equations.

$36x + 42y + 12 = 0$

$\underline{49x - 42y - 182 = 0}$

$85x \qquad\;\; -170 = 0 \Rightarrow 85x = 170 \Rightarrow x = 2$

Substitute 2 for x in equation (1).

$6(2) + 7y + 2 = 0$

$\quad 12 + 7y + 2 = 0$

$\qquad\;\; 7y + 14 = 0$

$\qquad\qquad\;\; 7y = -14 \Rightarrow y = -2$

Check: $6x + 7y + 2 = 0$ (1)

$\qquad\quad 6(2) + 7(-2) + 2 = 0$?

$\qquad\qquad\quad 12 - 14 + 2 = 0 \Rightarrow 0 = 0$ True

$\qquad\qquad\quad 7x - 6y - 26 = 0$ (2)

$\qquad\quad 7(2) - 6(-2) - 26 = 0$?

$\qquad\qquad 14 + 12 - 26 = 0 \Rightarrow 0 = 0$ True

Solution set: $\{(2, -2)\}$

27. $\frac{x}{2} + \frac{y}{3} = 4$ (1)

$\quad \frac{3x}{2} + \frac{3y}{2} = 15$ (2)

To clear denominators, multiply equation (1) by 6 and equation (2) by 2.

$6\left(\frac{x}{2} + \frac{y}{3}\right) = 6(4)$

$2\left(\frac{3x}{2} + \frac{3y}{2}\right) = 2(15)$

This gives the system

$3x + 2y = 24$ (3)

$3x + 3y = 30$ (4).

Multiply equation (3) by -1 and add to equation (4).

$-3x - 2y = -24$

$\underline{\;\;3x + 3y = \;\;30\;}$

$\qquad\quad y = \;\; 6$

Substitute 6 for y in equation (1).

$\frac{x}{2} + \frac{6}{3} = 4 \Rightarrow \frac{x}{2} + 2 = 4 \Rightarrow \frac{x}{2} = 2 \Rightarrow x = 4$

Check: $\frac{x}{2} + \frac{y}{3} = 4$ (1) $\Big|$ $\frac{3x}{2} + \frac{3y}{2} = 15$ (2)

$\qquad\;\; \frac{4}{2} + \frac{6}{3} = 4$? $\Big|$ $\frac{3(4)}{2} + \frac{3(6)}{2} = 15$?

$\qquad\qquad 2 + 2 = 4$ $\Big|$ $6 + 9 = 15$

$\qquad\qquad\quad 4 = 4$ True $\Big|$ $15 = 15$ True

Solution set: $\{(4, 6)\}$

29. $\frac{2x-1}{3} + \frac{y+2}{4} = 4$ (1)

$\quad \frac{x+3}{2} - \frac{x-y}{3} = 3$ (2)

Multiply equation (1) by 12 and equation (2) by 6 to clear denominators. Also, remove parentheses and combine like terms.

$12\left(\frac{2x-1}{3} + \frac{y+2}{4}\right) = 12(4)$

$4(2x - 1) + 3(y + 2) = 48$

$\quad 8x - 4 + 3y + 6 = 48 \Rightarrow 8x + 3y = 46$ (3)

$\qquad 6\left(\frac{x+3}{2} - \frac{x-y}{3}\right) = 6(3)$

$3(x + 3) - 2(x - y) = 18$

$\quad 3x + 9 - 2x + 2y = 18 \Rightarrow x + 2y = 9$ (4)

Multiply equation (4) by -8 and then add the result to equation (3).

$8x + \;\; 3y = \;\; 46$

$\underline{-8x - 16y = -72\;}$

$\qquad -13y = -26 \Rightarrow y = 2$

Substitute 2 for y into equation (4) and solve for x: $x + 2(2) = 9 \Rightarrow x + 4 = 9 \Rightarrow x = 5$

31. $\quad 9x - \;\; 5y = 1$ (1)

$\;\; -18x + 10y = 1$ (2)

Multiply equation (1) by 2 and add the result to equation (2).

$\quad 18x - 10y = 2$

$\underline{-18x + 10y = 1\;}$

$\qquad\qquad\quad 0 = 3$

This is a false statement. The solution set is $\varnothing$, and the system is inconsistent.

33. $\quad 4x - \;\; y = \;\;\;\; 9$ (1)

$\;\; -8x + 2y = -18$ (2)

Multiply equation (1) by 2 and add the result to equation (2).

$\quad 8x - 2y = \;\; 18$

$\underline{-8x + 2y = -18\;}$

$\qquad\qquad 0 = \;\;\; 0$

This is a true statement. There are infinitely many solutions. We will now express the solution set with y as the arbitrary variable. Solve equation (1) for x.

$4x - y = 9 \Rightarrow 4x = y + 9 \Rightarrow x = \frac{y+9}{4}$

Solution set: $\left\{\left(\frac{y+9}{4}, y\right)\right\}$

Copyright © 2013 Pearson Education, Inc.

35. $5x - 5y - 3 = 0$ (1)
$x - y - 12 = 0$ (2)

Multiply equation (2) by −5 and then add the resulting equations.
$$5x - 5y - 3 = 0$$
$$\underline{-5x + 5y + 60 = 0}$$
$$57 = 0$$

This is a false statement. The solution set is $\varnothing$, and the system is inconsistent.

37. $7x + 2y = 6$ (1)
$14x + 4y = 12$ (2)

Multiply equation (1) by −2 and add the result to equation (2).
$$-14x - 4y = -12$$
$$\underline{14x + 4y = 12}$$
$$0 = 0$$

This is a true statement. There are infinitely many solutions. We will express the solution set with y as the arbitrary variable. Solve equation (1) for x.
$$7x + 2y = 6 \Rightarrow 7x = 6 - 2y \Rightarrow x = \tfrac{6-2y}{7}$$

Solution set: $\left\{ \left(\tfrac{6-2y}{7}, \ y \right) \right\}$

39. $2x - 6y = 0$ (1)
$-7x + 21y = 10$ (2)

Multiply equation (1) by $\tfrac{7}{2}$ and then add the result to equation (1).
$$7x - 21y = 0$$
$$\underline{-7x + 21y = 10}$$
$$0 = 10$$

This is a false statement. The solution set is $\varnothing$, and the system is inconsistent.

41. Since $y = ax + b$ and the line passes through $(2, 0)$ and $(0, 3)$, we have the equations
$$0 = a(2) + b \text{ and } 3 = a(0) + b$$
These becomes the following system.
$$2a + b = 0 \quad (1)$$
$$b = 3 \quad (2)$$
Substitute $b = 3$ into equation (1) to solve for a: $2a + 3 = 0 \Rightarrow 2a = -3 \Rightarrow a = -\tfrac{3}{2}$. Thus, the equation is $y = -\tfrac{3}{2}x + 3 \Rightarrow 3x + 2y = 6$.
The other line goes through the points $(0, 1)$ and $(-3, 0)$, so we have the equations
$1 = a(0) + b$ and $0 = a(-3) + b$. These become the following system:

$$b = 1 \quad (3)$$
$$-3a + b = 0 \quad (4)$$

Substitute $b = 1$ into equation (4), then solve for a:
$$-3a + 1 = 0 \Rightarrow a = \tfrac{1}{3}$$
Thus, the equation of this line is
$$y = \tfrac{1}{3}x + 1 \Rightarrow 3y = x + 3 \Rightarrow x - 3y = -3$$

43. Solve each equation for y.
$$\tfrac{11}{3}x + y = 0.5 \Rightarrow y = 0.5 - \tfrac{11}{3}x \text{ and}$$
$$0.6x - y = 3 \Rightarrow -y = 3 - 0.6x \Rightarrow y = 0.6x - 3$$

Solution set: $\{(0.820, -2.508)\}$

45. Solve each equation for y.
$$\sqrt{7}x + \sqrt{2}y - 3 = 0 \Rightarrow \sqrt{2}y = 3 - \sqrt{7}x \Rightarrow$$
$$y = \frac{3 - \sqrt{7}x}{\sqrt{2}} \text{ and } \sqrt{6}x - y - \sqrt{3} = 0 \Rightarrow$$
$$y = \sqrt{6}x - \sqrt{3}$$

Solution set: $\{0.892, 0.453)\}$

Copyright © 2013 Pearson Education, Inc.

47. $x + y + z = 2$ (1)
$2x + y - z = 5$ (2)
$x - y + z = -2$ (3)

Eliminate z by adding equations (1) and (2) to get $3x + 2y = 7$ (4).

Eliminate z by adding equations (2) and (3) to get $3x = 3 \Rightarrow x = 1$ (5).

Using $x = 1$, find y from equation (4) by substitution.

$3(1) + 2y = 7 \Rightarrow 3 + 2y = 7 \Rightarrow 2y = 4 \Rightarrow y = 2$

Substitute 1 for x and 2 for y in equation (1) to find z.

$1 + 2 + z = 2 \Rightarrow 3 + z = 2 \Rightarrow z = -1$

Verify that the ordered triple $(1, 2, -1)$ satisfies all three equations.

Check:

$x + y + z = 2$ (1) $\quad\quad 2x + y - z = 5$ (2)
$1 + 2 + (-1) = 2$? $\quad 2(1) + 2 - (-1) = 5$?
$\quad\quad 2 = 2$ True $\quad\quad\quad 2 + 2 + 1 = 5$
$\quad\quad\quad\quad\quad\quad\quad\quad\quad\quad 5 = 5$ True

$x - y + z = -2$ (3)
$1 - 2 + (-1) = -2$?
$\quad\quad -1 - 1 = -2$
$\quad\quad\quad -2 = -2$ True

Solution set: $\{(1, 2, -1)\}$

49. $x + 3y + 4z = 14$ (1)
$2x - 3y + 2z = 10$ (2)
$3x - y + z = 9$ (3)

Eliminate y by adding equations (1) and (2) to get $3x + 6z = 24$ (4).

Multiply equation (3) by 3 and add the result to equation (1).

$x + 3y + 4z = 14$
$9x - 3y + 3z = 27$
$\overline{10x + \quad 7z = 41}$ (5)

Multiply equation (4) by 10 and equation (5) by −3 and add in order to eliminate y.

$30x + 60z = 240$
$-30x - 21z = -123$
$\overline{\quad\quad 39z = 117} \Rightarrow z = 3$

Using $z = 3$, find x from equation (5) by substitution.

$10x + 7(3) = 41 \Rightarrow 10x + 21 = 41 \Rightarrow$
$\quad 10x = 20 \Rightarrow x = 2$

Substitute 2 for x and 3 for z in equation (1) to find y.

$2 + 3y + 4(3) = 14 \Rightarrow 3y = 0 \Rightarrow y = 0$

Verify that the ordered triple $(2, 0, 3)$ satisfies all three equations.

Check:

$x + 3y + 4z = 14$ (1)
$2 + 3(0) + 4(3) = 14$?
$\quad 2 + 0 + 12 = 14$
$\quad\quad\quad 14 = 14$ True

$2x - 3y + 2z = 10$ (2)
$2(2) - 3(0) + 2(3) = 10$?
$\quad\quad 4 - 0 + 6 = 10$
$\quad\quad\quad\quad 10 = 10$ True

$3x - y + z = 9$ (3)
$3(2) - 0 + 3 = 9$?
$\quad 6 - 0 + 3 = 9$
$\quad\quad\quad 9 = 9$ True

Solution set: $\{(2, 0, 3)\}$

51. $x + 4y - z = 6$ (1)
$2x - y + z = 3$ (2)
$3x + 2y + 3z = 16$ (3)

Eliminate z by adding equations (1) and (2) to get $3x + 3y = 9$ or $x + y = 3$ (4).

Multiply equation (1) by 3 and add the result to equation (3).

$3x + 12y - 3z = 18$
$3x + 2y + 3z = 16$
$\overline{6x + 14y \quad\quad = 34}$ (5)

Multiply equation (4) by −6 and then add the result to equation (5) in order to eliminate x.

$-6x - 6y = -18$
$6x + 14y = 34$
$\overline{\quad 8y = 16} \Rightarrow y = 2$

Using $y = 2$, find x from equation (4) by substitution: $x + 2 = 3 \Rightarrow x = 1$

Substitute 1 for x and 2 for y in equation (1) to find z.

$1 + 4(2) - z = 6 \Rightarrow 1 + 8 - z = 6 \Rightarrow$
$\quad\quad 9 - z = 6 \Rightarrow -z = -3 \Rightarrow z = 3$

Verify that the ordered triple $(1, 2, 3)$ satisfies all three equations.

Check:

$x + 4y - z = 6$ (1) $\quad\quad 2x - y + z = 3$ (2)
$1 + 4(2) - 3 = 6$? $\quad\quad 2(1) - 2 + 3 = 3$?
$\quad 1 + 8 - 3 = 6$ $\quad\quad\quad\quad 2 - 2 + 3 = 3$
$\quad\quad\quad 6 = 6$ True $\quad\quad\quad\quad\quad 3 = 3$ True

$3x + 2y + 3z = 16$ (3)
$3(1) + 2(2) + 3(3) = 16$?
$\quad\quad 3 + 4 + 9 = 16$
$\quad\quad\quad\quad 16 = 16$ True

Solution set: $\{(1, 2, 3)\}$

Copyright © 2013 Pearson Education, Inc.

53. $x - 3y - 2z = -3$ (1)
$3x + 2y - z = 12$ (2)
$-x - y + 4z = 3$ (3)

Eliminate x by adding equations (1) and (3) to get $-4y + 2z = 0$ (4).

Eliminate x by multiplying equation (3) by 3 and add to equation (2).
$-3x - 3y + 12z = 9$
$\underline{3x + 2y - z = 12}$
$-y + 11z = 21$ (5)

Multiply equation (5) by -4 and add to equation (4) in order to eliminate y.
$-4y + 2z = 0$
$\underline{4y - 44z = -84}$
$-42z = -84 \Rightarrow z = 2$

Using $z = 2$, find y from equation (4) by substitution.
$-4y + 2(2) = 0 \Rightarrow -4y + 4 = 0 \Rightarrow$
$-4y = -4 \Rightarrow y = 1$

Substitute 1 for y and 2 for z in equation (1) to find x.
$x - 3(1) - 2(2) = -3 \Rightarrow x - 3 - 4 = -3 \Rightarrow$
$x - 7 = -3 \Rightarrow x = 4$

Verify that the ordered triple $(4,1,2)$ satisfies all three equations.
Check: $x - 3y - 2z = -3$ (1)
$4 - 3(1) - 2(2) = -3$?
$4 - 3 - 4 = -3$
$-3 = -3$ True

$3x + 2y - z = 12$ (2)
$3(4) + 2(1) - 2 = 12$?
$12 + 2 - 2 = 12$
$12 = 12$ True

$-x - y + 4z = 3$ (3)
$-4 - 1 + 4(2) = 3$?
$-4 - 1 + 8 = 3$
$3 = 3$ True

Solution set: $\{(4,1,2)\}$

55. $2x + 6y - z = 6$ (1)
$4x - 3y + 5z = -5$ (2)
$6x + 9y - 2z = 11$ (3)

Eliminate y by multiplying equation (2) by 2 and add to equation (1).
$2x + 6y - z = 6$
$\underline{8x - 6y + 10z = -10}$
$10x + 9z = -4$ (4)

Eliminate y by multiplying equation (2) by 3 and add to equation (3).
$6x + 9y - 2z = 11$
$\underline{12x - 9y + 15z = -15}$
$18x + 13z = -4$ (5)

Multiply equation (4) by 9 and equation (5) by -5 and add in order to eliminate x.
$90x + 81z = -36$
$\underline{-90x - 65z = 20}$
$16z = -16 \Rightarrow z = -1$

Using $z = -1$, find x from equation (4) by substitution.
$10x + 9(-1) = -4 \Rightarrow 10x - 9 = -4$
$10x = 5 \Rightarrow x = \frac{1}{2}$

Substitute $\frac{1}{2}$ for x and -1 for z in equation (1) to find y.
$2\left(\frac{1}{2}\right) + 6y - (-1) = 6 \Rightarrow 1 + 6y + 1 = 6 \Rightarrow$
$6y + 2 = 6 \Rightarrow 6y = 4 \Rightarrow y = \frac{2}{3}$

Verify that the ordered triple $\left(\frac{1}{2}, \frac{2}{3}, -1\right)$ satisfies all three equations.
Check:
$2x + 6y - z = 6$ (1)
$2\left(\frac{1}{2}\right) + 6\left(\frac{2}{3}\right) - (-1) = 6$?
$1 + 4 + 1 = 6$
$6 = 6$ True

$4x - 3y + 5z = -5$ (2)
$4\left(\frac{1}{2}\right) - 3\left(\frac{2}{3}\right) + 5(-1) = -5$?
$2 - 2 - 5 = -5$
$-5 = -5$ True

$6x + 9y - 2z = 11$ (3)
$6\left(\frac{1}{2}\right) + 9\left(\frac{2}{3}\right) - 2(-1) = 11$?
$3 + 6 + 2 = 11$
$11 = 11$ True

Solution set: $\left\{\left(\frac{1}{2}, \frac{2}{3}, -1\right)\right\}$

Copyright © 2013 Pearson Education, Inc.

57. $2x - 3y + 2z - 3 = 0$ (1)
$\quad 4x + 8y + z - 2 = 0$ (2)
$\quad -x - 7y + 3z - 14 = 0$ (3)

Eliminate x by multiplying equation (3) by 2 and add to equation (1).

$\quad 2x - 3y + 2z - 3 = 0$
$\quad \underline{-2x - 14y + 6z - 28 = 0}$
$\qquad -17y + 8z - 31 = 0$ (4)

Eliminate x by multiplying equation (3) by 4 and add to equation (2).

$\quad 4x + 8y + z - 2 = 0$
$\quad \underline{-4x - 28y + 12z - 56 = 0}$
$\qquad -20y + 13z - 58 = 0$ (5)

Multiply equation (5) by 17, equation (4) by -20, and add in order to eliminate y.

$\quad 340y - 160z + 620 = 0$
$\quad \underline{-340y + 221z - 986 = 0}$
$\qquad 61z - 366 = 0 \Rightarrow z = 6$

Using $z = 6$, find y from equation (5) by substitution.

$-20y + 13(6) - 58 = 0 \Rightarrow -20y + 78 - 58 = 0$
$\quad -20y + 20 = 0 \Rightarrow -20y = -20 \Rightarrow y = 1$

Substitute 1 for y and 6 for z in equation (1) to find x.

$2x - 3(1) + 2(6) - 3 = 0 \Rightarrow 2x - 3 + 12 - 3 = 0 \Rightarrow$
$\qquad 2x + 6 = 0 \Rightarrow 2x = -6 \Rightarrow x = -3$

Verify that the ordered triple $(-3, 1, 6)$ satisfies all three equations.

Check: $\quad 2x - 3y + 2z - 3 = 0$ (1)
$\quad\quad 2(-3) - 3(1) + 2(6) - 3 = 0$?
$\quad\quad\quad\quad -6 - 3 + 12 - 3 = 0$
$\quad\quad\quad\quad\quad\quad\quad\quad 0 = 0$ True

$\quad\quad\quad 4x + 8y + z - 2 = 0$ (2)
$\quad\quad 4(-3) + 8(1) + 6 - 2 = 0$
$\quad\quad\quad\quad -12 + 8 + 6 - 2 = 0$?
$\quad\quad\quad\quad\quad\quad\quad\quad 0 = 0$ True

$\quad\quad\quad -x - 7y + 3z - 14 = 0$ (3)
$\quad\quad -(-3) - 7(1) + 3(6) - 14 = 0$?
$\quad\quad\quad\quad\quad 3 - 7 + 18 - 14 = 0$
$\quad\quad\quad\quad\quad\quad\quad\quad 0 = 0$ True

Solution set: $\{(-3, 1, 6)\}$

59. $\quad x - 2y + 3z = 6$ (1)
$\quad\quad 2x - y + 2z = 5$ (2)

Multiply equation (2) by -2 and add to equation (1) in order to eliminate y.

$\quad x - 2y + 3z = 6$
$\quad \underline{-4x + 2y - 4z = -10}$
$\quad -3x \quad\quad - z = -4$ (3)

Solve equation (3) for z.

$-3x - z = -4 \Rightarrow -z = 3x - 4 \Rightarrow z = -3x + 4$

Express y in terms of x by solving equation (1) for y and substituting $-3x + 4$ for z.

$x - 2y + 3z = 6 \Rightarrow -2y = -x - 3z + 6 \Rightarrow$

$y = \dfrac{-x - 3z + 6}{-2} = \dfrac{x + 3z - 6}{2} \Rightarrow$

$y = \dfrac{x + 3(-3x + 4) - 6}{2} = \dfrac{x - 9x + 12 - 6}{2}$

$\quad = \dfrac{-8x + 6}{2} = -4x + 3$

With x arbitrary, the solution set is of the form $\{(x, -4x + 3, -3x + 4)\}$.

61. $5x - 4y + z = 9$ (1)
$\quad\quad\quad\quad y + z = 15$ (2)

Multiply equation (2) by -1 and add to equation (1) in order to eliminate z.

$\quad 5x - 4y + z = 9$
$\quad \underline{\quad\quad -y - z = -15}$
$\quad 5x - 5y \quad = -6$ (3)

Solve equation (3) for y.

$5x - 5y = -6 \Rightarrow -5y = -5x - 6 \Rightarrow$

$y = \dfrac{-5x - 6}{-5} = \dfrac{5x + 6}{5}$

Express z in terms of x by solving equation (1) for z and substituting $\frac{5x+6}{5}$ for y.

$5x - 4y + z = 9 \Rightarrow z = -5x + 4y + 9 \Rightarrow$

$z = -5x + 4\left(\dfrac{5x + 6}{5}\right) + 9$

$\quad = \dfrac{-25x + 20x + 24 + 45}{5} = \dfrac{-5x + 69}{5}$

With x arbitrary, the solution set is of the form $\left\{\left(x, \dfrac{5x + 6}{5}, \dfrac{-5x + 69}{5}\right)\right\}$.

Copyright © 2013 Pearson Education, Inc.

63. $3x + 4y - z = 13$ (1)
$\quad x + y + 2z = 15$ (2)

Multiply equation (2) by -4 and add to equation (1) in order to eliminate y.

$\quad 3x + 4y - z = 13$
$\underline{-4x - 4y - 8z = -60}$
$\quad -x \quad -9z = -47$ (3)

Solve equation (3) for z.

$-x - 9z = -47 \Rightarrow -9z = -47 + x \Rightarrow$

$$z = \frac{-47 + x}{-9} = \frac{47 - x}{9}$$

Express y in terms of x by solving equation (2) for y and substituting $\frac{47-x}{9}$ for z.

$x + y + 2z = 15 \Rightarrow y = -x - 2z + 15 \Rightarrow$

$$y = -x - 2\left(\frac{47 - x}{9}\right) + 15$$

$$= \frac{-9x - 2(47 - x) + 135}{9}$$

$$= \frac{-9x - 94 + 2x + 135}{9} = \frac{41 - 7x}{9}$$

With x arbitrary, the solution set is of the form

$$\left\{\left(x, \frac{41 - 7x}{9}, \frac{47 - x}{9}\right)\right\}.$$

65. $\quad 3x + 5y - z = -2$ (1)
$\quad 4x - y + 2z = 1$ (2)
$\quad -6x - 10y + 2z = 0$ (3)

Multiply equation (1) by 2 and add the result to equation (3).

$\quad 6x + 10y - 2z = -4$
$\underline{-6x - 10y + 2z = 0}$
$\quad\quad\quad\quad\quad 0 = -4$

We obtain a false statement. The solution set is $\varnothing$, and the system is inconsistent.

67. $\quad 5x - 4y + z = 0$ (1)
$\quad\quad x + y \quad = 0$ (2)
$\quad -10x + 8y - 2z = 0$ (3)

Multiply equation (2) by 4 and add the result to equation (1) in order to eliminate y.

$\quad 5x - 4y + z = 0$
$\underline{\quad 4x + 4y \quad = 0}$
$\quad 9x \quad\quad + z = 0$ (4)

Multiply equation (2) by -8 and add the result to equation (3) in order to eliminate y.

$\quad -8x - 8y \quad = 0$
$\underline{-10x + 8y - 2z = 0}$
$\quad -18x \quad\quad - 2z = 0$ (5)

Multiply equation (4) by 2 and add the result to equation (5).

$\quad 18x + 2z = 0$
$\underline{-18x - 2z = 0}$
$\quad\quad\quad\quad 0 = 0$

This is a true statement and the system has infinitely many solutions.

Solve equation (4) for x.

$9x + z = 0 \Rightarrow 9x = -z \Rightarrow x = -\frac{z}{9}$

Express y in terms of z by solving equation (1) for y and substituting $-\frac{z}{9}$ for x.

$5x - 4y + z = 0 \Rightarrow -4y = -5x - z \Rightarrow$

$$y = \frac{-5x - z}{-4} = \frac{5x + z}{4} = \frac{5\left(-\frac{z}{9}\right) + z}{4}$$

$$= \frac{-5z + 9z}{36} = \frac{4z}{36} = \frac{z}{9}$$

With z arbitrary, the solution set is of the form

$$\left\{\left(-\frac{z}{9}, \frac{z}{9}, z\right)\right\}.$$

69. $\dfrac{2}{x} + \dfrac{1}{y} = \dfrac{3}{2}$ (1)

$\dfrac{3}{x} - \dfrac{1}{y} = 1$ (2)

Let $\frac{1}{x} = t$ and $\frac{1}{y} = u$. With these substitutions, the system becomes

$2t + u = \frac{3}{2}$ (3)
$3t - u = 1$ (4)

Add these equations, eliminate u, and solve for t.

$5t = \frac{5}{2} \Rightarrow t = \frac{1}{2}$

Substitute $\frac{1}{2}$ for t in equation (3) and solve for u.

$2\left(\frac{1}{2}\right) + u = \frac{3}{2} \Rightarrow 1 + u = \frac{3}{2} \Rightarrow u = \frac{1}{2}$

Now find the values of x and y, the variables in the original system. So, $\frac{1}{x} = t, tx = 1$, and $x = \frac{1}{t}$. Likewise $y = \frac{1}{u}$.

$x = \dfrac{1}{t} = \dfrac{1}{\frac{1}{2}} = 2$ and $y = \dfrac{1}{u} = \dfrac{1}{\frac{1}{2}} = 2$

Solution set: $\{(2, 2)\}$

Copyright © 2013 Pearson Education, Inc.

71. $\dfrac{2}{x}+\dfrac{1}{y}=11$ (1)

$\dfrac{3}{x}-\dfrac{5}{y}=10$ (2)

Let $\dfrac{1}{x}=t$ and $\dfrac{1}{y}=u$. With these substitutions,

the system becomes

$2t+u=11$ (3)

$3t-5u=10$ (4)

Multiply equation (3) by 5 and add to equation (4), eliminating u, and solve for t.

$10t+5u=55$

$\underline{3t-5u=10}$

$13t=65 \Rightarrow t=5$

Substitute 5 for t in equation (3) and solve for u: $2(5)+u=11 \Rightarrow 10+u=11 \Rightarrow u=1$

Now find the values of x and y, the variables in the original system.

$x=\dfrac{1}{t}=\dfrac{1}{5}$ and $y=\dfrac{1}{u}=\dfrac{1}{1}=1$

Solution set: $\left\{\left(\dfrac{1}{5},1\right)\right\}$

73. $\dfrac{2}{x}+\dfrac{3}{y}-\dfrac{2}{z}=-1$ (1)

$\dfrac{8}{x}-\dfrac{12}{y}+\dfrac{5}{z}=5$ (2)

$\dfrac{6}{x}+\dfrac{3}{y}-\dfrac{1}{z}=1$ (3)

Let $\dfrac{1}{x}=t$, $\dfrac{1}{y}=u$, and $\dfrac{1}{z}=v$. With these

substitutions, the system becomes

$2t+3u-2v=-1$ (4)

$8t-12u+5v=5$ (5)

$6t+3u-v=1$ (6)

Eliminate v by multiplying equation (6) by 5 and add to equation (5).

$8t-12u+5v=5$

$\underline{30t+15u-5v=5}$

$38t+3u=10$ (7)

Eliminate v by multiplying equation (6) by -2 and add to equation (4).

$2t+3u-2v=-1$

$\underline{-12t-6u+2v=-2}$

$-10t-3u=-3$ (8)

Add equations (7) and (8) in order to eliminate u.

$38t+3u=10$

$\underline{-10t-3u=-3}$

$28t=7 \Rightarrow t=\dfrac{1}{4}$

Using $t=\dfrac{1}{4}$, find u from equation (7) by substitution.

$38\left(\dfrac{1}{4}\right)+3u=10 \Rightarrow \dfrac{19}{2}+3u=10 \Rightarrow$

$3u=\dfrac{20}{2}-\dfrac{19}{2}=\dfrac{1}{2} \Rightarrow u=\dfrac{1}{6}$

Substitute $\dfrac{1}{4}$ for t and $\dfrac{1}{6}$ for u in equation (4) to find v.

$2\left(\dfrac{1}{4}\right)+3\left(\dfrac{1}{6}\right)-2v=-1 \Rightarrow \dfrac{1}{2}+\dfrac{1}{2}-2v=-1 \Rightarrow$

$1-2v=-1 \Rightarrow -2v=-2 \Rightarrow v=1$

Now find the values of x, y, and z, the variables in the original system.

$x=\dfrac{1}{t}=\dfrac{1}{\frac{1}{4}}=4$, $y=\dfrac{1}{u}=\dfrac{1}{\frac{1}{6}}=6$, and

$z=\dfrac{1}{v}=\dfrac{1}{1}=1$

Solution set: $\left\{(4,6,1)\right\}$

75. $x-2y=3$ (1)

$-2x+4y=k$ (2)

Multiply equation (1) by 2 and add the result to equation (2).

$2x-4y=6$

$\underline{-2x+4y=k}$

$0=6+k$

The system will have no solution when $6+k\neq 0 \Rightarrow k\neq -6$.

The system will have infinitely many solutions when $6+k=0 \Rightarrow k=-6$.

77. Since $y=ax+b$ and the line passes through $(-2, 1)$ and $(-1, -2)$, we have the equations $1=a(-2)+b$ and $-2=a(-1)+b$.

This becomes the following system.

$-2a+b=1$ (1)

$-a+b=-2$ (2)

Multiply equation (1) by -1 and add the result to equation (2) in order to eliminate b.

$2a-b=-1$

$\underline{-a+b=-2}$

$a=-3$

Substitute this value into equation (1).

$-2(-3)+b=1 \Rightarrow 6+b=1 \Rightarrow b=-5$

The equation is $y=-3x-5$.

Copyright © 2013 Pearson Education, Inc.

79. Since $y = ax^2 + bx + c$ and the parabola passes through the points (2, 3), (–1, 0), and (–2, 2), we have the following equations.

$$3 = a(2)^2 + b(2) + c \implies 4a + 2b + c = 3 \quad (1)$$
$$0 = a(-1)^2 + b(-1) + c \implies a - b + c = 0 \quad (2)$$
$$2 = a(-2)^2 + b(-2) + c \implies 4a - 2b + c = 2 \quad (3)$$

Multiply equation (2) by –1 and add the result to equation (1) in order to eliminate c.

$$4a + 2b + c = 3$$
$$\underline{-a + b - c = 0}$$
$$3a + 3b \quad = 3 \quad (4)$$

Multiply equation (2) by –1 and add the result to equation (3) in order to eliminate c.

$$-a + b - c = 0$$
$$\underline{4a - 2b + c = 2}$$
$$3a - b \quad = 2 \quad (5)$$

Multiply equation (4) by –1 and then add the result to equation (5) and solve for b.

$$-3a - 3b = -3$$
$$\underline{3a - b = 2}$$
$$-4b = -1 \implies b = \tfrac{1}{4}$$

Substitute this value into equation (5).

$$3a - \tfrac{1}{4} = 2 \implies 3a = \tfrac{8}{4} + \tfrac{1}{4} = \tfrac{9}{4} \implies a = \tfrac{3}{4}$$

Substitute $a = \tfrac{3}{4}$ and $b = \tfrac{1}{4}$ into equation (1) in order to solve for c.

$$4\left(\tfrac{3}{4}\right) + 2\left(\tfrac{1}{4}\right) + c = 3$$
$$3 + \tfrac{1}{2} + c = 3 \implies c = -\tfrac{1}{2}$$

The equation of the parabola is
$$y = \tfrac{3}{4}x^2 + \tfrac{1}{4}x - \tfrac{1}{2}.$$

81. Since $y = ax + b$ and the line passes through the points (2, 5) and (–1, –4), we have the equations $5 = a(2) + b$ and $-4 = a(-1) + b$.

This becomes the following system.

$$2a + b = 5 \quad (1)$$
$$-a + b = -4 \quad (2)$$

Multiply equation (2) by –1 and add the result to equation (1) in order to eliminate b.

$$2a + b = 5$$
$$\underline{a - b = 4}$$
$$3a \quad = 9 \implies a = 3$$

Substitute this value into equation (1).

$$2(3) + b = 5 \implies 6 + b = 5 \implies b = -1$$

The equation is $y = 3x - 1$.

83. Since $y = ax^2 + bx + c$ and the parabola passes through the points $(-2, -3.75)$, $(4, -3.75)$, and $(-1, -1.25)$, we have the following equations.

$$-3.75 = a(-2)^2 + b(-2) + c$$
$$4a - 2b + c = -3.75 \quad (1)$$
$$-3.75 = a(4)^2 + b(4) + c$$
$$16a + 4b + c = -3.75 \quad (2)$$
$$-1.25 = a(-1)^2 + b(-1) + c$$
$$a - b + c = -1.25 \quad (3)$$

Multiply equation (2) by –1 and add the result to equation (1) in order to eliminate c.

$$4a - 2b + c = -3.75$$
$$\underline{-16a - 4b - c = 3.75}$$
$$-12a - 6b \quad = 0 \text{ or } 2a + b = 0 \quad (4)$$

Multiply equation (2) by –1 and add the result to equation (3) in order to eliminate c.

$$-16a - 4b - c = 3.75$$
$$\underline{a - b + c = -1.25}$$
$$-15a - 5b \quad = 2.50 \text{ or } -3a - b = 0.5 \quad (5)$$

Add equations (4) and (5) in order to solve for b.

$$2a + b = 0$$
$$\underline{-3a - b = 0.5}$$
$$-a \quad = 0.5 \implies a = -0.5$$

Substitute $a = -0.5$ into equation (4) in order to solve for b.

$$2(-0.5) + b = 0 \implies -1 + b = 0 \implies b = 1$$

Substitute $a = -0.5$ and $b = 1$ into equation (3) in order to solve for c.

$$-.5 - 1 + c = -1.25$$
$$-1.5 + c = -1.25 \implies c = 0.25$$

The equation of the parabola is
$$y = -0.5x^2 + x + 0.25 \text{ or } y = -\tfrac{1}{2}x^2 + x + \tfrac{1}{4}.$$

Copyright © 2013 Pearson Education, Inc.

85. Since $x^2 + y^2 + ax + by + c = 0$ and the circle passes through the points $(-1,3)$, $(6,2)$, and $(-2,-4)$, we have the following equations.

$$(-1)^2 + 3^2 + a(-1) + b(3) + c = 0$$
$$-a + 3b + c = -10 \quad (1)$$
$$6^2 + 2^2 + a(6) + b(2) + c = 0 \Rightarrow$$
$$6a + 2b + c = -40 \quad (2)$$
$$(-2)^2 + (-4)^2 + a(-2) + b(-4) + c = 0$$
$$-2a - 4b + c = -20 \quad (3)$$

Multiply equation (1) by –2 and add the result to equation (3) in order to eliminate a.

$$2a - 6b - 2c = 20$$
$$\underline{-2a - 4b + c = -20}$$
$$-10b - c = 0 \, (4)$$

Multiply equation (1) by 6 and add the result to equation (2) in order to eliminate a.

$$-6a + 18b + 6c = -60$$
$$\underline{6a + 2b + c = -40}$$
$$20b + 7c = -100 \, (5)$$

Multiply equation (4) by 7 and add the result to equation (5) in order to eliminate c.

$$-70b - 7c = 0$$
$$\underline{20b + 7c = -100}$$
$$-50b = -100 \Rightarrow b = 2$$

We substitute this value into equation (4) in order to solve for c.

$$-10(2) - c = 0 \Rightarrow -20 - c = 0 \Rightarrow -20 = c$$

Substitute $b = 2$ and $c = -20$ into equation (1).

$$-a + 3(2) - 20 = -10 \Rightarrow -a + 6 - 20 = -10 \Rightarrow$$
$$-a - 14 = -10 \Rightarrow -a = 4 \Rightarrow a = -4$$

The equation of the circle is
$$x^2 + y^2 - 4x + 2y - 20 = 0.$$

87. Since $x^2 + y^2 + ax + by + c = 0$ and the circle passes through the points $(2, 1)$, $(-1, 0)$, and $(3, 3)$, we have the following equations.

$$2^2 + 1^2 + a(2) + b(1) + c = 0$$
$$2a + b + c = -5 \quad (1)$$
$$(-1)^2 + 0^2 + a(-1) + b(0) + c = 0$$
$$-a + c = -1 \quad (2)$$
$$3^2 + 3^2 + a(3) + b(3) + c = 0$$
$$3a + 3b + c = -18 \quad (3)$$

Multiply equation (1) by –3 and add the result to equation (3) in order to eliminate b.

$$-6a - 3b - 3c = 15$$
$$\underline{3a + 3b + c = -18}$$
$$-3a - 2c = -3 \quad (4)$$

Multiply equation (2) by 2 and add the result to equation (4) in order to eliminate c.

$$-2a + 2c = -2$$
$$\underline{-3a - 2c = -3}$$
$$-5a = -5 \Rightarrow a = 1$$

We substitute this value into equation (2) in order to solve for c.

$$-1 + c = -1 \Rightarrow c = 0$$

Substitute $a = 1$ and $c = 0$ into equation (1).

$$2(1) + b + 0 = -5 \Rightarrow 2 + b = -5 \Rightarrow b = -7$$

The equation of the circle is
$$x^2 + y^2 + x - 7y = 0.$$

89. Since $x^2 + y^2 + ax + by + c = 0$ and the circle passes through points $(0, 3)$, $(4, 2)$, and $(-5, -2)$, we have the following equations.

$$0^2 + 3^2 + a(0) + b(3) + c = 0$$
$$3b + c = -9 \quad (1)$$
$$4^2 + 2^2 + a(4) + b(2) + c = 0$$
$$4a + 2b + c = -20 \quad (2)$$
$$(-5)^2 + (-2)^2 + a(-5) + b(-2) + c = 0$$
$$-5a - 2b + c = -29 \quad (3)$$

Multiply equation (2) by 5 and multiply equation (3) by 4, and then add the results in order to eliminate a.

$$20a + 10b + 5c = -100$$
$$\underline{-20a - 8b + 4c = -116}$$
$$2b + 9c = -216 \quad (4)$$

Multiply equation (1) by –9 and add the result to equation (4) in order to eliminate c.

$$-27b - 9c = 81$$
$$\underline{2b + 9c = -216}$$
$$-25b = -135 \Rightarrow b = \tfrac{27}{5}$$

Substitute $b = \tfrac{27}{5}$ into equation (1) in order to solve for c.

$$3\left(\tfrac{27}{5}\right) + c = -9 \Rightarrow \tfrac{81}{5} + c = -9 \Rightarrow$$
$$c = -\tfrac{81}{5} - \tfrac{45}{5} = -\tfrac{126}{5}$$

Substitute $b = \tfrac{27}{5}$ and $c = -\tfrac{126}{5}$ into equation (2).

$$4a + 2\left(\tfrac{27}{5}\right) + \left(-\tfrac{126}{5}\right) = -20 \Rightarrow$$
$$4a + \tfrac{54}{5} - \tfrac{126}{5} = -20$$
$$4a - \tfrac{72}{5} = -20 \Rightarrow 4a = \tfrac{72}{5} - \tfrac{100}{5} \Rightarrow$$
$$4a = -\tfrac{28}{5} \Rightarrow a = -\tfrac{7}{5}$$

The equation of the circle is
$$x^2 + y^2 - \tfrac{7}{5}x + \tfrac{27}{5}y - \tfrac{126}{5} = 0.$$

Copyright © 2013 Pearson Education, Inc.

91. (a) Since $C = at^2 + bt + c$ and we have the ordered pairs (0, 317), (20, 339), and (49, 385), we have the following equations.

$$317 = a(0)^2 + b(0) + c$$
$$c = 317 \qquad (1)$$
$$339 = a(20)^2 + b(20) + c$$
$$400a + 20b + c = 339 \qquad (2)$$
$$385 = a(49)^2 + b(49) + c$$
$$2401a + 49b + c = 385 \qquad (3)$$

Since $c = 317$, we substitute this value into equations (2) and (3) to obtain the following system.

$$400a + 20b + 317 = 339$$
$$400a + 20b = 22 \qquad (4)$$
$$2401a + 49b + 317 = 385$$
$$2401a + 49b = 68 \qquad (5)$$

Multiply equation (4) by –49, multiply equation (5) by 20, then add the results in order to eliminate b.

$$-19600a - 9800b = -1078$$
$$\underline{48020a + 9800b = 1360}$$
$$28420a = 282$$
$$a = \frac{141}{14210} \approx 0.009922$$

Substitute $a = \frac{141}{14210}$ into equation (4) in order to solve for b.

$$400\left(\tfrac{141}{14210}\right) + 20b = 22$$
$$20b = 22 - 400\left(\tfrac{141}{14210}\right)$$
$$b = \frac{22 - 400\left(\tfrac{141}{14210}\right)}{20} \Rightarrow$$
$$b = \frac{12811}{14210} \approx 0.9015$$

The constants are $a = \frac{141}{14210} \approx 0.009922$, $b = \frac{12811}{14210} \approx 0.9015$, and $c = 317$.
The equation is
$$C = \tfrac{141}{14210}t^2 + \tfrac{12811}{14210}t + 317 \text{ or}$$
$$C = 0.009922t^2 + 0.9015t + 317.$$

(b) Since $t = 0$ corresponds to 1960, the amount of carbon dioxide will be double its 1960 level when
$$2 \cdot 317 = \tfrac{141}{14210}t^2 + \tfrac{12811}{14210}t + 317$$
We will solve this equation graphically.

```
WINDOW
 Xmin=0
 Xmax=200
 Xscl=20
 Ymin=100
 Ymax=900
 Yscl=50
↓Xres=1
```

```
Plot1 Plot2 Plot3
\Y1◳141/14210X²+
12811/14210X+317
\Y2◳2*317
\Y3=
\Y4=
\Y5=
```

```
Intersection
X=138.99192  .Y=634 . . .
```

According to the model, the amount of carbon dioxide in the atmosphere will double at the very end of 2098, 138 years after 1960.

93. Let x = one number; let y = the other number
We have the following system of equations.
$$x + y = 47 \quad (1)$$
$$x - y = 1 \quad (2)$$
Add the two equations to eliminate y.
$2x = 48 \Rightarrow x = 24$. Substitute this value into equation (2) and solve for y:
$24 + y = 47 \Rightarrow y = 23$.
The two numbers are 23 and 24.

95. Let x = the FCI for football; let y = the FCI for baseball. We have the following system of equations.
$$\frac{x + y}{2} = 307.76 \Rightarrow x + y = 615.52 \quad (1)$$
$$x - y = 225.56 \quad (2)$$
Add the two equations to eliminate y.
$2x = 841.08 \Rightarrow x = 420.54$
Substitute this value into equation (2) and solve for y.
$420.54 - y = 225.56 \Rightarrow y = 194.98$
Verify that (420.54, 194.98) satisfies both equations in the original system.
The FCI for football was \$420.54, and the FCI for baseball was \$194.98.

97. Let x = the number of \$3.00 gallons;
y = the number of \$4.50 gallons;
z = the number of \$9.00 gallons.
We have the following equations.
$$x + y + z = 300 \qquad (1)$$
$$2x = y \Rightarrow$$
$$2x - y = 0 \qquad (2)$$
$$3.00x + 4.50y + 9.00z = 6.00(300) \Rightarrow$$
$$3x + 4.5y + 9z = 1800 \qquad (3)$$

Multiply equation (1) by –9 and add to equation (3) in order to eliminate z.
$$-9x - 9y - 9z = -2700$$
$$\underline{3x + 4.5y + 9z = 1800}$$
$$-6x - 4.5y = -900 \,(4)$$

Multiply equation (2) by 3 and add to equation (4) in order to eliminate x.

(continued on next page)

Copyright © 2013 Pearson Education, Inc.

(*continued*)

$$6x - 3y = 0$$
$$\underline{-6x - 4.5y = -900}$$
$$-7.5y = -900 \Rightarrow y = 120$$

Substitute this value into equation (2) to solve for x.

$2x - 120 = 0 \Rightarrow 2x = 120 \Rightarrow x = 60$

Substitute $x = 60$ and $y = 120$ into equation (1) and solve for z.

$60 + 120 + z = 300 \Rightarrow 180 + z = 300 \Rightarrow z = 120$

She should use 60 gal of the $3.00 water, 120 gal of the $4.50 water, and 120 gal of the $9.00 water.

99. Let x = the length of the shortest side;
y = the length of the medium side;
z = the length of the longest side.
We have the following equations.

$$x + y + z = 59 \quad (1)$$
$$z = y + 11 \Rightarrow \quad -y + z = 11 \quad (2)$$
$$y = x + 3 \Rightarrow -x + y \quad = 3 \quad (3)$$

Add equations (1) and (3) together in order to eliminate x.

$$x + y + z = 59$$
$$\underline{-x + y \quad = 3}$$
$$2y + z = 62 \ (4)$$

Multiply equation (2) by 2 and then add the result to equation (4) to solve for z.

$$-2y + 2z = 22$$
$$\underline{2y + \quad z = 62}$$
$$3z = 84 \Rightarrow z = 28$$

Substitute this value into equation (2) in order to solve for y.

$-y + 28 = 11 \Rightarrow -y = -17 \Rightarrow y = 17$

Substitute $y = 17$ into equation (3) in order to solve for x: $-x + 17 = 3 \Rightarrow -x = -14 \Rightarrow x = 14$

The lengths of the sides of the triangle are 14 inches, 17 inches, and 28 inches.
Note: $14 + 17 + 28 = 59$

101. Let x = the amount invested in real estate (at 3%); y = the amount invested in a money market account (at 2.5%); z = the amount invested in CDs (at 1.5%).
Completing the table we have the following.

	Amount Invested	Rate (as a decimal)	Annual Interest
Real Estate	x	0.03	$0.03x$
Money Market	y	0.025	$0.025y$
CDs	z	0.015	$0.015z$

We have the following equations.
$$x + y + z = 200,000 \quad (1)$$
$$z = x + y - 80,000 \Rightarrow$$
$$x + y - z = 80,000 \quad (2)$$
$$0.03x + .025y + 0.015z = 4900 \Rightarrow$$
$$30x + 25y + 15z = 4,900,000 \quad (3)$$

Add equations (1) and (2) in order to eliminate z.
$$x + \quad y + z = 200,000$$
$$\underline{x + \quad y - z = \quad 80,000}$$
$$2x + 2y \quad = 280,000 \text{ or } x + y = 140,000 \ (4)$$

Multiply equation (2) by 15 and add the result to equation (3) in order to eliminate z.
$$15x + 15y - 15z = 1,200,000$$
$$\underline{30x + 25y + 15z = 4,900,000}$$
$$45x + 40y \quad = 6,100,000 \text{ or }$$
$$9x + 8y = 1,220,000 \quad (5)$$

Multiply equation (4) by -8 and add the result to equation (5).
$$-8x - 8y = -1,120,000$$
$$\underline{9x + 8y = \quad 1,220,000}$$
$$x \quad = \quad 100,000$$

Substitute this value into equation (4) in order to solve for y.
$100,000 + y = 140,000 \Rightarrow y = 40,000$

Substitute $x = 100,000$ and $y = 40,000$ into equation (1) in order to solve for z.
$$100,000 + 40,000 + z = 200,000$$
$$140,000 + z = 200,000 \Rightarrow z = 60,000$$

The amounts invested were $100,000 at 3% (real estate), $40,000 at 2.5% (money market), and $60,000 at 1.5% (CDs).

103. $p = 16 - \frac{5}{4}q$

(a) $p = 16 - \frac{5}{4}(0) = 16 - 0 = 16$
The price is $16.

(b) $p = 16 - \frac{5}{4}(4) = 16 - 5 = 16 - 5 = 11$
The price is $11.

(c) $p = 16 - \frac{5}{4}(8) = 16 - 10 = 6$
The price is $6.

105. See answer to Exercise 107.

107.

Copyright © 2013 Pearson Education, Inc.

109.
$$25x + 40y + 20z = 2200 \ (4)$$
$$4x + 2y + 3z = 280 \ (5)$$
$$3x + 2y + z = 180 \ (6)$$

Eliminate z by multiplying equation (6) by -3 and add to equation (5).
$$-9x - 6y - 3z = -540$$
$$\underline{4x + 2y + 3z = \ 280}$$
$$-5x - 4y = -260 \ (7)$$

Eliminate z by multiplying equation (6) by -20 and add to equation (4).
$$-60x - 40y - 20z = -3600$$
$$\underline{25x + 40y + 20z = \ 2200}$$
$$-35x = -1400 \Rightarrow x = 40$$

Substitute this value in equation (7) in order to solve for y.
$$-5(40) - 4y = -260 \Rightarrow -200 - 4y = -260 \Rightarrow$$
$$-4y = -60 \Rightarrow y = 15$$

Substitute $x = 40$ and $y = 15$ into equation (6) to solve for z.
$$3(40) + 2(15) + z = 180 \Rightarrow$$
$$120 + 30 + z = 180 \Rightarrow$$
$$150 + z = 180 \Rightarrow z = 30$$

Solution set: $\{(40, 15, 30)\}$

111. Let x = the number of pounds of Arabian Mocha Sanani; y = the number of pounds of Organic Shade-Grown Mexico; z = the number of pounds of Guatemala Antigua. Completing the table we have the following.

	Number of Pounds	Cost per Pound	Total Cost
Arabian Mocha	x	15.99	$15.99x$
Organic Mexico	y	12.99	$12.99y$
Guatemala Antigua	z	10.19	$10.19z$
Total	50	12.37	$50(12.37) = 618.50$

We have the following equations.
$$x + y + z = 50 \quad (1)$$
$$z = 2x \quad (2)$$
$$15.99x + 12.99y + 10.19z = 618.50 \quad (3)$$

Substitute $z = 2x$ into equations (1) and (3):
$$x + y + 2x = 50 \Rightarrow$$
$$3x + y = 50 \quad (4)$$
$$15.99x + 12.99y + 10.19(2x) = 618.50 \Rightarrow$$
$$36.37x + 12.99y = 618.50 \quad (5)$$

Solve equation (4) for y, then substitute that value into equation (5) and solve for x:
$$3x + y = 50 \Rightarrow y = -3x + 50$$

$$36.37x + 12.99(-3x + 50) = 618.50 \Rightarrow$$
$$36.37x - 38.97x + 649.50 = 618.50 \Rightarrow$$
$$-2.6x = -31 \Rightarrow x \approx 11.9231 \approx 11.92$$

Substitute $x = 11.9231$ into equation (2) to solve for z: $z \approx 2(11.9231) \approx 23.85$

Substitute the values for x and z into equation (1) to solve for y:
$$11.92 + y + 23.85 = 50 \Rightarrow y \approx 14.23$$

11.92 pounds of Arabian Mocha Sanani, 14.23 pounds of Organic Shade-Grown Mexico, and 23.85 pounds of Guatemala Antigua are needed. (Answers are approximations.)

Section 5.2 Matrix Solution of Linear Systems

1. $\begin{bmatrix} 1 & 4 \\ 4 & 7 \end{bmatrix}$; -4 times row 1 added to row 2

$$\begin{bmatrix} 1 & 4 \\ 4+(-4)(1) & 7+(-4)(4) \end{bmatrix} = \begin{bmatrix} 1 & 4 \\ 0 & -9 \end{bmatrix}$$

3. $\begin{bmatrix} 1 & 5 & 6 \\ -2 & 3 & -1 \\ 4 & 7 & 0 \end{bmatrix}$; 2 times row 1 added to row 2

$$\begin{bmatrix} 1 & 5 & 6 \\ -2+2(1) & 3+2(5) & -1+2(6) \\ 4 & 7 & 0 \end{bmatrix} = \begin{bmatrix} 1 & 5 & 6 \\ 0 & 13 & 11 \\ 4 & 7 & 0 \end{bmatrix}$$

5.
$$2x + 3y = 11$$
$$x + 2y = 8$$

The augmented matrix is $\left[\begin{array}{cc|c} 2 & 3 & 11 \\ 1 & 2 & 8 \end{array}\right]$. The size is 2×3.

7.
$$2x + y + z - 3 = 0$$
$$3x - 4y + 2z + 7 = 0$$
$$x + y + z - 2 = 0$$

Each equation of a linear system must have the constant term isolated on one side of the equal sign. Rewriting this system, we have
$$2x + y + z = 3$$
$$3x - 4y + 2z = -7$$
$$x + y + z = 2$$

The augmented matrix is $\left[\begin{array}{ccc|c} 2 & 1 & 1 & 3 \\ 3 & -4 & 2 & -7 \\ 1 & 1 & 1 & 2 \end{array}\right]$.

The size is 3×4.

Copyright © 2013 Pearson Education, Inc.

9. $\begin{bmatrix} 3 & 2 & 1 & | & 1 \\ 0 & 2 & 4 & | & 22 \\ -1 & -2 & 3 & | & 15 \end{bmatrix}$ is associated with the

system $\begin{aligned} 3x+2y+z &= 1 \\ 2y+4z &= 22 \\ -x-2y+3z &= 15. \end{aligned}$

11. $\begin{bmatrix} 1 & 0 & 0 & | & 2 \\ 0 & 1 & 0 & | & 3 \\ 0 & 0 & 1 & | & -2 \end{bmatrix}$ is associated with the system

$x = 2$
$y = 3$
$z = -2.$

13. $\begin{bmatrix} 1 & 1 & 0 & | & 3 \\ 0 & 2 & 1 & | & -4 \\ 1 & 0 & -1 & | & 5 \end{bmatrix}$ is associated with the

system $\begin{aligned} x+y &= 3 \\ 2y+z &= -4 \\ x-z &= 5. \end{aligned}$

15. $x+y=5$
$x-y=-1$

This system has the augmented matrix
$\begin{bmatrix} 1 & 1 & | & 5 \\ 1 & -1 & | & -1 \end{bmatrix}.$

$\begin{bmatrix} 1 & 1 & | & 5 \\ 0 & -2 & | & -6 \end{bmatrix} -1R1+R2 \Rightarrow$

$\begin{bmatrix} 1 & 1 & | & 5 \\ 0 & 1 & | & 3 \end{bmatrix} -\frac{1}{2}R2 \Rightarrow \begin{bmatrix} 1 & 0 & | & 2 \\ 0 & 1 & | & 3 \end{bmatrix} -1R2+R1$

Solution set: $\{(2,3)\}$

17. $3x+2y=-9$
$2x-5y=-6$

This system has the augmented matrix
$\begin{bmatrix} 3 & 2 & | & -9 \\ 2 & -5 & | & -6 \end{bmatrix}.$

$\begin{bmatrix} 1 & 7 & | & -3 \\ 2 & -5 & | & -6 \end{bmatrix} -1R2+R1 \Rightarrow$

$\begin{bmatrix} 1 & 7 & | & -3 \\ 0 & -19 & | & 0 \end{bmatrix} -2R1+R2 \Rightarrow$

$\begin{bmatrix} 1 & 7 & | & -3 \\ 0 & 1 & | & 0 \end{bmatrix} -\frac{1}{19}R2 \Rightarrow$

$\begin{bmatrix} 1 & 0 & | & -3 \\ 0 & 1 & | & 0 \end{bmatrix} -7R2+R1$

Solution set: $\{(-3,0)\}$

19. $6x-3y-4=0$
$3x+6y-7=0$

Rewrite the system as $\begin{aligned} 6x-3y &= 4 \\ 3x+6y &= 7 \end{aligned}$

The augmented matrix is $\begin{bmatrix} 6 & -3 & | & 4 \\ 3 & 6 & | & 7 \end{bmatrix}.$

$\begin{bmatrix} 1 & -\frac{1}{2} & | & \frac{2}{3} \\ 3 & 6 & | & 7 \end{bmatrix} \frac{1}{6}R1 \Rightarrow \begin{bmatrix} 1 & -\frac{1}{2} & | & \frac{2}{3} \\ 0 & \frac{15}{2} & | & 5 \end{bmatrix} -3R1+R2 \Rightarrow$

$\begin{bmatrix} 1 & -\frac{1}{2} & | & \frac{2}{3} \\ 0 & 1 & | & \frac{2}{3} \end{bmatrix} \frac{2}{15}R2 \Rightarrow \begin{bmatrix} 1 & 0 & | & 1 \\ 0 & 1 & | & \frac{2}{3} \end{bmatrix} \frac{1}{2}R2+R1$

Solution set: $\left\{\left(1, \frac{2}{3}\right)\right\}$

21. $2x-y=6$
$4x-2y=0$

This system has the augmented matrix
$\begin{bmatrix} 2 & -1 & | & 6 \\ 4 & -2 & | & 0 \end{bmatrix}.$

$\begin{bmatrix} 1 & -\frac{1}{2} & | & 3 \\ 4 & -2 & | & 0 \end{bmatrix} \frac{1}{2}R1 \Rightarrow$

$\begin{bmatrix} 1 & -\frac{1}{2} & | & 3 \\ 0 & 0 & | & -12 \end{bmatrix} -4R1+R2$

The second row of the augmented matrix corresponds to the equation $0x+0y=-12,$ which has no solution. Thus, the solution set is $\varnothing.$

23. $\frac{3}{8}x-\frac{1}{2}y=\frac{7}{8}$
$-6x+8y=-14$

This system has the augmented matrix
$\begin{bmatrix} \frac{3}{8} & -\frac{1}{2} & | & \frac{7}{8} \\ -6 & 8 & | & -14 \end{bmatrix}$

$\begin{bmatrix} 3 & -4 & | & 7 \\ -6 & 8 & | & -14 \end{bmatrix} 8R1$

$\begin{bmatrix} 1 & -\frac{4}{3} & | & \frac{7}{3} \\ -6 & 8 & | & -14 \end{bmatrix} \frac{1}{3}R1$

$\begin{bmatrix} 1 & -\frac{4}{3} & | & \frac{7}{3} \\ 0 & 0 & | & 0 \end{bmatrix} 6R1+R2$

It is impossible to go further. The equation that corresponds to the first row in the final matrix is $x-\frac{4}{3}y=\frac{7}{3} \Rightarrow x=\frac{4}{3}y+\frac{7}{3}=\frac{4y+7}{3}$

Solution set: $\left\{\frac{4y+7}{3}, y\right\}$

Copyright © 2013 Pearson Education, Inc.

25. $\begin{aligned} x+\ y-5z &= -18 \\ 3x-3y+\ z &=\ \ 6 \\ x+3y-2z &= -13 \end{aligned}$; This system has the augmented matrix $\begin{bmatrix} 1 & 1 & -5 & -18 \\ 3 & -3 & 1 & 6 \\ 1 & 3 & -2 & -13 \end{bmatrix}$.

$\begin{bmatrix} 1 & 1 & -5 & -18 \\ 0 & -6 & 16 & 60 \\ 1 & 3 & -2 & -13 \end{bmatrix} \begin{matrix} \\ -3R1+R2 \end{matrix} \Rightarrow \begin{bmatrix} 1 & 1 & -5 & -18 \\ 0 & -6 & 16 & 60 \\ 0 & 2 & 3 & 5 \end{bmatrix} \begin{matrix} \\ \\ -1R1+R3 \end{matrix} \Rightarrow \begin{bmatrix} 1 & 1 & -5 & -18 \\ 0 & 1 & -\frac{8}{3} & -10 \\ 0 & 2 & 3 & 5 \end{bmatrix} \begin{matrix} \\ -\frac{1}{6}R2 \\ \end{matrix}$

$\begin{bmatrix} 1 & 0 & -\frac{7}{3} & -8 \\ 0 & 1 & -\frac{8}{3} & -10 \\ 0 & 2 & 3 & 5 \end{bmatrix} \begin{matrix} -1R2+R1 \\ \\ \end{matrix} \Rightarrow \begin{bmatrix} 1 & 0 & -\frac{7}{3} & -8 \\ 0 & 1 & -\frac{8}{3} & -10 \\ 0 & 0 & \frac{25}{3} & 25 \end{bmatrix} \begin{matrix} \\ \\ -2R2+R3 \end{matrix} \Rightarrow \begin{bmatrix} 1 & 0 & -\frac{7}{3} & -8 \\ 0 & 1 & -\frac{8}{3} & -10 \\ 0 & 0 & 1 & 3 \end{bmatrix} \begin{matrix} \\ \\ \frac{3}{25}R3 \end{matrix}$

$\begin{bmatrix} 1 & 0 & 0 & -1 \\ 0 & 1 & -\frac{8}{3} & -10 \\ 0 & 0 & 1 & 3 \end{bmatrix} \begin{matrix} \frac{7}{3}R3+R1 \\ \\ \end{matrix} \Rightarrow \begin{bmatrix} 1 & 0 & 0 & -1 \\ 0 & 1 & 0 & -2 \\ 0 & 0 & 1 & 3 \end{bmatrix} \begin{matrix} \\ \frac{8}{3}R3+R2 \\ \end{matrix}$

Solution set: $\{(-1,-2,3)\}$

27. $\begin{aligned} x+\ y-\ z &=\ 6 \\ 2x-\ y+\ z &= -9 \\ x-2y+3z &=\ 1 \end{aligned}$; This system has the augmented matrix $\begin{bmatrix} 1 & 1 & -1 & 6 \\ 2 & -1 & 1 & -9 \\ 1 & -2 & 3 & 1 \end{bmatrix}$.

$\begin{bmatrix} 1 & 1 & -1 & 6 \\ 0 & -3 & 3 & -21 \\ 1 & -2 & 3 & 1 \end{bmatrix} \begin{matrix} \\ -2R1+R2 \\ \end{matrix} \Rightarrow \begin{bmatrix} 1 & 1 & -1 & 6 \\ 0 & -3 & 3 & -21 \\ 0 & -3 & 4 & -5 \end{bmatrix} \begin{matrix} \\ \\ -1R1+R3 \end{matrix} \Rightarrow \begin{bmatrix} 1 & 1 & -1 & 6 \\ 0 & 1 & -1 & 7 \\ 0 & -3 & 4 & -5 \end{bmatrix} \begin{matrix} \\ -\frac{1}{3}R2 \\ \end{matrix}$

$\begin{bmatrix} 1 & 0 & 0 & -1 \\ 0 & 1 & -1 & 7 \\ 0 & -3 & 4 & -5 \end{bmatrix} \begin{matrix} -1R2+R1 \\ \\ \end{matrix} \Rightarrow \begin{bmatrix} 1 & 0 & 0 & -1 \\ 0 & 1 & -1 & 7 \\ 0 & 0 & 1 & 16 \end{bmatrix} \begin{matrix} \\ \\ 3R2+R3 \end{matrix} \Rightarrow \begin{bmatrix} 1 & 0 & 0 & -1 \\ 0 & 1 & 0 & 23 \\ 0 & 0 & 1 & 16 \end{bmatrix} \begin{matrix} \\ R3+R2 \\ \end{matrix}$

Solution set: $\{(-1,23,16)\}$

29. $\begin{aligned} x-z &= -3 \\ y+z &= 9 \\ x+z &= 7 \end{aligned}$; This system has the augmented matrix $\begin{bmatrix} 1 & 0 & -1 & -3 \\ 0 & 1 & 1 & 9 \\ 1 & 0 & 1 & 7 \end{bmatrix}$.

$\begin{bmatrix} 1 & 0 & -1 & -3 \\ 0 & 1 & 1 & 9 \\ 0 & 0 & 2 & 10 \end{bmatrix} \begin{matrix} \\ \\ -1R1+R3 \end{matrix} \Rightarrow \begin{bmatrix} 1 & 0 & -1 & -3 \\ 0 & 1 & 1 & 9 \\ 0 & 0 & 1 & 5 \end{bmatrix} \begin{matrix} \\ \\ \frac{1}{2}R3 \end{matrix}$

$\begin{bmatrix} 1 & 0 & 0 & 2 \\ 0 & 1 & 1 & 9 \\ 0 & 0 & 1 & 5 \end{bmatrix} \begin{matrix} R3+R1 \\ \\ \end{matrix} \Rightarrow \begin{bmatrix} 1 & 0 & 0 & 2 \\ 0 & 1 & 0 & 4 \\ 0 & 0 & 1 & 5 \end{bmatrix} \begin{matrix} \\ -1R3+R2 \\ \end{matrix}$

Solution set: $\{(2,4,5)\}$

Copyright © 2013 Pearson Education, Inc.

31. $\begin{array}{l} y = -2x - 2z + 1 \\ x = -2y - z + 2 \\ z = x - y \end{array}$; Rewrite the system as $\begin{array}{l} 2x + y + 2z = 1 \\ x + 2y + z = 2 \\ x - y - z = 0 \end{array}$. The augmented matrix is $\begin{bmatrix} 2 & 1 & 2 & | & 1 \\ 1 & 2 & 1 & | & 2 \\ 1 & -1 & -1 & | & 0 \end{bmatrix}$.

$\begin{bmatrix} 1 & 2 & 1 & | & 2 \\ 2 & 1 & 2 & | & 1 \\ 1 & -1 & -1 & | & 0 \end{bmatrix} R1 \leftrightarrow R2 \Rightarrow \begin{bmatrix} 1 & 2 & 1 & | & 2 \\ 0 & -3 & 0 & | & -3 \\ 1 & -1 & -1 & | & 0 \end{bmatrix} -2R1 + R2 \Rightarrow \begin{bmatrix} 1 & 2 & 1 & | & 2 \\ 0 & -3 & 0 & | & -3 \\ 0 & -3 & -2 & | & -2 \end{bmatrix} -1R1 + R3$

$\begin{bmatrix} 1 & 2 & 1 & | & 2 \\ 0 & 1 & 0 & | & 1 \\ 0 & -3 & -2 & | & -2 \end{bmatrix} -\frac{1}{3}R2 \Rightarrow \begin{bmatrix} 1 & 0 & 1 & | & 0 \\ 0 & 1 & 0 & | & 1 \\ 0 & -3 & -2 & | & -2 \end{bmatrix} -2R2 + R1 \Rightarrow \begin{bmatrix} 1 & 0 & 1 & | & 0 \\ 0 & 1 & 0 & | & 1 \\ 0 & 0 & -2 & | & 1 \end{bmatrix} 3R2 + R3$

$\begin{bmatrix} 1 & 0 & 1 & | & 0 \\ 0 & 1 & 0 & | & 1 \\ 0 & 0 & 1 & | & -\frac{1}{2} \end{bmatrix} -\frac{1}{2}R3 \Rightarrow \begin{bmatrix} 1 & 0 & 0 & | & \frac{1}{2} \\ 0 & 1 & 0 & | & 1 \\ 0 & 0 & 1 & | & -\frac{1}{2} \end{bmatrix} -1R3 + R1$

Solution set: $\left\{\left(\frac{1}{2}, 1, -\frac{1}{2}\right)\right\}$

33. $\begin{array}{l} 2x - y + 3z = 0 \\ x + 2y - z = 5 \\ 2y + z = 1 \end{array}$; This system has the augmented matrix $\begin{bmatrix} 2 & -1 & 3 & | & 0 \\ 1 & 2 & -1 & | & 5 \\ 0 & 2 & 1 & | & 1 \end{bmatrix}$.

$\begin{bmatrix} 1 & 2 & -1 & | & 5 \\ 2 & -1 & 3 & | & 0 \\ 0 & 2 & 1 & | & 1 \end{bmatrix} R1 \leftrightarrow R2 \Rightarrow \begin{bmatrix} 1 & 2 & -1 & | & 5 \\ 0 & -5 & 5 & | & -10 \\ 0 & 2 & 1 & | & 1 \end{bmatrix} -2R1 + R2 \Rightarrow \begin{bmatrix} 1 & 2 & -1 & | & 5 \\ 0 & 1 & -1 & | & 2 \\ 0 & 2 & 1 & | & 1 \end{bmatrix} -\frac{1}{5}R2$

$\begin{bmatrix} 1 & 0 & 1 & | & 1 \\ 0 & 1 & -1 & | & 2 \\ 0 & 2 & 1 & | & 1 \end{bmatrix} -2R2 + R1 \Rightarrow \begin{bmatrix} 1 & 0 & 1 & | & 1 \\ 0 & 1 & -1 & | & 2 \\ 0 & 0 & 3 & | & -3 \end{bmatrix} -2R2 + R3 \Rightarrow \begin{bmatrix} 1 & 0 & 1 & | & 1 \\ 0 & 1 & -1 & | & 2 \\ 0 & 0 & 1 & | & -1 \end{bmatrix} \frac{1}{3}R3$

$\begin{bmatrix} 1 & 0 & 0 & | & 2 \\ 0 & 1 & -1 & | & 2 \\ 0 & 0 & 1 & | & -1 \end{bmatrix} -1R3 + R1 \Rightarrow \begin{bmatrix} 1 & 0 & 0 & | & 2 \\ 0 & 1 & 0 & | & 1 \\ 0 & 0 & 1 & | & -1 \end{bmatrix} R3 + R2$

Solution set: $\left\{(2, 1, -1)\right\}$

35. $\begin{array}{l} 3x + 5y - z + 2 = 0 \\ 4x - y + 2z - 1 = 0 \\ -6x - 10y + 2z = 0 \end{array}$; Rewrite the system as $\begin{array}{l} 3x + 5y - z = -2 \\ 4x - y + 2z = 1 \\ -6x - 10y + 2z = 0 \end{array}$.

The augmented matrix is $\begin{bmatrix} 3 & 5 & -1 & | & -2 \\ 4 & -1 & 2 & | & 1 \\ -6 & -10 & 2 & | & 0 \end{bmatrix}$.

$\begin{bmatrix} 1 & \frac{5}{3} & -\frac{1}{3} & | & -\frac{2}{3} \\ 4 & -1 & 2 & | & 1 \\ -6 & -10 & 2 & | & 0 \end{bmatrix} \frac{1}{3}R1 \Rightarrow \begin{bmatrix} 1 & \frac{5}{3} & -\frac{1}{3} & | & -\frac{2}{3} \\ 0 & -\frac{23}{3} & \frac{10}{3} & | & \frac{11}{3} \\ -6 & -10 & 2 & | & 0 \end{bmatrix} -4R1 + R2 \Rightarrow \begin{bmatrix} 1 & \frac{5}{3} & -\frac{1}{3} & | & -\frac{2}{3} \\ 0 & -\frac{23}{3} & \frac{10}{3} & | & \frac{11}{3} \\ 0 & 0 & 0 & | & -4 \end{bmatrix} 6R1 + R3$

The last row indicates that there is no solution. The solution set is $\varnothing$.

Copyright © 2013 Pearson Education, Inc.

37. $x - 8y + z = 4$
$3x - y + 2z = -1$

This system has the augmented matrix $\begin{bmatrix} 1 & -8 & 1 & | & 4 \\ 3 & -1 & 2 & | & -1 \end{bmatrix}$.

$\begin{bmatrix} 1 & -8 & 1 & | & 4 \\ 0 & 23 & -1 & | & -13 \end{bmatrix}{}_{-3R1+R2} \Rightarrow \begin{bmatrix} 1 & -8 & 1 & | & 4 \\ 0 & 1 & -\frac{1}{23} & | & -\frac{13}{23} \end{bmatrix}{}_{\frac{1}{23}R2} \Rightarrow \begin{bmatrix} 1 & 0 & \frac{15}{23} & | & -\frac{12}{23} \\ 0 & 1 & -\frac{1}{23} & | & -\frac{13}{23} \end{bmatrix}{}^{8R2+R1}$

The equations that correspond to the final matrix are $x + \frac{15}{23}z = -\frac{12}{23}$ and $y - \frac{1}{23}z = -\frac{13}{23}$.

This system has infinitely many solutions. We will express the solution set with z as the arbitrary variable.

Therefore, $x = -\frac{15}{23}z - \frac{12}{23} = -\frac{15z-12}{23}$ and $y = \frac{1}{23}z - \frac{13}{23} = \frac{z-13}{23}$.

Solution set: $\left\{ \left(-\frac{15z-12}{23}, \frac{z-13}{23}, z \right) \right\}$

39. $x - y + 2z + w = 4$
$y + z = 3$
$z - w = 2$
$x - y = 0$

This system has the augmented matrix $\begin{bmatrix} 1 & -1 & 2 & 1 & | & 4 \\ 0 & 1 & 1 & 0 & | & 3 \\ 0 & 0 & 1 & -1 & | & 2 \\ 1 & -1 & 0 & 0 & | & 0 \end{bmatrix}$

$\begin{bmatrix} 1 & -1 & 2 & 1 & | & 4 \\ 0 & 1 & 1 & 0 & | & 3 \\ 0 & 0 & 1 & -1 & | & 2 \\ 0 & 0 & -2 & -1 & | & -4 \end{bmatrix}{}_{-1R1+R4} \Rightarrow \begin{bmatrix} 1 & 0 & 3 & 1 & | & 7 \\ 0 & 1 & 1 & 0 & | & 3 \\ 0 & 0 & 1 & -1 & | & 2 \\ 0 & 0 & -2 & -1 & | & -4 \end{bmatrix}{}^{R2+R1} \Rightarrow$

$\begin{bmatrix} 1 & 0 & 0 & 4 & | & 1 \\ 0 & 1 & 1 & 0 & | & 3 \\ 0 & 0 & 1 & -1 & | & 2 \\ 0 & 0 & -2 & -1 & | & -4 \end{bmatrix}{}^{-3R3+R1} \Rightarrow \begin{bmatrix} 1 & 0 & 0 & 4 & | & 1 \\ 0 & 1 & 0 & 1 & | & 1 \\ 0 & 0 & 1 & -1 & | & 2 \\ 0 & 0 & -2 & -1 & | & -4 \end{bmatrix}{}^{-1R3+R2} \Rightarrow$

$\begin{bmatrix} 1 & 0 & 0 & 4 & | & 1 \\ 0 & 1 & 0 & 1 & | & 1 \\ 0 & 0 & 1 & -1 & | & 2 \\ 0 & 0 & 0 & -3 & | & 0 \end{bmatrix}{}_{2R3+R4} \Rightarrow \begin{bmatrix} 1 & 0 & 0 & 4 & | & 1 \\ 0 & 1 & 0 & 1 & | & 1 \\ 0 & 0 & 1 & -1 & | & 2 \\ 0 & 0 & 0 & 1 & | & 0 \end{bmatrix}{}_{-\frac{1}{3}R4} \Rightarrow$

$\begin{bmatrix} 1 & 0 & 0 & 0 & | & 1 \\ 0 & 1 & 0 & 1 & | & 1 \\ 0 & 0 & 1 & -1 & | & 2 \\ 0 & 0 & 0 & 1 & | & 0 \end{bmatrix}{}^{-4R4+R1} \Rightarrow \begin{bmatrix} 1 & 0 & 0 & 0 & | & 1 \\ 0 & 1 & 0 & 0 & | & 1 \\ 0 & 0 & 1 & -1 & | & 2 \\ 0 & 0 & 0 & 1 & | & 0 \end{bmatrix}{}^{-1R4+R2} \Rightarrow \begin{bmatrix} 1 & 0 & 0 & 0 & | & 1 \\ 0 & 1 & 0 & 0 & | & 1 \\ 0 & 0 & 1 & 0 & | & 2 \\ 0 & 0 & 0 & 1 & | & 0 \end{bmatrix}{}_{R4+R3}$

Solution set: $\{(1, 1, 2, 0)\}$

Copyright © 2013 Pearson Education, Inc.

41.
$$\begin{aligned} x+3y-2z-\ w&=9 \\ 4x+\ y+\ z+2w&=2 \\ -3x-\ y+\ z-\ w&=-5 \\ x-\ y-3z-2w&=2 \end{aligned}$$
; This system has the augmented matrix
$\begin{bmatrix} 1 & 3 & -2 & -1 & 9 \\ 4 & 1 & 1 & 2 & 2 \\ -3 & -1 & 1 & -1 & -5 \\ 1 & -1 & -3 & -2 & 2 \end{bmatrix}$.

$\begin{bmatrix} 1 & 3 & -2 & -1 & 9 \\ 0 & -11 & 9 & 6 & -34 \\ -3 & -1 & 1 & -1 & -5 \\ 1 & -1 & -3 & -2 & 2 \end{bmatrix}$ $-4R1+R2$ $\Rightarrow$ $\begin{bmatrix} 1 & 3 & -2 & -1 & 9 \\ 0 & -11 & 9 & 6 & -34 \\ 0 & 8 & -5 & -4 & 22 \\ 1 & -1 & -3 & -2 & 2 \end{bmatrix}$ $3R1+R3$ $\Rightarrow$

$\begin{bmatrix} 1 & 3 & -2 & -1 & 9 \\ 0 & -11 & 9 & 6 & -34 \\ 0 & 8 & -5 & -4 & 22 \\ 0 & -4 & -1 & -1 & -7 \end{bmatrix}$ $-1R1+R4$ $\Rightarrow$ $\begin{bmatrix} 1 & 3 & -2 & -1 & 9 \\ 0 & 1 & -\frac{9}{11} & -\frac{6}{11} & \frac{34}{11} \\ 0 & 8 & -5 & -4 & 22 \\ 0 & -4 & -1 & -1 & -7 \end{bmatrix}$ $-\frac{1}{11}R2$ $\Rightarrow$

$\begin{bmatrix} 1 & 0 & \frac{5}{11} & \frac{7}{11} & -\frac{3}{11} \\ 0 & 1 & -\frac{9}{11} & -\frac{6}{11} & \frac{34}{11} \\ 0 & 8 & -5 & -4 & 22 \\ 0 & -4 & -1 & -1 & -7 \end{bmatrix}$ $-3R2+R1$ $\Rightarrow$ $\begin{bmatrix} 1 & 0 & \frac{5}{11} & \frac{7}{11} & -\frac{3}{11} \\ 0 & 1 & -\frac{9}{11} & -\frac{6}{11} & \frac{34}{11} \\ 0 & 0 & \frac{17}{11} & \frac{4}{11} & -\frac{30}{11} \\ 0 & -4 & -1 & -1 & -7 \end{bmatrix}$ $-8R2+R3$ $\Rightarrow$

$\begin{bmatrix} 1 & 0 & \frac{5}{11} & \frac{7}{11} & -\frac{3}{11} \\ 0 & 1 & -\frac{9}{11} & -\frac{6}{11} & \frac{34}{11} \\ 0 & 0 & \frac{17}{11} & \frac{4}{11} & -\frac{30}{11} \\ 0 & 0 & -\frac{47}{11} & -\frac{35}{11} & \frac{59}{11} \end{bmatrix}$ $4R2+R4$ $\Rightarrow$ $\begin{bmatrix} 1 & 0 & \frac{5}{11} & \frac{7}{11} & -\frac{3}{11} \\ 0 & 1 & -\frac{9}{11} & -\frac{6}{11} & \frac{34}{11} \\ 0 & 0 & 1 & \frac{4}{17} & -\frac{30}{17} \\ 0 & 0 & -\frac{47}{11} & -\frac{35}{11} & \frac{59}{11} \end{bmatrix}$ $\frac{11}{17}R3$ $\Rightarrow$

$\begin{bmatrix} 1 & 0 & 0 & \frac{9}{17} & \frac{9}{17} \\ 0 & 1 & -\frac{9}{11} & -\frac{6}{11} & \frac{34}{11} \\ 0 & 0 & 1 & \frac{4}{17} & -\frac{30}{17} \\ 0 & 0 & -\frac{47}{11} & -\frac{35}{11} & \frac{59}{11} \end{bmatrix}$ $-\frac{5}{11}R3+R1$ $\Rightarrow$ $\begin{bmatrix} 1 & 0 & 0 & \frac{9}{17} & \frac{9}{17} \\ 0 & 1 & 0 & -\frac{6}{17} & \frac{28}{71} \\ 0 & 0 & 1 & \frac{4}{17} & -\frac{30}{17} \\ 0 & 0 & -\frac{47}{11} & -\frac{35}{11} & \frac{59}{11} \end{bmatrix}$ $\frac{9}{11}R3+R2$ $\Rightarrow$

$\begin{bmatrix} 1 & 0 & 0 & \frac{9}{17} & \frac{9}{17} \\ 0 & 1 & 0 & -\frac{6}{17} & \frac{28}{17} \\ 0 & 0 & 1 & \frac{4}{17} & -\frac{30}{17} \\ 0 & 0 & 0 & -\frac{37}{17} & -\frac{37}{17} \end{bmatrix}$ $\frac{47}{11}R3+R4$ $\Rightarrow$ $\begin{bmatrix} 1 & 0 & 0 & \frac{9}{17} & \frac{9}{17} \\ 0 & 1 & 0 & -\frac{6}{17} & \frac{28}{17} \\ 0 & 0 & 1 & \frac{4}{17} & -\frac{30}{17} \\ 0 & 0 & 0 & 1 & 1 \end{bmatrix}$ $-\frac{17}{37}R4$ $\Rightarrow$

$\begin{bmatrix} 1 & 0 & 0 & 0 & 0 \\ 0 & 1 & 0 & -\frac{6}{17} & \frac{28}{17} \\ 0 & 0 & 1 & \frac{4}{17} & -\frac{30}{17} \\ 0 & 0 & 0 & 1 & 1 \end{bmatrix}$ $-\frac{9}{17}R4+R1$ $\Rightarrow$ $\begin{bmatrix} 1 & 0 & 0 & 0 & 0 \\ 0 & 1 & 0 & 0 & 2 \\ 0 & 0 & 1 & \frac{4}{17} & -\frac{30}{17} \\ 0 & 0 & 0 & 1 & 1 \end{bmatrix}$ $\frac{6}{17}R4+R2$ $\Rightarrow$

$\begin{bmatrix} 1 & 0 & 0 & 0 & 0 \\ 0 & 1 & 0 & 0 & 2 \\ 0 & 0 & 1 & 0 & -2 \\ 0 & 0 & 0 & 1 & 1 \end{bmatrix}$ $-\frac{4}{17}R4+R3$

Solution set: $\{(0,2,-2,1)\}$

Copyright © 2013 Pearson Education, Inc.

43. $0.3x + 2.7y - \sqrt{2}z = 3$
$\sqrt{7}x - 20y + 12z = -2$
$4x + \sqrt{3}y - 1.2z = \frac{3}{4}$

This system has the augmented matrix

$$\begin{bmatrix} 0.3 & 2.7 & -\sqrt{2} & | & 3 \\ \sqrt{7} & -20 & 12 & | & -2 \\ 4 & \sqrt{3} & -1.2 & | & \frac{3}{4} \end{bmatrix}.$$

```
rref([A])
[[1 0 0 .571337…
 [0 1 0 7.04055…
 [0 0 1 11.4416…
```

Using a graphing calculator. we obtain the approximate solution set.
Solution set: $\{(0.571, 7.041, 11.442)\}$.

45. $2x + 3y = 5 \Rightarrow 3y = 5 - 2x \Rightarrow y = \frac{5-2x}{3}$
$-3x + 5y = 22 \Rightarrow 5y = 3x + 22 \Rightarrow y = \frac{3x+22}{5}$
$2x + y = -1 \Rightarrow y = -2x - 1$

```
Plot1 Plot2 Plot3        WINDOW
\Y1B(5-2X)/3             Xmin=-3
\Y2B(3X+22)/5            Xmax=-1
\Y3B-2X-1               Xscl=1
\Y4=                    Ymin=2
\Y5=                    Ymax=4
\Y6=                    Yscl=1
\Y7=                    Xres=1
```

There are no solutions since the three lines do not intersect at one point.

47. $\dfrac{1}{(x-1)(x+1)} = \dfrac{A}{x-1} + \dfrac{B}{x+1}$

If we add the rational expression on the right, we get the following.

$$\frac{1}{(x-1)(x+1)} = \frac{A(x+1)}{(x-1)(x+1)} + \frac{B(x-1)}{(x-1)(x+1)}$$

$$\frac{1}{(x-1)(x+1)} = \frac{A(x+1) + B(x-1)}{(x-1)(x+1)}$$

Since the denominators are equal, the numerators must be equal. Thus we have the equation $1 = A(x+1) + B(x-1)$.

$1 = A(x+1) + B(x-1) \Rightarrow$
$1 = Ax + A + Bx - B$
$1 = (A - B) + (Ax + Bx) \Rightarrow$
$1 = (A - B) + (A + B)x$

Equating the coefficients of like powers of x gives the following system of equations.
$A + B = 0$
$A - B = 1$

Solve this system by the Gauss-Jordan method.

$$\begin{bmatrix} 1 & 1 & | & 0 \\ 1 & -1 & | & 1 \end{bmatrix} \Rightarrow \begin{bmatrix} 1 & 1 & | & 0 \\ 0 & -2 & | & 1 \end{bmatrix} -R1 + R2 \Rightarrow$$

$$\begin{bmatrix} 1 & 1 & | & 0 \\ 0 & 1 & | & -\frac{1}{2} \end{bmatrix} -\frac{1}{2}R2 \Rightarrow$$

$$\begin{bmatrix} 1 & 0 & | & \frac{1}{2} \\ 0 & 1 & | & -\frac{1}{2} \end{bmatrix} -1R2 + R1$$

Thus, $A = \frac{1}{2}$ and $B = -\frac{1}{2}$.

49. $\dfrac{x}{(x-a)(x+a)} = \dfrac{A}{x-a} + \dfrac{B}{x+a}$

If we add the rational expression on the right, we get the following.

$$\frac{x}{(x-a)(x+a)} = \frac{A(x+a)}{(x-a)(x+a)} + \frac{B(x-a)}{(x-a)(x+a)}$$

$$\frac{x}{(x-a)(x+a)} = \frac{A(x+a) + B(x-a)}{(x-a)(x+a)}$$

Since the denominators are equal, the numerators must be equal. Thus we have the equation $x = A(x+a) + B(x-a)$.

$x = A(x+a) + B(x-a) \Rightarrow$
$x = Ax + Aa + Bx - Ba \Rightarrow$
$x = (A + B)x + (A - B)a$

Equating the coefficients of like powers of x gives the following system of equations.
$A + B = 1$
$A - B = 0$

Solve this system by the Gauss-Jordan method.

$$\begin{bmatrix} 1 & 1 & | & 1 \\ 1 & -1 & | & 0 \end{bmatrix} \Rightarrow \begin{bmatrix} 1 & 1 & | & 1 \\ 0 & -2 & | & -1 \end{bmatrix} -R1 + R2 \Rightarrow$$

$$\begin{bmatrix} 1 & 1 & | & 1 \\ 0 & 1 & | & \frac{1}{2} \end{bmatrix} -\frac{1}{2}R2 \Rightarrow \begin{bmatrix} 1 & 0 & | & \frac{1}{2} \\ 0 & 1 & | & \frac{1}{2} \end{bmatrix} -1R2 + R1$$

Thus, $A = \frac{1}{2}$ and $B = \frac{1}{2}$.

Copyright © 2013 Pearson Education, Inc.

51. Let x = the daily wage for a day laborer
y = the daily wage for a concrete finisher
From the information, we can write the system
$$7x + 2y = 1384$$
$$x + 5y = 952$$

Solve this system by the Gauss-Jordan method:

$$\begin{bmatrix} 7 & 2 & | & 1384 \\ 1 & 5 & | & 952 \end{bmatrix} \Rightarrow$$

$$\begin{bmatrix} 1 & 5 & | & 952 \\ 7 & 2 & | & 1384 \end{bmatrix} \text{R1} \leftrightarrow \text{R2} \Rightarrow$$

$$\begin{bmatrix} 1 & 5 & | & 952 \\ 0 & -33 & | & -5280 \end{bmatrix} {-7}\text{R1}+\text{R2} \Rightarrow$$

$$\begin{bmatrix} 1 & 5 & | & 952 \\ 0 & 1 & | & 160 \end{bmatrix} -\frac{1}{33}\text{R2} \Rightarrow$$

$$\begin{bmatrix} 1 & 0 & | & 152 \\ 0 & 1 & | & 160 \end{bmatrix} {-5}\text{R2}+\text{R1}$$

From the final matrix, $x = 152$ and $y = 160$, so the day laborers earn \$152 per day and the concreter finishers earn \$160 per day.

53. Let x = the first number; y = the second number; z = the third number.
From the information, we can write the system
$$x + y + z = 20$$
$$x = 3(y - z) \Rightarrow x = 3y - 3z \Rightarrow x - 3y + 3z = 0$$
$$y = 2 + 2z \Rightarrow \qquad\qquad y - 2z = 2$$

Solve this system by the Gauss-Jordan method:

$$\begin{bmatrix} 1 & 1 & 1 & | & 20 \\ 1 & -3 & 3 & | & 0 \\ 0 & 1 & -2 & | & 2 \end{bmatrix} \Rightarrow$$

$$\begin{bmatrix} 1 & 1 & 1 & | & 20 \\ 0 & 1 & -\frac{1}{2} & | & 5 \\ 0 & 1 & -2 & | & 2 \end{bmatrix} \frac{1}{4}(\text{R1}-\text{R2}) \Rightarrow$$

$$\begin{bmatrix} 1 & 1 & 1 & | & 20 \\ 0 & 1 & -\frac{1}{2} & | & 5 \\ 0 & 0 & 1 & | & 2 \end{bmatrix} \frac{2}{3}(\text{R2}-\text{R3}) \Rightarrow$$

$$\begin{bmatrix} 1 & 1 & 1 & | & 20 \\ 0 & 1 & 0 & | & 6 \\ 0 & 0 & 1 & | & 2 \end{bmatrix} \text{R2}+\frac{1}{2}\text{R3} \Rightarrow$$

$$\begin{bmatrix} 1 & 0 & 1 & | & 14 \\ 0 & 1 & 0 & | & 6 \\ 0 & 0 & 1 & | & 2 \end{bmatrix} \text{R1}-\text{R2} \Rightarrow$$

$$\begin{bmatrix} 1 & 0 & 0 & | & 12 \\ 0 & 1 & 0 & | & 6 \\ 0 & 0 & 1 & | & 2 \end{bmatrix} \text{R1}-\text{R3}$$

From the final matrix, $x = 12$, $y = 6$, and $z = 2$. Thus, the numbers are 12, 6, and 2.

55. Let x = number of cubic centimeters of the 2% solution; y = number of cubic centimeters of the 7% solution.
From the information we can write the system
$$x + y = 40 \qquad\Rightarrow\qquad x + y = 40$$
$$0.02x + 0.07y = 0.032(40) \Rightarrow 0.02x + 0.07y = 1.28$$

Solve this system by the Gauss-Jordan method.

$$\begin{bmatrix} 1 & 1 & | & 40 \\ 0.02 & 0.07 & | & 1.28 \end{bmatrix} \Rightarrow$$

$$\begin{bmatrix} 1 & 1 & | & 40 \\ 0 & 0.05 & | & 0.48 \end{bmatrix} {-0.02}\text{R1}+\text{R2}$$

$$\begin{bmatrix} 1 & 1 & | & 40 \\ 0 & 1 & | & 9.6 \end{bmatrix} \frac{1}{0.05}\text{R2} \Rightarrow \begin{bmatrix} 1 & 0 & | & 30.4 \\ 0 & 1 & | & 9.6 \end{bmatrix} {-1}\text{R2}+\text{R1}$$

From the final matrix, $x = 30.4$ and $y = 9.6$, so the chemist should mix 30.4 cm^3 of the 2% solution with 9.6 cm^3 of the 7% solution.

57. Answers will vary.
If the condition that correlates to the second row in the augmented matrix,

$$\begin{bmatrix} 1 & 1 & 1 & | & 25{,}000 \\ -1 & 2 & 0 & | & 4{,}000 \\ 4 & 6 & 5 & | & 122{,}000 \end{bmatrix} \text{ was dropped, then that}$$

row would be dropped. We would have

$$\begin{bmatrix} 1 & 1 & 1 & | & 25{,}000 \\ 4 & 6 & 5 & | & 122{,}000 \end{bmatrix}.$$

Solve by the Gauss-Jordan method.

$$\begin{bmatrix} 1 & 1 & 1 & | & 25{,}000 \\ 4 & 6 & 5 & | & 122{,}000 \end{bmatrix} \Rightarrow$$

$$\begin{bmatrix} 1 & 1 & 1 & | & 25{,}000 \\ 0 & 2 & 1 & | & 22{,}000 \end{bmatrix} {-4}\text{R1}+\text{R2} \Rightarrow$$

$$\begin{bmatrix} 1 & 1 & 1 & | & 25{,}000 \\ 0 & 1 & \frac{1}{2} & | & 11{,}000 \end{bmatrix} \frac{1}{2}\text{R2} \Rightarrow$$

$$\begin{bmatrix} 1 & 0 & \frac{1}{2} & | & 14{,}000 \\ 0 & 1 & \frac{1}{2} & | & 11{,}000 \end{bmatrix} {-1}\text{R2}+\text{R1}$$

We can see it is not possible to find a unique solution.

Copyright © 2013 Pearson Education, Inc.

59. Let x = number of grams of food A;
y = number of grams of food B;
z = number of grams of food C.
Completing the table we have the following.

Food Group	A	B	C	Total
Grams/Meal	x	y	z	400

From the information given in the exercise, we can write the system.

$$x + y + z = 400$$
$$x = \tfrac{1}{3}y \Rightarrow 3x - y = 0$$
$$x + z = 2y \Rightarrow x - 2y + z = 0$$

This system has the augmented matrix

$$\begin{bmatrix} 1 & 1 & 1 & | & 400 \\ 3 & -1 & 0 & | & 0 \\ 1 & -2 & 1 & | & 0 \end{bmatrix}$$

Solve by the Gauss-Jordan method.

$$\begin{bmatrix} 1 & 1 & 1 & | & 400 \\ 0 & -4 & -3 & | & -1200 \\ 1 & -2 & 1 & | & 0 \end{bmatrix} \begin{matrix} \\ -3R1+R2 \Rightarrow \\ \\ \end{matrix}$$

$$\begin{bmatrix} 1 & 1 & 1 & | & 400 \\ 0 & -4 & -3 & | & -1200 \\ 0 & -3 & 0 & | & -400 \end{bmatrix} \begin{matrix} \\ \Rightarrow \\ -1R1+R3 \end{matrix}$$

$$\begin{bmatrix} 1 & 1 & 1 & | & 400 \\ 0 & 1 & \tfrac{3}{4} & | & 300 \\ 0 & -3 & 0 & | & -400 \end{bmatrix} \begin{matrix} \\ -\tfrac{1}{4}R2 \quad \Rightarrow \\ \\ \end{matrix}$$

$$\begin{bmatrix} 1 & 0 & \tfrac{1}{4} & | & 100 \\ 0 & 1 & \tfrac{3}{4} & | & 300 \\ 0 & -3 & 0 & | & -400 \end{bmatrix} \begin{matrix} -1R2+R1 \\ \Rightarrow \\ \end{matrix}$$

$$\begin{bmatrix} 1 & 0 & \tfrac{1}{4} & | & 100 \\ 0 & 1 & \tfrac{3}{4} & | & 300 \\ 0 & 0 & \tfrac{9}{4} & | & 500 \end{bmatrix} \begin{matrix} \\ \Rightarrow \\ 3R2+R3 \end{matrix}$$

$$\begin{bmatrix} 1 & 0 & \tfrac{1}{4} & | & 100 \\ 0 & 1 & \tfrac{3}{4} & | & 300 \\ 0 & 0 & 1 & | & \tfrac{2000}{9} \end{bmatrix} \begin{matrix} \\ \\ \tfrac{4}{9}R3 \end{matrix}$$

$$\begin{bmatrix} 1 & 0 & 0 & | & \tfrac{400}{9} \\ 0 & 1 & \tfrac{3}{4} & | & 300 \\ 0 & 0 & 1 & | & \tfrac{2000}{9} \end{bmatrix} \begin{matrix} -\tfrac{1}{4}R3+R1 \\ \Rightarrow \\ \end{matrix}$$

$$\begin{bmatrix} 1 & 0 & 0 & | & \tfrac{400}{9} \\ 0 & 1 & 0 & | & \tfrac{400}{3} \\ 0 & 0 & 1 & | & \tfrac{2000}{9} \end{bmatrix} \begin{matrix} \\ -\tfrac{3}{4}R3+R2 \\ \end{matrix}$$

From the final matrix, we have
$x = \tfrac{400}{9} \approx 44.4,\ y = \tfrac{400}{3} \approx 133.3,$ and

$z = \tfrac{2000}{9} \approx 222.2.$

The diet should include 44.4 g of food A, 133.3 g of food B, and 222.2 g of food C.

61. (a) For the 65 or older group:
With $x = 0$ representing 2010 and $x = 40$ representing 2050, we have information that correlates to the points (0, 0.130) and (40, 0.202).

$$m = \frac{0.202 - 0.130}{40 - 0} = \frac{0.072}{40} = 0.0018$$

Since we have the point (0, 0.130), the equation in slope-intercept form is
$y = 0.0018x + 0.13$

For the 25–34 group:
With $x = 0$ representing 2010 and $x = 40$ representing 2050, we have information that correlates to the points (0, 0.135) and (40, 0.128).

$$m = \frac{0.128 - 0.135}{40 - 0} = \frac{-0.007}{40} = -0.000175$$

(three significant figures)
Since we have the point (0, 0.135), the equation in slope-intercept form is
$y = -0.000175x + 0.135$

(b) We need to solve the following system.

$$\begin{cases} y = 0.0018x + 0.13 \\ y = -0.000175x + 0.135 \end{cases} \Rightarrow$$

$$\begin{cases} -0.0018x + y = 0.13 \\ 0.000175x + y = 0.135 \end{cases} \Rightarrow$$

$$-1800x + 1,000,000y = 130,000$$
$$175x + 1,000,000y = 135,000$$

This system has the augmented matrix

$$\begin{bmatrix} -1800 & 1,000,000 & | & 130,000 \\ 175 & 1,000,000 & | & 135,000 \end{bmatrix}$$

Solve by the Gauss-Jordan method.

$$\begin{bmatrix} 1 & -\tfrac{1,000,000}{1800} & | & -\tfrac{130,000}{1800} \\ 1 & \tfrac{1,000,000}{175} & | & \tfrac{135,000}{175} \end{bmatrix} \begin{matrix} -\tfrac{1}{1800}R1 \\ \tfrac{1}{175}R2 \end{matrix}$$

Reducing the fractions, we have

$$\begin{bmatrix} 1 & -\tfrac{5000}{9} & | & -\tfrac{650}{9} \\ 1 & \tfrac{40000}{7} & | & \tfrac{5400}{7} \end{bmatrix} \Rightarrow$$

$$\begin{bmatrix} 1 & -\tfrac{5000}{9} & | & -\tfrac{650}{9} \\ 0 & -\tfrac{395,000}{63} & | & -\tfrac{53,150}{63} \end{bmatrix} \begin{matrix} \\ R1-R2 \end{matrix} \Rightarrow$$

(continued)

Copyright © 2013 Pearson Education, Inc.

(*continued*)

$$\begin{bmatrix} 1 & -\frac{5000}{9} & \Big| & -\frac{650}{9} \\ 0 & 1 & \Big| & \frac{1063}{7900} \end{bmatrix} -\frac{63}{395,000}\text{R2} \Rightarrow$$

$$\begin{bmatrix} 1 & 0 & \Big| & \frac{200}{79} \\ 0 & 1 & \Big| & \frac{1063}{7900} \end{bmatrix} \frac{5000}{9}\text{R2}+\text{R1}$$

The solution set is

$$\left\{\left(\tfrac{200}{79},\tfrac{1063}{7900}\right)\right\}\approx\left\{(2.5316,0.1346)\right\}$$

$x=\frac{200}{79}\approx 2.5616$ represents the year

2012 and $y=\frac{1063}{7900}\approx 0.1346\approx 13.46\%$.

The two groups will have the same percentage of the population during 2012, about 13.5%.

(c) Answers will vary.

63. (a) A height of 6'11" is 83".
If $W=7.46H-374$, then
$W=7.46(83)-374=245.18$.
Using the first equation, the predicted weight is approximately 245 pounds.
If $W=7.93H-405$, then
$W=7.93(83)-405=253.19$.
Using the second equation, the predicted weight is approximately 253 pounds.

(b) For the first model $W=7.46H-374$, a 1-inch increase in height results in a 7.46-pound increase in weight.
For the second model $W=7.93H-405$, a 1-inch increase in height results in a 7.93-pound increase in weight. In each case, the change is given by the slope of the line that is the graph of the given equation.

(c) $W-7.46H=-374$
$W-7.93H=-405$
This system has the augmented matrix
$$\begin{bmatrix} 1 & -7.46 & \Big| & -374 \\ 1 & -7.93 & \Big| & -405 \end{bmatrix}.$$
Solve this system by the Gauss-Jordan method.
$$\begin{bmatrix} 1 & -7.46 & \Big| & -374 \\ 0 & -0.47 & \Big| & -31 \end{bmatrix} -1\text{R1}+\text{R2} \Rightarrow$$
$$\begin{bmatrix} 1 & -7.46 & \Big| & -374 \\ 0 & 1 & \Big| & 65.957 \end{bmatrix} -\frac{1}{0.47}\text{R2} \Rightarrow$$
$$\begin{bmatrix} 1 & 0 & \Big| & 118.043 \\ 0 & 1 & \Big| & 65.957 \end{bmatrix} 7.46\text{R2}+\text{R1}$$

From the last matrix, we have $W\approx 118$ and $H\approx 66$. The two models agree at a height of 66 inches and a weight of 118 pounds.

65. $F=a+bA+cP+dW$
Substituting the values, we have the following system of equations.
$a+871b+11.5c+3d=239$
$a+847b+12.2c+2d=234$
$a+685b+10.6c+5d=192$
$a+969b+14.2c+1d=343$.

67. Using these values,
$F=-714.457+0.34756A+48.6585P$
$\qquad\qquad\qquad\qquad +30.71951W$

69. $T(n)=\frac{2}{3}n^3+\frac{3}{2}n^2-\frac{7}{6}n$
Continuing in this manner we have the following.

n	T
3	28
6	191
10	805
29	17,487
100	681,550
200	5,393,100
400	42,906,200
1000	668,165,500
5000	8.3×10^{10}
10,000	6.7×10^{11}
100,000	6.7×10^{14}

71. If the number of variables doubles, the number of operations increases by a factor of 8. 100 variables require 681,550 operations and 200 variables requires 5,393,100 operations.
The ratio $\frac{5,393,100}{681,550}\approx 7.91\approx 8$.

Section 5.3 Determinant Solution of Linear Systems

1. $\begin{vmatrix} -5 & 9 \\ 4 & -1 \end{vmatrix}=-5(-1)-4\cdot 9=5-36=-31$

```
[A]
        [[-5 9 ]
         [4  -1]]
det([A])
              -31
```

Copyright © 2013 Pearson Education, Inc.

3. $\begin{vmatrix} -1 & -2 \\ 5 & 3 \end{vmatrix} = -1 \cdot 3 - 5(-2) = -3 - (-10) = 7$

```
[A]
        [[-1 -2]
         [5  3 ]]
det([A])
                7
```

5. $\begin{vmatrix} 9 & 3 \\ -3 & -1 \end{vmatrix} = 9(-1) - (-3) \cdot 3 = -9 - (-9) = 0$

```
[A]
        [[9  3 ]
         [-3 -1]]
det([A])
                0
```

7. $\begin{vmatrix} 3 & 4 \\ 5 & -2 \end{vmatrix} = 3(-2) - 5 \cdot 4 = -6 - 20 = -26$

```
[A]
        [[3  4 ]
         [5 -2]]
det([A])
              -26
```

9. $\begin{vmatrix} -7 & 0 \\ 3 & 0 \end{vmatrix} = -7 \cdot 0 - 3 \cdot 0 = 0 - 0 = 0$

```
[A]
        [[-7 0]
         [3  0]]
det([A])
               0
```

11. $\begin{vmatrix} -2 & 0 & 1 \\ 1 & 2 & 0 \\ 4 & 2 & 1 \end{vmatrix}$

To find the cofactor of 1, we have
$i = 2, j = 1,$

$M_{21} = \begin{vmatrix} 0 & 1 \\ 2 & 1 \end{vmatrix} = 0 \cdot 1 - 2 \cdot 1 = 0 - 2 = -2$

Thus, the cofactor is

$(-1)^{2+1}(-2) = (-1)^3(-2) = (-1)(-2) = 2.$

To find the cofactor of 2, we have
$i = 2, j = 2,$

$M_{22} = \begin{vmatrix} -2 & 1 \\ 4 & 1 \end{vmatrix} = -2 \cdot 1 - 4 \cdot 1 = -2 - 4 = -6$

Thus, the cofactor is

$(-1)^{2+2}(-6) = (-1)^4(-6) = 1(-6) = -6.$

To find the cofactor of 0, we have
$i = 2, j = 3,$

$M_{23} = \begin{vmatrix} -2 & 0 \\ 4 & 2 \end{vmatrix} = -2 \cdot 2 - 4 \cdot 0 = -4 - 0 = -4$

Thus, the cofactor is

$(-1)^{2+3}(-4) = (-1)^5(-4) = (-1)(-4) = 4.$

13. $\begin{vmatrix} 1 & 2 & -1 \\ 2 & 3 & -2 \\ -1 & 4 & 1 \end{vmatrix}$

To find the cofactor of 2, we have $i = 2, j = 1,$

$M_{21} = \begin{vmatrix} 2 & -1 \\ 4 & 1 \end{vmatrix} = 2 \cdot 1 - 4(-1) = 2 - (-4) = 6$

Thus, the cofactor is

$(-1)^{2+1}(6) = (-1)^3(6) = (-1)(6) = -6.$

To find the cofactor of 3, we have $i = 2, j = 2,$

$M_{22} = \begin{vmatrix} 1 & -1 \\ -1 & 1 \end{vmatrix} = 1 \cdot 1 - (-1)(-1) = 1 - 1 = 0$

Thus, the cofactor is

$(-1)^{2+2}(0) = (-1)^4(0) = 1(0) = 0.$

To find the cofactor of $-2,$ we have
$i = 2, j = 3,$

$M_{23} = \begin{vmatrix} 1 & 2 \\ -1 & 4 \end{vmatrix} = 1 \cdot 4 - (-1) \cdot 2 = 4 - (-2) = 6$

Thus, the cofactor is

$(-1)^{2+3}(6) = (-1)^5(6) = (-1)(6) = -6.$

For Exercises 15–29, an answer can be arrived at by expanding on any row or column. In the solutions, we will expand on a row or column that allows a minimum number of calculations. Any row or column containing zero will reduce the number of calculations. You may want to refer to the Determinant Theorems on page 508 of your text.

15. $\begin{vmatrix} 4 & -7 & 8 \\ 2 & 1 & 3 \\ -6 & 3 & 0 \end{vmatrix}$

If we expand by the third row, we will need to find $a_{31} \cdot A_{31} + a_{32} \cdot A_{32} + a_{33} \cdot A_{33}.$ However, we do not need to calculate $A_{33},$ since $a_{33} = 0.$

(continued on next page)

Copyright © 2013 Pearson Education, Inc.

(*continued*)

$$A_{31} = (-1)^{3+1} \begin{vmatrix} -7 & 8 \\ 1 & 3 \end{vmatrix} = (-1)^4 (-7 \cdot 3 - 1 \cdot 8)$$
$$= 1(-21 - 8) = 1(-29) = -29$$

$$A_{32} = (-1)^{3+2} \begin{vmatrix} 4 & 8 \\ 2 & 3 \end{vmatrix} = (-1)^5 (4 \cdot 3 - 2 \cdot 8)$$
$$= -1(12 - 16) = -1(-4) = 4$$

$$a_{31} \cdot A_{31} + a_{32} \cdot A_{32} + a_{33} \cdot A_{33}$$
$$= a_{31} \cdot A_{31} + a_{32} \cdot A_{32} + 0 \cdot A_{33}$$
$$= -6(-29) + 3(4) + 0 = 174 + 12 = 186$$

17. $\begin{vmatrix} 1 & 2 & 0 \\ -1 & 2 & -1 \\ 0 & 1 & 4 \end{vmatrix}$

If we expand by the third row, we will need to find $a_{31} \cdot A_{31} + a_{32} \cdot A_{32} + a_{33} \cdot A_{33}$. However, we do not need to calculate A_{31}, since $a_{31} = 0$.

$$A_{32} = (-1)^{3+2} \begin{vmatrix} 1 & 0 \\ -1 & -1 \end{vmatrix}$$
$$= (-1)^5 [1(-1) - (-1) \cdot 0]$$
$$= -1(-1 - 0) = -1(-1) = 1$$

$$A_{33} = (-1)^{3+3} \begin{vmatrix} 1 & 2 \\ -1 & 2 \end{vmatrix}$$
$$= (-1)^6 [1 \cdot 2 - (-1) \cdot 2]$$
$$= 1[2 - (-2)] = 1(4) = 4$$

$$a_{31} \cdot A_{31} + a_{32} \cdot A_{32} + a_{33} \cdot A_{33}$$
$$= 0 \cdot A_{31} + a_{32} \cdot A_{32} + a_{33} \cdot A_{33}$$
$$= 0 + 1(1) + 4(4) = 1 + 16 = 17$$

For Exercises 19–28, we will use the sign checkerboard shown on page 507 in calculations.

19. $\begin{vmatrix} 10 & 2 & 1 \\ -1 & 4 & 3 \\ -3 & 8 & 10 \end{vmatrix}$

If we expand by the first row, we will need to find $a_{11} \cdot M_{11} - a_{12} \cdot M_{12} + a_{13} \cdot M_{13}$.

$$M_{11} = \begin{vmatrix} 4 & 3 \\ 8 & 10 \end{vmatrix} = 4 \cdot 10 - 8 \cdot 3 = 40 - 24 = 16,$$

$$M_{12} = \begin{vmatrix} -1 & 3 \\ -3 & 10 \end{vmatrix} = -1 \cdot 10 - (-3) \cdot 3$$
$$= -10 - (-9) = -1, \text{ and}$$

$$M_{13} = \begin{vmatrix} -1 & 4 \\ -3 & 8 \end{vmatrix} = -1 \cdot 8 - (-3) \cdot 4 = -8 - (-12) = 4$$

$$a_{11} \cdot M_{11} - a_{12} \cdot M_{12} + a_{13} \cdot M_{13}$$
$$= 10 \cdot 16 - 2(-1) + 1 \cdot 4$$
$$= 160 - (-2) + 4 = 166$$

21. $\begin{vmatrix} 1 & -2 & 3 \\ 0 & 0 & 0 \\ 1 & 10 & -12 \end{vmatrix}$

If we expand by the second row, we will need to find $-a_{21} \cdot M_{21} + a_{22} \cdot M_{22} - a_{23} \cdot M_{23}$. Since $a_{21} = a_{22} = a_{23} = 0$, the result will be zero.

23. $\begin{vmatrix} 3 & 3 & -1 \\ 2 & 6 & 0 \\ -6 & -6 & 2 \end{vmatrix}$

If we expand by the second row, we will need to find $-a_{21} \cdot M_{21} + a_{22} \cdot M_{22} - a_{23} \cdot M_{23}$. However, we do not need to calculate M_{23}, since $a_{23} = 0$.

$$M_{21} = \begin{vmatrix} 3 & -1 \\ -6 & 2 \end{vmatrix} = 3 \cdot 2 - (-6)(-1) = 6 - 6 = 0$$

and

$$M_{22} = \begin{vmatrix} 3 & -1 \\ -6 & 2 \end{vmatrix} = 3 \cdot 2 - (-6)(-1) = 6 - 6 = 0$$

$$a_{21} \cdot M_{21} - a_{22} \cdot M_{22} + a_{23} \cdot M_{23}$$
$$= a_{21} \cdot M_{21} - a_{22} \cdot M_{22} + 0 \cdot M_{23}$$
$$= 2 \cdot 0 - 6 \cdot 0 + 0 = 0 - 0 = 0$$

25. $\begin{vmatrix} 1 & 0 & 0 \\ 0 & 1 & 0 \\ 0 & 0 & 1 \end{vmatrix}$

If we expand by the first row, we will need to find $a_{11} \cdot M_{11} - a_{12} \cdot M_{12} + a_{13} \cdot M_{13}$. However, we do not need to calculate M_{12} or M_{13}, since $a_{12} = a_{13} = 0$.

$$M_{11} = \begin{vmatrix} 1 & 0 \\ 0 & 1 \end{vmatrix} = 1 \cdot 1 - 0 \cdot 0 = 1 - 0 = 1$$

$$a_{11} \cdot M_{11} - a_{12} \cdot M_{12} + a_{13} \cdot M_{13}$$
$$= a_{11} \cdot M_{11} - 0 \cdot M_{12} + 0 \cdot M_{13}$$
$$= 1 \cdot 1 - 0 + 0 = 1$$

Copyright © 2013 Pearson Education, Inc.

27. $\begin{vmatrix} -2 & 0 & 1 \\ 0 & 1 & 0 \\ 0 & 0 & -1 \end{vmatrix}$

If we expand by the first column, we will need to find $a_{11} \cdot M_{11} - a_{21} \cdot M_{21} + a_{31} \cdot M_{31}$. However, we do not need to calculate M_{21} or M_{31}, since $a_{21} = a_{31} = 0$.

$M_{11} = \begin{vmatrix} 1 & 0 \\ 0 & -1 \end{vmatrix} = 1(-1) - 0 \cdot 0 = -1 - 0 = -1$

$a_{11} \cdot M_{11} - a_{21} \cdot M_{21} + a_{31} \cdot M_{31}$
$= a_{11} \cdot M_{11} - 0 \cdot M_{21} + 0 \cdot M_{31}$
$= -2(-1) - 0 + 0 = 2$

29. $\begin{vmatrix} \sqrt{2} & 4 & 0 \\ 1 & -\sqrt{5} & 7 \\ -5 & \sqrt{5} & 1 \end{vmatrix}$

If we expand by the third column, we will need to find $a_{13} \cdot M_{13} - a_{23} \cdot M_{23} + a_{33} \cdot M_{33}$. However, we do not need to calculate M_{13} since $a_{13} = 0$.

$M_{23} = \begin{vmatrix} \sqrt{2} & 4 \\ -5 & \sqrt{5} \end{vmatrix} = \sqrt{2} \cdot \sqrt{5} - (4)(-5)$
$= \sqrt{10} + 20$

$M_{33} = \begin{vmatrix} \sqrt{2} & 4 \\ 1 & -\sqrt{5} \end{vmatrix} = \sqrt{2}(-\sqrt{5}) - (1)(4)$
$= -\sqrt{10} - 4$

$a_{13} \cdot M_{13} - a_{23} \cdot M_{23} + a_{33} \cdot M_{33}$.
$= 0 \cdot M_{13} - 7(\sqrt{10} + 20) + 1(-\sqrt{10} - 4)$
$= -7\sqrt{10} - 140 - \sqrt{10} - 4 = -144 - 8\sqrt{10}$

31. $\begin{vmatrix} 0.4 & -0.8 & 0.6 \\ 0.3 & 0.9 & 0.7 \\ 3.1 & 4.1 & -2.8 \end{vmatrix}$

The determinant is -5.5.

```
[A]
[[.4  -.8  .6 ]
 [.3   .9  .7 ]
 [3.1 4.1 -2.8]]
det([A]
            -5.5
```

33. $\begin{vmatrix} a_{11} & a_{12} & a_{13} \\ a_{21} & a_{22} & a_{23} \\ a_{31} & a_{32} & a_{33} \end{vmatrix} \begin{matrix} a_{11} & a_{12} \\ a_{21} & a_{22} \\ a_{31} & a_{32} \end{matrix}$

$d_1 = a_{11}a_{22}a_{33}; d_2 = a_{12}a_{23}a_{31}; d_3 = a_{13}a_{21}a_{32}$

$d_4 = a_{13}a_{22}a_{31}; d_5 = a_{11}a_{23}a_{32}; d_6 = a_{12}a_{21}a_{33}$

$(d_1 + d_2 + d_3) - (d_4 + d_5 + d_6)$
$= (a_{11}a_{22}a_{33} + a_{12}a_{23}a_{31} + a_{13}a_{21}a_{32})$
$\quad - (a_{13}a_{22}a_{31} + a_{11}a_{23}a_{32} + a_{12}a_{21}a_{33})$
$= (a_{11}a_{22}a_{33} + a_{12}a_{23}a_{31} + a_{13}a_{21}a_{32})$
$\quad - (a_{31}a_{22}a_{13} + a_{32}a_{23}a_{11} + a_{33}a_{21}a_{12})$

35. $\begin{vmatrix} 1 & 3 & 2 \\ 0 & 2 & 6 \\ 7 & 1 & 5 \end{vmatrix}$

If we expand by the first column, we will need to find $a_{11} \cdot M_{11} - a_{21} \cdot M_{21} + a_{31} \cdot M_{31}$.

$(-1)^{1+1} \cdot 1 \cdot \begin{vmatrix} 2 & 6 \\ 1 & 5 \end{vmatrix} - (-1)^{1+2} \cdot 0 \cdot \begin{vmatrix} 3 & 2 \\ 1 & 5 \end{vmatrix}$
$\quad + (-1)^{1+3} \cdot 7 \cdot \begin{vmatrix} 3 & 2 \\ 2 & 6 \end{vmatrix}$

$= 1(10 - 6) - 0 + 7(18 - 4) = 4 + 98 = 102$

Both methods give the same determinant.

37. To solve the equation $\begin{vmatrix} 5 & x \\ -3 & 2 \end{vmatrix} = 6$, we need to solve $5 \cdot 2 - (-3) \cdot x = 6$.

$5 \cdot 2 - (-3) \cdot x = 6 \Rightarrow 10 + 3x = 6 \Rightarrow$
$3x = -4 \Rightarrow x = -\frac{4}{3}$

Verifying $\begin{vmatrix} 5 & -\frac{4}{3} \\ -3 & 2 \end{vmatrix} = 6$, we have

$5 \cdot 2 - (-3)(-\frac{4}{3}) = 10 - 4 = 6$.

Solution set: $\{-\frac{4}{3}\}$

39. To solve the equation $\begin{vmatrix} x & 3 \\ x & x \end{vmatrix} = 4$, we need to solve

$x \cdot x - 3x = 4 \Rightarrow x^2 - 3x = 4 \Rightarrow$
$x^2 - 3x - 4 = 0 \Rightarrow (x+1)(x-4) = 0$
$x + 1 = 0 \Rightarrow x = -1$ or $x - 4 = 0 \Rightarrow x = 4$

Verify $x = -1$.

$\begin{vmatrix} -1 & 3 \\ -1 & -1 \end{vmatrix} = -1(-1) - (-1) \cdot 3 = 1 - (-3) = 4$

Verify $x = 4$.

$\begin{vmatrix} 4 & 3 \\ 4 & 4 \end{vmatrix} = 4 \cdot 4 - 4 \cdot 3 = 16 - 12 = 4$

Solution set: $\{-1, 4\}$

Copyright © 2013 Pearson Education, Inc.

41. To solve the equation $\begin{vmatrix} -2 & 0 & 1 \\ -1 & 3 & x \\ 5 & -2 & 0 \end{vmatrix} = 3,$ expand

by the first row. In order to do this, we will need to find $a_{11} \cdot M_{11} - a_{12} \cdot M_{12} + a_{13} \cdot M_{13}.$ However, we do not need to calculate $M_{12},$ since $a_{12} = 0.$

$$M_{11} = \begin{vmatrix} 3 & x \\ -2 & 0 \end{vmatrix} = 3 \cdot 0 - (-2) \cdot x = 0 - (-2x) = 2x$$

$$M_{13} = \begin{vmatrix} -1 & 3 \\ 5 & -2 \end{vmatrix} = (-1)(-2) - 5 \cdot 3 = 2 - 15 = -13$$

$a_{11} \cdot M_{11} - a_{12} \cdot M_{12} + a_{13} \cdot M_{13}$
$= a_{11} \cdot M_{11} - 0 \cdot M_{12} + a_{13} \cdot M_{13}$
$= -2(2x) - 0 + 1(-13)$
$= -4x + (-13) = -4x - 13$

Set this equal to 3 and solve to get
$-4x - 13 = 3 \Rightarrow -4x = 16 \Rightarrow x = -4.$

Verify $\begin{vmatrix} -2 & 0 & 1 \\ -1 & 3 & -4 \\ 5 & -2 & 0 \end{vmatrix} = 3.$

Since

$$M_{11} = \begin{vmatrix} 3 & -4 \\ -2 & 0 \end{vmatrix} = 3 \cdot 0 - (-2)(-4) = 0 - 8 = -8$$

and $M_{13} = \begin{vmatrix} -1 & 3 \\ 5 & -2 \end{vmatrix} = -13,$ we have

$a_{11} \cdot M_{11} - a_{12} \cdot M_{12} + a_{13} \cdot M_{13}$
$= a_{11} \cdot M_{11} - 0 \cdot M_{12} + a_{13} \cdot M_{13}$
$= -2(-8) - 0 + 1(-13) = 16 + (-13) = 3$

Solution set: $\{-4\}$

43. To solve the equation $\begin{vmatrix} 5 & 3x & -3 \\ 0 & 2 & -1 \\ 4 & -1 & x \end{vmatrix} = -7,$

expand by the second row. We will need to find $-a_{21} \cdot M_{21} + a_{22} \cdot M_{22} - a_{23} \cdot M_{23}.$ However, we do not need to calculate $M_{21},$ since $a_{21} = 0.$

$$M_{22} = \begin{vmatrix} 5 & -3 \\ 4 & x \end{vmatrix} = 5x - 4(-3) = 5x - (-12) = 5x + 12$$

and $M_{23} = \begin{vmatrix} 5 & 3x \\ 4 & -1 \end{vmatrix} = 5(-1) - 4 \cdot 3x = -5 - 12x$

$-a_{21} \cdot M_{21} + a_{22} \cdot M_{22} - a_{23} \cdot M_{23}$
$= -0 \cdot M_{21} + a_{22} \cdot M_{22} - a_{23} \cdot M_{23}$
$= 0 + 2(5x + 12) - (-1)(-5 - 12x)$
$= 2(5x + 12) + (-5 - 12x)$
$= 10x + 24 - 5 - 12x = 19 - 2x$

Set this equal to -7 and solve to get
$19 - 2x = -7 \Rightarrow -2x = -26 \Rightarrow x = 13.$

Verify $\begin{vmatrix} 5 & 3(13) & -3 \\ 0 & 2 & -1 \\ 4 & -1 & 13 \end{vmatrix} = \begin{vmatrix} 5 & 39 & -3 \\ 0 & 2 & -1 \\ 4 & -1 & 13 \end{vmatrix} = -7.$

Since

$$M_{22} = \begin{vmatrix} 5 & -3 \\ 4 & 13 \end{vmatrix} = 5 \cdot 13 - 4(-3) = 65 - (-12) = 77$$

and

$$M_{23} = \begin{vmatrix} 5 & 39 \\ 4 & -1 \end{vmatrix} = 5(-1) - 4 \cdot 39 = -5 - 156 = -161,$$

we have the following.

$-a_{21} \cdot M_{21} + a_{22} \cdot M_{22} - a_{23} \cdot M_{23}$
$= -0 \cdot M_{21} + a_{22} \cdot M_{22} - a_{23} \cdot M_{23}$
$= 0 + 2 \cdot 77 - (-1)(-161) = 154 - 161 = -7$

Solution set: $\{13\}$

45. To solve the equation $\begin{vmatrix} x & 0 & -1 \\ 2 & -3 & x \\ x & 0 & 7 \end{vmatrix} = 12,$ expand

by the second column. We will need to find $-a_{12} \cdot M_{12} + a_{22} \cdot M_{22} - a_{32} \cdot M_{32}.$ However, we do not need to calculate M_{12} or M_{32} since $a_{12} = 0$ and $a_{32} = 0.$

$$M_{22} = \begin{vmatrix} x & -1 \\ x & 7 \end{vmatrix} = 7x + x = 8x$$

$-a_{12} \cdot M_{12} + a_{22} \cdot M_{22} - a_{32} \cdot M_{32}.$
$= 0 \cdot M_{12} + (-3)8x + 0 \cdot M_{32} = -24x$

Set this equal to 12 and solve to get
$-24x = 12 \Rightarrow x = -\frac{1}{2}$

Verify $\begin{vmatrix} -\frac{1}{2} & 0 & -1 \\ 2 & -3 & -\frac{1}{2} \\ -\frac{1}{2} & 0 & 7 \end{vmatrix} = 12.$

$$M_{22} = \begin{vmatrix} -\frac{1}{2} & -1 \\ -\frac{1}{2} & 7 \end{vmatrix} = -\frac{1}{2} \cdot 7 - \left(-\frac{1}{2}\right)(-1)$$
$$= -\frac{7}{2} - \frac{1}{2} = -\frac{8}{2} = -4$$

$-a_{12} \cdot M_{12} + a_{22} \cdot M_{22} - a_{32} \cdot M_{32}.$
$= 0 \cdot M_{12} + (-3)(-4) + 0 \cdot M_{32} = 12$

Solution set: $\left\{-\frac{1}{2}\right\}$

Copyright © 2013 Pearson Education, Inc.

47. $P(0,0), Q(0,2), R(1,4)$

Find $D = \dfrac{1}{2}\begin{vmatrix} x_1 & y_1 & 1 \\ x_2 & y_2 & 1 \\ x_3 & y_3 & 1 \end{vmatrix}$, where

$P = (x_1, y_1) = (0,0), Q(x_2, y_2) = (0,2)$, and $R = (x_3, y_3) = (1,4)$

Expanding by the first row, we have the following

$D = \dfrac{1}{2}\begin{vmatrix} 0 & 0 & 1 \\ 0 & 2 & 1 \\ 1 & 4 & 1 \end{vmatrix} = \dfrac{1}{2}\left[0\begin{vmatrix} 2 & 1 \\ 4 & 1 \end{vmatrix} - 0\begin{vmatrix} 0 & 1 \\ 1 & 1 \end{vmatrix} + 1\begin{vmatrix} 0 & 2 \\ 1 & 4 \end{vmatrix} \right]$

$= \tfrac{1}{2}\left[0(2-4) - 0(0-1) + 1(0-2) \right]$

$= \tfrac{1}{2}\left[0(-2) - 0(-1) + 1(-2) \right] = \tfrac{1}{2}(-2) = -1$

Area of triangle $= |D| = |-1| = 1 \text{ unit}^2$.

49. $P(2,5), Q(-1,3), R(4,0)$

Find $D = \dfrac{1}{2}\begin{vmatrix} x_1 & y_1 & 1 \\ x_2 & y_2 & 1 \\ x_3 & y_3 & 1 \end{vmatrix}$, where

$P = (x_1, y_1) = (2,5), Q(x_2, y_2) = (-1,3)$, and $R = (x_3, y_3) = (4,0)$.

Expanding by the third row, we have the following.

$D = \dfrac{1}{2}\begin{vmatrix} 2 & 5 & 1 \\ -1 & 3 & 1 \\ 4 & 0 & 1 \end{vmatrix}$

$= \dfrac{1}{2}\left[4\begin{vmatrix} 5 & 1 \\ 3 & 1 \end{vmatrix} - 0\begin{vmatrix} 2 & 1 \\ -1 & 1 \end{vmatrix} + 1\begin{vmatrix} 2 & 5 \\ -1 & 3 \end{vmatrix} \right]$

$= \tfrac{1}{2}\left[4(5-3) - 0(2+1) + 1(6+5) \right]$

$= \tfrac{1}{2}\left[4(2) - 0(3) + 1(11) \right] = \tfrac{1}{2}(8-0+11)$

$= \tfrac{1}{2}(19) = \tfrac{19}{2} = 9.5$

Area of triangle $= |D| = |9.5| = 9.5 \text{ units}^2$.

51. $(101.3, 52.7), (117.2, 253.9), (313.1, 301.6)$

Label the points as follows.

$P = (x_1, y_1) = (101.3, 52.7)$,

$Q(x_2, y_2) = (117.2, 253.9)$, and

$R = (x_3, y_3) = (313.1, 301.6)$

Since

$D = \dfrac{1}{2}\begin{vmatrix} x_1 & y_1 & 1 \\ x_2 & y_2 & 1 \\ x_3 & y_3 & 1 \end{vmatrix} = \dfrac{1}{2}\begin{vmatrix} 101.3 & 52.7 & 1 \\ 117.2 & 253.9 & 1 \\ 313.1 & 301.6 & 1 \end{vmatrix}$,

we will enter the 3×3 as $\begin{bmatrix} 101.3 & 52.7 & 1 \\ 117.2 & 253.9 & 1 \\ 313.1 & 301.6 & 1 \end{bmatrix}$

and perform the calculations as shown below.

```
[A]
[[101.3 52.7  1...
 [117.2 253.9 1...
 [313.1 301.6 1...
(1/2)det([A])
        -19328.325
```

Area of triangular lot is $|-19,328.325|$ ft^2 or approximately 19,328.3 ft^2.

For exercises 53–73, refer to the determinant theorems on page 508 in the text.

Let $|A| = \begin{vmatrix} 1 & 2 & 3 \\ 4 & 5 & 6 \\ 7 & 9 & 10 \end{vmatrix}$.

53. $\begin{vmatrix} 4 & 5 & 6 \\ 1 & 2 & 3 \\ 7 & 9 & 10 \end{vmatrix}$

Interchange rows 1 and 2.

$\begin{vmatrix} 4 & 5 & 6 \\ 1 & 2 & 3 \\ 7 & 9 & 10 \end{vmatrix} = -\begin{vmatrix} 1 & 2 & 3 \\ 4 & 5 & 6 \\ 7 & 9 & 10 \end{vmatrix} = -3$

55. If each element of a row are multiplied by a real number k, then the determinant of the new matrix equals the determinant of the original matrix.

$\begin{vmatrix} 5 & 10 & 15 \\ 4 & 5 & 6 \\ 7 & 9 & 10 \end{vmatrix} = \begin{vmatrix} 5 \cdot 1 & 5 \cdot 2 & 5 \cdot 3 \\ 4 & 5 & 6 \\ 7 & 9 & 10 \end{vmatrix} = 5 \cdot 3 = 15$

57. Adding a row to another row does not change the value of the determinant.

$\begin{vmatrix} 1 & 2 & 3 \\ 4 & 5 & 6 \\ 8 & 11 & 13 \end{vmatrix} = \begin{vmatrix} 1 & 2 & 3 \\ 4 & 5 & 6 \\ 7+1 & 9+2 & 10+3 \end{vmatrix} = 3$

Copyright © 2013 Pearson Education, Inc.

59. $\begin{vmatrix} 1 & 0 & 0 \\ 1 & 0 & 1 \\ 3 & 0 & 0 \end{vmatrix}$

By Theorem 1, the determinant is 0 since every entry in column 2 is 0.

61. $\begin{vmatrix} 6 & 8 & -12 \\ -1 & 0 & 2 \\ 4 & 0 & -8 \end{vmatrix}$

Given $\begin{bmatrix} 6 & 8 & -12 \\ -1 & 0 & 2 \\ 4 & 0 & -8 \end{bmatrix}$, add 2 times column 1

to column 3 to obtain $\begin{bmatrix} 6 & 8 & 0 \\ -1 & 0 & 0 \\ 4 & 0 & 0 \end{bmatrix}$.

By Theorem 6 the following statement is true.

$\begin{vmatrix} 6 & 8 & -12 \\ -1 & 0 & 2 \\ 4 & 0 & -8 \end{vmatrix} = \begin{vmatrix} 6 & 8 & 0 \\ -1 & 0 & 0 \\ 4 & 0 & 0 \end{vmatrix}$

By Theorem 1, the determinant is 0 since every entry in column 3 is 0.

63. $\begin{vmatrix} -4 & 1 & 4 \\ 2 & 0 & 1 \\ 0 & 2 & 4 \end{vmatrix}$

Since $\begin{bmatrix} -4 & 1 & 4 \\ 2 & 0 & 1 \\ 0 & 2 & 4 \end{bmatrix} \Rightarrow \begin{bmatrix} 0 & 1 & 6 \\ 2 & 0 & 1 \\ 0 & 2 & 4 \end{bmatrix} 2R2 + R1$,

by Theorem 6 we have the following.

$\begin{vmatrix} -4 & 1 & 4 \\ 2 & 0 & 1 \\ 0 & 2 & 4 \end{vmatrix} = \begin{vmatrix} 0 & 1 & 6 \\ 2 & 0 & 1 \\ 0 & 2 & 4 \end{vmatrix}$

Expanding by the first column, we have the following.

$\begin{vmatrix} 0 & 1 & 6 \\ 2 & 0 & 1 \\ 0 & 2 & 4 \end{vmatrix} = 0\begin{vmatrix} 0 & 1 \\ 2 & 4 \end{vmatrix} - 2\begin{vmatrix} 1 & 6 \\ 2 & 4 \end{vmatrix} + 0\begin{vmatrix} 1 & 6 \\ 0 & 1 \end{vmatrix}$

$= 0 - 2(4 - 12) + 0 = -2(-8) = 16$

65. $\begin{vmatrix} 0 & 1 & -3 \\ 7 & 5 & 2 \\ 1 & -2 & 6 \end{vmatrix}$

Add 2 times row 1 to row 3 to obtain

$\begin{vmatrix} 0 & 1 & -3 \\ 7 & 5 & 2 \\ 1 & 0 & 0 \end{vmatrix}$.

By Theorem 6 the following statement is true.

$\begin{vmatrix} 0 & 1 & -3 \\ 7 & 5 & 2 \\ 1 & -2 & 6 \end{vmatrix} = \begin{vmatrix} 0 & 1 & -3 \\ 7 & 5 & 2 \\ 1 & 0 & 0 \end{vmatrix}$

Expanding by the third row, we have the following.

$\begin{vmatrix} 0 & 1 & -3 \\ 7 & 5 & 2 \\ 1 & 0 & 0 \end{vmatrix} = 1\begin{vmatrix} 1 & -3 \\ 5 & 2 \end{vmatrix} - 0\begin{vmatrix} 0 & -3 \\ 7 & 2 \end{vmatrix} + 0\begin{vmatrix} 0 & 1 \\ 7 & 5 \end{vmatrix}$

$= 1(2 + 15) - 0 + 0 = 1(17) = 17$

67. $\begin{vmatrix} 1 & 6 & 7 \\ 0 & 6 & 7 \\ 0 & 0 & 9 \end{vmatrix}$

Add −1 times row 2 to row 1 to obtain

$\begin{vmatrix} 1 & 0 & 0 \\ 0 & 6 & 7 \\ 0 & 0 & 9 \end{vmatrix}$.

By Theorem 6 the following statement is true.

$\begin{vmatrix} 1 & 6 & 7 \\ 0 & 6 & 7 \\ 0 & 0 & 9 \end{vmatrix} = \begin{vmatrix} 1 & 0 & 0 \\ 0 & 6 & 7 \\ 0 & 0 & 9 \end{vmatrix}$

Expanding by the third row, we have the following.

$\begin{vmatrix} 1 & 0 & 0 \\ 0 & 6 & 7 \\ 0 & 0 & 9 \end{vmatrix} = 0\begin{vmatrix} 0 & 0 \\ 6 & 7 \end{vmatrix} - 0\begin{vmatrix} 1 & 0 \\ 0 & 7 \end{vmatrix} + 9\begin{vmatrix} 1 & 0 \\ 0 & 6 \end{vmatrix}$

$= 0 - 0 + 9(6) = 54$

Copyright © 2013 Pearson Education, Inc.

69. $\begin{vmatrix} 2 & -1 & 3 \\ 6 & 4 & 10 \\ 4 & 5 & 7 \end{vmatrix}$

Add 4 times row 1 to row 2 to obtain

$\begin{vmatrix} 2 & -1 & 3 \\ 14 & 0 & 22 \\ 4 & 5 & 7 \end{vmatrix}$

Now add 5 times row 1 to row 3 to obtain

$\begin{vmatrix} 2 & -1 & 3 \\ 14 & 0 & 22 \\ 14 & 0 & 22 \end{vmatrix}$

By Theorem 6 the following statement is true.

$\begin{vmatrix} 2 & -1 & 3 \\ 6 & 4 & 10 \\ 4 & 5 & 7 \end{vmatrix} = \begin{vmatrix} 2 & -1 & 3 \\ 14 & 0 & 22 \\ 4 & 5 & 7 \end{vmatrix} = \begin{vmatrix} 2 & -1 & 3 \\ 14 & 0 & 22 \\ 14 & 0 & 22 \end{vmatrix}$

Theorem 5 states that if any two rows of a matrix are identical, then the determinant of the matrix is 0. Thus,

$\begin{vmatrix} 2 & -1 & 3 \\ 6 & 4 & 10 \\ 4 & 5 & 7 \end{vmatrix} = \begin{vmatrix} 2 & -1 & 3 \\ 14 & 0 & 22 \\ 4 & 5 & 7 \end{vmatrix} = \begin{vmatrix} 2 & -1 & 3 \\ 14 & 0 & 22 \\ 14 & 0 & 22 \end{vmatrix} = 0$

71. $\begin{vmatrix} -1 & 0 & 2 & 3 \\ 5 & 4 & -3 & 7 \\ 8 & 2 & 9 & -5 \\ 4 & 4 & -1 & 10 \end{vmatrix}$

Add 5 times row 1 to row 2 to obtain

$\begin{vmatrix} -1 & 0 & 2 & 3 \\ 0 & 4 & 7 & 22 \\ 8 & 2 & 9 & -5 \\ 4 & 4 & -1 & 10 \end{vmatrix}$

Add 4 times row 1 to row 4 to obtain

$\begin{vmatrix} -1 & 0 & 2 & 3 \\ 0 & 4 & 7 & 22 \\ 8 & 2 & 9 & -5 \\ 0 & 4 & 7 & 22 \end{vmatrix}$

Theorem 5 states that if any two rows of a matrix are identical, then the determinant of the matrix is 0. Thus,

$\begin{vmatrix} -1 & 0 & 2 & 3 \\ 5 & 4 & -3 & 7 \\ 8 & 2 & 9 & -5 \\ 4 & 4 & -1 & 10 \end{vmatrix} = 0$

From exercise 46, we have the following array of signs. Refer to this array for exercises 73–75.

For 4×4 matrices

+	−	+	−
−	+	−	+
+	−	+	−
−	+	−	+

Like previous exercises, you can arrive at a solution by expanding on any row or column. We will also use determinant theorems to reduce the number of calculations.

73. $\begin{vmatrix} 4 & 0 & 0 & 2 \\ -1 & 0 & 3 & 0 \\ 2 & 4 & 0 & 1 \\ 0 & 0 & 1 & 2 \end{vmatrix}$

Expanding by the second column, we have

$\begin{vmatrix} 4 & 0 & 0 & 2 \\ -1 & 0 & 3 & 0 \\ 2 & 4 & 0 & 1 \\ 0 & 0 & 1 & 2 \end{vmatrix} = -4 \begin{vmatrix} 4 & 0 & 2 \\ -1 & 3 & 0 \\ 0 & 1 & 2 \end{vmatrix}$

Using the definition of the determinant in the text, we have the following.

$-4 \begin{vmatrix} 4 & 0 & 2 \\ -1 & 3 & 0 \\ 0 & 1 & 2 \end{vmatrix}$

$= -4 \left(\begin{bmatrix} 4(3)(2) + 0(0)(0) + 2(-1)(1) \end{bmatrix} \right.$
$\left. - \begin{bmatrix} 0(3)(2) + 1(0)(4) + 2(-1)(0) \end{bmatrix} \right)$

$= -4 \begin{bmatrix} (24 + 0 - 2) - (0 + 0 + 0) \end{bmatrix}$

$= -4(22 - 0) = -4(22) = -88$

75. $\begin{vmatrix} 3 & -6 & 5 & -1 \\ 0 & 2 & -1 & 3 \\ -6 & 4 & 2 & 0 \\ -7 & 3 & 1 & 1 \end{vmatrix}$

Add 3 times row 1 to row 2 and add row 1 to row 4 to obtain

$\begin{vmatrix} 3 & -6 & 5 & -1 \\ 0 & 2 & -1 & 3 \\ -6 & 4 & 2 & 0 \\ -7 & 3 & 1 & 1 \end{vmatrix} = \begin{vmatrix} 3 & -6 & 5 & -1 \\ 9 & -16 & 14 & 0 \\ -6 & 4 & 2 & 0 \\ -4 & -3 & 6 & 0 \end{vmatrix}.$

(continued on next page)

Copyright © 2013 Pearson Education, Inc.

(*continued*)

Expanding by the fourth column, we have the following.

$$\begin{vmatrix} 3 & -6 & 5 & -1 \\ 9 & -16 & 14 & 0 \\ -6 & 4 & 2 & 0 \\ -4 & -3 & 6 & 0 \end{vmatrix} = -(-1)\begin{vmatrix} 9 & -16 & 14 \\ -6 & 4 & 2 \\ -4 & -3 & 6 \end{vmatrix}$$

$$= \begin{vmatrix} 9 & -16 & 14 \\ -6 & 4 & 2 \\ -4 & -3 & 6 \end{vmatrix}$$

Adding −7 times row 2 to row 1 and −3 times row 2 added to row 3 we have the following.

$$\begin{vmatrix} 9 & -16 & 14 \\ -6 & 4 & 2 \\ -4 & -3 & 6 \end{vmatrix} = \begin{vmatrix} 51 & -44 & 0 \\ -6 & 4 & 2 \\ 14 & -15 & 0 \end{vmatrix}$$

Expanding by the third column, we have

$$-2\begin{vmatrix} 51 & -44 \\ 14 & -15 \end{vmatrix} = -2(-765 + 616)$$
$$= -2(-149) = 298$$

77. For
$$\begin{aligned} 4x + 3y - 2z &= 1 \\ 7x - 4y + 3z &= 2 \\ -2x + y - 8z &= 0 \end{aligned}$$
we have the following.

$$D = \begin{vmatrix} 4 & 3 & -2 \\ 7 & -4 & 3 \\ -2 & 1 & -8 \end{vmatrix}, D_x = \begin{vmatrix} 1 & 3 & -2 \\ 2 & -4 & 3 \\ 0 & 1 & -8 \end{vmatrix},$$

$$D_y = \begin{vmatrix} 4 & 1 & -2 \\ 7 & 2 & 3 \\ -2 & 0 & -8 \end{vmatrix}, \text{ and } D_z = \begin{vmatrix} 4 & 3 & 1 \\ 7 & -4 & 2 \\ -2 & 1 & 0 \end{vmatrix}$$

   (a)   D   **(b)**   A

   (c)   C   **(d)**   B

79. $\begin{aligned} x + y &= 4 \\ 2x - y &= 2 \end{aligned}$

$$D = \begin{vmatrix} 1 & 1 \\ 2 & -1 \end{vmatrix} = 1(-1) - 2(1) = -1 - 2 = -3,$$

$$D_x = \begin{vmatrix} 4 & 1 \\ 2 & -1 \end{vmatrix} = 4(-1) - 2(1) = -4 - 2 = -6,$$

$$D_y = \begin{vmatrix} 1 & 4 \\ 2 & 2 \end{vmatrix} = 1(2) - 2(4) = 2 - 8 = -6 \Rightarrow$$

$$x = \frac{D_x}{D} = \frac{-6}{-3} = 2 \text{ and } y = \frac{D_y}{D} = \frac{-6}{-3} = 2.$$

Solution set: $\{(2, 2)\}$

81. $\begin{aligned} 4x + 3y &= -7 \\ 2x + 3y &= -11 \end{aligned}$

$$D = \begin{vmatrix} 4 & 3 \\ 2 & 3 \end{vmatrix} = 4(3) - 2(3) = 12 - 6 = 6,$$

$$D_x = \begin{vmatrix} -7 & 3 \\ -11 & 3 \end{vmatrix} = -7(3) - (-11)(3)$$
$$= -21 + 33 = 12,$$

$$D_y = \begin{vmatrix} 4 & -7 \\ 2 & -11 \end{vmatrix} = 4(-11) - 2(-7)$$
$$= -44 + 14 = -30 \Rightarrow$$

$$x = \frac{D_x}{D} = \frac{12}{6} = 2 \text{ and } y = \frac{D_y}{D} = \frac{-30}{6} = -5.$$

Solution set: $\{(2, -5)\}$

83. $\begin{aligned} 5x + 4y &= 10 \\ 3x - 7y &= 6 \end{aligned}$

$$D = \begin{vmatrix} 5 & 4 \\ 3 & -7 \end{vmatrix} = 5(-7) - 3(4) = -35 - 12 = -47,$$

$$D_x = \begin{vmatrix} 10 & 4 \\ 6 & -7 \end{vmatrix} = 10(-7) - 6(4) = -70 - 24 = -94,$$

$$D_y = \begin{vmatrix} 5 & 10 \\ 3 & 6 \end{vmatrix} = 5(6) - 3(10) = 30 - 30 = 0 \Rightarrow$$

$$x = \frac{D_x}{D} = \frac{-94}{-47} = 2 \text{ and } y = \frac{D_y}{D} = \frac{0}{-47} = 0.$$

Solution set: $\{(2, 0)\}$

85. $\begin{aligned} 1.5x + 3y &= 5 \quad (1) \\ 2x + 4y &= 3 \quad (2) \end{aligned}$

$$D = \begin{vmatrix} 1.5 & 3 \\ 2 & 4 \end{vmatrix} = (1.5)(4) - 2(3) = 6 - 6 = 0$$

Because $D = 0$, Cramer's rule does not apply. To determine whether the system is inconsistent or has infinitely many solutions, use the elimination method.

$\begin{aligned} 6x + 12y &= 20 \quad \text{Multiply equation (1) by 4.} \\ -6x - 12y &= -9 \quad \text{Multiply equation (2) by} -3. \\ \hline 0 &= 11 \quad \text{False} \end{aligned}$

The system is inconsistent.
Solution set: $\varnothing$

Copyright © 2013 Pearson Education, Inc.

87. $3x + 2y = 4$ (1)
$6x + 4y = 8$ (2)

$$D = \begin{vmatrix} 3 & 2 \\ 6 & 4 \end{vmatrix} = 3(4) - 6(2) = 12 - 12 = 0$$

Because $D = 0$, Cramer's rule does not apply. To determine whether the system is inconsistent or has infinitely many solutions, use the elimination method.

$$\begin{array}{l} -6x - 4y = -8 \quad \text{Multiply equation (1) by } -2. \\ \underline{6x + 4y = 8} \\ 0 = 0 \quad \text{True} \end{array}$$

This shows that equations (1) and (2) are dependent. To write the solution set with y as the arbitrary variable, solve equation (1) for x in terms of y.

$$3x + 2y = 4 \Rightarrow 3x = 4 - 2y \Rightarrow x = \tfrac{4-2y}{3}$$

Solution set: $\left\{ \left(\tfrac{4-2y}{3}, y \right) \right\}$

89. $\frac{1}{2}x + \frac{1}{3}y = 2$
$\frac{3}{2}x - \frac{1}{2}y = -12$

$$D = \begin{vmatrix} \frac{1}{2} & \frac{1}{3} \\ \frac{3}{2} & -\frac{1}{2} \end{vmatrix} = \tfrac{1}{2}\left(-\tfrac{1}{2}\right) - \tfrac{3}{2}\left(\tfrac{1}{3}\right)$$

$$= -\tfrac{1}{4} - \tfrac{1}{2} = -\tfrac{1}{4} - \tfrac{2}{4} = -\tfrac{3}{4}$$

$$D_x = \begin{vmatrix} 2 & \frac{1}{3} \\ -12 & -\frac{1}{2} \end{vmatrix} = 2\left(-\tfrac{1}{2}\right) - (-12)\left(\tfrac{1}{3}\right) \text{ and}$$

$$= -1 + 4 = 3$$

$$D_y = \begin{vmatrix} \frac{1}{2} & 2 \\ \frac{3}{2} & -12 \end{vmatrix} = \tfrac{1}{2}(-12) - \tfrac{3}{2}(2) = -6 - 3 = -9$$

Thus, we have $x = \frac{D_x}{D} = \frac{3}{-\frac{3}{4}} = 3\left(-\tfrac{4}{3}\right) = -4$ and

$$y = \frac{D_y}{D} = \frac{-9}{-\frac{3}{4}} = -9\left(-\tfrac{4}{3}\right) = 12$$

Solution set: $\{(-4, 12)\}$

In Exercises 91–101, we will be using the Determinant Theorems on page 508 of the text to reduce the number of calculations necessary in simplifying determinants.

91. $2x - y + 4z = -2$
$3x + 2y - z = -3$
$x + 4y + 2z = 17$

Adding 2 times row 1 to row 2 and 4 times row 1 to row 3 we have

$$D = \begin{vmatrix} 2 & -1 & 4 \\ 3 & 2 & -1 \\ 1 & 4 & 2 \end{vmatrix} = \begin{vmatrix} 2 & -1 & 4 \\ 7 & 0 & 7 \\ 9 & 0 & 18 \end{vmatrix}.$$

Expanding by column two, we have

$$D = -(-1) \begin{vmatrix} 7 & 7 \\ 9 & 18 \end{vmatrix} = 126 - 63 = 63.$$

Adding 2 times row 1 to row 2 and 4 times row 1 to row 3, we have

$$D_x = \begin{vmatrix} -2 & -1 & 4 \\ -3 & 2 & -1 \\ 17 & 4 & 2 \end{vmatrix} = \begin{vmatrix} -2 & -1 & 4 \\ -7 & 0 & 7 \\ 9 & 0 & 18 \end{vmatrix}.$$

Expanding about column two, we have

$$D_x = -(-1) \begin{vmatrix} -7 & 7 \\ 9 & 18 \end{vmatrix} = -126 - 63 = -189.$$

Adding column 2 to column 1, we have

$$D_y = \begin{vmatrix} 2 & -2 & 4 \\ 3 & -3 & -1 \\ 1 & 17 & 2 \end{vmatrix} = \begin{vmatrix} 0 & -2 & 4 \\ 0 & -3 & -1 \\ 18 & 17 & 2 \end{vmatrix}.$$

Expanding by column one, we have

$$D_y = 18 \begin{vmatrix} -2 & 4 \\ -3 & -1 \end{vmatrix} = 18(2 + 12) = 18(14) = 252.$$

Adding column 3 to column 1, we have

$$D_z = \begin{vmatrix} 2 & -1 & -2 \\ 3 & 2 & -3 \\ 1 & 4 & 17 \end{vmatrix} = \begin{vmatrix} 0 & -1 & -2 \\ 0 & 2 & -3 \\ 18 & 4 & 17 \end{vmatrix}.$$

Expanding by column one we have

$$D_z = 18 \begin{vmatrix} -1 & -2 \\ 2 & -3 \end{vmatrix} = 18(3 + 4) = 18(7) = 126.$$

Thus, we have $x = \frac{D_x}{D} = \frac{-189}{63} = -3$,

$y = \frac{D_y}{D} = \frac{252}{63} = 4$, and $z = \frac{D_z}{D} = \frac{126}{63} = 2$.

Solution set: $\{(-3, 4, 2)\}$

93. $\begin{array}{ll} 4x - 3y + z - 1 = 0 & 4x - 3y + z = -1 \\ 5x + 7y + 2z - 2 = 0 \Rightarrow & 5x + 7y + 2z = -2 \\ 3x - 5y - z - 1 = 0 & 3x - 5y - z = 1 \end{array}$

Adding row 3 to row 1 and 2 times row 3 to row 2, we have

$$D = \begin{vmatrix} 4 & -3 & 1 \\ 5 & 7 & 2 \\ 3 & -5 & -1 \end{vmatrix} = \begin{vmatrix} 7 & -8 & 0 \\ 11 & -3 & 0 \\ 3 & -5 & -1 \end{vmatrix}.$$

Expanding by column three, we have

$$D = -1 \begin{vmatrix} 7 & -8 \\ 11 & -3 \end{vmatrix} = -1(-21 + 88) = -1(67) = -67$$

(*continued on next page*)

Copyright © 2013 Pearson Education, Inc.

(continued)

Adding column 1 to column 3 we have

$$D_x = \begin{vmatrix} -1 & -3 & 1 \\ -2 & 7 & 2 \\ 1 & -5 & -1 \end{vmatrix} = \begin{vmatrix} -1 & 3 & 0 \\ -2 & 7 & 0 \\ 1 & -5 & 0 \end{vmatrix}.$$

Since we have a column of zeros, $D_x = 0$.
Adding column 2 to column 3, we have

$$D_y = \begin{vmatrix} 4 & -1 & 1 \\ 5 & -2 & 2 \\ 3 & 1 & -1 \end{vmatrix} = \begin{vmatrix} 4 & -1 & 0 \\ 5 & -2 & 0 \\ 3 & 1 & 0 \end{vmatrix}.$$

Since we have a column of zeros, $D_y = 0$.

Adding row 3 to row 1 and 2 times row 3 to row 2, we have

$$D_z = \begin{vmatrix} 4 & -3 & -1 \\ 5 & 7 & -2 \\ 3 & -5 & 1 \end{vmatrix} = \begin{vmatrix} 7 & -8 & 0 \\ 11 & -3 & 0 \\ 3 & -5 & 1 \end{vmatrix}.$$

Expanding by the third column, we have

$$D_z = 1\begin{vmatrix} 7 & -8 \\ 11 & -3 \end{vmatrix} = -21 + 88 = 67.$$

Thus, we have $x = \dfrac{D_x}{D} = \dfrac{0}{-67} = 0$,

$y = \dfrac{D_y}{D} = \dfrac{0}{-67} = 0$, and $z = \dfrac{D_z}{D} = \dfrac{67}{-67} = -1$.

Solution set: $\{(0,0,-1)\}$

95.
$$\begin{aligned} x + 2y + 3z &= 4 \quad (1) \\ 4x + 3y + 2z &= 1 \quad (2) \\ -x - 2y - 3z &= 0 \quad (3) \end{aligned}$$

Adding row 1 to row 3, we have

$$D = \begin{vmatrix} 1 & 2 & 3 \\ 4 & 3 & 2 \\ -1 & -2 & -3 \end{vmatrix} = \begin{vmatrix} 1 & 2 & 3 \\ 4 & 3 & 2 \\ 0 & 0 & 0 \end{vmatrix}.$$

Since we have a row of zeros, $D = 0$ and we cannot use Cramer's rule.
Using the elimination method, we can add equations (1) and (3).

$$\begin{aligned} x + 2y + 3z &= 4 \\ -x - 2y - 3z &= 0 \\ \hline 0 &= 4 \quad \text{False} \end{aligned}$$

The system is inconsistent.
Solution set: $\varnothing$

97.
$$\begin{aligned} -2x - 2y + 3z &= 4 \quad (1) \\ 5x + 7y - z &= 2 \quad (2) \\ 2x + 2y - 3z &= -4 \quad (3) \end{aligned}$$

Adding row 1 to row 3, we have

$$D = \begin{vmatrix} -2 & -2 & 3 \\ 5 & 7 & -1 \\ 2 & 2 & -3 \end{vmatrix} = \begin{vmatrix} -2 & -2 & 3 \\ 5 & 7 & -1 \\ 0 & 0 & 0 \end{vmatrix}.$$

Since we have a row of zeros, $D = 0$ and we cannot use Cramer's rule. Using the elimination method, we can add equations (1) and (3).

$$\begin{aligned} -2x - 2y + 3z &= 4 \\ 2x + 2y - 3z &= -4 \\ \hline 0 &= 0 \quad \text{True} \end{aligned}$$

This system will have infinitely many solutions.
Solve the system made up of equations (2) and (3) in terms of the arbitrary variable z. To eliminate x, multiply equation (2) by -2 and equation (3) by 5 and add the results

$$\begin{aligned} -10x - 14y + 2z &= -4 \\ 10x + 10y - 15z &= -20 \\ \hline -4y - 13z &= -24 \end{aligned}$$

Solve for y in terms of z.
$$-4y - 13z = -24 \Rightarrow -4y = -24 + 13z \Rightarrow$$
$$y = \frac{-24 + 13z}{-4} = \frac{24 - 13z}{4}$$

Express x also in terms of z by solving equation (3) for x and substituting $\frac{24-13z}{4}$ for y.
$$2x + 2y - 3z = -4 \Rightarrow 2x = -2y + 3z - 4 \Rightarrow$$
$$x = \frac{-2y + 3z - 4}{2}$$
$$x = \frac{-2\left(\frac{24-13z}{4}\right) + 3z - 4}{2} = \frac{\frac{-24+13z}{2} + 3z - 4}{2}$$
$$= \frac{-24 + 13z + 6z - 8}{4} = \frac{-32 + 19z}{4}$$

Solution set (with z arbitrary):
$$\left\{ \left(\frac{-32 + 19z}{4}, \frac{24 - 13z}{4}, z \right) \right\}$$

99.
$$\begin{aligned} 5x - y &= -4 \\ 3x + 2z &= 4 \\ 4y + 3z &= 22 \end{aligned}$$

Adding 4 times row 1 to row 3, we have

$$D = \begin{vmatrix} 5 & -1 & 0 \\ 3 & 0 & 2 \\ 0 & 4 & 3 \end{vmatrix} = \begin{vmatrix} 5 & -1 & 0 \\ 3 & 0 & 2 \\ 20 & 0 & 3 \end{vmatrix}.$$

(continued on next page)

Copyright © 2013 Pearson Education, Inc.

(*continued*)

Expanding by column two, we have

$$D = -(-1)\begin{vmatrix} 3 & 2 \\ 20 & 3 \end{vmatrix} = 9 - 40 = -31.$$

Adding 4 times row 1 to row 3, we have

$$D_x = \begin{vmatrix} -4 & -1 & 0 \\ 4 & 0 & 2 \\ 22 & 4 & 3 \end{vmatrix} = \begin{vmatrix} -4 & -1 & 0 \\ 4 & 0 & 2 \\ 6 & 0 & 3 \end{vmatrix}.$$

Expand by column two, we have

$$D_x = -(-1)\begin{vmatrix} 4 & 2 \\ 6 & 3 \end{vmatrix} = 12 - 12 = 0.$$

Adding column 2 to column 1, we have

$$D_y = \begin{vmatrix} 5 & -4 & 0 \\ 3 & 4 & 2 \\ 0 & 22 & 3 \end{vmatrix} = \begin{vmatrix} 1 & -4 & 0 \\ 7 & 4 & 2 \\ 22 & 22 & 3 \end{vmatrix}.$$

Adding 4 times column 1 to column 2, we now

have $D_y = \begin{vmatrix} 1 & 0 & 0 \\ 7 & 32 & 2 \\ 22 & 110 & 3 \end{vmatrix}.$

Expanding by row one, we have

$$D_y = 1\begin{vmatrix} 32 & 2 \\ 110 & 3 \end{vmatrix} = 96 - 220 = -124.$$

Adding 4 times row 1 to row 3, we have

$$D_z = \begin{vmatrix} 5 & -1 & -4 \\ 3 & 0 & 4 \\ 0 & 4 & 22 \end{vmatrix} = \begin{vmatrix} 5 & -1 & -4 \\ 3 & 0 & 4 \\ 20 & 0 & 6 \end{vmatrix}.$$

Expanding by column two, we have

$$D_z = -(-1)\begin{vmatrix} 3 & 4 \\ 20 & 6 \end{vmatrix} = 18 - 80 = -62.$$

Thus, we have

$$x = \frac{D_x}{D} = \frac{0}{-31} = 0, \; y = \frac{D_y}{D} = \frac{-124}{-31} = 4, \text{ and}$$

$$z = \frac{D_z}{D} = \frac{-62}{-31} = 2.$$

Solution set: $\{(0,4,2)\}$

101. $x + 2y = 10$
$3x + 4z = 7$
$-y - z = 1$

Adding -3 times row 1 to row 2, we have

$$D = \begin{vmatrix} 1 & 2 & 0 \\ 3 & 0 & 4 \\ 0 & -1 & -1 \end{vmatrix} = \begin{vmatrix} 1 & 2 & 0 \\ 0 & -6 & 4 \\ 0 & -1 & -1 \end{vmatrix}.$$

Expanding by column one, we have

$$D = 1\begin{vmatrix} -6 & 4 \\ -1 & -1 \end{vmatrix} = 6 + 4 = 10.$$

Adding column 1 to column 2 and column 1 to column 3, we have

$$D_x = \begin{vmatrix} 10 & 2 & 0 \\ 7 & 0 & 4 \\ 1 & -1 & -1 \end{vmatrix} = \begin{vmatrix} 10 & 12 & 10 \\ 7 & 7 & 11 \\ 1 & 0 & 0 \end{vmatrix}.$$

Expanding by row three, we have

$$D_x = 1\begin{vmatrix} 12 & 10 \\ 7 & 11 \end{vmatrix} = 132 - 70 = 62.$$

Adding column 2 to column 3, we have

$$D_y = \begin{vmatrix} 1 & 10 & 0 \\ 3 & 7 & 4 \\ 0 & 1 & -1 \end{vmatrix} = \begin{vmatrix} 1 & 10 & 10 \\ 3 & 7 & 11 \\ 0 & 1 & 0 \end{vmatrix}.$$

Expanding by row three, we have

$$D_y = -1\begin{vmatrix} 1 & 10 \\ 3 & 11 \end{vmatrix} = -1(11 - 30) = -1(-19) = 19.$$

Adding column 3 to column 2, we have

$$D_z = \begin{vmatrix} 1 & 2 & 10 \\ 3 & 0 & 7 \\ 0 & -1 & 1 \end{vmatrix} = \begin{vmatrix} 1 & 12 & 10 \\ 3 & 7 & 7 \\ 0 & 0 & 1 \end{vmatrix}.$$

Expanding by row three, we have

$$D_z = 1\begin{vmatrix} 1 & 12 \\ 3 & 7 \end{vmatrix} = 7 - 36 = -29.$$

Thus, we have $x = \frac{D_x}{D} = \frac{62}{10} = \frac{31}{5}$, $y = \frac{D_y}{D} = \frac{19}{10}$,

and $z = \frac{D_z}{D} = \frac{-29}{10} = -\frac{29}{10}.$

Solution set: $\left\{\left(\frac{31}{5}, \frac{19}{10}, -\frac{29}{10}\right)\right\}$

103. $\frac{\sqrt{3}}{2}(W_1 + W_2) = 100$
$W_1 - W_2 = 0$

Using the distributive property, we have the following system.

$\frac{\sqrt{3}}{2}W_1 + \frac{\sqrt{3}}{2}W_2 = 100$
$W_1 - W_2 = 0$

Using Cramer's rule, we have

$$D = \begin{vmatrix} \frac{\sqrt{3}}{2} & \frac{\sqrt{3}}{2} \\ 1 & -1 \end{vmatrix} = -\frac{\sqrt{3}}{2} - \frac{\sqrt{3}}{2} = -\sqrt{3},$$

$$D_{W_1} = \begin{vmatrix} 100 & \frac{\sqrt{3}}{2} \\ 0 & -1 \end{vmatrix} = -100 - 0 = -100, \text{ and}$$

$$D_{W_2} = \begin{vmatrix} \frac{\sqrt{3}}{2} & 100 \\ 1 & 0 \end{vmatrix} = 0 - 100 = -100.$$

(*continued on next page*)

Copyright © 2013 Pearson Education, Inc.

(*continued*)

This yields the following solution.

$$W_1 = \frac{D_{W_1}}{D} = \frac{-100}{-\sqrt{3}} = \frac{100}{\sqrt{3}} = \frac{100\sqrt{3}}{3} \approx 58 \text{ and}$$

$$W_2 = \frac{D_{W_2}}{D} = \frac{-100}{-\sqrt{3}} \approx 58$$

Both W_1 and W_2 are approximately 58 lb.

105. $bx + y = a^2$

$ax + y = b^2$

Using Cramer's rule, we have

$$D = \begin{vmatrix} b & 1 \\ a & 1 \end{vmatrix} = b - a, \quad D_x = \begin{vmatrix} a^2 & 1 \\ b^2 & 1 \end{vmatrix} = a^2 - b^2,$$

and $D_y = \begin{vmatrix} b & a^2 \\ a & b^2 \end{vmatrix} = b^3 - a^3.$

$$x = \frac{D_x}{D} = \frac{a^2 - b^2}{b - a} = \frac{(a+b)(a-b)}{b-a}$$

$$= \frac{(a+b)(a-b)}{-(a-b)} = -(a+b) = -a - b$$

$$y = \frac{D_y}{D} = \frac{b^3 - a^3}{b - a} = \frac{(b-a)(b^2 + ab + a^2)}{b - a}$$

$$= b^2 + ab + a^2$$

Solution set: $\left\{ \left(-a - b, a^2 + ab + b^2 \right) \right\}$

107. $b^2 x + a^2 y = b^2$

$ax + by = a$

Using Cramer's rule, we have

$$D = \begin{vmatrix} b^2 & a^2 \\ a & b \end{vmatrix} = b^3 - a^3,$$

$$D_x = \begin{vmatrix} b^2 & a^2 \\ a & b \end{vmatrix} = b^3 - a^3, \text{ and}$$

$$D_y = \begin{vmatrix} b^2 & b^2 \\ a & a \end{vmatrix} = ab^2 - ab^2 = 0.$$

$$x = \frac{D_x}{D} = \frac{b^3 - a^3}{b^3 - a^3} = 1$$

$$y = \frac{D_y}{D} = \frac{0}{b^3 - a^3} = 0$$

Note: In order for $D \neq 0$, we also assumed $a \neq b$.

Solution set: $\left\{ (1, 0) \right\}$

109. $ax + by = c$

$dx + ey = f$

Since a, b, c, d, e, and f are consecutive integers, we can rewrite any integer in terms of any other.

Using Cramer's rule, we have

$$D = \begin{vmatrix} a & b \\ d & e \end{vmatrix} = \begin{vmatrix} a & a+1 \\ a+3 & a+4 \end{vmatrix}$$

$$= a^2 + 4a - \left(a^2 + 4a + 3 \right) = -3$$

$$D_x = \begin{vmatrix} c & b \\ f & e \end{vmatrix} = \begin{vmatrix} b+1 & b \\ b+4 & b+3 \end{vmatrix} \text{ and}$$

$$= \left(b^2 + 4b + 3 \right) - \left(b^2 + 4b \right) = 3$$

$$D_y = \begin{vmatrix} a & c \\ d & f \end{vmatrix} = \begin{vmatrix} a & a+2 \\ a+3 & a+5 \end{vmatrix}$$

$$= \left(a^2 + 5a \right) - \left(a^2 + 5a + 6 \right) = -6$$

$$x = \frac{D_x}{D} = \frac{3}{-3} = -1$$

$$y = \frac{D_y}{D} = \frac{-6}{-3} = 2$$

Solution set: $\{(-1, 2)\}$

Section 5.4   Partial Fractions

1. $\dfrac{5}{3x(2x+1)} = \dfrac{A}{3x} + \dfrac{B}{2x+1}$

Multiply both sides by $3x(2x+1)$ to get

$5 = A(2x+1) + B(3x).$ (1)

First substitute 0 for x to get

$5 = A(2 \cdot 0 + 1) + B(3 \cdot 0) \Rightarrow A = 5.$

Replace A with 5 in equation (1) and substitute $-\frac{1}{2}$ for x to get the following.

$$5 = 5\left[2\left(-\tfrac{1}{2} \right) + 1 \right] + B\left[3\left(-\tfrac{1}{2} \right) \right]$$

$$= 5(-1+1) - \tfrac{3}{2}B = 5(0) - \tfrac{3}{2}B \Rightarrow$$

$$5 = -\tfrac{3}{2}B \Rightarrow -\tfrac{10}{3} = B$$

Thus, we have $\dfrac{5}{3x(2x+1)} = \dfrac{5}{3x} + \dfrac{-\tfrac{10}{3}}{2x+1} \Rightarrow$

$$\dfrac{5}{3x(2x+1)} = \dfrac{5}{3x} + \dfrac{-10}{3(2x+1)}$$

Copyright © 2013 Pearson Education, Inc.

3. $\dfrac{4x+2}{(x+2)(2x-1)} = \dfrac{A}{x+2} + \dfrac{B}{2x-1}$

Multiply both sides by $(x+2)(2x-1)$ to get

$4x+2 = A(2x-1)+B(x+2).$ (1)

First substitute -2 for x to get the following.

$4(-2)+2 = A[2(-2)-1]+B(-2+2)$

$-8+2 = A(-4-1)+B(0) = -6 = -5A$

$A = \frac{6}{5}$

Replace A with $\frac{6}{5}$ in equation (1) and

substitute $\frac{1}{2}$ for x to get the following.

$4 \cdot \frac{1}{2} + 2 = \frac{6}{5}\left(2 \cdot \frac{1}{2} - 1\right) + B\left(\frac{1}{2}+2\right)$

$2+2 = \frac{6}{5}(1-1) + B \cdot \frac{5}{2} \Rightarrow 4 = \frac{6}{5}(0) + \frac{5}{2}B \Rightarrow$

$4 = \frac{5}{2}B \Rightarrow B = \frac{8}{5}$

Thus, we have

$\dfrac{4x+2}{(x+2)(2x-1)} = \dfrac{\frac{6}{5}}{(x+2)} + \dfrac{\frac{8}{5}}{(2x-1)} \Rightarrow$

$\dfrac{4x+2}{(x+2)(2x-1)} = \dfrac{6}{5(x+2)} + \dfrac{8}{5(2x-1)}$

5. $\dfrac{x}{x^2+4x-5} = \dfrac{x}{(x+5)(x-1)} = \dfrac{A}{x+5} + \dfrac{B}{x-1}$

Multiply both sides by $(x+5)(x-1)$ to get

$x = A(x-1)+B(x+5).$ (1)

First substitute -5 for x to get

$-5 = A(-5-1)+B(-5+5) \Rightarrow$

$-5 = A(-6)+B(0) \Rightarrow -5 = -6A \Rightarrow A = \frac{5}{6}$

Replace A with $\frac{5}{6}$ in equation (1) and

substitute 1 for x to get

$1 = \frac{5}{6}(1-1)+B(1+5) \Rightarrow 1 = \frac{5}{6}(0)+B(6) \Rightarrow$

$1 = 6B \Rightarrow B = \frac{1}{6}$

Thus, we have

$\dfrac{x}{(x+5)(x-1)} = \dfrac{\frac{5}{6}}{x+5} + \dfrac{\frac{1}{6}}{x-1}$

$= \dfrac{5}{6(x+5)} + \dfrac{1}{6(x-1)}$

7. $\dfrac{4}{x(1-x)} = \dfrac{A}{x} + \dfrac{B}{1-x}$

Multiply both sides by $x(1-x)$ to get

$4 = A(1-x)+Bx.$ (1)

First substitute 0 for x to get the following.

$4 = A(1-0)+B(0) \Rightarrow 4 = A+0 \Rightarrow A = 4$

Replace A with 4 in equation (1) and substitute
1 for x to get the following.

$4 = 4(1-1)+B(1) \Rightarrow 4 = 0+B \Rightarrow B = 4$

Thus, we have $\dfrac{4}{x(1-x)} = \dfrac{4}{x} + \dfrac{4}{1-x}.$

9. $\dfrac{4x^2-x-15}{x(x+1)(x-1)} = \dfrac{A}{x} + \dfrac{B}{x+1} + \dfrac{C}{x-1}$

Multiply both sides by $x(x+1)(x-1)$ to get

$4x^2 - x - 15$
$= A(x+1)(x-1) + Bx(x-1) + Cx(x+1)$ (1)

First substitute 0 for x to get the following.

$4(0)^2 - 0 - 15$
$= A(0+1)(0-1) + B(0)(0-1) + C(0)(0+1)$

$0-0-15 = -A+0+0 \Rightarrow -15 = -A \Rightarrow A = 15$

Replace A with 15 in equation (1) and substitute
-1 for x to get the following.

$4(-1)^2 - (-1) - 15$
$= 15(-1+1)(-1-1) + B(-1)(-1-1)$
$\qquad\qquad\qquad + C(-1)(-1+1)$

$4+1-15$
$= 15(0)(-2) + B(-1)(-2) + C(-1)(0)$

$-10 = 0+2B+0 \Rightarrow -10 = 2B \Rightarrow B = -5$

$4x^2 - x - 15$
$= 15(x+1)(x-1) - 5x(x-1) + Cx(x+1)$ (2)

Now substitute 1 for x in equation (2) to get
the following.

$4(1)^2 - 1 - 15 = 15(1+1)(1-1) - 5(1)(1-1)$
$\qquad\qquad\qquad + C(1)(1+1)$

$4-1-15 = 15(2)(0) - 5(1)(0) + C(1)(2)$

$-12 = 0 - 0 + 2C$

$-12 = 2C \Rightarrow C = -6$

Thus, we have

$\dfrac{4x^2-x-15}{x(x+1)(x-1)} = \dfrac{15}{x} + \dfrac{-5}{x+1} + \dfrac{-6}{x-1}.$

Copyright © 2013 Pearson Education, Inc.

11. $\dfrac{2x+1}{(x+2)^3} = \dfrac{A}{x+2} + \dfrac{B}{(x+2)^2} + \dfrac{C}{(x+2)^3}$

Multiply both sides by $(x+2)^3$ to get

$2x+1 = A(x+2)^2 + B(x+2) + C.$ (1)

First substitute -2 for x to get the following.

$2(-2)+1 = A(-2+2)^2 + B(-2+2) + C$
$-4+1 = 0+0+C \Rightarrow C = -3$

Replace C with -3 in equation (1) and substitute 0 (arbitrary choice) for x to get the following.

$2(0)+1 = A(0+2)^2 + B(0+2) - 3$
$1 = 4A + 2B - 3 \Rightarrow 4A + 2B = 4$
$2A + B = 2$ (3)

Replace C with -3 in equation (1) and substitute -1 (arbitrary choice) for x to get the following.

$2(-1)+1 = A(-1+2)^2 + B(-1+2) - 3$
$-1 = A + B - 3 \Rightarrow A + B = 2$ (4)

Solve the system of equations by multiplying equation (4) by -1 and adding to equation 3.

$2A + B = 2$
$\underline{-A - B = -2}$
$A = 0$

Substituting 0 for A in equation (4) we obtain $0 + B = 2 \Rightarrow B = 2$.

Thus, we have

$\dfrac{2x+1}{(x+2)^3} = \dfrac{0}{x+2} + \dfrac{2}{(x+2)^2} + \dfrac{-3}{(x+2)^3}$

$= \dfrac{2}{(x+2)^2} + \dfrac{-3}{(x+2)^3}$

13. $\dfrac{x^2}{x^2+2x+1}$

This is not a proper fraction; the numerator has degree greater than or equal to that of the denominator. Divide the numerator by the denominator.

$$\begin{array}{r} 1 \\ x^2+2x+1\overline{)\,x^2} \\ \underline{x^2+2x+1} \\ -2x-1 \end{array}$$

Find the partial fraction decomposition for

$\dfrac{-2x-1}{x^2+2x+1} = \dfrac{-2x-1}{(x+1)^2}$.

$\dfrac{-2x-1}{(x+1)^2} = \dfrac{A}{x+1} + \dfrac{B}{(x+1)^2}$

Find the partial fraction decomposition for

$\dfrac{-2x-1}{x^2+2x+1} = \dfrac{-2x-1}{(x+1)^2}$.

$\dfrac{-2x-1}{(x+1)^2} = \dfrac{A}{x+1} + \dfrac{B}{(x+1)^2}$

Multiply both sides by $(x+1)^2$ to get

$-2x-1 = A(x+1) + B.$ (1)

First substitute -1 for x to get the following.

$-2(-1)-1 = A(-1+1) + B$
$2-1 = 0 + B \Rightarrow 1 = B$

Replace B with 1 in equation (1) and substitute 2 (arbitrary choice) for x to get the following.

$-2(2)-1 = A(2+1) + 1 \Rightarrow -5 = 3A + 1$
$-6 = 3A \Rightarrow A = -2$

Thus, we have

$\dfrac{x^2}{x^2+2x+1} = 1 + \dfrac{-2}{x+1} + \dfrac{1}{(x+1)^2}$.

15. $\dfrac{2x^5 + 3x^4 - 3x^3 - 2x^2 + x}{2x^2 + 5x + 2}$

The degree of the numerator is greater than the degree of the denominator, so first find the quotient.

$$\begin{array}{r} x^3 - x^2 \\ 2x^2+5x+2\overline{)\,2x^5+3x^4-3x^3-2x^2+x} \\ \underline{2x^5+5x^4+2x^3} \\ -2x^4-5x^3-2x^2+x \\ \underline{-2x^4-5x^3-2x^2} \\ x \end{array}$$

Find the partial fraction decomposition for

$\dfrac{x}{2x^2+5x+2} = \dfrac{x}{(2x+1)(x+2)}$.

$\dfrac{x}{(2x+1)(x+2)} = \dfrac{A}{2x+1} + \dfrac{B}{x+2}$

Multiply both sides by $(2x+1)(x+2)$ to get

$x = A(x+2) + B(2x+1).$ (1)

First substitute $-\frac{1}{2}$ for x to get the following.

$-\frac{1}{2} = A\left(-\frac{1}{2}+2\right) + B\left[2\left(-\frac{1}{2}\right)+1\right]$
$-\frac{1}{2} = A\left(\frac{3}{2}\right) + B(0) \Rightarrow -\frac{1}{2} = \frac{3}{2}A \Rightarrow A = -\frac{1}{3}$

Replace A with $-\frac{1}{3}$ in equation (1) and substitute -2 for x to get the following.

$-2 = -\frac{1}{3}(-2+2) + B[2(-2)+1]$
$-2 = -\frac{1}{3}(0) + B(-3) \Rightarrow -2 = -3B \Rightarrow B = \frac{2}{3}$

(continued on next page)

Copyright © 2013 Pearson Education, Inc.

(continued)

Thus, we have

$$\frac{2x^5 + 3x^4 - 3x^3 - 2x^2 + x}{2x^2 + 5x + 2}$$

$$= x^3 - x^2 + \frac{-1}{3(2x+1)} + \frac{2}{3(x+2)}$$

17. $\dfrac{x^3 + 4}{9x^3 - 4x}$

Find the quotient since the degrees of the numerator and denominator are the same.

$$9x^3 - 4x \overline{\smash{\big)}\, x^3 + 0x^2 + 0x + 4} \quad \tfrac{1}{9}$$
$$\underline{x^3 \qquad\quad -\tfrac{4}{9}x}$$
$$\tfrac{4}{9}x + 4$$

Find the partial fraction decomposition for

$$\frac{\frac{4}{9}x + 4}{9x^3 - 4x}.$$

$$\frac{\frac{4}{9}x + 4}{9x^3 - 4x} = \frac{\frac{4}{9}x + 4}{x(9x^2 - 4)} = \frac{\frac{4}{9}x + 4}{x(3x+2)(3x-2)}$$

$$= \frac{A}{x} + \frac{B}{3x+2} + \frac{C}{3x-2}$$

Multiply both sides of

$$\frac{\frac{4}{9}x + 4}{9x^3 - 4x} = \frac{A}{x} + \frac{B}{3x+2} + \frac{C}{3x-2} \quad \text{by}$$

$x(3x+2)(3x-2)$ to get the following.

$$\tfrac{4}{9}x + 4 = A(3x+2)(3x-2) + Bx(3x-2)$$
$$+ Cx(3x+2) \quad (1)$$

First substitute 0 for x to get the following.
$$\tfrac{4}{9}(0) + 4$$
$$= A[3(0) + 2][3(0) - 2] + B(0)[3(0) - 2]$$
$$+ C(0)[3(0) + 2]$$

$$0 + 4 = A(2)(-2) + 0 + 0 \Rightarrow 4 = -4A \Rightarrow A = -1$$

Replace A with -1 in equation (1) and substitute $-\tfrac{2}{3}$ for x to get the following.

$$\tfrac{4}{9}\left(-\tfrac{2}{3}\right) + 4 = -\left[3\left(-\tfrac{2}{3}\right) + 2\right]\left[3\left(-\tfrac{2}{3}\right) - 2\right]$$
$$+ B\left(-\tfrac{2}{3}\right)\left[3\left(-\tfrac{2}{3}\right) - 2\right]$$
$$+ C\left(-\tfrac{2}{3}\right)\left[3\left(-\tfrac{2}{3}\right) + 2\right]$$

$$-\tfrac{8}{27} + 4 = 0 + B\left(-\tfrac{2}{3}\right)(-4) + 0$$
$$\tfrac{100}{27} = \tfrac{8}{3}B \Rightarrow B = \tfrac{25}{18}$$

$$\tfrac{4}{9}x + 4 = -(3x+2)(3x-2)$$
$$+ \tfrac{25}{18}x(3x-2) + Cx(3x+2) \quad (2)$$

Substitute $\tfrac{2}{3}$ in equation (2) for x to get the following.

$$\tfrac{4}{9}\left(\tfrac{2}{3}\right) + 4 = -\left[3\left(\tfrac{2}{3}\right) + 2\right]\left[3\left(\tfrac{2}{3}\right) - 2\right]$$
$$+ \tfrac{25}{18}\left(\tfrac{2}{3}\right)\left[3\left(\tfrac{2}{3}\right) - 2\right] + C\left(\tfrac{2}{3}\right)\left[3\left(\tfrac{2}{3}\right) + 2\right]$$

$$\tfrac{8}{27} + 4 = 0 + 0 + C\left(\tfrac{2}{3}\right)(4)$$
$$\tfrac{116}{27} = \tfrac{8}{3}C \Rightarrow C = \tfrac{29}{18}$$

Thus, we have

$$\frac{x^3 + 4}{9x^3 - 4x} = \frac{1}{9} + \frac{-1}{x} + \frac{25}{18(3x+2)} + \frac{29}{18(3x-2)}.$$

19. $\dfrac{-3}{x^2\left(x^2 + 5\right)} = \dfrac{A}{x} + \dfrac{B}{x^2} + \dfrac{Cx + D}{x^2 + 5}$

Multiply both sides by $x^2\left(x^2 + 5\right)$ to get the following.

$$-3 = Ax\left(x^2 + 5\right) + B\left(x^2 + 5\right) + (Cx + D)x^2$$

Distributing and combining like terms on the right side of the equation, we have the following.

$$-3 = Ax\left(x^2 + 5\right) + B\left(x^2 + 5\right) + (Cx + D)x^2$$
$$-3 = Ax^3 + 5Ax + Bx^2 + 5B + Cx^3 + Dx^2$$
$$-3 = (A + C)x^3 + (B + D)x^2 + (5A)x + 5B$$

Equate the coefficients of like powers of x on the two sides of the equation.

For the x^3- term, we have $0 = A + C$.

For the x^2- term, we have $0 = B + D$.

For the x-term, we have $0 = 5A \Rightarrow A = 0$.

For the constant term, we have

$$-3 = 5B \Rightarrow B = -\tfrac{3}{5}.$$

Since $A = 0$ and $0 = A + C$, we have $C = 0$.

Since $B = -\tfrac{3}{5}$ and $0 = B + D$, we have

$$D = \tfrac{3}{5}.$$

Thus, we have $\dfrac{-3}{x^2\left(x^2 + 5\right)} = \dfrac{-3}{5x^2} + \dfrac{3}{5\left(x^2 + 5\right)}.$

Copyright © 2013 Pearson Education, Inc.

21. $\dfrac{3x-2}{(x+4)(3x^2+1)} = \dfrac{A}{x+4} + \dfrac{Bx+C}{3x^2+1}$

Multiply both sides by $(x+4)(3x^2+1)$ to get the following.

$3x-2 = A(3x^2+1) + (Bx+C)(x+4)$ (1)

First substitute -4 for x to get the following.

$3(-4)-2 = A\left[3(-4)^2+1\right]$
$\qquad\qquad + \left[B(-4)+C\right](-4+4)$

$-12-2 = A(48+1)+0$

$-14 = 49A \Rightarrow -\dfrac{14}{49} = A \Rightarrow A = -\dfrac{2}{7}$

Replace A with $-\dfrac{2}{7}$ in equation (1) and substitute 0 for x to get the following.

$3(0)-2 = -\dfrac{2}{7}\left[3(0)^2+1\right] + \left[B(0)+C\right](0+4)$

$-2 = -\dfrac{2}{7}+4C \Rightarrow -\dfrac{12}{7}=4C \Rightarrow C=-\dfrac{3}{7}$

$3x-2 = -\dfrac{2}{7}(3x^2+1) + \left(Bx-\dfrac{3}{7}\right)(x+4)$ (2)

Substitute 1 (arbitrary value) in equation (2) for x to get the following.

$3(1)-2 = -\dfrac{2}{7}\left[3(1)^2+1\right] + \left[B(1)-\dfrac{3}{7}\right](1+4)$

$3-2 = -\dfrac{2}{7}(4) + \left[B-\dfrac{3}{7}\right](5)$

$1 = -\dfrac{8}{7}+5B-\dfrac{15}{7}$

$1 = 5B-\dfrac{23}{7} \Rightarrow \dfrac{30}{7}=5B \Rightarrow B=\dfrac{6}{7}$

Thus, we have

$\dfrac{3x-2}{(x+4)(3x^2+1)} = \dfrac{-\dfrac{2}{7}}{x+4} + \dfrac{\dfrac{6}{7}x+\left(-\dfrac{3}{7}\right)}{3x^2+1}$

$\qquad\qquad = \dfrac{-2}{7(x+4)} + \dfrac{6x-3}{7(3x^2+1)}$

23. $\dfrac{1}{x(2x+1)(3x^2+4)} = \dfrac{A}{x} + \dfrac{B}{2x+1} + \dfrac{Cx+D}{3x^2+4}$

Multiply both sides by $x(2x+1)(3x^2+4)$ to get the following.

$1 = A(2x+1)(3x^2+4) + Bx(3x^2+4)$
$\qquad\qquad + (Cx+D)(x)(2x+1)$

Expanding and combining like terms on the right side of the equation, we have the following.

$1 = A(2x+1)(3x^2+4) + Bx(3x^2+4)$
$\qquad\qquad + (Cx+D)(x)(2x+1)$

$1 = A(6x^3+3x^2+8x+4) + B(3x^3+4x)$
$\qquad\qquad + C(2x^3+x^2)$
$\qquad\qquad + D(2x^2+x)$

$1 = 6Ax^3+3Ax^2+8Ax+4A+3Bx^3$
$\qquad\qquad + 4Bx+2Cx^3+Cx^2+2Dx^2+Dx$

$1 = (6A+3B+2C)x^3 + (3A+C+2D)x^2$
$\qquad\qquad + (8A+4B+D)x+4A$

Equate the coefficients of like powers of x on the two sides of the equation.

For the x^3- term, we have $0 = 6A+3B+2C$.

For the x^2- term, we have $0 = 3A+C+2D$.

For the x-term, we have $0 = 8A+4B+D$.

For the constant term, we have $1 = 4A$.

If we use the Gauss-Jordan method, we begin with the following augmented matrix.

$\begin{bmatrix} 6 & 3 & 2 & 0 & 0 \\ 3 & 0 & 1 & 2 & 0 \\ 8 & 4 & 0 & 1 & 0 \\ 4 & 0 & 0 & 0 & 1 \end{bmatrix}$

$\begin{bmatrix} 6 & 3 & 2 & 0 & 0 \\ 3 & 0 & 1 & 2 & 0 \\ 8 & 4 & 0 & 1 & 0 \\ 1 & 0 & 0 & 0 & \frac{1}{4} \end{bmatrix} \frac{1}{4}R4 \Rightarrow$

$\begin{bmatrix} 1 & 0 & 0 & 0 & \frac{1}{4} \\ 3 & 0 & 1 & 2 & 0 \\ 8 & 4 & 0 & 1 & 0 \\ 6 & 3 & 2 & 0 & 0 \end{bmatrix} R1 \leftrightarrow R4 \Rightarrow$

$\begin{bmatrix} 1 & 0 & 0 & 0 & \frac{1}{4} \\ 0 & 0 & 1 & 2 & -\frac{3}{4} \\ 0 & 4 & 0 & 1 & -2 \\ 0 & 3 & 2 & 0 & -\frac{3}{2} \end{bmatrix} \begin{matrix} \\ -3R1+R2 \\ -8R1+R3 \\ -6R1+R4 \end{matrix} \Rightarrow$

$\begin{bmatrix} 1 & 0 & 0 & 0 & \frac{1}{4} \\ 0 & 0 & 1 & 2 & -\frac{3}{4} \\ 0 & 1 & 0 & \frac{1}{4} & -\frac{1}{2} \\ 0 & 3 & 2 & 0 & -\frac{3}{2} \end{bmatrix} \begin{matrix} \\ \\ \frac{1}{4}R3 \\ \end{matrix} \Rightarrow$

(continued on next page)

Copyright © 2013 Pearson Education, Inc.

(continued)

$$\begin{bmatrix} 1 & 0 & 0 & 0 & | & \frac{1}{4} \\ 0 & 1 & 0 & \frac{1}{4} & | & -\frac{1}{2} \\ 0 & 0 & 1 & 2 & | & -\frac{3}{4} \\ 0 & 3 & 2 & 0 & | & -\frac{3}{2} \end{bmatrix} \text{R2} \leftrightarrow \text{R3} \quad \Rightarrow$$

$$\begin{bmatrix} 1 & 0 & 0 & 0 & | & \frac{1}{4} \\ 0 & 1 & 0 & \frac{1}{4} & | & -\frac{1}{2} \\ 0 & 0 & 1 & 2 & | & -\frac{3}{4} \\ 0 & 0 & 2 & -\frac{3}{4} & | & 0 \end{bmatrix} \quad \Rightarrow$$
$$-3\text{R2} + \text{R4}$$

$$\begin{bmatrix} 1 & 0 & 0 & 0 & | & \frac{1}{4} \\ 0 & 1 & 0 & \frac{1}{4} & | & -\frac{1}{2} \\ 0 & 0 & 1 & 2 & | & -\frac{3}{4} \\ 0 & 0 & 0 & -\frac{19}{4} & | & \frac{3}{2} \end{bmatrix} \quad \Rightarrow$$
$$-2\text{R3} + \text{R4}$$

$$\begin{bmatrix} 1 & 0 & 0 & 0 & | & \frac{1}{4} \\ 0 & 1 & 0 & \frac{1}{4} & | & -\frac{1}{2} \\ 0 & 0 & 1 & 2 & | & -\frac{3}{4} \\ 0 & 0 & 0 & 1 & | & -\frac{6}{19} \end{bmatrix} \quad \Rightarrow$$
$$-\frac{4}{19}\text{R4}$$

$$\begin{bmatrix} 1 & 0 & 0 & 0 & | & \frac{1}{4} \\ 0 & 1 & 0 & 0 & | & -\frac{8}{19} \\ 0 & 0 & 1 & 0 & | & -\frac{9}{76} \\ 0 & 0 & 0 & 1 & | & -\frac{6}{19} \end{bmatrix} \begin{array}{l} \\ -\frac{1}{4}\text{R4} + \text{R2} \\ -2\text{R4} + \text{R3} \\ \end{array}$$

Thus, we have

$$\frac{1}{x(2x+1)(3x^2+4)}$$

$$= \frac{\frac{1}{4}}{x} + \frac{-\frac{8}{19}}{2x+1} + \frac{-\frac{9}{76}x + \left(-\frac{6}{19}\right)}{3x^2+4}$$

$$= \frac{\frac{1}{4}}{x} + \frac{-\frac{8}{19}}{2x+1} + \frac{-\frac{9}{76}x - \frac{24}{76}}{3x^2+4}$$

$$= \frac{1}{4x} + \frac{-8}{19(2x+1)} + \frac{-9x-24}{76(3x^2+4)}$$

25. $\dfrac{3x-1}{x(2x^2+1)^2} = \dfrac{A}{x} + \dfrac{Bx+C}{2x^2+1} + \dfrac{Dx+E}{(2x^2+1)^2}$

Multiply both sides by $x(2x^2+1)^2$ to get the following.

$$3x-1 = A(2x^2+1)^2 + (Bx+C)(x)(2x^2+1) + (Dx+E)x$$

Expanding and combining like terms on the right side of the equation, we have the following.

$$3x-1 = A(2x^2+1)^2 + (Bx+C)(x)(2x^2+1)$$
$$+ (Dx+E)x$$
$$= A(4x^4 + 4x^2 + 1) + B(2x^4 + x^2)$$
$$+ C(2x^3 + x)$$
$$+ Dx^2 + Ex$$
$$= 4Ax^4 + 4Ax^2 + A + 2Bx^4 + Bx^2$$
$$+ 2Cx^3 + Cx + Dx^2 + Ex$$
$$3x-1 = (4A+2B)x^4 + 2Cx^3 + (4A+B+D)x^2$$
$$+ (C+E)x + A$$

Equate the coefficients of like powers of x on the two sides of the equation.

For the x^4-term, we have $0 = 4A + 2B$.

For the x^3-term, we have $0 = 2C \Rightarrow C = 0$.

For the x^2-term, we have $0 = 4A + B + D$.

For the x-term, we have $3 = C + E$.

For the constant term, we have $-1 = A$.

Since $A = -1$ and $0 = 4A + 2B$, we have $B = 2$.

Since $A = -1$, $B = 2$ and $0 = 4A + B + D$, we have $D = 2$. Since $C = 0$ and $3 = C + E$, we have $E = 3$.

Thus, we have

$$\frac{3x-1}{x(2x^2+1)^2} = \frac{-1}{x} + \frac{2x+0}{2x^2+1} + \frac{2x+3}{(2x^2+1)^2}$$

$$= \frac{-1}{x} + \frac{2x}{2x^2+1} + \frac{2x+3}{(2x^2+1)^2}$$

27. $\dfrac{-x^4 - 8x^2 + 3x - 10}{(x+2)(x^2+4)^2}$

$$= \frac{A}{x+2} + \frac{Bx+C}{x^2+4} + \frac{Dx+E}{(x^2+4)^2}$$

Multiply both sides by $(x+2)(x^2+4)^2$ to get the following.

$$-x^4 - 8x^2 + 3x - 10$$
$$= A(x^2+4)^2 + (Bx+C)(x+2)(x^2+4)$$
$$+ (Dx+E)(x+2) \quad (1)$$

Substitute then expand and combine like terms on the right side of the equation.

Substitute -2 in equation (1) for x to get the following.

(continued on next page)

Copyright © 2013 Pearson Education, Inc.

(*continued*)

$$-(-2)^4 - 8(-2)^2 + 3(-2) - 10$$
$$= A\left[(-2)^2 + 4\right]^2$$
$$\quad + \left[B(-2) + C\right](-2+2)\left[(-2)^2 + 4\right]$$
$$\quad + \left[D(-2) + E\right](-2+2)$$
$$-16 - 32 + (-6) - 10 = A(4+4)^2 + 0 + 0$$
$$-64 = 64A \Rightarrow A = -1$$

Substituting -1 for A in equation (1) and expanding and combining like terms, we have the following.

$$-x^4 - 8x^2 + 3x - 10$$
$$= -1(x^2+4)^2 + (Bx+C)(x+2)(x^2+4)$$
$$\quad + (Dx+E)(x+2)$$
$$= -1(x^4 + 8x^2 + 16)$$
$$\quad + (Bx+C)(x^3 + 2x^2 + 4x + 8)$$
$$\quad + D(x^2 + 2x) + E(x+2)$$
$$= -x^4 - 8x^2 - 16 + Bx^4 + 2Bx^3 + 4Bx^2 + 8Bx$$
$$\quad + Cx^3 + 2Cx^2 + 4Cx + 8C + Dx^2 + 2Dx$$
$$\quad + Ex + 2E$$
$$= (-1+B)x^4 + (2B+C)x^3$$
$$\quad + (-8 + 4B + 2C + D)x^2$$
$$\quad + (8B + 4C + 2D + E)x$$
$$\quad + (-16 + 8C + 2E)$$

Equate the coefficients of like powers of x on the two sides of the equation.

For the x^4-term, we have
$-1 = -1 + B \Rightarrow B = 0$.

For the x^3-term, we have $0 = 2B + C$.

For the x^2-term, we have
$-8 = -8 + 4B + 2C + D \Rightarrow 0 = 4B + 2C + D$.

For the x-term, we have
$3 = 8B + 4C + 2D + E$.

For the constant term, we have
$-10 = -16 + 8C + 2E \Rightarrow 6 = 8C + 2E$.

Since $B = 0$ and $0 = 2B + C$, we have $C = 0$.
Since $B = 0$, $C = 0$, and $0 = 4B + 2C + D$, we have $D = 0$. Since $C = 0$ and $6 = 8C + 2E$, we have $E = 3$.

Thus we have

$$\frac{-x^4 - 8x^2 + 3x - 10}{(x+2)(x^2+4)^2}$$
$$= \frac{-1}{x+2} + \frac{0 \cdot x + 0}{x^2+4} + \frac{0 \cdot x + 3}{(x^2+4)^2}$$
$$= \frac{-1}{x+2} + \frac{3}{(x^2+4)^2}$$

29. $\dfrac{5x^5 + 10x^4 - 15x^3 + 4x^2 + 13x - 9}{x^3 + 2x^2 - 3x}$

Since the degree of the numerator is higher than the degree of the denominator, first find the quotient.

$$\begin{array}{r} 5x^2 \\ x^3 + 2x^2 - 3x \overline{) 5x^5 + 10x^4 - 15x^3 + 4x^2 + 13x - 9} \\ \underline{5x^5 + 10x^4 - 15x^3 } \\ 4x^2 + 13x - 9 \end{array}$$

Find the partial fraction decomposition of
$\dfrac{4x^2 + 13x - 9}{x^3 + 2x^2 - 3x}$.

$$\frac{4x^2 + 13x - 9}{x^3 + 2x^2 - 3x} = \frac{4x^2 + 13x - 9}{x(x^2 + 2x - 3)}$$
$$= \frac{4x^2 + 13x - 9}{x(x+3)(x-1)}$$
$$= \frac{A}{x} + \frac{B}{x+3} + \frac{C}{x-1}$$

Multiply by $x(x+3)(x-1)$ to obtain

$$4x^2 + 13x - 9$$
$$= A(x+3)(x-1) + Bx(x-1) + Cx(x+3) \quad (1)$$

First substitute 0 for x in equation (1) to get the following.

$$4(0)^2 + 13(0) - 9$$
$$= A(0+3)(0-1) + B(0)(0-1)$$
$$\qquad\qquad + C(0)(0+3)$$
$$-9 = -3A \Rightarrow A = 3$$

Replace A with 3 in equation (1) and substitute -3 for x to get the following.

$$4(-3)^2 + 13(-3) - 9$$
$$= 3(-3+3)(-3-1) + B(-3)(-3-1)$$
$$\qquad\qquad + C(-3)(-3+3)$$
$$-12 = 12B \Rightarrow B = -1$$

$$4x^2 + 13x - 9$$
$$= 3(x+3)(x-1) - x(x-1) + Cx(x+3) \quad (2)$$

(continued on next page)

Copyright © 2013 Pearson Education, Inc.

(continued)

Now substitute 1 (arbitrary choice) for x in equation (2) to get the following.

$4(1)^2 + 13(1) - 9$
$= 3(1+3)(1-1) - 1(1-1) + C(1)(1+3)$
$8 = 4C \Rightarrow C = 2$

Thus, we have

$\dfrac{5x^5 + 10x^4 - 15x^3 + 4x^2 + 13x - 9}{x^3 + 2x^2 - 3x}$
$= 5x^2 + \dfrac{3}{x} + \dfrac{-1}{x+3} + \dfrac{2}{x-1}$

31. $\dfrac{x^2}{x^4 - 1} = \dfrac{x^2}{(x^2+1)(x^2-1)}$

$= \dfrac{x^2}{(x^2+1)(x+1)(x-1)}$

$= \dfrac{A}{x+1} + \dfrac{B}{x-1} + \dfrac{C}{x^2+1}$

Multiply both sides by $(x^2+1)(x+1)(x-1)$ to get the following.

$x^2 = A(x^2+1)(x-1) + B(x^2+1)(x+1)$
$\qquad + C(x+1)(x-1)$ (1)

Substitute, then expand and combine like terms on the right side of the equation.

Substitute −1 in equation (1) for x to get the following.

$(-1)^2 = A\big[(-1)^2 + 1\big](-1-1)$
$\qquad + B\big[(-1)^2 + 1\big](-1+1)$
$\qquad + \big[C(-1) + D\big](-1+1)(-1-1)$
$1 = A(1+1)(-1-1) + 0 + 0$
$1 = -4A \Rightarrow A = -\frac{1}{4}$

Substituting $-\frac{1}{4}$ for A and 1 for x in equation (1), we have:

$1^2 = -\frac{1}{4}(1^2+1)(1-1) + B(1^2+1)(1+1)$
$\qquad + C(1+1)(1-1)$
$1 = 0 + 0 + 4B \Rightarrow 1 = 4B \Rightarrow B = \frac{1}{4}$

$x^2 = -\frac{1}{4}(x^2+1)(x-1) + \frac{1}{4}(x^2+1)(x+1)$
$\qquad + C(x+1)(x-1)$ (2)

Expanding and combining like terms, we have the following.

$x^2 = -\frac{1}{4}(x^2+1)(x-1) + \frac{1}{4}(x^2+1)(x+1)$
$\qquad + C(x+1)(x-1)$
$= -\frac{1}{4}(x^3 - x^2 + x - 1) + \frac{1}{4}(x^3 + x^2 + x + 1)$
$\qquad + C(x^2 - 1)$
$= Cx^2 + \frac{1}{2}x^2 - C + \frac{1}{2} = (C + \frac{1}{2})x^2 - C + \frac{1}{2}$

Equate the coefficients of like powers of x on the two sides of the equation.

For the x^2-term, we have $1 = C + \frac{1}{2} \Rightarrow C = \frac{1}{2}$.

For the constant term, we have

$0 = -C + \frac{1}{2} \Rightarrow C = \frac{1}{2}$ (as above)

Thus, we have

$\dfrac{x^2}{x^4 - 1} = \dfrac{-\frac{1}{4}}{x+1} + \dfrac{\frac{1}{4}}{x-1} + \dfrac{\frac{1}{2}}{x^2+1}$

$= \dfrac{-1}{4(x+1)} + \dfrac{1}{4(x-1)} + \dfrac{1}{2(x^2+1)}$

Copyright © 2013 Pearson Education, Inc.

33. $\dfrac{4x^2-3x-4}{x^3+x^2-2x}=\dfrac{4x^2-3x-4}{x\left(x^2+x-2\right)}=\dfrac{4x^2-3x-4}{x(x-1)(x+2)}=\dfrac{A}{x}+\dfrac{B}{x-1}+\dfrac{C}{x+2}$

Multiply both sides by $x(x-1)(x+2)$ to get the following.

$4x^2-3x-4=A(x-1)(x+2)+Bx(x+2)+Cx(x-1)$ (1)

Substitute, then expand and combine like terms on the right side of the equation.
Substitute 1 in equation (1) for x to get the following.

$4(1)^2-3(1)-4=A(1-1)(1+2)+B(1)(1+2)+C(1)(1-1)\Rightarrow-3=0+3B+0\Rightarrow B=-1$

Now substitute -1 for B and -2 for x in equation (1).

$4(-2)^2-3(-2)-4=A(-2-1)(-2+2)+(-1)(-2)(-2+2)+C(-2)(-2-1)\Rightarrow18=0+0+6C\Rightarrow C=3$

Substitute the values for B and C into equation (1), then expand and combine like terms.

$4x^2-3x-4=A(x-1)(x+2)-1x(x+2)+3x(x-1)=A\left(x^2+x-2\right)-x^2-2x+3x^2-3x$

$=Ax^2+Ax-2A-x^2-2x+3x^2-3x=Ax^2-x^2+3x^2+Ax-2x-3x-2A$

$=(A+2)x^2+(A-5)x-2A$

Equate the coefficients of like powers of x on the two sides of the equation.
For the x^2-term, we have $4=A+2\Rightarrow2=A$.
For the x-term, we have $-3=A-5\Rightarrow2=A$ (as above).
For the constant term, we have $-4=-2A\Rightarrow2=A$ (again).
Thus,

$\dfrac{4x^2-3x-4}{x^3+x^2-2x}=\dfrac{2}{x}+\dfrac{-1}{x-1}+\dfrac{3}{x+2}$

Chapter 5 Quiz
(Sections 5.1–5.4)

1. $2x+y=-4$ (1)
$-x+2y=2$ (2)
Solve equation (1) for y:
$2x+y=-4\Rightarrow y=-2x-4$ (3)
Replace y in equation (2) with $-2x-4$ and solve for x:
$-x+2y=2\Rightarrow-x+2(-2x-4)=2$
$-5x-8=2\Rightarrow x=-2$
Substitute -2 for x in equation (3) and solve
for y: $y=-2(-2)-4=0$
Verify that the ordered pair $(-2,0)$ satisfies both equations.
Check:
$\begin{array}{l|l}2x+y=-4 \text{ (1)} & -x+2y=2 \text{ (2)}\\2(-2)+0=-4 \quad? & -(-2)+2(0)=2 \quad?\\-4=-4 \text{ True} & 2=2 \text{ True}\end{array}$
Solution set: $\{(-2,0)\}$

3. $x-y=6$ (1)
$x-y=4$ (2)
Multiply equation (2) by -1, then add the result to equation (1):
$\begin{array}{r}x-y=6\\-x+y=-4\\\hline0=2\end{array}$
This is a false statement. The system is inconsistent and the solution set is $\varnothing$.

5. $3x+5y=-5$ (1)
$-2x+3y=16$ (2)
This system has the augmented matrix
$\begin{bmatrix}3&5&|&-5\\-2&3&|&16\end{bmatrix}.$
$\begin{bmatrix}1&8&|&11\\-2&3&|&16\end{bmatrix}\text{R1+R2}\quad\Rightarrow$
$\begin{bmatrix}1&8&|&11\\0&19&|&38\end{bmatrix}\text{2R1+R2}\quad\Rightarrow$
$\begin{bmatrix}1&8&|&11\\0&1&|&2\end{bmatrix}\frac{1}{19}\text{R2}\quad\Rightarrow$
$\begin{bmatrix}1&0&|&-5\\0&1&|&2\end{bmatrix}\text{R1-8R2}$

(continued on next page)

Copyright © 2013 Pearson Education, Inc.

(continued)

Verify that the ordered pair $(-5, 2)$ satisfies both equations.

Check:
$$3x + 5y = -5 \quad (1)$$
$$3(-5) + 5(2) = -5 \quad ?$$
$$-15 + 10 = -5$$
$$-5 = -5 \quad \text{True}$$

$$-2x + 3y = 16 \quad (2)$$
$$-2(-5) + 3(2) = 16 \quad ?$$
$$10 + 6 = 16$$
$$16 = 16 \quad \text{True}$$

Solution set: $\{(-5, 2)\}$

7.
$$\begin{aligned} x + y + z &= 1 \quad (1) \\ -x + y + z &= 5 \quad (2) \\ y + 2z &= 5 \quad (3) \end{aligned}$$

Eliminate x by adding equations (1) and (2) to get $2y + 2z = 6 \Rightarrow y + z = 3$ (4).

Now solve the system consisting of equations (3) and (4) by subtracting equation (4) from equation (3):
$$\begin{aligned} y + 2z &= 5 \quad (3) \\ y + z &= 3 \quad (4) \\ \hline z &= 2 \end{aligned}$$

Substitute the value for z into equation (4) to solve for y: $y + 2 = 3 \Rightarrow y = 1$

Now, substitute the values for y and z into equation (1) to solve for x:
$x + 1 + 2 = 1 \Rightarrow x = -2$

Verify that the ordered triple $(-2, 1, 2)$ satisfies all three equations.

Check:

$$\begin{array}{l|l} x + y + z = 1 \ (1) & -x + y + z = 5 \quad (2) \\ -2 + 1 + 2 = 1 \ ? & -(-2) + 1 + 2 = 5 \ ? \\ \quad\quad 1 = 1 \ \text{True} & \quad\quad 5 = 5 \quad \text{True} \end{array}$$

$$y + 2z = 5 \quad (3)$$
$$1 + 2(2) = 5 \quad ?$$
$$5 = 5 \quad \text{True}$$

Solution set: $\{(-2, 1, 2)\}$

9.
$$\begin{aligned} 7x + y - z &= 4 \quad (1) \\ 2x - 3y + z &= 2 \quad (2) \\ -6x + 9y - 3z &= -6 \quad (3) \end{aligned}$$

$$D = \begin{vmatrix} 7 & 1 & -1 \\ 2 & -3 & 1 \\ -6 & 9 & -3 \end{vmatrix}$$

Adding 3 times row 2 to row 3, we have

$$D = \begin{vmatrix} 7 & 1 & -1 \\ 2 & -3 & 1 \\ 0 & 0 & 0 \end{vmatrix}$$

Since we have a row of zeros, $D = 0$ and we cannot use Cramer's rule.

Using the elimination method, we can add 3 times equation (2) to equation (3).
$$\begin{aligned} 6x - 9y + 3z &= 6 \quad (2) \\ -6x + 9y - 3z &= -6 \quad (3) \\ \hline 0 &= 0 \quad \text{True} \end{aligned}$$

This system will have infinitely many solutions.

Solve the system made up of equations (1) and (3) in terms of the arbitrary variable y. To eliminate z, multiply equation (1) by -3 and add the result to equation (3):
$$\begin{aligned} -21x - 3y + 3z &= -12 \\ -6x + 9y - 3z &= -6 \\ \hline -27x + 6y &= -18 \Rightarrow -27x = -6y - 18 \Rightarrow \end{aligned}$$
$$x = \frac{6y + 18}{27} = \frac{2y + 6}{9}$$

Now, express z also in terms of y by solving equation (1) for y and substituting $\frac{2y+6}{9}$ for x in the result.
$$7x + y - z = 4 \Rightarrow z = 7x + y - 4 \Rightarrow$$
$$z = 7\left(\frac{2y + 6}{9}\right) + y - 4 = \frac{14y + 42}{9} + y - 4$$
$$= \frac{23y + 6}{9}$$

Solution set: $\left\{\left(\dfrac{2y + 6}{9}, y, \dfrac{23y + 6}{9}\right)\right\}$

11. Let x = the amount invested at 3%; y = the amount invested at 4%; z = the amount invested at 6%

The information in the problem gives the system:
$$\begin{cases} x + y + z = 5000 \\ z = x + y \Rightarrow \\ 0.03x + 0.04y + 0.06z = 240 \end{cases}$$
$$\begin{cases} x + y + z = 5000 & (1) \\ -x - y + z = 0 & (2) \\ 0.03x + 0.04y + 0.06 = 240 & (3) \end{cases}$$

This system has the augmented matrix
$$\begin{bmatrix} 1 & 1 & 1 & 5000 \\ -1 & -1 & 1 & 0 \\ 0.03 & 0.04 & 0.06 & 240 \end{bmatrix}$$

(continued on next page)

Copyright © 2013 Pearson Education, Inc.

(continued)

$$\begin{bmatrix} 1 & 1 & 1 & | & 5000 \\ 0 & 0 & 1 & | & 2500 \\ 0.03 & 0.04 & 0.06 & | & 240 \end{bmatrix} \tfrac{1}{2}(R1+R2) \Rightarrow$$

$$\begin{bmatrix} 1 & 1 & 1 & | & 5000 \\ 0.03 & 0.04 & 0.06 & | & 240 \\ 0 & 0 & 1 & | & 2500 \end{bmatrix} R2 \leftrightarrow R3 \Rightarrow$$

$$\begin{bmatrix} 1 & 1 & 1 & | & 5000 \\ 3 & 4 & 6 & | & 24,000 \\ 0 & 0 & 1 & | & 2500 \end{bmatrix} 100R2 \Rightarrow$$

$$\begin{bmatrix} 1 & 1 & 1 & | & 5000 \\ 0 & 1 & 3 & | & 9000 \\ 0 & 0 & 1 & | & 2500 \end{bmatrix} -3R1+R2 \Rightarrow$$

$$\begin{bmatrix} 1 & 1 & 1 & | & 5000 \\ 0 & 1 & 0 & | & 1500 \\ 0 & 0 & 1 & | & 2500 \end{bmatrix} R2-3R3 \Rightarrow$$

$$\begin{bmatrix} 1 & 0 & 1 & | & 3500 \\ 0 & 1 & 0 & | & 1500 \\ 0 & 0 & 1 & | & 2500 \end{bmatrix} R1-R2 \Rightarrow$$

$$\begin{bmatrix} 1 & 0 & 0 & | & 1000 \\ 0 & 1 & 0 & | & 1500 \\ 0 & 0 & 1 & | & 2500 \end{bmatrix} R1-R3 \Rightarrow$$

$1000 was invested at 3%, $1500 was invested at 4%, and $2500 was invested at 6%.

13. $\begin{vmatrix} -1 & 2 & 4 \\ -3 & -2 & -3 \\ 2 & -1 & 5 \end{vmatrix}$

To expand by the first row, we will need to find $a_{11} \cdot M_{11} - a_{12} \cdot M_{12} + a_{13} \cdot M_{13}$.

$$M_{11} = \begin{vmatrix} -2 & -3 \\ -1 & 5 \end{vmatrix} = (-2)(5) - (-1)(-3) = -13$$

$$M_{12} = \begin{vmatrix} -3 & -3 \\ 2 & 5 \end{vmatrix} = (-3)(5) - (2)(-3) = -9$$

$$M_{13} = \begin{vmatrix} -3 & -2 \\ 2 & -1 \end{vmatrix} = (-3)(-1) - (2)(-2) = 7$$

$a_{11} \cdot M_{11} - a_{12} \cdot M_{12} + a_{13} \cdot M_{13}$
$= (-1)(-13) - (2)(-9) + (4)(7) = 59$

15. $\dfrac{2x^2 - 15x - 32}{(x-1)(x^2 + 6x + 8)} = \dfrac{2x^2 - 15x - 32}{(x-1)(x+2)(x+4)}$

$\qquad = \dfrac{A}{x-1} + \dfrac{B}{x+2} + \dfrac{C}{x+4}$

Multiply both sides by $(x-1)(x+2)(x+4)$:

$2x^2 - 15x - 32$
$= A(x+2)(x+4) + B(x-1)(x+4)$
$\qquad\qquad + C(x-1)(x+2)$

Expanding and combining like terms on the right side of the equation, we have

$2x^2 - 15x - 32$
$= A(x+2)(x+4) + B(x-1)(x+4)$
$\qquad\qquad + C(x-1)(x+2)$
$= Ax^2 + 6Ax + 8A + Bx^2 + 3Bx - 4B$
$\qquad\qquad + Cx^2 + Cx - 2C$
$= x^2(A + B + C) + x(6A + 3B + C)$
$\qquad\qquad + (8A - 4B - 2C)$

Equate the coefficients of like powers of x on the two sides of the equation.

For the x^2-term, $2 = A + B + C$
For the x-term, $-15 = 6A + 3B + C$
For the constant term, $-32 = 8A - 4B - 2C$
Using the Gauss-Jordan method, we have

$$\begin{bmatrix} 1 & 1 & 1 & | & 2 \\ 6 & 3 & 1 & | & -15 \\ 8 & -4 & -2 & | & -32 \end{bmatrix} \Rightarrow$$

$$\begin{bmatrix} 1 & 1 & 1 & | & 2 \\ 0 & -3 & -5 & | & -27 \\ 0 & -12 & -10 & | & -48 \end{bmatrix} \begin{matrix} -6R1+R2 \\ -8R1+R2 \end{matrix} \Rightarrow$$

$$\begin{bmatrix} 1 & 1 & 1 & | & 2 \\ 0 & -3 & -5 & | & -27 \\ 0 & 0 & 1 & | & 6 \end{bmatrix} \tfrac{1}{10}(-4R2+R3) \Rightarrow$$

$$\begin{bmatrix} 1 & 1 & 1 & | & 2 \\ 0 & 1 & 0 & | & -1 \\ 0 & 0 & 1 & | & 6 \end{bmatrix} -\tfrac{1}{3}(5R3+R2) \Rightarrow$$

$$\begin{bmatrix} 1 & 0 & 1 & | & 3 \\ 0 & 1 & 0 & | & -1 \\ 0 & 0 & 1 & | & 6 \end{bmatrix} R1-R2 \Rightarrow$$

$$\begin{bmatrix} 1 & 0 & 0 & | & -3 \\ 0 & 1 & 0 & | & -1 \\ 0 & 0 & 1 & | & 6 \end{bmatrix} R1-R3$$

$A = -3$, $B = -1$, $C = 6$.
Thus

$$\frac{2x^2 - 15x - 32}{(x-1)(x^2 + 6x + 8)} = \frac{-3}{x-1} + \frac{-1}{x+2} + \frac{6}{x+4}$$

Copyright © 2013 Pearson Education, Inc.

Section 5.5 Nonlinear Systems of Equations

1. The system is $x^2 = y - 1$
$$y = 3x + 5$$

The proposed solution set is $\{(-1, 2), (4, 17)\}$.

Check $(-1, 2)$.

$$x^2 = y - 1 \qquad y = 3x + 5$$
$$(-1)^2 = 2 - 1 \,? \qquad 2 = 3(-1) + 5 \,?$$
$$1 = 1 \qquad \qquad 2 = -3 + 5$$
$$\qquad \qquad \qquad 2 = 2$$

Check $(4, 17)$.

$$x^2 = y - 1 \qquad y = 3x + 5$$
$$4^2 = 17 - 1 \,? \qquad 17 = 3(4) + 5 \,?$$
$$16 = 16 \qquad \quad 17 = 12 + 5$$
$$\qquad \qquad \qquad 17 = 17$$

Both solutions are valid.

3. The system is $x^2 + y^2 = 5$
$$-3x + 4y = 2$$

The proposed solution set is
$$\left\{(-2, -1), \left(\tfrac{38}{25}, \tfrac{41}{25}\right)\right\}.$$

Check $(-2, -1)$.

$$x^2 + y^2 = 5 \qquad -3x + 4y = 2$$
$$(-2)^2 + (-1)^2 = 5 \,? \qquad -3(-2) + 4(-1) = 2 \,?$$
$$4 + 1 = 5 \qquad \qquad 6 + (-4) = 2$$
$$5 = 5 \qquad \qquad \quad 2 = 2$$

Check $\left(\tfrac{38}{25}, \tfrac{41}{25}\right)$.

$$x^2 + y^2 = 5 \qquad -3x + 4y = 2$$
$$\left(\tfrac{38}{25}\right)^2 + \left(\tfrac{41}{25}\right)^2 = 5 \,? \quad -3\left(\tfrac{38}{25}\right) + 4\left(\tfrac{41}{25}\right) = 2 \,?$$
$$\tfrac{1444}{625} + \tfrac{1681}{625} = 5 \qquad -\tfrac{114}{25} + \tfrac{164}{25} = 2$$
$$\tfrac{3125}{625} = 5 \qquad \qquad \tfrac{50}{25} = 2$$
$$5 = 5 \qquad \qquad \quad 2 = 2$$

Both solutions are valid.

5. The system is $y = \log x$
$$x^2 - y^2 = 4$$

The proposed approximate solution set is
$$\{(2.0232821,\ 0.30605644)\}.$$

To show this, we will recreate the solution with the graphing calculator. In order to get $x^2 - y^2 = 4$ to be displayed, solve for y and enter the relation as two functions.

$$x^2 - y^2 = 4 \Rightarrow y^2 = x^2 - 4 \Rightarrow y = \pm\sqrt{x^2 - 4}$$

7. The system $\begin{array}{c} x^2 - y = 4 \\ x + y = -2 \end{array}$ cannot have more than two solutions because a parabola and a line cannot intersect in more than two points.

In the solutions to Exercises 9–13, we will include both the algebraic solution and graphing calculator solution (like Example 1). For Exercises 15–41, we will include just the algebraic solution. Also, in the solutions to Exercises 9–41, we will omit the checking of solutions. Recall, though, when checking elements of the solution set, substitute the ordered pair(s) into both equations of the system.

9. $x^2 - y = 0$ (1)
$\quad x + y = 2$ (2)

Algebraic Solution:
Solving equation (2) for y, we have $y = 2 - x$.
Substitute this result into equation (1).
$$2 - x = x^2 \Rightarrow x^2 + x - 2 = 0$$
$$(x + 2)(x - 1) = 0 \Rightarrow x = -2 \text{ or } x = 1$$
If $x = -2$, then $y = 2 - (-2) = 4$. If $x = 1$, then $y = 2 - 1 = 1$.

Graphing Calculator Solution
$$x^2 - y = 0 \Rightarrow y = x^2 \text{ and } x + y = 2 \Rightarrow y = 2 - x$$

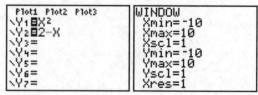

(continued on next page)

Copyright © 2013 Pearson Education, Inc.

(continued)

Solution set: $\{(1,1),(-2,4)\}$

11. $y = x^2 - 2x + 1$ (1)
$x - 3y = -1$ (2)

Algebraic Solution:

Solving equation (2) for y, we have $y = \frac{x+1}{3}$.

Substitute this result into equation (1).

$\frac{x+1}{3} = x^2 - 2x + 1 \Rightarrow x + 1 = 3x^2 - 6x + 3 \Rightarrow$

$3x^2 - 7x + 2 = 0 \Rightarrow (3x-1)(x-2) = 0 \Rightarrow$

$x = \frac{1}{3}$ or $x = 2$

If $x = 2$, then $y = \frac{2+1}{3} = 1$. If $x = \frac{1}{3}$, then

$y = \frac{\frac{1}{3}+1}{3} = \frac{1+3}{9} = \frac{4}{9}$.

Graphing Calculator Solution

Solution set: $\left\{(2,1),\left(\frac{1}{3},\frac{4}{9}\right)\right\}$

13. $y = x^2 + 4x$ (1)
$2x - y = -8$ (2)

Algebraic Solution:

Solving equation (2) for y, we have

$y = 2x + 8$. Substitute this result into equation

(1): $2x + 8 = x^2 + 4x \Rightarrow x^2 + 2x - 8 = 0 \Rightarrow$

$(x+4)(x-2) = 0 \Rightarrow x = -4$ or $x = 2$

If $x = -4$, then $y = 2(-4) + 8 = 0$.

If $x = 2$, then $y = 2(2) + 8 = 12$.

Graphing Calculator Solution

Solution set: $\{(2,12),(-4,0)\}$

15. $3x^2 + 2y^2 = 5$ (1)
$x - y = -2$ (2)

Solving equation (2) for y, we have $y = x + 2$.
Substitute this result into equation (1).

$3x^2 + 2(x+2)^2 = 5$

$3x^2 + 2(x^2 + 4x + 4) = 5$

$3x^2 + 2x^2 + 8x + 8 = 5$

$5x^2 + 8x + 3 = 0$

$(5x+3)(x+1) = 0 \Rightarrow x = -\frac{3}{5}$ or $x = -1$

If $x = -\frac{3}{5}$, then $y = -\frac{3}{5} + 2 = -\frac{3}{5} + \frac{10}{5} = \frac{7}{5}$. If

$x = -1$, then $y = -1 + 2 = 1$.

Solution set: $\left\{\left(-\frac{3}{5},\frac{7}{5}\right),(-1,1)\right\}$

17. $x^2 + y^2 = 8$ (1)
$x^2 - y^2 = 0$ (2)

Using the elimination method, we add
equations (1) and (2).

$\begin{array}{r} x^2 + y^2 = 8 \\ \underline{x^2 - y^2 = 0} \\ 2x^2 = 8 \end{array} \Rightarrow x^2 = 4 \Rightarrow x = \pm 2$

Find y by substituting back into equation (2).

If $x = 2$, then $2^2 - y^2 = 0 \Rightarrow 4 - y^2 = 0 \Rightarrow$

$y^2 = 4 \Rightarrow y = \pm 2$. If $x = -2$, then

$(-2)^2 - y^2 = 0 \Rightarrow 4 - y^2 = 0 \Rightarrow y^2 = 4 \Rightarrow$

$y = \pm 2$.

Solution set: $\{(2,2),(2,-2),(-2,2),(-2,-2)\}$

Copyright © 2013 Pearson Education, Inc.

19. $5x^2 - y^2 = 0 \ (1)$
$3x^2 + 4y^2 = 0 \ (2)$

Using the elimination method, we multiply equation (1) by 4 and add to equation (2).
$$20x^2 - 4y^2 = 0$$
$$\underline{3x^2 + 4y^2 = 0}$$
$$23x^2 \qquad = 0 \Rightarrow x^2 = 0 \Rightarrow x = 0$$

Find y by substituting back into equation (1). If $x = 0$, then
$$5(0)^2 - y^2 = 0 \Rightarrow 0 - y^2 = 0 \Rightarrow y^2 = 0 \Rightarrow y = 0.$$

Solution set: $\{(0,0)\}$

21. $3x^2 + y^2 = 3 \ (1)$
$4x^2 + 5y^2 = 26 \ (2)$

Using the elimination method, we multiply equation (1) by -5 and add to equation (2).
$$-15x^2 - 5y^2 = -15$$
$$\underline{4x^2 + 5y^2 = 26}$$
$$-11x^2 \qquad = 11 \Rightarrow x^2 = -1 \Rightarrow x = \pm i$$

Find y by substituting back into equation (1). If $x = i$, then $3(i)^2 + y^2 = 3 \Rightarrow$
$$3(-1) + y^2 = 3 \Rightarrow -3 + y^2 = 3 \Rightarrow y^2 = 6 \Rightarrow$$
$y = \pm\sqrt{6}$. If $x = -i$, then $3(-i)^2 + y^2 = 3 \Rightarrow$
$$3(-1) + y^2 = 3 \Rightarrow -3 + y^2 = 3 \Rightarrow y^2 = 6 \Rightarrow$$
$y = \pm\sqrt{6}$.

Solution set:
$$\left\{\left(i, \sqrt{6}\right), \left(-i, \sqrt{6}\right), \left(i, -\sqrt{6}\right), \left(-i, -\sqrt{6}\right)\right\}$$

23. $2x^2 + 3y^2 = 5 \ (1)$
$3x^2 - 4y^2 = -1 \ (2)$

Using the elimination method, we multiply equation (1) by 4 and equation (2) by 3 and then add the resulting equations.
$$8x^2 + 12y^2 = 20$$
$$\underline{9x^2 - 12y^2 = -3}$$
$$17x^2 \qquad = 17 \Rightarrow x^2 = 1 \Rightarrow x = \pm 1$$

Find y by substituting back into equation (1). If $x = 1$, then $2(1)^2 + 3y^2 = 5 \Rightarrow$
$$2 + 3y^2 = 5 \Rightarrow 3y^2 = 3 \Rightarrow y^2 = 1 \Rightarrow y = \pm 1.$$

If $x = -1$, then $2(-1)^2 + 3y^2 = 5 \Rightarrow$
$$2 + 3y^2 = 5 \Rightarrow 3y^2 = 3 \Rightarrow y^2 = 1 \Rightarrow y = \pm 1.$$

Solution set: $\{(1,-1),(-1,1),(1,1),(-1,-1)\}$

25. $2x^2 + 2y^2 = 20 \ (1)$
$4x^2 + 4y^2 = 30 \ (2)$

Using the elimination method, we multiply equation (1) by -2 and add to equation (2).
$$-4x^2 - 4y^2 = -40$$
$$\underline{4x^2 + 4y^2 = \ \ 30} \ |$$
$$0 = -10$$

This is a false statement.
Solution set: $\varnothing$

27. $2x^2 - 3y^2 = 12 \ (1)$
$6x^2 + 5y^2 = 36 \ (2)$

Using the elimination method, we multiply equation (1) by -3 and add to equation (2).
$$-6x^2 + 9y^2 = -36$$
$$\underline{6x^2 + 5y^2 = \ \ 36}$$
$$14y^2 = \ \ 0 \Rightarrow y^2 = 0 \Rightarrow y = 0$$

Find x by substituting back into equation (2).

If $y = 0$, then $6x^2 + 5(0)^2 = 36 \Rightarrow$
$$6x^2 + 0 = 36 \Rightarrow x^2 = 6 \Rightarrow x = \pm\sqrt{6}.$$

Solution set: $\left\{\left(-\sqrt{6}, 0\right), \left(-\sqrt{6}, 0\right)\right\}$

29. $xy = -15 \qquad (1)$
$4x + 3y = 3 \qquad (2)$

Solving equation (1) for y, we have $y = -\frac{15}{x}$.
Substitute this result into equation (2).
$$4x + 3\left(-\frac{15}{x}\right) = 3 \Rightarrow 4x^2 - 45 = 3x \Rightarrow$$
$$4x^2 - 3x - 45 = 0 \Rightarrow (x+3)(4x-15) = 0 \Rightarrow$$
$x = -3$ or $x = \frac{15}{4}$

If $x = -3$, then $y = -\frac{15}{-3} = 5$. If $x = \frac{15}{4}$, then
$$y = -\frac{15}{\frac{15}{4}} = -4.$$

Solution set: $\left\{(-3, 5), \left(\frac{15}{4}, -4\right)\right\}$

31. $2xy + 1 = 0 \ (1)$
$x + 16y = 2 \ (2)$

Solving equation (2) for x, we have
$x = 2 - 16y$.
Substitute this result into equation (1).
$$2(-16y + 2)y + 1 = 0 \Rightarrow -32y^2 + 4y + 1 = 0 \Rightarrow$$
$$32y^2 - 4y - 1 = 0 \Rightarrow (8y+1)(4y-1) = 0 \Rightarrow$$
$y = -\frac{1}{8}$ or $y = \frac{1}{4}$

(*continued on next page*)

Copyright © 2013 Pearson Education, Inc.

(continued)

If $y = -\frac{1}{8}$, then $x = 2 - 16\left(-\frac{1}{8}\right) = 2 + 2 = 4$. If

$y = \frac{1}{4}$, then $x = 2 - 16\left(\frac{1}{4}\right) = 2 - 4 = -2$.

Solution set: $\left\{\left(4, -\frac{1}{8}\right), \left(-2, \frac{1}{4}\right)\right\}$

33. $3x^2 - y^2 = 11$ (1)

        $xy = 12$ (2)

Solving equation (2) for x, we have $x = \frac{12}{y}$.

Substitute this result into equation (1).

$3\left(\frac{12}{y}\right)^2 - y^2 = 11 \Rightarrow \frac{432}{y^2} - y^2 = 11 \Rightarrow$

$432 - y^4 = 11y^2 \Rightarrow y^4 + 11y^2 - 432 = 0 \Rightarrow$

$\left(y^2 + 27\right)\left(y^2 - 16\right) = 0$

$y^2 + 27 = 0 \Rightarrow y = \pm\sqrt{-27} = \pm 3i\sqrt{3}$ or

$y^2 - 16 = 0 \Rightarrow y^2 = \sqrt{16} \Rightarrow y = \pm 4$

If $y = 3i\sqrt{3}$, then $x = \frac{12}{3i\sqrt{3}} = \frac{4}{i\sqrt{3}} = -\frac{4\sqrt{3}}{3}i$.

If $y = -3i\sqrt{3}$, then $x = \frac{12}{-3i\sqrt{3}} = -\frac{4}{i\sqrt{3}} = \frac{4\sqrt{3}}{3}i$.

If $y = -4$, then $x = \frac{12}{-4} = -3$. If $y = 4$, then

$x = \frac{12}{4} = 3$.

Solution set:

$\left\{\left(-\frac{4\sqrt{3}}{3}i,\ 3i\sqrt{3}\right), \left(\frac{4\sqrt{3}}{3}i,\ -3i\sqrt{3}\right), (3, 4),\right.$

$\left.(-3, -4)\right\}$

35. $x^2 - xy + y^2 = 5$ (1)

    $2x^2 + xy - y^2 = 10$ (2)

Using the elimination method, we add equations (1) and (2).

$\begin{array}{l} x^2 - xy + y^2 = 5 \\ 2x^2 + xy - y^2 = 10 \\ \hline 3x^2 \qquad\qquad = 15 \end{array} \Rightarrow x^2 = 5 \Rightarrow x = \pm\sqrt{5}$

Find y by substituting back into equation (1).

If $x = \sqrt{5}$, then

$\left(\sqrt{5}\right)^2 - \sqrt{5}y + y^2 = 5 \Rightarrow 5 - \sqrt{5}y + y^2 = 5 \Rightarrow$

$\qquad y^2 - \sqrt{5}y = 0 \Rightarrow y\left(y - \sqrt{5}\right) = 0$

Thus, we have $y = 0$ or $y = \sqrt{5}$.

If $x = -\sqrt{5}$, then $\left(-\sqrt{5}\right)^2 - \left(-\sqrt{5}\right)y + y^2 = 5$

$5 + \sqrt{5}y + y^2 = 5 \Rightarrow y^2 + \sqrt{5}y = 0 \Rightarrow$

$y\left(y + \sqrt{5}\right) = 0 \Rightarrow y = 0$ or $y = -\sqrt{5}$

Thus, we have $y = 0$ or $y = -\sqrt{5}$.

Solution set:

$\left\{\left(\sqrt{5}, 0\right), \left(-\sqrt{5}, 0\right), \left(\sqrt{5}, \sqrt{5}\right), \left(-\sqrt{5}, -\sqrt{5}\right)\right\}$

37. $x^2 + 2xy - y^2 = 14$ (1)

    $x^2 - y^2 = -16$ (2)

This system can be solved using a combination of the elimination and substitution methods. Using the elimination method, multiply equation (2) by -1 and add to equation (1).

$\begin{array}{l} \quad x^2 + 2xy - y^2 = 14 \\ -x^2 + \qquad\quad y^2 = 16 \\ \hline \qquad\quad 2xy \qquad = 30 \end{array} \Rightarrow xy = 15\,(3)$

Solve equation (3) for y.

$xy = 15 \Rightarrow y = \frac{15}{x}$ (4)

Find x by substituting equation (4) into equation (2).

$x^2 - \left(\frac{15}{x}\right)^2 = -16 \Rightarrow x^2 - \frac{225}{x^2} = -16 \Rightarrow$

$x^4 - 225 = -16x^2 \Rightarrow$

$x^4 + 16x^2 - 225 = 0 \Rightarrow \left(x^2 - 9\right)\left(x^2 + 25\right) = 0$

$x^2 - 9 = 0 \Rightarrow x^2 = 9 \Rightarrow x = \pm 3$

$x^2 + 25 = 0 \Rightarrow x^2 = -25 \Rightarrow x = \pm 5i$

If $x = 3$, then $y = \frac{15}{3} = 5$. If $x = -3$, then

$y = \frac{15}{-3} = -5$. If $x = 5i$, then $y = \frac{15}{5i} = \frac{3}{i} = -3i$.

If $x = -5i$, then $y = \frac{15}{-5i} = -\frac{3}{i} = 3i$.

Solution set:

$\left\{(3, 5), (-3, -5), (5i, -3i), (-5i, 3i)\right\}$

39. $x^2 + y^2 = 25$ (1)

    $|x| - y = 5$ (2)

Solve equation (2) for x, then substitute this value into equation (1) and solve for y.

$|x| = y + 5 \Rightarrow x = y + 5$ or $x = -(y + 5)$

If $x = y + 5$, then

$(y + 5)^2 + y^2 = 25 \Rightarrow$

$y^2 + 10y + 25 + y^2 = 25 \Rightarrow 2y^2 + 10y = 0 \Rightarrow$

$2y(y + 5) = 0 \Rightarrow 2y = 0$ or $y + 5 = 0 \Rightarrow$

$y = 0$ or $y = -5$

If $y = 0$, then $x = 0 + 5 = 5$.

If $y = -5$, then $x = -5 + 5 = 0$.

(continued on next page)

Copyright © 2013 Pearson Education, Inc.

(*continued*)

If $x = -(y + 5)$, then

$$\left[-(y+5)\right]^2 + y^2 = 25 \Rightarrow$$

$$y^2 + 10y + 25 + y^2 = 25 \Rightarrow 2y^2 + 10y = 0 \Rightarrow$$

$$2y(y+5) = 0 \Rightarrow 2y = 0 \text{ or } y + 5 = 0 \Rightarrow$$

$$y = 0 \text{ or } y = -5$$

If $y = 0$, then $x = -(0 + 5) = -5$.
If $y = -5$, then $x = -(-5 + 5) = 0$.
Solution set: $\{(5, 0), (-5, 0), (0, -5)\}$

41. $x = |y|$ (1)

$x^2 + y^2 = 18$ (2)

If $x = |y|$, then $x^2 = y^2$ since $|y|^2 = y^2$.

Substitute $x^2 = y^2$ into equation (2).

$$y^2 + y^2 = 18 \Rightarrow 2y^2 = 18 \Rightarrow y^2 = 9 \Rightarrow y = \pm 3$$

If $y = 3$, then $x = |y| = |3| = 3$. If $y = -3$, then

$x = |y| = |-3| = 3$.

Solution set: $\{(3, -3), (3, 3)\}$

43. $2x^2 - y^2 = 4$ (1)

$|x| = |y|$ (2)

Since $|x|^2 = x^2$ and $|y|^2 = y^2$, we substitute

$y^2 = x^2$ into equation (1) and solve for x.

$$2x^2 - x^2 = 4 \Rightarrow x^2 = 4 \Rightarrow x = \pm 2$$

By substituting these values back into equation (2) we have the following.

If $x = 2$, then $|2| = |y| \Rightarrow 2 = |y| \Rightarrow y = \pm 2$. If

$x = -2$, then $|-2| = |y| \Rightarrow 2 = |y| \Rightarrow y = \pm 2$.

Solution set: $\{(2,2),(-2,-2),(2,-2),(-2,2)\}$

45. $y = \log(x + 5)$

$y = x^2$

Solution set: $\{(-0.79, 0.62), (0.88, 0.77)\}$

47. $y = e^{x+1}$

$2x + y = 3 \Rightarrow y = 3 - 2x$

Solution set: $\{(0.06, 2.88)\}$

49. Let x and y represent the numbers. We obtain the following system.

$x + y = 17$ (1)

$xy = 42$ (2)

Solving equation (1) for y we have $y = 17 - x$. (3). Substituting this into equation (2) we have $x(17 - x) = 42$. Solving this equation for x we have the following.

$$x(17 - x) = 42 \Rightarrow 17x - x^2 = 42 \Rightarrow$$

$$0 = x^2 - 17x + 42$$

$$(x - 3)(x - 14) = 0 \Rightarrow x = 3 \text{ or } x = 14$$

Using equation (3), if $x = 3$ then
$3 + y = 17 \Rightarrow y = 14$. If $x = 14$, then
$14 + y = 17 \Rightarrow y = 3$.

The two numbers are 14 and 3.

51. Let x and y represent the numbers. We obtain the following system.

$x^2 + y^2 = 100$ (1)

$x^2 - y^2 = 28$ (2)

Adding equations (1) and (2) in order to eliminate y^2, we have $2x^2 = 128$.

Solving this equation for x we have

$$2x^2 = 128 \Rightarrow x^2 = 64 \Rightarrow x = \pm 8$$

Using equation (1), if $x = 8$ then

$$8^2 + y^2 = 100 \Rightarrow 64 + y^2 = 100 \Rightarrow$$

$$y^2 = 36 \Rightarrow y = \pm 6$$

If $x = -8$ then $(-8)^2 + y^2 = 100 \Rightarrow$

$$64 + y^2 = 100 \Rightarrow y^2 = 36 \Rightarrow y = \pm 6$$

The two numbers are 8 and 6, or 8 and −6, or −8 and 6, or −8 and −6.

Copyright © 2013 Pearson Education, Inc.

53. Let x and y represent the numbers.
We obtain the following system.

$$\frac{x}{y} = \frac{9}{2} \quad (1)$$
$$xy = 162 \, (2)$$

Solve equation (1) for x.

$$x = \frac{9}{2}y$$

Substitute $\frac{9}{2}y$ for x in equation (2) and solve for y.

$$\left(\frac{9}{2}y\right)y = 162 \Rightarrow \frac{9}{2}y^2 = 162 \Rightarrow$$
$$\frac{2}{9}\left(\frac{9}{2}y^2\right) = \frac{2}{9}(162) \Rightarrow y^2 = 36 \Rightarrow y = \pm 6$$

If $y = 6$, $x = \frac{9}{2}(6) = 27$. If $y = -6$,

$$x = \frac{9}{2}(-6) = -27.$$

The two numbers are either 27 and 6, or –27 and –6.

55. Let $x = $ the length of the second side and $y = $ the length of the third side.
We obtain the following system.

$$x^2 + y^2 = 13^2 = 169 \quad (1)$$
$$x = y + 7 \qquad (2)$$

Substitute $y + 7$ for x in equation (1) and solve for y.

$$(y+7)^2 + y^2 = 169$$
$$y^2 + 14y + 49 + y^2 = 169$$
$$2y^2 + 14y - 120 = 0$$
$$y^2 + 7y - 60 = 0$$
$$(y+12)(y-5) = 0 \Rightarrow y = -12 \text{ or } y = 5$$

We disregard the negative solution since a length cannot be negative. If $y = 5$, then $x = 5 + 7 = 12$.
The lengths of the two shorter sides are 5 m and 12 m.

57. If the system formed by the following equations has a solution, the line and the circle intersect.

$$3x - 2y = 9 \quad (1)$$
$$x^2 + y^2 = 25 \, (2)$$

Solving equation (1) for x we have

$$3x = 9 + 2y \Rightarrow x = \frac{9 + 2y}{3}. \quad (3)$$

Substituting into equation (3) into equation (2) we have $\left(\dfrac{9+2y}{3}\right)^2 + y^2 = 25$.

Solving this equation for y we have the following.

$$\left(\frac{9+2y}{3}\right)^2 + y^2 = 25 \Rightarrow \frac{(9+2y)^2}{9} + y^2 = 25 \Rightarrow$$
$$(9+2y)^2 + 9y^2 = 225 \Rightarrow$$
$$81 + 36y + 4y^2 + 9y^2 = 225 \Rightarrow$$
$$13y^2 + 36y + 81 = 225 \Rightarrow$$
$$13y^2 + 36y - 144 = 0$$

Using the quadratic formula where $a = 13$, $b = 36$, and $c = -144$, we have the following.

$$y = \frac{-36 \pm \sqrt{(36)^2 - 4(13)(-144)}}{2(13)}$$
$$= \frac{-36 \pm \sqrt{1296 + 7488}}{26} = \frac{-36 \pm \sqrt{8784}}{26}$$
$$y = \frac{-36 - \sqrt{8784}}{26} \approx -4.989 \text{ and}$$
$$y = \frac{-36 + \sqrt{8784}}{26} \approx 2.220$$

If $y = \frac{-36+\sqrt{8784}}{26}$, then

$$x = \frac{9 + 2\left(\frac{-36+\sqrt{8784}}{26}\right)}{3} = \frac{234 + 2\left(-36 + \sqrt{8784}\right)}{78}$$
$$= \frac{162 + 2\sqrt{8784}}{78} \approx 4.480$$

If $y = \frac{-36-\sqrt{8784}}{26}$, then

$$x = \frac{9 + 2\left(\frac{-36-\sqrt{8784}}{26}\right)}{3} = \frac{234 + 2\left(-36 - \sqrt{8784}\right)}{78}$$
$$= \frac{162 - 2\sqrt{8784}}{78} \approx -0.326$$

Thus, the circle and line do intersect, in fact twice, at approximately (4.48, 2.22) and (–0.326, –4.99).

59. We must first find the solution to the following system.

$$y = x^2 \qquad (1)$$
$$x^2 + y^2 = 90 \, (2)$$

Substitute y for x^2 in equation (2) and solve for y.

$$y + y^2 = 90 \Rightarrow y^2 + y - 90 = 0$$
$$(y+10)(y-9) = 0 \Rightarrow y = -10 \text{ or } y = 9$$

If $y = -10$, then $-10 = x^2 \Rightarrow x = \pm i\sqrt{10}$.
Since we are seeking points of intersection, we reject these nonreal solutions. If $y = 9$, then $9 = x^2 \Rightarrow x = \pm 3$. The two points of intersection of the two graphs are (3, 9) and (–3, 9). The slope of the line through these two points is $m = \frac{9-9}{-3-3} = 0$, so this is the horizontal line with equation $y = 9$.

Copyright © 2013 Pearson Education, Inc.

61. Let x represent the length and width of the square base, and let y represent the height. Using the formula for the volume of a box, $V = LWH$, we have the equation

$x^2 y = 360$ (1). Since the surface area consists of a square base and four rectangular sides we have the equation $x^2 + 4xy = 276$ (2). Solve equation (1) for y, then substitute this expression into equation (2) and solve for x.

$$x^2 y = 360 \Rightarrow y = \frac{360}{x^2}$$

$$x^2 + 4xy = 276 \Rightarrow x^2 + 4x\left(\frac{360}{x^2}\right) = 276 \Rightarrow$$

$$x^2 + \frac{1440}{x} = 276 \Rightarrow x^3 + 1440 = 276x \Rightarrow$$

$$x^3 - 276x + 1440 = 0$$

Since we are looking for the dimensions of the box, we are restricted to positive solutions. Also, the solutions should be relatively small. By the rational zeros theorem, factors of 1440 are the only possible rational solutions. Using synthetic division, we see that 6 is a solution.

```
6)1    0    -276    1440
       6     36    -1440
   1   6    -240       0
```

Therefore, one possible solution is $x = 6$ and $6^2 y = 360 \Rightarrow 36y = 360 \Rightarrow y = 10$.

To find any other possible solutions, solve $x^2 + 6x - 240 = 0$, the new quotient polynomial.

Using the quadratic formula, we have

$$x = \frac{-6 \pm \sqrt{6^2 - 4(1)(-240)}}{2(1)} = \frac{-6 \pm \sqrt{996}}{2}$$

$$\approx 12.780 \text{ or } -18.780$$

We disregard the negative solution.

If $x = 12.780$, then $12.780^2 y = 360 \Rightarrow$

$163.3284y = 360 \Rightarrow y \approx 2.204$.

The solutions are:
length = width = 6 ft; height = 10 ft or
length = width $\approx$ 12.780 ft; height $\approx$ 2.204 ft

63. The system is $\begin{array}{l} x^2 y = 75 \\ x^2 + 4xy = 85 \end{array}$

The first possible solution is $x = 5$ and $y = 3$.

$$\begin{array}{ll} x^2 y = 75 & x^2 + 4xy = 85 \\ 5^2 \cdot 3 = 75 \text{ ?} & 5^2 + 4 \cdot 5 \cdot 3 = 85 \text{ ?} \\ 25 \cdot 3 = 75 & 25 + 60 = 85 \\ \quad 75 = 75 \text{ True} & \quad 85 = 85 \text{ True} \end{array}$$

Thus, the solution length = width = 5 in. and height = 3 in. is valid.

Ideally we should check the exact values in the system in order to determine if the approximate dimensions are valid. From Example 6 we have

$$x = \frac{-5 + \sqrt{5^2 - 4(1)(-60)}}{2(1)} = \frac{-5 + \sqrt{25 + 240}}{2} = \frac{-5 + \sqrt{265}}{2}.$$

The corresponding y-value would be

$$y = \frac{75}{x^2} = \frac{75}{\left(\frac{-5 + \sqrt{265}}{2}\right)^2} = \frac{75}{\frac{\left(-5 + \sqrt{265}\right)^2}{4}} = \frac{300}{\left(-5 + \sqrt{265}\right)^2}.$$

The second possible solution is $x = \frac{-5 + \sqrt{265}}{2}$

and $y = \frac{300}{\left(-5 + \sqrt{265}\right)^2}$.

$$x^2 y = 75$$

$$\left(\frac{-5 + \sqrt{265}}{2}\right)^2 \cdot \frac{300}{\left(-5 + \sqrt{265}\right)^2} = 75 \text{ ?}$$

$$\frac{\left(-5 + \sqrt{265}\right)^2}{4} \cdot \frac{300}{\left(-5 + \sqrt{265}\right)^2} = 75$$

$$75 = 75 \text{ True}$$

$$x^2 + 4xy = 85$$

$$\frac{\left(-5 + \sqrt{265}\right)^2}{4} + \frac{600}{-5 + \sqrt{265}} = 85 \text{ ?}$$

$$\frac{\left(-5 + \sqrt{265}\right)^3 + 2400}{4\left(-5 + \sqrt{265}\right)} = 85$$

$$\frac{\left(-5 + \sqrt{265}\right)^2 \left(-5 + \sqrt{265}\right) + 2400}{4\left(-5 + \sqrt{265}\right)} = 85$$

$$\frac{\left(25 - 10\sqrt{265} + 265\right)\left(-5 + \sqrt{265}\right) + 2400}{4\left(-5 + \sqrt{265}\right)} = 85$$

$$\frac{\left(290 - 10\sqrt{265}\right)\left(-5 + \sqrt{265}\right) + 2400}{4\left(-5 + \sqrt{265}\right)} = 85$$

$$\frac{\left(-1450 + 340\sqrt{265} - 2650\right) + 2400}{4\left(-5 + \sqrt{265}\right)} = 85$$

$$\frac{-4100 + 340\sqrt{265} + 2400}{4\left(-5 + \sqrt{265}\right)} = 85$$

$$\frac{-1700 + 340\sqrt{265}}{4\left(-5 + \sqrt{265}\right)} = 85$$

$$\frac{340\left(-5 + \sqrt{265}\right)}{4\left(-5 + \sqrt{265}\right)} = 85$$

$$85 = 85 \text{ True}$$

Copyright © 2013 Pearson Education, Inc.

65. supply: $p = \sqrt{0.1q+9} - 2$
demand: $p = \sqrt{25-0.1q}$

  (a) Equilibrium occurs when supply equals demand, so solve the system formed by the supply and demand equations. This system can be solved by substitution. Substitute $\sqrt{0.1q+9} - 2$ for p in the demand equation and solve the resulting equation for q.

$$\sqrt{0.1q+9} - 2 = \sqrt{25-0.1q}$$
$$\left(\sqrt{0.1q+9} - 2\right)^2 = \left(\sqrt{25-0.1q}\right)^2$$
$$0.1q+9-4\sqrt{0.1q+9}+4 = 25-0.1q$$
$$0.2q-12 = 4\sqrt{0.1q+9}$$
$$(0.2q-12)^2 = \left(4\sqrt{0.1q+9}\right)^2$$
$$0.04q^2-4.8q+144 = 16(0.1q+9)$$
$$0.04q^2-4.8q+144 = 1.6q+144$$
$$0.04q^2-6.4q = 0$$
$$0.04q(q-160) = 0 \Rightarrow$$
$$q = 0 \text{ or } q = 160$$

Disregard an equilibrium demand of 0. The equilibrium demand is 160 units.

  (b) Substitute 160 for q in either equation and solve for p.
$$p = \sqrt{0.1(160)+9} - 2 = \sqrt{16+9} - 2$$
$$= \sqrt{25} - 2 = 5 - 2 = 3$$
The equilibrium price is \$3.

67. (a) The carbon dioxide emissions from consumption of fossil fuels are increasing in both countries, but they are increasing much more rapidly in China than in the United States.

  (b) Carbon dioxide emissions were decreasing in China from 1996 to 2000.

  (c) The emissions were equal in 2007, when the level was approximately 6000 million metric tons in each country.

  (d) $U = C \Rightarrow$
$$5092(1.0105)^{(x-1990)}$$
$$= 2082(1.0638)^{(x-1990)}$$
$$\ln\left[5092(1.0105)^{(x-1990)}\right]$$
$$= \ln\left[2082(1.0638)^{(x-1990)}\right]$$
$$\ln 5092 + (x-1990)\ln 1.0105$$
$$= \ln 2082 + (x-1990)\ln 1.0638$$

$$\ln 5092 - \ln 2082$$
$$= (x-1990)(\ln 1.0638 - \ln 1.0105)$$
$$\frac{\ln 5092 - \ln 2082}{\ln 1.0638 - \ln 1.0105} = x - 1990$$
$$\frac{\ln 5092 - \ln 2082}{\ln 1.0638 - \ln 1.0105} + 1990 = x$$
$$2007.4 \approx x$$
If x is approximately 2007.4,
$$U = 5092(1.0105)^{(2007.4-1990)} \approx 6106.9.$$
Emissions from both countries will be equal in 2007. The emissions level will be about 6107 million metric tons.

69. Shift the graph of $y = x^2$ four units down to obtain the graph of $y = x^2 - 4$.

71. $x^2-4 = x-1$ if $x \geq 1$ and $x^2-4 = 1-x$ if $x < 1$

73. If $y = |x-1|$ and $x = \frac{1+\sqrt{13}}{2}$, then
$$y = \left|\frac{1+\sqrt{13}}{2} - 1\right| = \left|\frac{1+\sqrt{13}}{2} - \frac{2}{2}\right| = \left|\frac{-1+\sqrt{13}}{2}\right| = \frac{-1+\sqrt{13}}{2}.$$
If $y = 1-x$ and $x = \frac{-1-\sqrt{21}}{2}$, then
$$y = 1 - \left(\frac{-1-\sqrt{21}}{2}\right) = \frac{2}{2} + \frac{1+\sqrt{21}}{2} = \frac{3+\sqrt{21}}{2}.$$
Solution set:
$$\left\{\left(\frac{1+\sqrt{13}}{2}, \frac{-1+\sqrt{13}}{2}\right), \left(\frac{-1-\sqrt{21}}{2}, \frac{3+\sqrt{21}}{2}\right)\right\}$$

Summary Exercises on Systems of Equations

As noted in the text, different methods of solving equations have been introduced. In the solutions to these exercises, we will present the solution using one method (or combination of methods). In general, the nonlinear systems cannot use the matrix, Gauss-Jordan, or Cramer's rule methods. Only the methods of substitution or elimination should be considered for such systems.

1. $2x+5y = 4$
$3x-2y = -13$

$$D = \begin{vmatrix} 2 & 5 \\ 3 & -2 \end{vmatrix} = 2(-2) - 3(5) = -4 - 15 = -19,$$

$$D_x = \begin{vmatrix} 4 & 5 \\ -13 & -2 \end{vmatrix} = 4(-2) - (-13)(5)$$
$$= -8 - (-65) = 57$$

(continued on next page)

Copyright © 2013 Pearson Education, Inc.

$$D_y = \begin{vmatrix} 2 & 4 \\ 3 & -13 \end{vmatrix} = 2(-13) - 3(4)$$
$$= -26 - 12 = -38 \Rightarrow$$

$x = \dfrac{D_x}{D} = \dfrac{57}{-19} = -3$ and $y = \dfrac{D_y}{D} = \dfrac{-38}{-19} = 2.$

Solution set: $\{(-3, 2)\}$

3. $2x^2 + y^2 = 5$ (1)
$3x^2 + 2y^2 = 10$ (2)

Using the elimination method, we add -2 times equation (1) to equation (2).

$$\begin{array}{rcl} -4x^2 - 2y^2 &=& -10 \\ 3x^2 + 2y^2 &=& 10 \\ \hline -x^2 &=& 0 \Rightarrow x^2 = 0 \Rightarrow x = 0 \end{array}$$

If $x = 0$, then $2(0)^2 + y^2 = 5 \Rightarrow$
$0 + y^2 = 5 \Rightarrow y^2 = 5 \Rightarrow y = \pm\sqrt{5}.$

Solution set: $\left\{(0, \sqrt{5}), (0, -\sqrt{5})\right\}$

5. $6x - y = 5$ (1)
$xy = 4$ (2)

Solving equation (2) for y, we have $y = \dfrac{4}{x}.$

Substitute this result into equation (1).

$$6x - \dfrac{4}{x} = 5 \Rightarrow 6x^2 - 4 = 5x \Rightarrow$$
$$6x^2 - 5x - 4 = 0 \Rightarrow (2x+1)(3x-4) = 0 \Rightarrow$$
$$x = -\tfrac{1}{2} \text{ or } x = \tfrac{4}{3}$$

If $x = \tfrac{4}{3}$, then $y = \dfrac{4}{\frac{4}{3}} = 3.$ If $x = -\tfrac{1}{2}$, then

$y = \dfrac{4}{-\frac{1}{2}} = -8.$

Solution set: $\left\{\left(\tfrac{4}{3}, 3\right), \left(-\tfrac{1}{2}, -8\right)\right\}$

7. $x + 2y + z = 5$ (1)
$y + 3z = 9$ (2)

This system has infinitely many solutions. We will express the solution set with z as the arbitrary variable. Solving equation (2) for y, we have $y = 9 - 3z.$ Substituting y (in terms of z) into equation (1) and solving for x we have the following.

$$x + 2(9 - 3z) + z = 5 \Rightarrow x + 18 - 6z + z = 5 \Rightarrow$$
$$x + 18 - 5z = 5 \Rightarrow x = -13 + 5z$$

Solution set: $\{(-13 + 5z, 9 - 3z, z)\}$

9. $3x + 6y - 9z = 1$
$2x + 4y - 6z = 1$
$3x + 4y + 5z = 0$

This system has the augmented matrix

$$\begin{bmatrix} 3 & 6 & -9 & | & 1 \\ 2 & 4 & -6 & | & 1 \\ 3 & 4 & 5 & | & 0 \end{bmatrix}.$$

$$\begin{bmatrix} 2 & 4 & -6 & | & 1 \\ 3 & 6 & -9 & | & 1 \\ 3 & 4 & 5 & | & 0 \end{bmatrix} \begin{matrix} \text{R1} \leftrightarrow \text{R2} \\ \\ \Rightarrow \\ \\ \end{matrix}$$

$$\begin{bmatrix} 1 & 2 & -3 & | & \frac{1}{2} \\ 3 & 6 & -9 & | & 1 \\ 3 & 4 & 5 & | & 0 \end{bmatrix} \begin{matrix} \frac{1}{2}\text{R1} \\ \\ \Rightarrow \\ \\ \end{matrix}$$

$$\begin{bmatrix} 1 & 2 & -3 & | & \frac{1}{2} \\ 0 & 0 & 0 & | & -\frac{1}{2} \\ 0 & -2 & 14 & | & -\frac{3}{2} \end{bmatrix} \begin{matrix} \\ -3\text{R1}+\text{R2} \\ -3\text{R1}+\text{R3} \end{matrix}$$

The second row of the augmented matrix corresponds to the statement $0 = -\tfrac{1}{2}.$ This is a false statement. Thus the solution set is $\varnothing.$

11. $x^2 + y^2 = 4$ (1)
$y = x + 6$ (2)

Substitute equation (1) into equation (2).

$$x^2 + (x+6)^2 = 4$$
$$x^2 + x^2 + 12x + 36 = 4$$
$$2x^2 + 12x + 36 = 4$$
$$2x^2 + 12x + 32 = 0 \Rightarrow x^2 + 6x + 16 = 0$$

Using the quadratic formula where $a = 1, b = 6,$ and $c = 16,$ we have the following.

$$x = \frac{-6 \pm \sqrt{6^2 - 4(1)(16)}}{2(1)} = \frac{-6 \pm \sqrt{36 - 64}}{2}$$
$$= \frac{-6 \pm \sqrt{-28}}{2} = \frac{-6 \pm 2i\sqrt{7}}{2} = -3 \pm i\sqrt{7}$$

If $x = -3 + i\sqrt{7},$ then
$y = -3 + i\sqrt{7} + 6 = 3 + i\sqrt{7}.$ If $x = -3 - i\sqrt{7},$
then $y = -3 - i\sqrt{7} + 6 = 3 - i\sqrt{7}.$
Solution set:

$$\left\{(-3 + i\sqrt{7}, 3 + i\sqrt{7}), (-3 - i\sqrt{7}, 3 - i\sqrt{7})\right\}$$

Copyright © 2013 Pearson Education, Inc.

13. $y+1 = x^2 + 2x$ (1)
$y + 2x = 4$ (2)

Solving equation (2) for y, we have $y = 4 - 2x$.
Substitute this result into equation (1).

$4 - 2x + 1 = x^2 + 2x \Rightarrow 5 - 2x = x^2 + 2x \Rightarrow$
$\qquad 0 = x^2 + 4x - 5 \Rightarrow (x+5)(x-1) = 0 \Rightarrow$
$\qquad x = -5 \text{ or } x = 1$

If $x = -5$, then $y = 4 - 2(-5) = 4 + 10 = 14$. If
$x = 1$, then $y = 4 - 2(1) = 4 - 2 = 2$.

Solution set: $\{(1,2),(-5,14)\}$

15. $\quad 2x + 3y + 4z = 3$
$\quad -4x + 2y - 6z = 2$
$\qquad\quad 4x + 3z = 0$

This system has the augmented matrix

$$\begin{bmatrix} 2 & 3 & 4 & | & 3 \\ -4 & 2 & -6 & | & 2 \\ 4 & 0 & 3 & | & 0 \end{bmatrix}.$$

$$\begin{bmatrix} 1 & \frac{3}{2} & 2 & | & \frac{3}{2} \\ -4 & 2 & -6 & | & 2 \\ 4 & 0 & 3 & | & 0 \end{bmatrix} \begin{matrix} \frac{1}{2}R1 \\ \\ \\ \end{matrix} \Rightarrow$$

$$\begin{bmatrix} 1 & \frac{3}{2} & 2 & | & \frac{3}{2} \\ 0 & 8 & 2 & | & 8 \\ 0 & -6 & -5 & | & -6 \end{bmatrix} \begin{matrix} \\ 4R1+R2 \\ -4R1+R3 \end{matrix} \Rightarrow$$

$$\begin{bmatrix} 1 & \frac{3}{2} & 2 & | & \frac{3}{2} \\ 0 & 1 & \frac{1}{4} & | & 1 \\ 0 & -6 & -5 & | & -6 \end{bmatrix} \begin{matrix} \\ \frac{1}{8}R2 \\ \\ \end{matrix} \Rightarrow$$

$$\begin{bmatrix} 1 & 0 & \frac{13}{8} & | & 0 \\ 0 & 1 & \frac{1}{4} & | & 1 \\ 0 & 0 & -\frac{7}{2} & | & 0 \end{bmatrix} \begin{matrix} -\frac{3}{2}R2+R1 \\ \\ 6R2+R3 \end{matrix} \Rightarrow$$

$$\begin{bmatrix} 1 & 0 & \frac{13}{8} & | & 0 \\ 0 & 1 & \frac{1}{4} & | & 1 \\ 0 & 0 & 1 & | & 0 \end{bmatrix} \begin{matrix} \\ \\ -\frac{2}{7}R3 \end{matrix} \Rightarrow$$

$$\begin{bmatrix} 1 & 0 & 0 & | & 0 \\ 0 & 1 & 0 & | & 1 \\ 0 & 0 & 1 & | & 0 \end{bmatrix} \begin{matrix} -\frac{13}{8}R3+R1 \\ -\frac{1}{4}R3+R2 \\ \\ \end{matrix}$$

Solution set: $\{(0, 1, 0)\}$

17. $-5x + 2y + z = 5$ (1)
$-3x - 2y - z = 3$ (2)
$-x + 6y \quad = 1$ (3)

Add equations (1) and (2) in order to eliminate
z to obtain $-8x = 8$ or $x = -1$.
Substitute $x = -1$ into equation (3) to solve
for y.

$-(-1) + 6y = 1 \Rightarrow 1 + 6y = 1 \Rightarrow 6y = 0 \Rightarrow y = 0$

Substitute $x = -1$ and $y = 0$ into equation (1)
to solve for z.

$-5(-1) + 2(0) + z = 5 \Rightarrow 5 + 0 + z = 5 \Rightarrow$
$\qquad\qquad 5 + z = 5 \Rightarrow z = 0$

Solution set: $\{(-1, 0, 0)\}$

19. $2x^2 + y^2 = 9$ (1)
$3x - 2y = -6$ (2)

Solving equation (2) for x, we have

$3x = 2y - 6 \Rightarrow x = \dfrac{2y-6}{3}.$

Substitute this result into equation (1).

$$2\left(\frac{2y-6}{3}\right)^2 + y^2 = 9$$

$$2 \cdot \frac{(2y-6)^2}{9} + y^2 = 9$$

$$2 \cdot (2y-6)^2 + 9y^2 = 81$$

$$2(4y^2 - 24y + 36) + 9y^2 = 81$$

$$8y^2 - 48y + 72 + 9y^2 = 81$$

$$17y^2 - 48y + 72 = 81$$

$$17y^2 - 48y - 9 = 0$$

$$(y - 3)(17y + 3) = 0 \Rightarrow y - 3 \text{ or } y = -\tfrac{3}{17}$$

If $y = 3$, then $x = \frac{2(3)-6}{3} = \frac{6-6}{3} = \frac{0}{3} = 0$.

If $y = -\frac{3}{17}$, then $x = \frac{2\left(-\frac{3}{17}\right)-6}{3} = \frac{2(-3)-102}{51}$

$= \frac{-6-102}{51} = \frac{-108}{51} = -\frac{36}{17}.$

Solution set: $\left\{(0,3),\left(-\frac{36}{17}, -\frac{3}{17}\right)\right\}$

21. $\quad x + y - z = 0$ (1)
$\qquad 2y - z = 1$ (2)
$\quad 2x + 3y - 4z = -4$ (3)

Add -2 times equation (1) to equation (3) in
order to eliminate x.

$\quad -2x - 2y + 2z = 0$
$\quad \underline{2x + 3y - 4z = -4}$
$\qquad\qquad y - 2z = -4$ (4)

Add -2 times equation (4) to equation (2) in
order to eliminate y and solve for z.

$\quad 2y - z = 1$
$\quad \underline{-2y + 4z = 8}$
$\qquad\quad 3z = 9 \Rightarrow z = 3$

Substitute 3 for z in equation (4) to obtain

$y - 2(3) = -4 \Rightarrow y - 6 = -4 \Rightarrow y = 2.$

Substitute 3 for z and 2 for y in equation (1) to
obtain $x + 2 - 3 = 0 \Rightarrow x - 1 = 0 \Rightarrow x = 1$.
Solution set: $\{(1, 2, 3)\}$

Copyright © 2013 Pearson Education, Inc.

23.
$$4x - z = -6 \ (1)$$
$$\tfrac{3}{5}y + \tfrac{1}{2}z = 0 \ (2)$$
$$\tfrac{1}{3}x + \tfrac{2}{3}z = -5 \ (3)$$

Before applying one of the methods, it would be helpful to clear the fractions in equations (2) and (3) by multiplying equation (2) by 10 and equation (3) by 3 to obtain the following.

$$6y + 5z = 0 \ (4)$$
$$x + 2z = -15 \ (5)$$

The system formed by equations (1), (4), and (5) can be represented by the following augmented matrix.

$$\begin{bmatrix} 4 & 0 & -1 & -6 \\ 0 & 6 & 5 & 0 \\ 1 & 0 & 2 & -15 \end{bmatrix}$$

$$\begin{bmatrix} 1 & 0 & 2 & -15 \\ 0 & 6 & 5 & 0 \\ 4 & 0 & -1 & -6 \end{bmatrix} \begin{matrix} R1 \leftrightarrow R3 \\ \\ \\ \end{matrix} \Rightarrow$$

$$\begin{bmatrix} 1 & 0 & 2 & -15 \\ 0 & 6 & 5 & 0 \\ 0 & 0 & -9 & 54 \end{bmatrix} \begin{matrix} \\ \\ -4R1+R3 \end{matrix} \Rightarrow$$

$$\begin{bmatrix} 1 & 0 & 2 & -15 \\ 0 & 1 & \tfrac{5}{6} & 0 \\ 0 & 0 & -9 & 54 \end{bmatrix} \begin{matrix} \\ \tfrac{1}{6}R2 \\ \\ \end{matrix} \Rightarrow$$

$$\begin{bmatrix} 1 & 0 & 2 & -15 \\ 0 & 1 & \tfrac{5}{6} & 0 \\ 0 & 0 & 1 & -6 \end{bmatrix} \begin{matrix} \\ \\ -\tfrac{1}{9}R3 \end{matrix} \Rightarrow$$

$$\begin{bmatrix} 1 & 0 & 0 & -3 \\ 0 & 1 & 0 & 5 \\ 0 & 0 & 1 & -6 \end{bmatrix} \begin{matrix} -2R3+R1 \\ -\tfrac{5}{6}R3+R2 \\ \\ \end{matrix}$$

Solution set: $\{(-3, 5, -6)\}$

25.
$$x^2 + 3y^2 = 28 \ (1)$$
$$y - x = -2 \ (2)$$

Solving equation (2) for y, we have $y = x - 2$.
Substitute this result into equation (1) and solve for y

$$x^2 + 3(x-2)^2 = 28$$
$$x^2 + 3(x^2 - 4x + 4) = 28$$
$$x^2 + 3x^2 - 12x + 12 = 28$$
$$4x^2 - 12x + 12 = 28$$
$$4x^2 - 12x - 16 = 0$$
$$x^2 - 3x - 4 = 0$$
$$(x+1)(x-4) = 0 \Rightarrow x = -1 \text{ or } x = 4$$

If $x = -1$, then $y = -1 - 2 = -3$. If $x = 4$, then $y = 4 - 2 = 2$.

Solution set: $\{(4, 2), (-1, -3)\}$

27.
$$2x^2 + 3y^2 = 20 \ (1)$$
$$x^2 + 4y^2 = 5 \ (2)$$

Using the elimination method, we add -2 times equation (2) to equation (1).

$$2x^2 + 3y^2 = 20$$
$$\underline{-2x^2 - 8y^2 = -10}$$
$$-5y^2 = 10 \Rightarrow y^2 = -2 \Rightarrow y = \pm i\sqrt{2}$$

Find x by substituting back into equation (2).

If $y = i\sqrt{2}$, then $x^2 + 4\left(i\sqrt{2}\right)^2 = 5 \Rightarrow$

$x^2 + 4(-1)(2) = 5 \Rightarrow x^2 - 8 = 5 \Rightarrow x^2 = 13 \Rightarrow$

$x = \pm\sqrt{13}$. If $y = -i\sqrt{2}$, then

$x^2 + 4\left(-i\sqrt{2}\right)^2 = 5 \Rightarrow x^2 + 4(-1)(2) = 5 \Rightarrow$

$x^2 - 8 = 5 \Rightarrow x^2 = 13 \Rightarrow x = \pm\sqrt{13}$

Solution set: $\left\{\left(\sqrt{13}, i\sqrt{2}\right), \left(-\sqrt{13}, i\sqrt{2}\right),\right.$
$\left.\left(\sqrt{13}, -i\sqrt{2}\right), \left(-\sqrt{13}, -i\sqrt{2}\right)\right\}$

29.
$$x + 2z = 9$$
$$y + z = 1$$
$$3x - 2y = 9$$

Adding -3 times row 1 to row 3, we have

$$D = \begin{vmatrix} 1 & 0 & 2 \\ 0 & 1 & 1 \\ 3 & -2 & 0 \end{vmatrix} = \begin{vmatrix} 1 & 0 & 2 \\ 0 & 1 & 1 \\ 0 & -2 & -6 \end{vmatrix}.$$

Expanding by column one, we have

$$D = 1 \cdot \begin{vmatrix} 1 & 1 \\ -2 & -6 \end{vmatrix} = -6 + 2 = -4.$$

Adding 2 times row 2 to row 3, we have

$$D_x = \begin{vmatrix} 9 & 0 & 2 \\ 1 & 1 & 1 \\ 9 & -2 & 0 \end{vmatrix} = \begin{vmatrix} 9 & 0 & 2 \\ 1 & 1 & 1 \\ 11 & 0 & 2 \end{vmatrix}.$$

Expanding by column two, we have

$$D_x = 1 \cdot \begin{vmatrix} 9 & 2 \\ 11 & 2 \end{vmatrix} = 18 - 22 = -4.$$

Adding -3 times row 1 to row 3 we have

$$D_y = \begin{vmatrix} 1 & 9 & 2 \\ 0 & 1 & 1 \\ 3 & 9 & 0 \end{vmatrix} = \begin{vmatrix} 1 & 9 & 2 \\ 0 & 1 & 1 \\ 0 & -18 & -6 \end{vmatrix}.$$

(continued on next page)

Copyright © 2013 Pearson Education, Inc.

(*continued*)

Expanding by column one, we have

$$D_y = 1 \cdot \begin{vmatrix} 1 & 1 \\ -18 & -6 \end{vmatrix} = -6 + 18 = 12.$$

Adding -3 times row 1 to row 3, we have

$$D_z = \begin{vmatrix} 1 & 0 & 9 \\ 0 & 1 & 1 \\ 3 & -2 & 9 \end{vmatrix} = \begin{vmatrix} 1 & 0 & 9 \\ 0 & 1 & 1 \\ 0 & -2 & -18 \end{vmatrix}.$$

Expanding by column one, we have

$$D_z = 1 \cdot \begin{vmatrix} 1 & 1 \\ -2 & -18 \end{vmatrix} = -18 + 2 = -16.$$

Thus, we have

$$x = \frac{D_x}{D} = \frac{-4}{-4} = 1, \; y = \frac{D_y}{D} = \frac{12}{-4} = -3,$$

and $z = \frac{D_z}{D} = \frac{-16}{-4} = 4.$

Solution set: $\{(1, -3, 4)\}$

31.
$$\begin{aligned} -x + y & = -1 \\ x + \quad z & = 4 \\ 6x - 3y + 2z & = 10 \end{aligned}$$

This system has the augmented matrix

$$\begin{bmatrix} -1 & 1 & 0 & | & -1 \\ 1 & 0 & 1 & | & 4 \\ 6 & -3 & 2 & | & 10 \end{bmatrix}.$$

$$\begin{bmatrix} 1 & -1 & 0 & | & 1 \\ 1 & 0 & 1 & | & 4 \\ 6 & -3 & 2 & | & 10 \end{bmatrix} \begin{matrix} -1R1 \\ \\ \\ \end{matrix} \Rightarrow$$

$$\begin{bmatrix} 1 & -1 & 0 & | & 1 \\ 0 & 1 & 1 & | & 3 \\ 0 & 3 & 2 & | & 4 \end{bmatrix} \begin{matrix} \\ -1R1+R2 \\ -6R1+R3 \end{matrix} \Rightarrow$$

$$\begin{bmatrix} 1 & 0 & 1 & | & 4 \\ 0 & 1 & 1 & | & 3 \\ 0 & 0 & -1 & | & -5 \end{bmatrix} \begin{matrix} R2+R1 \\ \\ -3R2+R3 \end{matrix} \Rightarrow$$

$$\begin{bmatrix} 1 & 0 & 1 & | & 4 \\ 0 & 1 & 1 & | & 3 \\ 0 & 0 & 1 & | & 5 \end{bmatrix} \begin{matrix} \\ \\ -1R3 \end{matrix} \Rightarrow$$

$$\begin{bmatrix} 1 & 0 & 0 & | & -1 \\ 0 & 1 & 0 & | & -2 \\ 0 & 0 & 1 & | & 5 \end{bmatrix} \begin{matrix} -1R3+R1 \\ -1R3+R2 \\ \end{matrix}$$

Solution set: $\{(-1, -2, 5)\}$

33.
$$\begin{aligned} xy & = -3 \quad (1) \\ x + y & = -2 \quad (2) \end{aligned}$$

Solving equation (1) for y, we have $y = -\frac{3}{x}$.

Substitute this result into equation (2).

$$x + \left(-\frac{3}{x} \right) = -2 \Rightarrow x^2 - 3 = -2x \Rightarrow$$

$$x^2 + 2x - 3 = 0 \Rightarrow (x+3)(x-1) = 0 \Rightarrow$$
$$x = -3 \text{ or } x = 1$$

If $x = -3$, then $y = -\frac{3}{-3} = 1$. If $x = 1$, then

$$y = -\frac{3}{1} = -3.$$

Solution set: $\{(1, -3), (-3, 1)\}$

35.
$$\begin{aligned} y & = x^2 + 6x + 9 \; (1) \\ x + y & = 3 \quad (2) \end{aligned}$$

Solve equation (2) for y to obtain

$$x + y = 3 \Rightarrow y = 3 - x. \; (3)$$

Substitute equation (3) into equation (1) and solve for x.

$$3 - x = x^2 + 6x + 9$$
$$0 = x^2 + 7x + 6$$
$$(x+6)(x+1) = 0 \Rightarrow x = -6 \text{ or } x = -1$$

If $x = -6$, then $y = 3 - (-6) = 9$. If $x = -1$, then $y = 3 - (-1) = 4$.

Solution set: $\{(-6, 9), (-1, 4)\}$

37.
$$\begin{aligned} 2x - 3y & = -2 \quad (1) \\ x + y - 4z & = -16 \quad (2) \\ 3x - 2y + z & = 7 \quad (3) \end{aligned}$$

Add 4 times equation (3) to equation (2) in order to eliminate z.

$$\begin{aligned} x + y - 4z & = -16 \\ 12x - 8y + 4z & = 28 \\ \hline 13x - 7y \quad\;\; & = 12 \quad (4) \end{aligned}$$

Add 3 times equation (4) to -7 times equation (1) in order to eliminate y and solve for x.

$$\begin{aligned} -14x + 21y & = 14 \\ 39x - 21y & = 36 \\ \hline 25x \quad\quad\;\; & = 50 \Rightarrow x = 2 \end{aligned}$$

Substitute 2 for x in equation (4) to obtain

$$13(2) - 7y = 12 \Rightarrow 26 - 7y = 12 \Rightarrow$$
$$-7y = -14 \Rightarrow y = 2$$

Substitute 2 for x and 2 for y in equation (2) to obtain the following.

$$2 + 2 - 4z = -16 \Rightarrow 4 - 4z = -16 \Rightarrow$$
$$-4z = -20 \Rightarrow z = 5$$

Solution set: $\{(2, 2, 5)\}$

Copyright © 2013 Pearson Education, Inc.

39. $y = (x-1)^2 + 2$ (1)

     $y = 2x - 1$ (2)

Substitute $2x - 1$ for y in equation (1) and solve for x:

$2x - 1 = (x-1)^2 + 2 \Rightarrow 2x - 1 = x^2 - 2x + 1 + 2$

$x^2 - 4x + 4 = 0 \Rightarrow (x-2)^2 = 0 \Rightarrow x = 2$

Substitute 2 for x in equation (2) and solve for y: $y = 2(2) - 1 = 3$

Solution set: $\{(2, 3)\}$

Section 5.6   Systems of Inequalities and Linear Programming

1. $x + 2y \le 6$

The boundary is the line $x + 2y = 6$, which can be graphed using the x-intercept 6 and y-intercept 3. The boundary is included in the graph, so draw a solid line. Solving for y, we have the following.

$x + 2y \le 6 \Rightarrow 2y \le -x + 6 \Rightarrow y \le -\frac{1}{2}x + 3$

Since y is *less than* or equal to $-\frac{1}{2}x + 3$, the graph of the solution set is the line and the half-plane *below* the boundary. We also can use (0, 0) as a test point. Since $0 + 2(0) \le 6 \Rightarrow 0 \le 6$ is a true statement, shade line and the side of the graph containing the test point (0, 0) (the half-plane below the boundary).

3. $2x + 3y \ge 4$

The boundary is the line $2x + 3y = 4$, which can be graphed using the x-intercept 2 and y-intercept $\frac{4}{3}$. The boundary is included in the graph, so draw a solid line. Solving for y, we have the following.

$2x + 3y \ge 4 \Rightarrow 3y \ge -2x + 4 \Rightarrow y \ge -\frac{2}{3}x + \frac{4}{3}$

Since y is *greater than* or equal to $-\frac{2}{3}x + \frac{4}{3}$, the graph of the solution set is the line and the half-plane *above* the boundary. We also can use (0, 0) as a test point.

Since $2(0) + 3(0) \ge 4 \Rightarrow 0 \ge 4$ is a false statement, shade the line and the side of the graph not containing the test point (0,0) (the half-plane above the boundary).

$2x + 3y \ge 4$

5. $3x - 5y > 6$

The boundary is the line $3x - 5y = 6$, which can be graphed using the x-intercept 2 and y-intercept $-\frac{6}{5}$. The boundary is not included in the graph, so draw a dashed line. Solving for y, we have the following Since y is *less than* $\frac{3}{5}x - \frac{6}{5}$, the graph of the solution set is the half-plane *below* the boundary. We also can use (0, 0) as a test point. Since $3(0) - 5(0) > 6 \Rightarrow 0 > 6$ is a false statement, shade the side of the graph not containing the test point (0, 0) (the half-plane below the boundary).

| Exercise 5 | Exercise 7 |

7. $5x < 4y - 2$

The boundary is the line $5x = 4y - 2$, which can be graphed using the x-intercept $-\frac{2}{5}$ and y-intercept $\frac{1}{2}$. The boundary is not included in the graph, so draw a dasshed line. Solving for y, we have the following.

$5x < 4y - 2 \Rightarrow -4y < -5x - 2 \Rightarrow y > \frac{5}{4}x + \frac{1}{2}$

Since y is greater than $\frac{5}{4}x + \frac{1}{2}$, the graph of the solution set is the half-plane *above* the boundary. We also can use (0, 0) as a test point. Since $5(0) < 4(0) - 2 \Rightarrow 0 < -2$ is a false statement, shade the side of the graph not containing the test point (0, 0) (the half-plane above the boundary).

Copyright © 2013 Pearson Education, Inc.

9. $x \le 3$

The boundary is the vertical line $x = 3$, which intersects the x-axis at 3. The boundary is included in the graph, so draw a solid line. Since $x \le 3$, it can easily be determined that we should shade to the left of the boundary. We also can use $(0, 0)$ as a test point. Since $0 \le 3$ is a true statement, shade the line and the side of the graph containing the test point $(0, 0)$ (the half-plane to the left of the boundary).

11. $y < 3x^2 + 2$

The boundary is the parabola $y = 3x^2 + 2$, which opens upwards. It has vertex $(0, 2)$, y-intercept 2, and no x-intercepts. Since the inequality symbol is <, draw a dashed curve. Since y is *less than* $3x^2 + 2$, the graph of the solution set is the half-plane *below* the boundary. We also can use $(0, 0)$ as a test point. Since $0 < 3(0)^2 + 2 \Rightarrow 0 < 2$ is a true statement, shade the region of the graph containing the test point $(0, 0)$.

13. $y > (x - 1)^2 + 2$

The boundary is the parabola $y = (x - 1)^2 + 2$, which opens upwards. It has vertex $(1, 2)$, y-intercept 3, and no x-intercepts. Since the inequality symbol is >, draw a dashed curve.

Since y is *greater than* $(x - 1)^2 + 2$, the graph of the solution set is the half-plane *above* the boundary. We also can use $(0, 0)$ as a test point.

Since $0 > (0 - 1)^2 + 2 \Rightarrow 0 > 3$ is a false statement, shade the region of the graph not containing the test point $(0, 0)$.

15. $x^2 + (y + 3)^2 \le 16$

The boundary is a circle with center $(0, -3)$ and radius 4. Draw a solid circle to show that the boundary is included in the graph. Since $x^2 + (y + 3)^2 \le 16$, it can easily be determined that points in the interior or on the boundary would satisfy this relation. We also can use $(0, 0)$ as a test point. Since we have $0^2 + (0 + 3)^2 \le 16 \Rightarrow 9 \le 16$ is a true statement, shade the region of the graph containing the test point $(0, 0)$.

17. $y > 2^x + 1$

The boundary is an exponential function with y-intercept 2. The inequality symbol is >, so draw a dashed curve. Since y is *greater than* $2^x + 1$, the graph of the solution set is the half-plane *above* the boundary. We also can use $(0, 0)$ as a test point. Since $0 > 2^0 + 1 \Rightarrow 0 > 2$ is a false statement, shade the region of the graph not containing the test point $(0, 0)$.

Copyright © 2013 Pearson Education, Inc.

19. If an inequality is a "strict" inequality (< or >), then the boundary is not included and is represented by a dashed line. If an inequality is a "nonstrict" or "weak" inequality (≤ or ≥), then the boundary is included and is represented by a solid line.

21. $Ax + By \geq C$, $B > 0$

Solving for y we have

$$Ax + By \geq C \Rightarrow By \geq -Ax + C \Rightarrow y \geq -\frac{A}{B}x + \frac{C}{B}$$

Since $B > 0$, the inequality symbol was not reversed when both sides are divided by B. Since $y \geq -\frac{A}{B}x + \frac{C}{B}$, you would shade above the line.

23. The graph of $(x-5)^2 + (y-2)^2 = 4$ is a circle with center $(5, 2)$ and radius $r = \sqrt{4} = 2$. The graph of $(x-5)^2 + (y-2)^2 < 4$ is the region in the interior of this circle. The correct response is B.

25. The graph of $y \leq 3x - 6$ is the region below the line with slope 3 and y-intercept -6. This is graph C.

27. The graph of $y \leq -3x - 6$ is the region below the line with slope -3 and y-intercept -6. This is graph A.

29. $x + y \geq 0$
$2x - y \geq 3$

Graph $x + y = 0$ as a solid line through the origin with a slope of -1. Shade the region above this line. Graph $2x - y = 3$ as a solid line with x-intercept $\frac{3}{2}$ and y-intercept -3. Shade the region below this line. To find where the boundaries of the two lines intersect, solve the system $\begin{array}{l} x + y = 0 \ (1) \\ 2x - y = 3 \ (2) \end{array}$ by adding the two equations together to obtain $3x = 3 \Rightarrow x = 1$. Substituting this value into equation (1), we obtain $1 + y = 0 \Rightarrow y = -1$. Thus, the boundaries intersect at $(1, -1)$. The solution set is the common region, which is shaded in the final graph.

For Exercises 31–61, we will omit the steps to find the points in which the boundaries intersect.

31. $2x + y > 2$
$x - 3y < 6$

Graph $2x + y = 2$ as a dashed line with y-intercept 2 and x-intercept 1. Shade the region above this line. Graph $x - 3y = 6$ as a dashed line with y-intercept -2 and x-intercept 6. Shade the region above this line. The solution set is the common region, which is shaded in the final graph. The lines intersect at $\left(\frac{12}{7}, -\frac{10}{7}\right)$.

33. $3x + 5y \leq 15$
$x - 3y \geq 9$

Graph $3x + 5y = 15$ as a solid line with y-intercept 3 and x-intercept 5. Shade the region below this line. Graph $x - 3y = 9$ as a solid line with y-intercept -3 and x-intercept 9. Shade the region below this line. The solution set is the common region, which is shaded in the final graph. The lines intersect at $\left(\frac{45}{7}, -\frac{6}{7}\right)$.

Copyright © 2013 Pearson Education, Inc.

35. $4x - 3y \le 12$

$y \le x^2$

Graph $4x - 3y = 12$ as a solid line with y-intercept -4 and x-intercept 3. Shade the region above this line. Graph the solid parabola $y = x^2$. Shade the region outside of this parabola. The solution set is the intersection of these two regions, which is shaded in the final graph. The line and the parabola do not intersect.

37. $x + 2y \le 4$

$y \ge x^2 - 1$

Graph $x + 2y = 4$ as a solid line with y-intercept 2 and x-intercept 4. Shade the region below the line. Graph the solid parabola $y = x^2 - 1$. Shade the region inside of the parabola. The solution set is the common region, which is shaded in the final graph. The parabola and the line intersect at $\left(\frac{3}{2}, \frac{5}{4}\right)$ and $(-2, 3)$.

39. $y \le (x + 2)^2$

$y \ge -2x^2$

Graph $y = (x + 2)^2$ as a solid parabola opening up with a vertex at $(-2, 0)$. Shade the region below the parabola. Graph $y = -2x^2$ as a solid parabola opening down with a vertex at the origin. Shade the region above the parabola. The solution set is the intersection of these two regions, which is shaded in the final graph. The parabolas do not intersect.

$y \le (x + 2)^2$

$y \ge -2x^2$

41. $x + y \le 36$

$-4 \le x \le 4$

Graph $x + y = 36$ as a solid line with y-intercept 36 and x-intercept 36. Shade the region below this line. Graph the vertical lines $x = -4$ and $x = 4$ as solid lines. Shade the region between these lines. The solution set is the intersection of these two regions, which is shaded in the final graph. The lines intersect at $(-4, 40)$ and $(4, 32)$.

43. $y \ge x^2 + 4x + 4$

$y < -x^2$

Graph the solid parabola $y = x^2 + 4x + 4$ or $y = (x + 2)^2$, which has vertex $(-2, 0)$ and opens upward. Shade the region inside of this parabola. Graph the dashed parabola $y = -x^2$, which has vertex $(0, 0)$ and opens downward. Shade the region inside this parabola. These two regions have no points in common, so the system has no solution. The parabolas do not intersect.

$y \ge x^2 + 4x + 4$

$y < -x^2$

Copyright © 2013 Pearson Education, Inc.

45. $3x - 2y \geq 6$ (1)
$x + y \leq -5$ (2)
$y \leq 4$ (3)

Graph $3x - 2y = 6$ as a solid line and shade the region below it. Graph $x + y = -5$ as a solid line and shade the region below it. Graph $y = 4$ as a solid horizontal line and shade the region below it. The solution set is the intersection of these three regions, which is shaded in the final graph. Lines (1) and (2) intersect at $\left(-\frac{4}{5}, -\frac{21}{5}\right)$.

$3x - 2y \geq 6$
$x + y \leq -5$
$y \leq 4$

47. $-2 < x < 2$
$y > 1$
$x - y > 0$

Graph the vertical lines $x = -2$ and $x = 2$ as a dashed line. Shade the region between the two lines. Graph the horizontal line $y = 1$ as a dashed line. Shade the region above the line. Graph the line $x - y = 0$ as a dashed line through the origin with a slope of 1. Shade the region below this line. The solution set is the intersection of these three regions, which is shaded in the final graph.

$-2 < x < 2$
$y > 1$
$x - y > 0$

49. $x \leq 4$
$x \geq 0$
$y \geq 0$
$x + 2y \geq 2$

Graph $x = 4$ as a solid vertical line. Shade the region to the left of this line. Graph $x = 0$ as a solid vertical line. (This is the y-axis.) Shade the region to the right of this line. Graph $y = 0$ as a solid horizontal line. (This is the x-axis.) Shade the region above the line. Graph $x + 2y = 2$ as a solid line with x-intercept 2 and y-intercept 1.

Shade the region above the line. The solution set is the intersection of these four regions, which is shaded in the final graph.

$x \leq 4$
$x \geq 0$
$y \geq 0$
$x + 2y \geq 2$

51. $2x + 3y \leq 12$
$2x + 3y > -6$
$3x + y < 4$
$x \geq 0$
$y \geq 0$

Graph $2x + 3y = 12$ as a solid line and shade the region below it. Graph $2x + 3y = 6$ as a dashed line and shade the region above it. Graph $3x + y = 4$ as a dashed line and shade the region below it. $x = 0$ is the y-axis. Shade the region to the right of it. $y = 0$ is the x-axis. Shade the region above it. The solution set is the intersection of these five regions, which is shaded in the final graph. The open circles at $(0, 4)$ and $\left(\frac{4}{3}, 0\right)$ indicate that those points are not included in the solution (due to the fact that the boundary line on which they lie, $3x + y = 4$, is not included).

$2x + 3y \leq 12$
$2x + 3y > -6$
$3x + y < 4$
$x \geq 0, y \geq 0$

$y \leq \left(\frac{1}{2}\right)^x$
$y \geq 4$

Exercise 51   Exercise 53

53. $y \leq \left(\frac{1}{2}\right)^x$
$y \geq 4$

Graph $y = \left(\frac{1}{2}\right)^x$ using a solid curve passing through the points $(-2, 4)$, $(-1, 2)$, $(0, 1)$, $\left(1, \frac{1}{2}\right)$, and $\left(2, \frac{1}{4}\right)$. Shade the region below this curve. Graph the solid horizontal line $y = 4$ and shade the region above it. The solution set consists of the intersection of these two regions, which is shaded in the final graph. The curves intersect at $(-2, 4)$.

Copyright © 2013 Pearson Education, Inc.

55. $y \le \log x$

   $y \ge |x - 2|$

   Graph $y = \log x$ using a solid curve because $y \le \log x$ is a nonstrict inequality. (Recall that "log x" means $\log_{10} x$.) This graph contains the points $(0.1, -1)$, $(1, 0)$ and $(10, 1)$. Use a calculator to approximate other points on the graph, such as $(2, 0.30)$ and $(4, 0.60)$. Because the symbol is $\le$, shade the region *below* the curve. Now graph $y = |x - 2|$. Make this boundary solid because $y \ge |x - 2|$ is also a nonstrict inequality. This graph can be obtained by translating the graph of $y = |x|$ to the right 2 units. It contains points $(0, 2)$, $(2, 0)$, and $(4, 2)$. Because the symbol is $\ge$, shade the region *above* the absolute value graph. The solution set is the intersection of the two regions, which is shaded in the final graph.

57. $y > x^3 + 1$

   $y \ge -1$

   Graph $y > x^3 + 1$ as a dashed curve, which is the graph of $y = x^3$ translated up 1 unit. Shade the region above the curve. Graph $y = -1$ as a solid horizontal line through $(0, -1)$. Shade the area above the line. The solution set is the common region, which is shaded in the final graph.

59. $3x + 2y \ge 6$

   Solving the inequality for y we have

   $3x + 2y \ge 6 \Rightarrow 2y \ge -3x + 6 \Rightarrow y \ge -\frac{3}{2}x + 3$.

   Enter $Y_1 = (-3/2)x + 3$ and use a graphing calculator to shade the region above the line. Notice the icon to the left of Y_1.

61. $x + y \ge 2$; $x + y \le 6$

   Solving each inequality for y, we have $y = -x + 2$ and $y = -x + 6$. Enter $Y_1 = -x + 2$ and $Y_2 = -x + 6$ and use the graphing calculator to shade the region above the graph of $y_1 = -x + 2$ and below the graph of $y_2 = -x + 6$.

63. The upper line passes through $(0, 2)$ and $(4, 0)$, so $m = \frac{0-2}{4-0} = -\frac{1}{2}$, and the equation is

   $y = -\frac{1}{2}x + 2 \Rightarrow x + 2y = 4$. The line is solid, so it is included in the inequality. The lower line passes through $(0, -3)$ and $(4, 0)$, so $m = \frac{0-(-3)}{4-0} = \frac{3}{4}$ and the equation is

   $y = \frac{3}{4}x - 3 \Rightarrow 3x - 4y = 12$. The line is solid, so it is included in the inequality. The shaded region contains the point $(0, 0)$, so test $(0, 0)$ in each equation to determine the direction of the inequalities:

   $0 + 2(0) = 0 < 4 \Rightarrow x + 2y \le 4$,

   $3(0) + 4(0) = 0 < 12 \Rightarrow 3x - 4y \le 12$

   The system is $x + 2y \le 4$.

   $3x - 4y \le 12$

Copyright © 2013 Pearson Education, Inc.

65. The circle has center (0, 0) and radius 4, so its equation is $x^2 + y^2 = 16$. The curve is solid, so it is included in the inequality. The horizontal line passes through (0, 2), so its equation is $y = 2$. It is solid, so it is included in the inequality. The shaded region includes the point (−1, 3), so test (−1, 3) in each equation to determine the direction of the inequalities:

$(-1)^2 + 3^2 = 10 < 16 \Rightarrow x^2 + y^2 \leq 16$,

$3 > 2 \Rightarrow y > 2$. The system is $\begin{array}{l} x^2 + y^2 \leq 16. \\ y \geq 2 \end{array}$

67. The graph is in the first quadrant, so $x > 0$ and $y > 0$. A circle with radius 2 centered at the origin has the equation $x^2 + y^2 = 4$. Since the region includes and is inside the circle, the inequality is $x^2 + y^2 \leq 4$. A line that passes through (0, −1) and (2, 2) has slope

$m = \frac{2-(-1)}{2-0} = \frac{3}{2}$ and equation $y = \frac{3}{2}x - 1$.

Since the region is above the line, but does not include the line, the inequality is $y > \frac{3}{2}x - 1$.

The system is

$$\begin{array}{l} x > 0 \\ y > 0 \\ x^2 + y^2 \leq 4 \\ y > \frac{3}{2}x - 1 \end{array}$$

69.

Point	Value of $3x + 5y$	
$(1,1)$	$3(1)+5(1)=8$	← Minimum
$(2,7)$	$3(2)+5(7)=41$	
$(5,10)$	$3(5)+5(10)=65$	← Maximum
$(6,3)$	$3(6)+5(3)=33$	

The maximum value is 65 at (5, 10). The minimum value is 8 at (1, 1).

71.

Point	Value of $3x + 5y$	
$(1,0)$	$3(1)+5(0)=3$	← Minimum
$(1,10)$	$3(1)+5(10)=53$	
$(7,9)$	$3(7)+5(9)=66$	← Maximum
$(7,6)$	$3(7)+5(6)=51$	

The maximum value is 66 at (7, 9). The minimum value is 3 at (1, 0).

73.

Point	Value of $10y$	
$(1,0)$	$10(0)=0$	← Minimum
$(1,10)$	$10(10)=100$	← Maximum
$(7,9)$	$10(9)=90$	
$(7,6)$	$10(6)=60$	

The maximum value is 100 at (1, 10). The minimum value is 0 at (1, 0).

75. Let x = the number of Brand X pills; y = the number of Brand Y pills.
The following table is helpful in organizing the information.

	Number of Brand X pills (x)	Number of Brand Y pills (y)	Restrictions
Vitamin A	3000	1000	At least 6000
Vitamin C	45	50	At least 195
Vitamin D	75	200	At least 600

We have

$$\begin{array}{l} 3000x + 1000y \geq 6000 \\ 45x + 50y \geq 195 \\ 75x + 200y \geq 600 \\ x \geq 0, y \geq 0. \end{array}$$

Graph $3000x + 1000y = 6000$ as a solid line with x-intercept 2 and y-intercept 6. Shade the region above the line. Graph $45x + 50y = 195$ as a solid line with x-intercept $4.\overline{3}$ and y-intercept 3.9. Shade the region above the line. Graph $75x + 200y = 600$ as a solid line with x-intercept 8 and y-intercept 3. Shade the region above the line. Graph $x = 0$ (the y-axis) as a solid line and shade the region to the right of it. Graph $y = 0$ (the x-axis) as a solid line and shade the region above it. The region of feasible solutions is the intersection of these five regions.

Copyright © 2013 Pearson Education, Inc.

77. (a) Let x = number of cartons of food;
y = number of cartons of clothing.
The following table is helpful in organizing the information.

	Number of cartons of food (x)	Number of cartons of clothing (y)	Restrictions
Weight	40	10	Cannot Exceed 16,000
Volume	20	30	No more than 18,000

We have
$$40x + 10y \le 16,000$$
$$20x + 30y \le 18,000$$
$$x \ge 0, y \ge 0.$$

Maximize objective function, number of people = $10x + 8y$.

Find the region of feasible solutions by graphing the system of inequalities that is made up of the constraints. To graph $40x + 10y \le 16,000$, draw the line with x-intercept 400 and y-intercept 1600 as a solid line. Because the test point (0, 0) satisfies this inequality, shade the region *below* the line. To graph $20x + 30y \le 18,000$, draw the line with x-intercept 900 and y-intercept 600 as a solid line. Because the test point (0, 0) satisfies this inequality, shade the region *below* the line. The constraints $x \ge 0$ and $y \ge 0$ restrict the graph to the first quadrant. The graph of the feasible region is the intersection of the regions that are the graphs of the individual constraints.

From the graph, observe that three are the vertices are $(0,0), (0,600)$, and $(400,0)$.

The fourth vertex is the intersection point of the lines $40x + 10y = 16,000$ and $20x + 30y = 18,000$.

To find this point, solve the system
$$40x + 10y = 16,000$$
$$20x + 30y = 18,000.$$

The first equation can be written as
$4x + y = 1600 \Rightarrow y = 1600 - 4x.$

Substituting this equation into $20x + 30y = 18,000$, we have
$$20x + 30(1600 - 4x) = 18,000$$
$$20x + 48,000 - 120x = 18,000$$
$$48,000 - 100x = 18,000$$
$$-100x = -30,000$$
$$x = 300$$

Substituting $x = 300$ into $y = 1600 - 4x$, we have
$$y = 1600 - 4(300) = 1600 - 1200 = 400.$$

Thus, the fourth vertex is (300, 400). Next, evaluate the objective function at each vertex.

Point	Number of people = $10x + 8y$	
$(0,0)$	$10(0)+8(0)=0$	
$(0,600)$	$10(0)+8(600)=4800$	
$(300,400)$	$10(300)+8(400)=6200$	← Maximum
$(400,0)$	$10(400)+8(0)=4000$	

The maximum value of $10x + 8y$ occurs at $(300,400)$, so they should send 300 cartons of food and 400 cartons of clothes to maximize the number of people helped.

(b) The maximum number of people helped is 6200.

79. Let x = number of cabinet A;
y = number of cabinet B.
The cost constraint is $10x + 20y \le 140$. The space constraint is $6x + 8y \le 72$. Since the numbers of cabinets cannot be negative, we also have $x \ge 0$ and $y \ge 0$.

We want to maximize the objective function, storage capacity = $8x + 12y$. Find the region of feasible solutions by graphing the system of inequalities that is made up of the constraints. To graph $10x + 20y \le 140$, draw the line with x-intercept 14 and y-intercept 7 as a solid line. Because the test point (0, 0) satisfies this inequality, shade the region below the line.

(*continued on next page*)

Copyright © 2013 Pearson Education, Inc.

(*continued*)

To graph $6x + 8y \le 72$, draw the line with x-intercept 12 and y-intercept 9 as a solid line. Because the test point (0, 0) satisfies this inequality, shade the region below the line. The constraints $x \ge 0$ and $y \ge 0$ restrict the graph to the first quadrant. The graph of the feasible region is the intersection of the regions that are the graphs of the individual constraints.

From the graph, observe that three vertices are (0, 0), (0, 7), and (12, 0). The fourth vertex is the intersection point of the lines $10x + 20y = 140$ and $6x + 8y = 72$. To find this point, solve the system

$$10x + 20y = 140$$
$$6x + 8y = 72.$$

The first equation can be written as $x + 2y = 14 \Rightarrow x = 14 - 2y$. Substituting this equation into $6x + 8y = 72$, we have

$$6(14 - 2y) + 8y = 72 \Rightarrow 84 - 12y + 8y = 72 \Rightarrow$$
$$84 - 4y = 72 \Rightarrow -4y = -12 \Rightarrow y = 3$$

Substituting $y = 3$ into $x = 14 - 2y$, we have $x = 14 - 2(3) = 14 - 6 = 8$. Thus, the fourth vertex is (8, 3). Next, evaluate the objective function at each vertex.

Point	Storage Capacity $= 8x + 12y$	
$(0,0)$	$8(0) + 12(0) = 0$	
$(0,7)$	$8(0) + 12(7) = 84$	
$(8,3)$	$8(8) + 12(3) = 100$	← Maximum
$(12,0)$	$8(12) + 12(0) = 96$	

The maximum value of $8x + 12y$ occurs at (8, 3), so the office manager should buy 8 of cabinet A and 3 of cabinet B for a total storage capacity of 100 ft^3.

81. Let x = number of servings of product A; y = number of servings of product B. The Supplement I constraint is $3x + 2y \ge 15$. The Supplement II constraint is $2x + 4y \ge 15$.

Since the numbers of servings cannot be negative, we also have $x \ge 0$ and $y \ge 0$. We want to minimize the objective function, cost $= 0.25x + 0.40y$. Find the region of feasible solutions by graphing the system of inequalities that is made up of the constraints. To graph $3x + 2y \ge 15$, draw the line with x-intercept 5 and y-intercept $\frac{15}{2} = 7\frac{1}{2}$ as a solid line. Because the test point (0, 0) does not satisfy this inequality, shade the region *above* the line. To graph $2x + 4y \ge 15$, draw the line with x-intercept $\frac{15}{2} = 7\frac{1}{2}$ and y-intercept $\frac{15}{4} = 3\frac{3}{4}$ as a solid line. Because the test point (0, 0) does not satisfy this inequality, shade the region *above* the line.

The constraints $x \ge 0$ and $y \ge 0$ restrict the graph to the first quadrant. The graph of the feasible region is the intersection of the regions that are the graphs of the individual constraints.

From the graph, observe that two vertices are $\left(0, \frac{15}{2}\right)$ and $\left(\frac{15}{2}, 0\right)$. The third vertex is the intersection point of the lines $3x + 2y = 15$ and $2x + 4y = 15$. To find this point, solve the system

$$3x + 2y = 15$$
$$2x + 4y = 15.$$

Multiply the first equation by −2 and add it to the second equation.

$$-6x - 4y = -30$$
$$\underline{2x + 4y = 15}$$
$$-4x = -15 \Rightarrow x = \frac{15}{4}$$

Substituting this equation into $2x + 4y = 15$, we have $2\left(\frac{15}{4}\right) + 4y = 15 \Rightarrow \frac{15}{2} + 4y = 15 \Rightarrow$ $15 + 8y = 30 \Rightarrow 8y = 15 \Rightarrow y = \frac{15}{8}$.

(*continued on next page*)

Copyright © 2013 Pearson Education, Inc.

(*continued*)

Thus, the third vertex is $\left(\frac{15}{4}, \frac{15}{8}\right)$. Next, evaluate the objective function at each vertex.

Point	Cost = 0.25x + 0.40y
$\left(0, \frac{15}{2}\right)$	$0.25(0) + 0.40\left(\frac{15}{2}\right) = 3.00$
$\left(\frac{15}{4}, \frac{15}{8}\right)$	$0.25\left(\frac{15}{4}\right) + 0.40\left(\frac{15}{8}\right) = 1.69$ ← Minimum
$\left(\frac{15}{2}, 0\right)$	$0.25\left(\frac{15}{2}\right) + 0.40(0) = 1.88$

The minimum cost is \$1.69 for $\frac{15}{4} = 3\frac{3}{4}$

servings of *A* and $\frac{15}{8} = 1\frac{7}{8}$ servings of *B*.

Section 5.7   Properties of Matrices

1. $\begin{bmatrix} -3 & a \\ b & 5 \end{bmatrix} = \begin{bmatrix} c & 0 \\ 4 & d \end{bmatrix}$

Since corresponding elements are equal, we have the following.
$a = 0, b = 4, c = -3, d = 5$

3. $\begin{bmatrix} x+2 & y-6 \\ z-3 & w+5 \end{bmatrix} = \begin{bmatrix} -2 & 8 \\ 0 & 3 \end{bmatrix}$

Since corresponding elements are equal, we have the following.
$x + 2 = -2 \Rightarrow x = -4$
$y - 6 = 8 \Rightarrow y = 14$
$z - 3 = 0 \Rightarrow z = 3$
$w + 5 = 3 \Rightarrow w = -2$

5. Since $\begin{bmatrix} x & y & z \end{bmatrix}$ is a 1×3 matrix and $\begin{bmatrix} 21 & 6 \end{bmatrix}$ is a 1×2 matrix, the statement cannot be true, hence we cannot find values of *x*, *y*, and *z*.

7. $\begin{bmatrix} 0 & 5 & x \\ -1 & 3 & y+2 \\ 4 & 1 & z \end{bmatrix} = \begin{bmatrix} 0 & w & 6 \\ -1 & 3 & 0 \\ 4 & 1 & 8 \end{bmatrix}$

Since corresponding elements are equal, we have the following.
$x = 6$
$y + 2 = 0 \Rightarrow y = -2$
$z = 8$
$w = 5$

9. $\begin{bmatrix} -7+z & 4r & 8s \\ 6p & 2 & 5 \end{bmatrix} + \begin{bmatrix} -9 & 8r & 3 \\ 2 & 5 & 4 \end{bmatrix}$

$= \begin{bmatrix} 2 & 36 & 27 \\ 20 & 7 & 12a \end{bmatrix} \Rightarrow$

$\begin{bmatrix} -16+z & 12r & 8s+3 \\ 6p+2 & 7 & 9 \end{bmatrix} = \begin{bmatrix} 2 & 36 & 27 \\ 20 & 7 & 12a \end{bmatrix}$

Since corresponding elements are equal, we have
$-16 + z = 2 \Rightarrow z = 18,\ 12r = 36 \Rightarrow r = 3,$
$8s + 3 = 27 \Rightarrow s = 3,\ 6p + 2 = 20 \Rightarrow p = 3,$
$9 = 12a \Rightarrow a = \frac{3}{4}$

Thus, $z = 18, r = 3, s = 3, p = 3,$ and $a = \frac{3}{4}$.

11. Two matrices are equal if they have the same <u>dimension</u> and if corresponding elements are <u>equal</u>.

13. $\begin{bmatrix} -4 & 8 \\ 2 & 3 \end{bmatrix}$

This matrix has 2 rows and 2 columns, so it is a 2×2 square matrix.

15. $\begin{bmatrix} -6 & 8 & 0 & 0 \\ 4 & 1 & 9 & 2 \\ 3 & -5 & 7 & 1 \end{bmatrix}$

This matrix has 3 rows and 4 columns, so it is a 3×4 matrix.

17. $\begin{bmatrix} 2 \\ 4 \end{bmatrix}$

This matrix has 2 rows and 1 column, so it is a 2×1 column matrix.

19. Answers will vary.

21. $\begin{bmatrix} -4 & 3 \\ 12 & -6 \end{bmatrix} + \begin{bmatrix} 2 & -8 \\ 5 & 10 \end{bmatrix} = \begin{bmatrix} -4+2 & 3+(-8) \\ 12+5 & -6+10 \end{bmatrix}$

$= \begin{bmatrix} -2 & -5 \\ 17 & 4 \end{bmatrix}$

23. $\begin{bmatrix} 6 & -9 & 2 \\ 4 & 1 & 3 \end{bmatrix} + \begin{bmatrix} -8 & 2 & 5 \\ 6 & -3 & 4 \end{bmatrix}$

$= \begin{bmatrix} 6+(-8) & -9+2 & 2+5 \\ 4+6 & 1+(-3) & 3+4 \end{bmatrix}$

$= \begin{bmatrix} -2 & -7 & 7 \\ 10 & -2 & 7 \end{bmatrix}$

Copyright © 2013 Pearson Education, Inc.

25. Since $\begin{bmatrix} 2 & 4 & 6 \end{bmatrix}$ is a 1×3 matrix and $\begin{bmatrix} -2 \\ -4 \\ -6 \end{bmatrix}$ is

a 3×1 matrix, $\begin{bmatrix} 2 & 4 & 6 \end{bmatrix} + \begin{bmatrix} -2 \\ -4 \\ -6 \end{bmatrix}$ cannot be

added.

27.
$$\begin{bmatrix} -6 & 8 \\ 0 & 0 \end{bmatrix} - \begin{bmatrix} 0 & 0 \\ -4 & -2 \end{bmatrix} = \begin{bmatrix} -6-0 & 8-0 \\ 0-(-4) & 0-(-2) \end{bmatrix}$$
$$= \begin{bmatrix} -6 & 8 \\ 4 & 2 \end{bmatrix}$$

29.
$$\begin{bmatrix} 12 \\ -1 \\ 3 \end{bmatrix} - \begin{bmatrix} 8 \\ 4 \\ -1 \end{bmatrix} = \begin{bmatrix} 12-8 \\ -1-4 \\ 3-(-1) \end{bmatrix} = \begin{bmatrix} 4 \\ -5 \\ 4 \end{bmatrix}$$

31. Since $\begin{bmatrix} -4 & 3 \end{bmatrix}$ is a 1×2 matrix and

$\begin{bmatrix} 5 & 8 & 2 \end{bmatrix}$ is a 1×3 matrix,

$\begin{bmatrix} -4 & 3 \end{bmatrix} - \begin{bmatrix} 5 & 8 & 2 \end{bmatrix}$ cannot be subtracted.

33.
$$\begin{bmatrix} \sqrt{3} & -4 \\ 2 & -\sqrt{5} \\ -8 & \sqrt{8} \end{bmatrix} - \begin{bmatrix} 2\sqrt{3} & 9 \\ -2 & \sqrt{5} \\ -7 & 3\sqrt{2} \end{bmatrix}$$
$$= \begin{bmatrix} \sqrt{3}-2\sqrt{3} & -4-9 \\ 2-(-2) & -\sqrt{5}-\sqrt{5} \\ -8-(-7) & \sqrt{8}-3\sqrt{2} \end{bmatrix} = \begin{bmatrix} -\sqrt{3} & -13 \\ 4 & -2\sqrt{5} \\ -1 & -\sqrt{2} \end{bmatrix}$$

35.
$$\begin{bmatrix} 3x+y & -2y \\ x+2y & 3y \end{bmatrix} + \begin{bmatrix} 2x & 3y \\ 5x & x \end{bmatrix}$$
$$= \begin{bmatrix} (3x+y)+2x & -2y+3y \\ (x+2y)+5x & 3y+x \end{bmatrix}$$
$$= \begin{bmatrix} 5x+y & y \\ 6x+2y & x+3y \end{bmatrix}$$

In Exercises 37–39, $A = \begin{bmatrix} -2 & 4 \\ 0 & 3 \end{bmatrix}$ and $B = \begin{bmatrix} -6 & 2 \\ 4 & 0 \end{bmatrix}$.

37. $2A = 2\begin{bmatrix} -2 & 4 \\ 0 & 3 \end{bmatrix} = \begin{bmatrix} 2(-2) & 2(4) \\ 2(0) & 2(3) \end{bmatrix} = \begin{bmatrix} -4 & 8 \\ 0 & 6 \end{bmatrix}$

39. $\frac{3}{2}B = \frac{3}{2}\begin{bmatrix} -6 & 2 \\ 4 & 0 \end{bmatrix} = \begin{bmatrix} \frac{3}{2}(-6) & \frac{3}{2}(2) \\ \frac{3}{2}(4) & \frac{3}{2}(0) \end{bmatrix} = \begin{bmatrix} -9 & 3 \\ 6 & 0 \end{bmatrix}$

41. $2A - 3B = 2\begin{bmatrix} -2 & 4 \\ 0 & 3 \end{bmatrix} - 3\begin{bmatrix} -6 & 2 \\ 4 & 0 \end{bmatrix}$
$$= \begin{bmatrix} -4 & 8 \\ 0 & 6 \end{bmatrix} - \begin{bmatrix} -18 & 6 \\ 12 & 0 \end{bmatrix}$$
$$= \begin{bmatrix} -4-(-18) & 8-6 \\ 0-12 & 6-0 \end{bmatrix} = \begin{bmatrix} 14 & 2 \\ -12 & 6 \end{bmatrix}$$

43. $-A + \frac{1}{2}B = -\begin{bmatrix} -2 & 4 \\ 0 & 3 \end{bmatrix} + \frac{1}{2}\begin{bmatrix} -6 & 2 \\ 4 & 0 \end{bmatrix}$
$$= \begin{bmatrix} 2 & -4 \\ 0 & -3 \end{bmatrix} + \begin{bmatrix} -3 & 1 \\ 2 & 0 \end{bmatrix}$$
$$= \begin{bmatrix} 2+(-3) & -4+1 \\ 0+2 & -3+0 \end{bmatrix} = \begin{bmatrix} -1 & -3 \\ 2 & -3 \end{bmatrix}$$

45. AB can be calculated and the result will be a 2×5 matrix.

Matrix A Matrix B
  2×3 3×5
     matches
   size of AB
     2×5

47. BA cannot be calculated.

Matrix B Matrix A
  3×5 2×3
     different

49. BC can be calculated and the result will be a 3×2 matrix.

Matrix B Matrix C
  3×5 5×2
     matches
   size of CA
     3×2

51. A 2×2 matrix multiplied by a 2×1 matrix results in a 2×1 matrix.
$$\begin{bmatrix} 1 & 2 \\ 3 & 4 \end{bmatrix}\begin{bmatrix} -1 \\ 7 \end{bmatrix} = \begin{bmatrix} 1(-1)+2(7) \\ 3(-1)+4(7) \end{bmatrix}$$
$$= \begin{bmatrix} -1+14 \\ -3+28 \end{bmatrix} = \begin{bmatrix} 13 \\ 25 \end{bmatrix}$$

53. A 2×3 matrix multiplied by a 3×1 matrix results in a 2×1 matrix.
$$\begin{bmatrix} 3 & -4 & 1 \\ 5 & 0 & 2 \end{bmatrix}\begin{bmatrix} -1 \\ 4 \\ 2 \end{bmatrix} = \begin{bmatrix} 3(-1)+(-4)(4)+1(2) \\ 5(-1)+0(4)+2(2) \end{bmatrix}$$
$$= \begin{bmatrix} -3+(-16)+2 \\ -5+0+4 \end{bmatrix} = \begin{bmatrix} -17 \\ -1 \end{bmatrix}$$

Copyright © 2013 Pearson Education, Inc.

55. A 2×3 matrix multiplied by a 3×2 matrix results in a 2×2 matrix.

$$\begin{bmatrix} \sqrt{2} & \sqrt{2} & -\sqrt{18} \\ \sqrt{3} & \sqrt{27} & 0 \end{bmatrix}\begin{bmatrix} 8 & -10 \\ 9 & 12 \\ 0 & 2 \end{bmatrix} = \begin{bmatrix} \sqrt{2} & \sqrt{2} & -3\sqrt{2} \\ \sqrt{3} & 3\sqrt{3} & 0 \end{bmatrix}\begin{bmatrix} 8 & -10 \\ 9 & 12 \\ 0 & 2 \end{bmatrix}$$

$$= \begin{bmatrix} 8\sqrt{2}+9\sqrt{2} & -10\sqrt{2}+12\sqrt{2} \\ \quad -3\sqrt{2}(0) & \quad -3\sqrt{2}(2) \\ 8\sqrt{3}+3\sqrt{3}(9) & -10\sqrt{3}+3\sqrt{3}(12) \\ \quad +0(0) & \quad +0(2) \end{bmatrix} = \begin{bmatrix} 17\sqrt{2} & -4\sqrt{2} \\ 35\sqrt{3} & 26\sqrt{3} \end{bmatrix}$$

57. A 2×2 matrix multiplied by a 2×2 matrix results in a 2×2 matrix.

$$\begin{bmatrix} \sqrt{3} & 1 \\ 2\sqrt{5} & 3\sqrt{2} \end{bmatrix}\begin{bmatrix} \sqrt{3} & -\sqrt{6} \\ 4\sqrt{3} & 0 \end{bmatrix} = \begin{bmatrix} \sqrt{3}(\sqrt{3})+1(4\sqrt{3}) & \sqrt{3}(-\sqrt{6})+1(0) \\ 2\sqrt{5}(\sqrt{3})+3\sqrt{2}(4\sqrt{3}) & 2\sqrt{5}(-\sqrt{6})+3\sqrt{2}(0) \end{bmatrix}$$

$$= \begin{bmatrix} 3+4\sqrt{3} & -\sqrt{18} \\ 2\sqrt{15}+12\sqrt{6} & -2\sqrt{30} \end{bmatrix} = \begin{bmatrix} 3+4\sqrt{3} & -3\sqrt{2} \\ 2\sqrt{15}+12\sqrt{6} & -2\sqrt{30} \end{bmatrix}$$

59. $\begin{bmatrix} -3 & 0 & 2 & 1 \\ 4 & 0 & 2 & 6 \end{bmatrix}\begin{bmatrix} -4 & 2 \\ 0 & 1 \end{bmatrix}$

A 2×4 matrix cannot be multiplied by a 2×2 matrix because the number of columns of the first matrix (four) is not equal to the number of rows of the second matrix (two).

61. A 1×3 matrix multiplied by a 3×3 matrix results in a 1×3 matrix.

$$\begin{bmatrix} -2 & 4 & 1 \end{bmatrix}\begin{bmatrix} 3 & -2 & 4 \\ 2 & 1 & 0 \\ 0 & -1 & 4 \end{bmatrix} = \begin{bmatrix} -2(3)+4(2)+1(0) & -2(-2)+4(1)+1(-1) & -2(4)+4(0)+1(4) \end{bmatrix}$$

$$= \begin{bmatrix} -6+8+0 & 4+4+(-1) & -8+0+4 \end{bmatrix} = \begin{bmatrix} 2 & 7 & -4 \end{bmatrix}$$

63. A 3×3 matrix multiplied by a 3×3 matrix results in a 3×3 matrix.

$$\begin{bmatrix} -2 & -3 & -4 \\ 2 & -1 & 0 \\ 4 & -2 & 3 \end{bmatrix}\begin{bmatrix} 0 & 1 & 4 \\ 1 & 2 & -1 \\ 3 & 2 & -2 \end{bmatrix}$$

$$= \begin{bmatrix} -2(0)+(-3)(1)+(-4)(3) & -2(1)+(-3)(2)+(-4)(2) & -2(4)+(-3)(-1)+(-4)(-2) \\ 2(0)+(-1)(1)+0(3) & 2(1)+(-1)(2)+0(2) & 2(4)+(-1)(-1)+0(-2) \\ 4(0)+(-2)(1)+3(3) & 4(1)+(-2)(2)+3(2) & 4(4)+(-2)(-1)+3(-2) \end{bmatrix}$$

$$= \begin{bmatrix} 0+(-3)+(-12) & -2+(-6)+(-8) & -8+3+8 \\ 0+(-1)+0 & 2+(-2)+0 & 8+1+0 \\ 0+(-2)+9 & 4+(-4)+6 & 16+2+(-6) \end{bmatrix} = \begin{bmatrix} -15 & -16 & 3 \\ -1 & 0 & 9 \\ 7 & 6 & 12 \end{bmatrix}$$

In Exercises 65–71, $A = \begin{bmatrix} 4 & -2 \\ 3 & 1 \end{bmatrix}$, $B = \begin{bmatrix} 5 & 1 \\ 0 & -2 \\ 3 & 7 \end{bmatrix}$ and $C = \begin{bmatrix} -5 & 4 & 1 \\ 0 & 3 & 6 \end{bmatrix}$.

65. A 3×2 matrix multiplied by a 2×2 matrix results in a 3×2 matrix.

$$BA = \begin{bmatrix} 5 & 1 \\ 0 & -2 \\ 3 & 7 \end{bmatrix}\begin{bmatrix} 4 & -2 \\ 3 & 1 \end{bmatrix} = \begin{bmatrix} 5(4)+1(3) & 5(-2)+1(1) \\ 0(4)+(-2)(3) & 0(-2)+(-2)(1) \\ 3(4)+7(3) & 3(-2)+7(1) \end{bmatrix} = \begin{bmatrix} 20+3 & -10+1 \\ 0+(-6) & 0+(-2) \\ 12+21 & -6+7 \end{bmatrix} = \begin{bmatrix} 23 & -9 \\ -6 & -2 \\ 33 & 1 \end{bmatrix}$$

Copyright © 2013 Pearson Education, Inc.

67. A 3×2 matrix multiplied by a 2×3 matrix results in a 3×3 matrix.

$$BC = \begin{bmatrix} 5 & 1 \\ 0 & -2 \\ 3 & 7 \end{bmatrix}\begin{bmatrix} -5 & 4 & 1 \\ 0 & 3 & 6 \end{bmatrix} = \begin{bmatrix} 5(-5)+1(0) & 5(4)+1(3) & 5(1)+1(6) \\ 0(-5)+(-2)(0) & 0(4)+(-2)(3) & 0(1)+(-2)(6) \\ 3(-5)+7(0) & 3(4)+7(3) & 3(1)+7(6) \end{bmatrix}$$

$$= \begin{bmatrix} -25+0 & 20+3 & 5+6 \\ 0+0 & 0+(-6) & 0+(-12) \\ -15+0 & 12+21 & 3+42 \end{bmatrix} = \begin{bmatrix} -25 & 23 & 11 \\ 0 & -6 & -12 \\ -15 & 33 & 45 \end{bmatrix}$$

69. $AB = \begin{bmatrix} 4 & -2 \\ 3 & 1 \end{bmatrix}\begin{bmatrix} 5 & 1 \\ 0 & -2 \\ 3 & 7 \end{bmatrix}$ is the product of a 2×2 matrix multiplied by a 3×2 matrix, which is not possible.

71. Since $A^2 = AA$, we are finding the product of two 2×2 matrices, which results in a 2×2 matrix.

$$A^2 = \begin{bmatrix} 4 & -2 \\ 3 & 1 \end{bmatrix}\begin{bmatrix} 4 & -2 \\ 3 & 1 \end{bmatrix} = \begin{bmatrix} 4(4)+(-2)(3) & 4(-2)+(-2)(1) \\ 3(4)+1(3) & 3(-2)+1(1) \end{bmatrix} = \begin{bmatrix} 16+(-6) & -8+(-2) \\ 12+3 & -6+1 \end{bmatrix} = \begin{bmatrix} 10 & -10 \\ 15 & -5 \end{bmatrix}$$

73. Since the answers to 65 and 69 are not equal, $BA \ne AB$.
Since the answers to 67 and 68 are not equal, $BC \ne CB$.
Since the answers to 66 and 70 are not equal, $AC \ne CA$.

75. $A = \begin{bmatrix} 3 & 4 \\ -2 & 1 \end{bmatrix}$ and $B = \begin{bmatrix} 6 & 0 \\ 5 & -2 \end{bmatrix}$

(a) $AB = \begin{bmatrix} 3 & 4 \\ -2 & 1 \end{bmatrix}\begin{bmatrix} 6 & 0 \\ 5 & -2 \end{bmatrix} = \begin{bmatrix} 3(6)+4(5) & 3(0)+4(-2) \\ -2(6)+1(5) & -2(0)+1(-2) \end{bmatrix} = \begin{bmatrix} 18+20 & 0+(-8) \\ -12+5 & 0+(-2) \end{bmatrix} = \begin{bmatrix} 38 & -8 \\ -7 & -2 \end{bmatrix}$

(b) $BA = \begin{bmatrix} 6 & 0 \\ 5 & -2 \end{bmatrix}\begin{bmatrix} 3 & 4 \\ -2 & 1 \end{bmatrix} = \begin{bmatrix} 6(3)+0(-2) & 6(4)+0(1) \\ 5(3)+(-2)(-2) & 5(4)+(-2)(1) \end{bmatrix} = \begin{bmatrix} 18+0 & 24+0 \\ 15+4 & 20+(-2) \end{bmatrix} = \begin{bmatrix} 18 & 24 \\ 19 & 18 \end{bmatrix}$

Note: $AB \ne BA$

77. $A = \begin{bmatrix} 0 & 1 & -1 \\ 0 & 1 & 0 \\ 0 & 0 & 1 \end{bmatrix}$ and $B = \begin{bmatrix} 1 & 0 & 0 \\ 0 & 1 & 0 \\ 0 & 0 & 1 \end{bmatrix}$

(a) $AB = \begin{bmatrix} 0 & 1 & -1 \\ 0 & 1 & 0 \\ 0 & 0 & 1 \end{bmatrix}\begin{bmatrix} 1 & 0 & 0 \\ 0 & 1 & 0 \\ 0 & 0 & 1 \end{bmatrix} = \begin{bmatrix} 0(1)+1(0)+(-1)(0) & 0(0)+1(1)+(-1)(0) & 0(0)+1(0)+(-1)(1) \\ 0(1)+1(0)+0(0) & 0(0)+1(1)+0(0) & 0(0)+1(0)+0(1) \\ 0(1)+0(0)+1(0) & 0(0)+0(1)+1(0) & 0(0)+0(0)+1(1) \end{bmatrix}$

$$= \begin{bmatrix} 0+0+0 & 0+1+0 & 0+0+(-1) \\ 0+0+0 & 0+1+0 & 0+0+0 \\ 0+0+0 & 0+0+0 & 0+0+1 \end{bmatrix} = \begin{bmatrix} 0 & 1 & -1 \\ 0 & 1 & 0 \\ 0 & 0 & 1 \end{bmatrix}$$

(b) $BA = \begin{bmatrix} 1 & 0 & 0 \\ 0 & 1 & 0 \\ 0 & 0 & 1 \end{bmatrix}\begin{bmatrix} 0 & 1 & -1 \\ 0 & 1 & 0 \\ 0 & 0 & 1 \end{bmatrix} = \begin{bmatrix} 1(0)+0(0)+0(0) & 1(1)+0(1)+0(0) & 1(-1)+0(0)+0(1) \\ 0(0)+1(0)+0(0) & 0(1)+1(1)+0(0) & 0(-1)+1(0)+0(1) \\ 0(0)+0(0)+1(0) & 0(1)+0(1)+1(0) & 0(-1)+0(0)+1(1) \end{bmatrix}$

$$= \begin{bmatrix} 0+0+0 & 1+0+0 & -1+0+0 \\ 0+0+0 & 0+1+0 & 0+0+0 \\ 0+0+0 & 0+0+0 & 0+0+1 \end{bmatrix} = \begin{bmatrix} 0 & 1 & -1 \\ 0 & 1 & 0 \\ 0 & 0 & 1 \end{bmatrix}$$

Note: $AB = BA = A$

Copyright © 2013 Pearson Education, Inc.

79. In Exercise 77, $AB = A$ and $BA = A$. For this pair of matrices, B acts in the same way for multiplication as the number $\underline{1}$ acts for multiplication of real numbers.

81. (a) The sales figure information may be written as the 3×3 matrix where column 1 represents nonfat, column 2 represents regular, and column 3 represents supercreamy.

$$\begin{array}{l}\text{Location I} \\ \text{Location II} \\ \text{Location III}\end{array}\begin{bmatrix} 50 & 100 & 30 \\ 10 & 90 & 50 \\ 60 & 120 & 40 \end{bmatrix}.$$

(b) The income per gallon information may be written as the 3×1 matrix $\begin{bmatrix} 12 \\ 10 \\ 15 \end{bmatrix}$.

Note: If the matrix in part (a) had been written with its rows and columns interchanged, then this income per gallon information would be written instead as a 1×3 matrix.

(c) $\begin{bmatrix} 50 & 100 & 30 \\ 10 & 90 & 50 \\ 60 & 120 & 40 \end{bmatrix}\begin{bmatrix} 12 \\ 10 \\ 15 \end{bmatrix} = \begin{bmatrix} 2050 \\ 1770 \\ 2520 \end{bmatrix}$

Note: This result may be written as a 1×3 matrix instead.

(d) $2050 + 1770 + 2520 = 6340$; The total daily income from the three locations is $6340.

83. Answers will vary depending on when numbers are rounded.

(a) $j_1 = 690, s_p = 210, a_1 = 2100$

1st year:
$$j_2 = 0.33a_1 = 0.33(2100) = 693$$
$$s_2 = 0.18j_1 = 0.18(690)$$
$$\approx 124.2 \approx 124$$
$$a_2 = 0.71s_1 + 0.94a_1$$
$$= 0.71(210) + 0.94(2100)$$
$$\approx 2123.1 \approx 2123$$
$$j_2 + s_2 + a_2 = 693 + 124 + 2123 = 2940$$

2nd year:
$$j_3 = 0.33a_2 = 0.33(2123) \approx 700.6$$
$$s_3 = 0.18j_2 = 0.18(693) \approx 124.7$$

$$a_3 = 0.71s_2 + 0.94a_2$$
$$= 0.71(124) + 0.94(2123)$$
$$\approx 2083.7$$
$$j_3 + s_3 + a_3 = 700.6 + 124.7 + 2083.7$$
$$= 2909$$

3rd year:
$$j_4 = 0.33a_3 = 0.33(2084) \approx 687$$
$$s_4 = 0.18j_3 = 0.18(700) = 126$$
$$a_4 = 0.72s_3 + 0.94a_3$$
$$= 0.71(125) + 0.94(2084)$$
$$= 2048$$
$$j_3 + s_3 + a_3 = 687 + 126 + 2048 = 2861$$

4th year:
$$j_5 = 0.33a_4 = 0.33(2048) \approx 676$$
$$s_5 = 0.18j_4 = 0.18(687) \approx 124$$
$$a_5 = 0.72s_4 + 0.94a_4$$
$$= 0.71(126) + 0.94(2048)$$
$$= 2014$$
$$j_5 + s_5 + a_5 = 686 + 124 + 2048 = 2814$$

5th year:
$$j_6 = 0.33a_5 = 0.33(2014) \approx 664$$
$$s_6 = 0.18j_5 = 0.18(676) \approx 122$$
$$a_6 = 0.72s_5 + 0.94a_5$$
$$= 0.71(124) + 0.94(2014)$$
$$\approx 1981$$
$$j_6 + s_6 + a_6 = 664 + 122 + 1981 = 2767$$

(b) The northern spotted owl will become extinct.

(c) $j_1 = 690, s_1 = 210, a_1 = 2100$

1st year:
$$j_2 = 0.33a_1 = 0.33(2100) = 693$$
$$s_2 = 0.3j_1 = 0.3(690) = 207$$
$$a_2 = 0.71s_1 + 0.94a_1$$
$$= 0.71(210) + 0.94(2100)$$
$$\approx 2123$$
$$j_2 + s_2 + a_2 = 693 + 207 + 2123 = 3023$$

2nd year:
$$j_3 = 0.33a_2 = 0.33(2123) \approx 701$$
$$s_3 = 0.3j_2 = 0.3(693) \approx 208$$
$$a_3 = 0.71s_2 + 0.94a_2$$
$$= 0.71(207) = 0.94(2123)$$
$$\approx 2143$$
$$j_3 + s_3 + a_3 = 701 + 208 + 2143 = 3052$$

(*continued on next page*)

Copyright © 2013 Pearson Education, Inc.

(*continued*)

3rd year:
$$j_4 = 0.33a_3 = 0.33(2143) \approx 707$$
$$s_4 = 0.3j_3 = 0.3(701) = 210$$
$$a_4 = 0.71s_3 + 0.94a_3 = 0.71(208) + 0.94(2143) \approx 2162$$
$$j_4 + s_4 + a_4 = 707 + 210 + 2162 = 3079$$

4th year:
$$j_5 = 0.33a_4 = 0.33(2162) \approx 714$$
$$s_5 = 0.3j_4 = 0.3(707) \approx 212$$
$$a_5 = 0.72s_4 + 0.94a_4 = 0.71(210) + 0.94(2162) \approx 2181$$
$$j_5 + s_5 + a_5 = 714 + 212 + 2181 = 3107$$

5th year:
$$j_6 = 0.71s_5 = 0.33a_5 = 0.33(2181) \approx 720$$
$$s_6 = 0.3j_5 = 0.3(714) \approx 214$$
$$a_6 = 0.71s_5 + 0.94a_5 = 0.71(212) + 0.94(2181) = 2201$$
$$j_6 + s_6 + a_6 = 720 + 214 + 2201 = 3135$$

85. Expanding $\begin{vmatrix} -x & 0 & 0.33 \\ 0.18 & -x & 0 \\ 0 & 0.71 & 0.94-x \end{vmatrix}$ by row one, we have the following.

$$-x\begin{vmatrix} -x & 0 \\ 0.71 & 0.94-x \end{vmatrix} - 0 + 0.33\begin{vmatrix} 0.18 & -x \\ 0 & 0.71 \end{vmatrix} = -x\left[-x(0.94-x) - 0.71(0)\right] + 0.33\left[(0.18)(0.71) - 0(-x)\right]$$

$$= -x\left(-0.94x + x^2 - 0\right) + 0.33(0.1278 - 0)$$

$$= 0.94x^2 - x^3 + 0.042174$$

Thus, the polynomial is $-x^3 + 0.94x^2 + 0.042174$. Evaluating this polynomial with $x = 0.98359$, we have

$$-(0.98359)^3 + 0.94(0.98359)^2 + 0.042174 \approx 0.0000029.$$ Thus 0.98359 is an approximate zero.

In Exercises 87 – 93, $A = \begin{bmatrix} a_{11} & a_{12} \\ a_{21} & a_{22} \end{bmatrix}$, $B = \begin{bmatrix} b_{11} & b_{12} \\ b_{21} & b_{22} \end{bmatrix}$, and $C = \begin{bmatrix} c_{11} & c_{12} \\ c_{21} & c_{22} \end{bmatrix}$.

87. $A + B = B + A$

$$A + B = \begin{bmatrix} a_{11} & a_{12} \\ a_{21} & a_{22} \end{bmatrix} + \begin{bmatrix} b_{11} & b_{12} \\ b_{21} & b_{22} \end{bmatrix} = \begin{bmatrix} a_{11} + b_{11} & a_{12} + b_{12} \\ a_{21} + b_{21} & a_{22} + b_{22} \end{bmatrix} = \begin{bmatrix} b_{11} + a_{11} & b_{12} + a_{12} \\ b_{21} + a_{21} & b_{22} + a_{22} \end{bmatrix} = B + A$$

89. $(AB)C = \left(\begin{bmatrix} a_{11} & a_{12} \\ a_{21} & a_{22} \end{bmatrix}\begin{bmatrix} b_{11} & b_{12} \\ b_{21} & b_{22} \end{bmatrix}\right)C = \begin{bmatrix} a_{11}b_{11} + a_{12}b_{21} & a_{11}b_{12} + a_{12}b_{22} \\ a_{21}b_{11} + a_{22}b_{21} & a_{21}b_{12} + a_{22}b_{22} \end{bmatrix}\begin{bmatrix} c_{11} & c_{12} \\ c_{21} & c_{22} \end{bmatrix}$

$$= \begin{bmatrix} (a_{11}b_{11} + a_{12}b_{21})c_{11} + (a_{11}b_{12} + a_{12}b_{22})c_{21} & (a_{11}b_{11} + a_{12}b_{21})c_{12} + (a_{11}b_{12} + a_{12}b_{22})c_{22} \\ (a_{21}b_{11} + a_{22}b_{21})c_{11} + (a_{21}b_{12} + a_{22}b_{22})c_{21} & (a_{21}b_{11} + a_{22}b_{21})c_{12} + (a_{21}b_{12} + a_{22}b_{22})c_{22} \end{bmatrix}$$

$$= \begin{bmatrix} a_{11}b_{11}c_{11} + a_{12}b_{21}c_{11} + a_{11}b_{12}c_{21} + a_{12}b_{22}c_{21} & a_{11}b_{11}c_{12} + a_{12}b_{21}c_{12} + a_{11}b_{12}c_{22} + a_{12}b_{22}c_{22} \\ a_{21}b_{11}c_{11} + a_{22}b_{21}c_{11} + a_{21}b_{12}c_{21} + a_{22}b_{22}c_{21} & a_{21}b_{11}c_{12} + a_{22}b_{21}c_{12} + a_{21}b_{12}c_{22} + a_{22}b_{22}c_{22} \end{bmatrix}$$

(*continued on next page*)

Copyright © 2013 Pearson Education, Inc.

(*continued*)

$$A(BC) = A\left(\begin{bmatrix} b_{11} & b_{12} \\ b_{21} & b_{22} \end{bmatrix}\begin{bmatrix} c_{11} & c_{12} \\ c_{21} & c_{22} \end{bmatrix}\right) = \begin{bmatrix} a_{11} & a_{12} \\ a_{21} & a_{22} \end{bmatrix}\begin{bmatrix} b_{11}c_{11} + b_{12}c_{21} & b_{11}c_{12} + b_{12}c_{22} \\ b_{21}c_{11} + b_{22}c_{21} & b_{21}c_{12} + b_{22}c_{22} \end{bmatrix}$$

$$= \begin{bmatrix} a_{11}(b_{11}c_{11} + b_{12}c_{21}) + a_{12}(b_{21}c_{11} + b_{22}c_{21}) & a_{11}(b_{11}c_{12} + b_{12}c_{22}) + a_{12}(b_{21}c_{12} + b_{22}c_{22}) \\ a_{21}(b_{11}c_{11} + b_{12}c_{21}) + a_{22}(b_{21}c_{11} + b_{22}c_{21}) & a_{21}(b_{11}c_{12} + b_{12}c_{22}) + a_{22}(b_{21}c_{12} + b_{22}c_{22}) \end{bmatrix}$$

$$= \begin{bmatrix} a_{11}b_{11}c_{11} + a_{11}b_{12}c_{21} + a_{12}b_{21}c_{11} + a_{12}b_{22}c_{21} & a_{11}b_{11}c_{12} + a_{11}b_{12}c_{22} + a_{12}b_{21}c_{12} + a_{12}b_{22}c_{22} \\ a_{21}b_{11}c_{11} + a_{21}b_{12}c_{21} + a_{22}b_{21}c_{11} + a_{22}b_{22}c_{21} & a_{21}b_{11}c_{12} + a_{21}b_{12}c_{22} + a_{22}b_{21}c_{12} + a_{22}b_{22}c_{22} \end{bmatrix}$$

$$= \begin{bmatrix} a_{11}b_{11}c_{11} + a_{12}b_{21}c_{11} + a_{11}b_{12}c_{21} + a_{12}b_{22}c_{21} & a_{11}b_{11}c_{12} + a_{12}b_{21}c_{12} + a_{11}b_{12}c_{22} + a_{12}b_{22}c_{22} \\ a_{21}b_{11}c_{11} + a_{22}b_{21}c_{11} + a_{21}b_{12}c_{21} + a_{22}b_{22}c_{21} & a_{21}b_{11}c_{12} + a_{22}b_{21}c_{12} + a_{21}b_{12}c_{22} + a_{22}b_{22}c_{22} \end{bmatrix}$$

Since the final matrix for $(AB)C$ and $A(BC)$ are the same, we have obtained the desired results.

91. $c(A + B) = cA + cB$, for any real number c.

$$c(A + B) = c\left(\begin{bmatrix} a_{11} & a_{12} \\ a_{21} & a_{22} \end{bmatrix} + \begin{bmatrix} b_{11} & b_{12} \\ b_{21} & b_{22} \end{bmatrix}\right) = c\begin{bmatrix} a_{11} + b_{11} & a_{12} + b_{12} \\ a_{21} + b_{21} & a_{22} + b_{22} \end{bmatrix} = \begin{bmatrix} c(a_{11} + b_{11}) & c(a_{12} + b_{12}) \\ c(a_{21} + b_{21}) & c(a_{22} + b_{22}) \end{bmatrix}$$

$$= \begin{bmatrix} c \cdot a_{11} + c \cdot b_{11} & c \cdot a_{12} + c \cdot b_{12} \\ c \cdot a_{21} + c \cdot b_{21} & c \cdot a_{22} + c \cdot b_{22} \end{bmatrix} = \begin{bmatrix} c \cdot a_{11} & c \cdot a_{12} \\ c \cdot a_{21} & c \cdot a_{22} \end{bmatrix} + \begin{bmatrix} c \cdot b_{11} & c \cdot b_{12} \\ c \cdot b_{21} & c \cdot b_{22} \end{bmatrix}$$

$$= c\begin{bmatrix} a_{11} & a_{12} \\ a_{21} & a_{22} \end{bmatrix} + c\begin{bmatrix} b_{11} & b_{12} \\ b_{21} & b_{22} \end{bmatrix} = cA + cB$$

93. $c(A)d = (cd)A$, for any real numbers c and d.

$$c(A)d = \left(c\begin{bmatrix} a_{11} & a_{12} \\ a_{21} & a_{22} \end{bmatrix}\right)d = \begin{bmatrix} c \cdot a_{11} & c \cdot a_{12} \\ c \cdot a_{21} & c \cdot a_{22} \end{bmatrix}d = \begin{bmatrix} c \cdot a_{11} \cdot d & c \cdot a_{12} \cdot d \\ c \cdot a_{21} \cdot d & c \cdot a_{22} \cdot d \end{bmatrix} = \begin{bmatrix} c \cdot d \cdot a_{11} & c \cdot d \cdot a_{12} \\ c \cdot d \cdot a_{21} & c \cdot d \cdot a_{22} \end{bmatrix}$$

$$= \begin{bmatrix} (cd) \cdot a_{11} & (cd) \cdot a_{12} \\ (cd) \cdot a_{21} & (cd) \cdot a_{22} \end{bmatrix} = (cd)\begin{bmatrix} a_{11} & a_{12} \\ a_{21} & a_{22} \end{bmatrix} = (cd)A$$

Section 5.8   Matrix Inverses

1. $A = \begin{bmatrix} -2 & 4 & 0 \\ 3 & 5 & 9 \\ 0 & 8 & -6 \end{bmatrix}$ and $I_3 = \begin{bmatrix} 1 & 0 & 0 \\ 0 & 1 & 0 \\ 0 & 0 & 1 \end{bmatrix}$

$$I_3A = \begin{bmatrix} 1 & 0 & 0 \\ 0 & 1 & 0 \\ 0 & 0 & 1 \end{bmatrix}\begin{bmatrix} -2 & 4 & 0 \\ 3 & 5 & 9 \\ 0 & 8 & -6 \end{bmatrix} = \begin{bmatrix} 1(-2)+0(3)+0(0) & 1(4)+0(5)+0(8) & 1(0)+0(9)+0(-6) \\ 0(-2)+1(3)+0(0) & 0(4)+1(5)+0(8) & 0(0)+1(9)+0(-6) \\ 0(-2)+0(3)+1(0) & 0(4)+0(5)+1(8) & 0(0)+0(9)+1(-6) \end{bmatrix}$$

$$= \begin{bmatrix} -2+0+0 & 4+0+0 & 0+0+0 \\ 0+3+0 & 0+5+0 & 0+9+0 \\ 0+0+0 & 0+0+8 & 0+0+(-6) \end{bmatrix} = \begin{bmatrix} -2 & 4 & 0 \\ 3 & 5 & 9 \\ 0 & 8 & -6 \end{bmatrix}$$

3. $\begin{bmatrix} 5 & 7 \\ 2 & 3 \end{bmatrix}\begin{bmatrix} 3 & -7 \\ -2 & 5 \end{bmatrix} = \begin{bmatrix} 5(3)+7(-2) & 5(-7)+7(5) \\ 2(3)+3(-2) & 2(-7)+3(5) \end{bmatrix} = \begin{bmatrix} 15+(-14) & -35+35 \\ 6+(-6) & (-14)+15 \end{bmatrix} = \begin{bmatrix} 1 & 0 \\ 0 & 1 \end{bmatrix}$

$\begin{bmatrix} 3 & -7 \\ -2 & 5 \end{bmatrix}\begin{bmatrix} 5 & 7 \\ 2 & 3 \end{bmatrix} = \begin{bmatrix} 3(5)+(-7)(2) & 3(7)+(-7)(3) \\ (-2)(5)+5(2) & (-2)(7)+5(3) \end{bmatrix} = \begin{bmatrix} 15+(-14) & 21+(-21) \\ (-10)+10 & (-14)+15 \end{bmatrix} = \begin{bmatrix} 1 & 0 \\ 0 & 1 \end{bmatrix}$

Since the products obtained by multiplying the matrices in either order are both the 2×2 identity matrix, the given matrices are inverses of each other.

Copyright © 2013 Pearson Education, Inc.

5. $\begin{bmatrix} -1 & 2 \\ 3 & -5 \end{bmatrix}\begin{bmatrix} -5 & -2 \\ -3 & -1 \end{bmatrix} = \begin{bmatrix} -1(-5)+2(-3) & -1(-2)+2(-1) \\ 3(-5)+(-5)(-3) & 3(-2)+(-5)(-1) \end{bmatrix} = \begin{bmatrix} 5+(-6) & 2+(-2) \\ -15+15 & -6+5 \end{bmatrix} = \begin{bmatrix} -1 & 0 \\ 0 & -1 \end{bmatrix}$

Since this product is not the 2×2 identity matrix, the given matrices are not inverses of each other.

7. $\begin{bmatrix} 0 & 1 & 0 \\ 0 & 0 & -2 \\ 1 & -1 & 0 \end{bmatrix}\begin{bmatrix} 1 & 0 & 1 \\ 1 & 0 & 0 \\ 0 & -1 & 0 \end{bmatrix} = \begin{bmatrix} 0(1)+1(1)+0(0) & 0(0)+1(0)+0(-1) & 0(1)+1(0)+0(0) \\ 0(1)+0(1)+(-2)(0) & 0(0)+0(0)+(-2)(-1) & 0(1)+0(0)+(-2)(0) \\ 1(1)+(-1)(1)+0(0) & 1(0)+(-1)(0)+0(-1) & 1(1)+(-1)(0)+0(0) \end{bmatrix}$

$= \begin{bmatrix} 0+1+0 & 0+0+0 & 0+0+0 \\ 0+0+0 & 0+0+2 & 0+0+0 \\ 1+(-1)+0 & 0+0+0 & 1+0+0 \end{bmatrix} = \begin{bmatrix} 1 & 0 & 0 \\ 0 & 2 & 0 \\ 0 & 0 & 1 \end{bmatrix}$

Since this product is not the 3×3 identity matrix, the given matrices are not inverses of each other.

9. $\begin{bmatrix} 1 & 0 & 0 \\ 0 & -1 & 0 \\ 1 & 0 & 1 \end{bmatrix}\begin{bmatrix} 1 & 0 & 0 \\ 0 & -1 & 0 \\ -1 & 0 & 1 \end{bmatrix}$

$= \begin{bmatrix} 1(1)+0(0)+0(-1) & 1(0)+0(-1)+0(0) & 1(0)+0(0)+0(1) \\ 0(1)-1(0)+0(-1) & 0(0)-1(-1)+0(0) & 0(0)-1(0)+0(1) \\ 1(1)+0(0)+1(-1) & 1(0)+0(-1)+1(0) & 1(0)+0(0)+1(1) \end{bmatrix} = \begin{bmatrix} 1 & 0 & 0 \\ 0 & 1 & 0 \\ 0 & 0 & 1 \end{bmatrix} = I_3$

$\begin{bmatrix} 1 & 0 & 0 \\ 0 & -1 & 0 \\ -1 & 0 & 1 \end{bmatrix}\begin{bmatrix} 1 & 0 & 0 \\ 0 & -1 & 0 \\ 1 & 0 & 1 \end{bmatrix}$

$= \begin{bmatrix} 1(1)+0(0)+0(1) & 1(0)+0(-1)+0(0) & 1(0)+0(0)+0(1) \\ 0(1)-1(0)+0(1) & 0(0)-1(-1)+0(0) & 0(0)-1(0)+0(1) \\ -1(1)+0(0)+1(1) & -1(0)+0(-1)+1(0) & -1(0)+0(0)+1(1) \end{bmatrix} = \begin{bmatrix} 1 & 0 & 0 \\ 0 & 1 & 0 \\ 0 & 0 & 1 \end{bmatrix} = I_3$

The given matrices are inverses of each other.

11. Find the inverse of $A = \begin{bmatrix} -1 & 2 \\ -2 & -1 \end{bmatrix}$, if it exists. Since $[A \mid I_2] = \begin{bmatrix} -1 & 2 & 1 & 0 \\ -2 & -1 & 0 & 1 \end{bmatrix}$, we have

$\begin{bmatrix} 1 & -2 & -1 & 0 \\ -2 & -1 & 0 & 1 \end{bmatrix} -1\text{R}1 \Rightarrow \begin{bmatrix} 1 & -2 & -1 & 0 \\ 0 & -5 & -2 & 1 \end{bmatrix} 2\text{R}1+\text{R}2 \Rightarrow$

$\begin{bmatrix} 1 & -2 & -1 & 0 \\ 0 & 1 & \frac{2}{5} & -\frac{1}{5} \end{bmatrix} -\frac{1}{5}\text{R}2 \Rightarrow \begin{bmatrix} 1 & 0 & -\frac{1}{5} & -\frac{2}{5} \\ 0 & 1 & \frac{2}{5} & -\frac{1}{5} \end{bmatrix} 2\text{R}2+\text{R}1$

Thus, $A^{-1} = \begin{bmatrix} -\frac{1}{5} & -\frac{2}{5} \\ \frac{2}{5} & -\frac{1}{5} \end{bmatrix}$.

13. Find the inverse of $A = \begin{bmatrix} -1 & -2 \\ 3 & 4 \end{bmatrix}$, if it exists. Since $[A \mid I_2] = \begin{bmatrix} -1 & -2 & 1 & 0 \\ 3 & 4 & 0 & 1 \end{bmatrix}$, we have

$\begin{bmatrix} -1 & -2 & 1 & 0 \\ 0 & -2 & 3 & 1 \end{bmatrix} 3\text{R}1+\text{R}2 \Rightarrow \begin{bmatrix} 1 & 2 & -1 & 0 \\ 0 & -2 & 3 & 1 \end{bmatrix} -1\text{R}1 \Rightarrow \begin{bmatrix} 1 & 0 & 2 & 1 \\ 0 & -2 & 3 & 1 \end{bmatrix} \text{R}2+\text{R}1 \Rightarrow \begin{bmatrix} 1 & 0 & 2 & 1 \\ 0 & 1 & -\frac{3}{2} & -\frac{1}{2} \end{bmatrix} -\frac{1}{2}\text{R}2$

Thus, $A^{-1} = \begin{bmatrix} 2 & 1 \\ -\frac{3}{2} & -\frac{1}{2} \end{bmatrix}$.

Copyright © 2013 Pearson Education, Inc.

15. Find the inverse of $\begin{bmatrix} 5 & 10 \\ -3 & -6 \end{bmatrix}$, if it exists. Since $\begin{bmatrix} A\,|\,I_2 \end{bmatrix} = \begin{bmatrix} 5 & 10 & | & 1 & 0 \\ -3 & -6 & | & 0 & 1 \end{bmatrix}$, we have

$$\begin{bmatrix} 1 & 2 & | & \frac{1}{5} & 0 \\ -3 & -6 & | & 0 & 1 \end{bmatrix} \begin{smallmatrix}\frac{1}{5}R1\end{smallmatrix} \Rightarrow \begin{bmatrix} 1 & 2 & | & \frac{1}{5} & 0 \\ 0 & 0 & | & \frac{3}{5} & 1 \end{bmatrix} 3R1+R2$$

At this point, the matrix should be changed so that the second-row, second-column element will be 1. Since that element is now 0, the desired transformation cannot be completed. Therefore, the inverse of the given matrix does not exist.

17. Find the inverse of $A = \begin{bmatrix} 1 & 0 & 1 \\ 0 & -1 & 0 \\ 2 & 1 & 1 \end{bmatrix}$, if it exists. Since $\begin{bmatrix} A\,|\,I_3 \end{bmatrix} = \begin{bmatrix} 1 & 0 & 1 & | & 1 & 0 & 0 \\ 0 & -1 & 0 & | & 0 & 1 & 0 \\ 2 & 1 & 1 & | & 0 & 0 & 1 \end{bmatrix}$ we have

$$\begin{bmatrix} 1 & 0 & 1 & | & 1 & 0 & 0 \\ 0 & -1 & 0 & | & 0 & 1 & 0 \\ 0 & 1 & -1 & | & -2 & 0 & 1 \end{bmatrix} \begin{smallmatrix}-2R1+R3\end{smallmatrix} \Rightarrow \begin{bmatrix} 1 & 0 & 1 & | & 1 & 0 & 0 \\ 0 & 1 & 0 & | & 0 & -1 & 0 \\ 0 & 0 & -1 & | & -2 & 1 & 1 \end{bmatrix} \begin{smallmatrix}-1R2 \\ R2+R3\end{smallmatrix}$$

$$\begin{bmatrix} 1 & 0 & 1 & | & 1 & 0 & 0 \\ 0 & 1 & 0 & | & 0 & -1 & 0 \\ 0 & 0 & 1 & | & 2 & -1 & -1 \end{bmatrix} \begin{smallmatrix}-1R3\end{smallmatrix} \Rightarrow \begin{bmatrix} 1 & 0 & 0 & | & -1 & 1 & 1 \\ 0 & 1 & 0 & | & 0 & -1 & 0 \\ 0 & 0 & 1 & | & 2 & -1 & -1 \end{bmatrix} \begin{smallmatrix}-1R3+R1\end{smallmatrix}.$$

Thus, $A^{-1} = \begin{bmatrix} -1 & 1 & 1 \\ 0 & -1 & 0 \\ 2 & -1 & -1 \end{bmatrix}$.

19. Find the inverse of $A = \begin{bmatrix} 2 & 3 & 3 \\ 1 & 4 & 3 \\ 1 & 3 & 4 \end{bmatrix}$, if it exists. Since $\begin{bmatrix} A\,|\,I_3 \end{bmatrix} = \begin{bmatrix} 2 & 3 & 3 & | & 1 & 0 & 0 \\ 1 & 4 & 3 & | & 0 & 1 & 0 \\ 1 & 3 & 4 & | & 0 & 0 & 1 \end{bmatrix}$, we have

$$\begin{bmatrix} 2 & 3 & 3 & | & 1 & 0 & 0 \\ 1 & 4 & 3 & | & 0 & 1 & 0 \\ 0 & -1 & 1 & | & 0 & -1 & 1 \end{bmatrix} \begin{smallmatrix}-R2+R3\end{smallmatrix} \Rightarrow \begin{bmatrix} 1 & -1 & 0 & | & 1 & -1 & 0 \\ 1 & 4 & 3 & | & 0 & 1 & 0 \\ 0 & -1 & 1 & | & 0 & -1 & 1 \end{bmatrix} \begin{smallmatrix}-R2+R1\end{smallmatrix} \Rightarrow \begin{bmatrix} 1 & -1 & 0 & | & 1 & -1 & 0 \\ 1 & 7 & 0 & | & 0 & 4 & -3 \\ 0 & -1 & 1 & | & 0 & -1 & 1 \end{bmatrix} \begin{smallmatrix}-3R3+R2\end{smallmatrix} \Rightarrow$$

$$\begin{bmatrix} 8 & 0 & 0 & | & 7 & -3 & -3 \\ 1 & 7 & 0 & | & 0 & 4 & -3 \\ 0 & -1 & 1 & | & 0 & -1 & 1 \end{bmatrix} \begin{smallmatrix}7R1+R2\end{smallmatrix} \Rightarrow \begin{bmatrix} 1 & 0 & 0 & | & \frac{7}{8} & -\frac{3}{8} & -\frac{3}{8} \\ 1 & 7 & 0 & | & 0 & 4 & -3 \\ 0 & -1 & 1 & | & 0 & -1 & 1 \end{bmatrix} \begin{smallmatrix}\frac{1}{8}R1\end{smallmatrix} \Rightarrow \begin{bmatrix} 1 & 0 & 0 & | & \frac{7}{8} & -\frac{3}{8} & -\frac{3}{8} \\ 0 & 7 & 0 & | & -\frac{7}{8} & \frac{35}{8} & -\frac{21}{8} \\ 0 & -1 & 1 & | & 0 & -1 & 1 \end{bmatrix} \begin{smallmatrix}-R1+R2\end{smallmatrix} \Rightarrow$$

$$\begin{bmatrix} 1 & 0 & 0 & | & \frac{7}{8} & -\frac{3}{8} & -\frac{3}{8} \\ 0 & 1 & 0 & | & -\frac{1}{8} & \frac{5}{8} & -\frac{3}{8} \\ 0 & -1 & 1 & | & 0 & -1 & 1 \end{bmatrix} \begin{smallmatrix}\frac{1}{7}R2\end{smallmatrix} \Rightarrow \begin{bmatrix} 1 & 0 & 0 & | & \frac{7}{8} & -\frac{3}{8} & -\frac{3}{8} \\ 0 & 1 & 0 & | & -\frac{1}{8} & \frac{5}{8} & -\frac{3}{8} \\ 0 & 0 & 1 & | & -\frac{1}{8} & -\frac{3}{8} & \frac{5}{8} \end{bmatrix} \begin{smallmatrix}R2+R3\end{smallmatrix} \Rightarrow$$

Thus, $A^{-1} = \begin{bmatrix} \frac{7}{8} & -\frac{3}{8} & -\frac{3}{8} \\ -\frac{1}{8} & \frac{5}{8} & -\frac{3}{8} \\ -\frac{1}{8} & -\frac{3}{8} & \frac{5}{8} \end{bmatrix}$.

Copyright © 2013 Pearson Education, Inc.

21. Find the inverse of $A = \begin{bmatrix} 2 & 2 & -4 \\ 2 & 6 & 0 \\ -3 & -3 & 5 \end{bmatrix}$, if it exists. Since $\begin{bmatrix} A \mid I_3 \end{bmatrix} = \begin{bmatrix} 2 & 2 & -4 & | & 1 & 0 & 0 \\ 2 & 6 & 0 & | & 0 & 1 & 0 \\ -3 & -3 & 5 & | & 0 & 0 & 1 \end{bmatrix}$,

$$\begin{bmatrix} 1 & 1 & -2 & | & \frac{1}{2} & 0 & 0 \\ 2 & 6 & 0 & | & 0 & 1 & 0 \\ -3 & -3 & 5 & | & 0 & 0 & 1 \end{bmatrix} \begin{matrix} \frac{1}{2}R1 \\ \\ \end{matrix} \Rightarrow \begin{bmatrix} 1 & 1 & -2 & | & \frac{1}{2} & 0 & 0 \\ 0 & 4 & 4 & | & -1 & 1 & 0 \\ 0 & 0 & -1 & | & \frac{3}{2} & 0 & 1 \end{bmatrix} \begin{matrix} \\ -2R1+R2 \\ 3R1+R3 \end{matrix}$$

$$\begin{bmatrix} 1 & 1 & -2 & | & \frac{1}{2} & 0 & 0 \\ 0 & 1 & 1 & | & -\frac{1}{4} & \frac{1}{4} & 0 \\ 0 & 0 & -1 & | & \frac{3}{2} & 0 & 1 \end{bmatrix} \begin{matrix} \\ \frac{1}{4}R2 \\ \end{matrix} \Rightarrow \begin{bmatrix} 1 & 0 & -3 & | & \frac{3}{4} & -\frac{1}{4} & 0 \\ 0 & 1 & 1 & | & -\frac{1}{4} & \frac{1}{4} & 0 \\ 0 & 0 & -1 & | & \frac{3}{2} & 0 & 1 \end{bmatrix} \begin{matrix} -1R2+R1 \\ \\ \end{matrix}$$

$$\begin{bmatrix} 1 & 0 & -3 & | & \frac{3}{4} & -\frac{1}{4} & 0 \\ 0 & 1 & 1 & | & -\frac{1}{4} & \frac{1}{4} & 0 \\ 0 & 0 & 1 & | & -\frac{3}{2} & 0 & -1 \end{bmatrix} \begin{matrix} \\ \\ -1R3 \end{matrix} \Rightarrow \begin{bmatrix} 1 & 0 & 0 & | & -\frac{15}{4} & -\frac{1}{4} & -3 \\ 0 & 1 & 0 & | & \frac{5}{4} & \frac{1}{4} & 1 \\ 0 & 0 & 1 & | & -\frac{3}{2} & 0 & -1 \end{bmatrix} \begin{matrix} 3R3+R1 \\ -1R3+R2 \\ \end{matrix}.$$

Thus, $A^{-1} = \begin{bmatrix} -\frac{15}{4} & -\frac{1}{4} & -3 \\ \frac{5}{4} & \frac{1}{4} & 1 \\ -\frac{3}{2} & 0 & -1 \end{bmatrix}$.

23. Find the inverse of $A = \begin{bmatrix} 1 & 1 & 0 & 2 \\ 2 & -1 & 1 & -1 \\ 3 & 3 & 2 & -2 \\ 1 & 2 & 1 & 0 \end{bmatrix}$, if it exists.

$$\begin{bmatrix} A \mid I_4 \end{bmatrix} = \begin{bmatrix} 1 & 1 & 0 & 2 & | & 1 & 0 & 0 & 0 \\ 2 & -1 & 1 & -1 & | & 0 & 1 & 0 & 0 \\ 3 & 3 & 2 & -2 & | & 0 & 0 & 1 & 0 \\ 1 & 2 & 1 & 0 & | & 0 & 0 & 0 & 1 \end{bmatrix} \Rightarrow \begin{bmatrix} 1 & 1 & 0 & 2 & | & 1 & 0 & 0 & 0 \\ 0 & -3 & 1 & -5 & | & -2 & 1 & 0 & 0 \\ 0 & 0 & 2 & -8 & | & -3 & 0 & 1 & 0 \\ 0 & 1 & 1 & -2 & | & -1 & 0 & 0 & 1 \end{bmatrix} \begin{matrix} \\ -2R1+R2 \\ -3R1+R3 \\ -1R1+R4 \end{matrix}$$

$$\Rightarrow \begin{bmatrix} 1 & 1 & 0 & 2 & | & 1 & 0 & 0 & 0 \\ 0 & 1 & -\frac{1}{3} & \frac{5}{3} & | & \frac{2}{3} & -\frac{1}{3} & 0 & 0 \\ 0 & 0 & 2 & -8 & | & -3 & 0 & 1 & 0 \\ 0 & 1 & 1 & -2 & | & -1 & 0 & 0 & 1 \end{bmatrix} \begin{matrix} \\ -\frac{1}{3}R2 \\ \\ \end{matrix} \Rightarrow \begin{bmatrix} 1 & 0 & \frac{1}{3} & \frac{1}{3} & | & \frac{1}{3} & \frac{1}{3} & 0 & 0 \\ 0 & 1 & -\frac{1}{3} & \frac{5}{3} & | & \frac{2}{3} & -\frac{1}{3} & 0 & 0 \\ 0 & 0 & 2 & -8 & | & -3 & 0 & 1 & 0 \\ 0 & 0 & \frac{4}{3} & -\frac{11}{3} & | & -\frac{5}{3} & \frac{1}{3} & 0 & 1 \end{bmatrix} \begin{matrix} -1R2+R1 \\ \\ \\ -1R2+R4 \end{matrix}$$

$$\Rightarrow \begin{bmatrix} 1 & 0 & \frac{1}{3} & \frac{1}{3} & | & \frac{1}{3} & \frac{1}{3} & 0 & 0 \\ 0 & 1 & -\frac{1}{3} & \frac{5}{3} & | & \frac{2}{3} & -\frac{1}{3} & 0 & 0 \\ 0 & 0 & 1 & -4 & | & -\frac{3}{2} & 0 & \frac{1}{2} & 0 \\ 0 & 0 & \frac{4}{3} & -\frac{11}{3} & | & -\frac{5}{3} & \frac{1}{3} & 0 & 1 \end{bmatrix} \begin{matrix} \\ \\ \frac{1}{2}R3 \\ \end{matrix} \Rightarrow \begin{bmatrix} 1 & 0 & 0 & \frac{5}{3} & | & \frac{5}{6} & \frac{1}{3} & -\frac{1}{6} & 0 \\ 0 & 1 & 0 & \frac{1}{3} & | & \frac{1}{6} & -\frac{1}{3} & \frac{1}{6} & 0 \\ 0 & 0 & 1 & -4 & | & -\frac{3}{2} & 0 & \frac{1}{2} & 0 \\ 0 & 0 & 0 & \frac{5}{3} & | & \frac{1}{3} & \frac{1}{3} & -\frac{2}{3} & 1 \end{bmatrix} \begin{matrix} -\frac{1}{3}R3+R1 \\ \frac{1}{3}R3+R2 \\ \\ -\frac{4}{3}R3+R4 \end{matrix}$$

$$\Rightarrow \begin{bmatrix} 1 & 0 & 0 & \frac{5}{3} & | & \frac{5}{6} & \frac{1}{3} & -\frac{1}{6} & 0 \\ 0 & 1 & 0 & \frac{1}{3} & | & \frac{1}{6} & -\frac{1}{3} & \frac{1}{6} & 0 \\ 0 & 0 & 1 & -4 & | & -\frac{3}{2} & 0 & \frac{1}{2} & 0 \\ 0 & 0 & 0 & 1 & | & \frac{1}{5} & \frac{1}{5} & -\frac{2}{5} & \frac{3}{5} \end{bmatrix} \begin{matrix} \\ \\ \\ \frac{3}{5}R4 \end{matrix} \Rightarrow \begin{bmatrix} 1 & 0 & 0 & 0 & | & \frac{1}{2} & 0 & \frac{1}{2} & -1 \\ 0 & 1 & 0 & 0 & | & \frac{1}{10} & -\frac{2}{5} & \frac{3}{10} & -\frac{1}{5} \\ 0 & 0 & 1 & 0 & | & -\frac{7}{10} & \frac{4}{5} & -\frac{11}{10} & \frac{12}{5} \\ 0 & 0 & 0 & 1 & | & \frac{1}{5} & \frac{1}{5} & -\frac{2}{5} & \frac{3}{5} \end{bmatrix} \begin{matrix} -\frac{5}{3}R4+R1 \\ -\frac{1}{3}R4+R2 \\ 4R4+R3 \\ \end{matrix}$$

Thus, $A^{-1} = \begin{bmatrix} \frac{1}{2} & 0 & \frac{1}{2} & -1 \\ \frac{1}{10} & -\frac{2}{5} & \frac{3}{10} & -\frac{1}{5} \\ -\frac{7}{10} & \frac{4}{5} & -\frac{11}{10} & \frac{12}{5} \\ \frac{1}{5} & \frac{1}{5} & -\frac{2}{5} & \frac{3}{5} \end{bmatrix}$.

Copyright © 2013 Pearson Education, Inc.

25. Find the inverse of $A = \begin{bmatrix} 3 & 2 & 0 & -1 \\ 2 & 0 & 1 & 2 \\ 1 & 2 & -1 & 0 \\ 2 & -1 & 1 & 1 \end{bmatrix}$, if it exists.

$$[A \mid I_4] = \left[\begin{array}{cccc|cccc} 3 & 2 & 0 & -1 & 1 & 0 & 0 & 0 \\ 2 & 0 & 1 & 2 & 0 & 1 & 0 & 0 \\ 1 & 2 & -1 & 0 & 0 & 0 & 1 & 0 \\ 2 & -1 & 1 & 1 & 0 & 0 & 0 & 1 \end{array}\right] \Rightarrow \left[\begin{array}{cccc|cccc} 5 & 1 & 1 & 0 & 1 & 0 & 0 & 1 \\ 0 & 1 & 0 & 1 & 0 & 1 & 0 & -1 \\ 1 & 2 & -1 & 0 & 0 & 0 & 1 & 0 \\ 2 & -1 & 1 & 1 & 0 & 0 & 0 & 1 \end{array}\right] \begin{array}{l} R1+R4 \\ -R4+R2 \\ \\ \\ \end{array} \Rightarrow$$

$$\left[\begin{array}{cccc|cccc} 6 & 3 & 0 & 0 & 1 & 0 & 1 & 1 \\ 0 & 1 & 0 & 1 & 0 & 1 & 0 & -1 \\ 1 & 2 & -1 & 0 & 0 & 0 & 1 & 0 \\ 0 & -5 & 3 & 1 & 0 & 0 & -2 & 1 \end{array}\right] \begin{array}{l} R1+R3 \\ \\ \\ -2R3+R4 \end{array} \Rightarrow \left[\begin{array}{cccc|cccc} 6 & 3 & 0 & 0 & 1 & 0 & 1 & 1 \\ 0 & 1 & 0 & 1 & 0 & 1 & 0 & -1 \\ 0 & -9 & 6 & 0 & 1 & 0 & -5 & 1 \\ 0 & -5 & 3 & 1 & 0 & 0 & -2 & 1 \end{array}\right] \begin{array}{l} \\ \\ -6R3+R1 \\ \\ \end{array} \Rightarrow$$

$$\left[\begin{array}{cccc|cccc} 6 & 3 & 0 & 0 & 1 & 0 & 1 & 1 \\ 0 & 1 & 0 & 1 & 0 & 1 & 0 & -1 \\ 0 & 0 & 6 & 9 & 1 & 9 & -5 & -8 \\ 0 & 0 & 3 & 6 & 0 & 5 & -2 & -4 \end{array}\right] \begin{array}{l} \\ \\ 9R2+R3 \\ 5R2+R4 \end{array} \Rightarrow \left[\begin{array}{cccc|cccc} 6 & 3 & 0 & 0 & 1 & 0 & 1 & 1 \\ 0 & 1 & 0 & 1 & 0 & 1 & 0 & -1 \\ 0 & 0 & 6 & 9 & 1 & 9 & -5 & -8 \\ 0 & 0 & 0 & \frac{3}{2} & -\frac{1}{2} & \frac{1}{2} & \frac{1}{2} & 0 \end{array}\right] \begin{array}{l} \\ \\ \\ -\frac{1}{2}R3+R4 \end{array} \Rightarrow$$

$$\left[\begin{array}{cccc|cccc} 6 & 3 & 0 & 0 & 1 & 0 & 1 & 1 \\ 0 & 1 & 0 & 1 & 0 & 1 & 0 & -1 \\ 0 & 0 & 6 & 9 & 1 & 9 & -5 & -8 \\ 0 & 0 & 0 & 1 & -\frac{1}{3} & \frac{1}{3} & \frac{1}{3} & 0 \end{array}\right] \begin{array}{l} \\ \\ \\ \frac{2}{3}R4 \end{array} \Rightarrow \left[\begin{array}{cccc|cccc} 6 & 3 & 0 & 0 & 1 & 0 & 1 & 1 \\ 0 & 1 & 0 & 0 & \frac{1}{3} & \frac{2}{3} & -\frac{1}{3} & -1 \\ 0 & 0 & 6 & 0 & 4 & 6 & -8 & -8 \\ 0 & 0 & 0 & 1 & -\frac{1}{3} & \frac{1}{3} & \frac{1}{3} & 0 \end{array}\right] \begin{array}{l} \\ -R4+R2 \\ -9R4+R3 \\ \\ \end{array} \Rightarrow$$

$$\left[\begin{array}{cccc|cccc} 6 & 0 & 0 & 0 & 0 & -2 & 2 & 4 \\ 0 & 1 & 0 & 0 & \frac{1}{3} & \frac{2}{3} & -\frac{1}{3} & -1 \\ 0 & 0 & 1 & 0 & \frac{2}{3} & 1 & -\frac{4}{3} & -\frac{4}{3} \\ 0 & 0 & 0 & 1 & -\frac{1}{3} & \frac{1}{3} & \frac{1}{3} & 0 \end{array}\right] \begin{array}{l} -3R2+R1 \\ \\ \frac{1}{6}R3 \\ \\ \end{array} \Rightarrow \left[\begin{array}{cccc|cccc} 1 & 0 & 0 & 0 & 0 & -\frac{1}{3} & \frac{1}{3} & \frac{2}{3} \\ 0 & 1 & 0 & 0 & \frac{1}{3} & \frac{2}{3} & -\frac{1}{3} & -1 \\ 0 & 0 & 1 & 0 & \frac{2}{3} & 1 & -\frac{4}{3} & -\frac{4}{3} \\ 0 & 0 & 0 & 1 & -\frac{1}{3} & \frac{1}{3} & \frac{1}{3} & 0 \end{array}\right] \begin{array}{l} \frac{1}{6}R1 \\ \\ \\ \\ \end{array}$$

Thus, $A^{-1} = \begin{bmatrix} 0 & -\frac{1}{3} & \frac{1}{3} & \frac{2}{3} \\ \frac{1}{3} & \frac{2}{3} & -\frac{1}{3} & -1 \\ \frac{2}{3} & 1 & -\frac{4}{3} & -\frac{4}{3} \\ -\frac{1}{3} & \frac{1}{3} & \frac{1}{3} & 0 \end{bmatrix}$.

27. Since $A^{-1} = \begin{bmatrix} \frac{1}{11} & -\frac{7}{22} \\ \frac{1}{11} & \frac{2}{11} \end{bmatrix}$, we need to find A, where $A = \left(A^{-1}\right)^{-1}$.

$$[A^{-1} \mid I_2] = \left[\begin{array}{cc|cc} \frac{1}{11} & -\frac{7}{22} & 1 & 0 \\ \frac{1}{11} & \frac{2}{11} & 0 & 1 \end{array}\right] \Rightarrow \left[\begin{array}{cc|cc} 1 & -\frac{7}{2} & 11 & 0 \\ 1 & 2 & 0 & 11 \end{array}\right] \begin{array}{l} 11R1 \\ 11R2 \end{array} \Rightarrow \left[\begin{array}{cc|cc} 1 & -\frac{7}{2} & 11 & 0 \\ 0 & \frac{11}{2} & -11 & 11 \end{array}\right] \begin{array}{l} \\ -R1+R2 \end{array}$$

$$\left[\begin{array}{cc|cc} 1 & -\frac{7}{2} & 11 & 0 \\ 0 & 1 & -2 & 2 \end{array}\right] \begin{array}{l} \\ \frac{2}{11}R2 \end{array} \Rightarrow \left[\begin{array}{cc|cc} 1 & 0 & 4 & 7 \\ 0 & 1 & -2 & 2 \end{array}\right] \begin{array}{l} \frac{7}{2}R2+R1 \\ \\ \end{array}$$

Thus, $A = \begin{bmatrix} 4 & 7 \\ -2 & 2 \end{bmatrix}$.

29. If $A = \begin{bmatrix} a & b \\ c & d \end{bmatrix}$, then $ad - bc = ad - cb$ is the determinant of matrix A.

Copyright © 2013 Pearson Education, Inc.

31. $A^{-1} = \begin{bmatrix} \dfrac{d}{|A|} & \dfrac{-b}{|A|} \\ \dfrac{-c}{|A|} & \dfrac{a}{|A|} \end{bmatrix} = \begin{bmatrix} \dfrac{1}{|A|}d & \dfrac{1}{|A|}(-b) \\ \dfrac{1}{|A|}(-c) & \dfrac{a}{|A|}(a) \end{bmatrix}$

$\quad = \dfrac{1}{|A|}\begin{bmatrix} d & -b \\ -c & a \end{bmatrix}$

33. Given $A = \begin{bmatrix} 4 & 2 \\ 7 & 3 \end{bmatrix}$, $|A| = 12 - 14 = -2$. Thus

$\quad A^{-1} = \dfrac{1}{-2}\begin{bmatrix} 3 & -2 \\ -7 & 4 \end{bmatrix} = \begin{bmatrix} -\frac{3}{2} & 1 \\ \frac{7}{2} & -2 \end{bmatrix}.$

35. $-x + y = 1$
$\quad 2x - y = 1$

Given $A = \begin{bmatrix} -1 & 1 \\ 2 & -1 \end{bmatrix}$, $X = \begin{bmatrix} x \\ y \end{bmatrix}$, and $B = \begin{bmatrix} 1 \\ 1 \end{bmatrix}$,

find A^{-1}. Since $\left[A \mid I_2 \right] = \begin{bmatrix} -1 & 1 & 1 & 0 \\ 2 & -1 & 0 & 1 \end{bmatrix}$,

$\begin{bmatrix} 1 & -1 & -1 & 0 \\ 2 & -1 & 0 & 1 \end{bmatrix} \begin{matrix} -1R1 \\ \\ \end{matrix} \Rightarrow$

$\begin{bmatrix} 1 & -1 & -1 & 0 \\ 0 & 1 & 2 & 1 \end{bmatrix} \begin{matrix} \\ -2R1 + R2 \end{matrix} \Rightarrow$

$\begin{bmatrix} 1 & 0 & 1 & 1 \\ 0 & 1 & 2 & 1 \end{bmatrix} R2 + R1$

Thus, $A^{-1} = \begin{bmatrix} 1 & 1 \\ 2 & 1 \end{bmatrix}$.

$X = A^{-1}B = \begin{bmatrix} 1 & 1 \\ 2 & 1 \end{bmatrix}\begin{bmatrix} 1 \\ 1 \end{bmatrix} = \begin{bmatrix} 1(1)+1(1) \\ 2(1)+1(1) \end{bmatrix}$

$\quad = \begin{bmatrix} 1+1 \\ 2+1 \end{bmatrix} = \begin{bmatrix} 2 \\ 3 \end{bmatrix}$

Solution set: $\{(2,3)\}$

37. $2x - y = -8$
$\quad 3x + y = -2$

Given $A = \begin{bmatrix} 2 & -1 \\ 3 & 1 \end{bmatrix}$, $X = \begin{bmatrix} x \\ y \end{bmatrix}$, and $B = \begin{bmatrix} -8 \\ -2 \end{bmatrix}$,

find A^{-1}. We will use the relation given on page 588 to calculate A^{-1} in the solution to this exercise.

If $A = \begin{bmatrix} a & b \\ c & d \end{bmatrix}$, then $A^{-1} = \begin{bmatrix} \dfrac{d}{ad-bc} & \dfrac{-b}{ad-bc} \\ \dfrac{-c}{ad-bc} & \dfrac{a}{ad-bc} \end{bmatrix}$.

Thus we have:

$A^{-1} = \begin{bmatrix} \dfrac{1}{2(1)-(-1)(3)} & \dfrac{1}{2(1)-(-1)(3)} \\ \dfrac{-3}{2(1)-(-1)(3)} & \dfrac{2}{2(1)-(-1)(3)} \end{bmatrix}$

$\quad = \begin{bmatrix} \dfrac{1}{2-(-3)} & \dfrac{1}{2-(-3)} \\ \dfrac{-3}{2-(-3)} & \dfrac{2}{2-(-3)} \end{bmatrix} = \begin{bmatrix} \frac{1}{5} & \frac{1}{5} \\ -\frac{3}{5} & \frac{2}{5} \end{bmatrix}$

$X = A^{-1}B = \begin{bmatrix} \frac{1}{5} & \frac{1}{5} \\ -\frac{3}{5} & \frac{2}{5} \end{bmatrix}\begin{bmatrix} -8 \\ -2 \end{bmatrix}$

$\quad = \begin{bmatrix} \frac{1}{5}(-8)+\frac{1}{5}(-2) \\ -\frac{3}{5}(-8)+\frac{2}{5}(-2) \end{bmatrix} = \begin{bmatrix} -\frac{8}{5}+\left(-\frac{2}{5}\right) \\ \frac{24}{5}+\left(-\frac{4}{5}\right) \end{bmatrix}$

$\quad = \begin{bmatrix} -\frac{10}{5} \\ \frac{20}{5} \end{bmatrix} = \begin{bmatrix} -2 \\ 4 \end{bmatrix}$

Solution set: $\{(-2,4)\}$

39. $3x + 4y = -3$
$\quad -5x + 8y = 16$

Given $A = \begin{bmatrix} 3 & 4 \\ -5 & 8 \end{bmatrix}$, $X = \begin{bmatrix} x \\ y \end{bmatrix}$, and $B = \begin{bmatrix} -3 \\ 16 \end{bmatrix}$,

find A^{-1}. Since $\left[A \mid I_2 \right] = \begin{bmatrix} 3 & 4 & 1 & 0 \\ -5 & 8 & 0 & 1 \end{bmatrix}$,

$\begin{bmatrix} 3 & 4 & 1 & 0 \\ 0 & 44 & 5 & 3 \end{bmatrix} 3R2 + 5R1 \Rightarrow$

$\begin{bmatrix} 3 & 4 & 1 & 0 \\ 0 & 1 & \frac{5}{44} & \frac{3}{44} \end{bmatrix} \frac{1}{44}R2 \Rightarrow$

$\begin{bmatrix} 3 & 0 & \frac{6}{11} & -\frac{3}{11} \\ 0 & 1 & \frac{5}{44} & \frac{3}{44} \end{bmatrix} -4R2 + R1 \Rightarrow$

$\begin{bmatrix} 1 & 0 & \frac{2}{11} & -\frac{1}{11} \\ 0 & 1 & \frac{5}{44} & \frac{3}{44} \end{bmatrix} \frac{1}{3}R1$

Thus, $A^{-1} = \begin{bmatrix} \frac{2}{11} & -\frac{1}{11} \\ \frac{5}{44} & \frac{3}{44} \end{bmatrix}$.

$X = A^{-1}B = \begin{bmatrix} \frac{2}{11} & -\frac{1}{11} \\ \frac{5}{44} & \frac{3}{44} \end{bmatrix}\begin{bmatrix} -3 \\ 16 \end{bmatrix}$

$\quad = \begin{bmatrix} \frac{2}{11}(-3)-\frac{1}{11}(16) \\ \frac{5}{44}(-3)+\frac{3}{44}(16) \end{bmatrix}$

$\quad = \begin{bmatrix} -\frac{6}{11}-\frac{16}{11} \\ -\frac{15}{44}+\frac{48}{44} \end{bmatrix} = \begin{bmatrix} -\frac{22}{11} \\ \frac{33}{44} \end{bmatrix} = \begin{bmatrix} -2 \\ \frac{3}{4} \end{bmatrix}$

Solution set: $\left\{\left(-2, \frac{3}{4}\right)\right\}$

Copyright © 2013 Pearson Education, Inc.

41. $6x + 9y = 3$
$-8x + 3y = 6$

Given $A = \begin{bmatrix} 6 & 9 \\ -8 & 3 \end{bmatrix}$, $X = \begin{bmatrix} x \\ y \end{bmatrix}$, and $B = \begin{bmatrix} 3 \\ 6 \end{bmatrix}$,

find A^{-1}. We will use the results of the Relating Concepts exercises to calculate A^{-1} in the solution to this exercise.

If $A = \begin{bmatrix} a & b \\ c & d \end{bmatrix}$, then

$$A^{-1} = \frac{1}{ad - bc}\begin{bmatrix} d & -b \\ -c & a \end{bmatrix}.$$

Thus we have the following.

$$A^{-1} = \frac{1}{6(3)-9(-8)}\begin{bmatrix} 3 & -9 \\ 8 & 6 \end{bmatrix} = \frac{1}{18-(-72)}\begin{bmatrix} 3 & -9 \\ 8 & 6 \end{bmatrix}$$

$$= \frac{1}{90}\begin{bmatrix} 3 & -9 \\ 8 & 6 \end{bmatrix} = \begin{bmatrix} \frac{3}{90} & \frac{-9}{90} \\ \frac{8}{90} & \frac{6}{90} \end{bmatrix} = \begin{bmatrix} \frac{1}{30} & -\frac{1}{10} \\ \frac{4}{45} & \frac{1}{15} \end{bmatrix}$$

$$X = A^{-1}B = \begin{bmatrix} \frac{1}{30} & -\frac{1}{10} \\ \frac{4}{45} & \frac{1}{15} \end{bmatrix}\begin{bmatrix} 3 \\ 6 \end{bmatrix}$$

$$= \begin{bmatrix} \frac{1}{30}(3) + \left(-\frac{1}{10}\right)(6) \\ \frac{4}{45}(3) + \frac{1}{15}(6) \end{bmatrix} = \begin{bmatrix} \frac{1}{10} + \left(-\frac{6}{10}\right) \\ \frac{4}{15} + \frac{2}{5} \end{bmatrix}$$

$$= \begin{bmatrix} -\frac{5}{10} \\ \frac{4}{15} + \frac{6}{15} \end{bmatrix} = \begin{bmatrix} -\frac{1}{2} \\ \frac{10}{15} \end{bmatrix} = \begin{bmatrix} -\frac{1}{2} \\ \frac{2}{3} \end{bmatrix}$$

Solution set: $\left\{\left(-\frac{1}{2}, \frac{2}{3}\right)\right\}$

43. $0.2x + 0.3y = -1.9$
$0.7x - 0.2y = 4.6$

Given

$$A = \begin{bmatrix} 0.2 & 0.3 \\ 0.7 & -0.2 \end{bmatrix}, X = \begin{bmatrix} x \\ y \end{bmatrix}, \text{ and } B = \begin{bmatrix} -1.9 \\ 4.6 \end{bmatrix},$$

find A^{-1}. Since $[A \mid I_2] = \begin{bmatrix} 0.2 & 0.3 & 1 & 0 \\ 0.7 & -0.2 & 0 & 1 \end{bmatrix}$,

$$\begin{bmatrix} 1 & 1.5 & 5 & 0 \\ 0.7 & -0.2 & 0 & 1 \end{bmatrix} 5R1 \Rightarrow$$

$$\begin{bmatrix} 1 & 1.5 & 5 & 0 \\ 0 & -1.25 & -3.5 & 1 \end{bmatrix} -0.7R1 + R2 \Rightarrow$$

$$\begin{bmatrix} 1 & 1.5 & 5 & 0 \\ 0 & 1 & 2.8 & -0.8 \end{bmatrix} \frac{1}{-1.25}R2 \Rightarrow$$

$$\begin{bmatrix} 1 & 0 & 0.8 & 1.2 \\ 0 & 1 & 2.8 & -0.8 \end{bmatrix} -1.5R2 + R1$$

Thus, $A^{-1} = \begin{bmatrix} 0.8 & 1.2 \\ 2.8 & -0.8 \end{bmatrix}$.

$$X = A^{-1}B = \begin{bmatrix} 0.8 & 1.2 \\ 2.8 & -0.8 \end{bmatrix}\begin{bmatrix} -1.9 \\ 4.6 \end{bmatrix}$$

$$= \begin{bmatrix} 0.8(-1.9) + 1.2(4.6) \\ 2.8(-1.9) + (-0.8)(4.6) \end{bmatrix}$$

$$= \begin{bmatrix} -1.52 + 5.52 \\ -5.32 + (-3.68) \end{bmatrix} = \begin{bmatrix} 4 \\ -9 \end{bmatrix}$$

Solution set: $\{(4, -9)\}$

45. $\frac{1}{2}x + \frac{1}{3}y = \frac{49}{18}$
$\frac{1}{2}x + 2y = \frac{4}{3}$

Given $A = \begin{bmatrix} \frac{1}{2} & \frac{1}{3} \\ \frac{1}{2} & 2 \end{bmatrix}$, $X = \begin{bmatrix} x \\ y \end{bmatrix}$, and $B = \begin{bmatrix} \frac{49}{18} \\ \frac{4}{3} \end{bmatrix}$,

find A^{-1}. We will use the results of the Relating Concepts exercises to calculate A^{-1} in the solution to this exercise.

If $A = \begin{bmatrix} a & b \\ c & d \end{bmatrix}$, then $A^{-1} = \frac{1}{ad - bc}\begin{bmatrix} d & -b \\ -c & a \end{bmatrix}$.

Thus we have

$$A^{-1} = \frac{1}{\frac{1}{2}\cdot 2 - \frac{1}{2}\cdot\frac{1}{3}}\begin{bmatrix} 2 & -\frac{1}{3} \\ -\frac{1}{2} & \frac{1}{2} \end{bmatrix} = \frac{1}{\frac{5}{6}}\begin{bmatrix} 2 & -\frac{1}{3} \\ -\frac{1}{2} & \frac{1}{2} \end{bmatrix}$$

$$= \frac{6}{5}\begin{bmatrix} 2 & -\frac{1}{3} \\ -\frac{1}{2} & \frac{1}{2} \end{bmatrix} = \begin{bmatrix} \frac{12}{5} & -\frac{6}{15} \\ -\frac{6}{10} & \frac{6}{10} \end{bmatrix} = \begin{bmatrix} \frac{12}{5} & -\frac{2}{5} \\ -\frac{3}{5} & \frac{3}{5} \end{bmatrix}$$

$$X = A^{-1}B = \begin{bmatrix} \frac{12}{5} & -\frac{2}{5} \\ -\frac{3}{5} & \frac{3}{5} \end{bmatrix}\begin{bmatrix} \frac{49}{18} \\ \frac{4}{3} \end{bmatrix}$$

$$= \begin{bmatrix} \frac{12}{5}\left(\frac{49}{18}\right) - \frac{2}{5}\left(\frac{4}{3}\right) \\ -\frac{3}{5}\left(\frac{49}{18}\right) + \frac{3}{5}\left(\frac{4}{3}\right) \end{bmatrix}$$

$$= \begin{bmatrix} \frac{98}{15} - \frac{8}{15} \\ -\frac{49}{30} + \frac{4}{5} \end{bmatrix} = \begin{bmatrix} \frac{90}{15} \\ -\frac{5}{6} \end{bmatrix} = \begin{bmatrix} 6 \\ -\frac{5}{6} \end{bmatrix}$$

Solution set: $\left\{\left(6, -\frac{5}{6}\right)\right\}$

Copyright © 2013 Pearson Education, Inc.

47. In order to solve $\begin{array}{l} 7x-2y=3 \\ 14x-4y=1 \end{array}$ by the matrix inverse method, we must first find A^{-1} given that

$A = \begin{bmatrix} 7 & -2 \\ 14 & -4 \end{bmatrix}$. Since $[A|I_2] = \begin{bmatrix} 7 & -2 & | & 1 & 0 \\ 14 & -4 & | & 0 & 1 \end{bmatrix}$, we have the following.

$\begin{bmatrix} 1 & -\frac{2}{7} & | & \frac{1}{7} & 0 \\ 14 & -4 & | & 0 & 1 \end{bmatrix} \begin{array}{l} \frac{1}{7}R1 \\ \end{array} \Rightarrow \begin{bmatrix} 1 & -\frac{2}{7} & | & \frac{1}{7} & 0 \\ 0 & 0 & | & -2 & 1 \end{bmatrix} -14R1+R2$

Since it is not possible to get a 1 in the second entry of the second column, the inverse of A does not exist, hence the matrix inverse method cannot be used to solve the system.

49.
$x+ y+z=6$
$2x+3y-z=7$
$3x- y-z=6$

Given $A = \begin{bmatrix} 1 & 1 & 1 \\ 2 & 3 & -1 \\ 3 & -1 & -1 \end{bmatrix}$, $X = \begin{bmatrix} x \\ y \\ z \end{bmatrix}$, and $B = \begin{bmatrix} 6 \\ 7 \\ 6 \end{bmatrix}$, find A^{-1}. Since $[A|I_3] = \begin{bmatrix} 1 & 1 & 1 & | & 1 & 0 & 0 \\ 2 & 3 & -1 & | & 0 & 1 & 0 \\ 3 & -1 & -1 & | & 0 & 0 & 1 \end{bmatrix}$,

$\begin{bmatrix} 1 & 1 & 1 & | & 1 & 0 & 0 \\ 0 & 1 & -3 & | & -2 & 1 & 0 \\ 0 & -4 & -4 & | & -3 & 0 & 1 \end{bmatrix} \begin{array}{l} \\ -2R1+R2 \\ -3R1+R3 \end{array} \Rightarrow \begin{bmatrix} 1 & 0 & 4 & | & 3 & -1 & 0 \\ 0 & 1 & -3 & | & -2 & 1 & 0 \\ 0 & 0 & -16 & | & -11 & 4 & 1 \end{bmatrix} \begin{array}{l} -R2+R1 \\ \\ 4R2+R3 \end{array} \Rightarrow$

$\begin{bmatrix} 1 & 0 & 4 & | & 3 & -1 & 0 \\ 0 & 1 & -3 & | & -2 & 1 & 0 \\ 0 & 0 & 1 & | & \frac{11}{16} & -\frac{1}{4} & -\frac{1}{16} \end{bmatrix} \begin{array}{l} \\ \\ -\frac{1}{16}R3 \end{array} \Rightarrow \begin{bmatrix} 1 & 0 & 0 & | & \frac{1}{4} & 0 & \frac{1}{4} \\ 0 & 1 & 0 & | & \frac{1}{16} & \frac{1}{4} & -\frac{3}{16} \\ 0 & 0 & 1 & | & \frac{11}{16} & -\frac{1}{4} & -\frac{1}{16} \end{bmatrix} \begin{array}{l} -4R3+R1 \\ 3R3+R2. \\ \end{array}$

Thus, $A^{-1} = \begin{bmatrix} \frac{1}{4} & 0 & \frac{1}{4} \\ \frac{1}{16} & \frac{1}{4} & -\frac{3}{16} \\ \frac{11}{16} & -\frac{1}{4} & -\frac{1}{16} \end{bmatrix}$.

$X = A^{-1}B = \begin{bmatrix} \frac{1}{4} & 0 & \frac{1}{4} \\ \frac{1}{16} & \frac{1}{4} & -\frac{3}{16} \\ \frac{11}{16} & -\frac{1}{4} & -\frac{1}{16} \end{bmatrix} \begin{bmatrix} 6 \\ 7 \\ 6 \end{bmatrix} = \begin{bmatrix} \frac{1}{4}(6)+0(7)+\frac{1}{4}(6) \\ \frac{1}{16}(6)+\frac{1}{4}(7)+\left(-\frac{3}{16}\right)(6) \\ \frac{11}{16}(6)+\left(-\frac{1}{4}\right)(7)+\left(-\frac{1}{16}\right)(6) \end{bmatrix} = \begin{bmatrix} \frac{3}{2}+0+\frac{3}{2} \\ \frac{3}{8}+\frac{7}{4}+\left(-\frac{9}{8}\right) \\ \frac{33}{8}+\left(-\frac{7}{4}\right)+\left(-\frac{3}{8}\right) \end{bmatrix} = \begin{bmatrix} \frac{6}{2} \\ \frac{7}{4}+\left(-\frac{6}{8}\right) \\ \frac{30}{8}+\left(-\frac{7}{4}\right) \end{bmatrix}$

$= \begin{bmatrix} 3 \\ \frac{7}{4}+\left(-\frac{3}{4}\right) \\ \frac{15}{4}+\left(-\frac{7}{4}\right) \end{bmatrix} = \begin{bmatrix} 3 \\ \frac{4}{4} \\ \frac{8}{4} \end{bmatrix} = \begin{bmatrix} 3 \\ 1 \\ 2 \end{bmatrix}$

Solution set: $\{(3, 1, 2)\}$

51.
$2x+3y+3z=1$
$x+4y+3z=0$
$x+3y+4z=-1$

We have $A = \begin{bmatrix} 2 & 3 & 3 \\ 1 & 4 & 3 \\ 1 & 3 & 4 \end{bmatrix}$, $X = \begin{bmatrix} x \\ y \\ z \end{bmatrix}$, $B = \begin{bmatrix} 1 \\ 0 \\ -1 \end{bmatrix}$, and $A^{-1} = \begin{bmatrix} \frac{7}{8} & -\frac{3}{8} & -\frac{3}{8} \\ -\frac{1}{8} & \frac{5}{8} & -\frac{3}{8} \\ -\frac{1}{8} & -\frac{3}{8} & \frac{5}{8} \end{bmatrix}$. (from Exercise 19).

$X = A^{-1}B = \begin{bmatrix} \frac{7}{8} & -\frac{3}{8} & -\frac{3}{8} \\ -\frac{1}{8} & \frac{5}{8} & -\frac{3}{8} \\ -\frac{1}{8} & -\frac{3}{8} & \frac{5}{8} \end{bmatrix} \cdot \begin{bmatrix} 1 \\ 0 \\ -1 \end{bmatrix} = \begin{bmatrix} \frac{7}{8}(1)-\frac{3}{8}(0)-\frac{3}{8}(-1) \\ -\frac{1}{8}(1)+\frac{5}{8}(0)-\frac{3}{8}(-1) \\ -\frac{1}{8}(1)-\frac{3}{8}(0)+\frac{5}{8}(-1) \end{bmatrix} = \begin{bmatrix} \frac{7}{8}+0+\frac{3}{8} \\ -\frac{1}{8}+0+\frac{3}{8} \\ -\frac{1}{8}+0-\frac{5}{8} \end{bmatrix} = \begin{bmatrix} \frac{10}{8} \\ \frac{2}{8} \\ -\frac{6}{8} \end{bmatrix} = \begin{bmatrix} \frac{5}{4} \\ \frac{1}{4} \\ -\frac{3}{4} \end{bmatrix}$

Solution set: $\left\{\left(\frac{5}{4}, \frac{1}{4}, -\frac{3}{4}\right)\right\}$

Copyright © 2013 Pearson Education, Inc.

53. $2x + 2y - 4z = 12$
$2x + 6y = 16$
$-3x - 3y + 5z = -20$

We have $A = \begin{bmatrix} 2 & 2 & -4 \\ 2 & 6 & 0 \\ -3 & -3 & 5 \end{bmatrix}$, $X = \begin{bmatrix} x \\ y \\ z \end{bmatrix}$, $B = \begin{bmatrix} 12 \\ 16 \\ -20 \end{bmatrix}$, and $A^{-1} = \begin{bmatrix} -\frac{15}{4} & -\frac{1}{4} & -3 \\ \frac{5}{4} & \frac{1}{4} & 1 \\ -\frac{3}{2} & 0 & -1 \end{bmatrix}$ (from Exercise 21).

$X = A^{-1}B = \begin{bmatrix} -\frac{15}{4} & -\frac{1}{4} & -3 \\ \frac{5}{4} & \frac{1}{4} & 1 \\ -\frac{3}{2} & 0 & -1 \end{bmatrix} \begin{bmatrix} 12 \\ 16 \\ -20 \end{bmatrix} = \begin{bmatrix} -\frac{15}{4}(12) + \left(-\frac{1}{4}\right)(16) + (-3)(-20) \\ \frac{5}{4}(12) + \frac{1}{4}(16) + 1(-20) \\ -\frac{3}{2}(12) + 0(16) + (-1)(-20) \end{bmatrix} = \begin{bmatrix} -45 + (-4) + 60 \\ 15 + 4 + (-20) \\ -18 + 0 + 20 \end{bmatrix} = \begin{bmatrix} 11 \\ -1 \\ 2 \end{bmatrix}$

Solution set: $\{(11, -1, 2)\}$

55. $x + y + 2w = 3$
$2x - y + z - w = 3$
$3x + 3y + 2z - 2w = 5$
$x + 2y + z = 3$

We have $A = \begin{bmatrix} 1 & 1 & 0 & 2 \\ 2 & -1 & 1 & -1 \\ 3 & 3 & 2 & -2 \\ 1 & 2 & 1 & 0 \end{bmatrix}$, $X = \begin{bmatrix} x \\ y \\ z \\ w \end{bmatrix}$, $B = \begin{bmatrix} 3 \\ 3 \\ 5 \\ 3 \end{bmatrix}$, and $A^{-1} = \begin{bmatrix} \frac{1}{2} & 0 & \frac{1}{2} & -1 \\ \frac{1}{10} & -\frac{2}{5} & \frac{3}{10} & -\frac{1}{5} \\ -\frac{7}{10} & \frac{4}{5} & -\frac{11}{10} & \frac{12}{5} \\ \frac{1}{5} & \frac{1}{5} & -\frac{2}{5} & \frac{3}{5} \end{bmatrix}$ (from Exercise 23).

$X = A^{-1}B = \begin{bmatrix} \frac{1}{2} & 0 & \frac{1}{2} & -1 \\ \frac{1}{10} & -\frac{2}{5} & \frac{3}{10} & -\frac{1}{5} \\ -\frac{7}{10} & \frac{4}{5} & -\frac{11}{10} & \frac{12}{5} \\ \frac{1}{5} & \frac{1}{5} & -\frac{2}{5} & \frac{3}{5} \end{bmatrix} \begin{bmatrix} 3 \\ 3 \\ 5 \\ 3 \end{bmatrix} = \begin{bmatrix} \frac{1}{2}(3) + 0(3) + \frac{1}{2}(5) + (-1)(3) \\ \frac{1}{10}(3) + \left(-\frac{2}{5}\right)(3) + \frac{3}{10}(5) + \left(-\frac{1}{5}\right)(3) \\ -\frac{7}{10}(3) + \frac{4}{5}(3) + \left(-\frac{11}{10}\right)(5) + \frac{12}{5}(3) \\ \frac{1}{5}(3) + \frac{1}{5}(3) + \left(-\frac{2}{5}\right)(5) + \frac{3}{5}(3) \end{bmatrix}$

$= \begin{bmatrix} \frac{3}{2} + 0 + \frac{5}{2} + (-3) \\ \frac{3}{10} + \left(-\frac{6}{5}\right) + \frac{3}{2} + \left(-\frac{3}{5}\right) \\ -\frac{21}{10} + \frac{12}{5} + \left(-\frac{11}{2}\right) + \frac{36}{5} \\ \frac{3}{5} + \frac{3}{5} + (-2) + \frac{9}{5} \end{bmatrix} = \begin{bmatrix} \frac{8}{2} + (-3) \\ \frac{3}{10} + \left(-\frac{9}{5}\right) + \frac{3}{2} \\ -\frac{21}{10} + \left(-\frac{11}{2}\right) + \frac{48}{5} \\ (-2) + \frac{15}{5} \end{bmatrix} = \begin{bmatrix} 4 + (-3) \\ \frac{3}{10} + \left(-\frac{18}{10}\right) + \frac{15}{10} \\ -\frac{21}{10} + \left(-\frac{55}{10}\right) + \frac{96}{10} \\ (-2) + 3 \end{bmatrix} = \begin{bmatrix} 1 \\ \frac{0}{10} \\ \frac{20}{10} \\ 1 \end{bmatrix} = \begin{bmatrix} 1 \\ 0 \\ 2 \\ 1 \end{bmatrix}$

Solution set: $\{(1, 0, 2, 1)\}$

57. (a) $602.7 = a + 5.543b + 37.14c$
$656.7 = a + 6.933b + 41.30c$
$778.5 = a + 7.638b + 45.62c$

(b) $A = \begin{bmatrix} 1 & 5.543 & 37.14 \\ 1 & 6.933 & 41.30 \\ 1 & 7.638 & 45.62 \end{bmatrix}$, $X = \begin{bmatrix} a \\ b \\ c \end{bmatrix}$, and $B = \begin{bmatrix} 602.7 \\ 656.7 \\ 778.5 \end{bmatrix}$

Using a graphing calculator with matrix capabilities, we obtain the following.

```
[A]⁻¹[B]
[[-490.547375]
 [-89        ]
 [42.71875   ]]
```

$X = A^{-1}B$ or $\begin{bmatrix} a \\ b \\ c \end{bmatrix} = \begin{bmatrix} -490.547375 \\ -89 \\ 42.71875 \end{bmatrix}$.

Thus, $a \approx -490.547$, $b = -89$, $c = 42.71875$.

(c) $S = -490.547 - 89A + 42.71875B$

Copyright © 2013 Pearson Education, Inc.

(d) If $A = 7.752$ and $B = 47.38$, the predicted value of S is given by the following.
$$S = -490.547 - 89(7.752) + 42.71875(47.38) = 843.539375 \approx 843.5$$
The predicted value is approximately 843.5.

(e) If $A = 8.9$ and $B = 66.25$, the predicted value of S is given by the following.
$$S = -490.547 - 89(8.9) + 42.71875(66.25) = 1547.470188 \approx 1547.5$$
The predicted value is approximately 1547.5.
Using only three consecutive years to forecast six years into the future, it is probably not very accurate.

59. Answers will vary.

61. Given $A = \begin{bmatrix} \frac{2}{3} & 0.7 \\ 22 & \sqrt{3} \end{bmatrix}$, we obtain $A^{-1} \approx \begin{bmatrix} -0.1215875322 & 0.0491390161 \\ 1.544369078 & -0.046799063 \end{bmatrix}$ using a graphing calculator.

63. Given $A = \begin{bmatrix} \frac{1}{2} & \frac{1}{4} & \frac{1}{3} \\ 0 & \frac{1}{4} & \frac{1}{3} \\ \frac{1}{2} & \frac{1}{2} & \frac{1}{3} \end{bmatrix}$, we obtain $A^{-1} = \begin{bmatrix} 2 & -2 & 0 \\ -4 & 0 & 4 \\ 3 & 3 & -3 \end{bmatrix}$ using a graphing calculator.

65. $\begin{aligned} 2.1x + y &= \sqrt{5} \\ \sqrt{2}x - 2y &= 5 \end{aligned}$

Given $A = \begin{bmatrix} 2.1 & 1 \\ \sqrt{2} & -2 \end{bmatrix}$, $X = \begin{bmatrix} x \\ y \end{bmatrix}$, and $B = \begin{bmatrix} \sqrt{5} \\ 5 \end{bmatrix}$, using a graphing calculator, we have

$X = A^{-1}B = \begin{bmatrix} 1.68717058 \\ -1.306990242 \end{bmatrix}$. Thus, the solution set is $\{(1.68717058, -1.306990242)\}$.

67. $\begin{aligned} (\log 2)x + (\ln 3)y + (\ln 4)z &= 1 \\ (\ln 3)x + (\log 2)y + (\ln 8)z &= 5 \\ (\log 12)x + (\ln 4)y + (\ln 8)z &= 9 \end{aligned}$

Given $A = \begin{bmatrix} \log 2 & \ln 3 & \ln 4 \\ \ln 3 & \log 2 & \ln 8 \\ \log 12 & \ln 4 & \ln 8 \end{bmatrix}$, $X = \begin{bmatrix} x \\ y \\ z \end{bmatrix}$, and $B = \begin{bmatrix} 1 \\ 5 \\ 9 \end{bmatrix}$, using a graphing calculator, we have

$X = A^{-1}B = \begin{bmatrix} 13.58736702 \\ 3.929011993 \\ -5.342780076 \end{bmatrix}$.

Solution set: $\{(13.58736702, 3.929011993, -5.342780076)\}$

69. Suppose that A and B are square matrices of the same size, and A^{-1} exists. If $AB = O$, we can multiply both sides by A^{-1} to obtain $A^{-1}AB = A^{-1}O \Rightarrow IB = O \Rightarrow B = O$.

Copyright © 2013 Pearson Education, Inc.

71. Let $A = \begin{bmatrix} 1 & 0 \\ 1 & 1 \end{bmatrix}$ and $B = \begin{bmatrix} 1 & 1 \\ 0 & 1 \end{bmatrix}$. (Other answers are possible.)

$$AB = \begin{bmatrix} 1 & 0 \\ 1 & 1 \end{bmatrix} \begin{bmatrix} 1 & 1 \\ 0 & 1 \end{bmatrix} = \begin{bmatrix} 1(1)+0(0) & 1(1)+0(1) \\ 1(1)+1(0) & 1(1)+1(1) \end{bmatrix} = \begin{bmatrix} 1+0 & 1+0 \\ 1+0 & 1+1 \end{bmatrix} = \begin{bmatrix} 1 & 1 \\ 1 & 2 \end{bmatrix}$$

Find $(AB)^{-1}$. Since $\left[AB | I_2 \right] = \begin{bmatrix} 1 & 1 & | & 1 & 0 \\ 1 & 2 & | & 0 & 1 \end{bmatrix}$, $\begin{bmatrix} 1 & 1 & | & 1 & 0 \\ 0 & 1 & | & -1 & 1 \end{bmatrix}_{-R1+R2} \Rightarrow \begin{bmatrix} 1 & 0 & | & 2 & -1 \\ 0 & 1 & | & -1 & 1 \end{bmatrix}_{-R2+R1}$

Thus, $(AB)^{-1} = \begin{bmatrix} 2 & -1 \\ -1 & 1 \end{bmatrix}$.

Now find A^{-1}. Since $\left[A | I_2 \right] = \begin{bmatrix} 1 & 0 & | & 1 & 0 \\ 1 & 1 & | & 0 & 1 \end{bmatrix} \Rightarrow \begin{bmatrix} 1 & 0 & | & 1 & 0 \\ 0 & 1 & | & -1 & 1 \end{bmatrix}_{-R1+R2}$

Thus, $A^{-1} = \begin{bmatrix} 1 & 0 \\ -1 & 1 \end{bmatrix}$.

Now find B^{-1}. Since $\left[B | I_2 \right] = \begin{bmatrix} 1 & 1 & | & 1 & 0 \\ 0 & 1 & | & 0 & 1 \end{bmatrix} \Rightarrow \begin{bmatrix} 1 & 0 & | & 1 & -1 \\ 0 & 1 & | & 0 & 1 \end{bmatrix}_{-R2+R1}$

Thus, $B^{-1} = \begin{bmatrix} 1 & -1 \\ 0 & 1 \end{bmatrix}$.

Now find $A^{-1}B^{-1}$.

$$A^{-1}B^{-1} = \begin{bmatrix} 1 & 0 \\ -1 & 1 \end{bmatrix} \begin{bmatrix} 1 & -1 \\ 0 & 1 \end{bmatrix} = \begin{bmatrix} 1(1)+0(0) & 1(-1)+0(1) \\ -1(1)+1(0) & -1(-1)+1(1) \end{bmatrix} = \begin{bmatrix} 1+0 & -1+0 \\ -1+0 & 1+1 \end{bmatrix} = \begin{bmatrix} 1 & -1 \\ -1 & 2 \end{bmatrix}$$

Thus $(AB)^{-1} \neq A^{-1}B^{-1}$.

73. Given $A = \begin{bmatrix} a & 0 & 0 \\ 0 & b & 0 \\ 0 & 0 & c \end{bmatrix}$, and since a, b, and c are all nonzero, $\frac{1}{a}$, $\frac{1}{b}$, and $\frac{1}{c}$ all exist.

Thus, $\left[A | I \right] = \begin{bmatrix} a & 0 & 0 & | & 1 & 0 & 0 \\ 0 & b & 0 & | & 0 & 1 & 0 \\ 0 & 0 & c & | & 0 & 0 & 1 \end{bmatrix} \Rightarrow \begin{bmatrix} 1 & 0 & 0 & | & \frac{1}{a} & 0 & 0 \\ 0 & 1 & 0 & | & 0 & \frac{1}{b} & 0 \\ 0 & 0 & 1 & | & 0 & 0 & \frac{1}{c} \end{bmatrix} \begin{matrix} \frac{1}{a}R1 \\ \frac{1}{b}R2 \\ \frac{1}{c}R3 \end{matrix}$ and $A^{-1} = \begin{bmatrix} \frac{1}{a} & 0 & 0 \\ 0 & \frac{1}{b} & 0 \\ 0 & 0 & \frac{1}{c} \end{bmatrix}$.

75. Find the inverses of I_n, $-A$, and kA.

I_n is its own inverse, since $I_n \cdot I_n = I_n$.

The inverse of $-A$ is $-\left(A^{-1}\right)$, since $(-A)\left[-\left(A^{-1}\right)\right] = (-1)(-1)\left(A \cdot A^{-1}\right) = I_n$.

The inverse of kA (k a scalar) is $\frac{1}{k}A^{-1}$, since $(kA)\left(\frac{1}{k}A^{-1}\right) = \left(k \cdot \frac{1}{k}\right)\left(A \cdot A^{-1}\right) = I_n$.

Copyright © 2013 Pearson Education, Inc.

Chapter 5   Review Exercises

1. $2x + 6y = 6$ (1)
  $5x + 9y = 9$ (2)

Solving equation (1) for x, we have
$2x + 6y = 6 \Rightarrow 2x = 6 - 6y \Rightarrow x = 3 - 3y$ (3).

Substituting equation (3) into equation (2) and solving for y, we have the following.
$5(3 - 3y) + 9y = 9 \Rightarrow 15 - 15y + 9y = 9 \Rightarrow$
  $15 - 6y = 9 \Rightarrow -6y = -6 \Rightarrow y = 1$

Substituting 1 for y in equation (3), we have
$x = 3 - 3(1) = 3 - 3 = 0.$

Solution set: $\{(0,1)\}$

3. $x + 5y = 9$ (1)
  $2x + 10y = 18$ (2)

Multiply equation (1) by -2 and add the resulting equation to equation (2).
$\begin{array}{r} -2x - 10y = -18 \\ 2x + 10y = 18 \\ \hline 0 = 0 \end{array}$

This is a true statement and implies that the system has infinitely many solutions. Solving equation (1) for x we have $x = -5y + 9$. Given y as an arbitrary value, the solution set is $\{(-5y + 9, y)\}.$

5. $y = -x + 3$ (1)
  $2x + 2y = 1$ (2)

Substituting equation (1) into equation (2), we have $2x + 2(-x + 3) = 1$. Solving for x we have $2x + 2(-x + 3) = 1 \Rightarrow$

$2x + (-2x) + 6 = 1 \Rightarrow 6 = 1.$ This is a false statement and the solution is inconsistent. Thus, the solution set is $\varnothing$.

7. $3x - 2y = 0$ (1)
  $9x + 8y = 7$ (2)

Multiply equation (1) by -3 and add the result to equation (2).
$\begin{array}{r} -9x + 6y = 0 \\ 9x + 8y = 7 \\ \hline 14y = 7 \Rightarrow y = \frac{1}{2} \end{array}$

Substituting $\frac{1}{2}$ for y in equation (1) and solving for x, we have the following.
$3x - 2\left(\frac{1}{2}\right) = 0 \Rightarrow 3x - 1 = 0$
  $3x = 1 \Rightarrow x = \frac{1}{3}$

Solution set: $\left\{\left(\frac{1}{3}, \frac{1}{2}\right)\right\}$

9. $2x - 5y + 3z = -1$ (1)
  $x + 4y - 2z = 9$ (2)
  $-x + 2y + 4z = 5$ (3)

First, we eliminate x. Multiply equation (2) by -2 and add the result to equation (1).
$\begin{array}{r} 2x - 5y + 3z = -1 \\ -2x - 8y + 4z = -18 \\ \hline -13y + 7z = -19 \text{ (4)} \end{array}$

Next, add equations (2) and (3) to obtain $6y + 2z = 14$ (5). Now, we solve the system
$\begin{array}{r} -13y + 7z = -19 \,(4) \\ 6y + 2z = 14 \,(5) \end{array}$

Multiply equation (4) by 2, multiply equation (5) by -7, and add the resulting equations.
$\begin{array}{r} -26y + 14z = -38 \\ -42y - 14z = -98 \\ \hline -68y = -136 \Rightarrow y = 2 \end{array}$

Substituting this value into equation (4), we have $-13(2) + 7z = -19 \Rightarrow$

$-26 + 7z = -19 \Rightarrow 7z = 7 \Rightarrow z = 1.$
Substituting 2 for y and 1 for z in equation (2), we have $x + 4(2) - 2(1) = 9 \Rightarrow$
$x + 8 - 2 = 9 \Rightarrow x = 3.$

Solution set: $\{(3,2,1)\}$

11. $5x - y = 26$ (1)
  $4y + 3z = -4$ (2)
  $3x + 3z = 15$ (3)

Multiply equation (2) by -1 and add the result to equation (3) in order to eliminate z.
$\begin{array}{r} -4y - 3z = 4 \\ 3x + 3z = 15 \\ \hline 3x - 4y = 19 \,(4) \end{array}$

Multiply equation (1) by -4 and add the result to equation (4).
$\begin{array}{r} -20x + 4y = -104 \\ 3x - 4y = 19 \\ \hline -17x = -85 \Rightarrow x = 5 \end{array}$

Substitute 5 for x in equation (1) and solve for y to obtain
$5(5) - y = 26 \Rightarrow 25 - y = 26 \Rightarrow y = -1.$

Substitute 5 for x equation (3) and solve for z:
$3(5) + 3z = 15 \Rightarrow 15 + 3z = 15 \Rightarrow$
$3z = 0 \Rightarrow z = 0$

Solution set: $\{(5,-1,0)\}$

13. Answers will vary. One possible answer is
  $x + y = 2$
  $x + y = 3.$

Copyright © 2013 Pearson Education, Inc.

15. Let x = amount of rice; y = amount of soybeans.

We get the following system of equations. The first equation relates protein and the second relates calories.

$$15x + 22.5y = 9.5 \quad (1)$$
$$810x + 270y = 324 \quad (2)$$

Multiply equation (1) by -12 and add the result to equation (2).

$$
\begin{array}{r}
-180x - 270y = -114 \\
\underline{810x + 270y = 324} \\
630x = 210 \Rightarrow x = \tfrac{1}{3}
\end{array}
$$

Substitute $\tfrac{1}{3}$ for x in equation (1) and solve for y.

$$15\left(\tfrac{1}{3}\right) + 22.5y = 9.5 \Rightarrow 5 + 22.5y = 9.5 \Rightarrow$$
$$22.5y = 4.5 \Rightarrow y = 0.20 = \tfrac{1}{5}$$

$\tfrac{1}{3}$ cup of rice and $\tfrac{1}{5}$ cup of soybeans should be used.

17. Let x = the number of blankets, y = the number of rugs, z = the number of skirts.

The following table is helpful in organizing the information.

	Number of blankets (x)	Number of rugs (y)	Number of skirts (z)	Available
Spinning yarn	24	30	12	306
Dying	4	5	3	59
Weaving	15	18	9	201

We have the following system of equations.

$$24x + 30y + 12z = 306 \Rightarrow 4x + 5y + 2z = 51 \quad (1)$$
$$4x + 5y + 3z = 59 \quad (2)$$
$$15x + 18y + 9z = 201 \Rightarrow 5x + 6y + 3z = 67 \quad (3)$$

Multiply equation (1) by -1 and add the result to equation (2).

$$
\begin{array}{r}
-4x - 5y - 2z = -51 \\
\underline{4x + 5y + 3z = 59} \\
z = 8
\end{array}
$$

Substitute 8 for z in equations (1) and (3) and simplify.

$$4x + 5y + 2(8) = 51 \Rightarrow 4x + 5y + 16 = 51 \Rightarrow$$
$$4x + 5y = 35 \quad (4)$$
$$5x + 6y + 3(8) = 67 \Rightarrow 5x + 6y + 24 = 67 \Rightarrow$$
$$5x + 6y = 43 \quad (5)$$

Multiply equation (4) by 5 and equation (5) by -4 and add the results.

$$
\begin{array}{r}
20x + 25y = 175 \\
\underline{-20x - 24y = -172} \\
y = 3
\end{array}
$$

Substituting 3 for y in equation (4) and solving for x, we obtain $4x + 5(3) = 35 \Rightarrow$

$$4x + 15 = 35 \Rightarrow 4x = 20 \Rightarrow x = 5$$

5 blankets, 3 rugs, and 8 skirts can be made.

19. $H = 0.491x + 0.468y + 11.2$
$H = -0.981x + 1.872y + 26.4$

Substitute $H = 180$ and solve each equation for y so they can be entered into the graphing calculator.

$$180 = 0.491x + 0.468y + 11.2$$
$$168.8 - 0.491x = 0.468y \Rightarrow y = \tfrac{168.8 - 0.491x}{0.468}$$
$$180 = -0.981x + 1.872y + 26.4$$
$$153.6 + 0.981x = 1.872y \Rightarrow y = \tfrac{153.6 + 0.981x}{1.872}$$

Thus, we have $x \approx 177.1$ and $y \approx 174.9$.

If an athlete's maximum heart rate is 180 beats per minute, then it will be about 177 beats per minute 5 seconds after stopping and 175 beats per minute 10 seconds after stopping.

21. Since $y = ax^2 + bx + c$ and the points $(1, -2.3)$, $(2, -1.3)$, and $(3, 4.5)$ are on the parabola, we have the following system of equations.

$$-2.3 = a(1)^2 + b(1) + c \Rightarrow a + b + c = -2.3 \quad (1)$$
$$-1.3 = a(2)^2 + b(2) + c \Rightarrow 4a + 2b + c = -1.3 \quad (2)$$
$$4.5 = a(3)^2 + b(3) + c \Rightarrow 9a + 3b + c = 4.5 \quad (3)$$

First, eliminate c by multiplying equation (1) by -1 and adding the result to equation (2).

$$
\begin{array}{r}
-a - b - c = 2.3 \\
\underline{4a + 2b + c = -1.3} \\
3a + b = 1 \quad (4)
\end{array}
$$

(*continued on next page*)

Copyright © 2013 Pearson Education, Inc.

(*continued*)

Next, multiply equation (2) by –1 and add to equation (3).

$$-4a - 2b - c = 1.3$$
$$\underline{9a + 3b + c = 4.5}$$
$$5a + b \quad = 5.8 \quad (5)$$

Next, eliminate b by multiplying equation (4) by –1 and adding the result to equation (5).

$$-3a - b = -1$$
$$\underline{5a + b = 5.8}$$
$$2a \quad = 4.8 \Rightarrow a = 2.4$$

Substitute this value into equation (4) to obtain
$$3(2.4) + b = 1 \Rightarrow 7.2 + b = 1 \Rightarrow b = -6.2.$$

Substitute 2.4 for a and –6.2 for b into equation (1) in order to solve for c.

$$(2.4) + (-6.2) + c = -2.3$$
$$-3.8 + c = -2.3 \Rightarrow c = 1.5$$

The equation of the parabola is
$$y = 2.4x^2 - 6.2x + 1.5 \text{ or } y = \tfrac{12}{5}x^2 - \tfrac{31}{5}x + \tfrac{3}{2}.$$

23. $3x - 4y + z = 2$ (1)
$\quad\quad 2x + y = 1$ (2)

Solving equation (2) for y, we obtain $y = 1 - 2x$. Substitute $1 - 2x$ for y in equation (1) and solve for z.

$$3x - 4(1 - 2x) + z = 2 \Rightarrow 3x - 4 + 8x + z = 2 \Rightarrow$$
$$11x + z = 6 \Rightarrow z = 6 - 11x$$

Solution set: $\{(x, 1 - 2x, 6 - 11x)\}$

25. Writing
$$5x + 2y = -10$$
$$3x - 5y = -6$$

as an augmented matrix, we have

$$\begin{bmatrix} 5 & 2 & | & -10 \\ 3 & -5 & | & -6 \end{bmatrix}.$$

$$\begin{bmatrix} 1 & \tfrac{2}{5} & | & -2 \\ 3 & -5 & | & -6 \end{bmatrix} \tfrac{1}{5}\text{R1} \Rightarrow$$

$$\begin{bmatrix} 1 & \tfrac{2}{5} & | & -2 \\ 0 & -\tfrac{31}{5} & | & 0 \end{bmatrix} -3\text{R1}+\text{R2} \Rightarrow$$

$$\begin{bmatrix} 1 & \tfrac{2}{5} & | & -2 \\ 0 & 1 & | & 0 \end{bmatrix} -\tfrac{5}{31}\text{R2} \Rightarrow$$

$$\begin{bmatrix} 1 & 0 & | & -2 \\ 0 & 1 & | & 0 \end{bmatrix} -\tfrac{2}{5}\text{R2}+\text{R1}$$

Solution set: $\{(-2, 0)\}$

27. Writing
$$3x + y = -7$$
$$x - y = -5$$

as an augmented matrix, we have

$$\begin{bmatrix} 3 & 1 & | & -7 \\ 1 & -1 & | & -5 \end{bmatrix}.$$

$$\begin{bmatrix} 1 & -1 & | & -5 \\ 3 & 1 & | & -7 \end{bmatrix} \text{R1} \leftrightarrow \text{R2} \Rightarrow$$

$$\begin{bmatrix} 1 & -1 & | & -5 \\ 0 & 4 & | & 8 \end{bmatrix} -3\text{R1}+\text{R2} \Rightarrow$$

$$\begin{bmatrix} 1 & -1 & | & -5 \\ 0 & 1 & | & 2 \end{bmatrix} \tfrac{1}{4}\text{R2} \Rightarrow$$

$$\begin{bmatrix} 1 & 0 & | & -3 \\ 0 & 1 & | & 2 \end{bmatrix} \text{R2}+\text{R1}$$

Solution set: $\{(-3, 2)\}$

29. Writing
$$x - z = -3$$
$$y + z = 6$$
$$2x - 3z = -9$$

as an augmented matrix, we have

$$\begin{bmatrix} 1 & 0 & -1 & | & -3 \\ 0 & 1 & 1 & | & 6 \\ 2 & 0 & -3 & | & -9 \end{bmatrix}.$$

$$\begin{bmatrix} 1 & 0 & -1 & | & -3 \\ 0 & 1 & 1 & | & 6 \\ 0 & 0 & -1 & | & -3 \end{bmatrix} -2\text{R1}+\text{R3} \Rightarrow$$

$$\begin{bmatrix} 1 & 0 & -1 & | & -3 \\ 0 & 1 & 1 & | & 6 \\ 0 & 0 & 1 & | & 3 \end{bmatrix} -1\text{R3} \Rightarrow$$

$$\begin{bmatrix} 1 & 0 & -1 & | & -3 \\ 0 & 1 & 0 & | & 3 \\ 0 & 0 & 1 & | & 3 \end{bmatrix} -1\text{R3}+\text{R2} \Rightarrow$$

$$\begin{bmatrix} 1 & 0 & 0 & | & 0 \\ 0 & 1 & 0 & | & 3 \\ 0 & 0 & 1 & | & 3 \end{bmatrix} \text{R3}+\text{R1}$$

Solution set: $\{(0, 3, 3)\}$

Copyright © 2013 Pearson Education, Inc.

31. Let x = number of pounds of \$4.60 tea;
y = the number of pounds of \$5.75 tea;
z = the number of pounds of \$6.50 tea.
From the information in the exercise, we obtain the system

$$x + y + z = 20$$
$$4.6x + 5.75y + 6.5z = 20(5.25).$$
$$x = y + z$$

Rewriting the system, we have

$$x + y + z = 20$$
$$4.6x + 5.75y + 6.5z = 105.$$
$$x - y - z = 0$$

With an augmented matrix of

$$\begin{bmatrix} 1 & 1 & 1 & | & 20 \\ 4.6 & 5.75 & 6.5 & | & 105 \\ 1 & -1 & -1 & | & 0 \end{bmatrix},$$

we solve by the Gauss-Jordan method.

$$\begin{bmatrix} 1 & 1 & 1 & | & 20 \\ 0 & 1.15 & 1.9 & | & 13 \\ 0 & -2 & -2 & | & -20 \end{bmatrix} \begin{matrix} \\ -4.6R1+R2 \Rightarrow \\ -1R1+R3 \end{matrix}$$

$$\begin{bmatrix} 1 & 1 & 1 & | & 20 \\ 0 & 1 & \frac{38}{23} & | & \frac{260}{23} \\ 0 & -2 & -2 & | & -20 \end{bmatrix} \begin{matrix} \\ \frac{20}{23}R2 \\ \\ \end{matrix} \Rightarrow$$

$$\begin{bmatrix} 1 & 0 & -\frac{15}{23} & | & \frac{200}{23} \\ 0 & 1 & \frac{38}{23} & | & \frac{260}{23} \\ 0 & 0 & \frac{30}{23} & | & \frac{60}{23} \end{bmatrix} \begin{matrix} -1R2+R1 \\ \\ 2R2+R3 \end{matrix} \Rightarrow$$

$$\begin{bmatrix} 1 & 0 & -\frac{15}{23} & | & \frac{200}{23} \\ 0 & 1 & \frac{38}{23} & | & \frac{260}{23} \\ 0 & 0 & 1 & | & 2 \end{bmatrix} \begin{matrix} \\ \\ \frac{23}{30}R3 \end{matrix} \Rightarrow$$

$$\begin{bmatrix} 1 & 0 & 0 & | & 10 \\ 0 & 1 & 0 & | & 8 \\ 0 & 0 & 1 & | & 2 \end{bmatrix} \begin{matrix} \frac{15}{23}R3+R1 \\ -\frac{38}{23}R3+R2 \\ \\ \end{matrix}$$

From the final matrix, $x = 10$, $y = 8$, and $z = 2$. Therefore, 10 lb of \$4.60 tea, 8 lb of \$5.75 tea, and 2 lb of \$6.50 tea should be used.

33. $y = 2.81x + 121$ (1)
$y = 7.25x + 65.5$ (2)

The given system can be solved by substitution. Substitute equation (1) into equation (2) to obtain
$2.81x + 121 = 7.25x + 65.5$. Solving this equation, we have $55.5 = 4.44x \Rightarrow x = 12.5$.

Since $x = 0$ corresponds 1970, $x = 12.5$ implies that during 1982 males and females earned the same number of master's degrees.
The number of masters degrees earned by males in this year is
$y = 2.81 \cdot 12.5 + 121 = 156.125$ thousand. The masters degrees earned by females in this year (which is the same as the males) is
$y = 7.25 \cdot 12.5 + 65.5 = 156.125$ million. The total number of master's degrees was therefore approximately 312 thousand.

35. $\begin{vmatrix} -1 & 8 \\ 2 & 9 \end{vmatrix} = -1(9) - 2(8) = -9 - 16 = -25$

37. $\begin{vmatrix} x & 4x \\ 2x & 8x \end{vmatrix} = x(8x) - 2x(4x) = 8x^2 - 8x^2 = 0$

39. Expanding
$$\begin{vmatrix} -1 & 2 & 3 \\ 4 & 0 & 3 \\ 5 & -1 & 2 \end{vmatrix}$$
by the second column, we have
$$-2\begin{vmatrix} 4 & 3 \\ 5 & 2 \end{vmatrix} + 0\begin{vmatrix} -1 & 3 \\ 5 & 2 \end{vmatrix} - (-1)\begin{vmatrix} -1 & 3 \\ 4 & 3 \end{vmatrix}$$
$$= -2[4(2) - 5(3)] + 0 + [(-1)(3) - (4)(3)]$$
$$= -2(8 - 15) + (-3 - 12) = -2(-7) + (-15)$$
$$= 14 - 15 = -1$$

41. $\begin{vmatrix} 3x & 7 \\ -x & 4 \end{vmatrix} = 8 \Rightarrow$

$12x - (-7x) = 8 \Rightarrow 19x = 8 \Rightarrow x = \frac{8}{19}$

Solution set: $\left\{\frac{8}{19}\right\}$

43. $3x + 7y = 2$
$5x - y = -22$

$$D = \begin{vmatrix} 3 & 7 \\ 5 & -1 \end{vmatrix} = 3(-1) - 5(7) = -3 - 35 = -38$$

$$D_x = \begin{vmatrix} 2 & 7 \\ -22 & -1 \end{vmatrix} = 2(-1) - (-22)(7)$$
$$= -2 - (-154) = 152$$

$$D_y = \begin{vmatrix} 3 & 2 \\ 5 & -22 \end{vmatrix} = 3(-22) - 5(2)$$
$$= -66 - 10 = -76$$

$x = \frac{D_x}{D} = \frac{152}{-38} = -4$ and $y = \frac{D_y}{D} = \frac{-76}{-38} = 2$.
Solution set: $\{(-4, 2)\}$

Copyright © 2013 Pearson Education, Inc.

45. Given $\begin{array}{l} 6x+\ y=-3 \quad (1) \\ 12x+2y=1 \quad (2) \end{array}$, we have

$$D = \begin{vmatrix} 6 & 1 \\ 12 & 2 \end{vmatrix} = 6(2) - 12(1) = 12 - 12 = 0.$$

Because $D = 0$, Cramer's rule does not apply. To determine whether the system is inconsistent or has infinitely many solutions, use the elimination method.

$$\begin{array}{l} -12x - 2y = 6 \quad \text{Multiply equation (1) by} - 2. \\ \underline{12x + 2y = 1} \\ 0 = 7 \quad \text{False} \end{array}$$

The system is inconsistent. Thus the solution set is $\varnothing$.

47.
$$\begin{array}{l} x+\ y=-1 \\ 2y+\ z=5 \\ 3x-2z=-28 \end{array}$$

Adding -3 times row 1 to row 3, we have

$$D = \begin{vmatrix} 1 & 1 & 0 \\ 0 & 2 & 1 \\ 3 & 0 & -2 \end{vmatrix} = \begin{vmatrix} 1 & 1 & 0 \\ 0 & 2 & 1 \\ 0 & -3 & -2 \end{vmatrix}$$

Expanding by column one, we have

$$D = \begin{vmatrix} 2 & 1 \\ -3 & -2 \end{vmatrix} = -4 - (-3) = -1$$

Adding -2 times row 1 to row 2, we have

$$D_x = \begin{vmatrix} -1 & 1 & 0 \\ 5 & 2 & 1 \\ -28 & 0 & -2 \end{vmatrix} = \begin{vmatrix} -1 & 1 & 0 \\ 7 & 0 & 1 \\ -28 & 0 & -2 \end{vmatrix}$$

Expanding about column two, we have

$$D_x = -(1) \begin{vmatrix} 7 & 1 \\ -28 & -2 \end{vmatrix} = -[-14 - (-28)] = -14$$

Adding 2 times row 2 to row 3, we have

$$D_y = \begin{vmatrix} 1 & -1 & 0 \\ 0 & 5 & 1 \\ 3 & -28 & -2 \end{vmatrix} = \begin{vmatrix} 1 & -1 & 0 \\ 0 & 5 & 1 \\ 3 & -18 & 0 \end{vmatrix}$$

Expanding by column three, we have

$$D_y = -(1) \begin{vmatrix} 1 & -1 \\ 3 & -18 \end{vmatrix}$$
$$= -1[-18 - (-3)] = -(-15) = 15$$

Adding -3 times row 1 to row 3, we have

$$D_z = \begin{vmatrix} 1 & 1 & -1 \\ 0 & 2 & 5 \\ 3 & 0 & -28 \end{vmatrix} = \begin{vmatrix} 1 & 1 & -1 \\ 0 & 2 & 5 \\ 0 & -3 & -25 \end{vmatrix}$$

Expanding by column one, we have

$$D_z = 1 \begin{vmatrix} 2 & 5 \\ -3 & -25 \end{vmatrix} = -50 - (-15) = -35.$$

Thus, we have

$$x = \frac{D_x}{D} = \frac{-14}{-1} = 14, \ y = \frac{D_y}{D} = \frac{15}{-1} = -15,$$
$$z = \frac{D_z}{D} = \frac{-35}{-1} = 35$$

Solution set: $\{(14, -15, 35)\}$

49. $\dfrac{2}{3x^2 - 5x + 2} = \dfrac{2}{(x-1)(3x-2)} = \dfrac{A}{x-1} + \dfrac{B}{3x-2}$

Multiply both sides by $(x-1)(3x-2)$ to get

$$2 = A(3x - 2) + B(x - 1). \ (1)$$

First substitute 1 for x to get

$$2 = A[3(1) - 2] + B(1 - 1)$$
$$2 = A(1) + B(0) \Rightarrow A = 2$$

Replace A with 2 in equation (1) and substitute $\frac{2}{3}$ for x to get

$$2 = 2\left[3\left(\tfrac{2}{3}\right) - 2\right] + B\left(\tfrac{2}{3} - 1\right)$$
$$2 = 2(0) + B\left(-\tfrac{1}{3}\right) \Rightarrow 2 = -\tfrac{1}{3}B \Rightarrow B = -6$$

Thus, we have

$$\frac{2}{3x^2 - 5x + 2} = \frac{2}{x-1} + \frac{-6}{3x-2} \ \text{or} \ \frac{2}{x-1} - \frac{6}{3x-2}$$

51. $\dfrac{5 - 2x}{(x^2 + 2)(x - 1)} = \dfrac{A}{x-1} + \dfrac{Bx + C}{x^2 + 2}$

Multiply both sides by $(x^2 + 2)(x - 1)$ to get the following.

$$5 - 2x = A(x^2 + 2) + (Bx + C)(x - 1) \ (1)$$

First substitute 1 for x to get the following.

$$5 - 2(1) = A(1^2 + 2) + [B(1) + C](1 - 1)$$
$$5 - 2 = A(1 + 2) + 0 \Rightarrow 3 = 3A \Rightarrow A = 1$$

Replace A with 1 in equation (1) and substitute 0 for x to get the following.

$$5 - 2(0) = (0^2 + 2) + [B(0) + C](0 - 1)$$
$$5 - 0 = (0 + 2) + C(-1) \Rightarrow 5 = 2 - C \Rightarrow$$
$$3 = -C \Rightarrow C = -3$$

$$5 - 2x = (x^2 + 2) + (Bx - 3)(x - 1) \ (2)$$

Substitute 2 (arbitrary value) in equation (2) for x to get the following.

$$5 - 2(2) = (2^2 + 2) + [B(2) - 3](2 - 1)$$
$$5 - 4 = (4 + 2) + (2B - 3)(1)$$
$$1 = 6 + 2B - 3 \Rightarrow 1 = 3 + 2B$$
$$-2 = 2B \Rightarrow B = -1$$

(continued on next page)

Copyright © 2013 Pearson Education, Inc.

(*continued*)

Thus, we have

$$\frac{5-2x}{\left(x^2+2\right)\left(x-1\right)} = \frac{1}{x-1} + \frac{\left(-1\right)x+\left(-3\right)}{x^2+2} \text{ or}$$

$$\frac{1}{x-1} - \frac{x+3}{x^2+2}$$

53. $y = 2x + 10$ (1)

$x^2 + y = 13$ (2)

Substituting equation (1) into equation (2), we have $x^2 + \left(2x+10\right) = 13$. Solving for x we get the following.

$$x^2 + \left(2x+10\right) = 13$$
$$x^2 + 2x - 3 = 0$$
$$\left(x+3\right)\left(x-1\right) = 0 \Rightarrow x = -3 \text{ or } x = 1$$

For each value of x, use equation (1) to find the corresponding value of y.

If $x = -3$, $y = 2\left(-3\right)+10 = -6+10 = 4$.

If $x = 1$, $y = 2\left(1\right)+10 = 2+10 = 12$.

Solution set: $\left\{\left(-3,4\right),\left(1,12\right)\right\}$

55. $x^2 + y^2 = 17$ (1)

$2x^2 - y^2 = 31$ (2)

Add equations (1) and (2) to obtain

$3x^2 = 48 \Rightarrow x^2 = 16 \Rightarrow x = \pm 4$.

For each value of x, use equation (1) to find the corresponding value of y.

If $x = -4$, $\left(-4\right)^2 + y^2 = 17 \Rightarrow 16 + y^2 = 17 \Rightarrow$

$y^2 = 1 \Rightarrow y = \pm 1$. If $x = 4$, $4^2 + y^2 = 17 \Rightarrow$

$16 + y^2 = 17 \Rightarrow y^2 = 1 \Rightarrow y = \pm 1$.

Solution set: $\left\{\left(-4,1\right),\left(-4,-1\right),\left(4,-1\right),\left(4,1\right)\right\}$

57. $xy = -10$ (1)

$x + 2y = 1$ (2)

Solving equation (2) for x gives

$x = 1 - 2y$ (3). Substituting this expression into equation (1), we obtain $\left(1-2y\right)y = -10$.

Solving for y we have

$$\left(1-2y\right)y = -10 \Rightarrow y - 2y^2 = -10$$
$$0 = 2y^2 - y - 10$$
$$\left(y+2\right)\left(2y-5\right) = 0 \Rightarrow y = -2 \text{ or } y = \tfrac{5}{2}$$

For each value of y, use equation (3) to find the corresponding value of x.

If $y = -2$, $x = 1 - 2\left(-2\right) = 1 - \left(-4\right) = 5$.

If $y = \tfrac{5}{2}$, $x = 1 - 2\left(\tfrac{5}{2}\right) = 1 - 5 = -4$.

Solution set: $\left\{\left(5,-2\right),\left(-4,\tfrac{5}{2}\right)\right\}$

59. $x^2 + 2xy + y^2 = 4$ (1)

$x - 3y = -2$ (2)

Solve equation (2) for x: $x = 3y - 2$ (3).

Substituting this expression into equation (1), we obtain $\left(3y-2\right)^2 + 2\left(3y-2\right)y + y^2 = 4$.

Solving for x we have the following.

$$\left(3y-2\right)^2 + 2\left(3y-2\right)y + y^2 = 4$$
$$9y^2 - 12y + 4 + 6y^2 - 4y + y^2 = 4$$
$$16y^2 - 16y + 4 = 4$$
$$16y^2 - 16y = 0$$
$$y^2 - y = 0$$
$$y\left(y-1\right) = 0$$
$$y = 0 \text{ or } y = 1$$

For each value of y, use equation (3) to find the corresponding value of x.

If $y = 0$, $x = 3\left(0\right) - 2 = 0 - 2 = -2$.

If $y = 1$, $x = 3\left(1\right) - 2 = 3 - 2 = 1$.

Solution set: $\left\{\left(-2,0\right),\left(1,1\right)\right\}$

61. $2x^2 - 3y^2 = 18$ (1)

$2x^2 - 2y^2 = 14$ (2)

Multiply equation (2) by -1, then add the result to equation (1) and then solve for y.

$$2x^2 - 3y^2 = 18$$
$$\underline{-2x^2 + 2y^2 = -14}$$
$$-y^2 = 4 \Rightarrow y^2 = -4 \Rightarrow y = \pm 2i$$

If $y = -2i$, then, using equation (1), we have

$$2x^2 - 3\left(-2i\right)^2 = 18 \Rightarrow 2x^2 - 3\left(-4\right) = 18 \Rightarrow$$
$$2x^2 + 12 = 18 \Rightarrow 2x^2 = 6 \to x^2 - 3 \to$$
$$x = \pm\sqrt{3}$$

If $y = 2i$, then, using equation (1), we have

$$2x^2 - 3\left(2i\right)^2 = 18 \Rightarrow 2x^2 - 3\left(-4\right) = 18 \Rightarrow$$
$$2x^2 + 12 = 18 \Rightarrow 2x^2 = 6 \Rightarrow x^2 = 3 \Rightarrow$$
$$x = \pm\sqrt{3}$$

Solution set: $\left\{\left(-\sqrt{3}, -2i\right), \left(-\sqrt{3}, 2i\right),\right.$

$\left.\left(\sqrt{3}, -2i\right), \left(\sqrt{3}, 2i\right)\right\}$

Copyright © 2013 Pearson Education, Inc.

63. $x^2 + y^2 = 144$ (1)

$x + 2y = 8$ (2)

Solving equation (2) for x, we have $x = 8 - 2y$ (3). Substitute $-2y + 8$ for x in equation (1) to obtain

$$(8 - 2y)^2 + y^2 = 144$$
$$64 - 32y + 4y^2 + y^2 = 144$$
$$5y^2 - 32y - 80 = 0$$

We solve this quadratic equation for x by using the quadratic formula, with $a = 5, b = -32,$ and $c = -80.$

$$y = \frac{-(-32) \pm \sqrt{(-32)^2 - 4(5)(-80)}}{2(5)}$$

$$= \frac{32 \pm \sqrt{1,024 + 1,600}}{10} = \frac{32 + \sqrt{2,624}}{10}$$

$$= \frac{32 \pm 8\sqrt{41}}{10} = \frac{16 \pm 4\sqrt{41}}{5}$$

For each value of y, use equation (3) to find the corresponding value of x.

If $y = \frac{16 + 4\sqrt{41}}{5}$

$$x = 8 - 2\left(\frac{16 + 4\sqrt{41}}{5}\right) = \frac{40}{5} + \frac{-32 - 8\sqrt{41}}{5} = \frac{8 - 8\sqrt{41}}{5}$$

If $y = \frac{16 - 4\sqrt{41}}{5}$,

$$x = 8 - 2\left(\frac{16 - 4\sqrt{41}}{5}\right) = \frac{40}{5} + \frac{-32 + 8\sqrt{41}}{5} = \frac{8 + 8\sqrt{41}}{5}$$

Yes, the circle and the line have two points in common,

$$\left(\frac{8 - 8\sqrt{41}}{5}, \frac{16 + 4\sqrt{41}}{5}\right) \text{ and } \left(\frac{8 + 8\sqrt{41}}{5}, \frac{16 - 4\sqrt{41}}{5}\right).$$

65. $x + y \le 6$

$2x - y \ge 3$

Graph the solid line $x + y = 6$, which has x-intercept 6 and y-intercept 6. Since $x + y \le 6 \Rightarrow y \le -x + 6$, shade the region below this line. Graph the solid line

$2x - y = 3$, which has x-intercept $\frac{3}{2}$ and y-intercept -3. Since $2x - y \ge 3 \Rightarrow$

$-y \ge -2x + 3 \Rightarrow y \le 2x - 3$, shade the region below this line. The solution set is the intersection of these two regions, which is shaded in the final graph.

67. Find $x \ge 0$ and $y \ge 0$ such that $\begin{matrix} 3x + 2y \le 12 \\ 5x + y \ge 5 \end{matrix}$

and $2x + 4y$ is maximized. To graph $3x + 2y \le 12$, draw the line with x-intercept 4 and y-intercept 6 as a solid line. Because the test point $(0, 0)$ satisfies this inequality, shade the region *below* the line. To graph $5x + y \ge 5$, draw the line with x-intercept 1 and y-intercept 5 as a solid line. Because the test point $(0, 0)$ does not satisfy this inequality, shade the region *above* the line. The constraints $x \ge 0$ and $y \ge 0$ restrict the graph to the first quadrant. The graph of the feasible region is the intersection of the regions that are the graphs of the individual constraints. The four vertices are $(1, 0)$, $(4, 0)$, $(0, 5)$ and $(0, 6)$.

Point	Value of $2x + 4y$	
$(1, 0)$	$2(1) + 4(0) = 2$	
$(4, 0)$	$2(4) + 4(0) = 8$	
$(0, 5)$	$2(0) + 4(5) = 20$	
$(0, 6)$	$2(0) + 4(6) = 24$	$\leftarrow$ Maximum

The maximum value is 24, which occurs at $(0, 6)$.

69. Let x = number of servings of food A; y = number of servings of food B. The protein constraint is $2x + 6y \ge 30$. The fat constraint is $4x + 2y \ge 20$. Also, $y \ge 2$ and since the numbers of servings cannot be negative, we also have $x \ge 0$. We want to minimize the objective function, cost $= .18x + .12y$. Find the region of feasible solutions by graphing the system of inequalities that is made up of the constraints. To graph $2x + 6y \ge 30$, draw the line with x-intercept 15 and y-intercept 5 as a solid line. Because the test point $(0, 0)$ does not satisfy this inequality, shade the region above the line.

(continued on next page)

Copyright © 2013 Pearson Education, Inc.

(*continued*)

To graph $4x + 2y \geq 20$, draw the line with x-intercept 5 and y-intercept 10 as a solid line. Because the test point $(0,0)$ does not satisfy this inequality, shade the region above the line. To graph $y \geq 2$, draw the horizontal line with y-intercept 2 as a solid line and shade the region above the line. The graph of the feasible region is the intersection of the regions that are the graphs of the individual constraints.

From the graph, observe that one vertex is $(0,10)$. The second occurs when $y = 2$ and

$2x + 6y = 30 \Rightarrow 2x + 6(2) = 30 \Rightarrow 2x = 18 \Rightarrow x = 9.$

Thus, $(9,2)$ is a vertex. The third vertex is the intersection point of the lines $2x + 6y = 30$ and $4x + 2y = 20$. To find this point, solve the system $2x + 6y = 30$

$\qquad\qquad\qquad\qquad 4x + 2y = 20.$

Multiply the first equation by –2 and add it to the second equation.

$-4x - 12y = -60$
$\underline{4x + 2y = 20}$
$-10y = -40 \Rightarrow y = 4$

Substituting 4 for y into $2x + 6y = 30$, we have the following.

$2x + 6(4) = 30 \Rightarrow 2x + 24 = 30 \Rightarrow$
$\qquad 2x = 6 \Rightarrow x = 3$

Thus, the third vertex is (3, 4). Next, evaluate the objective function at each vertex.

Point	Cost $= 0.18x + 0.12y$
$(0,10)$	$0.18(0) + 0.12(10) = 1.20$
$(3,4)$	$0.18(3) + 0.12(4) = 1.02$ ← Minimum
$(9,2)$	$0.18(9) + 0.12(2) = 1.86$

The minimum cost of $1.02 per serving will be produced by 3 units of food A and 4 units of food B.

71. $\begin{bmatrix} 5 & x+2 \\ -6y & z \end{bmatrix} = \begin{bmatrix} a & 3x-1 \\ 5y & 9 \end{bmatrix}$

$a = 5; x+2 = 3x-1 \Rightarrow 3 = 2x \Rightarrow x = \frac{3}{2};$
$-6y = 5y \Rightarrow 0 = 11y \Rightarrow y = 0; z = 9$

Thus, $a = 5$, $x = \frac{3}{2}$, $y = 0$, and $z = 9$.

73. $\begin{bmatrix} 3 \\ 2 \\ 5 \end{bmatrix} - \begin{bmatrix} 8 \\ -4 \\ 6 \end{bmatrix} + \begin{bmatrix} 1 \\ 0 \\ 2 \end{bmatrix} = \begin{bmatrix} 3-8 \\ 2-(-4) \\ 5-6 \end{bmatrix} + \begin{bmatrix} 1 \\ 0 \\ 2 \end{bmatrix}$

$= \begin{bmatrix} -5 \\ 6 \\ -1 \end{bmatrix} + \begin{bmatrix} 1 \\ 0 \\ 2 \end{bmatrix} = \begin{bmatrix} -5+1 \\ 6+0 \\ -1+2 \end{bmatrix}$

$= \begin{bmatrix} -4 \\ 6 \\ 1 \end{bmatrix}$

75. $\begin{bmatrix} 2 & 5 & 8 \\ 1 & 9 & 2 \end{bmatrix} - \begin{bmatrix} 3 & 4 \\ 7 & 1 \end{bmatrix}$

This operation is not possible because one matrix is 2×3 and the other 2×2. Matrices of different sizes cannot be added or subtracted.

77. $\begin{bmatrix} -1 & 0 \\ 2 & 5 \end{bmatrix}\begin{bmatrix} -3 & 4 \\ 2 & 8 \end{bmatrix}$

$= \begin{bmatrix} -1(-3)+0(2) & -1(4)+0(8) \\ 2(-3)+5(2) & 2(4)+5(8) \end{bmatrix}$

$= \begin{bmatrix} 3+0 & -4+0 \\ -6+10 & 8+40 \end{bmatrix} = \begin{bmatrix} 3 & -4 \\ 4 & 48 \end{bmatrix}$

79. The product of a 2×3 matrix and a 3×2 matrix is a 2×2 matrix.

$\begin{bmatrix} 3 & 2 & -1 \\ 4 & 0 & 6 \end{bmatrix}\begin{bmatrix} -2 & 0 \\ 0 & 2 \\ 3 & 1 \end{bmatrix}$

$= \begin{bmatrix} 3(-2)+2(0)+(-1)(3) & 3(0)+2(2)+(-1)(1) \\ 4(-2)+0(0)+6(3) & 4(0)+0(2)+6(1) \end{bmatrix}$

$= \begin{bmatrix} -6+0-3 & 0+4-1 \\ -8+0+18 & 0+0+6 \end{bmatrix} = \begin{bmatrix} -9 & 3 \\ 10 & 6 \end{bmatrix}$

Copyright © 2013 Pearson Education, Inc.

81. The product of a 3×3 matrix and a 3×3 matrix is a 3×3 matrix.

$$\begin{bmatrix} -2 & 5 & 5 \\ 0 & 1 & 4 \\ 3 & -4 & -1 \end{bmatrix}\begin{bmatrix} 1 & 0 & -1 \\ -1 & 0 & 0 \\ 1 & 1 & -1 \end{bmatrix}$$

$$=\begin{bmatrix} -2(1)+5(-1)+5(1) & -2(0)+5(0)+5(1) & -2(-1)+5(0)+5(-1) \\ 0(1)+1(-1)+4(1) & 0(0)+1(0)+4(1) & 0(-1)+1(0)+4(-1) \\ 3(1)+(-4)(-1)+(-1)(1) & 3(0)+(-4)(0)+(-1)(1) & 3(-1)+(-4)(0)+(-1)(-1) \end{bmatrix}$$

$$=\begin{bmatrix} -2+(-5)+5 & 0+0+5 & 2+0+(-5) \\ 0+(-1)+4 & 0+0+4 & 0+0+(-4) \\ 3+4+(-1) & 0+0+(-1) & -3+0+1 \end{bmatrix}=\begin{bmatrix} -2 & 5 & -3 \\ 3 & 4 & -4 \\ 6 & -1 & -2 \end{bmatrix}$$

83. Find the inverse of $A=\begin{bmatrix} 2 & 1 \\ 5 & 3 \end{bmatrix}$, if it exists. We have $[A|I_2]=\begin{bmatrix} 2 & 1 & | & 1 & 0 \\ 5 & 3 & | & 0 & 1 \end{bmatrix}$, which yields

$$\begin{bmatrix} 1 & \frac{1}{2} & | & \frac{1}{2} & 0 \\ 5 & 3 & | & 0 & 1 \end{bmatrix}\frac{1}{2}R1 \Rightarrow \begin{bmatrix} 1 & \frac{1}{2} & | & \frac{1}{2} & 0 \\ 0 & \frac{1}{2} & | & -\frac{5}{2} & 1 \end{bmatrix}_{-5R1+R2} \Rightarrow \begin{bmatrix} 1 & \frac{1}{2} & | & \frac{1}{2} & 0 \\ 0 & 1 & | & -5 & 2 \end{bmatrix}2R2 \Rightarrow \begin{bmatrix} 1 & 0 & | & 3 & -1 \\ 0 & 1 & | & -5 & 2 \end{bmatrix}-\frac{1}{2}R2+R1$$

Thus, $A^{-1}=\begin{bmatrix} 3 & -1 \\ -5 & 2 \end{bmatrix}$.

85. Find the inverse of $A=\begin{bmatrix} 2 & -1 & 0 \\ 1 & 0 & 1 \\ 1 & -2 & 0 \end{bmatrix}$ if it exists. We have $\begin{bmatrix} A|I_3 \end{bmatrix}=\begin{bmatrix} 2 & -1 & 0 & | & 1 & 0 & 0 \\ 1 & 0 & 1 & | & 0 & 1 & 0 \\ 1 & -2 & 0 & | & 0 & 0 & 1 \end{bmatrix}$.

$$\begin{bmatrix} 1 & 0 & 1 & | & 0 & 1 & 0 \\ 2 & -1 & 0 & | & 1 & 0 & 0 \\ 1 & -2 & 0 & | & 0 & 0 & 1 \end{bmatrix}R1\leftrightarrow R2 \Rightarrow \begin{bmatrix} 1 & 0 & 1 & | & 0 & 1 & 0 \\ 0 & -1 & -2 & | & 1 & -2 & 0 \\ 0 & -2 & -1 & | & 0 & -1 & 1 \end{bmatrix}\begin{matrix} -2R1+R2 \\ -1R1+R3 \end{matrix} \Rightarrow \begin{bmatrix} 1 & 0 & 1 & | & 0 & 1 & 0 \\ 0 & 1 & 2 & | & -1 & 2 & 0 \\ 0 & -2 & -1 & | & 0 & -1 & 1 \end{bmatrix}-1R2 \Rightarrow$$

$$\begin{bmatrix} 1 & 0 & 1 & | & 0 & 1 & 0 \\ 0 & 1 & 2 & | & -1 & 2 & 0 \\ 0 & 0 & 3 & | & -2 & 3 & 1 \end{bmatrix}2R2+R3 \Rightarrow \begin{bmatrix} 1 & 0 & 1 & | & 0 & 1 & 0 \\ 0 & 1 & 2 & | & -1 & 2 & 0 \\ 0 & 0 & 1 & | & -\frac{2}{3} & 1 & \frac{1}{3} \end{bmatrix}\frac{1}{3}R3 \Rightarrow \begin{bmatrix} 1 & 0 & 0 & | & \frac{2}{3} & 0 & -\frac{1}{3} \\ 0 & 1 & 0 & | & \frac{1}{3} & 0 & -\frac{2}{3} \\ 0 & 0 & 1 & | & -\frac{2}{3} & 1 & \frac{1}{3} \end{bmatrix}\begin{matrix} -1R3+R1 \\ -2R3+R2 \end{matrix}$$

Thus, $A^{-1}=\begin{bmatrix} \frac{2}{3} & 0 & -\frac{1}{3} \\ \frac{1}{3} & 0 & -\frac{2}{3} \\ -\frac{2}{3} & 1 & \frac{1}{3} \end{bmatrix}$.

87. $\begin{matrix} 5x-4y=1 \\ x+4y=3 \end{matrix} \Rightarrow A=\begin{bmatrix} 5 & -4 \\ 1 & 4 \end{bmatrix}$, $X=\begin{bmatrix} x \\ y \end{bmatrix}$, and $B=\begin{bmatrix} 1 \\ 3 \end{bmatrix}$.

$$\begin{bmatrix} A|I_2 \end{bmatrix}=\begin{bmatrix} 5 & -4 & | & 1 & 0 \\ 1 & 4 & | & 0 & 1 \end{bmatrix} \Rightarrow \begin{bmatrix} 6 & 0 & | & 1 & 1 \\ 1 & 4 & | & 0 & 1 \end{bmatrix}R1+R2 \Rightarrow \begin{bmatrix} 1 & 0 & | & \frac{1}{6} & \frac{1}{6} \\ 1 & 4 & | & 0 & 1 \end{bmatrix}\frac{1}{6}R1 \Rightarrow$$

$$\begin{bmatrix} 1 & 0 & | & \frac{1}{6} & \frac{1}{6} \\ 0 & 4 & | & -\frac{1}{6} & \frac{5}{6} \end{bmatrix}-R1+R2 \Rightarrow \begin{bmatrix} 1 & 0 & | & \frac{1}{6} & \frac{1}{6} \\ 0 & 1 & | & -\frac{1}{24} & \frac{5}{24} \end{bmatrix}\frac{1}{4}R2$$

Thus, $A^{-1}=\begin{bmatrix} \frac{1}{6} & \frac{1}{6} \\ -\frac{1}{24} & \frac{5}{24} \end{bmatrix}$.

Now, solve. $X=A^{-1}B=\begin{bmatrix} \frac{1}{6} & \frac{1}{6} \\ -\frac{1}{24} & \frac{5}{24} \end{bmatrix}\begin{bmatrix} 1 \\ 3 \end{bmatrix}=\begin{bmatrix} \frac{1}{6}(1)+\frac{1}{6}(3) \\ -\frac{1}{24}(1)+\frac{5}{24}(3) \end{bmatrix}=\begin{bmatrix} \frac{4}{6} \\ \frac{14}{24} \end{bmatrix}=\begin{bmatrix} \frac{2}{3} \\ \frac{7}{12} \end{bmatrix}$

Solution set: $\left\{\left(\frac{2}{3},\frac{7}{12}\right)\right\}$

Copyright © 2013 Pearson Education, Inc.

89. Given

$$3x + 2y + z = -5$$
$$x - y + 3z = -5$$
$$2x + 3y + z = 0$$

we have

$$A = \begin{bmatrix} 3 & 2 & 1 \\ 1 & -1 & 3 \\ 2 & 3 & 1 \end{bmatrix}, \; X = \begin{bmatrix} x \\ y \\ z \end{bmatrix}, \text{ and } B = \begin{bmatrix} -5 \\ -5 \\ 0 \end{bmatrix}.$$

Finding A^{-1}, we have

$$\left[A \mid I_3 \right] = \begin{bmatrix} 3 & 2 & 1 & 1 & 0 & 0 \\ 1 & -1 & 3 & 0 & 1 & 0 \\ 2 & 3 & 1 & 0 & 0 & 1 \end{bmatrix} \Rightarrow$$

$$\begin{bmatrix} 1 & -1 & 3 & 0 & 1 & 0 \\ 3 & 2 & 1 & 1 & 0 & 0 \\ 2 & 3 & 1 & 0 & 0 & 1 \end{bmatrix} \begin{matrix} R1 \leftrightarrow R2 \\ \\ \\ \end{matrix} \Rightarrow$$

$$\begin{bmatrix} 1 & -1 & 3 & 0 & 1 & 0 \\ 0 & 5 & -8 & 1 & -3 & 0 \\ 0 & 5 & -5 & 0 & -2 & 1 \end{bmatrix} \begin{matrix} \\ -3R1 + R2 \Rightarrow \\ -2R1 + R3 \end{matrix}$$

$$\begin{bmatrix} 1 & -1 & 3 & 0 & 1 & 0 \\ 0 & 1 & -\frac{8}{5} & \frac{1}{5} & -\frac{3}{5} & 0 \\ 0 & 5 & -5 & 0 & -2 & 1 \end{bmatrix} \begin{matrix} \\ \frac{1}{5}R2 \\ \\ \end{matrix} \Rightarrow$$

$$\begin{bmatrix} 1 & 0 & \frac{7}{5} & \frac{1}{5} & \frac{2}{5} & 0 \\ 0 & 1 & -\frac{8}{5} & \frac{1}{5} & -\frac{3}{5} & 0 \\ 0 & 0 & 3 & -1 & 1 & 1 \end{bmatrix} \begin{matrix} R2 + R1 \\ \\ -5R2 + R3 \end{matrix} \Rightarrow$$

$$\begin{bmatrix} 1 & 0 & \frac{7}{5} & \frac{1}{5} & \frac{2}{5} & 0 \\ 0 & 1 & -\frac{8}{5} & \frac{1}{5} & -\frac{3}{5} & 0 \\ 0 & 0 & 1 & -\frac{1}{3} & \frac{1}{3} & \frac{1}{3} \end{bmatrix} \begin{matrix} \\ \\ \frac{1}{3}R3 \end{matrix} \Rightarrow$$

$$\begin{bmatrix} 1 & 0 & 0 & \frac{2}{3} & -\frac{1}{15} & -\frac{7}{15} \\ 0 & 1 & 0 & -\frac{1}{3} & -\frac{1}{15} & \frac{8}{15} \\ 0 & 0 & 1 & -\frac{1}{3} & \frac{1}{3} & \frac{1}{3} \end{bmatrix} \begin{matrix} -\frac{7}{5}R3 + R1 \\ \frac{8}{5}R3 + R2 \\ \\ \end{matrix}$$

Thus, $A^{-1} = \begin{bmatrix} \frac{2}{3} & -\frac{1}{15} & -\frac{7}{15} \\ -\frac{1}{3} & -\frac{1}{15} & \frac{8}{15} \\ -\frac{1}{3} & \frac{1}{3} & \frac{1}{3} \end{bmatrix}$

Finally, we need to find $X = A^{-1}B$.

$$X = A^{-1}B = \begin{bmatrix} \frac{2}{3} & -\frac{1}{15} & -\frac{7}{15} \\ -\frac{1}{3} & -\frac{1}{15} & \frac{8}{15} \\ -\frac{1}{3} & \frac{1}{3} & \frac{1}{3} \end{bmatrix} \begin{bmatrix} -5 \\ -5 \\ 0 \end{bmatrix}$$

$$= \begin{bmatrix} \frac{2}{3}(-5) + \left(-\frac{1}{15}\right)(-5) + \left(-\frac{7}{15}\right)(0) \\ -\frac{1}{3}(-5) + \left(-\frac{1}{15}\right)(-5) + \frac{8}{15}(0) \\ -\frac{1}{3}(-5) + \frac{1}{3}(-5) + \frac{1}{3}(0) \end{bmatrix}$$

$$= \begin{bmatrix} -\frac{10}{3} + \frac{1}{3} + 0 \\ \frac{5}{3} + \frac{1}{3} + 0 \\ \frac{5}{3} + \left(-\frac{5}{3}\right) + 0 \end{bmatrix} = \begin{bmatrix} -\frac{9}{3} \\ \frac{6}{3} \\ 0 \end{bmatrix} = \begin{bmatrix} -3 \\ 2 \\ 0 \end{bmatrix}$$

Solution set: $\{(-3, 2, 0)\}$

Chapter 5   Test

1. $3x - y = 9$ (1)

$x + 2y = 10$ (2)

Solving equation (2) for x, we obtain $x = 10 - 2y$ (3). Substituting this result into equation (1) and solving for y, we obtain

$3(10 - 2y) - y = 9 \Rightarrow 30 - 6y - y = 9 \Rightarrow$
$$-7y = -21 \Rightarrow y = 3$$

Substituting $y = 3$ back into equation (3) in order to find x, we obtain

$x = 10 - 2(3) = 10 - 6 = 4.$

Solution set: $\{(4, 3)\}$

2. $6x + 9y = -21$ (1)

$4x + 6y = -14$ (2)

Solving equation (1) for x, we obtain

$$6x = -9y - 21 \Rightarrow x = \frac{-9y - 21}{6} \Rightarrow$$

$$x = \frac{-3y - 7}{2} \text{ (3)}.$$

Substituting this result into equation (2) and solving for y, we obtain the following.

$$4\left(\frac{-3y - 7}{2}\right) + 6y = -14$$
$$2(-3y - 7) + 6y = -14$$
$$-6y - 14 + 6y = -14 \Rightarrow -14 = -14$$

This true statement implies there are infinitely many solutions. We can represent the solution set with y as the arbitrary variable.

Solution set: $\left\{\left(\dfrac{-3y - 7}{2}, \; y\right)\right\}$

Copyright © 2013 Pearson Education, Inc.

3. $\frac{1}{4}x - \frac{1}{3}y = -\frac{5}{12}$ (1)

$\frac{1}{10}x + \frac{1}{5}y = \frac{1}{2}$ (2)

To eliminate fractions, multiply equation (1) by 12 and equation (2) by 10.

$3x - 4y = -5$ (3)

$x + 2y = 5$ (4)

Multiply equation (4) by 2 and add the result ;to equation (3).

$3x - 4y = -5$

$2x + 4y = 10$

$\overline{5x \quad\quad = 5} \Rightarrow x = 1$

Substituting $x = 1$ in equation (4) to find y, we obtain $1 + 2y = 5 \Rightarrow 2y = 4 \Rightarrow y = 2$.

Solution set: $\{(1, 2)\}$

4. $x - 2y = 4$ (1)

$-2x + 4y = 6$ (2)

Multiply equation (1) by 2 and add the result to equation (2).

$2x - 4y = 8$

$-2x + 4y = 6$

$\overline{\quad\quad 0 = 14}$

The system is inconsistent. The solution set is $\varnothing$.

5. $2x + y + z = 3$ (1)

$x + 2y - z = 3$ (2)

$3x - y + z = 5$ (3)

Eliminate z first by adding equations (1) and (2) to obtain $3x + 3y = 6$ (4). Add equations (2) and (3) to eliminate z and obtain $4x + y = 8$ (5). Multiply equation (5) by -3 and add the results to equation (4).

$3x + 3y = 6$

$-12x - 3y = -24$

$\overline{-9x \quad\quad = -18} \Rightarrow x = 2$

Substituting $x = 2$ in equation (5) to find y, we have $4(2) + y = 8 \Rightarrow 8 + y = 8 \Rightarrow y = 0$.

Substituting $x = 2$ and $y = 0$ in equation (1) to find z, we have $2(2) + 0 + z = 3 \Rightarrow z = -1$.

Solution set: $\{(2, 0, -1)\}$

6. Writing

$3x - 2y = 13$

$4x - y = 19$

as an augmented matrix, we have

$\begin{bmatrix} 3 & -2 & | & 13 \\ 4 & -1 & | & 19 \end{bmatrix}.$

$\begin{bmatrix} 1 & -\frac{2}{3} & | & \frac{13}{3} \\ 4 & -1 & | & 19 \end{bmatrix} \frac{1}{3}R1 \quad \Rightarrow$

$\begin{bmatrix} 1 & -\frac{2}{3} & | & \frac{13}{3} \\ 0 & \frac{5}{3} & | & \frac{5}{3} \end{bmatrix} -4R1 + R2 \quad \Rightarrow$

$\begin{bmatrix} 1 & -\frac{2}{3} & | & \frac{13}{3} \\ 0 & 1 & | & 1 \end{bmatrix} \frac{3}{5}R2 \Rightarrow \begin{bmatrix} 1 & 0 & | & 5 \\ 0 & 1 & | & 1 \end{bmatrix} \frac{2}{3}R2 + R1$

Solution set $\{(5, 1)\}$

7. Writing

$3x - 4y + 2z = 15$

$2x - y + z = 13$

$x + 2y - z = 5$

as an augmented matrix, we have

$\begin{bmatrix} 3 & -4 & 2 & | & 15 \\ 2 & -1 & 1 & | & 13 \\ 1 & 2 & -1 & | & 5 \end{bmatrix}.$

$\begin{bmatrix} 1 & 2 & -1 & | & 5 \\ 2 & -1 & 1 & | & 13 \\ 3 & -4 & 2 & | & 15 \end{bmatrix} R1 \leftrightarrow R3 \quad \Rightarrow$

$\begin{bmatrix} 1 & 2 & -1 & | & 5 \\ 0 & -5 & 3 & | & 3 \\ 0 & -10 & 5 & | & 0 \end{bmatrix} \begin{matrix} \\ -2R1 + R2 \\ -3R1 + R3 \end{matrix} \Rightarrow$

$\begin{bmatrix} 1 & 2 & -1 & | & 5 \\ 0 & 1 & -\frac{3}{5} & | & -\frac{3}{5} \\ 0 & -10 & 5 & | & 0 \end{bmatrix} -\frac{1}{5}R2 \quad \Rightarrow$

$\begin{bmatrix} 1 & 0 & \frac{1}{5} & | & \frac{31}{5} \\ 0 & 1 & -\frac{3}{5} & | & -\frac{3}{5} \\ 0 & 0 & -1 & | & -6 \end{bmatrix} \begin{matrix} -2R2 + R1 \\ \\ 10R2 + R3 \end{matrix} \Rightarrow$

$\begin{bmatrix} 1 & 0 & \frac{1}{5} & | & \frac{31}{5} \\ 0 & 1 & -\frac{3}{5} & | & -\frac{3}{5} \\ 0 & 0 & 1 & | & 6 \end{bmatrix} \begin{matrix} \\ \\ -1R3 \end{matrix} \Rightarrow$

$\begin{bmatrix} 1 & 0 & 0 & | & 5 \\ 0 & 1 & 0 & | & 3 \\ 0 & 0 & 1 & | & 6 \end{bmatrix} \begin{matrix} -\frac{1}{5}R3 + R1 \\ \frac{3}{5}R3 + R2 \\ \end{matrix}$

Solution set: $\{(5, 3, 6)\}$

8. Since $y = ax^2 + bx + c$, and the points (1, 5), (2, 3), and (4, 11) are on the graph, we have the following equations which are then simplified.

$5 = a(1)^2 + b(1) + c \Rightarrow a + b + c = 5$ (1)

$3 = a(2)^2 + b(2) + c \Rightarrow 4a + 2b + c = 3$ (2)

$11 = a(4)^2 + b(4) + c \Rightarrow 16a + 4b + c = 11$ (3)

(*continued on next page*)

Copyright © 2013 Pearson Education, Inc.

(continued)

First, eliminate c by adding -1 times equation (1) to equation (2).

$$\begin{array}{r} -a-\ b-c=-5 \\ 4a+2b+c=\ \ 3 \\ \hline 3a+\ b\ \ \ \ \ \ =-2\ (4) \end{array}$$

Eliminating c again by adding -1 times equation (1) to equation (3), we have the following.

$$\begin{array}{r} -a-\ b-c=-5 \\ 16a+4b+c=11 \\ \hline 15a+3b\ \ \ \ \ =\ 6\ (5) \end{array}$$

Next, eliminate b by adding -3 times equation (4) to equation (5).

$$\begin{array}{r} -9a-3b=6 \\ 15a+3b=6 \\ \hline 6a\ \ \ \ \ \ \ \ =12 \Rightarrow a=2 \end{array}$$

Substituting this value into equation (4), we obtain $3(2)+b=-2 \Rightarrow 6+b=-2 \Rightarrow b=-8$.

Substituting 2 for a and -8 for b into equation (1), we have $2+(-8)+c=5 \Rightarrow$

$-6+c=5 \Rightarrow c=11$.

The equation of the parabola is

$y=2x^2-8x+11$.

9. Let x = number of units from Toronto;
y = number of units from Montreal;
z = number of units from Ottawa.
The information in the problem gives the system

$$\begin{array}{c} x+y+z=100 \\ 80x+50y+65z=5990 \\ x=z \end{array}$$

Multiply the first equation by -50 and add to the second equation.

$$\begin{array}{r} -50x-50y-50z=-5000 \\ 80x+50y+65z=\ \ \ 5990 \\ \hline 30x\ \ \ \ \ \ \ \ \ +15z=\ \ \ \ \ 990 \end{array}$$

Substitute x for z in this equation to obtain
$30x+15x=990 \Rightarrow 45x=990 \Rightarrow x=22$.
If $x=22$, then $z=22$.

Substitute 22 for x and for z in the first equation and solve for y:
$22+y+22=100 \Rightarrow y=56$

22 units from Toronto, 56 units from Montreal, and 22 units from Ottawa were ordered.

10. $\begin{vmatrix} 6 & 8 \\ 2 & -7 \end{vmatrix} = 6(-7)-2(8)=-58$

11. $\begin{vmatrix} 2 & 0 & 8 \\ -1 & 7 & 9 \\ 12 & 5 & -3 \end{vmatrix}$

This determinant may be evaluated by expanding about any row or any column. If we expand by the first row, we have the following.

$$\begin{vmatrix} 2 & 0 & 8 \\ -1 & 7 & 9 \\ 12 & 5 & -3 \end{vmatrix} = 2\begin{vmatrix} 7 & 9 \\ 5 & -3 \end{vmatrix} - 0\begin{vmatrix} -1 & 9 \\ 12 & -3 \end{vmatrix} + 8\begin{vmatrix} -1 & 7 \\ 12 & 5 \end{vmatrix}$$

$$= 2\big[7(-3)-5(9)\big]-0$$
$$+8\big[(-1)(5)-12(7)\big]$$
$$= 2(-21-45)+8(-5-84)$$
$$= 2(-66)+8(-89)$$
$$= -132-712=-844$$

12. $2x-3y=-33$
$4x+5y=11$

$$D=\begin{vmatrix} 2 & -3 \\ 4 & 5 \end{vmatrix}=2(5)-4(-3)=10-(-12)=22$$

$$D_x=\begin{vmatrix} -33 & -3 \\ 11 & 5 \end{vmatrix}=-33(5)-11(-3)$$
$$=-165-(-33)=-132$$

$$D_y=\begin{vmatrix} 2 & -33 \\ 4 & 11 \end{vmatrix}=2(11)-4(-33)$$
$$=22-(-132)=154$$

$$x=\frac{D_x}{D}=\frac{-132}{22}=-6;\ y=\frac{D_y}{D}=\frac{154}{22}=7$$

Solution set: $\{(-6,7)\}$

13. $\ \ \ x+y-z=-4$
$2x-3y-z=5$
$x+2y+2z=3$

Adding column 3 to columns 1 and 2, we have

$$D=\begin{vmatrix} 1 & 1 & -1 \\ 2 & -3 & -1 \\ 1 & 2 & 2 \end{vmatrix}=\begin{vmatrix} 0 & 0 & -1 \\ 1 & -4 & -1 \\ 3 & 4 & 2 \end{vmatrix}.$$

Expanding by row one, we have

$$D=-1\begin{vmatrix} 1 & -4 \\ 3 & 4 \end{vmatrix}=-\big[4-(-12)\big]=-16.$$

Adding -1 times row 1 to row 2 and 2 times row 1 to row 3, we have the following.

$$D_x=\begin{vmatrix} -4 & 1 & -1 \\ 5 & -3 & -1 \\ 3 & 2 & 2 \end{vmatrix}=\begin{vmatrix} -4 & 1 & -1 \\ 9 & -4 & 0 \\ -5 & 4 & 0 \end{vmatrix}$$

(continued on next page)

Copyright © 2013 Pearson Education, Inc.

(*continued*)

Expanding by column three, we have

$$D_x = -1 \begin{vmatrix} 9 & -4 \\ -5 & 4 \end{vmatrix} = -(36-20) = -16.$$

Adding -1 times row 1 to row 2 and 2 times row 1 to row 3, we have the following.

$$D_y = \begin{vmatrix} 1 & -4 & -1 \\ 2 & 5 & -1 \\ 1 & 3 & 2 \end{vmatrix} = \begin{vmatrix} 1 & -4 & -1 \\ 1 & 9 & 0 \\ 3 & -5 & 0 \end{vmatrix}.$$

Expanding by column three, we have

$$D_y = -1 \begin{vmatrix} 1 & 9 \\ 3 & -5 \end{vmatrix} = -1(-5-27) = -(-32) = 32.$$

Adding -2 times row 1 to row 2 and -1 times row 1 to row 3, we have

$$D_z = \begin{vmatrix} 1 & 1 & -4 \\ 2 & -3 & 5 \\ 1 & 2 & 3 \end{vmatrix} = \begin{vmatrix} 1 & 1 & -4 \\ 0 & -5 & 13 \\ 0 & 1 & 7 \end{vmatrix}.$$

Expanding by column one, we have

$$D_z = 1 \begin{vmatrix} -5 & 13 \\ 1 & 7 \end{vmatrix} = 1(-35-13) = -48.$$

Thus we have

$$x = \frac{D_x}{D} = \frac{-16}{-16} = 1, \quad y = \frac{D_y}{D} = \frac{32}{-16} = -2,$$

and $z = \frac{D_z}{D} = \frac{-48}{-16} = 3.$

Solution set: $\{(1, -2, 3)\}$

14. $\dfrac{x+2}{x^3+x^2+x} = \dfrac{x+2}{x(x^2+x+1)} = \dfrac{x+2}{x(x+1)^2}$

$$= \frac{A}{x} + \frac{B}{x+1} + \frac{C}{(x+1)^2}$$

Multiply both sides by $x(x+1)^2$ to get

$$x+2 = A(x+1)^2 + Bx(x+1) + Cx \ (1).$$

Substituting 0 for x in equation (1), we get

$$0+2 = A(0+1)^2 + B(0)(0+1) + C(0) \Rightarrow A = 2$$

Substituting 2 for A and -1 for x in equation (1), we get the following.

$$-1+2 = 2(-1+1)^2 + B(-1)(-1+1) + C(-1)$$
$$1 = -C \Rightarrow C = -1$$

We now have

$$x+2 = 2(x+1)^2 + Bx(x+1) + (-1)x \ (2).$$

Now substitute 1 for x (arbitrary choice) in equation (2) to obtain the following.

$$1+2 = 2(1+1)^2 + B(1)(1+1) + (-1)(1)$$
$$3 = 2(4) + 2B - 1$$
$$3 = 8 + 2B - 1 \Rightarrow 3 = 7 + 2B$$
$$-4 = 2B \Rightarrow B = -2$$

Thus,

$$\frac{x+2}{x^3+x^2+x} = \frac{x+2}{x(x^2+x+1)} = \frac{x+2}{x(x+1)^2}$$

$$= \frac{2}{x} + \frac{-2}{x+1} + \frac{-1}{(x+1)^2}.$$

15. The parabola passes through $(0, 4)$, $(1, 1)$, and $(2, 0)$, so those points satisfy the equation $y = ax^2 + bx + c$. Substituting each ordered pair into the equation gives the system

$$4 = a(0)^2 + b(0) + c$$
$$1 = a(1)^2 + b(1) + c$$
$$0 = a(2)^2 + b(2) + c$$

which simplifies to

$$4 = \qquad\quad c \quad (1)$$
$$1 = a + b + c \quad (2)$$
$$0 = 4a + 2b + c \quad (3)$$

Substitute $c = 4$ into equations (2) and (3), then solve the system consisting of those two equations for a and b.

$$\begin{aligned}1 &= a + b + 4 \\ 0 &= 4a + 2b + 4\end{aligned} \Rightarrow$$
$$-3 = a + b \quad (4)$$
$$-4 = 4a + 2b \quad (5)$$

From equation (4), we have $a = -b - 3$, so

$$-4 = 4(-b-3) + 2b$$
$$-4 = -4b - 12 + 2b \Rightarrow 8 = -2b \Rightarrow -4 = b$$
$$a = -b - 3 \Rightarrow a = -(-4) - 3 \Rightarrow a = 1$$

Thus, the equation of the parabola is

$$y = x^2 - 4x + 4$$

The line passes through $(-2, 0)$ and $(1, 1)$, so

$m = \frac{1-0}{1-(-2)} = \frac{1}{3}$, and the equation is

$$y - 0 = \frac{1}{3}(x+2) \Rightarrow 3y = x + 2 \Rightarrow x - 3y = -2.$$

The system is

$$y = x^2 - 4x + 4.$$
$$x - 3y = -2$$

Copyright © 2013 Pearson Education, Inc.

16. $2x^2 + y^2 = 6$ (1)

$x^2 - 4y^2 = -15$ (2)

Multiply equation (1) by 4 and add to equation (2), then solve for x.

$$8x^2 + 4y^2 = 24$$
$$\underline{x^2 - 4y^2 = -15}$$
$$9x^2 \qquad = \quad 9 \Rightarrow x^2 = 1 \Rightarrow x = \pm 1$$

Substitute these values into equation (1) and solve for y.

If $x = 1$, then $2(1)^2 + y^2 = 6 \Rightarrow 2 + y^2 = 6 \Rightarrow$

$y^2 = 4 \Rightarrow y = \pm 2$.

If $x = -1$, then $2(-1)^2 + y^2 = 6 \Rightarrow$

$2 + y^2 = 6 \Rightarrow y^2 = 4 \Rightarrow y = \pm 2$.

Solution set: $\{(1,2), (-1,2), (1,-2), (-1,-2)\}$

17. $x^2 + y^2 = 25$ (1)

$x + y = 7$ (2)

Solving equation (2) for x, we have $x = 7 - y$ (3). Substituting this result into equation (1), we have

$$(7 - y)^2 + y^2 = 25$$
$$49 - 14y + y^2 + y^2 = 25$$
$$2y^2 - 14y + 24 = 0$$
$$y^2 - 7y + 12 = 0$$
$$(y - 3)(y - 4) = 0 \Rightarrow y = 3 \text{ or } y = 4$$

Substitute these values into equation (3) and solve for x. If $y = 3$, $x = 7 - 3 = 4$.

If $y = 4$, $x = 7 - 4 = 3$.

Solution set: $\{(3, 4), (4, 3)\}$

18. Let x and y be the numbers.

$x + y = -1$ (1)

$x^2 + y^2 = 61$ (2)

Solving equation (1) for y, we have $y = -x - 1$ (3). Substituting this result into equation (2), we have

$$x^2 + (-x - 1)^2 = 61 \Rightarrow x^2 + x^2 + 2x + 1 = 61$$
$$2x^2 + 2x - 60 = 0 \Rightarrow x^2 + x - 30 = 0$$
$$(x + 6)(x - 5) = 0 \Rightarrow x = -6 \text{ or } x = 5$$

Substitute these values in equation (3) to find the corresponding values of y.

If $x = -6$, $y = -(-6) - 1 = 6 - 1 = 5$.

If $x = 5$, $y = -5 - 1 = -6$.

The same pair of numbers results from both cases. The numbers are 5 and –6.

19. $x - 3y \geq 6$

$y^2 \leq 16 - x^2$

Graph $x - 3y = 6$ as a solid line with x-intercept 6 and y-intercept of –2. Because the test point $(0,0)$ does not satisfies this inequality, shade the region *below* the line. Graph $y^2 = 16 - x^2$ or $x^2 + y^2 = 16$ as a solid circle with a center at the origin and radius 4. Shade the region, which is the interior of the circle. The solution set is the intersection of these two regions, which is the region shaded in the final graph.

20. Find $x \geq 0$ and $y \geq 0$ such that

$x + 2y \leq 24$

$3x + 4y \leq 60$

and $2x + 3y$ is maximized.

Find the region of feasible solutions by graphing the system of inequalities that is made up of the constraints. To graph $x + 2y \leq 24$, draw the line with x-intercept 24 and y-intercept 12 as a solid line. Because the test point (0, 0) satisfies this inequality, shade the region *below* the line. To graph $3x + 4y \leq 60$, draw the line with x-intercept 20 and y-intercept 15 as a solid line. Because the test point (0, 0) satisfies this inequality, shade the region *below* the line. The constraints $x \geq 0$ and $y \geq 0$ restrict the graph to the first quadrant. The graph of the feasible region is the intersection of the regions that are the graphs of the individual constraints.

(continued on next page)

Copyright © 2013 Pearson Education, Inc.

(continued)

From the graph, observe that three vertices are $(0, 0)$, $(0, 12)$, and $(20, 0)$. A fourth vertex is the intersection point of the lines $x + y = 24$ and $3x + 4y = 60$. To find this point, solve the system

$$x + 2y = 24 \quad (1)$$
$$3x + 4y = 60 \quad (2)$$

To eliminate x, multiply equation (1) by -3 and add the result to equation (2).

$$-3x - 6y = -72$$
$$\underline{3x + 4y = 60}$$
$$-2y = -12 \Rightarrow y = 6$$

Substituting this value into equation (1), we obtain $x + 2(6) = 24 \Rightarrow x + 12 = 24 \Rightarrow x = 12$.

The fourth vertex is $(12, 6)$. Next, evaluate the objective function at each vertex.

Point	Profit $= 2x + 3y$.
$(0, 0)$	$2(0) + 3(0) = 0$
$(0, 12)$	$2(0) + 3(12) = 36$
$(20, 0)$	$2(20) + 3(0) = 40$
$(12, 6)$	$2(12) + 3(6) = 42$ ← Maximum

The maximum value is 42 at $(12, 6)$.

21. Let $x =$ the number of VIP rings; $y =$ the number of SST rings.
The constraints are $x + y \le 24$
$$3x + 2y \le 60$$
$$x \ge 0, \ y \ge 0.$$

Maximize profit function, profit $= 30x + 40y$. Find the region of feasible solutions by graphing the system of inequalities that is made up of the constraints. To graph $x + y \le 24$, draw the line with x-intercept 24 and y-intercept 24 as a solid line. Because the test point $(0, 0)$ satisfies this inequality, shade the region *below* the line. To graph $3x + 2y \le 60$, draw the line with x-intercept 20 and y-intercept 30 as a solid line. Because the test point $(0, 0)$ satisfies this inequality, shade the region *below* the line. The constraints $x \ge 0$ and $y \ge 0$ restrict the graph to the first quadrant. The graph of the feasible region is the intersection of the regions that are the graphs of the individual constraints.

From the graph, observe that three vertices are $(0, 0)$, $(20, 0)$, and $(0, 24)$. A fourth vertex is the intersection point of the lines $x + y = 24$ and $3x + 2y = 60$. To find this point, solve the system

$$x + y = 24$$
$$3x + 2y = 60.$$

The first equation can be written as $x = 24 - y$. Substituting this equation into $3x + 2y = 60$, we have the following.

$$3(24 - y) + y = 60 \Rightarrow 72 - 3y + 2y = 60 \Rightarrow$$
$$72 - y = 60 \Rightarrow -y = -12 \Rightarrow y = 12$$

Substituting $y = 12$ into $x = 24 - y$, we have $x = 24 - 12 = 12$. Thus, the four vertex is $(12, 12)$. Next, evaluate the objective function at each vertex.

Point	Profit $= 30x + 40y$.
$(0, 0)$	$30(0) + 40(0) = 0$
$(20, 0)$	$30(20) + 40(0) = 600$
$(0, 24)$	$30(0) + 40(24) = 960$ ← Maximum
$(12, 12)$	$30(12) + 40(12) = 840$

0 VIP rings and 24 SST rings should be made daily for a daily profit of \$960.

22. $\begin{bmatrix} 5 & x+6 \\ 0 & 4 \end{bmatrix} = \begin{bmatrix} y-2 & 4-x \\ 0 & w+7 \end{bmatrix}$

All corresponding elements, position by position, of the two matrices must be equal.
$x + 6 = 4 - x \Rightarrow 2x = -2 \Rightarrow x = -1$;
$5 = y - 2 \Rightarrow y = 7; 4 = w + 7 \Rightarrow w = -3$
Thus, $x = -1$, $y = 7$, and $w = -3$.

Copyright © 2013 Pearson Education, Inc.

23. $3\begin{bmatrix} 2 & 3 \\ 1 & -4 \\ 5 & 9 \end{bmatrix} - \begin{bmatrix} -2 & 6 \\ 3 & -1 \\ 0 & 8 \end{bmatrix}$

$= \begin{bmatrix} 3(2) & 3(3) \\ 3(1) & 3(-4) \\ 3(5) & 3(9) \end{bmatrix} - \begin{bmatrix} -2 & 6 \\ 3 & -1 \\ 0 & 8 \end{bmatrix}$

$= \begin{bmatrix} 6 & 9 \\ 3 & -12 \\ 15 & 27 \end{bmatrix} + \begin{bmatrix} 2 & -6 \\ -3 & 1 \\ 0 & -8 \end{bmatrix}$

$= \begin{bmatrix} 6+2 & 9+(-6) \\ 3+(-3) & -12+1 \\ 15+0 & 27+(-8) \end{bmatrix} = \begin{bmatrix} 8 & 3 \\ 0 & -11 \\ 15 & 19 \end{bmatrix}$

24. $\begin{bmatrix} 1 \\ 2 \end{bmatrix} + \begin{bmatrix} 4 \\ -6 \end{bmatrix} + \begin{bmatrix} 2 & 8 \\ -7 & 5 \end{bmatrix}$

The first two matrices are 2×1 and the third is 2×2. Only matrices of the same size can be added, so it is not possible to find this sum.

25. The product of a 2×3 matrix and a 3×2 matrix is a 2×2 matrix.

$\begin{bmatrix} 2 & 1 & -3 \\ 4 & 0 & 5 \end{bmatrix} \begin{bmatrix} 1 & 3 \\ 2 & 4 \\ 3 & -2 \end{bmatrix}$

$= \begin{bmatrix} 2(1)+1(2)+(-3)(3) & 2(3)+1(4) \\ & +(-3)(-2) \\ 4(1)+0(2)+5(3) & 4(3)+0(4) \\ & +5(-2) \end{bmatrix}$

$= \begin{bmatrix} 2+2+(-9) & 6+4+6 \\ 4+0+15 & 12+0+(-10) \end{bmatrix} = \begin{bmatrix} -5 & 16 \\ 19 & 2 \end{bmatrix}$

26. $\begin{bmatrix} 2 & -4 \\ 3 & 5 \end{bmatrix} \begin{bmatrix} 4 \\ 2 \\ 7 \end{bmatrix}$

The first matrix is 2×2 and the second is 3×1. The first matrix has two columns and the second has three rows, so it is not possible to find this product.

27. There are associative, distributive, and identity properties that apply to multiplication of matrices, but matrix multiplication is not commutative. The correct choice is A.

28. Find the inverse of $A = \begin{bmatrix} -8 & 5 \\ 3 & -2 \end{bmatrix}$, if it exists.

The augmented matrix is

$[A|I_2] = \begin{bmatrix} -8 & 5 & | & 1 & 0 \\ 3 & -2 & | & 0 & 1 \end{bmatrix}$.

$\begin{bmatrix} 1 & -\frac{5}{8} & | & -\frac{1}{8} & 0 \\ 3 & -2 & | & 0 & 1 \end{bmatrix} -\frac{1}{8}R1 \Rightarrow$

$\begin{bmatrix} 1 & -\frac{5}{8} & | & -\frac{1}{8} & 0 \\ 0 & -\frac{1}{8} & | & \frac{3}{8} & 1 \end{bmatrix} -3R1+R2 \Rightarrow$

$\begin{bmatrix} 1 & -\frac{5}{8} & | & -\frac{1}{8} & 0 \\ 0 & 1 & | & -3 & -8 \end{bmatrix} -8R2 \Rightarrow$

$\begin{bmatrix} 1 & 0 & | & -2 & -5 \\ 0 & 1 & | & -3 & -8 \end{bmatrix} \frac{5}{8}R2+R1$

Thus,

$A^{-1} = \begin{bmatrix} -2 & -5 \\ -3 & -8 \end{bmatrix}$.

29. Find the inverse of $A = \begin{bmatrix} 4 & 12 \\ 2 & 6 \end{bmatrix}$, if it exists.

The augmented matrix is

$[A|I_2] = \begin{bmatrix} 4 & 12 & | & 1 & 0 \\ 2 & 6 & | & 0 & 1 \end{bmatrix}$.

$\begin{bmatrix} 1 & 3 & | & \frac{1}{4} & 0 \\ 2 & 6 & | & 0 & 1 \end{bmatrix} \frac{1}{4}R1 \Rightarrow$

$\begin{bmatrix} 1 & 3 & | & \frac{1}{4} & 0 \\ 0 & 0 & | & -\frac{1}{2} & 1 \end{bmatrix} -2R1+R2$

The second row, second column element is now 0, so the desired transformation cannot be completed. Therefore, the inverse of the given matrix does not exist.

30. Find the inverse of $A = \begin{bmatrix} 1 & 3 & 4 \\ 2 & 7 & 8 \\ -2 & -5 & -7 \end{bmatrix}$, if it exists. Performing row operations on the augmented matrix, we have the following.

$[A|I_3] = \begin{bmatrix} 1 & 3 & 4 & | & 1 & 0 & 0 \\ 2 & 7 & 8 & | & 0 & 1 & 0 \\ -2 & -5 & -7 & | & 0 & 0 & 1 \end{bmatrix} \Rightarrow$

$\begin{bmatrix} 1 & 3 & 4 & | & 1 & 0 & 0 \\ 0 & 1 & 0 & | & -2 & 1 & 0 \\ 0 & 1 & 1 & | & 2 & 0 & 1 \end{bmatrix} \begin{matrix} \\ -2R1+R2 \Rightarrow \\ 2R1+R3 \end{matrix}$

(continued on next page)

Copyright © 2013 Pearson Education, Inc.

(*continued*)

$$\begin{bmatrix} 1 & 0 & 4 & | & 7 & -3 & 0 \\ 0 & 1 & 0 & | & -2 & 1 & 0 \\ 0 & 0 & 1 & | & 4 & -1 & 1 \end{bmatrix} \begin{matrix} -3R2+R1 \\ \\ -1R2+R3 \end{matrix} \Rightarrow$$

$$\begin{bmatrix} 1 & 0 & 0 & | & -9 & 1 & -4 \\ 0 & 1 & 0 & | & -2 & 1 & 0 \\ 0 & 0 & 1 & | & 4 & -1 & 1 \end{bmatrix} \begin{matrix} -4R3+R1 \\ \\ \end{matrix}$$

Thus,

$$A^{-1} = \begin{bmatrix} -9 & 1 & -4 \\ -2 & 1 & 0 \\ 4 & -1 & 1 \end{bmatrix}.$$

31. The system $\begin{matrix} 2x+y=-6 \\ 3x-y=-29 \end{matrix}$ yields the matrix

equation $AX = B$ where

$$A = \begin{bmatrix} 2 & 1 \\ 3 & -1 \end{bmatrix}, \ X = \begin{bmatrix} x \\ y \end{bmatrix}, \ \text{and } B = \begin{bmatrix} -6 \\ -29 \end{bmatrix}. \ \text{Find}$$

A^{-1}. The augmented matrix is

$$\begin{bmatrix} A | I_2 \end{bmatrix} = \begin{bmatrix} 2 & 1 & | & 1 & 0 \\ 3 & -1 & | & 0 & 1 \end{bmatrix}.$$

$$\begin{bmatrix} 1 & \frac{1}{2} & | & \frac{1}{2} & 0 \\ 3 & -1 & | & 0 & 1 \end{bmatrix} \begin{matrix} \frac{1}{2}R1 \\ \\ \end{matrix} \Rightarrow$$

$$\begin{bmatrix} 1 & \frac{1}{2} & | & \frac{1}{2} & 0 \\ 0 & -\frac{5}{2} & | & -\frac{3}{2} & 1 \end{bmatrix} \begin{matrix} \\ -3R1+R2 \end{matrix} \Rightarrow$$

$$\begin{bmatrix} 1 & 0 & | & \frac{1}{5} & \frac{1}{5} \\ 0 & -\frac{5}{2} & | & -\frac{3}{2} & 1 \end{bmatrix} \begin{matrix} \frac{1}{5}R2+R1 \\ \\ \end{matrix} \Rightarrow$$

$$\begin{bmatrix} 1 & 0 & | & \frac{1}{5} & \frac{1}{5} \\ 0 & 1 & | & \frac{3}{5} & -\frac{2}{5} \end{bmatrix} \begin{matrix} \\ -\frac{2}{5}R2 \end{matrix}$$

Thus,

$$A^{-1} = \begin{bmatrix} \frac{1}{5} & \frac{1}{5} \\ \frac{3}{5} & -\frac{2}{5} \end{bmatrix}.$$

Since $X = A^{-1}B$, we have the following.

$$A^{-1}B = \begin{bmatrix} \frac{1}{5} & \frac{1}{5} \\ \frac{3}{5} & -\frac{2}{5} \end{bmatrix} \begin{bmatrix} -6 \\ -29 \end{bmatrix}$$

$$= \begin{bmatrix} \frac{1}{5}(-6)+\frac{1}{5}(-29) \\ \frac{3}{5}(-6)+\left(-\frac{2}{5}\right)(-29) \end{bmatrix}$$

$$= \begin{bmatrix} -\frac{6}{5}+\left(-\frac{29}{5}\right) \\ -\frac{18}{5}+\frac{58}{5} \end{bmatrix} = \begin{bmatrix} -\frac{35}{5} \\ \frac{40}{5} \end{bmatrix} = \begin{bmatrix} -7 \\ 8 \end{bmatrix}$$

Solution set: $\{(-7, 8)\}$

32. The system $\begin{matrix} x+y=5 \\ y-2z=23 \\ x+3z=-27 \end{matrix}$ yields the matrix

equation $AX = B$ where $A = \begin{bmatrix} 1 & 1 & 0 \\ 0 & 1 & -2 \\ 1 & 0 & 3 \end{bmatrix}$,

$X = \begin{bmatrix} x \\ y \\ z \end{bmatrix}$, and $B = \begin{bmatrix} 5 \\ 23 \\ -27 \end{bmatrix}$. Find A^{-1}. The

augmented matrix is

$$\begin{bmatrix} A | I_3 \end{bmatrix} = \begin{bmatrix} 1 & 1 & 0 & | & 1 & 0 & 0 \\ 0 & 1 & -2 & | & 0 & 1 & 0 \\ 1 & 0 & 3 & | & 0 & 0 & 1 \end{bmatrix}.$$

$$\begin{bmatrix} 1 & 1 & 0 & | & 1 & 0 & 0 \\ 0 & 1 & -2 & | & 0 & 1 & 0 \\ 0 & -1 & 3 & | & -1 & 0 & 1 \end{bmatrix} \begin{matrix} \\ \\ -R1+R3 \end{matrix} \Rightarrow$$

$$\begin{bmatrix} 1 & 0 & 2 & | & 1 & -1 & 0 \\ 0 & 1 & -2 & | & 0 & 1 & 0 \\ 0 & 0 & 1 & | & -1 & 1 & 1 \end{bmatrix} \begin{matrix} -1R2+R1 \\ \\ R2+R3 \end{matrix} \Rightarrow$$

$$\begin{bmatrix} 1 & 0 & 0 & | & 3 & -3 & -2 \\ 0 & 1 & 0 & | & -2 & 3 & 2 \\ 0 & 0 & 1 & | & -1 & 1 & 1 \end{bmatrix} \begin{matrix} -2R3+R1 \\ 2R3+R2 \\ \end{matrix}$$

Thus, $A^{-1} = \begin{bmatrix} 3 & -3 & -2 \\ -2 & 3 & 2 \\ -1 & 1 & 1 \end{bmatrix}$. Since

$X = A^{-1}B$, we have the following.

$$A^{-1}B = \begin{bmatrix} 3 & -3 & -2 \\ -2 & 3 & 2 \\ -1 & 1 & 1 \end{bmatrix} \begin{bmatrix} 5 \\ 23 \\ -27 \end{bmatrix}$$

$$= \begin{bmatrix} 3(5)+(-3)(23)+(-2)(-27) \\ -2(5)+3(23)+2(-27) \\ -1(5)+1(23)+1(-27) \end{bmatrix}$$

$$= \begin{bmatrix} 15+(-69)+54 \\ -10+(69)+(-54) \\ -5+23+(-27) \end{bmatrix} = \begin{bmatrix} 0 \\ 5 \\ -9 \end{bmatrix} = X$$

Solution set: $\{(0, 5, -9)\}$

Copyright © 2013 Pearson Education, Inc.

Chapter 6

ANALYTIC GEOMETRY

Section 6.1   Parabolas

1. (a) The relation $y - 2 = (x + 4)^2$ has vertex $(-4, 2)$ and opens up, so the correct choice is D.

(b) The relation $y - 4 = (x + 2)^2$ has vertex $(-2, 4)$ and opens up, so the correct choice is B.

(c) The relation $y - 2 = -(x + 4)^2$ has vertex $(-4, 2)$ and opens down, so the correct choice is C.

(d) The relation $y - 4 = -(x + 2)^2$ has vertex $(-2, 4)$ and opens down, so the correct choice is A.

(e) The relation $x - 2 = (y + 4)^2$ has vertex $(2, -4)$ and opens to the right, so the correct choice is F.

(f) The relation $x - 4 = (y + 2)^2$ has vertex $(4, -2)$ and opens to the right, so the correct choice is H.

(g) The relation $x - 2 = -(y + 4)^2$ has vertex $(2, -4)$ and opens to the left, so the correct choice is E.

(h) The relation $x - 4 = -(y + 2)^2$ has vertex $(4, -2)$ and opens to the left, so the correct choice is G.

3. $x + 4 = y^2 \Rightarrow x = y^2 - 4 \Rightarrow x - 0 = y^2 - 4$
The vertex is $(0, -4)$. The graph opens to the right and has the same shape as $x = y^2$. It is a translation 4 units to the left of the graph of $x = y^2$. The domain is $[-4, \infty)$. The range is $(-\infty, \infty)$. The graph is symmetric about its axis, the horizontal line $y = 0$ (the x-axis). Use the vertex and axis and plot a few additional points.

x	y
-4	0
-3	± 1
0	± 2

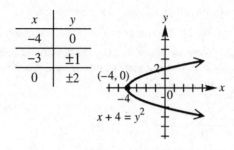

$x + 4 = y^2$

5. $x = (y - 3)^2 \Rightarrow x - 0 = (y - 3)^2$
The vertex is $(0, 3)$. The graph opens to the right and has the same shape as $x = y^2$. It is a translation 3 units up of the graph of $x = y^2$. The domain is $[0, \infty)$. The range is $(-\infty, \infty)$. The graph is symmetric about its axis, the horizontal line $y = 3$. Use the vertex and axis and plot a few additional points.

x	y
4	1
1	2
0	3
1	4
4	5

$x = (y - 3)^2$

7. $x - 2 = (y - 4)^2$
The vertex is $(2, 4)$. The graph opens to the right and has the same shape as $x = y^2$. It is a translation 2 units to the right and 4 units up of the graph of $x = y^2$. The domain is $[2, \infty)$. The range is $(-\infty, \infty)$. The graph is symmetric about its axis, the horizontal line $y = 4$. Use the vertex and axis and plot a few additional points.

x	y
6	6
3	5
2	4
3	3
6	2

$x - 2 = (y - 4)^2$

Copyright © 2013 Pearson Education, Inc.

9. $x - 2 = -3(y-1)^2$

The vertex is $(2,1)$. The graph opens to the left and has the same shape as $x = -3y^2$. It is a translation 1 unit up and 2 units to the right of the graph of $x = -3y^2$. The domain is $(-\infty, 2]$. The range is $(-\infty, \infty)$. The graph is symmetric about its axis, the horizontal line $y = 1$. Use the vertex and axis and plot a few additional points.

x	y
-10	3
-1	2
2	1
-1	0
-10	-1

$x - 2 = -3(y-1)^2$

11. $-\frac{1}{2}x = (y+3)^2 \Rightarrow x = -2(y+3)^2 \Rightarrow$
$x - 0 = -2[y - (-3)]^2$

The vertex is $(0,-3)$. The graph opens to the left and has the same shape as $x = -2y^2$. It is a translation 3 units down of the graph of $x = -2y^2$. The domain is $(-\infty, 0]$. The range is $(-\infty, \infty)$. The graph is symmetric about its axis, the horizontal line $y = -3$. Use the vertex and axis and plot a few additional points.

x	y
-8	-1
-2	-2
0	-3
-2	-4
-8	-5

$-\frac{1}{2}x = (y+3)^2$
$(0, -3)$

13. $x = y^2 + 4y + 2$

Complete the square on y to find the vertex and the axis.

$x = y^2 + 4y + 2 \Rightarrow x = \left(y^2 + 4y \quad \right) + 2$
$x = \left(y^2 + 4y + 4 - 4\right) + 2$
$x = \left(y^2 + 4y + 4\right) - 4 + 2$
$x = (y+2)^2 - 2 \Rightarrow x - (-2) = [y - (-2)]^2$

The vertex is $(-2,-2)$. The graph opens to the right and has the same shape as $x = y^2$, translated 2 units to the left and 2 units down. The domain is $[-2, \infty)$. The range is $(-\infty, \infty)$. The graph is symmetric about its axis, the horizontal line $y = -2$. Use the vertex and axis and plot a few additional points.

x	y
2	0
-1	-1
-2	-2
-1	-3
2	-4

$(-2, -2)$
$x = y^2 + 4y + 2$

15. $x = -4y^2 - 4y + 3$

Complete the square on y to find the vertex and the axis.

$x = -4y^2 - 4y + 3 \Rightarrow x = -4\left(y^2 + y \quad \right) + 3$
$x = -4\left(y^2 + y + \frac{1}{4} - \frac{1}{4}\right) + 3$
$x = -4\left(y^2 + y + \frac{1}{4}\right) - 4\left(-\frac{1}{4}\right) + 3$
$x = -4\left(y^2 + y + \frac{1}{4}\right) + 1 + 3$
$x = -4\left(y + \frac{1}{2}\right)^2 + 4 \Rightarrow x - 4 = -4\left[y - \left(-\frac{1}{2}\right)\right]^2$

The vertex is $\left(4, -\frac{1}{2}\right)$. The graph opens to the left and has the same shape as $x = -4y^2$, translated 4 units to the right and $\frac{1}{2}$ unit down. The domain is $(-\infty, 4]$. The range is $(-\infty, \infty)$. The graph is symmetric about its axis, the horizontal line $y = -\frac{1}{2}$. Use the vertex and axis and plot a few additional points.

x	y
-12	$1\frac{1}{2}$
0	$\frac{1}{2}$
4	$-\frac{1}{2}$
0	$-1\frac{1}{2}$
-12	$-2\frac{1}{2}$

$\left(4, -\frac{1}{2}\right)$
$x = -4y^2 - 4y + 3$

Copyright © 2013 Pearson Education, Inc.

17. $2x - y^2 + 4y - 6 = 0$

Solve for x, then complete the square on y to find the vertex and the axis.

$2x - y^2 + 4y - 6 = 0 \Rightarrow 2x = y^2 - 4y + 6 \Rightarrow$

$x = \tfrac{1}{2}y^2 - 2y + 3$

$x = \tfrac{1}{2}\left(y^2 - 4y \quad\right) + 3$

$x = \tfrac{1}{2}\left(y^2 - 4y + 4 - 4\right) + 3$

$x = \tfrac{1}{2}\left(y^2 - 4y + 4\right) + \tfrac{1}{2}(-4) + 3$

$x = \tfrac{1}{2}\left(y^2 - 4y + 4\right) - 2 + 3$

$x = \tfrac{1}{2}(y - 2)^2 + 1 \Rightarrow x - 1 = \tfrac{1}{2}(y - 2)^2$

The vertex is (1, 2). The graph opens to the right and has the same shape as $x = \tfrac{1}{2}y^2$, translated 1 unit to the right and 2 units up. The domain is $[1, \infty)$. The range is $(-\infty, \infty)$. The graph is symmetric about its axis, the horizontal line $y = 2$. Use the vertex and axis and plot a few additional points.

x	y
3	4
1.5	3
1	2
1.5	1
3	0

(1, 2)

$2x - y^2 + 4y - 6 = 0$

19. $-x = 3y^2 + 6y + 2 \Rightarrow x = -3y^2 - 6y - 2$

Complete the square on y to find the vertex and the axis.

$x = -3y^2 - 6y - 2$

$x = -3\left(y^2 + 2y \quad\right) - 2$

$x = -3\left(y^2 + 2y + 1 - 1\right) - 2$

$x = -3\left(y^2 + 2y + 1\right) - 3(-1) - 2$

$x = -3\left(y^2 + 2y + 1\right) + 3 - 2$

$x = -3(y + 1)^2 + 1$

The vertex is (1, –1) The graph opens to the left and has the same shape as $x = -3y^2$, translated 1 unit to the right and 1 unit down. The domain is $(-\infty, 1]$. The range is $(-\infty, \infty)$.

The graph is symmetric about its axis, the horizontal line $y = -1$. Use the vertex and axis and plot a few additional points.

x	y
-11	-3
-2	-2
1	-1
-2	0
-11	1

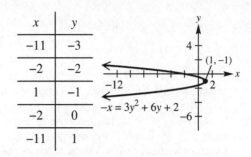

$-x = 3y^2 + 6y + 2$

21. The equation $x^2 = 24y$ has the form $x^2 = 4py$, so $4p = 24$, from which $p = 6$. Since the x-term is squared, the parabola is vertical, with focus $(0, p) = (0, 6)$ and directrix, $y = -p$, is $y = -6$. The vertex is $(0, 0)$, and the axis of the parabola is the y-axis.

23. The equation $y = -4x^2 \Rightarrow -\tfrac{1}{4}y = x^2$ has the form $x^2 = 4py$, so $4p = -\tfrac{1}{4}$, from which $p = -\tfrac{1}{16}$. Since the x-term is squared, the parabola is vertical, with focus $(0, p) = \left(0, -\tfrac{1}{16}\right)$ and directrix, $y = -p$, is $y = \tfrac{1}{16}$. The vertex is $(0, 0)$, and the axis of the parabola is the y-axis.

25. The equation $y^2 = -4x$ has the form $y^2 = 4px$, so $4p = -4$, from which $p = -1$. Since the y-term is squared, the parabola is horizontal, with focus $(p, 0) = (-1, 0)$ and directrix, $x = -p$, is $x = 1$. The vertex is (0, 0), and the axis of the parabola is the x-axis.

27. The equation $x = -32y^2 \Rightarrow -\tfrac{1}{32}x = y^2$ has the form $y^2 = 4px$, so $4p = -\tfrac{1}{32}$, from which $p = -\tfrac{1}{128}$. Since the y-term is squared, the parabola is horizontal, with focus $(p, 0) = \left(-\tfrac{1}{128}, 0\right)$ and directrix, $x = -p$, is $x = \tfrac{1}{128}$. The vertex is (0, 0), and the axis of the parabola is the x-axis.

Copyright © 2013 Pearson Education, Inc.

29. $(y-3)^2 = 12(x-1)$ has the form

$(y-k)^2 = 4p(x-h)$, with

$h=1, k=3,$ and $4p=12,$ so $p=3.$ The graph of the given equation is a parabola with vertical axis. The vertex (h, k) is (1, 3). Because this parabola has a horizontal axis and $p > 0$, the parabola opens right, so the focus is the distance $p = 3$ units to the right of the vertex. Thus the focus is (4, 3) The directrix is the vertical line $p = 3$ units to the left of the vertex, so the directrix is the line $x = -2$. The axis is the horizontal line through the vertex, so the equation of the axis is $y = 3$.

31. $(x-7)^2 = 16(y+5)$ can be written as

$(x-7)^2 = 16[y-(-5)]$. Thus, the parabola

has the form $(x-h)^2 = 4p(y-k)$, with $h = 7$, $k = -5$, and $4p = 16$, so $p = 4$. The graph of the given equation is a parabola with vertical axis. The vertex (h, k) is (7, -5). Because this parabola has a vertical axis and $p > 0$, the parabola opens up, so the focus is the distance $p = 4$ units above the vertex. Thus, the focus is (7, -1). The directrix is the horizontal line $p = 4$ units below the vertex, so the directrix is the line $y = -9$. The axis is the vertical line through the vertex, so the equation of the axis is $x = 7$.

33. A parabola with focus $(5,0)$ and vertex at the origin is a horizontal parabola. The equation has the form $y^2 = 4px$. Since $p = 5$ is positive, it opens to the right. Substituting 5 for p, we find that an equation for this parabola is $y^2 = 4(5)x \Rightarrow y^2 = 20x.$

35. A parabola with directrix $y = -\frac{1}{4}$ and vertex at the origin is a vertical parabola with focus $\left(0, \frac{1}{4}\right)$. The equation has the form $x^2 = 4py$. Since $p = \frac{1}{4}$ is positive, it opens up. Substituting $\frac{1}{4}$ for p, we find that an equation for this parabola is $x^2 = 4\left(\frac{1}{4}\right)y \Rightarrow x^2 = y.$

37. A parabola passing through $\left(\sqrt{3},3\right)$, opening up, and vertex at the origin has an equation of the form $x^2 = 4py$. Use this equation with the coordinates of the point $\left(\sqrt{3},3\right)$ to find the value of p. $x^2 = 4py \Rightarrow \left(\sqrt{3}\right)^2 = 4p\cdot 3 \Rightarrow$
$3 = 12p \Rightarrow \frac{1}{4} = p$
Thus, an equation of the parabola is
$x^2 = 4\left(\frac{1}{4}\right)y \Rightarrow x^2 = y.$

39. A parabola through (3, 2), symmetric with respect to the x-axis, and vertex at the origin has a horizontal axis (x-axis) and the equation is of the form $y^2 = 4px$. Use this equation with the coordinates of the point (3, 2) to find the value of p.
$y^2 = 4px \Rightarrow 2^2 = 4p\cdot 3 \Rightarrow 4 = 12p \Rightarrow p = \frac{1}{3}$
Thus, an equation for the parabola is
$y^2 = 4\left(\frac{1}{3}\right)x \Rightarrow y^2 = \frac{4}{3}x.$

41. The vertex is (4, 3) and the focus is (4, 5). Since the focus is above the vertex, the axis is vertical and the parabola opens upward. The distance between the vertex and the focus is $5 - 3 = 2$. Since the parabola opens upward, choose $p = 2$. The equation will have the form $(x-h)^2 = 4p(y-k)$. Substitute $p = 2$, $h = 4$, and $k = 3$ to find the required equation.
$(x-h)^2 = 4p(y-k)$
$(x-4)^2 = 4(2)(y-3) \Rightarrow (x-4)^2 = 8(y-3)$

43. The vertex is (-5, 6) and the directrix is $x = -12$. Since the directrix is 7 units to the left of the vertex, the focus is 7 units to the right of the vertex, at (2, 6) and $p = 7$. The focus is to the right of the vertex, so the axis is horizontal and the parabola opens to the right. The equation will have the form
$(y-k)^2 = 4p(x-h)$. Substitute $p = 7$, $h = -5$, and $k = 6$.
$(y-k)^2 = 4p(x-h)$
$(y-6)^2 = 4(7)[x-(-5)]$
$(y-6)^2 = 28(x+5)$

Copyright © 2013 Pearson Education, Inc.

45. Complete the square on y.

$$x = 3y^2 + 6y - 4 \Rightarrow 3\left(y^2 + 2y \quad \right) - 4 = x \Rightarrow$$

$$3\left(y^2 + 2y + 1 - 1\right) - 4 = x \Rightarrow$$

$$3\left(y^2 + 2y + 1\right) + 3\left(-1\right) - 4 = x \Rightarrow$$

$$3\left(y+1\right)^2 - 7 = x \Rightarrow 3\left(y+1\right)^2 = x + 7 \Rightarrow$$

$$\left(y+1\right)^2 = \frac{x+7}{3} \Rightarrow y + 1 = \pm\sqrt{\frac{x+7}{3}} \Rightarrow$$

$$y = -1 \pm \sqrt{\frac{x+7}{3}}$$

47. Solving for y we have

$$-\left(y+1\right)^2 = x + 2 \Rightarrow \left(y+1\right)^2 = -x - 2 \Rightarrow$$

$$y + 1 = \pm\sqrt{-x-2} \Rightarrow y = -1 \pm \sqrt{-x-2}$$

49. The equation is of the form $x = ay^2 + by + c$.

Substituting $x = -5$, $y = 1$, we get

$$-5 = a\left(1\right)^2 + b\left(1\right) + c \Rightarrow -5 = a + b + c \quad (1)$$

Substituting $x = -14$, $y = -2$, we get

$$-14 = a\left(-2\right)^2 + b\left(-2\right) + c \Rightarrow$$
$$-14 = 4a - 2b + c \quad (2)$$

Substituting $x = -10$, $y = 2$, we get

$$-10 = a\left(2\right)^2 + b\left(2\right) + c \Rightarrow$$
$$-10 = 4a + 2b + c. \quad (3)$$

51. Since $a = -2 < 0$, the parabola opens to the left.

53. (a) Sketch a cross-section of the dish. Place this parabola on a coordinate system with the vertex at the origin. (This solution will show that the focus lies outside of the dish.)

Since the parabola has vertex $(0, 0)$ and a vertical axis (the y-axis), it has an equation of the form $x^2 = 4py$. Substitute $x = 150$ and $y = 44$ to find the value of p.

$$x^2 = 4py \Rightarrow 150^2 = 4p \cdot 44$$

$$22{,}500 = 176p \Rightarrow p = \frac{22{,}500}{176} = \frac{5625}{44}$$

The required equation is

$$x^2 = 4\left(\frac{5625}{44}\right)y \Rightarrow x^2 = \frac{5625}{11}y$$

$$y = \frac{11}{5625}x^2$$

(b) Since $p = \frac{5625}{44} \approx 127.8$, the receiver should be placed approximately 127.8 ft from the vertex.

55. Place the parabola that represents the arch on a coordinate system with the center of the bottom of the arch at the origin. Then the vertex will be $(0, 12)$ and the points $(-6, 0)$ and $(6, 0)$ will also be on the parabola. Because the axis of the parabola is the y-axis and the vertex is $(0, 12)$, the equation will have the form $x^2 = 4p\left(y-12\right)$. Use the coordinates of the point $(6, 0)$ to find the value of p.

$$x^2 = 4p\left(y-12\right) \Rightarrow 6^2 = 4p\left(0-12\right) \Rightarrow$$

$$36 = -48p \Rightarrow p = \frac{36}{-48} \Rightarrow p = -\frac{3}{4}$$

Thus, the equation of the parabola is

$$x^2 = 4\left(-\frac{3}{4}\right)\left(y-12\right) \text{ or } x^2 = -3\left(y-12\right).$$

Now find the x-coordinate of point whose y-coordinate is 9 and whose x-coordinate is positive: $x^2 = -3\left(y-12\right) \Rightarrow$

$$x^2 = -3\left(9-12\right) \Rightarrow x^2 = 9 \Rightarrow x = \sqrt{9} = 3$$

From the symmetry of the parabola, we see that the width of the arch 9 ft up is 2(3 ft) = 6 ft.

Copyright © 2013 Pearson Education, Inc.

57. (a) Locate the cannon at the origin. With $v = 252.982$, the equation becomes

$$y = x - \frac{32}{v^2}x^2 \Rightarrow y = x - \frac{32}{252.982^2}x^2 \Rightarrow$$

$$y \approx x - \frac{1}{2000}x^2$$

Complete the square on x.

$$y = x - \frac{1}{2000}x^2$$

$$y = -\frac{1}{2000}\left(x^2 - 2000x \quad\quad\right)$$

$$y = -\frac{1}{2000}\left(x^2 - 2000x + 1,000,000 - 1,000,000\right)$$

$$y = -\frac{1}{2000}\left(x^2 - 2000x + 1,000,000\right)$$

$$\quad\quad\quad +\left(-\frac{1}{2000}\right)(-1,000,000)$$

$$y = -\frac{1}{2000}(x - 1000)^2 + 500$$

$$y - 500 = -\frac{1}{2000}(x - 1000)^2$$

Thus, the vertex of the parabola is located at (1000, 500). Because of symmetry, the shell then travels an additional 1000 feet for a maximum distance of 2000 feet.

(b) The envelope parabola has x-intercepts located at (–2000, 0) and (2000, 0). The vertex is easily found to be located at (0, 1000). Because the axis of the parabola is the y-axis and it opens down, the equation is of the form

$$x^2 = 4p(y - 1000).$$ Use the coordinates of the point (2000, 0) to find the value of p. $2000^2 = 4p(0 - 1000) \Rightarrow$

$$4,000,000 = -4000p \Rightarrow p = -1000$$

The equation of the parabola is

$$x^2 = 4(-1000)(y - 1000)$$

$$x^2 = -4000(y - 1000)$$

$$y - 1000 = -0.00025x^2$$

$$y = 1000 - 0.00025x^2$$

(c) Using the equation of the envelope parabola in part (b), we calculate the maximum possible height of a shell when x is 1500 feet.

$$y = 1000 - 0.00025x^2$$

$$y = 1000 - 0.00025(1500)^2$$

$$y = 1000 - 0.00025(2,250,000)$$

$$y = 1000 - 562.5 \Rightarrow y = 437.5$$

If the helicopter flies at a height of 450 feet, a shell fired by the cannon would never reach the helicopter.

59. (a) $y = \frac{19}{11}x - \frac{g}{3872}x^2$

For the moon, $g = 5.2$, so the equation is

$$y = \frac{19}{11}x - \frac{5.2}{3872}x^2.$$

For Mars, $g = 12.6$, so the equation is

$$y = \frac{19}{11}x - \frac{12.6}{3872}x^2.$$

Graph both equations on the same screen.

moon $Y_1 = \frac{19}{11}x - \frac{5.2}{3872}x^2$

Mars $Y_2 = \frac{19}{11}x - \frac{12.6}{3872}x^2$

(b) Using the "maximum" option in the CALC menu, we see that the ball reaches a maximum height of $y \approx 229$ ft on Mars and $y \approx 555$ ft on the moon.

61. Answers will vary. The proof is similar to the proof given on page 586 in the text for a parabola that opens up (which also applied to parabolas opening down).

(continued on next page)

Copyright © 2013 Pearson Education, Inc.

(*continued*)

$$d(P,F) = d(P,D)$$

$$\sqrt{(x-p)^2 + (y-0)^2} = \sqrt{(x-(-p))^2 + (y-y)^2}$$

$$\sqrt{(x-p)^2 + y^2} = \sqrt{(x+p)^2}$$

$$x^2 - 2xp + p^2 + y^2 = x^2 + 2xp + p^2$$

$$-2xp + y^2 = 2xp$$

$$y^2 = 4px$$

Section 6.2 Ellipses

1. (a) $36x^2 + 9y^2 = 324 \Rightarrow \dfrac{x^2}{9} + \dfrac{y^2}{36} = 1 \Rightarrow$

$\dfrac{x^2}{3^2} + \dfrac{y^2}{6^2} = 1$

This ellipse has endpoints of the minor axis of $(\pm 3, 0)$, which are also *x*-intercepts. It also has vertices of $(0, \pm 6)$, which are also *y*-intercepts. The correct choice is A.

(b) $9x^2 + 36y^2 = 324 \Rightarrow \dfrac{x^2}{36} + \dfrac{y^2}{9} = 1 \Rightarrow$

$\dfrac{x^2}{6^2} + \dfrac{y^2}{3^2} = 1$

This ellipse has vertices of $(\pm 6, 0)$, which are also *x*-intercepts. It also has endpoints of the minor axis of $(0, \pm 3)$, which are also *y*-intercepts. The correct choice is C.

(c) $\dfrac{x^2}{25} + \dfrac{y^2}{16} = 1 \Rightarrow \dfrac{x^2}{5^2} + \dfrac{y^2}{4^2} = 1$

This ellipse has vertices of $(\pm 5, 0)$, which are also *x*-intercepts. It also has endpoints of the minor axis of $(0, \pm 4)$, which are also *y*-intercepts. The correct choice is D.

(d) $\dfrac{x^2}{16} + \dfrac{y^2}{25} = 1 \Rightarrow \dfrac{x^2}{4^2} + \dfrac{y^2}{5^2} = 1$

This ellipse has endpoints of the minor axis of $(\pm 4, 0)$, which are also *x*-intercepts. It also has vertices of $(0, \pm 5)$, which are also *y*-intercepts. The correct choice is B.

3. $\dfrac{x^2}{25} + \dfrac{y^2}{9} = 1$

The graph is an ellipse with center (0, 0). Rewriting the given equation, we have

$\dfrac{x^2}{5^2} + \dfrac{y^2}{3^2} = 1.$ Since $5 > 3$, we have $a = 5$ and $b = 3$, and the major axis is horizontal. Thus, the vertices are (–5, 0) and (5, 0). The endpoints of the minor axis are (0, –3) and (0, 3). The domain is [–5, 5]. The range is [–3, 3]. To find the foci we need to find *c* such that $c^2 = a^2 - b^2$.

$c^2 = a^2 - b^2 \Rightarrow c^2 = 25 - 9 \Rightarrow c^2 = 16 \Rightarrow c = 4$

Since the major axis lies on the *x*-axis, the foci are (–4, 0) and (4, 0).

5. $\dfrac{x^2}{9} + y^2 = 1$

Rewriting the equation, we have $\dfrac{x^2}{3^2} + \dfrac{y^2}{1^2} = 1.$

The center is (0, 0). Since $3 > 1$, we have $a = 3$ and $b = 1$, and the major axis is horizontal. The vertices are (–3, 0) and (3, 0). The endpoints of the minor axis are (0, –1) and (0, 1). The domain is [–3, 3]. The range is [–1, 1]. To find the foci, we need to find *c* such that $c^2 = a^2 - b^2$.

$c^2 = a^2 - b^2 \Rightarrow c^2 = 9 - 1 \Rightarrow c^2 = 8 \Rightarrow$

$c = \sqrt{8} = 2\sqrt{2}$

Since the major axis lies on the *x*-axis, the foci are $(-2\sqrt{2}, 0)$ and $(2\sqrt{2}, 0)$.

Copyright © 2013 Pearson Education, Inc.

7. $9x^2 + y^2 = 81$

Rewriting the equation, we have

$\frac{x^2}{9} + \frac{y^2}{81} = 1$ or $\frac{x^2}{3^2} + \frac{y^2}{9^2} = 1$. The center is

(0, 0). Since 9 > 3, we have $a = 9$ and $b = 3$, and the major axis is vertical. The vertices are (0, –9) and (0, 9). The endpoints of the minor axis are (–3, 0) and (3, 0). The domain is [–3, 3]. The range is [–9, 9]. To find the foci, we need to find c such that $c^2 = a^2 - b^2$.

$c^2 = a^2 - b^2 \Rightarrow c^2 = 81 - 9 \Rightarrow c^2 = 72 \Rightarrow$
$c = \sqrt{72} = 6\sqrt{2}$

Since the major axis lies on the y-axis, the foci are $\left(0, -6\sqrt{2}\right)$ and $\left(0, 6\sqrt{2}\right)$.

9. $4x^2 = 100 - 25y^2 \Rightarrow 4x^2 + 25y^2 = 100$

Rewriting the equation, we have $\frac{x^2}{25} + \frac{y^2}{4} = 1$

or $\frac{x^2}{5^2} + \frac{y^2}{2^2} = 1$. The center is (0, 0). Since

5 > 2, we have $a = 5$ and $b = 2$, and the major axis is horizontal. The vertices are (–5, 0) and (5, 0). The endpoints of the minor axis are (0, –2) and (0, 2). The domain is [–5, 5]. The range is [–2, 2]. To find the foci, we need to find c such that $c^2 = a^2 - b^2$.

$c^2 = a^2 - b^2 \Rightarrow c^2 = 25 - 4 \Rightarrow$
$c^2 = 21 \Rightarrow c = \sqrt{21}$

Since the major axis lies on the x-axis, the foci are $\left(-\sqrt{21}, 0\right)$ and $\left(\sqrt{21}, 0\right)$.

11. $\frac{(x-2)^2}{25} + \frac{(y-1)^2}{4} = 1$

Rewriting the equation, we have

$\frac{(x-2)^2}{5^2} + \frac{(y-1)^2}{2^2} = 1$. The center is (2, 1). Since

$a = 5$ is associated with x^2, the major axis of the ellipse is horizontal.

The vertices are on a horizontal line through (2, 1), while the endpoints of the minor axis are on the vertical line through (2, 1). The vertices are 5 units to the left and right of the center at (–3, 1) and (7, 1). The endpoints of the minor axis are 2 units below and 2 units above the center at (2, –1) and (2, 3). The domain is [–3, 7]. The range is [–1, 3]. To find the foci, we need to find c such that

$c^2 = a^2 - b^2$.

$c^2 = a^2 - b^2 \Rightarrow c^2 = 25 - 4 \Rightarrow$
$c^2 = 21 \Rightarrow c = \sqrt{21}$

Since the major axis lies on $y = 1$, the foci are $\left(2 - \sqrt{21}, 1\right)$ and $\left(2 + \sqrt{21}, 1\right)$.

$\frac{(x-2)^2}{25} + \frac{(y-1)^2}{4} = 1$ $\frac{(x+3)^2}{16} + \frac{(y-2)^2}{36} = 1$
Exercise 11   Exercise 13

13. $\frac{(x+3)^2}{16} + \frac{(y-2)^2}{36} = 1$

Rewriting the equation, we have

$\frac{(x+3)^2}{4^2} + \frac{(y-2)^2}{6^2} = 1$. The center is (–3, 2). We

have $a = 6$ and $b = 4$. Since $a = 6$ is associated with y^2, the major axis of the ellipse is vertical. The vertices are on the vertical line through (–3, 2), and the endpoints of the minor axis are on the horizontal line through (–3, 2). The vertices are 6 units below and 6 units above the center at (–3, –4) and (–3, 8). The endpoints of the minor axis are 4 units to the left and 4 units to the right of the center at (–7, 2) and (1, 2). The domain is [–7, 1]. The range is [–4, 8]. To find the foci, we need to find c such that $c^2 = a^2 - b^2$.

$c^2 = a^2 - b^2 \Rightarrow c^2 = 36 - 16 \Rightarrow c^2 = 20 \Rightarrow$
$c = \sqrt{20} = 2\sqrt{5}$

Since the major axis lies on $x = -3$, the foci are $\left(-3, 2 - 2\sqrt{5}\right)$ and $\left(-3, 2 + 2\sqrt{5}\right)$.

Copyright © 2013 Pearson Education, Inc.

15. x-intercepts ± 5; y-intercepts ± 4
From the given information, $a = 5$ and $b = 4$.
Since $5 > 4$, the x-intercepts represent vertices.
The equation has the form $\frac{x^2}{a^2} + \frac{y^2}{b^2} = 1$. Since
$a^2 = 25$ and $b^2 = 16$, the equation of the
ellipse is $\frac{x^2}{25} + \frac{y^2}{16} = 1$.

17. Major axis with length 6, foci at $(0, 2)$, $(0, -2)$.
The length of the major axis is $2a$. Thus, we
have $2a = 6 \Rightarrow a = 3$. From the foci, we have
$c = 2$. Solving for b^2 we have
$c^2 = a^2 - b^2 \Rightarrow 4 = 9 - b^2 \Rightarrow b^2 = 5$.
Since the foci are on the y-axis and the ellipse
is centered at the origin, the equation has the
form $\frac{x^2}{b^2} + \frac{y^2}{a^2} = 1$. Since $a = 3 \Rightarrow a^2 = 9$ and
and $b^2 = 5$, the equation of the ellipse is
$\frac{x^2}{5} + \frac{y^2}{9} = 1$.

19. Center at $(3, 1)$; minor axis vertical, with
length 8; $c = 3$.
Since the center is $(3, 1)$ and the minor axis is
vertical, the equation has the form
$\frac{(x-3)^2}{a^2} + \frac{(y-1)^2}{b^2} = 1$. The length of the minor
axis is $2b$, so $b = 4$. Solving for a^2, we have
$c^2 = a^2 - b^2 \Rightarrow 9 = a^2 - 16 \Rightarrow a^2 = 25$.
Since $b = 4 \Rightarrow b^2 = 16$ and $a^2 = 25$, the
equation of the ellipse is $\frac{(x-3)^2}{25} + \frac{(y-1)^2}{16} = 1$.

21. Foci at $(0, 4)$, $(0, -4)$; sum of distances from
foci to point on ellipse is 10.
The center is halfway between the foci, so the
center is $(0, 0)$. The distance between the foci
is $4 - (-4) = 8$, so $2c = 8$ and thus $c = 4$. The
sum of the distances from the foci to any point
on the ellipse is 10, so $2a = 10 \Rightarrow a = 5$. Since
the foci lie on the y-axis, the equation is of the
form $\frac{x^2}{b^2} + \frac{y^2}{a^2} = 1$. Solving for b^2, we have
$c^2 = a^2 - b^2 \Rightarrow 4^2 = 5^2 - b^2 \Rightarrow 16 = 25 - b^2 \Rightarrow b^2 = 9$.
Since $a = 5 \Rightarrow a^2 = 25$ and $b^2 = 9$, the
equation of the ellipse is $\frac{x^2}{9} + \frac{y^2}{25} = 1$.

23. Foci at $(0, -3)$, $(0, 3)$; $(8, 3)$ on the ellipse.
The distance between the foci is $3 - (-3) = 6$,
so $2c = 6$ and thus $c = 3$. The center is the
midpoint between the foci, so the center is
$(0, 0)$.

Since the foci lie on the y-axis, the major axis
is vertical, so the equation is of the form
$\frac{x^2}{b^2} + \frac{y^2}{a^2} = 1$. Now recall from Figure 15 on
page 593 of the text, $d(P, F) + d(P, F') = 2a$.
If we let $P(8, 3)$, $F'(0, -3)$, and $F(0, 3)$
represent the point on the ellipse and the foci,
respectively, we have the following.
$d(P, F) + d(P, F')$
$= \sqrt{(8-0)^2 + (3+3)^2} + \sqrt{(8-0)^2 + (3-3)^2}$
$= 10 + 8 = 18 = 2a \Rightarrow a = 9$
Solving for b^2, we have
$c^2 = a^2 - b^2 \Rightarrow 9 = 81 - b^2 \Rightarrow b^2 = 72$. Since
$a = 9 \Rightarrow a^2 = 81$ and $b^2 = 72$, the equation of
the ellipse is $\frac{x^2}{72} + \frac{y^2}{81} = 1$.

25. Eccentricity $\frac{4}{5}$; vertices at $(-5, 0)$, $(5, 0)$.
From the vertices, we have $a = 5$ and $a^2 = 25$.
Since the vertices lie on the x-axis, the
equation is of the form $\frac{x^2}{a^2} + \frac{y^2}{b^2} = 1$. Thus, the
equation has the form $\frac{x^2}{25} + \frac{y^2}{b^2} = 1$. Use the
eccentricity and a to find c.
$e = \frac{c}{a} \Rightarrow \frac{4}{5} = \frac{c}{5} \Rightarrow 4 = c$
Now solving for b^2, we have
$c^2 = a^2 - b^2 \Rightarrow 4^2 = 5^2 - b^2 \Rightarrow$
$16 = 25 - b^2 \Rightarrow b^2 = 9 \Rightarrow b = 3$
The equation is therefore $\frac{x^2}{25} + \frac{y^2}{9} = 1$.

27. Eccentricity $\frac{3}{4}$, foci at $(0, -2)$, $(0, 2)$.
The foci are on the y-axis, so the equation has
the form $\frac{x^2}{b^2} + \frac{y^2}{a^2} = 1$. From the foci, we have
$c = 2$. Use the eccentricity and c to find a.
$e = \frac{c}{a} \Rightarrow \frac{3}{4} = \frac{2}{a} \Rightarrow 3a = 8 \Rightarrow a = \frac{8}{3}$
Solving for b^2, we have
$c^2 = a^2 - b^2 \Rightarrow 2^2 = \left(\frac{8}{3}\right)^2 - b^2 \Rightarrow$
$4 = \frac{64}{9} - b^2 \Rightarrow b^2 = \frac{64}{9} - \frac{36}{9} = \frac{28}{9}$
Since $a = \frac{8}{3} \Rightarrow a^2 = \frac{64}{9}$ and $b^2 = \frac{28}{9}$, the
equation is $\frac{x^2}{\frac{28}{9}} + \frac{y^2}{\frac{64}{9}} = 1$ or $\frac{9x^2}{28} + \frac{9y^2}{64} = 1$.

Copyright © 2013 Pearson Education, Inc.

29. $\dfrac{y}{2} = \sqrt{1 - \dfrac{x^2}{25}}$

Square both sides to get

$\dfrac{y^2}{4} = 1 - \dfrac{x^2}{25} \Rightarrow \dfrac{x^2}{25} + \dfrac{y^2}{4} = 1 \Rightarrow \dfrac{x^2}{5^2} + \dfrac{y^2}{2^2} = 1,$

which is the equation of an ellipse centered at the origin with x-intercepts ± 5 (vertices) and y-intercepts ± 2 (endpoints of minor axis).

Since $\sqrt{1 - \dfrac{x^2}{25}} \ge 0,$ the only possible values of

y are those making $\dfrac{y}{2} \ge 0 \Rightarrow y \ge 0.$ The

domain is $[-5, 5]$. The range is $[0, 2]$. The graph of the original equation is the upper half of the ellipse. By applying the vertical line test, we see that this is the graph of a function.

$\dfrac{y}{2} = \sqrt{1 - \dfrac{x^2}{25}}$

31. $x = -\sqrt{1 - \dfrac{y^2}{64}}$

Square both sides to get

$x^2 = 1 - \dfrac{y^2}{64} \Rightarrow \dfrac{x^2}{1} + \dfrac{y^2}{64} = 1 \Rightarrow \dfrac{x^2}{1^2} + \dfrac{y^2}{8^2} = 1,$ the

equation of an ellipse centered at the origin with x-intercepts ± 1 (endpoints of minor axis) and y-intercepts ± 8 (vertices). Since

$-\sqrt{1 - \dfrac{y^2}{64}} \le 0,$ we must have $x \le 0,$ so the

graph of the original equation is the left half of the ellipse. The domain is $[-1, 0]$. The range is $[-8, 8]$. The vertical line test shows that this is not the graph of a function.

$x = -\sqrt{1 - \dfrac{y^2}{64}}$

33. Solve for y in the equation of the ellipse.

$\dfrac{x^2}{16} + \dfrac{y^2}{4} = 1 \Rightarrow \dfrac{y^2}{4} = 1 - \dfrac{x^2}{16} \Rightarrow$

$y^2 = 4\left(1 - \dfrac{x^2}{16}\right) \Rightarrow y = \pm\sqrt{4\left(1 - \dfrac{x^2}{16}\right)} \Rightarrow$

$y = \pm 2\sqrt{1 - \dfrac{x^2}{16}}$

$Y_1 = 2\sqrt{1 - \dfrac{x^2}{16}}$

$Y_2 = -2\sqrt{1 - \dfrac{x^2}{16}}$

35. Solve for y in the equation of the ellipse.

$\dfrac{(x-3)^2}{25} + \dfrac{y^2}{9} = 1 \Rightarrow \dfrac{y^2}{9} = 1 - \dfrac{(x-3)^2}{25} \Rightarrow$

$y^2 = 9\left(1 - \dfrac{(x-3)^2}{25}\right) \Rightarrow$

$y = \pm\sqrt{9\left(1 - \dfrac{(x-3)^2}{25}\right)} = \pm 3\sqrt{1 - \dfrac{(x-3)^2}{25}}$

$Y_1 = 3\sqrt{1 - \dfrac{(x-3)^2}{25}}$

$Y_2 = -3\sqrt{1 - \dfrac{(x-3)^2}{25}}$

Copyright © 2013 Pearson Education, Inc.

37. $\dfrac{x^2}{3}+\dfrac{y^2}{4}=1$

Since $4>3$, $a^2=4$, which gives $a=2$. Also

$c=\sqrt{a^2-b^2}=\sqrt{4-3}=\sqrt{1}=1$. Thus,

$e=\dfrac{c}{a}=\dfrac{1}{2}$.

39. $4x^2+7y^2=28 \Rightarrow \dfrac{x^2}{7}+\dfrac{y^2}{4}=1$

Since $7>4$, $a^2=7$, which gives $a=\sqrt{7}$.

Also $c=\sqrt{a^2-b^2}=\sqrt{7-4}=\sqrt{3}$. Thus,

$e=\dfrac{c}{a}=\dfrac{\sqrt{7}}{\sqrt{3}}=\dfrac{\sqrt{21}}{3}\approx 0.65$.

41. Answers will vary.

43. Place the half-ellipse that represents the overpass on a coordinate system with the center of the bottom of the overpass at the origin. If the complete ellipse were drawn, the center of the ellipse would be $(0, 0)$. Then the half-ellipse will include the points $(0, 15)$, $(-10, 0)$, and $(10, 0)$. The equation is of the form $\dfrac{x^2}{b^2}+\dfrac{y^2}{a^2}=1$.

Thus, for the complete ellipse, we have $a=15$ and $b=10$. Thus, we have

$\dfrac{x^2}{10^2}+\dfrac{y^2}{15^2}=1 \Rightarrow \dfrac{x^2}{100}+\dfrac{y^2}{225}=1$.

To find the equation of the half-ellipse, solve this equation for y and use the positive square root since the overpass is represented by the upper half of the ellipse.

$\dfrac{x^2}{100}+\dfrac{y^2}{225}=1 \Rightarrow \dfrac{y^2}{225}=1-\dfrac{x^2}{100} \Rightarrow$

$y^2=225\left(1-\dfrac{x^2}{100}\right) \Rightarrow y=\sqrt{225\left(1-\dfrac{x^2}{100}\right)} \Rightarrow$

$y=15\sqrt{1-\dfrac{x^2}{100}}$

Find the y-coordinate of the point whose x-coordinate is $\dfrac{1}{2}(12)=6$.

$y=15\sqrt{1-\dfrac{x^2}{100}} \Rightarrow y=15\sqrt{1-\dfrac{6^2}{100}} \Rightarrow$

$y=15\sqrt{1-\dfrac{36}{100}}=15\sqrt{\dfrac{64}{100}}=15\left(\dfrac{4}{5}\right)=12$

The tallest truck that can pass under the overpass is 12 feet tall.

45. This problem is similar to Example 6 on page 600 of the text. The greatest distance between the comet and the sun is $a+c=3281$ million miles. If $a+c=3281 \Rightarrow c=3281-a$ and $e=\dfrac{c}{a}$, we have $e=\dfrac{3281-a}{a}$. Since $e=0.9673$, we have

$0.9673=\dfrac{3281-a}{a} \Rightarrow 0.9673a=3281-a \Rightarrow$

$1.9673a=3281 \Rightarrow a=\dfrac{3281}{1.9673}\approx 1668$.

We then have $c=3281-1668=1613$, which implies $a-c=1668-1613=55$. Thus, the shortest distance between Halley's Comet and the sun is about 55 million miles.

47. (a) Use the given values of e and a to find the value of c for each planet. Then use the values of a and c to find the value of b. Neptune:

$e=\dfrac{c}{a} \Rightarrow c=ea \Rightarrow c=(0.009)(30.1) \Rightarrow$

$c=0.2709$

$b^2=a^2-c^2 \Rightarrow b^2=(30.1)^2-(0.2709)^2 \Rightarrow$

$b^2\approx 905.9366 \Rightarrow b\approx 30.1$

Since $c=0.2709$, the graph should be translated 0.2709 units to the right so that the sun will be located at the origin. It's essentially circular with equation

$\dfrac{(x-0.2709)^2}{30.1^2}+\dfrac{y^2}{30.1^2}=1$.

Pluto:

$e=\dfrac{c}{a} \Rightarrow c=ea \Rightarrow c=(.249)(39.4) \Rightarrow$

$c=9.8106$

$b^2=a^2-c^2 \Rightarrow b^2=(39.4)^2-(9.8106)^2 \Rightarrow$

$b^2\approx 1456.1121 \Rightarrow b\approx 38.16$

As with Neptune, we translate the graph by c units to the right so that the sun will be located at the origin. The equation is

$\dfrac{(x-9.8106)^2}{39.4^2}+\dfrac{y^2}{38.16^2}=1$.

(b) In order to graph these equations on a graphing calculator, we must solve each equation for y. Each equation will be broken down into two functions, so we will need to graph four functions.
Neptune:

$\dfrac{(x-0.2709)^2}{30.1^2}+\dfrac{y^2}{30.1^2}=1 \Rightarrow$

$(x-0.2709)^2+y^2=30.1^2 \Rightarrow$

$y=\pm\sqrt{30.1^2-(x-0.2709)^2}$

(*continued on next page*)

Copyright © 2013 Pearson Education, Inc.

(*continued*)

Pluto:

$$\frac{(x-9.8106)^2}{39.4^2} + \frac{y^2}{38.16^2} = 1 \Rightarrow$$

$$\frac{y^2}{38.16^2} = 1 - \frac{(x-9.8106)^2}{39.4^2} \Rightarrow$$

$$y^2 = 38.16^2\left(1 - \frac{(x-9.8106)^2}{39.4^2}\right) \Rightarrow$$

$$y = \pm\sqrt{38.16^2\left(1 - \frac{(x-9.8106)^2}{39.4^2}\right)}$$

$$= \pm38.16\sqrt{1 - \frac{(x-9.8106)^2}{39.4^2}}$$

Graph the following four functions on the same screen.

$$Y_1 = \sqrt{30.1^2 - (x-0.2709)^2}$$

$$Y_2 = -\sqrt{30.1^2 - (x-0.2709)^2} = -Y_1$$

$$Y_3 = 38.16\sqrt{1 - \frac{(x-9.8106)^2}{39.4^2}}$$

$$Y_4 = -38.16\sqrt{1 - \frac{(x-9.8106)^2}{39.4^2}} = -Y_3$$

$$Y_3 = 38.16\sqrt{1 - \frac{(x-9.8106)^2}{39.4^2}}$$

$$Y_1 = \sqrt{30.1^2 - (x-0.2709)^2}$$

$$Y_2 = -Y_1 \qquad Y_4 = -Y_3$$

49. The stone and the wave source must be placed at the foci, (c, 0) and (–c, 0). Here $a^2 = 36$ and $b^2 = 9$, so $c^2 = a^2 - b^2 \Rightarrow c^2 = 36 - 9 \Rightarrow c^2 = 27 \Rightarrow c = \sqrt{27} = 3\sqrt{3}$. Thus, the kidney stone and source of the beam must be placed $3\sqrt{3}$ units from the center.

Chapter 6 Quiz
(Sections 6.1–6.2)

1. **(a)** $x + 3 = 4(y-1)^2$

   This is the equation of a parabola with vertex (–3, 1). The parabola opens right. The correct choices are B and D.

   (b) $(x+3)^2 + (y-1)^2 = 81$

   This is the equation of a circle with center (–3, 1). The correct choice is A.

   (c) $25(x-2)^2 + (y-1)^2 = 100 \Rightarrow$

   $$\frac{(x-2)^2}{4} + \frac{(y-1)^2}{100} = 1$$

   This is the equation of an ellipse with center (2, 1) and major axis vertical. The correct choice is E.

   (d) $\dfrac{(x-2)^2}{16} + \dfrac{(y-1)^2}{9} = 1$

   This is the equation of an ellipse with center (2, 1) and major axis horizontal. The correct choices are C and E.

   (e) $-2(x+3)^2 + 1 = y \Rightarrow -2(x+3)^2 = y - 1$

   This is the equation of a parabola with vertex (–3, 1). The parabola opens down. The correct choice is D.

3. A parabola passing through $\left(\sqrt{10}, -5\right)$, opening downward, and vertex at the origin has an equation of the form $x^2 = 4py$. Use this equation with the coordinates of the point $\left(\sqrt{10}, -5\right)$ to find the value of p.

   $$x^2 = 4py \Rightarrow \left(\sqrt{10}\right)^2 = 4p(-5) \Rightarrow$$

   $$10 = -20p \Rightarrow p = -\tfrac{1}{2}$$

   Thus, an equation of the parabola is

   $$x^2 = 4\left(-\tfrac{1}{2}\right)y = -2y \Rightarrow y = -\tfrac{1}{2}x^2.$$

Copyright © 2013 Pearson Education, Inc.

5. Foci at (–3, 3), (–3, 11), major axis with length 10. The length of the major axis is 2*a*. Thus, we have $2a = 10 \Rightarrow a = 5$. From the foci, we have $c = \frac{11-3}{2} = 4$. Solving for b^2 we have

$$c^2 = a^2 - b^2 \Rightarrow 16 = 25 - b^2 \Rightarrow b^2 = 9.$$ The center is the midpoint between the foci. Thus, the center is located at

$\left(\frac{(-3)+(-3)}{2}, \frac{3+11}{2}\right) = \left(\frac{-6}{2}, \frac{14}{2}\right) = (-3, 7).$ Since the foci lie on the vertical line $x = -3$, the major axis is vertical, so the equation is of the form

$$\frac{(x+3)^2}{b^2} + \frac{(y-7)^2}{a^2} = 1.$$

Since $a = 5 \Rightarrow a^2 = 25$ and and $b^2 = 9$, the equation of the ellipse is

$$\frac{(x+3)^2}{9} + \frac{(y-7)^2}{25} = 1.$$

7. $4x^2 + 9y^2 = 36$

Rewriting the equation, we have $\frac{4x^2}{36} + \frac{9y^2}{36} = 1$

or $\frac{x^2}{9} + \frac{y^2}{4} = 1 \Rightarrow \frac{x^2}{3^2} + \frac{y^2}{2^2} = 1.$ This is an ellipse with center (0, 0). Since 3 > 2, we have $a = 3$ and $b = 2$, and the major axis is horizontal. The vertices are $(-3, 0)$ and $(3, 0)$. The endpoints of the minor axis are $(0, -2)$ and $(0, 2)$. To find the foci, we need to find c such that $c^2 = a^2 - b^2.$

$$c^2 = a^2 - b^2 \Rightarrow c^2 = 9 - 4 \Rightarrow c^2 = 5 \Rightarrow c = \sqrt{5}$$

Since the major axis lies on the *x*-axis, the foci are $\left(-\sqrt{5}, 0\right)$ and $\left(\sqrt{5}, 0\right)$.

$4x^2 + 9y^2 = 36$

9. $\frac{(x+3)^2}{25} + \frac{(y+2)^2}{36} = 1$

Rewriting the equation, we have

$\frac{(x+3)^2}{5^2} + \frac{(y+2)^2}{6^2} = 1.$ This is an ellipse with center $(-3, -2)$. We have $a = 6$ and $b = 5$. Since $a = 6$ is associated with y^2, the major axis of the ellipse is vertical.

The vertices are on the vertical line through (–3, –2), and the endpoints of the minor axis are on the horizontal line through (–3, –2). The vertices are 6 units below and 6 units above the center at (–3, 4) and (–3, –8). The endpoints of the minor axis are 5 units to the left and 5 units to the right of the center at (–8, –2) and (2, –2). To find the foci, we need to find *c* such that $c^2 = a^2 - b^2.$

$$c^2 = a^2 - b^2 \Rightarrow c^2 = 36 - 25 \Rightarrow c^2 = 11 \Rightarrow$$
$$c = \sqrt{11}$$

Since the major axis lies on $x = -3$, the foci are $\left(-3, -2 + \sqrt{11}\right)$ and $\left(-3, -2 - \sqrt{11}\right)$.

$\frac{(x+3)^2}{25} + \frac{(y+2)^2}{36} = 1$

Section 6.3   Hyperbolas

1. $\dfrac{x^2}{25} + \dfrac{y^2}{9} = 1$

This is an equation of an ellipse with *x*-intercepts ± 5 and *y*-intercepts ± 3. The correct graph is C.

3. $\dfrac{x^2}{9} - \dfrac{y^2}{25} = 1$

This is the graph of a hyperbola with a horizontal transverse axis and *x*-intercepts ± 3 (no *y*-intercepts). The correct graph is D.

5. $\dfrac{x^2}{16} - \dfrac{y^2}{9} = 1,$

This equation may be written as $\frac{x^2}{4^2} - \frac{y^2}{3^2} = 1,$

which has the form $\frac{x^2}{a^2} - \frac{y^2}{b^2} = 1.$ The hyperbola is centered at (0, 0) with branches opening to the left and right. The graph has vertices at (–4, 0) and (4, 0) (also *x*-intercepts ± 4). There are no *y*-intercepts. The domain is $(-\infty, -4] \cup [4, \infty)$. The range is $(-\infty, \infty)$.

(continued on next page)

Copyright © 2013 Pearson Education, Inc.

(*continued*)

The foci are on the x-axis. Since $c^2 = a^2 + b^2 \Rightarrow c^2 = 16 + 9 \Rightarrow$ $c^2 = 25 \Rightarrow c = 5$, the foci are $(-5, 0)$ and $(5, 0)$. Since $a = 4$ and $b = 3$, the asymptotes are $y = \pm\frac{b}{a}x = \pm\frac{3}{4}x$.

$$\frac{x^2}{16} - \frac{y^2}{9} = 1$$

7. $\dfrac{y^2}{25} - \dfrac{x^2}{49} = 1$

This equation may be written as $\frac{y^2}{5^2} - \frac{x^2}{7^2} = 1$, which has the form $\frac{y^2}{a^2} - \frac{x^2}{b^2} = 1$. The hyperbola is centered at $(0, 0)$ with branches opening upward and downward. The graph has vertices at $(0, -5)$ and $(0, 5)$ (also y-intercepts ± 5). There are no x-intercepts. The domain is $(-\infty, \infty)$. The range is $(-\infty, -5] \cup [5, \infty)$. The foci are on the y-axis. Since $c^2 = a^2 + b^2$, we have

$$c^2 = a^2 + b^2 \Rightarrow c^2 = 25 + 49$$
$$c^2 = 74 \Rightarrow c = \sqrt{74}$$

The foci are $\left(0, -\sqrt{74}\right)$ and $\left(0, \sqrt{74}\right)$. Since $a = 5$ and $b = 7$, the asymptotes are $y = \pm\frac{a}{b}x = \pm\frac{5}{7}x$.

$$\frac{y^2}{25} - \frac{x^2}{49} = 1$$

9. $x^2 - y^2 = 9$

This equation may be written as

$\frac{x^2}{9} - \frac{y^2}{9} = 1 \Rightarrow \frac{x^2}{3^2} - \frac{y^2}{3^2} = 1$, which has the form

$\frac{x^2}{a^2} - \frac{y^2}{b^2} = 1$.

The hyperbola is centered at $(0, 0)$ with branches opening to the left and right. The graph has vertices at $(-3, 0)$ and $(3, 0)$ (also x-intercepts ± 3). There are no y-intercepts. The domain is $(-\infty, -3] \cup [3, \infty)$. The range is $(-\infty, \infty)$. The foci are on the x-axis. Since $c^2 = a^2 + b^2$, we have $c^2 = a^2 + b^2 \Rightarrow$ $c^2 = 9 + 9 \Rightarrow c^2 = 18 \Rightarrow c = \sqrt{18} = 3\sqrt{2}$. The foci are $\left(-3\sqrt{2}, 0\right)$ and $\left(3\sqrt{2}, 0\right)$. Since $a = 3$ and $b = 3$, the asymptotes are

$$y = \pm\frac{b}{a}x = \pm\frac{3}{3}x = \pm x$$

$$x^2 - y^2 = 9$$

11. $9x^2 - 25y^2 = 225$

This equation may be written as

$\frac{x^2}{25} - \frac{y^2}{9} = 1 \Rightarrow \frac{x^2}{5^2} - \frac{y^2}{3^2} = 1$, which has the form

$\frac{x^2}{a^2} - \frac{y^2}{b^2} = 1$. The hyperbola is centered at $(0, 0)$ with branches opening to the left and right. The graph has vertices at $(-5, 0)$ and $(5, 0)$ (also x-intercepts ± 5). There are no y-intercepts. The domain is $(-\infty, -5] \cup [5, \infty)$. The range is $(-\infty, \infty)$. The foci are on the x-axis. Since $c^2 = a^2 + b^2$, we have

$$c^2 = a^2 + b^2 \Rightarrow c^2 = 25 + 9 \Rightarrow$$
$$c^2 = 34 \Rightarrow c = \sqrt{34}$$

The foci are $\left(-\sqrt{34}, 0\right)$ and $\left(\sqrt{34}, 0\right)$. Since $a = 5$ and $b = 3$, the asymptotes are

$$y = \pm\frac{b}{a}x = \pm\frac{3}{5}x.$$

$$9x^2 - 25y^2 = 225$$

Copyright © 2013 Pearson Education, Inc.

13. $4y^2 - 25x^2 = 100$

This equation may be written as

$\frac{y^2}{25} - \frac{x^2}{4} = 1 \Rightarrow \frac{y^2}{5^2} - \frac{x^2}{2^2} = 1$, which has the form

$\frac{y^2}{a^2} - \frac{x^2}{b^2} = 1$. The hyperbola is centered at

$(0, 0)$ with branches opening upward and downward. The graph has vertices at $(0, -5)$ and $(0, 5)$ (also y-intercepts ± 5). There are no x-intercepts. The domain is $(-\infty, \infty)$. The range is $(-\infty, -5] \cup [5, \infty)$. The foci are on

the y-axis. Since $c^2 = a^2 + b^2$, we have

$c^2 = a^2 + b^2 \Rightarrow c^2 = 25 + 4$
$\quad c^2 = 29 \Rightarrow c = \sqrt{29}$

The foci are $\left(0, -\sqrt{29}\right)$ and $\left(0, \sqrt{29}\right)$. Since

$a = 5$ and $b = 2$, the asymptotes are

$y = \pm \frac{a}{b} x = \pm \frac{5}{2} x.$

4y² − 25x² = 100
Exercise 13

9x² − 4y² = 1
Exercise 15

15. $9x^2 - 4y^2 = 1$

This equation may be written as

$\frac{x^2}{\frac{1}{9}} - \frac{y^2}{\frac{1}{4}} = 1 \Rightarrow \frac{x^2}{\left(\frac{1}{3}\right)^2} - \frac{y^2}{\left(\frac{1}{2}\right)^2} = 1$, which has the

form $\frac{x^2}{a^2} - \frac{y^2}{b^2} = 1$. The hyperbola is centered at

$(0, 0)$ with branches opening to the left and right. The graph has vertices at $\left(-\frac{1}{3}, 0\right)$ and

$\left(\frac{1}{3}, 0\right)$ (also x-intercepts $\pm \frac{1}{3}$). There are no

y-intercepts. The domain is

$\left(-\infty, -\frac{1}{3}\right] \cup \left[\frac{1}{3}, \infty\right)$. The range is $(-\infty, \infty)$.

The foci are on the x-axis. Since $c^2 = a^2 + b^2$,

we have $c^2 = a^2 + b^2 \Rightarrow c^2 = \frac{1}{9} + \frac{1}{4} \Rightarrow$

$c^2 = \frac{4+9}{36} \Rightarrow c^2 = \frac{13}{36} \Rightarrow c = \sqrt{\frac{13}{36}} = \frac{\sqrt{13}}{6}$. The

foci are $\left(-\frac{\sqrt{13}}{6}, 0\right)$ and $\left(\frac{\sqrt{13}}{6}, 0\right)$. Since $a = \frac{1}{3}$

and $b = \frac{1}{2}$, the asymptotes are

$y = \pm \frac{b}{a} x = \pm \frac{\frac{1}{2}}{\frac{1}{3}} x = \pm \frac{3}{2} x.$

17. $\frac{(y-7)^2}{36} - \frac{(x-4)^2}{64} = 1$

Since this equation can be written as

$\frac{(y-7)^2}{6^2} - \frac{(x-4)^2}{8^2} = 1$, it has the form

$\frac{(y-k)^2}{a^2} - \frac{(x-h)^2}{b^2} = 1$ where $h = 4$, $k = 7$, $a = 6$,

and $b = 8$. The center is $(4, 7)$. The vertices are 6 units above and below the center $(4, 7)$. These points are $(4, 1)$ and $(4, 13)$. The domain is $(-\infty, \infty)$. The range is $(-\infty, 1] \cup$

$[13, \infty)$. Since $c^2 = a^2 + b^2 \Rightarrow$

$c^2 = 36 + 64 \Rightarrow c^2 = 100 \Rightarrow c = 10$, the foci

are 10 units below and above the center $(4, 7)$. Thus, the foci are $(4, -3)$ and $(4, 17)$. Since

$h = 4$, $k = 7$, $a = 6$, $b = 8$, and

$y = k \pm \frac{a}{b}(x - h)$, the asymptotes are

$y = 7 \pm \frac{6}{8}(x - 4) \Rightarrow y = 7 \pm \frac{3}{4}(x - 4)$

(y − 7)²/36 − (x − 4)²/64 = 1
Exercise 17

(x + 3)²/16 − (y − 2)²/9 = 1
Exercise 19

19. $\frac{(x+3)^2}{16} - \frac{(y-2)^2}{9} = 1,$

Since this equation can be written as

$\frac{[x-(-3)]^2}{4^2} - \frac{(y-2)^2}{3^2} = 1$, it has the form

$\frac{(x-h)^2}{a^2} - \frac{(y-k)^2}{b^2} = 1$ where $h = -3$, $k = 2$, $a = 4$,

and $b = 3$. The center is $(-3, 2)$. The vertices are 4 units left and right of the center $(-3, 2)$. These points are $(-7, 2)$ and $(1, 2)$ The domain is $(-\infty, -7] \cup [1, \infty)$. The range is $(-\infty, \infty)$.

Since $c^2 = a^2 + b^2 \Rightarrow c^2 = 16 + 9 \Rightarrow$

$c^2 = 25 \Rightarrow c = 5$, the foci are 5 units left and right of the center $(-3, 2)$. Thus, the foci are $(-8, 2)$ and $(2, 2)$. Since $h = -3$, $k = 2$, $a = 4$,

$b = 3$, and $y = k \pm \frac{b}{a}(x - h)$, the asymptotes

are $y = 2 \pm \frac{3}{4}\left[x - (-3)\right] \Rightarrow y = 2 \pm \frac{3}{4}(x + 3).$

Copyright © 2013 Pearson Education, Inc.

21. $16(x+5)^2 - (y-3)^2 = 1$

Since this equation can be written as

$\dfrac{[x-(-5)]^2}{\left(\frac{1}{4}\right)^2} - \dfrac{(y-3)^2}{1^2} = 1$, it has the form

$\dfrac{(x-h)^2}{a^2} - \dfrac{(y-k)^2}{b^2} = 1$ where $h = -5$, $k = 3$,

$a = \frac{1}{4}$, and $b = 1$. The center is $(-5, 3)$.

The vertices are $\frac{1}{4}$ unit left and right of the

center $(-5, 3)$. These points are $\left(-\frac{21}{4}, 3\right)$ and

$\left(-\frac{19}{4}, 3\right)$. The domain is

$\left(-\infty, -\frac{21}{4}\right] \cup \left[-\frac{19}{4}, \infty\right)$. The range is

$(-\infty, \infty)$. Since $c^2 = a^2 + b^2 \Rightarrow$

$c^2 = \frac{1}{16} + 1 \Rightarrow c^2 = \frac{1+16}{16} \Rightarrow c^2 = \frac{17}{16} \Rightarrow$

$c = \sqrt{\frac{17}{16}} = \frac{\sqrt{17}}{4}$, the foci are $\frac{\sqrt{17}}{4}$ units left and

right of the center $(-5, 3)$. Thus, the foci are

$\left(-5 - \frac{\sqrt{17}}{4}, 3\right)$ and $\left(-5 + \frac{\sqrt{17}}{4}, 3\right)$. Since

$h = -5$, $k = 3$, $a = \frac{1}{4}$, $b = 1$, and

$y = k \pm \frac{b}{a}(x-h)$, the asymptotes are

$y = 3 \pm \frac{1}{\frac{1}{4}}\left[x + (-5)\right] \Rightarrow y = 3 \pm 4(x+5)$

$16(x+5)^2 - (y-3)^2 = 1$

23. $\dfrac{y}{3} = \sqrt{1 + \dfrac{x^2}{16}}$

Square both sides to get $\frac{y^2}{9} = 1 + \frac{x^2}{16}$

or $\frac{y^2}{9} - \frac{x^2}{16} = 1$ or $\frac{y^2}{3^2} - \frac{x^2}{4^2} = 1$ This is the

equation of a hyperbola with center $(0, 0)$ and

vertices $(0, -3)$ and $(0, 3)$. Since $a = 3$ and

$b = 4$ and $y = \pm \frac{a}{b}x$, we have asymptotes

$y = \pm \frac{3}{4}x$. The original equation is the top half

of the hyperbola. The domain is $(-\infty, \infty)$. The

range is $[3, \infty)$. The vertical line test shows

this is a graph of a function.

$\dfrac{y}{3} = \sqrt{1 + \dfrac{x^2}{16}}$

25. $5x = -\sqrt{1 + 4y^2}$

Square both sides to get $25x^2 = 1 + 4y^2$ or

$25x^2 - 4y^2 = 1$ or $\dfrac{x^2}{\left(\frac{1}{5}\right)^2} - \dfrac{y^2}{\left(\frac{1}{2}\right)^2} = 1$. This is the

equation of a hyperbola with center $(0, 0)$ and

vertices $\left(-\frac{1}{5}, 0\right)$ and $\left(\frac{1}{5}, 0\right)$. Since

$a = \frac{1}{5}$ and $b = \frac{1}{2}$ and $y = \pm \frac{b}{a}x$, we have

asymptotes $y = \pm \frac{\frac{1}{2}}{\frac{1}{5}}x$ or $y = \pm \frac{5}{2}x$. The

original equation is the left half of the

hyperbola. The domain is $\left(-\infty, -\frac{1}{5}\right]$. The

range is $(-\infty, \infty)$. The vertical line test shows

that this is not the graph of a function.

$5x = -\sqrt{1 + 4y^2}$

27. $\dfrac{x^2}{8} - \dfrac{y^2}{8} = 1$

Since $a^2 = 8$, we have $a = 2\sqrt{2}$. Also,

$c = \sqrt{a^2 + b^2} = \sqrt{8+8} = \sqrt{16} = 4$. Thus,

$e = \frac{c}{a} = \frac{4}{2\sqrt{2}} = \frac{4\sqrt{2}}{4} = \sqrt{2} \approx 1.4$.

29. $16y^2 - 8x^2 = 16 \Rightarrow \dfrac{y^2}{1} - \dfrac{x^2}{2} = 1$

Since $a^2 = 1$, we have $a = 1$. Also,

$c = \sqrt{a^2 + b^2} = \sqrt{1+2} = \sqrt{3}$. Thus,

$e = \frac{c}{a} = \frac{\sqrt{3}}{1} = \sqrt{3} \approx 1.7$.

Copyright © 2013 Pearson Education, Inc.

31. x-intercepts ± 3; foci at $(-5, 0)$, $(5, 0)$

Since the center is halfway between the foci, the center is $(0,0)$. Since the foci are on a horizontal transverse axis, the equation has the form $\frac{x^2}{a^2} - \frac{y^2}{b^2} = 1$. The x-intercepts are also vertices, so $a = 3$ and thus $a^2 = 9$. With the given foci, we have $c = 5$. So given $c^2 = a^2 + b^2$, we have $c^2 = a^2 + b^2 \Rightarrow$ $5^2 = 3^2 + b^2 \Rightarrow 25 = 9 + b^2 \Rightarrow b^2 = 16$ Thus, the equation of the hyperbola is $\frac{x^2}{9} - \frac{y^2}{16} = 1$.

33. Vertices at $(0, 6)$, $(0, -6)$; asymptotes $y = \pm \frac{1}{2} x$

Since the center is halfway between the vertices, the center is $(0, 0)$. Since the vertices are on a vertical transverse axis, the equation has the form $\frac{y^2}{a^2} - \frac{x^2}{b^2} = 1$ and $a = 6$, which implies $a^2 = 36$. Also, since $y = \pm \frac{1}{2} x = \pm \frac{a}{b} x$, we have $\frac{a}{b} = \frac{1}{2} \Rightarrow \frac{6}{b} = \frac{1}{2} \Rightarrow b = 12 \Rightarrow b^2 = 144$. Thus, the equation of the hyperbola is $\frac{y^2}{36} - \frac{x^2}{144} = 1$

35. Vertices at $(-3, 0)$, $(3, 0)$; passing through $(-6, -1)$

Since the center is halfway between the vertices, the center is $(0,0)$. Since the vertices are on a horizontal transverse axis, the equation has the form $\frac{x^2}{a^2} - \frac{y^2}{b^2} = 1$ and $a = 3$, which implies $a^2 = 9$. Thus, we have $\frac{x^2}{9} - \frac{y^2}{b^2} = 1$. Since the hyperbola goes through the point $(-6, -1)$, substitute $x = -6$ and $y = -1$ into the equation and solve for b^2.

$\frac{x^2}{9} - \frac{y^2}{b^2} = 1 \rightarrow \frac{(-6)^2}{9} - \frac{(-1)^2}{b^2} = 1 \rightarrow \frac{36}{9} - \frac{1}{b^2} = 1 \Rightarrow$ $4 - \frac{1}{b^2} = 1 \Rightarrow -\frac{1}{b^2} = -3 \Rightarrow b^2 = \frac{1}{3}$

Thus, the equation is

$\frac{x^2}{9} - \frac{y^2}{\frac{1}{3}} = 1$ or $\frac{x^2}{9} - 3y^2 = 1$.

37. Foci at $\left(0, \sqrt{13}\right)$, $\left(0, -\sqrt{13}\right)$; asymptotes $y = \pm 5x$

Since the center is halfway between the foci, the center is $(0, 0)$. Since the foci are on a vertical transverse axis, the equation has the form $\frac{y^2}{a^2} - \frac{x^2}{b^2} = 1$ and $c = \sqrt{13}$, which implies $c^2 = 13$. Also, since $y = \pm 5x = \pm \frac{5}{1} x = \pm \frac{a}{b} x$, we have $\frac{a}{b} = \frac{5}{1} \Rightarrow a = 5b$ (1). Since $c^2 = a^2 + b^2$, we have $13 = a^2 + b^2$ (2). Substituting equation (1) into equation (2) and solving for b^2, we have

$13 = a^2 + b^2 \Rightarrow 13 = (5b)^2 + b^2 \Rightarrow$ $13 = 25b^2 + b^2 \Rightarrow 13 = 26b^2 \Rightarrow b^2 = \frac{1}{2}$. Since $13 = a^2 + b^2$ and $b^2 = \frac{1}{2}$, we have $13 = a^2 + \frac{1}{2} \Rightarrow 13 - \frac{1}{2} = a^2 \Rightarrow a^2 = \frac{26}{2} - \frac{1}{2} = \frac{25}{2}$. Thus, the equation of the hyperbola is $\frac{y^2}{\frac{25}{2}} - \frac{x^2}{\frac{1}{2}} = 1$ or $\frac{2y^2}{25} - 2x^2 = 1$.

39. Vertices at $(4, 5)$, $(4, 1)$; asymptotes $y = \pm 7(x - 4) + 3$

Since the center is halfway between the vertices, the center is located at $\left(\frac{4+4}{2}, \frac{5+1}{2}\right) = \left(\frac{8}{2}, \frac{6}{2}\right) = (4, 3)$. (This could have also been determined from the equation of the asymptotes.) Since the vertices are on a vertical transverse axis, the equation has the form $\frac{(y-k)^2}{a^2} - \frac{(x-h)^2}{b^2} = 1$. Since the distance between the vertices is 4, we have $2a = 4 \Rightarrow a = 2$. The slopes of the asymptotes $\pm 7 = \pm \frac{a}{b}$. This yields $\frac{7}{1} = \frac{a}{b} \Rightarrow 7b = a \Rightarrow b = \frac{a}{7}$. Since $a = 2 \Rightarrow b = \frac{2}{7}$. With $a = 2 \Rightarrow a^2 = 4$, $b = \frac{2}{7} \rightarrow b^2 = \frac{4}{49}$, $h = 4$, and $k = 3$ we have $\frac{(y-3)^2}{4} - \frac{(x-4)^2}{\frac{4}{49}} = 1$ or $\frac{(y-3)^2}{4} - \frac{49(x-4)^2}{4} = 1$.

Copyright © 2013 Pearson Education, Inc.

41. Center at $(1, -2)$; focus at $(-2, -2)$; vertex at $(-1, -2)$

Since the center, focus, and vertices are on a horizontal transverse axis, the equation has the

form $\frac{(x-h)^2}{a^2} - \frac{(y-k)^2}{b^2} = 1$. The distance from the

center to the vertex being 2 implies $a = 2$. The focus, $(-2, -2)$, is 3 units from the center, so

$c = 3$. Given that $c^2 = a^2 + b^2$, $a = 2$, and $c = 3$, we have

$c^2 = a^2 + b^2 \Rightarrow 3^2 = 2^2 + b^2 \Rightarrow 9 = 4 + b^2 \Rightarrow b^2 = 5$.

With $a = 2 \Rightarrow a^2 = 4$, $b^2 = 5$, $h = 1$, and

$k = -2$, we have $\frac{(x-1)^2}{4} - \frac{(y+2)^2}{5} = 1$.

43. Eccentricity 3; center at $(0, 0)$; vertex at $(0, 7)$

Since the center and vertex lie on a vertical transverse axis (the y-axis), the equation is of

the form $\frac{y^2}{a^2} - \frac{x^2}{b^2} = 1$. The distance between

the center and a vertex is 7, so $a = 7$. Use the

eccentricity to find c: $e = \frac{c}{a} \Rightarrow 3 = \frac{c}{7} \Rightarrow c = 21$

Now find the value of b^2 given $c^2 = a^2 + b^2$, $a = 7$, and $c = 21$.

$c^2 = a^2 + b^2 \Rightarrow 21^2 = 7^2 + b^2 \Rightarrow$

$441 = 49 + b^2 \Rightarrow b^2 = 392$

Since $a = 7 \Rightarrow a^2 = 49$ and $b^2 = 392$, the

equation of the hyperbola is $\frac{y^2}{49} - \frac{x^2}{392} = 1$.

45. Foci at $(9, -1)$, $(-11, -1)$; eccentricity $\frac{25}{9}$

Since the foci lie on a horizontal transverse axis, the equation is of the form

$\frac{(x-h)^2}{a^2} - \frac{(y-k)^2}{b^2} = 1$. Since the center is halfway

between the foci, the center is located at

$\left(\frac{9+(-11)}{2}, \frac{(-1)+(-1)}{2}\right) = \left(\frac{-2}{2}, \frac{-2}{2}\right) = (-1, -1)$. The

distance from the center to each foci is 10, so $c = 10$. Use the eccentricity to find a.

$e = \frac{c}{a} \Rightarrow \frac{25}{9} = \frac{10}{a} \Rightarrow 25a = 90 \Rightarrow a = \frac{90}{25} = \frac{18}{5}$

Now find the value of b^2 given $c^2 = a^2 + b^2$, $a = \frac{18}{5}$, and $c = 10$.

$c^2 = a^2 + b^2 \Rightarrow 10^2 = \left(\frac{18}{5}\right)^2 + b^2 \Rightarrow$

$100 = \frac{324}{25} + b^2 \Rightarrow$

$b^2 = 100 - \frac{324}{25} = \frac{2500}{25} - \frac{324}{25} = \frac{2176}{25}$

Since $a = \frac{18}{5} \Rightarrow a^2 = \frac{324}{25}$, $b^2 = \frac{2176}{25}$, $h = -1$,

and $k = 2$, we have $\frac{(x+1)^2}{\frac{324}{25}} - \frac{(y+1)^2}{\frac{2176}{25}} = 1$ or

$\frac{25(x+1)^2}{324} - \frac{25(y+1)^2}{2176} = 1$.

47. We solve $\frac{x^2}{4} - \frac{y^2}{16} = 1$ for y.

$\frac{x^2}{4} - \frac{y^2}{16} = 1 \Rightarrow \frac{x^2}{4} - 1 = \frac{y^2}{16} \Rightarrow$

$y^2 = 16\left(\frac{x^2}{4} - 1\right) \Rightarrow y^2 = 4x^2 - 16 \Rightarrow$

$y = \sqrt{4x^2 - 16} \Rightarrow y = \pm\sqrt{4\left(x^2 - 4\right)} \Rightarrow$

$y = \pm 2\sqrt{x^2 - 4}$

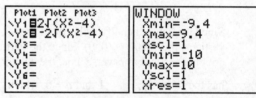

$Y_2 = -2\sqrt{x^2 - 4}$ $Y_1 = 2\sqrt{x^2 - 4}$

49. We solve $4y^2 - 36x^2 = 144$ for y.

$4y^2 - 36x^2 = 144 \Rightarrow 4y^2 = 36x^2 + 144 \Rightarrow$

$y^2 = 9x^2 + 36 \Rightarrow y = \pm\sqrt{9x^2 + 36}$

$y = \pm\sqrt{9\left(x^2 + 4\right)} \Rightarrow y = \pm 3\sqrt{x^2 + 4}$

$Y_1 = 3\sqrt{x^2 + 4}$ $Y_2 = -3\sqrt{x^2 + 4}$

Copyright © 2013 Pearson Education, Inc.

51. Solving $\frac{x^2}{4} - y^2 = 1$ for y, we have

$\frac{x^2}{4} - y^2 = 1 \Rightarrow y = \pm\sqrt{\frac{x^2}{4} - 1}$. The positive

square root is $y = \sqrt{\frac{x^2}{4} - 1} \Rightarrow y = \sqrt{\frac{x^2}{4} - \frac{4}{4}} \Rightarrow$

$y = \sqrt{\frac{x^2 - 4}{4}} \Rightarrow y = \frac{1}{2}\sqrt{x^2 - 4}$.

53. $y = \frac{1}{2}\sqrt{x^2 - 4}$

At $x = 50$, $y \approx 24.98$.

55. Because $24.98 < 25$, the graph of

$y = \frac{1}{2}\sqrt{x^2 - 4}$ lies below the graph of

$y = \frac{1}{2}x$ when $x = 50$.

57. (a) We must determine a and b in the

equation $\frac{x^2}{a^2} - \frac{y^2}{b^2} = 1$. The asymptotes are

$y = x$ and $y = -x$, which have slopes of 1
and -1, respectively, so $a = b$. Look at the
small right triangle that is shown in
quadrant III. The line $y = x$ intersects the
x-axis to form 45° angles. Since the right
angle vertex of the triangle lies on the line
$y = x$, we know that this triangle is a
45°–45°–90° triangle (an isosceles right
triangle). Thus, both legs of the triangle
have length d, and by the Pythagorean

theorem, $c^2 = d^2 + d^2 \Rightarrow c^2 = 2d^2 \Rightarrow$

$c = d\sqrt{2}$ (1). Thus, the coordinates of N

are $\left(-d\sqrt{2}, 0\right)$. Since $a = b$, we have

$c^2 = a^2 + a^2 \Rightarrow c^2 = 2a^2 \Rightarrow c = a\sqrt{2}$ (2).

From equations (1) and (2), we have

$d\sqrt{2} = a\sqrt{2} \Rightarrow d = a$. Thus,

$a = b = d = 5 \times 10^{-14}$. The equation of the

trajectory of A is given by $\frac{x^2}{a^2} - \frac{y^2}{b^2} = 1$.

$$\frac{x^2}{a^2} - \frac{y^2}{b^2} = 1$$

$$\frac{x^2}{\left(5 \times 10^{-14}\right)^2} - \frac{y^2}{\left(5 \times 10^{-14}\right)^2} = 1$$

$$x^2 - y^2 = \left(5 \times 10^{-14}\right)^2$$

$$x^2 - y^2 = 25 \times 10^{-28}$$

$$x^2 - y^2 = 2.5 \times 10^{-27}$$

$$x^2 = y^2 + 2.5 \times 10^{-27}$$

$$x = \sqrt{y^2 + 2.5 \times 10^{-27}}$$

(We choose the positive square root since
the trajectory occurs only where $x > 0$.
This equation represents the right half of
the hyperbola, as shown in the figure on
page 612.)

(b) The minimum distance between their

centers is $c + a = d\sqrt{2} + d$.

$c + a = d\sqrt{2} + d$

$c + a = \left(5 \times 10^{-14}\right)\sqrt{2} + \left(5 \times 10^{-14}\right)$

$c + a = \left(\sqrt{2} + 1\right)\left(5 \times 10^{-14}\right)$

$c + a \approx 12.07 \times 10^{-14} \approx 1.2 \times 10^{-13}\,\text{m}$

59. Since the center is the origin and the foci lie on
a horizontal transverse axis, the equation has

the form $\frac{x^2}{a^2} - \frac{y^2}{b^2} = 1$. We know from pages

605 of the text that $\left|d\left(P, F'\right) - d\left(P, F\right)\right| = 2a$

where F and F' are the locations of the foci,
in this case $(-c, 0)$ and $(c, 0)$. Since $d = rt$,

the difference between $d\left(P, F'\right)$ and $d\left(P, F\right)$

is $330t$. Thus, $2a = 330t \Rightarrow a = \frac{330t}{2}$. Since

$c^2 = a^2 + b^2$, we have the following.

$c^2 = a^2 + b^2 \Rightarrow c^2 = \left(\frac{330t}{2}\right)^2 + b^2 \Rightarrow$

$b^2 = c^2 - \frac{330^2 t^2}{4} \Rightarrow b^2 = \frac{4c^2 - 330^2 t^2}{4}$

Thus, the equation of the hyperbola is

$\frac{x^2}{\left(\frac{330t}{2}\right)^2} - \frac{y^2}{\frac{4c^2 - 330^2 t^2}{4}} = 1$ or $\frac{x^2}{330^2 t^2} - \frac{y^2}{4c^2 - 330^2 t^2} = \frac{1}{4}$.

Copyright © 2013 Pearson Education, Inc.

61. (a) Rewriting the equation of the hyperbola we have the following.

$$400x^2 - 625y^2 = 250,000$$

$$\frac{x^2}{625} - \frac{y^2}{400} = 1 \Rightarrow \frac{x^2}{25^2} - \frac{y^2}{20^2} = 1$$

The two branches of a hyperbola are closest at the vertices and this distance is $2a$. Since $a = 25$, the buildings are 50 m apart at their closest point.

(b) First, solve for y when $x = 50$.

$$\frac{x^2}{25^2} - \frac{y^2}{20^2} = 1 \Rightarrow \frac{50^2}{25^2} - \frac{y^2}{20^2} = 1 \Rightarrow$$

$$\left(\frac{50^2}{25^2}\right)^2 - \frac{y^2}{20^2} = 1 \Rightarrow 2^2 - \frac{y^2}{20^2} = 1 \Rightarrow$$

$$4 - \frac{y^2}{20^2} = 1 \Rightarrow \frac{y^2}{20^2} = 3 \Rightarrow$$

$$y^2 = 20^2 \cdot 3 \Rightarrow y = 20\sqrt{3}$$

Thus, $d = 2y = 2 \cdot 20\sqrt{3} = 40\sqrt{3} \approx 69.3$ m.

Section 6.4 Summary of the Conic Sections

1. $x^2 + y^2 = 144 \Rightarrow (x - 0)^2 + (y - 0)^2 = 12^2$

The graph of this equation is a circle with center $(0, 0)$ and radius 12. Also, note in our original equation, the x^2- and y^2-terms have the same positive coefficient.

3. $y = 2x^2 + 3x - 4 \Rightarrow y = 2\left(x^2 + \frac{3}{2}x\right) - 4$

$$y = 2\left(x^2 + \frac{3}{2}x + \frac{9}{16} - \frac{9}{16}\right) - 4$$

$$y = 2\left(x + \frac{3}{4}\right)^2 + 2\left(-\frac{9}{16}\right) - 4$$

$$y = 2\left(x + \frac{3}{4}\right)^2 - \frac{9}{8} - 4$$

$$y = 2\left(x + \frac{3}{4}\right)^2 - \frac{41}{8} \Rightarrow y - \left(-\frac{41}{8}\right) = 2\left[x - \left(-\frac{3}{4}\right)\right]^2$$

The graph of this equation is a parabola opening upwards with a vertex of $\left(-\frac{3}{4}, -\frac{41}{8}\right)$.

Also, note our original equation has an x^2-term, but no y^2-term.

5. $x - 1 = -3(y - 4)^2$

The graph of this equation is a parabola opening to the left with a vertex of $(1, 4)$. Also, note when expanded, our original equation has a y^2-term, but no x^2-term.

7. $\dfrac{x^2}{49} + \dfrac{y^2}{100} = 1 \Rightarrow \dfrac{x^2}{7^2} + \dfrac{y^2}{10^2} = 1$

The graph of this equation is an ellipse centered at the origin and x-intercepts of 7 and –7, and y-intercepts of 10 and –10. Also, note in our original equation, the x^2- and y^2-terms both have different positive coefficients.

9. $\dfrac{x^2}{4} - \dfrac{y^2}{16} = 1 \Rightarrow \dfrac{x^2}{2^2} - \dfrac{y^2}{4^2} = 1$

The graph of this equation is a hyperbola centered at the origin with x-intercepts of 2 and –2, and asymptotes of $y = \pm\frac{4}{2}x = \pm 2x$.

Also, note in our original equation, the x^2- and y^2-terms have coefficient that are opposite in sign.

11. $\dfrac{x^2}{25} - \dfrac{y^2}{25} = 1 \Rightarrow \dfrac{x^2}{5^2} - \dfrac{y^2}{5^2} = 1$

The graph of this equation is a hyperbola centered at the origin with x-intercepts of 5 and –5, and asymptotes of $y = \pm\frac{5}{5}x = \pm x$.

Also, note in our original equation, the x^2- and y^2-terms have coefficients that are opposite in sign.

13. $\dfrac{x^2}{4} = 1 - \dfrac{y^2}{9} \Rightarrow \dfrac{x^2}{4} + \dfrac{y^2}{9} = 1 \Rightarrow$

$$\frac{(x - 0)^2}{2^2} + \frac{(y - 0)^2}{3^2} = 1$$

The equation is of the form $\frac{(x-h)^2}{b^2} + \frac{(y-k)^2}{a^2} = 1$ with $a = 3$, $b = 2$, $h = 0$, and $k = 0$, so the graph of the given equation is an ellipse.

15. $\dfrac{(x + 3)^2}{16} + \dfrac{(y - 2)^2}{16} = 1$

$$(x + 3)^2 + (y - 2)^2 = 16$$

$$\left[x - (-3)\right]^2 + (y - 2)^2 = 4^2$$

The equation is of the form $(x - h)^2 + (y - k)^2 = r^2$ with $r = 4$, $h = -3$, and $k = 2$, so the graph of the given equation is a circle.

Copyright © 2013 Pearson Education, Inc.

17. $x^2 - 6x + y = 0 \Rightarrow y = -x^2 + 6x$

$$y = -\left(x^2 - 6x + 9 - 9\right)$$
$$y = -(x-3)^2 + 9$$
$$y - 9 = -(x-3)^2$$

The equation is of the form $y - k = a(x-h)^2$ with $a = -1$, $h = 3$, and $k = 9$, so the graph of the given equation is a parabola.

19. $4(x-3)^2 + 3(y+4)^2 = 0$

$$\frac{4(x-3)^2}{12} + \frac{3(y+4)^2}{12} = 0$$
$$\frac{(x-3)^2}{3} + \frac{[y-(-4)]^2}{4} = 0$$

The graph is the point $(3, -4)$.

21. $x - 4y^2 - 8y = 0 \Rightarrow x = 4y^2 + 8y$

$$x = 4\left(y^2 + 2y + 1 - 1\right)$$
$$x = 4(y+1)^2 - 4$$
$$x - (-4) = 4\left[y - (-1)\right]^2$$

The equation is of the form $x - h = a(y-k)^2$ with $a = 4$, $h = -4$, and $k = -1$, so the graph of the given equation is a parabola.

23.
$$4x^2 - 24x + 5y^2 + 10y + 41 = 0$$
$$4\left(x^2 - 6x + 9 - 9\right) + 5\left(y^2 + 2y + 1 - 1\right) = -41$$
$$4(x-3)^2 - 36 + 5(y+1)^2 - 5 = -41$$
$$4(x-3)^2 + 5(y+1)^2 = -41 + 36 + 5$$
$$4(x-3)^2 + 5(y+1)^2 = 0$$
$$\frac{4(x-3)^2}{20} + \frac{5(y+1)^2}{20} = 0$$
$$\frac{(x-3)^2}{5} + \frac{[y-(-1)]^2}{4} = 0$$

The graph is the point $(3, -1)$.

25. $\dfrac{x^2}{4} + \dfrac{y^2}{4} = -1 \Rightarrow x^2 + y^2 = -4$

A sum of squares can never be negative. This equation has no graph.

27. $x^2 = 25 + y^2 \Rightarrow x^2 - y^2 = 25$

$$\frac{x^2}{25} - \frac{y^2}{25} = 1 \Rightarrow \frac{(x-0)^2}{5^2} - \frac{(y-0)^2}{5^2} = 1$$

The equation is of the form $\dfrac{(x-h)^2}{a^2} - \dfrac{(y-k)^2}{b^2} = 1$ with $a = 5$, $b = 5$, $h = 0$, and $k = 0$, so the graph of the given equation is a hyperbola with center $(0, 0)$, vertices $(-5, 0)$ and $(5, 0)$, and asymptotes $y = \pm x$.

$$x^2 = 25 + y^2$$
$$\frac{x^2}{25} - \frac{y^2}{25} = 1$$

29. $x^2 = 4y - 8 \Rightarrow x^2 = 4(y-2) \Rightarrow$

$$y - 2 = \tfrac{1}{4}(x-0)^2$$

The equation is of the form $y - k = a(x-h)^2$ with $a = \tfrac{1}{4}$, $h = 0$, and $k = 2$, so the graph of the given equation is a parabola with vertex $(0, 2)$ and vertical axis $x = 0$ (the y-axis). Use the vertex and axis and plot a few additional points.

x	y
-4	6
-2	3
0	2
2	3
4	6

$$x^2 = 4y - 8$$
$$y = \tfrac{1}{4}x^2 + 2$$

31. $y^2 - 4y = x + 4 \Rightarrow y^2 - 4y + 4 - 4 = x + 4$

$$(y-2)^2 - 4 = x + 4$$
$$x + 8 = (y-2)^2 \Rightarrow x - (-8) = (y-2)^2$$

The equation is of the form $x - h = a(y-k)^2$ with $a = 1$, $h = -8$, and $k = 2$, so the graph of the given equation is a parabola with vertex $(-8, 2)$ and horizontal axis $y = 2$. Use the vertex and axis and plot a few additional points.

(*continued on next page*)

Copyright © 2013 Pearson Education, Inc.

(*continued*)

x	y
0	−0.8
−4	0
−8	2
−4	4
0	4.8

$$y^2 - 4y = x + 4$$
$$(y-2)^2 = x + 8$$

33.
$$3x^2 + 6x + 3y^2 - 12y = 12$$
$$x^2 + 2x + y^2 - 4y = 4$$
$$\left(x^2 + 2x + 1 - 1\right) + \left(y^2 - 4y + 4 - 4\right) = 4$$
$$(x+1)^2 - 1 + (y-2)^2 - 4 = 4$$
$$(x-1)^2 + (y-2)^2 = 4 + 1 + 4$$
$$(x+1)^2 + (y-2)^2 = 9$$
$$\left[x - (-1)\right]^2 + (y-2)^2 = 3^2$$

The equation is of the form
$$(x-h)^2 + (y-k)^2 = r^2 \text{ with}$$
$r = 3$, $h = -1$, and $k = 2$, so the graph of the given equation is a circle with center $(-1, 2)$ and radius 3.

$$3x^2 + 6x + 3y^2 - 12y = 12$$
$$(x+1)^2 + (y-2)^2 = 9$$

35.
$$4x^2 - 8x + 9y^2 - 36y = -4$$
$$4\left(x^2 - 2x + 1 - 1\right) + 9\left(y^2 - 4y + 4 - 4\right) = -4$$
$$4(x-1)^2 - 4 + 9(y-2)^2 - 36 = -4$$
$$4(x-1)^2 + 9(y-2)^2 = 36$$
$$\frac{4(x-1)^2}{36} + \frac{9(y-2)^2}{36} = 1$$
$$\frac{(x-1)^2}{9} + \frac{(y-2)^2}{4} = 1$$
$$\frac{(x-1)^2}{3^2} + \frac{(y-2)^2}{2^2} = 1$$

The equation is of the form $\frac{(x-h)^2}{a^2} + \frac{(y-k)^2}{b^2} = 1$ with $a = 3$, $b = 2$, $h = 1$, and $k = 2$, so the graph of the given equation is an ellipse with center $(1, 2)$ and vertices $(-2, 2)$, $(4, 2)$, $(1, 0)$ and $(1, 4)$.

$$4x^2 - 8x + 9y^2 - 36y = -4$$
$$\frac{(x-1)^2}{9} + \frac{(y-2)^2}{4} = 1$$

37. The definition of an ellipse states "an ellipse is the set of all points in a plane the sum of whose distances from two fixed points is constant." Therefore, the set of all points in a plane for which the sum of the distances from the points $(5, 0)$ and $(-5, 0)$ is 14 is an ellipse with foci $(5, 0)$ and $(-5, 0)$.

39. Refer to the Geometric Definition of a Conic Section on page 618 of the text. It gives the relation
[distance of *P* from *F*] $= e \cdot$ [distance of *P* from *L*].
In this exercise, $e = 1\frac{1}{2} = \frac{3}{2}$. Since $e > 1$, this is a hyperbola.

41. From the graph, the coordinates of *P* (a point on the graph) are $(-3, 8)$, the coordinates of *F* (a focus) are $(3, 0)$, the equation of *L* (the directrix) is $x = 27$. By the distance formula, the distance from *P* to *F* is
$$\sqrt{(x_2 - x_1)^2 + (y_2 - y_1)^2}$$
$$= \sqrt{\left[3 - (-3)\right]^2 + (0-8)^2} = \sqrt{6^2 + (-8)^2}$$
$$= \sqrt{36 + 64} = \sqrt{100} = 10$$
The distance from a point to a line is defined as the perpendicular distance, so the distance from *P* to *L* is $|27 - (-3)| = 30$. Thus,
$$e = \frac{\text{Distance of } P \text{ from } F}{\text{Distance of } P \text{ from } L} = \frac{10}{30} = \frac{1}{3}.$$

Copyright © 2013 Pearson Education, Inc.

43. From the graph, we see that $F = \left(\sqrt{2}, 0\right)$ and

L is the vertical line $x = -\sqrt{2}$. Choose $(0,0)$, the vertex of the parabola, as P.

Distance of P from $F = \sqrt{2}$, and distance of P from $L = \sqrt{2}$. Thus, we have

$$e = \frac{\text{Distance of } P \text{ from } F}{\text{Distance of } P \text{ from } L} = \frac{\sqrt{2}}{\sqrt{2}} = 1.$$

45. From the graph, we see that $P = (9, -7.5)$, $F = (9, 0)$ and L is the vertical line $x = 4$. Distance of P from $F = 7.5$, and distance of P from $L = 5$. Thus,

$$e = \frac{\text{Distance of } P \text{ from } F}{\text{Distance of } P \text{ from } L} = \frac{7.5}{5} = 1.5.$$

47. If $\dfrac{k}{\sqrt{D}} = \dfrac{2.82 \times 10^7}{\sqrt{42.5 \times 10^6}} = \dfrac{2.82 \times 10^7}{\sqrt{42.5} \times 10^3}$

$\approx 0.432568 \times 10^4 \approx 4326$ and $V = 2090$, then

we have $V < \dfrac{k}{\sqrt{D}}$. Thus, the shape of the

satellite's trajectory was elliptical.

49. Answers will vary.

51. To enter the relation $\dfrac{x^2}{16} + \dfrac{y^2}{12} = 1$ into the

graphing calculator, we must solve for y.

$$\frac{x^2}{16} + \frac{y^2}{12} = 1 \Rightarrow \frac{y^2}{12} = 1 - \frac{x^2}{16}$$

$$y^2 = 12\left(1 - \frac{x^2}{16}\right) \Rightarrow y = \pm\sqrt{12\left(1 - \frac{x^2}{16}\right)}$$

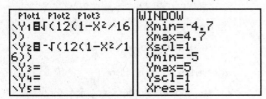

A sampling of points is as follows.

The distance of P from

$$(2,0) \approx \sqrt{(2-0)^2 + (0-3.4641016)^2}$$
$$= \sqrt{4 + (-3.4641016)^2} \approx 3.999999987$$

The distance of P from the line $x = 8$ is

$|0 - 8| = 8$. We have $3.999999987 \approx \frac{1}{2}(8) = 4$.

The distance of P from

$$(2,0) \approx \sqrt{(2-3)^2 + (0-2.2912878)^2}$$
$$= \sqrt{1 + (-2.2912878)^2} \approx 2.499999956$$

The distance of P from the line $x = 8$ is

$|3 - 8| = 5$. We have

$2.499999956 \approx \frac{1}{2}(5) = 2.5$.

The distance of P from

$$(2,0) \approx \sqrt{\left[2-(-1.8)\right]^2 + \left[0-(-3.093542)\right]^2}$$
$$= \sqrt{14.44 + (3.093542)^2} \approx 4.900000215$$

The distance of P from the line $x = 8$ is

$|-1.8 - 8| = 9.8$. We have

$4.900000215 \approx \frac{1}{2}(9.8) = 4.9$.

The distance of P from

$$(2,0) \approx \sqrt{\left[2-(-4)\right]^2 + (0-0)^2} = \sqrt{36} = 6$$

The distance of P from the line $x = 8$ is

$|-4 - 8| = 12$. We have $6 = \frac{1}{2}(12)$.

Copyright © 2013 Pearson Education, Inc.

Chapter 6   Review Exercises

1. $x = 4(y-5)^2 + 2 \Rightarrow x - 2 = 4(y-5)^2$

The vertex is $(2,5)$. The graph opens to the right and has the same shape as $x = 4y^2$. It is a translation 5 units up and 2 units to the right of the graph of $x = 4y^2$. The domain is $[2,\infty)$. The range is $(-\infty,\infty)$. The graph is symmetric about its axis, the horizontal line $y = 5$. Use the vertex and axis and plot a few additional points.

x	y
18	7
6	6
2	5
6	4
18	3

3. $x = 5y^2 - 5y + 3$

Complete the square on y to find the vertex and the axis.

$x = 5y^2 - 5y + 3$

$x = 5\left(y^2 - y \quad \right) + 3$

$x = 5\left(y^2 - y + \frac{1}{4} - \frac{1}{4}\right) + 3$

$x = 5\left(y - \frac{1}{2}\right)^2 + 5\left(-\frac{1}{4}\right) + 3$

$x = 5\left(y - \frac{1}{2}\right)^2 + \left(-\frac{5}{4}\right) + \frac{12}{4}$

$x = 5\left(y - \frac{1}{2}\right)^2 + \frac{7}{4} \Rightarrow x - \frac{7}{4} = 5\left(y - \frac{1}{2}\right)^2$

The vertex is $\left(\frac{7}{4}, \frac{1}{2}\right)$. The graph opens to the right and has the same shape as $x = 5y^2$, translated $\frac{7}{4}$ unit to the right and $\frac{1}{2}$ unit up. The domain is $\left[\frac{7}{4}, \infty\right)$. The range is $(-\infty,\infty)$. The graph is symmetric about its axis, the horizontal line $y = \frac{1}{2}$. Use the vertex and axis and plot a few additional points.

x	y
$\frac{27}{4}$	$\frac{3}{2}$
3	1
$\frac{7}{4}$	$\frac{1}{2}$
3	0
$\frac{27}{4}$	$-\frac{1}{2}$

5. The equation $y^2 = -\frac{2}{3}x$ has the form $y^2 = 4px$, so $4p = -\frac{2}{3}$, from which $p = -\frac{1}{6}$. Since the y-term is squared, the parabola is horizontal, with focus $(p,0) = \left(-\frac{1}{6}, 0\right)$ and directrix, $x = -p$, is $x = \frac{1}{6}$. The vertex is $(0,0)$, and the axis of the parabola is the x-axis. The domain is $(-\infty,0]$. The range is $(-\infty,\infty)$. Use this information and plot a few additional points.

x	y
-6	2
$-\frac{3}{2}$	1
0	0
$-\frac{3}{2}$	-1
-6	-2

7. The equation $3x^2 = y \Rightarrow x^2 = \frac{1}{3}y$ has the form $x^2 = 4py$, so $4p = \frac{1}{3}$, from which $p = \frac{1}{12}$. Since the x-term is squared, the parabola is vertical, with focus $(0, p) = \left(0, \frac{1}{12}\right)$ and directrix, $y = -p$, is $y = -\frac{1}{12}$. The vertex is $(0, 0)$, and the axis of the parabola is the y-axis. The domain is $(-\infty,\infty)$. The range is $[0, \infty)$. Use this information and plot a few additional points.

x	y
2	12
1	3
0	0
-1	3
-2	12

Copyright © 2013 Pearson Education, Inc.

9. A parabola with focus (4, 0) and vertex at the origin is a horizontal parabola. The equation has the form $y^2 = 4px$. Since $p = 4$ is positive, it opens to the right. Substituting 4 for p, we find that an equation for this parabola is $y^2 = 4(4)x \Rightarrow y^2 = 16x$.

11. A parabola passing through $(-3, 4)$, opening up, and vertex at the origin has an equation of the form $x^2 = 4py$. Use this equation with the coordinates of the point $(-3, 4)$ to find the value of p.

$$x^2 = 4py \Rightarrow (-3)^2 = 4p \cdot 4 \Rightarrow 9 = 16p \Rightarrow \tfrac{9}{16} = p$$

Thus, an equation of the parabola is

$$x^2 = 4\left(\tfrac{9}{16}\right)y \Rightarrow x^2 = \tfrac{9}{4}y.$$

13. $y^2 + 9x^2 = 9 \Rightarrow \dfrac{x^2}{1} + \dfrac{y^2}{9} = 1 \Rightarrow$

$$\dfrac{(x-0)^2}{1^2} + \dfrac{(y-0)^2}{3^2} = 1$$

The equation is of the form $\dfrac{(x-h)^2}{b^2} + \dfrac{(y-k)^2}{a^2} = 1$ with $a = 3$, $b = 1$, $h = 0$, and $k = 0$, so the graph of the given equation is an ellipse.

15. $3y^2 - 5x^2 = 30 \Rightarrow \dfrac{y^2}{10} - \dfrac{x^2}{6} = 1 \Rightarrow$

$$\dfrac{(y-0)^2}{\left(\sqrt{10}\right)^2} - \dfrac{(x-0)^2}{\left(\sqrt{6}\right)^2} = 1$$

The equation is of the form $\dfrac{(y-k)^2}{a^2} - \dfrac{(x-h)^2}{b^2} = 1$ with $a = \sqrt{10}$, $b = \sqrt{6}$, $h = 0$, and $k = 0$, so the graph of the given equation is a hyperbola.

17. $4x^2 - y = 0 \Rightarrow y = 4x^2 \Rightarrow y - 0 = 4(x - 0)^2$

The equation is of the form $y - k = a(x - h)^2$ with $a = 4$, $h = 0$, and $k = 0$, so the graph of the given equation is a parabola.

19.
$$4x^2 - 8x + 9y^2 + 36y = -4$$
$$4\left(x^2 - 2x + 1 - 1\right) + 9\left(y^2 + 4y + 4 - 4\right) = -4$$
$$4(x-1)^2 - 4 + 9(y+2)^2 - 36 = -4$$
$$4(x-1)^2 + 9(y+2)^2 = -4 + 4 + 36$$
$$4(x-1)^2 + 9(y+2)^2 = 36$$
$$\dfrac{(x-1)^2}{9} + \dfrac{(y+2)^2}{4} = 1$$
$$\dfrac{(x-1)^2}{3^2} + \dfrac{[y-(-2)]^2}{2^2} = 1$$

The equation is of the form $\dfrac{(x-h)^2}{a^2} + \dfrac{(y-k)^2}{b^2} = 1$ with $a = 3$, $b = 2$, $h = 1$, and $k = -2$, so the graph of the given equation is an ellipse.

21. $4x^2 + y^2 = 36 \Rightarrow \dfrac{x^2}{9} + \dfrac{y^2}{36} = 1 \Rightarrow$

$$\dfrac{(x-0)^2}{3^2} + \dfrac{(y-0)^2}{6^2} = 1$$

The equation is of the form $\dfrac{(x-h)^2}{b^2} + \dfrac{(y-k)^2}{a^2} = 1$ with $a = 6$, $b = 3$, $h = 0$, and $k = 0$, so the graph of the given equation is an ellipse. It is centered at the origin with vertices at $(0, -6)$ and $(0, 6)$ and endpoints of the minor axis at $(-3, 0)$ and $(3, 0)$. The correct graph is F.

23.
$$(x-2)^2 + (y+3)^2 = 36$$
$$(x-2)^2 + [y-(-3)]^2 = 6^2$$

The equation is of the form $(x-h)^2 + (y-k)^2 = r^2$ with $r = 6$, $h = 2$, and $k = -3$, so the graph of the given equation is a circle. This circle is centered at $(2, -3)$ and has radius 6. The correct graph is A.

25.
$$(y-1)^2 - (x-2)^2 = 36$$
$$\dfrac{(y-1)^2}{36} - \dfrac{(x-2)^2}{36} = 1$$
$$\dfrac{(y-1)^2}{6^2} - \dfrac{(x-2)^2}{6^2} = 1$$

The equation is of the form $\dfrac{(y-k)^2}{a^2} - \dfrac{(x-h)^2}{b^2} = 1$ with $a = 6$, $b = 6$, $h = 2$, and $k = 1$, so the graph of the given equation is a hyperbola. This hyperbola is centered at $(2, 1)$, opening upward and downward, with vertices at $(2, -5)$ and $(2, 7)$. The correct graph is B.

Copyright © 2013 Pearson Education, Inc.

27. $\dfrac{x^2}{4}+\dfrac{y^2}{9}=1\Rightarrow\dfrac{(x-0)^2}{2^2}+\dfrac{(y-0)^2}{3^2}=1$

The graph is an ellipse centered at the origin with domain [–2, 2], range [–3, 3], and vertices at (0, –3) and (0, 3). The endpoints of the minor axis are (–2, 0) and (2, 0).

29. $\dfrac{x^2}{64}-\dfrac{y^2}{36}=1\Rightarrow\dfrac{(x-0)^2}{8^2}-\dfrac{(y-0)^2}{6^2}=1$

The graph is a hyperbola centered at the origin with domain $(-\infty,-8]\cup[8,\infty)$, range $(-\infty,\infty)$, and vertices (–8, 0) and (8, 0). The asymptotes are the lines $y=\pm\dfrac{6}{8}x\Rightarrow$

$y=\pm\dfrac{3}{4}x.$

31. $\dfrac{(x+1)^2}{16}+\dfrac{(y-1)^2}{16}=1$

$(x+1)^2+(y-1)^2=16$

$\left[x-(-1)\right]^2+(y-1)^2=4^2$

The graph is a circle centered at $(-1,1)$ and radius 4. Its domain is $[-1-4,-1+4]=[-5,3].$ The range is $[1-4,1+4]=[-3,5].$

33. $4x^2+9y^2=36\Rightarrow\dfrac{x^2}{9}+\dfrac{y^2}{4}=1\Rightarrow$

$\dfrac{(x-0)^2}{3^2}+\dfrac{(y-0)^2}{2^2}=1$

The graph is an ellipse centered at the origin with domain [–3, 3], range [–2, 2], vertices at (–3, 0) and (3, 0). The endpoints of the minor axis at (0, –2) and (0, 2).

$4x^2+9y^2=36$

35. $\dfrac{(x-3)^2}{4}+(y+1)^2=1\Rightarrow$

$\dfrac{(x-3)^2}{2^2}+\dfrac{\left[y-(-1)\right]^2}{1^2}=1$

The graph is an ellipse centered at (3, –1). The vertices are (3 – 2, –1) = (1, –1) and (3 + 2, –1) = (5, –1). The endpoints of the minor axis are (3, –1 – 1) = (3, –2) and (3, –1 + 1) = (3, 0). The domain is [1, 5] and the range is [–2, 0].

$\dfrac{(x-3)^2}{4}+(y+1)^2=1$	$\dfrac{(y+2)^2}{4}-\dfrac{(x+3)^2}{9}=1$
Exercise 35	Exercise 37

37. $\dfrac{(y+2)^2}{4}-\dfrac{(x+3)^2}{9}=1\Rightarrow$

$\dfrac{\left[y-(-2)\right]^2}{2^2}-\dfrac{\left[x-(-3)\right]^2}{3^2}=1$

The graph is a hyperbola centered at $(-3,-2)$. The vertices are located at $(-3,-2-2)=(-3,-4)$ and $(-3,-2+2)=(-3,0)$. The domain is $(-\infty,\infty)$ and the range is $(-\infty,-4]\cup[0,\infty)$. The asymptotes are the lines

$y-(-2)=\pm\dfrac{2}{3}\left[x-(-3)\right]\Rightarrow$

$y+2=\pm\dfrac{2}{3}(x+3)\Rightarrow y=-2\pm\dfrac{2}{3}(x+3).$

Copyright © 2013 Pearson Education, Inc.

39. $x^2 - 4x + y^2 + 6y = -12$

Complete the squares on x and y:

$$x^2 - 4x + y^2 + 6y = -12$$
$$\left(x^2 - 4x + \quad\right) + \left(y^2 + 6y + \quad\right) = -12$$
$$\left(x^2 - 4x + 4\right) + \left(y^2 + 6y + 9\right) = -12 + 4 + 9$$
$$(x - 2)^2 + (y + 3)^2 = 1$$

This is the graph of a circle with center $(2, -3)$ and radius 1. The domain is $[1, 3]$, and the range is $[-4, -2]$.

$$x^2 - 4x + y^2 + 6y = -12$$
$$(x - 2)^2 + (y + 3)^2 = 1$$
Exercise 39

$$5x^2 + 20x + 2y^2 - 8y = -18$$
$$\frac{(x + 2)^2}{2} + \frac{(y - 2)^2}{5} = 1$$
Exercise 41

41. $5x^2 + 20x + 2y^2 - 8y = -18$

Complete the squares on x and y :

$$5x^2 + 20x + 2y^2 - 8y = -18$$
$$5\left(x^2 + 4x + \quad\right) + 2\left(y^2 - 4y + \quad\right) = -18$$
$$5\left(x^2 + 4x + 4\right) + 2\left(y^2 - 4y + 4\right) = -18 + 20 + 8$$
$$5(x + 2)^2 + 2(y - 2)^2 = 10$$
$$\frac{(x + 2)^2}{2} + \frac{(y - 2)^2}{5} = 1$$

$a^2 = 2 \Rightarrow a = \sqrt{2}$ and $b^2 = 5 \Rightarrow b = \sqrt{5}$. The graph is an ellipse with center $(-2, 2)$ and vertices $\left(-2, 2 + \sqrt{5}\right)$ and $\left(-2, 2 - \sqrt{5}\right)$. The domain is $\left[-2 - \sqrt{2}, -2 + \sqrt{2}\right]$, and the range is $\left[2 - \sqrt{5}, 2 + \sqrt{5}\right]$.

43. $\dfrac{x}{3} = -\sqrt{1 - \dfrac{y^2}{16}}$

Square both sides to get $\frac{x^2}{9} = 1 - \frac{y^2}{16}$ or

$\frac{x^2}{9} + \frac{y^2}{16} = 1$ or $\frac{x^2}{3^2} + \frac{y^2}{4^2} = 1$. This is the equation of an ellipse with center $(0, 0)$ and vertices $(0, -4)$ and $(0, 4)$. It has endpoints of the minor axis of $(-3, 0)$ and $(3, 0)$. The graph of the original equation is the left half of this ellipse.

The domain is $[-3, 0]$ and the range is $[-4, 4]$. The vertical line test shows that this relation is not a function.

$$\frac{x}{3} = -\sqrt{1 - \frac{y^2}{16}}$$
Exercise 43

$$y = -\sqrt{1 + x^2}$$
Exercise 45

45. $y = -\sqrt{1 + x^2}$

Square both sides of the equation to get

$y^2 = 1 + x^2$ or $\frac{y^2}{1} - \frac{x^2}{1} = 1$ or $\frac{y^2}{1^2} - \frac{x^2}{1^2} = 1$. This is the equation of a hyperbola with center $(0, 0)$ and vertices $(0, -1)$ and $(0, 1)$. Since $a = 1$, $b = 1$, and $y = \pm\frac{a}{b}x$, we have asymptotes $y = \pm\frac{1}{1}x \Rightarrow y = \pm x$. The original equation is the bottom half of the hyperbola. The domain is $(-\infty, \infty)$. The range is $\left(-\infty, -1\right]$. The vertical line test shows this is a graph of a function.

47. Ellipse; vertex at $(0, -4)$, focus at $(0, -2)$, center at the origin

Since the vertex is $(0, -4)$ and the focus is $(0, -2)$, we know that the major axis is vertical and thus, the equation is of the form $\frac{x^2}{b^2} + \frac{y^2}{a^2} = 1$. Since the ellipse is centered at the origin, we have $a = 4$ which implies $a^2 = 16$. Since the ellipse is centered at the origin and a focus is at $(0, 2)$, we have $c = 2$ which implies $c^2 = 4$. Since $c^2 = a^2 - b^2$, we have $c^2 = a^2 - b^2 \Rightarrow 4 = 16 - b^2 \Rightarrow b^2 = 12$. Thus, the equation of the ellipse is $\frac{x^2}{12} + \frac{y^2}{16} = 1$.

49. Hyperbola; focus at $(0, 5)$ transverse axis of length 8, center at the origin

Since the focus is $(0, 5)$ we know that the transverse axis is vertical and thus, the equation is of the form $\frac{y^2}{a^2} - \frac{x^2}{b^2} = 1$. Since the length of the transverse axis is 8, we have $2a = 8 \Rightarrow a = 4$. This implies that $a^2 = 16$.

(*continued on next page*)

Copyright © 2013 Pearson Education, Inc.

(*continued*)

Since the hyperbola is centered at the origin and a focus is at $(0, 5)$ we have $c = 5$ which implies $c^2 = 25$. Since $c^2 = a^2 + b^2$, we have $25 = 16 + b^2 \Rightarrow b^2 = 9$. Thus, the equation of the hyperbola is $\frac{y^2}{16} - \frac{x^2}{9} = 1$.

51. Parabola with focus at (3, 2) and directrix $x = -3$

Since the directrix is $x = -3$, a vertical line, the parabola is of the form $(y - k)^2 = 4p(x - h)$. Since the vertex is halfway between the focus and the directrix, we know that the vertex is $(0, 2) = (h, k)$. Because of the orientation of the directrix versus the focus, we know that the parabola opens to the right, so the p value must be positive. By examining the distance between the vertex and the focus or the directrix and the vertex, we know that $p = 3$. Thus, the equation of the parabola is

$$(y - 2)^2 = 4 \cdot 3(x - 0) \text{ or } (y - 2)^2 = 12x.$$

53. Ellipse with foci at (–2, 0) and (2, 0) and major axis of length 10

Since the foci are at (–2, 0) and (2, 0), we know that major axis is horizontal and the ellipse is centered at the origin. Thus, the equation is of the form $\frac{x^2}{a^2} + \frac{y^2}{b^2} = 1$. Since the foci are at (–2, 0) and (2, 0), we know that $c = 2$, which implies $c^2 = 4$. Since the major axis is of length 10 we have $2a = 10$. This implies that $a = 5$ and $a^2 = 25$. Since $c^2 = a^2 - b^2$, we have

$c^2 = a^2 - b^2 \Rightarrow 4 = 25 - b^2 \Rightarrow b^2 = 21$. Thus, the equation of the ellipse is $\frac{x^2}{25} + \frac{y^2}{21} = 1$.

55. Hyperbola with x-intercepts ±3; foci at (–5, 0), (5, 0)

Since the center is halfway between the foci, the center is at the origin. Since the foci are on a horizontal transverse axis, the equation has the form $\frac{x^2}{a^2} - \frac{y^2}{b^2} = 1$. The x-intercepts are also vertices, so $a = 3$. This implies $a^2 = 9$. Since the foci at (–5, 0) and (5, 0), we have $c = 5$.

This implies $c^2 = 25$. Since $c^2 = a^2 + b^2$, we have $c^2 = a^2 + b^2 \Rightarrow 25 = 9 + b^2 \Rightarrow b^2 = 16$. Thus, the equation of the hyperbola is

$\frac{x^2}{9} - \frac{y^2}{16} = 1$.

57. The points $F'(0,0)$ and $F(4,0)$ are the foci, so the center of the ellipse is $(2,0) = (h,k)$. Since the foci are on a horizontal axis, the equation is of the form $\frac{(x-h)^2}{a^2} + \frac{(y-k)^2}{b^2} = 1$. Since the distance from the center to each of the foci is 2, we have $c = 2 \Rightarrow c^2 = 4$. For any point P on the ellipse, $d(P, F) + d(P, F') = 2a = 8$, so $a = 4$. Thus, we have $a^2 = 16$. Since $c^2 = a^2 - b^2$, we have $c^2 = a^2 - b^2 \Rightarrow 4 = 16 - b^2 \Rightarrow b^2 = 12$. Thus, the equation of the ellipse is $\frac{(x-2)^2}{16} + \frac{y^2}{12} = 1$.

59. Refer to the chart on page 615 of the text and the Geometric Definition of a Conic Section on page 618. Graph A is an ellipse, so $0 < e < 1$. Graph B is a parabola, so $e = 1$. Graph C is a circle, so $e = 0$. Graph D is a hyperbola, so $e > 1$. Thus, in increasing order of eccentricity, the graphs are C, A, B, D.

61. Since the eccentricity is 0.964 and $e = \frac{c}{a}$, we have $0.964 = \frac{c}{a} \Rightarrow c = 0.964a$. Since the closest distance to the sun is 89 million mi, we have $a - c = 89$. Substituting $c = 0.964a$ into this equation, we have the following.

$a - c = 89 \Rightarrow a - 0.964a = 89 \Rightarrow$

$0.036a = 89 \Rightarrow a = \frac{89}{0.036} = \frac{22,250}{9}$

Since $c = 0.964a$, we have

$c = 0.964\left(\frac{22,250}{9}\right) = \frac{21,449}{9}$. Since

$c^2 = a^2 - b^2$, we have $c^2 = a^2 - b^2 \Rightarrow$

$\left(\frac{21,449}{9}\right)^2 = \left(\frac{22,250}{9}\right)^2 - b^2 \Rightarrow$

$\frac{460,059,601}{81} - \frac{495,062,500}{81} = -b^2 \Rightarrow b^2 = \frac{35,002,899}{81}$

Since $a^2 = \left(\frac{22,250}{9}\right)^2 = \frac{495,062,500}{81} \approx 6,111,883$

and $b^2 = \frac{35,002,899}{81} \approx 432,135$, the equation is

$\frac{x^2}{6,111,883} + \frac{y^2}{432,135} = 1$.

Copyright © 2013 Pearson Education, Inc.

Chapter 6 Test

1. $y = -x^2 + 6x$

   Completing the square on x, we have

   $y = -x^2 + 6x \Rightarrow y = -(x^2 - 6x + 9 - 9) \Rightarrow$

   $y = -(x-3)^2 + 9 \Rightarrow y - 9 = -(x-3)^2$

   The equation has the form $x - h = a(y-k)^2$.

   The vertex, (h, k), is (3, 9) Since the x-term is squared, the axis of symmetry is the vertical line $x = h$, or $x = 3$. The domain is $(-\infty, \infty)$.

   The range is $(-\infty, 9]$. Using this information and plot a few additional points.

x	y
5	5
4	8
3	9
2	8
1	5

2. $x = 4y^2 + 8y$

   Completing the square on y, we have

   $x = 4y^2 + 8y \Rightarrow x = 4(y^2 + 2y + 1 - 1) \Rightarrow$

   $x = 4(y+1)^2 - 4 \Rightarrow x - (-4) = 4[y - (-1)]^2$

   The equation has the form $y - k = a(x-h)^2$.

   The vertex, (h, k), is $(-4, -1)$. Since the y-term is squared, the axis of symmetry is the horizontal line $y = k$, or $y = -1$. The domain is $[-4, \infty)$. The range is $(-\infty, \infty)$. Using this information and plot a few additional points.

x	y
12	1
0	0
-4	-1
0	-2
12	-3

3. Since $x = 8y^2 \Rightarrow y^2 = \frac{1}{8}x$, the parabola is of the form $y^2 = 4px$. Thus, we have

   $4p = \frac{1}{8} \Rightarrow p = \frac{1}{32}$. Since the y-term is squared, the parabola is horizontal, with focus $(p, 0) = \left(\frac{1}{32}, 0\right)$ and directrix, $x = -p$, is

   $x = -\frac{1}{32}$.

4. Parabola; vertex at (2, 3) passing through $(-18, 1)$ opening to the left

   Since the parabola opens to the left, it is of the form $(y-k)^2 = 4p(x-h)$. Since the vertex is located at (2, 3), we have $(y-3)^2 = 4p(x-2)$.

   We can now substitute $x = -18$ and $y = 1$, and solve for p.

   $(1-3)^2 = 4p(-18-2) \Rightarrow$

   $(-2)^2 = 4p(-20) \Rightarrow 4 = -80p \Rightarrow p = -\frac{1}{20}$

   Thus, the equation is

   $(y-3)^2 = 4\left(-\frac{1}{20}\right)(x-2) \Rightarrow (y-3)^2 = -\frac{1}{5}(x-2)$

   or alternatively written as $x - 2 = -5(y-3)^2$.

   We could have also stated that since the parabola opens to the left, it is of the form $x - h = a(y-k)^2$. With $(h, k) = (2, 3)$, we have $x - 2 = a(y-3)^2$. Substituting x = −18 and y = 1, we have the following.

   $-18 - 2 = a(1-3)^2 \Rightarrow -20 = a(-2)^2 \Rightarrow$
   $-20 = 4a \Rightarrow a = 5$

   Thus, $x - 2 = -5(y-3)^2$.

5. Answers will vary.

6. $\dfrac{(x-8)^2}{100} + \dfrac{(y-5)^2}{49} = 1 \Rightarrow \dfrac{(x-8)^2}{10^2} + \dfrac{(y-5)^2}{7^2} = 1$

   The graph is an ellipse centered at (8, 5). We also have the vertices located at $(8-10, 5) = (-2, 5)$ and $(8+10, 5) = (18, 5)$. The endpoints of the minor axis are located at $(8, 5-7) = (8, -2)$ and $(8, 5+7) = (8, 12)$. The domain is [−2, 18] and the range is [−2, 12].

   $\dfrac{(x-8)^2}{100} + \dfrac{(y-5)^2}{49} = 1$

Copyright © 2013 Pearson Education, Inc.

7. $16x^2 + 4y^2 = 64 \Rightarrow \dfrac{x^2}{4} + \dfrac{y^2}{16} = 1 \Rightarrow$

$\dfrac{x^2}{2^2} + \dfrac{y^2}{4^2} = 1$

The graph is an ellipse centered at the origin The vertices are located at (0, –4) and (0, 4). The endpoints of the minor axis are located at (–2, 0) and (2, 0). The domain is [–2, 2] and the range is [–4, 4].

16x² + 4y² = 64

Exercise 7   Exercise 8

8. $y = -\sqrt{1 - \dfrac{x^2}{36}}$

Square both sides to get

$y = -\sqrt{1 - \dfrac{x^2}{36}} \Rightarrow y^2 = 1 - \dfrac{x^2}{36} \Rightarrow \dfrac{x^2}{36} + y^2 = 1 \Rightarrow$

$\dfrac{x^2}{6^2} + \dfrac{y^2}{1^2} = 1.$

This is the equation of an ellipse with vertices (–6, 0) and (6, 0). The endpoints of the minor axis are at (0, –1) and (0, 1). The graph of the original equation is the bottom half of the ellipse. The vertical line test shows that this relation is a function.

9. Ellipse; centered at the origin, horizontal major axis with length 6, minor axis with length 4 Since the ellipse is centered at the origin and major axis is horizontal, the ellipse has the

form $\dfrac{x^2}{a^2} + \dfrac{y^2}{b^2} = 1$. Since the major axis has

length 6, we have $2a = 6 \Rightarrow a = 3 \Rightarrow a^2 = 9$. Since the minor axis has length 4, we have

$2b = 4 \Rightarrow b = 2 \Rightarrow b^2 = 4$. Thus, the equation

of the ellipse is $\dfrac{x^2}{9} + \dfrac{y^2}{4} = 1.$

10. Place the half-ellipse that represents the overpass on a coordinate system with the center of the bottom of the overpass at the origin. If the complete ellipse were drawn, the center of the ellipse would be (0, 0). Then the half-ellipse will include the points (0, 12), (–20, 0), and (20, 0). The equation is of the

form $\dfrac{x^2}{a^2} + \dfrac{y^2}{b^2} = 1$. Thus, for the complete

ellipse, we have $a = 20$ and $b = 12$.

Thus, we have $\dfrac{x^2}{20^2} + \dfrac{y^2}{12^2} = 1 \Rightarrow \dfrac{x^2}{400} + \dfrac{y^2}{144} = 1.$

At a distance of 10 ft from the center of the bottom, $x = 10$. Find the positive y-coordinate of the point on the ellipse with x-coordinate 10.

$\dfrac{x^2}{400} + \dfrac{y^2}{144} = 1 \Rightarrow \dfrac{10^2}{400} + \dfrac{y^2}{144} = 1 \Rightarrow$

$\dfrac{100}{400} + \dfrac{y^2}{144} = 1 \Rightarrow \dfrac{1}{4} + \dfrac{y^2}{144} = 1 \Rightarrow \dfrac{y^2}{144} = \dfrac{3}{4} \Rightarrow$

$y^2 = 108 \Rightarrow y = \sqrt{108} \approx 10.39$

The arch is approximately 10.39 ft high 10 ft from the center of the bottom.

11. $\dfrac{x^2}{4} - \dfrac{y^2}{4} = 1 \Rightarrow \dfrac{x^2}{2^2} - \dfrac{y^2}{2^2} = 1$

The graph of the hyperbola is centered at the

origin. Since the x^2-term comes first, the branches open to the right and left, with vertices (–2, 0) and (2, 0). The domain is $(-\infty, -2] \cup [2, \infty)$. The range is $(-\infty, \infty)$. The

asymptotes are the lines $y = \pm \dfrac{2}{2} x = \pm x.$

$\dfrac{x^2}{4} - \dfrac{y^2}{4} = 1$

12. $9x^2 - 4y^2 = 36 \Rightarrow \dfrac{x^2}{4} - \dfrac{y^2}{9} = 1 \Rightarrow \dfrac{x^2}{2^2} - \dfrac{y^2}{3^2} = 1$

The graph of the hyperbola is centered at the origin. Since the x^2-term comes first, the branches open to the right and left, with vertices (–2, 0) and (2, 0). The domain is $(-\infty, -2] \cup [2, \infty)$. The range is $(-\infty, \infty)$. The

asymptotes are the lines $y = \pm \dfrac{3}{2} x.$

9x² – 4y² = 36

Copyright © 2013 Pearson Education, Inc.

13. Hyperbola; y-intercepts ± 5, foci at $(0, -6)$ and $(0, 6)$

Since the center is halfway between the foci, the center is $(0, 0)$. Since the foci are on a vertical transverse axis, the equation has the

form $\frac{y^2}{a^2} - \frac{x^2}{b^2} = 1$. The y-intercepts are also

vertices, so $a = 5$, and thus $a^2 = 25$. With the given foci, we have $c = 6$. So given

$c^2 = a^2 + b^2$, we have $c^2 = a^2 + b^2 \Rightarrow$

$6^2 = 5^2 + b^2 \Rightarrow 36 = 25 + b^2 \Rightarrow b^2 = 11$

Thus, the equation of the hyperbola is

$\frac{y^2}{25} - \frac{x^2}{11} = 1$.

14.
$$x^2 + 8x + y^2 - 4y + 2 = 0$$
$$\left(x^2 + 8x + 16 - 16\right) + \left(y^2 - 4y + 4 - 4\right) = -2$$
$$(x+4)^2 - 16 + (y-2)^2 - 4 = -2$$
$$(x+4)^2 + (y-2)^2 = -2 + 16 + 4$$
$$(x+4)^2 + (y-2)^2 = 18$$
$$\left[x-(-4)\right]^2 + (y-2)^2 = \left(\sqrt{18}\right)^2$$

The graph of this equation is a circle with center $(-4, 2)$ and radius $\sqrt{18}$. Also note in our original equation, the x^2- and y^2- terms have the same positive coefficient.

15.
$$5x^2 + 10x - 2y^2 - 12y - 23 = 0$$
$$5\left(x^2 + 2x + 1 - 1\right) - 2\left(y^2 + 6y + 9 - 9\right) = 23$$
$$5(x+1)^2 - 5 - 2(y+3)^2 + 18 = 23$$
$$5(x+1)^2 - 2(y+3)^2 = 23 + 5 - 18$$
$$5(x+1)^2 - 2(y+3)^2 = 10$$
$$\frac{(x+1)^2}{2} - \frac{(y+3)^2}{5} = 1 \Rightarrow \frac{\left[x-(-1)\right]^2}{\left(\sqrt{2}\right)^2} - \frac{\left[y-(-3)\right]^2}{\left(\sqrt{5}\right)^2} = 1$$

The graph of this equation is a hyperbola centered at $(-1, -3)$ and vertices

$\left(-1 \pm \sqrt{2}, -3\right)$. The asymptotes are

$y - (-3) = \pm \frac{\sqrt{5}}{\sqrt{2}}\left[x - (-2)\right] \Rightarrow y + 3 = \pm \frac{\sqrt{10}}{2}(x+2)$.

Also note in our original equation, the x^2- and y^2-terms have coefficients that are opposite in sign.

16. $3x^2 + 10y^2 - 30 = 0 \Rightarrow 3x^2 + 10y^2 = 30 \Rightarrow$
$$\frac{x^2}{10} + \frac{y^2}{3} = 1 \Rightarrow \frac{(x-0)^2}{\left(\sqrt{10}\right)^2} + \frac{(y-0)^2}{\left(\sqrt{3}\right)^2} = 1$$

The graph of this equation is an ellipse centered at the origin and x-intercepts (vertices) of $\pm\sqrt{10}$, and y-intercepts (endpoints of minor axis) of $\pm\sqrt{3}$. Also note in our original equation, the x^2- and y^2- terms both have different positive coefficients.

17. $x^2 - 4y = 0 \Rightarrow 4y = x^2 \Rightarrow y = \frac{1}{4}x^2$

This is a parabola with its vertex at the origin, opening upward. Also, note our original equation has an x^2-term, but no y^2-term.

18.
$$(x+9)^2 + (y-3)^2 = 0$$
$$\left[x-(-9)\right]^2 + (y-3)^2 = 0$$

This is the equation of a "circle" centered at $(-9, 3)$ with radius 0. The graph of this equation is a point.

19.
$$x^2 + 4x + y^2 - 6y + 30 = 0$$
$$\left(x^2 + 4x + 4 - 4\right) + \left(y^2 - 6y + 9 - 9\right) = -30$$
$$\left(x^2 + 4x + 4\right) - 4 + \left(y^2 - 6y + 9\right) - 9 = -30$$
$$(x+2)^2 + (y-3)^2 = -30 + 4 + 9$$
$$\left[x-(-2)\right]^2 + (y-3)^2 = -17$$

This equation has the form of the equation of a circle. However, since r^2 cannot be negative, there is no graph of this equation.

20. Solve the equation $\frac{x^2}{25} - \frac{y^2}{49} = 1$ for y.

$$\frac{x^2}{25} - \frac{y^2}{49} = 1 \Rightarrow -\frac{y^2}{49} = 1 - \frac{x^2}{25} \Rightarrow \frac{y^2}{49} = \frac{x^2}{25} - 1 \Rightarrow$$
$$y^2 = 49\left(\frac{x^2}{25} - 1\right) \Rightarrow y = \pm\sqrt{49\left(\frac{x^2}{25} - 1\right)} \Rightarrow$$
$$y = \pm 7\sqrt{\frac{x^2}{25} - 1}$$

Thus, the functions

$Y_1 = 7\sqrt{\frac{x^2}{25} - 1}$ and $Y_2 = 7\sqrt{\frac{x^2}{25} - 1}$ were used

to obtain the graph.

Copyright © 2013 Pearson Education, Inc.

Chapter 7

FURTHER TOPICS IN ALGEBRA

Section 7.1   Sequences and Series

1. $a_n = 4n + 10$
Replace n with 1, 2, 3, 4, and 5.
$n = 1$: $a_1 = 4(1) + 10 = 14$
$n = 2$: $a_2 = 4(2) + 10 = 18$
$n = 3$: $a_3 = 4(3) + 10 = 22$
$n = 4$: $a_4 = 4(4) + 10 = 26$
$n = 5$: $a_5 = 4(5) + 10 = 30$
The first five terms are 14, 18, 22, 26, and 30.

3. $a_n = \dfrac{n+5}{n+4}$
Replace n with 1, 2, 3, 4, and 5.
$n = 1$: $a_1 = \frac{1+5}{1+4} = \frac{6}{5}$
$n = 2$: $a_2 = \frac{2+5}{2+4} = \frac{7}{6}$
$n = 3$: $a_3 = \frac{3+5}{3+4} = \frac{8}{7}$
$n = 4$: $a_4 = \frac{4+5}{4+4} = \frac{9}{8}$
$n = 5$: $a_5 = \frac{5+5}{5+4} = \frac{10}{9}$
The first five terms are $\frac{6}{5}, \frac{7}{6}, \frac{8}{7}, \frac{9}{8}$, and $\frac{10}{9}$.

5. $a_n = \left(\frac{1}{3}\right)^n (n-1)$
Replace n with 1, 2, 3, 4, and 5.
$n = 1$: $a_1 = \left(\frac{1}{3}\right)^1 (1-1) = \left(\frac{1}{3}\right)(0) = 0$
$n = 2$: $a_2 = \left(\frac{1}{3}\right)^2 (2-1) = \left(\frac{1}{9}\right)(1) = \frac{1}{9}$
$n = 3$: $a_3 = \left(\frac{1}{3}\right)^3 (3-1) = \left(\frac{1}{27}\right)(2) = \frac{2}{27}$
$n = 4$: $a_4 = \left(\frac{1}{3}\right)^4 (4-1) = \left(\frac{1}{81}\right)(3) = \frac{1}{27}$
$n = 5$: $a_5 = \left(\frac{1}{3}\right)^5 (5-1) = \left(\frac{1}{243}\right)(4) = \frac{4}{243}$
The first five terms are $0, \frac{1}{9}, \frac{2}{27}, \frac{1}{27}$, and $\frac{4}{243}$.

7. $a_n = (-1)^n (2n)$
Replace n with 1, 2, 3, 4, and 5.
$n = 1$: $a_1 = (-1)^1 \left[2(1)\right] = -1(2) = -2$
$n = 2$: $a_2 = (-1)^2 \left[2(2)\right] = 1(4) = 4$
$n = 3$: $a_3 = (-1)^3 \left[2(3)\right] = -1(6) = -6$

$n = 4$: $a_4 = (-1)^4 \left[2(4)\right] = 1(8) = 8$
$n = 5$: $a_5 = (-1)^5 \left[2(5)\right] = -1(10) = -10$
The first five terms are $-2, 4, -6, 8$, and -10.

9. $a_n = \dfrac{4n-1}{n^2 + 2}$
Replace n with 1, 2, 3, 4, and 5.
$n = 1$: $a_1 = \dfrac{4(1)-1}{(1)^2 + 2} = \dfrac{4-1}{1+2} = \dfrac{3}{3} = 1$

$n = 2$: $a_2 = \dfrac{4(2)-1}{(2)^2 + 2} = \dfrac{8-1}{4+2} = \dfrac{7}{6}$

$n = 3$: $a_3 = \dfrac{4(3)-1}{(3)^2 + 2} = \dfrac{12-1}{9+2} = \dfrac{11}{11} = 1$

$n = 4$: $a_4 = \dfrac{4(4)-1}{(4)^2 + 2} = \dfrac{16-1}{16+2} = \dfrac{15}{18} = \dfrac{5}{6}$

$n = 5$: $a_5 = \dfrac{4(5)-1}{(5)^2 + 2} = \dfrac{20-1}{25+2} = \dfrac{19}{27}$

The first five terms are $1, \frac{7}{6}, 1, \frac{5}{6}$, and $\frac{19}{27}$.

11. $a_n = \dfrac{n^3 + 8}{n+2}$
Replace n with 1, 2, 3, 4, and 5.
$n = 1$: $a_1 = \dfrac{1^3 + 8}{1+2} = \dfrac{1+8}{3} = \dfrac{9}{3} = 3$

$n = 2$: $a_2 = \dfrac{2^3 + 8}{2+2} = \dfrac{8+8}{4} = \dfrac{16}{4} = 4$

$n = 3$: $a_3 = \dfrac{3^3 + 8}{3+2} = \dfrac{27+8}{5} = \dfrac{35}{5} = 7$

$n = 4$: $a_4 = \dfrac{4^3 + 8}{4+2} = \dfrac{64+8}{6} = \dfrac{72}{6} = 12$

$n = 5$: $a_5 = \dfrac{5^3 + 8}{5+2} = \dfrac{125+8}{7} = \dfrac{133}{7} = 19$

The first five terms are 3, 4, 7, 12, and 19.

13. Answers will vary.

15. The sequence of the days of the week has as its domain $\{1, 2, 3, 4, 5, 6, 7\}$. Therefore, it is a finite sequence.

Copyright © 2013 Pearson Education, Inc.

17. The sequence 1, 2, 3, 4, 5 has as its domain $\{1, 2, 3, 4, 5\}$. Therefore, it is a finite sequence.

19. The sequence 1, 2, 3, 4, 5, …has as its domain $\{1, 2, 3, 4, 5, …\}$. Therefore, the sequence is infinite.

21. The sequence $a_1 = 4$ and for $n \geq 2$,

$a_n = 4 \cdot a_{n-1}$ has as its domain $\{2, 3, 4, …\}$. Therefore, the sequence is infinite.

23. $a_1 = -2$, $a_n = a_{n-1} + 3$, for $n > 1$

This is an example of a recursive definition.
We know $a_1 = -2$. Since $a_n = a_{n-1} + 3$,

$n = 2: a_2 = a_1 + 3 = -2 + 3 = 1$
$n = 3: a_3 = a_2 + 3 = 1 + 3 = 4$
$n = 4: a_4 = a_3 + 3 = 4 + 3 = 7$.

The first four terms are –2, 1, 4, and 7.

25. $a_1 = 1$, $a_2 = 1$, $a_n = a_{n-1} + a_{n-2}$, for $n \geq 3$

$n = 3: a_3 = a_2 + a_1 = 1 + 1 = 2$
$n = 4: a_4 = a_3 + a_2 = 2 + 1 = 3$

The first four terms are 1, 1, 2, and 3.

27. $a_1 = 2$, $a_n = n \cdot a_{n-1}$, for $n > 1$
$a_2 = 2 \cdot a_1 = 2 \cdot 2 = 4$
$a_3 = 3 \cdot a_2 = 3 \cdot 4 = 12$
$a_4 = 4 \cdot a_3 = 4 \cdot 12 = 48$
The first four terms are 2, 4, 12, and 48.

29. $\displaystyle\sum_{i=1}^{5} (2i + 1)$

$= \left[2(1) + 1\right] + \left[2(2) + 1\right] + \left[2(3) + 1\right]$
$\qquad + \left[2(4) + 1\right] + \left[2(5) + 1\right]$
$= (2 + 1) + (4 + 1) + (6 + 1) + (8 + 1) + (10 + 1)$
$= 3 + 5 + 7 + 9 + 11 = 35$

31. $\displaystyle\sum_{j=1}^{4} j^{-1} = \sum_{j=1}^{4} \frac{1}{j} = \frac{1}{1} + \frac{1}{2} + \frac{1}{3} + \frac{1}{4}$

$\qquad = \frac{12}{12} + \frac{6}{12} + \frac{4}{12} + \frac{3}{12} = \frac{25}{12}$

33. $\displaystyle\sum_{i=1}^{4} i^i = 1^1 + 2^2 + 3^3 + 4^4 = 1 + 4 + 27 + 256 = 288$

35. $\displaystyle\sum_{k=1}^{6} (-1)^k \cdot k$

$= (-1)^1 \cdot 1 + (-1)^2 \cdot 2 + (-1)^3 \cdot 3 + (-1)^4 \cdot 4$
$\qquad + (-1)^5 \cdot 5 + (-1)^6 \cdot 6$
$= -1 \cdot 1 + 1 \cdot 2 + (-1) \cdot 3 + 1 \cdot 4 + (-1) \cdot 5 + 1 \cdot 6$
$= -1 + 2 - 3 + 4 - 5 + 6 = 3$

37. $\displaystyle\sum_{i=2}^{5} (6 - 3i)$

$= \left[6 - 3(2)\right] + \left[6 - 3(3)\right]$
$\qquad + \left[6 - 3(4)\right] + \left[6 - 3(5)\right]$
$= (6 - 6) + (6 - 9) + (6 - 12) + (6 - 15)$
$= 0 + (-3) + (-6) + (-9) = -18$

39. $\displaystyle\sum_{i=-2}^{3} 2(3)^i$

$= 2(3)^{-2} + 2(3)^{-1} + 2(3)^0$
$\qquad + 2(3)^1 + 2(3)^2 + 2(3)^3$
$= 2\left(\frac{1}{9}\right) + 2\left(\frac{1}{3}\right) + 2(1) + 2(3) + 2(9) + 2(27)$
$= \frac{2}{9} + \frac{2}{3} + 2 + 6 + 18 + 54$
$= \frac{2}{9} + \frac{6}{9} + 80 = 80\frac{8}{9} = \frac{728}{9}$

41. $\displaystyle\sum_{i=-1}^{5} (i^2 - 2i)$

$= \left[(-1)^2 - 2(-1)\right] + \left[0^2 - 2(0)\right] + \left[1^2 - 2(1)\right]$
$\qquad + \left[2^2 - 2(2)\right] + \left[3^2 - 2(3)\right]$
$\qquad + \left[4^2 - 2(4)\right] + \left[5^2 - 2(5)\right]$
$= \left[1 - (-2)\right] + (0 - 0) + (1 - 2) + (4 - 4)$
$\qquad + (9 - 6) + (16 - 8) + (25 - 10)$
$= 3 + 0 + (-1) + 0 + 3 + 8 + 15 = 28$

43. $\displaystyle\sum_{i=1}^{5} (3^i - 4) = (3^1 - 4) + (3^2 - 4) + (3^3 - 4)$

$\qquad + (3^4 - 4) + (3^5 - 4)$
$= (3 - 4) + (9 - 4) + (27 - 4)$
$\qquad + (81 - 4) + (243 - 4)$
$= (-1) + 5 + 23 + 77 + 239 = 343$

45. $\displaystyle\sum_{i=1}^{3} (i^3 - i) = (1^3 - 1) + (2^3 - 2) + (3^3 - 3)$

$\qquad = (1 - 1) + (8 - 2) + (27 - 3)$
$\qquad = 0 + 6 + 24 = 30$

Copyright © 2013 Pearson Education, Inc.

47. $\displaystyle\sum_{i=1}^{10}(4i^2-5)=1490$

To find the seq(feature, go to the LIST menu (2nd STAT), go to OPS and choose option 5.

```
NAMES OPS MATH      seq(4I²-5,I,1,10
1:SortA(            )→L₁
2:SortD(            {-1 11 31 59 95…
3:dim(
4:Fill(
5:seq(
6:cumSum(
7↓ΔList(
```

To find the sum(feature, go to the LIST menu (2nd STAT) go to MATH and choose option 5.

```
NAMES OPS MATH      seq(4I²-5,I,1,10
1:min(              )→L₁
2:max(              {-1 11 31 59 95…
3:mean(             sum(L₁)
4:median(                        1490
5:sum(
6:prod(
7↓stdDev(
```

49. $\displaystyle\sum_{j=3}^{9}(3j-j^2)=-154$

```
seq(3J-J²,J,3,9)
→L₁
{0 -4 -10 -18 -…
sum(L₁)
              -154
```

In Exercises 51–59, $x_1=-2$, $x_2=-1$, $x_3=0$, $x_4=1$, and $x_5=2$.

51. $\displaystyle\sum_{i=1}^{5}x_i=x_1+x_2+x_3+x_4+x_5=-2+(-1)+0+1+2=0$

53. $\displaystyle\sum_{i=1}^{5}(2x_i+3)=(2x_1+3)+(2x_2+3)+(2x_3+3)+(2x_4+3)+(2x_5+3)$

$$=\left[2(-2)+3\right]+\left[2(-1)+3\right]+\left[2(0)+3\right]+\left[2(1)+3\right]+\left[2(2)+3\right]$$
$$=(-4+3)+(-2+3)+(0+3)+(2+3)+(4+3)$$
$$=-1+1+3+5+7=15$$

55. $\displaystyle\sum_{i=1}^{3}\left(3x_i-x_i^2\right)$

$$=\left(3x_1-x_1^2\right)+\left(3x_2-x_2^2\right)+\left(3x_3-x_3^2\right)$$
$$=\left[3(-2)-(-2)^2\right]+\left[3(-1)-(-1)^2\right]$$
$$\qquad\qquad+\left[3(0)-0^2\right]$$
$$=(-6-4)+(-3-1)+(0-0)$$
$$=-10+(-4)+0=-14$$

57. $\displaystyle\sum_{i=2}^{5}\frac{x_i+1}{x_i+2}=\frac{x_2+1}{x_2+2}+\frac{x_3+1}{x_3+2}+\frac{x_4+1}{x_4+2}+\frac{x_5+1}{x_5+2}=\frac{-1+1}{-1+2}+\frac{0+1}{0+2}+\frac{1+1}{1+2}+\frac{2+1}{2+2}$

$$=\frac{0}{1}+\frac{1}{2}+\frac{2}{3}+\frac{3}{4}=0+\frac{6}{12}+\frac{8}{12}+\frac{9}{12}=\frac{23}{12}$$

59. $\displaystyle\sum_{i=1}^{4}\frac{x_i^3+1000}{x_i+10}=\frac{x_1^3+1000}{x_1+10}+\frac{x_2^3+1000}{x_2+10}+\frac{x_3^3+1000}{x_3+10}+\frac{x_4^3+1000}{x_4+10}$

$$=\frac{(-2)^3+1000}{(-2)+10}+\frac{(-1)^3+1000}{(-1)+10}+\frac{0^3+1000}{0+10}+\frac{1^3+1000}{1+10}$$
$$=\frac{-8+1000}{8}+\frac{-1+1000}{9}+\frac{1000}{10}+\frac{1+1000}{11}=\frac{992}{8}+\frac{999}{9}+\frac{1000}{10}+\frac{1001}{11}$$
$$=124+111+100+91=426$$

Copyright © 2013 Pearson Education, Inc.

In Exercises 61–65, $x_1 = 0$, $x_2 = 2$, $x_3 = 4$, $x_4 = 6$, and $\Delta x = 0.5$.

61. $f(x) = 4x - 7 \Rightarrow f(x_i) = 4x_i - 7$

$$\sum_{i=1}^{4} f(x_i)\Delta x = f(x_1)\Delta x + f(x_2)\Delta x + f(x_3)\Delta x + f(x_4)\Delta x$$

$$= (4x_1 - 7)\Delta x + (4x_2 - 7)\Delta x + (4x_3 - 7)\Delta x + (4x_4 - 7)\Delta x$$
$$= [4(0) - 7](0.5) + [4(2) - 7](0.5) + [4(4) - 7](0.5) + [4(6) - 7](0.5)$$
$$= (0 - 7)(0.5) + (8 - 7)(0.5) + (16 - 7)(0.5) + (24 - 7)(0.5)$$
$$= -7(0.5) + 1(0.5) + 9(0.5) + 17(0.5)$$
$$= -3.5 + 0.5 + 4.5 + 8.5 = 10$$

63. $f(x) = 2x^2 \Rightarrow f(x_i) = 2x_i^2$

$$\sum_{i=1}^{4} f(x_i)\Delta x = f(x_1)\Delta x + f(x_2)\Delta x + f(x_3)\Delta x + f(x_4)\Delta x$$

$$= 2(x_1^2)\Delta x + 2(x_2^2)\Delta x + 2(x_3^2)\Delta x + 2(x_4^2)\Delta x$$
$$= 2(0^2)(0.5) + 2(2^2)(0.5) + 2(4^2)(0.5) + 2(6^2)(0.5)$$
$$= 2(0)(0.5) + 2(4)(0.5) + 2(16)(0.5) + 2(36)(0.5)$$
$$= 0 + 4 + 16 + 36 = 56$$

65. $f(x) = \dfrac{-2}{x+1} \Rightarrow f(x_i) = \dfrac{-2}{x_i + 1}$

$$\sum_{i=1}^{4} f(x_i)\Delta x = f(x_1)\Delta x + f(x_2)\Delta x + f(x_3)\Delta x + f(x_4)\Delta x$$

$$= \left(\frac{-2}{x_1 + 1}\right)(0.5) + \left(\frac{-2}{x_2 + 1}\right)(0.5) + \left(\frac{-2}{x_3 + 1}\right)(0.5) + \left(\frac{-2}{x_4 + 1}\right)(0.5)$$

$$= \left(\frac{-2}{0+1}\right)(0.5) + \left(\frac{-2}{2+1}\right)(0.5) + \left(\frac{-2}{4+1}\right)(0.5) + \left(\frac{-2}{6+1}\right)(0.5)$$

$$= \left(\frac{-2}{1}\right)(0.5) + \left(-\frac{2}{3}\right)(0.5) + \left(-\frac{2}{5}\right)(0.5) + \left(-\frac{2}{7}\right)(0.5)$$

$$= -1 + \left(-\frac{1}{3}\right) + \left(-\frac{1}{5}\right) + \left(-\frac{1}{7}\right)$$

$$= \frac{-105}{105} + \left(\frac{-35}{105}\right) + \left(\frac{-21}{105}\right) + \left(\frac{-15}{105}\right) = -\frac{176}{105}$$

67. $\displaystyle\sum_{i=1}^{100} 6 = 100(6) = 600$

69. $\displaystyle\sum_{i=1}^{15} i^2 = \frac{15(15+1)[2(15)+1]}{6} = \frac{15(16)(30+1)}{6} = \frac{15(16)(31)}{6} = \frac{7440}{6} = 1240$

71. $\displaystyle\sum_{i=1}^{5}(5i + 3) = \sum_{i=1}^{5} 5i + \sum_{i=1}^{5} 3 = 5\sum_{i=1}^{5} i + 5(3) = 5 \cdot \frac{5(5+1)}{2} + 15 = 5 \cdot \frac{5(6)}{2} + 15$

$$= 5 \cdot \frac{30}{2} + 15 = 5(15) + 15 = 75 + 15 = 90$$

Copyright © 2013 Pearson Education, Inc.

73. $\displaystyle\sum_{i=1}^{5}\left(4i^2-2i+6\right)=\sum_{i=1}^{5}4i^2-\sum_{i=1}^{5}2i+\sum_{i=1}^{5}6=4\sum_{i=1}^{5}i^2-2\sum_{i=1}^{5}i+5(6)=4\cdot\dfrac{5(5+1)(10+1)}{6}-2\cdot\dfrac{5(5+1)}{2}+30$

$\qquad=4\cdot\dfrac{5(6)(11)}{6}-2\cdot\dfrac{5(6)}{2}+30=4\cdot\left[5(11)\right]-2\cdot\left[5(3)\right]+30=220-30+30=220$

75. $\displaystyle\sum_{i=1}^{4}\left(3i^3+2i-4\right)=\sum_{i=1}^{4}3i^3+\sum_{i=1}^{4}2i-\sum_{i=1}^{4}4=3\sum_{i=1}^{4}i^3+2\sum_{i=1}^{4}i-4(4)=3\cdot\dfrac{4^2\left(4+1\right)^2}{4}+2\cdot\dfrac{4(4+1)}{2}-16$

$\qquad=3\cdot\dfrac{16(5)^2}{4}+2\cdot\dfrac{4(5)}{2}-16=3\cdot\left[4(25)\right]+2\cdot\left[2(5)\right]-16=300+20-16=304$

In Exercises 77–79, there are other acceptable forms of the answers.

77. $\dfrac{1}{3(1)}+\dfrac{1}{3(2)}+\dfrac{1}{3(3)}+\cdots+\dfrac{1}{3(9)}=\displaystyle\sum_{i=1}^{9}\dfrac{1}{3i}$

79. $1-\dfrac{1}{2}+\dfrac{1}{4}-\dfrac{1}{8}+\cdots-\dfrac{1}{128}=\dfrac{1}{2^0}-\dfrac{1}{2^1}+\dfrac{1}{2^2}-\dfrac{1}{2^3}+\cdots-\dfrac{1}{2^7}$

$\qquad=\left(\dfrac{1}{2}\right)^0-\left(\dfrac{1}{2}\right)^1+\left(\dfrac{1}{2}\right)^2-\left(\dfrac{1}{2}\right)^3+\cdots-\left(\dfrac{1}{2}\right)^7$

$\qquad=\left(\dfrac{1}{2}\right)^{1-1}-\left(\dfrac{1}{2}\right)^{2-1}+\left(\dfrac{1}{2}\right)^{3-1}-\left(\dfrac{1}{2}\right)^{4-1}+\cdots-\left(\dfrac{1}{2}\right)^{8-1}$

$\qquad=\left(-\dfrac{1}{2}\right)^{1-1}+\left(-\dfrac{1}{2}\right)^{2-1}+\left(-\dfrac{1}{2}\right)^{3-1}+\left(-\dfrac{1}{2}\right)^{4-1}\qquad+\cdots+\left(-\dfrac{1}{2}\right)^{8-1}$

$\qquad=\displaystyle\sum_{k=1}^{8}\left(-\dfrac{1}{2}\right)^{k-1}$

For Exercises 81–85, enter the set $\{1, 2, 3, \ldots, 10\}$ into L_1. Turn the STAT PLOTS on and clear any functions you have in the $Y=$ screen.

81. $a_n=\dfrac{n+4}{2n}$

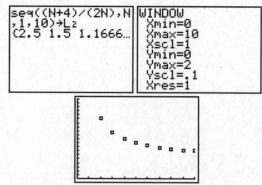

It appears that the sequence converges to $\frac{1}{2}$.

The line $y=\frac{1}{2}$ is graphed in the same screen as the plot.

Copyright © 2013 Pearson Education, Inc.

83. $a_n = 2e^n$

It appears that the sequence diverges.

85. $a_n = \left(1 + \dfrac{1}{n}\right)^n$

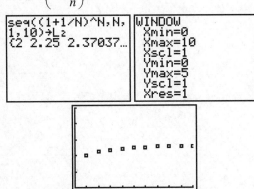

It appears that the sequence converges to $e \approx 2.71828$ The line $y = e$ is graphed in the same screen as the plot.

87. (a) $a_1 = 8; a_n = 2.9a_{n-1} - 0.2a_{n-1}^{\;2}$, for $n > 1$

$a_1 = 8$ thousand per acre

$a_2 = 2.9a_1 - 0.2a_1^{\;2} = 2.9(8) - 0.2(8)^2$
$\quad - 23.2 - 12.8 = 10.4$ thousand per acre

$a_3 = 2.9a_2 - 0.2a_2^{\;2} = 2.9(10.4) - 0.2(10.4)^2$
$\quad = 30.16 - 21.632 = 8.528$ thousand per acre

(b) One way to enter the required data into a list is to first enter the set $\{1, 2, 3, \ldots, 20\}$ into L_1. Turn the STAT PLOTS on and clear any functions you have in the $Y =$ screen.

Next, to enter the data from the recursive sequence, we can enter the values in individually as follows.

Hit enter, then edit the element in the list until you have completed all 20 elements.

Your lists should look like the following screen shots.

Plotting the points we have the following.

(*continued on next page*)

Copyright © 2013 Pearson Education, Inc.

(*continued*)

The population density oscillates above and below 9.5 thousand per acre (approximately). The line $y = 9.5$ is graphed in the same screen as the plot.

89. **(a)** Since the number of bacteria doubles every 40 minutes, it follows that $N_{j+1} = 2N_j$ for $j \geq 1$.

(b) Two hours is 120 minutes. If $120 = 40(j-1)$, then $3 = j-1 \Rightarrow j = 4$. Since $N_1 = 230$, $N_2 = 460$, $N_3 = 920$, and $N_4 = 1840$, there will be 1840 bacteria after two hours.

(c) We must graph the sequence $N_{j+1} = 2N_j$ for $j = 1, 2, 3, \ldots, 7$ if $N_1 = 230$. You should clear lists that you may have used in previous exercises and enter the set $\{1, 2, 3, \ldots, 7\}$ into L_1.

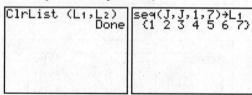

Next, to enter the data from the recursive sequence, we can enter the values in individually as follows.

Hit enter, then edit the element in the list until you have completed all 7 elements.

Your lists should look like the following screen shots.

Plotting the points, we have the following.

(d) The growth is very rapid. Since there is a doubling of the bacteria at equal intervals, their growth is exponential.

91. **(a)** $\ln 1.02 = \ln(1 + 0.02)$

$$\approx 0.02 - \frac{0.02^2}{2} + \frac{0.02^3}{3} - \frac{0.02^4}{4} + \frac{0.02^5}{5} - \frac{0.02^6}{6}$$

$$\approx 0.0198026273$$

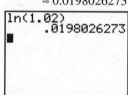

(b) $\ln 1.97 = \ln(1 - 0.03)$

$$\approx (-0.03) - \frac{(-0.03)^2}{2} + \frac{(-0.03)^3}{3} - \frac{(-0.03)^4}{4} + \frac{(-0.03)^5}{5} - \frac{(-0.03)^6}{6}$$

$$\approx -0.0304592075$$

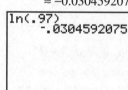

Copyright © 2013 Pearson Education, Inc.

93. $e^a \approx 1 + a + \dfrac{a^2}{2!} + \dfrac{a^3}{3!} + \cdots + \dfrac{a^n}{n!}$ where

$n! = 1 \cdot 2 \cdot 3 \cdots \cdot n$

(a) Calculate

$e^1 \approx 1 + 1 + \dfrac{1^2}{2!} + \dfrac{1^3}{3!} + \dfrac{1^4}{4!} + \dfrac{1^5}{5!} + \dfrac{1^6}{6!}$
$+ \dfrac{1^7}{7!} + \dfrac{1^8}{8!}$

The approximation is accurate to four decimal places.

(b) Calculate

$e^{-1} \approx 1 + (-1) + \dfrac{(-1)^2}{2!} + \dfrac{(-1)^3}{3!} + \dfrac{(-1)^4}{4!}$
$+ \dfrac{(-1)^5}{5!} + \dfrac{(-1)^6}{6!} + \dfrac{(-1)^7}{7!} + \dfrac{(-1)^8}{8!}$

The approximation is accurate to four decimal places.

Section 7.2   Arithmetic Sequences and Series

1. $2, 5, 8, 11, \ldots$
$d = a_2 - a_1 = 5 - 2 = 3$

3. $3, -2, -7, -12, \ldots$
$d = a_2 - a_1 = -2 - 3 = -5$

5. $x + 3y, 2x + 5y, 3x + 7y, \ldots$
$d = a_2 - a_1 = (2x + 5y) - (x + 3y)$
$= 2x + 5y - x - 3y = x + 2y$

7. $a_1 = 8$ and $d = 6$
Starting with $a_1 = 8$, add $d = 6$ to each term to get the next term.
$a_2 = 8 + 6 = 14;$ $a_3 = 14 + 6 = 20$
$a_4 = 20 + 6 = 26;$ $a_5 = 26 + 6 = 32$
The first five terms are 8, 14, 20, 26, and 32.

9. $a_1 = 5, d = -2$
$a_2 = 5 + (-2) = 3; a_3 = 3 + (-2) = 1$
$a_4 = 1 + (-2) = -1$
$a_5 = -1 + (-2) = -3$
The first five terms are 5, 3, 1, –1, and –3.

11. $a_1 = 10 + \sqrt{7}, a_2 = 10$
$d = a_2 - a_1 = 10 - \left(10 + \sqrt{7}\right) = -\sqrt{7}$
$a_3 = a_2 + d = 10 + \left(-\sqrt{7}\right) = 10 - \sqrt{7}$
$a_4 = a_3 + d = \left(10 - \sqrt{7}\right) - \sqrt{7} = 10 - 2\sqrt{7}$
$a_5 = a_4 + d = \left(10 - 2\sqrt{7}\right) - \sqrt{7} = 10 - 3\sqrt{7}$
The first five terms are $10 + \sqrt{7}$, 10, $10 - \sqrt{7}$, $10 - 2\sqrt{7}$, and $10 - 3\sqrt{7}$.

13. $5, 7, 9, \ldots$
$d = a_2 - a_1 = 7 - 5 = 2$
Since $a_n = a_1 + (n - 1)d$, we have
$a_8 = 5 + (8 - 1)2 = 5 + 7 \cdot 2 = 5 + 14 = 19$. Also, we have
$a_n = 5 + (n - 1)2 = 5 + 2n - 2 = 2n + 3$.

15. $a_1 = 5, a_4 = 15$
Since $a_n = a_1 + (n - 1)d$, we have
$a_4 = 5 + (4 - 1)d = 15 \Rightarrow 5 + 3d = 15 \Rightarrow$
$3d = 10 \Rightarrow d = \frac{10}{3}$
We now have
$a_8 = 5 + (8 - 1)\left(\frac{10}{3}\right) = 5 + 7\left(\frac{10}{3}\right) = \frac{15}{3} + \frac{70}{3} = \frac{85}{3}$.
Also, we have
$a_n = 5 + (n - 1)\left(\frac{10}{3}\right)$
$= 5 + \frac{10}{3}n - \frac{10}{3} = \frac{15}{3} + \frac{10}{3}n - \frac{10}{3} = \frac{10}{3}n + \frac{5}{3}$

17. $a_{10} = 6, a_{11} = 10.5$
Since a_{10} and a_{11} are successive terms in the sequence, we have
$d = a_{11} - a_{10} = 10.5 - 6 = 4.5$ Now find a_1
using the formula $a_n = a_1 + (n - 1)d$.
$a_{11} = a_1 + (11 - 1)d \Rightarrow 10.5 = a_1 + 10(4.5) \Rightarrow$
$10.5 = a_1 + 45 \Rightarrow a_1 = -34.5$
Now find a_8 and a_n. We have
$a_8 = -34.5 + (8 - 1)(4.5) = -34.5 + 7(4.5)$
$= -34.5 + 31.5 = -3$
$a_n = -34.5 + (n - 1)(4.5) = -34.5 + 4.5n - 4.5$
$= 4.5n - 39$

Copyright © 2013 Pearson Education, Inc.

19. $a_1 = x, \; a_2 = x+3$

$d = a_2 - a_1 = (x+3) - x = 3$

Since $a_n = a_1 + (n-1)d$, we have

$a_8 = x + (8-1)3 = x + 7 \cdot 3 = x + 21$. Also, we

have $a_n = x + (n-1)3 = x + 3n - 3$.

21. $a_4 = s + 6p, d = 2p$

First, find a_1. Since $a_n = a_1 + (n-1)d$, we

have $a_4 = a_1 + 3d \Rightarrow a_1 + 3(2p) = s + 6p \Rightarrow$

$a_1 + 6p = s + 6p \Rightarrow a_1 = s$. Thus,

$a_8 = s + 7(2p) = s + 14p$ and

$a_n = s + (n-1)(2p) = s + 2pn - 2p$

23. $a_5 = 27, \; a_{15} = 87$

Since $a_{15} = a_5 + 10d$, it follows that

$10d = a_{15} - a_5$. This implies

$10d = 87 - 27 \Rightarrow 10d = 60 \Rightarrow d = 6$.

Since $a_5 = a_1 + 4d$, we have

$27 = a_1 + 4(6) \Rightarrow 27 = a_1 + 24 \Rightarrow a_1 = 3$.

25. $S_{16} = -160, \; a_{16} = -25$

Since $S_n = \frac{n}{2}(a_1 + a_n)$, we have

$S_{16} = \frac{16}{2}(a_1 + a_{16}) \Rightarrow -160 = 8[a_1 + (-25)] \Rightarrow$

$-20 = a_1 - 25 \Rightarrow a_1 = 5$

27. The sequence is comprised of the points

$\{(1,-2),(2,-1),(3,0),(4,1),(5,2),(6,3)\}$. If

the points were connected, they would form a

line with slope $m = \frac{y_2 - y_1}{x_2 - x_1} = \frac{-1-(-2)}{2-1} = \frac{1}{1} = 1$.

Using the point-slope form of a line, we have

$y - y_1 = m(x - x_1) \Rightarrow y - (-2) = 1(x-1) \Rightarrow$

$y + 2 = x - 1 \Rightarrow y = x - 3$

The nth term of the sequence is determined by

$a_n = n - 3$. The domain is $\{1,2,3,4,5,6\}$ and

the range is $\{-2,-1,0,1,2,3\}$.

29. The sequence is comprised of the points

$\{(1,\frac{5}{2}),(2,2),(3,\frac{3}{2}),(4,1),(5,\frac{1}{2}),(6,0)\}$. If the

points were connected, they would form a line

with slope $m = \frac{y_2 - y_1}{x_2 - x_1} = \frac{2 - \frac{5}{2}}{2-1} = \frac{-\frac{1}{2}}{1} = -\frac{1}{2}$. Using

the point-slope form of a line, we have

$y - y_1 = m(x - x_1) \Rightarrow y - \frac{5}{2} = -\frac{1}{2}(x-1) \Rightarrow$

$y - \frac{5}{2} = -\frac{1}{2}x + \frac{1}{2} \Rightarrow y = -\frac{1}{2}x + 3$

The nth term of the sequence is determined by

$a_n = -\frac{1}{2}n + 3$ or $a_n = 3 - \frac{1}{2}n$. The domain is

$\{1,2,3,4,5,6\}$ and the range is

$\{0, \frac{1}{2}, 1, \frac{3}{2}, 2, \frac{5}{2}\}$ or $\{0, 0.5, 1, 1.5, 2, 2.5\}$.

31. The sequence is comprised of the points

$\{(1,10),(2,-10),(3,-30),(4,-50),(5,-70)\}$.

If the points were connected, they would form

a line with slope

$m = \frac{y_2 - y_1}{x_2 - x_1} = \frac{-10-10}{2-1} = \frac{-20}{1} = -20$. Using the

point-slope form of a line, we have

$y - y_1 = m(x - x_1) \Rightarrow y - 10 = -20(x-1) \Rightarrow$

$y - 10 = -20x + 20 \Rightarrow y = -20x + 30$

The nth term of the sequence is determined by

$a_n = -20n + 30$ or $a_n = 30 - 20n$. The domain

is $\{1, 2, 3, 4, 5\}$ and the range is

$\{-70, -50, -30, -10, 10\}$.

33. 8, 11, 14, …

Since $a_1 = 8, \; d = a_2 - a_1 = 11 - 8 = 3$, and

$S_n = \frac{n}{2}[2a_1 + (n-1)d]$, we have

$S_{10} = \frac{10}{2}[2(8) + (10-1)(3)] = 5[16 + 9(3)]$

$= 5(16 + 27) = 5(43) = 215$

35. 5, 9, 13, …

Since $a_1 = 5, \; d = a_2 - a_1 = 9 - 5 = 4$, and

$S_n = \frac{n}{2}[2a_1 + (n-1)d]$, we have the

following.

$S_{10} = \frac{10}{2}[2(5) + (10-1)(4)] = 5[10 + 9(4)]$

$= 5(10 + 36) = 5(46) = 230$

37. $a_2 = 9, \; a_4 = 13$

To find S_{10}, we need a_1 and d. Since

$a_4 = a_2 + 2d$, we have

$13 = 9 + 2d \Rightarrow 4 = 2d \Rightarrow d = 2$. Also, since

$d = a_2 - a_1$, we have $2 = 9 - a_1 \Rightarrow a_1 = 7$.

Thus, since $S_n = \frac{n}{2}[2a_1 + (n-1)d]$, we have

$S_{10} = \frac{10}{2}[2(7) + (10-1)(2)] = 5[14 + 9(2)]$

$= 5(14 + 18) = 5(32) = 160$

Copyright © 2013 Pearson Education, Inc.

39. $a_1 = 10$, $a_{10} = 5.5$

To find S_{10}, we need a_1 and d. Since
$a_{10} = a_1 + 9d$, we have
$5.5 = 10 + 9d \Rightarrow 9d = -4.5 \Rightarrow d = -0.5$. Thus,
since $S_n = \frac{n}{2}\left[2a_1 + (n-1)d\right]$, we have

$$S_{10} = \frac{10}{2}\left[2(10) + (10-1)(-0.5)\right]$$
$$= 5\left[20 + 9(-0.5)\right] = 5\left[20 + (-4.5)\right] = 77.5$$

41. $a_1 = \pi$, $a_{10} = 10\pi$

To find S_{10}, we need a_1 and d. Since
$a_{10} = a_1 + 9d$, we have
$10\pi = \pi + 9d \Rightarrow 9\pi = 9d \Rightarrow d = \pi$. Thus,
since $S_n = \frac{n}{2}\left[2a_1 + (n-1)d\right]$, we have

$$S_{10} = \frac{10}{2}\left[2\pi + (10-1)\pi\right]$$
$$= 5\left[2\pi + 9\pi\right] = 5(11\pi) = 55\pi$$

43. The first 80 positive integers form the
arithmetic sequence 1, 2, 3, 4, ..., 80. Thus, in
the formula $S_n = \frac{n}{2}(a_1 + a_n)$, we have
$n = 80, a_1 = 1$, and $a_{80} = 80$. Thus, the sum of
the first 80 positive integers is
$S_{80} = \frac{80}{2}(1 + 80) = 40(81) = 3240$.

45. The positive odd integers form an arithmetic
sequence 1, 3, 5, 7, ... with $a_1 = 1$ and $d = 2$.
Find the sum of the first 50 terms of this
sequence. Since $S_n = \frac{n}{2}\left[2a_1 + (n-1)d\right]$, we
have

$$S_{50} = \frac{50}{2}\left[2(1) + (50-1)(2)\right] = 25\left[2 + 49(2)\right]$$
$$= 25(2 + 98) = 25(100) = 2500$$

47. The positive even integers form an arithmetic
sequence 2, 4, 6, 8, ... with $a_1 = 2$ and $d = 2$.
Find the sum of the first 60 terms of this
sequence. Since $S_n = \frac{n}{2}\left[2a_1 + (n-1)d\right]$, we
have

$$S_{60} = \frac{60}{2}\left[2(2) + (60-1)(2)\right] = 30\left[4 + 59(2)\right]$$
$$= 30(4 + 118) = 30(122) = 3660$$

49. $S_{20} = 1090$, $a_{20} = 102$

Since the formula for the sum is
$S_n = \frac{n}{2}(a_1 + a_n)$, we have $S_{20} = \frac{20}{2}(a_1 + a_{20})$.
This yields
$1090 = 10(a_1 + 102) \Rightarrow 109 = a_1 + 102 \Rightarrow a_1 = 7$.
To solve for d, we use the fact that
$a_n = a_1 + (n-1)d$.

$a_n = a_1 + (n-1)d \Rightarrow a_{20} = a_1 + (20-1)d \Rightarrow$
$102 = 7 + 19d \Rightarrow 95 = 19d \Rightarrow d = 5$
Thus, $a_1 = 7$ and $d = 5$.

51. $S_{12} = -108$, $a_{12} = -19$

Since the formula for the sum is
$S_n = \frac{n}{2}(a_1 + a_n)$, we have $S_{12} = \frac{12}{2}(a_1 + a_{12})$.
This yields $-108 = 6\left[a_1 + (-19)\right] \Rightarrow$
$-18 = a_1 + (-19) \Rightarrow a_1 = 1$. To solve for d, we
use the fact that $a_n = a_1 + (n-1)d$.

$a_n = a_1 + (n-1)d \Rightarrow a_{12} = a_1 + (12-1)d \Rightarrow$
$-19 = 1 + 11d \Rightarrow -20 = 11d \Rightarrow d = -\frac{20}{11}$
Thus, $a_1 = 1$ and $d = -\frac{20}{11}$.

53. $\displaystyle\sum_{i=1}^{3}(i+4)$

This is a sum of three terms having a common
difference of 1. Thus, in the formula
$S_n = \frac{n}{2}(a_1 + a_n)$, we have $a_1 = 1 + 4 = 5$,
$n = 3$, and $a_n = a_3 = 3 + 4 = 7$. The sum can,
therefore, be determined to be

$$\sum_{i=1}^{3}(i+4) = S_3 = \frac{3}{2}(a_1 + a_3) = \frac{3}{2}(5+7) = 18$$

55. $\displaystyle\sum_{j=1}^{10}(2j+3)$

This is the sum of the arithmetic sequence with
$a_1 = 2(1) + 3 = 5$, $d = 2$, and $n = 10$. Using the
formula $S_n = \frac{n}{2}\left[2a_1 + (n-1)d\right]$, we have

$$\sum_{j=1}^{10}(2j+3) = S_{10} = \frac{10}{2}\left[2(5) + 9(2)\right]$$
$$= 5(10 + 18) = 5(28) = 140$$

57. $\displaystyle\sum_{i=4}^{12}(-5-8i)$

This is the sum of an arithmetic sequence with
$d = -8$ and $a_1 = -5 - 8(4) = -37$. If i started
at 1, there would be 12 terms. Since three
terms are missing, $n = 9$. Using the formula
$S_n = \frac{n}{2}\left[2a_1 + (n-1)d\right]$, we have

$$\sum_{i=4}^{12}(-5-8i) = S_9 = \frac{9}{2}\left[2(-37) + 8(-8)\right]$$
$$= \frac{9}{2}\left[-74 + (-64)\right] = -621$$

Copyright © 2013 Pearson Education, Inc.

59. $\displaystyle\sum_{i=1}^{1000} i$

This is the sum of 1000 terms having a common difference of 1. Thus, in the formula $S_n = \frac{n}{2}(a_1 + a_n)$, we have $a_1 = 1$, $n = 1000$, and $a_n = a_{1000} = 1000$. The sum can, therefore, be determined as follows.

$$\sum_{i=1}^{1000} i = S_{1000} = \frac{1000}{2}(1 + 1000) = 500,500$$

61. $\displaystyle\sum_{k=1}^{100} (-k)$

This is the sum of 100 terms with $a_1 = -1$, $d = -1$ and $n = 100$. Using the formula $S_n = \frac{n}{2}\left[2a_1 + (n-1)d\right]$, we have

$$\sum_{k=1}^{100} (-k) = S_{100} = \frac{100}{2}\left[2(-1) + 99(-1)\right]$$
$$= 50(-2 - 99) = 50(-101) = -5050$$

For Exercises 63–65, you will need the sum(and seq(features of your graphing calculator. On the TI-84, the sum(feature is located under the LIST then MATH menu. The seq(feature is under the LIST then OPS menu.

```
NAMES OPS MATH       NAMES OPS MATH
1:min(               1:SortA(
2:max(               2:SortD(
3:mean(              3:dim(
4:median(            4:Fill(
5:sum(               5:seq(
6:prod(              6:cumSum(
7↓stdDev(            7↓ΔList(
```

63. $a_n = 4.2n + 9.73$

Using the sequence feature of a graphing calculator, we obtain $S_{10} = 328.3$.

```
sum(seq(4.2N+9.7
3,N,1,10,1))
              328.3
```

65. $a_n = \sqrt{8}n + \sqrt{3}$

Using the sequence feature of a graphing calculator, we obtain $S_{10} \approx 172.884$.

```
sum(seq(√(8)N+√(
3),N,1,10,1))
         172.8839999
```

67. Find the sum of all the integers from 51 to 71.
We need to find

$$\sum_{i=51}^{71} i = \sum_{i=1}^{71} i - \sum_{i=1}^{50} i = S_{71} - S_{50}.$$ Since we know

$a_1 = 1$, $d = 1$, $a_{50} = 50$, and $a_{71} = 71$, we

have $S_{71} = \frac{71}{2}(1 + 71) = 71(36) = 2556$ and

$S_{50} = \frac{50}{2}(1 + 50) = 25(51) = 1275$. Thus, the

sum is $S_{71} - S_{50} = 2556 - 1275 = 1281$.

69. In every 12-hour cycle, the clock will chime $1 + 2 + 3 + \cdots + 12$ times. Since $a_1 = 1$, $n = 12$, $a_{12} = 12$, and $S_n = \frac{n}{2}(a_1 + a_n)$, we have $S_{12} = \frac{12}{2}(1 + 12) = 6(13) = 78$. Since there are two 12-hour cycles in 1 day, every day, the clock will chime $2(78) = 156$ times. Since there are 30 days in this month, the clock will chime $156(30) = 4680$ times.

71. In this exercise, we have $a_1 = 49,000$, $d = 580$, and $n = 5$. Using $a_n = a_1 + (n-1)d$, we need to find a_{10}. Since

$$a_{11} = 49,000 + (11-1)580 = 49,000 + 10(580)$$
$$= 49,000 + 5800 = 54,800$$

we would expect that five years from now, the population will be 54,800.

73. The longest rung measures 28 in., and the shortest measures 18 in. Since the rungs are uniformly tapered, the sum of the lengths of these 31 supports can be thought of as the sum of an arithmetic sequence with $a_1 = 28$, $a_n = a_{31} = 18$, and $n = 31$. Since $S_n = \frac{n}{2}(a_1 + a_n)$, we have

$$S_{31} = \frac{31}{2}(a_1 + a_{31}) = \frac{31}{2}(28 + 18)$$
$$= \frac{31}{2}(46) = 713$$

Thus, a total of 713 in. of material would be needed.

Copyright © 2013 Pearson Education, Inc.

75. Consider the arithmetic sequence
$a_1, a_2, a_3, \ldots$. Then, consider the sequence
$a_1, a_3, a_5, \ldots$. We have
$a_3 - a_1 = (a_1 + 2d) - a_1 = 2d$ and
$a_5 - a_3 = (a_1 + 4d) - (a_1 + 2d) = 2d$. In
general, we have the following.

$a_n - a_{n-2}$
$= \left[a_1 + (n-1)d \right] - \left[a_1 + \left((n-2)-1 \right)d \right]$
$= a_1 + nd - d - \left[a_1 + (n-3)d \right]$
$= a_1 + nd - d - (a_1 + nd - 3d)$
$= a_1 + nd - d - a_1 - nd + 3d = 2d$

Thus, $a_1, a_3, a_5, \ldots$ is an arithmetic sequence
whose first term is a_1 and common difference
is $2d$.

Section 7.3 Geometric Sequences and Series

For Exercises 1–3, we are examining the geometric
sequence 1, 2, 4, 8, …. As stated on page 648 of the
text, the nth term of a geometric sequence is given by
$a_n = a_1 r^{n-1}$, where a_1 is the first term and r is the
common ratio. For this situation, $a_n = 1 \cdot 2^{n-1} = 2^{n-1}$.

1. day 10

(a) $a_{10} = 2^{10-1} = 2^9 = 512$ cents or \$5.12

(b) Since $S_n = \dfrac{a_1 \left(1 - r^n \right)}{1-r}$ for $r \neq 1$, we have

$S_{10} = \dfrac{1 \left(1 - 2^{10} \right)}{1-2} = \dfrac{1 - 1024}{-1} = \dfrac{-1023}{-1} = 1023$
cents or \$10.23.

3. day 15

(a) $a_{15} = 2^{15-1} = 2^{14} = 16,384$ cents or
\$163.84

(b) Since $S_n = \dfrac{a_1 \left(1 - r^n \right)}{1-r}$ for $r \neq 1$, we have

$S_{15} = \dfrac{1 \left(1 - 2^{15} \right)}{1-2} = \dfrac{1 - 32,768}{-1} = \dfrac{-32,767}{-1} = 32,767$
cents or \$327.67.

5. $a_1 = 5$, $r = -2$
$a_5 = a_1 r^{5-1} = 5(-2)^4 = 5(16) = 80$
$a_n = a_1 r^{n-1} = 5(-2)^{n-1}$

7. $a_2 = -4$, $r = 3$
We need to determine a_1. Since
$a_2 = a_1 r^{2-1} \Rightarrow -4 = a_1 (3)^1 \Rightarrow$
$-4 = a_1 (3) \Rightarrow a_1 = -\frac{4}{3}$, we have
$a_5 = a_1 r^{5-1} = -\frac{4}{3}(3)^4 = -\frac{4}{3}(81) = -108$ and
$a_n = a_1 r^{n-1} = -\frac{4}{3}(3)^{n-1}$

9. $a_4 = 243$, $r = -3$
We need to determine a_1. Since
$a_4 = a_1 r^{4-1} \Rightarrow 243 = a_1 (-3)^3 \Rightarrow$
$243 = a_1 (-27) \Rightarrow a_1 = -9$, we have
$a_5 = a_1 r^{5-1} = -9(-3)^4 = -9(81) = -729$
$a_n = a_1 r^{n-1} = -9(-3)^{n-1}$
Note that
$(-9)(-3)^{n-1} = -(-3)^2 (-3)^{n-1} = (-3)^{2+n-1}$
$= -(-3)^{n+1}$, so $a_n = -(-3)^{n+1}$ is an
equivalent formula for the nth term of this
sequence.

11. $-4, -12, -36, -108, \ldots$
First find r. Since $r = \frac{a_2}{a_1}$, we have
$r = \frac{-12}{-4} = 3$. Also, given that $a_1 = -4$, we have
$a_5 = a_1 r^{5-1} = -4(3)^4 = -4(81) = -324$
$a_n = a_1 r^{n-1} = -4(3)^{n-1}$

13. $\frac{4}{5}$, 2, 5, $\frac{25}{2}$, …
First find r. Since $r = \frac{a_2}{a_1}$, we have
$r = \frac{2}{\frac{4}{5}} = 2\left(\frac{5}{4}\right) = \frac{5}{2}$. Also, given that $a_1 = \frac{4}{5}$, we
have $a_5 = a_1 r^{5-1} = \frac{4}{5}\left(\frac{5}{2}\right)^4 = \frac{4}{5}\left(\frac{625}{16}\right) = \frac{125}{4}$ and
$a_n = a_1 r^{n-1} = \frac{4}{5}\left(\frac{5}{2}\right)^{n-1}$
Note that $\left(\frac{4}{5}\right)\left(\frac{5}{2}\right)^{n-1} = \frac{2^2}{5^1} \cdot \frac{5^{n-1}}{2^{n-1}} = \frac{5^{n-2}}{2^{n-3}}$, so
$a_n = \frac{5^{n-2}}{2^{n-3}}$ is an equivalent formula for the nth
term of this sequence.

Copyright © 2013 Pearson Education, Inc.

15. $10, -5, \frac{5}{2}, -\frac{5}{4}, \ldots$

First find r. Since $r = \frac{a_2}{a_1}$, we have

$r = \frac{-5}{10} = -\frac{1}{2}$. Also, given that $a_1 = 10$, we

have $a_5 = a_1 r^{5-1} = 10\left(-\frac{1}{2}\right)^4 = 10\left(\frac{1}{16}\right) = \frac{5}{8}$ and

$a_n = a_1 r^{n-1} = 10\left(-\frac{1}{2}\right)^{n-1}$

17. $a_2 = -6$, $a_7 = -192$

Using $a_n = a_1 r^{n-1}$, we have $a_2 = a_1 r = -6$.

This yields $a_1 = \frac{-6}{r}$. Also, $a_7 = a_1 r^6 = -192$.

Substituting $a_1 = \frac{-6}{r}$ into $a_1 r^6 = -192$

equation and solving for r, we have

$a_1 r^6 = -192 \Rightarrow \left(\frac{-6}{r}\right) r^6 = -192 \Rightarrow$

$-6r^5 = -192 \Rightarrow r^5 = 32 \Rightarrow r^5 = 2^5 \Rightarrow r = 2$

Since $a_1 = \frac{-6}{r}$, we have $a_1 = \frac{-6}{2} = -3$.

19. $a_3 = 50$, $a_7 = .005$

Using $a_n = a_1 r^{n-1}$, we have $a_3 = a_1 r^2 = 50$.

This yields $a_1 = \frac{50}{r^2}$. Also, $a_7 = a_1 r^6 = 0.005$.

Substituting $a_1 = \frac{50}{r^2}$ into $a_1 r^6 = 0.005$

equation and solving for r, we have

$a_1 r^6 = 0.005 \Rightarrow \left(\frac{50}{r^2}\right) r^6 = 0.005 \Rightarrow$

$50 r^4 = 0.005 \Rightarrow r^4 = 0.0001 \Rightarrow r = \pm 0.1$

$a_1 = \frac{50}{(\pm 0.1)^2} = \frac{50}{0.01} = 5000$.

21. $2, 8, 32, 128, \ldots$

First find r. Since $r = \frac{a_2}{a_1}$, we have $r = \frac{8}{2} = 4$.

Also, given that $a_1 = 2$, $n = 5$, and

$S_n = \frac{a_1(1-r^n)}{1-r}$, we have

$S_n = \frac{a_1(1-r^n)}{1-r} = \frac{2(1-4^5)}{1-4} = \frac{2(1-1024)}{-3}$

$= \frac{2(-1023)}{-3} = \frac{-2046}{-3} = 682$

23. $18, -9, \frac{9}{2}, -\frac{9}{4}, \ldots$

First find r. Since $r = \frac{a_2}{a_1}$, we have

$r = \frac{-9}{18} = -\frac{1}{2}$. Also, given that $a_1 = 18$, $n = 5$,

and $S_n = \frac{a_1(1-r^n)}{1-r}$, we have

$S_n = \frac{a_1(1-r^n)}{1-r} = \frac{18\left[1-\left(-\frac{1}{2}\right)^5\right]}{1-\left(-\frac{1}{2}\right)} = \frac{18\left[1-\left(-\frac{1}{32}\right)\right]}{\frac{3}{2}}$

$= \frac{18\left(\frac{32}{32}+\frac{1}{32}\right)}{\frac{3}{2}} = 18\left(\frac{33}{32}\right)\left(\frac{2}{3}\right) = 18\left(\frac{11}{16}\right) = \frac{99}{8}$

25. $a_1 = 8.423$, $r = 2.859$

Since $S_n = \frac{a_1(1-r^n)}{1-r}$, we have

$S_5 = \frac{8.423\left[1-(2.859)^5\right]}{1-2.859} \approx 860.95$.

27. $\displaystyle\sum_{i=1}^{5} (-3)^i$

For this geometric series, $a_1 = (-3)^1 = -3$,

$r = -3$, and $n = 5$. Since $S_n = \frac{a_1(1-r^n)}{1-r}$, we

have

$\displaystyle\sum_{i=1}^{5} (-3)^i = S_5 = \frac{-3\left(1-(-3)^5\right)}{1-(-3)} = \frac{-3(1+243)}{4}$

$= \frac{-3(244)}{4} = -183$

29. $\displaystyle\sum_{j=1}^{6} 48\left(\frac{1}{2}\right)^j$

For this geometric series,

$a_1 = 48\left(\frac{1}{2}\right)^1 = 48\left(\frac{1}{2}\right) = 24$, $r = \frac{1}{2}$, and $n = 6$.

Since $S_n = \frac{a_1(1-r^n)}{1-r}$, we have

$\displaystyle\sum_{j=1}^{6} 48\left(\frac{1}{2}\right)^j = S_6 = \frac{24\left[1-\left(\frac{1}{2}\right)^6\right]}{1-\frac{1}{2}} = \frac{24\left(1-\frac{1}{64}\right)}{\frac{1}{2}}$

$= \frac{24\left(\frac{64}{64}-\frac{1}{64}\right)}{\frac{1}{2}} = 24\left(\frac{63}{64}\right)(2)$

$= 24\left(\frac{63}{32}\right) = \frac{189}{4}$

Copyright © 2013 Pearson Education, Inc.

31. $\displaystyle\sum_{k=4}^{10} 2^k$

This series is the sum of the fourth through tenth terms of a geometric sequence with $a_1 = 2^1 = 2$ and $r = 2$. To find this sum, find the difference between the sum of the first ten terms and the sum of the first three terms.

$$\sum_{k=4}^{10} 2^k = \sum_{k=1}^{10} 2^k - \sum_{k=1}^{3} 2^k = \frac{2(1-2^{10})}{1-2} - \frac{2(1-2^3)}{1-2}$$

$$= \frac{2(1-1024)}{-1} - \frac{2(1-8)}{-1} = \frac{2(-1023)}{-1} - \frac{2(-7)}{-1}$$

$$= 2046 - 14 = 2032$$

Note: We could also consider $\displaystyle\sum_{k=4}^{10} 2^k$ as a geometric series with $a_1 = 2^4 = 16$ and $r = 2$. If the sequence started with $k = 1$, there would be 10 terms. Since it starts with 4, three of the terms are missing, so there are seven terms and $n = 7$. Thus we have

$$\sum_{k=4}^{10} 2^k = S_7 = \frac{16(1-2^7)}{1-2} = \frac{16(1-128)}{-1}$$

$$= \frac{16(-127)}{-1} = 2032$$

33. The sum of an infinite geometric series exists if $|r| < 1$.

35. 12, 24, 48, 96, ...

Since $r = \frac{a_2}{a_1}$, we have $r = \frac{24}{12} = 2$. The sum of the terms of this infinite geometric sequence would not converge since $r = 2$ is not between -1 and 1.

37. $-48, -24, -12, -6, ...$

Since $r = \frac{a_2}{a_1}$, we have $r = \frac{-24}{-48} = \frac{1}{2}$. Since $-1 < r < 1$, the sum converges.

39. $2 + 1 + \frac{1}{2} + \frac{1}{4} + \cdots$ is a geometric series with

$r = \frac{a_2}{a_1} = \frac{1}{2}$ and $a_1 = 2$. Since $S_n = \frac{a_1(1-r^n)}{1-r}$,

we have $S_n = \frac{2\left[1-\left(\frac{1}{2}\right)^n\right]}{1-\frac{1}{2}} = \frac{2\left[1-\left(\frac{1}{2}\right)^n\right]}{\frac{1}{2}}$.

Similar to Example 7, we have $\displaystyle\lim_{n\to\infty} \left(\frac{1}{2}\right)^n = 0$.

Thus, we obtain

$$\lim_{n\to\infty} S_n = \lim_{n\to\infty} \frac{2\left[1-\left(\frac{1}{2}\right)^n\right]}{\frac{1}{2}} = \frac{2(1-0)}{\frac{1}{2}} = \frac{2}{\frac{1}{2}} = 4.$$

41. $18 + 6 + 2 + \frac{2}{3} + \cdots$

For this geometric series, $a_1 = 18$ and $r = \frac{6}{18} = \frac{1}{3}$. Because $-1 < r < 1$, this series converges. We have

$$S_\infty = \frac{a_1}{1-r} = \frac{18}{1-\frac{1}{3}} = \frac{18}{\frac{2}{3}} = 18 \cdot \frac{3}{2} = 27.$$

43. $\frac{1}{4} - \frac{1}{6} + \frac{1}{9} - \frac{2}{27} + \cdots$

For this geometric series, $a_1 = \frac{1}{4}$ and

$r = \frac{-\frac{1}{6}}{\frac{1}{4}} = -\frac{1}{6} \cdot \frac{4}{1} = -\frac{2}{3}$. Because $-1 < r < 1$,

this series converges. We have

$$S_\infty = \frac{a_1}{1-r} = \frac{\frac{1}{4}}{1-\left(-\frac{2}{3}\right)} = \frac{\frac{1}{4}}{\frac{5}{3}} = \frac{1}{4} \cdot \frac{3}{5} = \frac{3}{20}.$$

45. $\displaystyle\sum_{i=1}^{\infty} 3\left(\frac{1}{4}\right)^{i-1}$

For this geometric series,

$a_1 = 3\left(\frac{1}{4}\right)^{1-1} = 3\left(\frac{1}{4}\right)^0 = 3 \cdot 1 = 3$ and $r = \frac{1}{4}$.

Because $-1 < r < 1$, this series converges. We

have $S_\infty = \frac{a_1}{1-r} = \frac{3}{1-\frac{1}{4}} = \frac{3}{\frac{3}{4}} = 3 \cdot \frac{4}{3} = 4.$

47. $\displaystyle\sum_{k=1}^{\infty} 3^{-k}$

For this geometric series, $a_1 = 3^{-1} = \frac{1}{3}$ and $r = \frac{1}{3}$. Because $-1 < r < 1$ this series converges. We have

$$S_\infty = \frac{a_1}{1-r} = \frac{\frac{1}{3}}{1-\frac{1}{3}} = \frac{\frac{1}{3}}{\frac{2}{3}} = \frac{1}{2}.$$

Copyright © 2013 Pearson Education, Inc.

49. $\displaystyle\sum_{i=1}^{\infty}\left(-\frac{2}{3}\right)\left(-\frac{1}{4}\right)^{i-1}$

For this geometric series,

$a_1 = \left(-\frac{2}{3}\right)\left(-\frac{1}{4}\right)^{1-1} = -\frac{2}{3}$ and $r = -\frac{1}{4}$.

$S_\infty = \dfrac{a_1}{1-r} = \dfrac{-\frac{2}{3}}{1-\left(-\frac{1}{4}\right)} = \dfrac{-\frac{2}{3}}{\frac{5}{4}} = -\dfrac{8}{15}$.

51. $\displaystyle\sum_{i=1}^{\infty}\left(\frac{3}{7}\right)^{i}$

For this geometric series, $a_1 = \left(\frac{3}{7}\right)^{1} = \frac{3}{7}$ and $r = \frac{3}{7}$.

$S_\infty = \dfrac{a_1}{1-r} = \dfrac{\frac{3}{7}}{1-\frac{3}{7}} = \dfrac{\frac{3}{7}}{\frac{4}{7}} = \dfrac{3}{4}$.

For Exercises 53–55, you will need the sum(and seq(features of your graphing calculator. On the TI-84, the sum(feature is located under the LIST then MATH menu. The seq(feature is under the LIST then OPS menu.

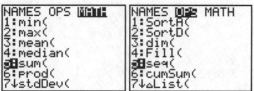

53. $\displaystyle\sum_{i=1}^{10}(1.4)^{i}$

Using the sequence feature of a graphing calculator, we obtain $S_{10} \approx 97.739$.

```
sum(seq(1.4^I,I,
1,10))
          97.73912924
```

55. $\displaystyle\sum_{j=3}^{8}2(0.4)^{j}$

Using the sequence feature of a graphing calculator, the sum is approximately 0.212.

```
sum(seq(2*.4^J,J
,3,8))
          .21245952
```

57. (a) $a_n = 1276(0.916)^{n}$

$a_1 = 1276(0.916)^{1} = 1168.816 \approx 1169$

and $r = 0.916$

(b) $a_n = 1276(0.916)^{n}$

$a_{10} = 1276(0.916)^{10} \approx 531$

$a_{20} = 1276(0.916)^{20} \approx 221$

This means that a person who is 10 years from retirement should have savings of 531% of his or her annual salary; a person 20 years from retirement should have savings of 221% of his or her annual salary.

59. (a) Since the first term is a_1 and the common ratio is r, we have $a_n = a_1 \cdot 2^{n-1}$.

(b) If $a_1 = 100$, we have $a_n = 100 \cdot 2^{n-1}$.

Since $100 = 10^2$ and $1{,}000{,}000 = 10^6$, we need to solve the equation $10^2 \cdot 2^{n-1} = 10^6$. Divide both sides by 10^2 to obtain $2^{n-1} = 10^4$. Take the common logarithm (base 10) of both sides and solve for n.

$\log 2^{n-1} = \log 10^4 \Rightarrow (n-1)\log 2 = 4 \Rightarrow$

$n-1 = \dfrac{4}{\log 2} \Rightarrow n = \dfrac{4}{\log 2} + 1 \approx 14.28$

Since the number of bacteria is increasing, the first value of n where $a_n > 1{,}000{,}000$ is 15.

(c) Since a_n represents the number of bacteria after 40(n – 1) minutes, a_{15} represents the number after $40(15-1) = 40 \cdot 14 = 560$ minutes or 9 hours, 20 minutes.

61. Since there are 200 insects in the first generation, $a_1 = 200$. The r value will be 1.25. Thus, $a_n = a_1 r^{n-1} \Rightarrow$

$a_5 = 200(1.25)^{5-1} = 200(1.25)^{4} = 488.28125$.

There are approximately 488 fruit flies in the fifth generation.

Copyright © 2013 Pearson Education, Inc.

63. Use the formula for the sum of an infinite geometric sequence with $a_1 = 1000$ and $r = 0.1$. Since

$$S_\infty = \frac{a_1}{1-r} = \frac{1000}{1-0.1} = \frac{1000}{0.9} = \frac{10,000}{9}, \text{ the}$$

manager should have ordered $\frac{10,000}{9}$ units of sugar.

65. Use the formula for the sum of the first n terms of a geometric sequence with $a_1 = 2$, $r = 2$, and $n = 5$.

$$S_n = \frac{a_1\left(1-r^n\right)}{1-r} \Rightarrow$$

$$S_5 = \frac{2\left(1-2^5\right)}{1-2} = \frac{2(1-32)}{-1} = \frac{2(-31)}{-1} = 62$$

Going back five generations, the total number of ancestors is 62. Next, use the same formula with $a_1 = 2$, $r = 2$, and $n = 10$.

$$S_{10} = \frac{2\left(1-2^{10}\right)}{1-2} = \frac{2(1-1024)}{-1} = 2046$$

Going back ten generations, the total number of ancestors is 2046.

67. When the midpoints of the sides of an equilateral triangle are connected, the length of a side of the new triangle is one-half the length of a side of the original triangle. Use the formula for the nth term of a geometric sequence with $a_1 = 2$, $r = \frac{1}{2}$, and $n = 8$.

$$a_n = a_1 r^{n-1} \Rightarrow$$

$$a_8 = (2)\left(\tfrac{1}{2}\right)^{8-1} = (2)\left(\tfrac{1}{2}\right)^7 = (2)\left(\tfrac{1}{128}\right) = \tfrac{1}{64}$$

The eighth triangle has sides of length $\frac{1}{64}$ m.

69. Option 1 is modeled by the arithmetic sequence $a_n = 5000 + 10,000(n-1)$ with the following sum.

$$S_n = \tfrac{n}{2}\left[2a_1 + (n-1)d\right]$$

$$S_{30} = \tfrac{30}{2}\left[2a_1 + (30-1)d\right] = 15\left[2a_1 + 29d\right]$$

$$= 15\left[2(5000) + 29 \cdot 10,000\right]$$

$$= 15\left[10,000 + 290,000\right]$$

$$S_{30} = 15(300,000) = 4,500,000$$

Thus, the first option will pay $4,500,000 for the month's work. Option 2 is modeled by the sequence $a_n = 0.01(2)^{n-1}$ with the following sum.

$$S_n = \frac{a_1\left(1-r^n\right)}{1-r} \Rightarrow S_{30} = \frac{a_1\left(1-r^{30}\right)}{1-r} \Rightarrow$$

$$S_{30} = \frac{0.01\left(1-2^{30}\right)}{1-2}$$

$$= \frac{0.01(1-1,073,741,824)}{-1} \Rightarrow$$

$$S_{30} = -0.01(-1,073,741,823) = 10,737,418.23$$

The second will pay $10,737,418.23. Option 2 pays better.

71. The future value of an annuity uses the formula

$$S_n = \frac{a_1(1-r^n)}{1-r}, \text{ where } r = 1 + \text{interest rate.}$$

The payments are $1000 for 9 yr at 3% compounded annually, so $a_1 = 1000$, $r = 1.03$, and $n = 9$.

$$S_9 = \frac{1000\left[1-(1.03)^9\right]}{1-1.03} \approx 10,159.11$$

The future value is $10,159.11.

73. The future value of an annuity uses the formula

$$S_n = \frac{a_1(1-r^n)}{1-r}, \text{ where } r = 1 + \text{interest rate.}$$

The payments are $2430 for 10 yr at 1% compounded annually, so $a_1 = 2430$, $r = 1.01$, and $n = 10$.

$$S_{10} = \frac{2430\left[1-(1.01)^{10}\right]}{1-1.01} \approx 25,423.18$$

The future value is $25,423.18.

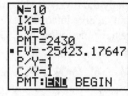

Copyright © 2013 Pearson Education, Inc.

75. $S = R\left[\dfrac{(1+i)^n - 1}{i}\right]$ where $R = 2403$, $i = 0.01$,

and $n = 11$.

$$S = 2430\left[\dfrac{(1.01)^{11} - 1}{0.01}\right] \approx 28,107.41$$

The balance after 11 years is $28,107.41

```
N=11
I%=1
PV=0
PMT=2430
■FV=-28107.40824
P/Y=1
C/Y=1
PMT:BEGIN BEGIN
```

77. For deposits made annually for twenty-five

years, we need to find $S = R\left[\dfrac{(1+i)^n - 1}{i}\right]$

where $R = 2000$, $i = 0.03$, and $n = 25$. Thus,

we have $S = 2000\left[\dfrac{(1.03)^{25} - 1}{0.03}\right] \approx 72,918.53.$

The total amount in Michael's IRA will be
$72,918.53.

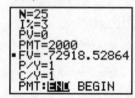

```
N=25
I%=3
PV=0
PMT=2000
■FV=-72918.52864
P/Y=1
C/Y=1
PMT:BEGIN BEGIN
```

79. In the formula $S_n = \dfrac{a_1(1 - r^n)}{1 - r}$, we are

considering the sum
$a_1 + a_1 r + a_1 r^2 + \cdots + a_1 r^{n-1}$. With an annuity
with deposits at the end of each year, this
summation is actually the reversed since the
last deposit will not receive interest. Reversing
the summation we have the following.
$$a_1 r^{n-1} + a_1 r^{n-2} + \cdots + a_1 r^2 + a_1 r + a_1$$
Now the first term of this series represents the
first amount deposited having gone through
$n - 1$ compound periods (since it is deposited
at the end of the first year). The value of r is
$1 + i$, where i is the interest rate in decimal
form. The value of a_1 is the equal amounts of
money that are deposited annually.

If we call this value R, then the formula

$$S_n = \dfrac{a_1(1 - r^n)}{1 - r} \quad \text{becomes} \quad S = \dfrac{R(1 - (1+i)^n)}{1 - (1+i)}.$$

This simplifies as follows.

$$S = \dfrac{R(1 - (1+i)^n)}{1 - (1+i)} \Rightarrow S = \dfrac{R(1 - (1+i)^n)}{1 - 1 - i} \Rightarrow$$

$$S = R\dfrac{1 - (1+i)^n}{-i} \Rightarrow S = R\dfrac{(1+i)^n - 1}{i}$$

81. If $a_1, a_2, a_3, \ldots$ is a geometric sequence, then
$a_n = a_1 r_a^{n-1}$ (where r_a is the common ratio
for this sequence).
$a_3 = a_1 \cdot r^2, a_5 = a_3 \cdot r^2, a_7 = a_5 \cdot r^2, \ldots$ So,
$a_1, a_3, a_5, \ldots$ is a geometric sequence, with
common ratio r^2.

Summary Exercises on Sequences and Series

1. 2, 4, 8, 16, 32, … is a geometric sequence
with $r = 2$.
Notice $2(2) = 4$, $4(2) = 8$, $8(2) = 16$,
and $16(2) = 32$.

3. $3, \frac{1}{2}, -2, -\frac{9}{2}, -7, \ldots$ is an arithmetic

sequence with $d = -\frac{5}{2}$.

Notice $3 + \left(-\frac{5}{2}\right) = \frac{1}{2}$, $\frac{1}{2} + \left(-\frac{5}{2}\right) = -2$,

$-2 + \left(-\frac{5}{2}\right) = -\frac{9}{2}$, and $-\frac{9}{2} + \left(-\frac{5}{2}\right) = -7$.

5. $\frac{3}{4}, 1, \frac{4}{3}, \frac{16}{9}, \frac{64}{27}, \ldots$ is a geometric sequence

with $r = \frac{4}{3}$.

Notice $\frac{3}{4}\left(\frac{4}{3}\right) = 1, 1\left(\frac{4}{3}\right) = \frac{4}{3}, \frac{4}{3}\left(\frac{4}{3}\right) = \frac{16}{9}$, and

$\frac{16}{9}\left(\frac{4}{3}\right) = \frac{64}{27}$.

7. $\frac{1}{2}, \frac{1}{3}, \frac{1}{4}, \frac{1}{5}, \frac{1}{6}, \ldots$ is neither an arithmetic

sequence nor a geometric sequence.
Notice $\frac{1}{2} + \left(-\frac{1}{6}\right) = \frac{1}{3}$ but $\frac{1}{3} + \left(-\frac{1}{6}\right) \neq \frac{1}{4}$, so the

sequence is not arithmetic. Also, $\frac{1}{2}\left(\frac{2}{3}\right) = \frac{1}{3}$ but

$\frac{1}{3}\left(\frac{2}{3}\right) \neq \frac{1}{4}$, so the sequence is not geometric.

Copyright © 2013 Pearson Education, Inc.

9. 1, 9, 10, 19, 29, ... is neither an arithmetic sequence nor a geometric sequence. Notice $1+8=9$, but $9+8 \neq 10$, so the sequence is not arithmetic. Also, $1 \cdot 9 = 9$, but $9 \cdot 9 \neq 10$, so the sequence is not geometric.

11. The sequence $3, 6, 12, 24, 48, ...$ is geometric with $a_1 = 3$ and $r = 2$. Thus, we have

$$a_n = a_1 r^{n-1} \Rightarrow a_n = 3(2)^{n-1}$$

Since $\sum_{i=1}^{n} a_1 r^{i-1} = S_n = \frac{a_1(1-r^n)}{1-r}$, we have

$$\sum_{i=1}^{10} 3(2)^{i-1} = S_{10} = \frac{3(1-2^{10})}{1-2} = \frac{3(1-1024)}{-1}$$
$$= -3(-1023) = 3069$$

13. The sequence $4, \frac{5}{2}, 1, -\frac{1}{2}, -2, ...$ is arithmetic with $a_1 = 4$ and $d = -\frac{3}{2}$. Thus, we have

$$a_n = a_1 + (n-1)d \Rightarrow a_n = 4 + (n-1)\left(-\frac{3}{2}\right) \Rightarrow$$
$$a_n = 4 - \frac{3}{2}n + \frac{3}{2} \Rightarrow a_n = \frac{11}{2} - \frac{3}{2}n$$

Since

$$\sum_{i=1}^{n}\left[a_1 + (i-1)d\right] = S_n = \frac{n}{2}\left[2a_1 + (n-1)d\right],$$

we have

$$\sum_{i=1}^{10} \frac{11}{2} - \frac{3}{2}i = S_{10} = \frac{10}{2}\left[2(4) + (10-1)\left(-\frac{3}{2}\right)\right]$$
$$= 5\left[8 + 9\left(-\frac{3}{2}\right)\right] = 5\left[8 + \left(-\frac{27}{2}\right)\right]$$
$$= 5\left(-\frac{11}{2}\right) = -\frac{55}{2}$$

15. The sequence $3, -6, 12, -24, 48, ...$ is geometric with $a_1 = 3$ and $r = -2$. Thus, we have $a_n = a_1 r^{n-1} \Rightarrow a_n = 3(-2)^{n-1}$

Since $\sum_{i=1}^{n} a_1 r^{i-1} = S_n = \frac{a_1(1-r^n)}{1-r}$, we have

$$\sum_{i=1}^{10} 3(2)^{i-1} = S_{10} = \frac{3\left[1-(-2)^{10}\right]}{1-(-2)}$$
$$= \frac{3(1-1024)}{3} = -1023$$

17. $\sum_{i=1}^{\infty} \frac{1}{3}(-2)^{i-1}$ is an infinite geometric series with $a_1 = \frac{1}{3}$ and $r = -2$. Since $|r| = 2 > 1$, this series diverges.

19. $\sum_{i=1}^{25}(4-6i)$ is an arithmetic series with $a_1 = -2$ and $d = -6$. Since the sum of the first n terms of an arithmetic series is

$$S_n = \frac{n}{2}\left[2a_1 + (n-1)d\right], \text{ we have}$$

$$\sum_{i=1}^{25}(4-6i) = S_{25} = \frac{25}{2}\left[2(-2) + (25-1)(-6)\right]$$
$$= \frac{25}{2}\left[-4 + 24(-6)\right]$$
$$= \frac{25}{2}\left[-4 + (-144)\right] = \frac{25}{2}(-148)$$
$$= -1850$$

21. $\sum_{i=1}^{\infty} 4\left(-\frac{1}{2}\right)^i$ is an infinite geometric series with $a_1 = -2$ and $r = -\frac{1}{2}$. Since $|r| = \frac{1}{2} < 1$, then the sum of this series is

$$S = \frac{a_1}{1-r} = \frac{-2}{1-\left(-\frac{1}{2}\right)} = \frac{-2}{\frac{3}{2}} = -\frac{4}{3}.$$

23. $\sum_{j=1}^{12}(2j-1)$ is an arithmetic series with $a_1 = 1$ and $d = 2$. To use the formula

$$S_n = \frac{n}{2}(a_1 + a_n), \text{ we need find } a_{12} \text{ (the last}$$

term). Since $a_{12} = 2(12) - 1 = 24 - 1 = 23$, we have

$$\sum_{j=1}^{12}(2j-1) = S_{12} = \frac{12}{2}(1+23) = 6(24) = 144$$

25. $\sum_{k=1}^{\infty} 1.0001^i$ is an infinite geometric series with $a_1 = 1.0001$ and $r = 1.0001$ Since $|r| = 1.0001 > 1$, this series diverges.

Section 7.4   The Binomial Theorem

1. $\frac{6!}{3!3!} = \frac{6 \cdot 5 \cdot 4 \cdot 3 \cdot 2 \cdot 1}{3 \cdot 2 \cdot 1 \cdot 3 \cdot 2 \cdot 1} = \frac{6 \cdot 5 \cdot 4}{3 \cdot 2 \cdot 1} = 20$

3. $\frac{7!}{3!4!} = \frac{7 \cdot 6 \cdot 5 \cdot 4 \cdot 3 \cdot 2 \cdot 1}{3 \cdot 2 \cdot 1 \cdot 4 \cdot 3 \cdot 2 \cdot 1} = \frac{7 \cdot 6 \cdot 5}{3 \cdot 2 \cdot 1} = 35$

5. $\binom{8}{5} = \frac{8!}{5!3!} = \frac{8 \cdot 7 \cdot 6 \cdot 5 \cdot 4 \cdot 3 \cdot 2 \cdot 1}{5 \cdot 4 \cdot 3 \cdot 2 \cdot 1 \cdot 3 \cdot 2 \cdot 1} = \frac{8 \cdot 7 \cdot 6}{3 \cdot 2 \cdot 1} = 56$

Copyright © 2013 Pearson Education, Inc.

7. $\dbinom{10}{2} = \dfrac{10!}{2!8!} = \dfrac{10 \cdot 9 \cdot 8!}{2 \cdot 1 \cdot 8!} = \dfrac{10 \cdot 9}{2 \cdot 1} = 45$

9. $\dbinom{14}{14} = \dfrac{14!}{14!(14-14)!} = \dfrac{14!}{14!0!} = \dfrac{14!}{14! \cdot 1} = 1$

11. $\dbinom{n}{n-1} = \dfrac{n!}{(n-1)![n-(n-1)]!} = \dfrac{n!}{(n-1)!(n-n+1)!} = \dfrac{n!}{(n-1)!1!} = \dfrac{n(n-1)!}{(n-1)!} = n$

13. $_8C_3 = \dfrac{8!}{3!(8-3)!} = \dfrac{8!}{3!5!} = \dfrac{8 \cdot 7 \cdot 6}{3 \cdot 2 \cdot 1} = 56$

15. $_{100}C_{98} = \dfrac{100!}{98!(100-98)!} = \dfrac{100!}{98!2!} = \dfrac{100 \cdot 99}{2 \cdot 1} = 4950$

17. $_9C_0 = \dfrac{9!}{0!(9-0)!} = \dfrac{9!}{1 \cdot 9!} = 1$

19. $_{12}C_1 = \dfrac{12!}{1!(12-1)!} = \dfrac{12 \cdot 11!}{1!11!} = \dfrac{12}{1} = 12$

21. The first term in the expansion of $(2x+3y)^4$ is $(2x)^4 = 16x^4$ and the last is $(3y)^4 = 81y^4$.

23. $(x+y)^6 = x^6 + \dbinom{6}{1}x^5y + \dbinom{6}{2}x^4y^2 + \dbinom{6}{3}x^3y^3 + \dbinom{6}{4}x^2y^4 + \dbinom{6}{5}xy^5 + y^6$

$= x^6 + \dfrac{6!}{1!5!}x^5y + \dfrac{6!}{2!4!}x^4y^2 + \dfrac{6!}{3!3!}x^3y^3 + \dfrac{6!}{4!2!}x^2y^4 + \dfrac{6!}{5!1!}xy^5 + y^6$

$= x^6 + 6x^5y + \dfrac{6 \cdot 5}{2 \cdot 1}x^4y^2 + \dfrac{6 \cdot 5 \cdot 4}{3 \cdot 2 \cdot 1}x^3y^3 + \dfrac{6 \cdot 5}{2 \cdot 1}x^2y^4 + 6xy^5 + y^6$

$= x^6 + 6x^5y + 15x^4y^2 + 20x^3y^3 + 15x^2y^4 + 6xy^5 + y^6$

25. $(p-q)^5 = \left[p+(-q)\right]^5$

$= p^5 + \dbinom{5}{1}p^4(-q) + \dbinom{5}{2}p^3(-q)^2 + \dbinom{5}{3}p^2(-q)^3 + \dbinom{5}{4}p(-q)^4 + (-q)^5$

$= p^5 + \dfrac{5!}{1!4!}p^4(-q) + \dfrac{5!}{2!3!}p^3q^2 + \dfrac{5!}{3!2!}p^2(-q)^3 + \dfrac{5!}{4!1!}pq^4 + (-q)^5$

$= p^5 + 5p^4(-q) + \dfrac{5 \cdot 4}{2 \cdot 1}p^3q^2 + \dfrac{5 \cdot 4}{2 \cdot 1}p^2(-q)^3 + 5pq^4 + (-q)^5$

$= p^5 - 5p^4q + 10p^3q^2 - 10p^2q^3 + 5pq^4 - q^5$

27. $(r^2+s)^5 = (r^2)^5 + \dbinom{5}{1}(r^2)^4s + \dbinom{5}{2}(r^2)^3s^2 + \dbinom{5}{3}(r^2)^2s^3 + \dbinom{5}{4}(r^2)s^4 + s^5$

$= r^{10} + \dfrac{5!}{1!4!}r^8s + \dfrac{5!}{2!3!}r^6s^2 + \dfrac{5!}{3!2!}r^4s^3 + \dfrac{5!}{4!1!}r^2s^4 + s^5$

$= r^{10} + 5r^8s + \dfrac{5 \cdot 4}{2 \cdot 1}r^6s^2 + \dfrac{5 \cdot 4}{2 \cdot 1}r^4s^3 + 5r^2s^4 + s^5$

$= r^{10} + 5r^8s + 10r^6s^2 + 10r^4s^3 + 5r^2s^4 + s^5$

Copyright © 2013 Pearson Education, Inc.

29. $(p + 2q)^4 = p^4 + \binom{4}{1} p^3 (2q) + \binom{4}{2} p^2 (2q)^2 + \binom{4}{3} p (2q)^3 + (2q)^4$

$\qquad = p^4 + \dfrac{4!}{1!3!} p^3 (2q) + \dfrac{4!}{2!2!} p^2 (4q^2) + \dfrac{4!}{3!1!} p (8q^3) + 16q^4$

$\qquad = p^4 + 4 p^3 (2q) + \dfrac{4 \cdot 3}{2 \cdot 1} p^2 (4q^2) + 4 p (8q^3) + 16q^4 = p^4 + 8 p^3 q + 24 p^2 q^2 + 32 pq^3 + 16q^4$

31. $(7p - 2q)^4 = (7p)^4 + \binom{4}{1}(7p)^3 (-2q) + \binom{4}{2}(7p)^2 (-2q)^2 + \binom{4}{3}(7p)(-2q)^3 + (-2q)^4$

$\qquad = 2401 p^4 + \dfrac{4!}{1!3!}(343 p^3)(-2q) + \dfrac{4!}{2!2!}(49 p^2)(4q^2) + \dfrac{4!}{3!1!}(7p)(-8q^3) + 16q^4$

$\qquad = 2401 p^4 - 4(686 p^3 q) + \dfrac{4 \cdot 3}{2 \cdot 1}(196 p^2 q^2) - 4(56 pq^3) + 16q^4$

$\qquad = 2401 p^4 - 2744 p^3 q + 6(196 p^2 q^2) - 224 pq^3 + 16q^4$

$\qquad = 2401 p^4 - 2744 p^3 q + 1176 p^2 q^2 - 224 pq^3 + 16q^4$

33. $(3x - 2y)^6 = (3x)^6 + \binom{6}{1}(3x)^5 (-2y) + \binom{6}{2}(3x)^4 (-2y)^2 + \binom{6}{3}(3x)^3 (-2y)^3$

$\qquad\qquad + \binom{6}{4}(3x)^2 (-2y)^4 + \binom{6}{5}(3x)(-2y)^5 + (-2y)^6$

$\qquad = 729 x^6 + \dfrac{6!}{1!5!}(243 x^5)(-2y) + \dfrac{6!}{2!4!}(81 x^4)(4y^2) + \dfrac{6!}{3!3!}(27 x^3)(-8y^3)$

$\qquad\qquad + \dfrac{6!}{4!2!}(9 x^2)(16 y^4) + \dfrac{6!}{5!1!}(3x)(-32 y^5) + 64 y^6$

$\qquad = 729 x^6 - 6(486 x^5 y) + \dfrac{6 \cdot 5}{2 \cdot 1}(324 x^4 y^2) - \dfrac{6 \cdot 5 \cdot 4}{3 \cdot 2 \cdot 1}(216 x^3 y^3)$

$\qquad\qquad + \dfrac{6 \cdot 5}{2 \cdot 1}(144 x^2 y^4) - 6(96 xy^5) + 64 y^6$

$\qquad = 729 x^6 - 2916 x^5 y + 15(324 x^4 y^2) - 20(216 x^3 y^3) + 15(144 x^2 y^4) - 576 xy^5 + 64 y^6$

$\qquad = 729 x^6 - 2916 x^5 y + 4860 x^4 y^2 - 4320 x^3 y^3 + 2160 x^2 y^4 - 576 xy^5 + 64 y^6$

35. $\left(\dfrac{m}{2} - 1\right)^6 = \left(\dfrac{m}{2}\right)^6 + \binom{6}{1}\left(\dfrac{m}{2}\right)^5 (-1) + \binom{6}{2}\left(\dfrac{m}{2}\right)^4 (-1)^2 + \binom{6}{3}\left(\dfrac{m}{2}\right)^3 (-1)^3$

$\qquad\qquad + \binom{6}{4}\left(\dfrac{m}{2}\right)^2 (-1)^4 + \binom{6}{5}\left(\dfrac{m}{2}\right)(-1)^5 + (-1)^6$

$\qquad = \dfrac{m^6}{64} - \dfrac{6!}{1!5!}\left(\dfrac{m^5}{32}\right) + \dfrac{6!}{2!4!}\left(\dfrac{m^4}{16}\right) - \dfrac{6!}{3!3!}\left(\dfrac{m^3}{8}\right) + \dfrac{6!}{4!2!}\left(\dfrac{m^2}{4}\right) - \dfrac{6!}{5!1!}\left(\dfrac{m}{2}\right) + 1$

$\qquad = \dfrac{m^6}{64} - 6\left(\dfrac{m^5}{32}\right) + \dfrac{6 \cdot 5}{2 \cdot 1}\left(\dfrac{m^4}{16}\right) - \dfrac{6 \cdot 5 \cdot 4}{3 \cdot 2 \cdot 1}\left(\dfrac{m^3}{8}\right) + \dfrac{6 \cdot 5}{2 \cdot 1}\left(\dfrac{m^2}{4}\right) - 6\left(\dfrac{m}{2}\right) + 1$

$\qquad = \dfrac{m^6}{64} - \dfrac{3m^5}{16} + 15\left(\dfrac{m^4}{16}\right) - 20\left(\dfrac{m^3}{8}\right) + 15\left(\dfrac{m^2}{4}\right) - 3m + 1$

$\qquad = \dfrac{m^6}{64} - \dfrac{3m^5}{16} + \dfrac{15 m^4}{16} - \dfrac{5 m^3}{2} + \dfrac{15 m^2}{4} - 3m + 1$

Copyright © 2013 Pearson Education, Inc.

37. $\left(\sqrt{2}r + \dfrac{1}{m}\right)^4 = \left(\sqrt{2}r\right)^4 + \dbinom{4}{1}\left(\sqrt{2}r\right)^3\left(\dfrac{1}{m}\right) + \dbinom{4}{2}\left(\sqrt{2}r\right)^2\left(\dfrac{1}{m}\right)^2 + \dbinom{4}{3}\left(\sqrt{2}r\right)\left(\dfrac{1}{m}\right)^3 + \left(\dfrac{1}{m}\right)^4$

$= \left(\sqrt{2}r\right)^4 + \dfrac{4!}{1!3!}\left(\sqrt{2}r\right)^3\left(\dfrac{1}{m}\right) + \dfrac{4!}{2!2!}\left(\sqrt{2}r\right)^2\dfrac{1}{m^2} + \dfrac{4!}{3!1!}\left(\sqrt{2}r\right)\dfrac{1}{m^3} + \dfrac{1}{m^4}$

$= 4r^4 + 4\left(2\sqrt{2}\right)r^3\left(\dfrac{1}{m}\right) + \dfrac{4 \cdot 3}{2 \cdot 1}\left(2r^2\right)\dfrac{1}{m^2} + 4\left(\sqrt{2}r\right)\dfrac{1}{m^3} + \dfrac{1}{m^4}$

$= 4r^4 + \dfrac{8\sqrt{2}r^3}{m} + (6)\dfrac{2r^2}{m^2} + \dfrac{4\sqrt{2}r}{m^3} + \dfrac{1}{m^4}$

$= 4r^4 + \dfrac{8\sqrt{2}r^3}{m} + \dfrac{12r^2}{m^2} + \dfrac{4\sqrt{2}r}{m^3} + \dfrac{1}{m^4}$

39. $\left(\dfrac{1}{x^4} + x^4\right)^4 = \left(\dfrac{1}{x^4}\right)^4 + \dbinom{4}{1}\left(\dfrac{1}{x^4}\right)^3\left(x^4\right) + \dbinom{4}{2}\left(\dfrac{1}{x^4}\right)^2\left(x^4\right)^2 + \dbinom{4}{3}\left(\dfrac{1}{x^4}\right)\left(x^4\right)^3 + \left(x^4\right)^4$

$= \dfrac{1}{x^{16}} + \dfrac{4!}{1!3!}\left(\dfrac{1}{x^{12}}\right)\left(x^4\right) + \dfrac{4!}{2!2!}\left(\dfrac{1}{x^8}\right)\left(x^8\right) + \dfrac{4!}{3!1!}\left(\dfrac{1}{x^4}\right)\left(x^{12}\right) + x^{16}$

$= \dfrac{1}{x^{16}} + 4\left(\dfrac{1}{x^8}\right) + 6 + 4x^8 + x^{16} = \dfrac{1}{x^{16}} + \dfrac{4}{x^8} + 6 + 4x^8 + x^{16}$

41. $\left(4h - j\right)^8$

Using the formula $\dbinom{n}{k-1}x^{n-(k-1)}y^{k-1}$ with

$n = 8$, $k - 1 = 5$, and $n - (k-1) = 3$, the sixth term of the expansion is

$\dbinom{8}{5}\left(4h\right)^3\left(-j\right)^5 = \dfrac{8!}{5!3!}\left(64h^3\right)\left(-j^5\right)$

$= \dfrac{8 \cdot 7 \cdot 6}{3 \cdot 2 \cdot 1}\left(64h^3\right)\left(-j^5\right)$

$= 56\left(-64h^3 j^5\right) = -3584h^3 j^5$

43. $\left(a^2 + b\right)^{22}$

Use the formula $\dbinom{n}{k-1}x^{n-(k-1)}y^{k-1}$ with

$n = 22$, $k = 17$, $k - 1 = 16$, and $n - (k-1) = 6$. The seventeenth term of the expansion is

$\dbinom{22}{16}\left(a^2\right)^6\left(b\right)^{16}$

$= \dfrac{22!}{16!6!}\left(a^{12}\right)\left(b^{16}\right)$

$= \dfrac{22 \cdot 21 \cdot 20 \cdot 19 \cdot 18 \cdot 17}{6 \cdot 5 \cdot 4 \cdot 3 \cdot 2 \cdot 1}\left(a^{12}b^{16}\right)$

$= 74,613a^{12}b^{16}$

45. $\left(x - y^3\right)^{20}$

Use the formula $\dbinom{n}{k-1}x^{n-(k-1)}y^{k-1}$ with

$n = 20$, $k = 15$, $k - 1 = 14$, and $n - (k-1) = 6$. The fifteenth term of the expansion is

$\dbinom{20}{14}\left(x\right)^6\left(-y^3\right)^{14}$

$= \dfrac{20!}{14!6!}\left(x^6\right)\left(y^{42}\right)$

$= \dfrac{20 \cdot 19 \cdot 18 \cdot 17 \cdot 16 \cdot 15}{6 \cdot 5 \cdot 4 \cdot 3 \cdot 2 \cdot 1}\left(x^6 y^{42}\right)$

$= 38,760x^6 y^{42}$

47. $\left(3x^7 + 2y^3\right)^8$

This expansion has nine terms, so the middle term is the fifth term. Using the formula

$\dbinom{n}{k-1}x^{n-(k-1)}y^{k-1}$ with $n = 8$, $k = 5$,

$k - 1 = 4$, and $n - (k-1) = 4$, the fifth term is

$\dbinom{8}{4}\left(3x^7\right)^4\left(2y^3\right)^4 = \dfrac{8!}{4!4!}\left(3^4 x^{28}\right)\left(2^4 y^{12}\right)$

$= \dfrac{8 \cdot 7 \cdot 6 \cdot 5}{4 \cdot 3 \cdot 2 \cdot 1}\left(81x^{28}\right)\left(16y^{12}\right)$

$= (70)\left(1296x^{28} y^{12}\right)$

$= 90,720x^{28} y^{12}$

Copyright © 2013 Pearson Education, Inc.

49. If the coefficients of the fifth and eighth terms in the expansion of $(x+y)^n$ are the same, then the symmetry of the expansion can be used to determine n. There are four terms before the fifth term, so there must be four terms after the eighth term. This means that the last term of the expansion is the twelfth term. This in turn means that $n=11$, since $(x+y)^{11}$ is the expansion that has twelve terms.

51. Using a calculator, we obtain the exact value $10! = 3,628,800$. Using Stirling's formula, we obtain
$$10! \approx \sqrt{2\pi(10)} \cdot 10^{10} \cdot e^{-10} \approx 3,598,695.619$$

53. Using a calculator, we obtain the exact value $12! = 479,001,600$. Using Stirling's formula, we obtain
$$12! \approx \sqrt{2\pi(12)} \cdot 12^{12} \cdot e^{-12} \approx 475,687,486.5$$

Since $\dfrac{479,001,600 - 475,687,486.5}{479,001,600} \approx 0.00692$, the percent error is approximately 0.692%.

For Exercises 55–58, we will be using the first four terms of the formula
$$(1+x)^n = 1 + nx + \frac{n(n-1)}{2!}x^2 + \frac{n(n-1)(n-2)}{3!}x^3 + \cdots$$

55. $(1.02)^{-3} = (1+0.02)^{-3} = 1 + (-3)(0.02) + \dfrac{(-3)(-3-1)}{2!}(0.02)^2 + \dfrac{(-3)(-3-1)(-3-2)}{3!}(0.02)^3$

$= 1 - 0.06 + \dfrac{(-3)(-4)}{2\cdot 1}(0.0004) + \dfrac{(-3)(-4)(-5)}{3\cdot 2\cdot 1}(0.000008)$

$= 1 - 0.06 + 6(0.0004) - 10(0.000008) = 1 - 0.06 + 0.0024 - 0.00008 = 0.94232 \approx 0.942$

57. $(1.01)^{1.5} = (1+0.01)^{1.5} = 1 + (1.5)(0.01) + \dfrac{(1.5)(1.5-1)}{2!}(0.01)^2 + \dfrac{(1.5)(1.5-1)(1.5-2)}{3!}(0.01)^3$

$= 1 + 0.015 + \dfrac{(1.5)(0.5)}{2\cdot 1}(0.0001) + \dfrac{(1.5)(0.5)(-0.5)}{3\cdot 2\cdot 1}(0.000001)$

$= 1 + 0.015 + 0.375(0.0001) - 0.0625(0.000001)$

$= 1 + 0.015 + 0.0000375 - 0.0000000625 = 1.0150374375 \approx 1.015$

Section 7.5   Mathematical Induction

1. Let S_n be the statement $1 + 3 + 5 + \cdots + (2n-1) = n^2$.

$S_1 : 1 = 1^2 \Rightarrow 1 = 1$, which is true.

$S_2 : 1 + 3 = 2^2 \Rightarrow 4 = 4$, which is true.

$S_3 : 1 + 3 + 5 = 3^2 \Rightarrow 9 = 9$, which is true.

$S_4 : 1 + 3 + 5 + 7 = 4^2 \Rightarrow 16 = 16$, which is true.

$S_5 : 1 + 3 + 5 + 7 + 9 = 5^2 \Rightarrow 25 = 25$, which is true.

Now prove that S_n is true for every positive integer n.

(a) *Verify the statement for $n=1$.*

S_1 is the statement $1 = 1^2 \Rightarrow 1 = 1$, which is true.

(b) Write the statement for $n=k$: $S_k = 1 + 3 + 5 + \cdots + (2k-1) = k^2$

Copyright © 2013 Pearson Education, Inc.

(c) *Write the statement for* $n = k+1$: $S_{k+1} = 1+3+5+\cdots+(2k-1)+\left[2(k+1)-1\right] = (k+1)^2$

(d) *Assume the statement is true for* $n = k$. *Use algebra to change the statement in part* (b) *to the statement in part* (c): *Start with* $S_k: 1+3+5+\cdots+(2k-1) = k^2$ *and add the* $(k+1)$ *st term,* $\left[2(k+1)-1\right]$, *to both sides of this equation.*

$$1+3+5+\ldots+(2k-1) = k^2$$
$$1+3+5+\cdots+(2k-1)+\left[2(k+1)-1\right] = k^2+\left[2(k+1)-1\right]$$
$$1+3+5+\cdots+(2k-1)+\left[2(k+1)-1\right] = k^2+(2k+2-1)$$
$$1+3+5+\cdots+(2k-1)+\left[2(k+1)-1\right] = k^2+2k+1$$
$$1+3+5+\cdots+(2k-1)+\left[2(k+1)-1\right] = (k+1)^2$$

(e) *Write a conclusion based on Steps* (a) – (d).
Since S_n is true for $n = 1$ and $n = k+1$ when it is true for $n = k$, S_n is true for every positive integer n.

3. S_n is the statement $3+6+9+\ldots+3n = \dfrac{3n(n+1)}{2}$.

(a) *Verify the statement for* $n = 1$.

$3(1) = \dfrac{3(1)(1+1)}{2} \Rightarrow 3 = \dfrac{6}{2} \Rightarrow 3 = 3$, which is true. Thus, S_n is true for $n = 1$.

(b) *Write the statement for* $n = k$: $3+6+9+\cdots+3k = \dfrac{3k(k+1)}{2}$

(c) *Write the statement for* $n = k+1$: $3+6+9+\cdots+3k+3(k+1) = \dfrac{3(k+1)\left[(k+1)+1\right]}{2}$

(d) *Assume the statement is true for* $n = k$. *Use algebra to change the statement in part* (b) *to the statement in part* (c): *Add the* $(k+1)$ *st term,* $3(k+1)$, *to both sides of this equation.*

$$3+6+9+\cdots+3k = \frac{3k(k+1)}{2} \Rightarrow 3+6+9+\cdots+3k+3(k+1) = \frac{3k(k+1)}{2}+3(k+1) \Rightarrow$$

$$3+6+9+\cdots+3k+3(k+1) = \frac{3k(k+1)+6(k+1)}{2} \Rightarrow 3+6+9+\cdots+3k+3(k+1) = \frac{3k^2+3k+6k+6}{2} \Rightarrow$$

$$3+6+9+\cdots+3k+3(k+1) = \frac{3k^2+9k+6}{2} \Rightarrow 3+6+9+\cdots+3k+3(k+1) = \frac{3(k^2+3k+2)}{2} \Rightarrow$$

$$3+6+9+\cdots+3k+3(k+1) = \frac{3(k+1)(k+2)}{2} \Rightarrow 3+6+9+\cdots+3k+3(k+1) = \frac{3(k+1)\left[(k+1)+1\right]}{2}$$

(e) *Write a conclusion based on Steps* (a) – (d).
Since S_n is true for $n = 1$ and $n = k+1$ when it is true for $n = k$, S_n is true for every positive integer n.

5. S_n is the statement $2+4+8+\ldots+2^n = 2^{n+1}-2$.

(a) *Verify the statement for* $n = 1$.
$2^1 = 2^{1+1}-2 \Rightarrow 2 = 2^2-2 \Rightarrow 2 = 4-2 \Rightarrow 2 = 2$, which is true. Thus, S_n is true for $n = 1$.

(b) *Write the statement for* $n = k$: $2+4+8+\ldots+2^k = 2^{k+1}-2$

(c) *Write the statement for* $n = k+1$: $2+4+8+\cdots+2^k+2^{k+1} = 2^{(k+1)+1}-2$

Copyright © 2013 Pearson Education, Inc.

(d) *Assume the statement is true for* $n = k$. *Use algebra to change the statement in part* (b) *to the statement in part* (c): *Add the* $(k+1)$ *st term,* 2^{k+1}, *to both sides of this equation.*

$$2 + 4 + 8 + \cdots + 2^k = 2^{k+1} - 2 \Rightarrow 2 + 4 + 8 + \cdots + 2^k + 2^{k+1} = 2^{k+1} - 2 + 2^{k+1} \Rightarrow$$
$$2 + 4 + 8 + \cdots + 2^k + 2^{k+1} = 2 \cdot 2^{k+1} - 2 \Rightarrow 2 + 4 + 8 + \cdots + 2^k + 2^{k+1} = 2^1 \cdot 2^{k+1} - 2 \Rightarrow$$
$$2 + 4 + 8 + \cdots + 2^k + 2^{k+1} = 2^{(k+1)+1} - 2$$

(e) *Write a conclusion based on Steps* (a) – (d).
Since S_n is true for $n = 1$ and $n = k+1$ when it is true for $n = k$, S_n is true for every positive integer n.

7. S_n is the statement $1^2 + 2^2 + 3^2 + \cdots + n^2 = \dfrac{n(n+1)(2n+1)}{6}$.

(a) *Verify the statement for* $n = 1$.

$$1^2 = \frac{1(1+1)(2 \cdot 1 + 1)}{6} \Rightarrow 1 = \frac{1(2)(2+1)}{6} \Rightarrow 1 = \frac{2(3)}{6} \Rightarrow 1 = 1, \text{ which is true. Thus, } S_n \text{ is true for } n = 1.$$

(b) *Write the statement for* $n = k$: $1^2 + 2^2 + 3^2 + \cdots + k^2 = \dfrac{k(k+1)(2k+1)}{6}$

(c) *Write the statement for* $n = k+1$: $1^2 + 2^2 + 3^2 + \cdots + k^2 + (k+1)^2 = \dfrac{(k+1)\big[(k+1)+1\big]\big[2(k+1)+1\big]}{6}$

(d) *Assume the statement is true for* $n = k$. *Use algebra to change the statement in part* (b) *to the statement in part* (c): *Add the* $(k+1)$ *st term,* $(k+1)^2$, *to both sides of this equation.*

$$1^2 + 2^2 + 3^2 + \cdots + k^2 = \frac{k(k+1)(2k+1)}{6}$$
$$1^2 + 2^2 + 3^2 + \cdots + k^2 + (k+1)^2 = \frac{k(k+1)(2k+1)}{6} + (k+1)^2$$
$$1^2 + 2^2 + 3^2 + \cdots + k^2 + (k+1)^2 = \frac{k(k+1)(2k+1) + 6(k+1)^2}{6}$$
$$1^2 + 2^2 + 3^2 + \cdots + k^2 + (k+1)^2 = \frac{(k+1)\big[k(2k+1) + 6(k+1)\big]}{6}$$
$$1^2 + 2^2 + 3^2 + \cdots + k^2 + (k+1)^2 = \frac{(k+1)(2k^2 + k + 6k + 6)}{6}$$
$$1^2 + 2^2 + 3^2 + \cdots + k^2 + (k+1)^2 = \frac{(k+1)(2k^2 + 7k + 6)}{6}$$
$$1^2 + 2^2 + 3^2 + \cdots + k^2 + (k+1)^2 = \frac{(k+1)(k+2)(2k+3)}{6}$$
$$1^2 + 2^2 + 3^2 + \cdots + k^2 + (k+1)^2 = \frac{(k+1)(k+2)(2k+2+1)}{6}$$
$$1^2 + 2^2 + 3^2 + \cdots + k^2 + (k+1)^2 = \frac{(k+1)\big[(k+1)+1\big]\big[2(k+1)+1\big]}{6}$$

(e) *Write a conclusion based on Steps* (a) – (d).
Since S_n is true for $n = 1$ and $n = k+1$ when it is true for $n = k$, S_n is true for every positive integer n.

9. S_n is the statement $5 \cdot 6 + 5 \cdot 6^2 + 5 \cdot 6^3 + \cdots + 5 \cdot 6^n = 6(6^n - 1)$.

(a) *Verify the statement for* $n = 1$.

$$5 \cdot 6^1 = 6(6^1 - 1) \Rightarrow 5 \cdot 6 = 6(6 - 1) \Rightarrow 30 = 6(5) \Rightarrow 30 = 30, \text{ which is true. Thus, } S_n \text{ is true for } n = 1.$$

Copyright © 2013 Pearson Education, Inc.

(b) *Write the statement for* $n = k$: $\;5 \cdot 6 + 5 \cdot 6^2 + 5 \cdot 6^3 + \cdots + 5 \cdot 6^k = 6\left(6^k - 1\right)$

(c) *Write the statement for* $n = k+1$: $\;5 \cdot 6 + 5 \cdot 6^2 + 5 \cdot 6^3 + \cdots + 5 \cdot 6^k + 5 \cdot 6^{k+1} = 6\left(6^{k+1} - 1\right)$

(d) *Assume the statement is true for* $n = k$. *Use algebra to change the statement in part (b) to the statement in part (c)*: *Add the* $(k+1)$ *st term,* $5 \cdot 6^{k+1}$, *to both sides of this equation.*

$$5 \cdot 6 + 5 \cdot 6^2 + 5 \cdot 6^3 + \cdots + 5 \cdot 6^k = 6\left(6^k - 1\right)$$

$$5 \cdot 6 + 5 \cdot 6^2 + 5 \cdot 6^3 + \cdots + 5 \cdot 6^k + 5 \cdot 6^{k+1} = 6\left(6^k - 1\right) + 5 \cdot 6^{k+1}$$

$$5 \cdot 6 + 5 \cdot 6^2 + 5 \cdot 6^3 + \cdots + 5 \cdot 6^k + 5 \cdot 6^{k+1} = 6\left[\left(6^k - 1\right) + 5 \cdot 6^k\right]$$

$$5 \cdot 6 + 5 \cdot 6^2 + 5 \cdot 6^3 + \cdots + 5 \cdot 6^k + 5 \cdot 6^{k+1} = 6\left[6 \cdot 6^k - 1\right]$$

$$5 \cdot 6 + 5 \cdot 6^2 + 5 \cdot 6^3 + \cdots + 5 \cdot 6^k + 5 \cdot 6^{k+1} = 6\left(6^{k+1} - 1\right)$$

(e) *Write a conclusion based on Steps (a) – (d).*
Since S_n is true for $n = 1$ and $n = k+1$ when it is true for $n = k$, S_n is true for every positive integer n.

11. S_n is the statement $\dfrac{1}{1 \cdot 2} + \dfrac{1}{2 \cdot 3} + \dfrac{1}{3 \cdot 4} + \cdots + \dfrac{1}{n(n+1)} = \dfrac{n}{n+1}$.

(a) *Verify the statement for* $n = 1$.

$$\frac{1}{(1)(1+1)} = \frac{1}{1+1} \Rightarrow \frac{1}{(1)(2)} = \frac{1}{2} \Rightarrow \frac{1}{2} = \frac{1}{2},$$ which is true. Thus, S_n is true for $n = 1$.

(b) *Write the statement for* $n = k$: $\;\dfrac{1}{1 \cdot 2} + \dfrac{1}{2 \cdot 3} + \dfrac{1}{3 \cdot 4} + \cdots + \dfrac{1}{k(k+1)} = \dfrac{k}{k+1}$

(c) *Write the statement for* $n = k+1$: $\;\dfrac{1}{1 \cdot 2} + \dfrac{1}{2 \cdot 3} + \dfrac{1}{3 \cdot 4} + \cdots + \dfrac{1}{k(k+1)} + \dfrac{1}{(k+1)\left[(k+1)+1\right]} = \dfrac{k+1}{(k+1)+1}$

(d) *Assume the statement is true for* $n = k$. *Use algebra to change the statement in part (b) to the statement in part (c)*: *Add the* $(k+1)$ *st term,* $\dfrac{1}{(k+1)\left[(k+1)+1\right]}$, *to both sides of this equation.*

$$\frac{1}{1 \cdot 2} + \frac{1}{2 \cdot 3} + \frac{1}{3 \cdot 4} + \cdots + \frac{1}{k(k+1)} = \frac{k}{k+1}$$

$$\frac{1}{1 \cdot 2} + \frac{1}{2 \cdot 3} + \frac{1}{3 \cdot 4} + \cdots + \frac{1}{k(k+1)} + \frac{1}{(k+1)\left[(k+1)+1\right]} = \frac{k}{k+1} + \frac{1}{(k+1)\left[(k+1)+1\right]}$$

$$\frac{1}{1 \cdot 2} + \frac{1}{2 \cdot 3} + \frac{1}{3 \cdot 4} + \cdots + \frac{1}{k(k+1)} + \frac{1}{(k+1)\left[(k+1)+1\right]} = \frac{k}{k+1} + \frac{1}{(k+1)(k+2)}$$

$$\frac{1}{1 \cdot 2} + \frac{1}{2 \cdot 3} + \frac{1}{3 \cdot 4} + \cdots + \frac{1}{k(k+1)} + \frac{1}{(k+1)\left[(k+1)+1\right]} = \frac{k(k+2)+1}{(k+1)(k+2)}$$

$$\frac{1}{1 \cdot 2} + \frac{1}{2 \cdot 3} + \frac{1}{3 \cdot 4} + \cdots + \frac{1}{k(k+1)} + \frac{1}{(k+1)\left[(k+1)+1\right]} = \frac{k^2 + 2k + 1}{(k+1)(k+2)}$$

$$\frac{1}{1 \cdot 2} + \frac{1}{2 \cdot 3} + \frac{1}{3 \cdot 4} + \cdots + \frac{1}{k(k+1)} + \frac{1}{(k+1)\left[(k+1)+1\right]} = \frac{(k+1)^2}{(k+1)(k+2)}$$

(*continued on next page*)

Copyright © 2013 Pearson Education, Inc.

(continued)

$$\frac{1}{1\cdot2}+\frac{1}{2\cdot3}+\frac{1}{3\cdot4}+\cdots+\frac{1}{k(k+1)}+\frac{1}{(k+1)\big[(k+1)+1\big]}=\frac{k+1}{k+2}$$

$$\frac{1}{1\cdot2}+\frac{1}{2\cdot3}+\frac{1}{3\cdot4}+\cdots+\frac{1}{k(k+1)}+\frac{1}{(k+1)\big[(k+1)+1\big]}=\frac{k+1}{(k+1)+1}$$

(e) *Write a conclusion based on Steps* $(a)-(d)$.

Since S_n is true for $n=1$ and $n=k+1$ when it is true for $n=k$, S_n is true for every positive integer n.

13. S_n is the statement $\dfrac{1}{2}+\dfrac{1}{2^2}+\dfrac{1}{2^3}+\cdots+\dfrac{1}{2^n}=1-\dfrac{1}{2^n}$.

(a) *Verify the statement for* $n=1$.

$\dfrac{1}{2^1}=1-\dfrac{1}{2^1}\Rightarrow\dfrac{1}{2}=1-\dfrac{1}{2}\Rightarrow\dfrac{1}{2}=\dfrac{1}{2}$, which is true. Thus, S_n is true for $n=1$.

(b) *Write the statement for* $n=k$: $\dfrac{1}{2}+\dfrac{1}{2^2}+\dfrac{1}{2^3}+\cdots+\dfrac{1}{2^k}=1-\dfrac{1}{2^k}$

(c) *Write the statement for* $n=k+1$: $\dfrac{1}{2}+\dfrac{1}{2^2}+\dfrac{1}{2^3}+\cdots+\dfrac{1}{2^k}+\dfrac{1}{2^{k+1}}=1-\dfrac{1}{2^{k+1}}$

(d) *Assume the statement is true for* $n=k$. *Use algebra to change the statement in part* (b) *to the statement in part* (c): Add the $(k+1)$ st term, $\dfrac{1}{2^{k+1}}$, to both sides of this equation.

$$\frac{1}{2}+\frac{1}{2^2}+\frac{1}{2^3}+\cdots+\frac{1}{2^k}=1-\frac{1}{2^k}\Rightarrow\frac{1}{2}+\frac{1}{2^2}+\frac{1}{2^3}+\cdots+\frac{1}{2^k}+\frac{1}{2^{k+1}}=1-\frac{1}{2^k}+\frac{1}{2^{k+1}}\Rightarrow$$

$$\frac{1}{2}+\frac{1}{2^2}+\frac{1}{2^3}+\cdots+\frac{1}{2^k}+\frac{1}{2^{k+1}}=1-\frac{1\cdot2}{2^k\cdot2}+\frac{1}{2^{k+1}}\Rightarrow\frac{1}{2}+\frac{1}{2^2}+\frac{1}{2^3}+\cdots+\frac{1}{2^k}+\frac{1}{2^{k+1}}=1-\frac{2}{2^{k+1}}+\frac{1}{2^{k+1}}\Rightarrow$$

$$\frac{1}{2}+\frac{1}{2^2}+\frac{1}{2^3}+\cdots+\frac{1}{2^k}+\frac{1}{2^{k+1}}=1-\frac{1}{2^{k+1}}$$

(e) *Write a conclusion based on Steps* $(a)-(d)$.

Since S_n is true for $n=1$ and $n=k+1$ when it is true for $n=k$, S_n is true for every positive integer n.

15. $2^n>2n$

If $n=1$, we have $2^1>2\cdot1$ or $2>2$, which is false.

If $n=2$, we have $2^2>2\cdot2$ or $4>4$, which is false.

If $n=3$, we have $2^3>2\cdot3$ or $8>6$, which is true.

For $n\geq3$, the statement is true. The statement is false for $n=1$ or 2.

17. $2^n>n^2$

If $n=1$, we have $2^1>1^2$ or $2>1$, which is true.

If $n=2$, we have $2^2>2^2$ or $4>4$, which is false.

If $n=3$, we have $2^3>3^2$ or $8>9$, which is false.

If $n=4$, we have $2^4>4^2$ or $16>16$, which is false.

If $n=5$, we have $2^5>5^2$ or $32>25$, which is true.

For $n=1$ and $n\geq5$, the statement is true. The statement is false for $n=2,3,$ or 4.

19. Let S_n be the statement $(a^m)^n=a^{mn}$.

(Assume that a and m are constant.)

Step 1: Show that the statement is true when $n=1$. S_1 is the statement

$$\big(a^m\big)^1=a^{m\cdot1}\Rightarrow a^m=a^m,\text{ which is true.}$$

Step 2: Show that S_k implies S_{k+1}, where S_k is the statement $\big(a^m\big)^k=a^{mk}$, and S_{k+1} is the statement $\big(a^m\big)^{k+1}=a^{m(k+1)}$.

(continued on next page)

Copyright © 2013 Pearson Education, Inc.

(continued)

$$\left(a^m\right)^k = a^{mk} \Rightarrow \left(a^m\right)^k \cdot \left(a^m\right)^1 = a^{mk} \cdot \left(a^m\right)^1 \Rightarrow$$

$$\left(a^m\right)^{k+1} = a^{mk} \cdot a^m \Rightarrow \left(a^m\right)^{k+1} = a^{mk+m} \Rightarrow$$

$$\left(a^m\right)^{k+1} = a^{m(k+1)}$$

It has been shown that S_k implies S_{k+1}.
Therefore, the statement S_n is true for every positive integer value of n.

21. Let S_n be the statement $2^n > 2n$, if $n \geq 3$.
Step 1:
Show that the statement is true when $n = 3$: S_3 is the statement $2^3 > 2 \cdot 3 \Rightarrow 8 > 6$, which is true.
Step 2: Show that S_k implies S_{k+1}, where S_k is the statement $2^k > 2k$, and S_{k+1} is the statement $2^{k+1} > 2(k+1)$.

$$2^k > 2k \Rightarrow 2^k \cdot 2 > 2k \cdot 2 \Rightarrow$$
$$2^k \cdot 2^1 > 4k \Rightarrow 2^{k+1} > 2k + 2k$$
Since $k \geq 3$ we have $2k > 2$.
$$2^{k+1} > 2k + 2k > 2k + 2 \Rightarrow 2^{k+1} > 2k + 2 \Rightarrow$$
$$2^{k+1} > 2(k+1)$$

It has been shown that S_k implies S_{k+1}.
Therefore, the statement S_n is true for every positive integer value of n greater than or equal to 3.

23. Let S_n be the statement: If $a > 1$, then $a^n > 1$.
Step 1: Show that the statement is true when $n = 1$: S_1 is the statement, If $a > 1$, then

$a^1 > 1$, which is obviously true since $a = a^1$.
Step 2: Show that S_k implies S_{k+1}, where S_k is the statement: If $a > 1$, then $a^k > 1$. S_{k+1} is the statement: If $a > 1$, then $a^{k+1} > 1$.
If $a > 1$, then $a^k > 1$.
$$a^k > 1 \Rightarrow a^k \cdot a > 1 \cdot a \Rightarrow a^k \cdot a^1 > a \Rightarrow a^{k+1} > a$$
Since $a > 1$ we have the following.
$$a^{k+1} > a > 1 \Rightarrow a^{k+1} > 1$$
Note: Since $a > 1 > 0$, we did not need to be concerned about changing the direction of the inequality.
It has been shown that S_k implies S_{k+1}.
Therefore, the statement S_n is true for every positive integer value of n.

25. Let S_n be the statement: If $0 < a < 1$, then $a^n < a^{n-1}$.
Step 1: Show that the statement is true when $n = 1$: S_1 is the statement: If $0 < a < 1$, then $a^1 < a^{1-1}$. Since $a^1 < a^{1-1} \Rightarrow a < a^0 \Rightarrow a < 1$, which is true.
Step 2: Show that S_k implies S_{k+1}, where S_k is the statement: If $0 < a < 1$, then $a^k < a^{k-1}$.
S_{k+1} is the statement: If $0 < a < 1$, then $a^{k+1} < a^{(k+1)-1}$.
If $0 < a < 1$, then $a^k < a^{k-1}$.
$$a^k < a^{k-1} \Rightarrow a^k \cdot a^1 < a^{k-1}a^1 \Rightarrow$$
$$a^{k+1} < a^{(k-1)+1} \Rightarrow a^{k+1} < a^{(k+1)-1}$$
Note: Since $0 < a < 1$, we did not need to be concerned about changing the direction of the inequality.
It has been shown that S_k implies S_{k+1}.
Therefore, the statement S_n is true for every positive integer value of n.

27. Let S_n be the statement: If $n \geq 4$, then $n! > 2^n$, where
$$n! = n(n-1)(n-2)\cdots(3)(2)(1).$$
Step 1: Show that the statement is true when $n = 4$: S_4 is the statement
$$4! > 2^4 \Rightarrow 4(3)(2)(1) > (2)(2)(2)(2) \Rightarrow 24 > 16,$$
which is true.
Step 2: Show that S_k implies S_{k+1}, where S_k is the statement $k! > 2^k$, and S_{k+1} is the statement $(k+1)! > 2^{k+1}$.
$$k! > 2^k \Rightarrow k!(k+1) > 2^k(k+1) \Rightarrow$$
$$(k+1)! > 2^k(k+1)$$
Since $k + 1 > 2$, we have
$$(k+1)! > 2^k(k+1) > 2 \cdot 2^k \Rightarrow$$
$$(k+1)! > 2 \cdot 2^k \Rightarrow (k+1)! > 2^1 \cdot 2^k$$
$$(k+1)! > 2^{1+k} \Rightarrow (k+1)! > 2^{k+1}$$
Note: Since $k + 1 > 0$, we did not need to be concerned about changing the direction of the inequality.
It has been shown that S_k implies S_{k+1}.
Therefore, the statement S_n is true for every positive integer value of n greater than or equal to 4.

Copyright © 2013 Pearson Education, Inc.

29. Let S_n be the number of handshakes of the n

people is $\dfrac{n^2-n}{2}$.

Since 2 is the smallest number of people who
can shake hands, we need to prove this
statement for every positive integer $n \geq 2$.
Step 1: Show that the statement is true for
$n = 2$. S_2 is the statement that for two people,
the number of handshakes is

$\dfrac{2^2-2}{2} = \dfrac{4-2}{2} = \dfrac{2}{2} = 1,$ which is true.

Step 2: Show that S_k implies S_{k+1}, where S_k
is the statement: k people shake hands in

$\dfrac{k^2-k}{2}$ ways and S_{k+1} is the statement: $k+1$

people shake hands in $\dfrac{(k+1)^2-(k+1)}{2}$ ways.

Start with S_k and assume it is a true

statement: For k people, there are $\dfrac{k^2-k}{2}$

handshakes.
If one more person joins the k people, this
$(k+1)$st person will shake hands with the
previous k people one time each. Thus, there
will be k additional handshakes. Thus, the
number of handshakes for $k + 1$ people is as
follows.

$\dfrac{k^2-k}{2} + k = \dfrac{k^2-k}{2} + \dfrac{2k}{2} = \dfrac{k^2+k}{2}$

$= \dfrac{(k^2+2k+1)-(k+1)}{2}$

$= \dfrac{(k+1)^2-(k+1)}{2}$

This shows that if S_k is true, S_{k+1} is true.
Since both steps for a proof by the generalized
principle of mathematical induction have been
completed, the given statement is true for
every positive integer $n \geq 2$.

31. The number of sides of the nth figure is
$3 \cdot 4^{k-1}$ (from Exercise 30). To find the
perimeter of each figure, multiply the number
of sides by the length of each side. In each
figure, the lengths of the sides are $\frac{1}{3}$ the
lengths of the sides in the preceding figure. To
find the perimeter of the nth figure, P_n, we try
to find a pattern.

$P_1 = 3(1) = 3$

$P_2 = 3 \cdot 4\left(\frac{1}{3}\right) = 4$

$P_3 = 3 \cdot 4^2\left(\frac{1}{9}\right) = \frac{16}{3}$, and so on.

This gives a geometric sequence with
$a_1 = 3$ and $r = \frac{4}{3}$. Thus,

$P_n = a_1 r^{n-1} \Rightarrow P_n = 3\left(\frac{4}{3}\right)^{n-1}$. The result may

also be written as

$P_n = \dfrac{3^1 \cdot 4^{n-1}}{3^{n-1}} \Rightarrow P_n = \dfrac{4^{n-1}}{3^{-1} \cdot 3^{n-1}} \Rightarrow P_n = \dfrac{4^{n-1}}{3^{n-2}}$.

33. With 1 ring, 1 move is required. With 2 rings,
3 moves are required. Note that
$2^2 - 1 = 3 = 2 + 1$. With 3 rings, 7 moves are
required. Note that $2^3 - 1 = 7 = 2^2 + 2 + 1$.
With n rings, the number of moves required
are $2^{n-1} + 2^{n-2} + \cdots + 2^1 + 1$. This is a
geometric series with $a_1 = 1$ and $r = 2$. The
sum would be

$\dfrac{a_1\left(1-r^n\right)}{1-r} = \dfrac{1\left(1-2^n\right)}{1-2} = \dfrac{1-2^n}{-1} = 2^n - 1.$ Thus,

S_n is the statement: For n rings, the number of
required moves is $2^n - 1$. To prove this
statement using mathematical induction, we
perform the following steps.
Step 1: Show that the statement is true when
$n = 1$.
S_1 is the statement: For 1 ring, the number of
required moves is $2^1 - 1 = 2 - 1 = 1$, which is
true.
Step 2: Show that S_k implies S_{k+1}, where S_k
is the statement: For k rings, the number of
required moves is $2^k - 1$. S_{k+1} is the
statement: For $k+1$ rings, the number of
required moves is $2^{k+1} - 1$.
Assume $k + 1$ rings are on the first peg. Since
S_k is true, the top k rings can be moved to the
second peg in $2^k - 1$ moves. Now move the
bottom ring to the third peg. Since S_k is true,
move the k rings on the second peg on top of
the largest ring on the third peg in $2^k - 1$
moves. The total number of moves is
$\left(2^k - 1\right) + 1 + \left(2^k - 1\right) = 2 \cdot 2^k - 1 = 2^1 \cdot 2^k - 1 = 2^{k+1} - 1.$

Since S_n is true for $n = 1$ and $n = k+1$
when is true for $n = k$, S_n is true for every
positive integer n.

Copyright © 2013 Pearson Education, Inc.

Chapter Quiz
(Sections 7.1–7.5)

1. $a_n = -4n + 2$

Replace n with 1, 2, 3, 4, and 5.

$n = 1: a_1 = -4(1) + 2 = -2$

$n = 2: a_2 = -4(2) + 2 = -6$

$n = 3: a_3 = -4(3) + 2 = -10$

$n = 4: a_4 = -4(4) + 2 = -14$

$n = 5: a_5 = -4(5) + 2 = -18$

The first five terms are $-2, -6, -10, -14$, and -18. There is a common difference of -4, so the sequence is arithmetic.

3. $a_1 = 5, a_2 = 3, a_n = a_{n-1} + 3a_{n-2}$, for $n \geq 3$

$a_3 = a_2 + 3a_1 = 3 + 3(5) = 18$; $\quad a_4 = a_3 + 3a_2 = 18 + 3(3) = 27$; $\quad a_5 = a_4 + 3a_3 = 27 + 3(18) = 81$

The first five terms are 5, 3, 18, 27, and 81. There is neither a common difference nor a common ratio, so the sequence is neither arithmetic nor geometric.

5. (a) $a_1 = -20, d = 14$

$S_n = \frac{n}{2}\left[2a_1 + (n-1)d\right]$, so we have $S_{10} = \frac{10}{2}\left[2(-20) + (10-1)(14)\right] = 5\left[-40 + 9(14)\right] = 5(86) = 430$

(b) $a_1 = -20, r = -\frac{1}{2}$

$S_n = \frac{a_1\left(1 - r^n\right)}{1 - r}$, so we have $S_{10} = \frac{-20\left[1 - \left(-\frac{1}{2}\right)^{10}\right]}{1 - \left(-\frac{1}{2}\right)} = \frac{-20\left[1 - \frac{1}{1024}\right]}{\frac{3}{2}} = \frac{-20\left(\frac{1023}{1024}\right)}{\frac{3}{2}} = -\frac{1705}{128}$

7. $(x - 3y)^5 = \left[x + (-3y)\right]^5$

$= x^5 + \binom{5}{1}x^4(-3y) + \binom{5}{2}x^3(-3y)^2 + \binom{5}{3}x^2(-3y)^3 + \binom{5}{4}x(-3y)^4 + (-3y)^5$

$= x^5 + \frac{5!}{1!4!}x^4(-3y) + \frac{5!}{2!3!}x^3(9)y^2 + \frac{5!}{3!2!}x^2(-27)y^3 + \frac{5!}{4!1!}x(81)y^4 + (-243)y^5$

$= x^5 + 5(-3)x^4y + \frac{5 \cdot 4}{2 \cdot 1}(9)x^3y^2 + \frac{5 \cdot 4}{2 \cdot 1}(-27)x^2y^3 + 5(81)xy^4 - 243y^5$

$= x^5 - 15x^4y + 90x^3y^2 - 270x^2y^3 + 405xy^4 - 243y^5$

9. (a) $9! = 9 \cdot 8 \cdot 7 \cdot 6 \cdot 5 \cdot 4 \cdot 3 \cdot 2 \cdot 1 = 362,880$

(b) $\binom{10}{4} = \frac{10!}{4!(10-4)!} = \frac{10!}{4!6!} = \frac{10 \cdot 9 \cdot 8 \cdot 7}{4 \cdot 3 \cdot 2 \cdot 1} = 210$

Section 7.6 Counting Theory

1. There are $3 \cdot 4 \cdot 1 \cdot 2 = 24$ possible outfits.

3. There are $2 \cdot 20 \cdot 4 = 160$ ways an attendee can choose.

5. There are $7 \cdot 6 \cdot 5 \cdot 4 \cdot 3 \cdot 2 \cdot 1 = 5040$ ways to arrange the portraits.

7. There are $16 \cdot 15 \cdot 14 = 3360$ ways to select a first-place, second-place, and third-place winner.

9. $P(12, 8) = \frac{12!}{(12-8)!} = \frac{12!}{4!}$

$= \frac{12 \cdot 11 \cdot 10 \cdot 9 \cdot 8 \cdot 7 \cdot 6 \cdot 5 \cdot 4 \cdot 3 \cdot 2 \cdot 1}{4 \cdot 3 \cdot 2 \cdot 1}$

$= 12 \cdot 11 \cdot 10 \cdot 9 \cdot 8 \cdot 7 \cdot 6 \cdot 5$

$= 19,958,400$

11. $P(9, 2) = \frac{9!}{(9-2)!} = \frac{9!}{7!} = \frac{9 \cdot 8 \cdot 7!}{7!} = 9 \cdot 8 = 72$

13. $P(5, 1) = \frac{5!}{(5-1)!} = \frac{5!}{4!} = \frac{5 \cdot 4!}{4!} = 5$

Copyright © 2013 Pearson Education, Inc.

15. $C(4,2) = \begin{pmatrix} 4 \\ 2 \end{pmatrix} = \dfrac{4!}{(4-2)!2!} = \dfrac{4!}{2!2!}$

$\qquad = \dfrac{4 \cdot 3 \cdot 2!}{2!2!} = \dfrac{4 \cdot 3}{2 \cdot 1} = 6$

17. $C(6,0) = \begin{pmatrix} 6 \\ 0 \end{pmatrix} = \dfrac{6!}{(6-0)!0!} = \dfrac{6!}{6! \cdot 1} = 1$

19. $\begin{pmatrix} 12 \\ 4 \end{pmatrix} = \dfrac{12!}{(12-4)!4!} = \dfrac{12!}{8!4!} = \dfrac{12 \cdot 11 \cdot 10 \cdot 9 \cdot 8!}{8!4!}$

$\qquad = \dfrac{12 \cdot 11 \cdot 10 \cdot 9}{4!} = \dfrac{12 \cdot 11 \cdot 10 \cdot 9}{4 \cdot 3 \cdot 2 \cdot 1} = 495$

21. $_{20}P_5 = 1,860,480$

23. $_{15}P_8 = 259,459,200$

25. $_{20}C_5 = 15,504$

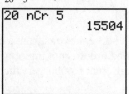

27. $\begin{pmatrix} 15 \\ 8 \end{pmatrix} = C(15,8) = 6435$

```
15 nCr 8
            6435
```

29. (a) Since the order of digits in a telephone number does matter, this involves a permutation.

(b) Since the order of digits in a Social Security number does matter, this involves a permutation.

(c) Since the order of the cards in a poker hand does not matter, this involves a combination.

(d) Since the order of members on a committee of politicians does not matter, this involves a combination.

(e) Since the order of numbers of the "combination" on a combination lock does matter, this involves a permutation.

(f) Since the order of digits and/or letters on a license plate does matter, this involves a permutation.

(g) Since the order does not matter, the lottery choice of six numbers involves a combination.

31. Using the fundamental principle of counting we have $5 \cdot 4 \cdot 2 = 40$. There are 40 different homes available if a builder offers a choice of 5 basic plans, 4 roof styles, and 2 exterior finishes.

33. (a) There are two choices for the first letter, K and W. The second letter can be any of the 26 letters of the alphabet except for the one chosen for the first letter, so there are 25 choices. The third letter can be any of the remaining 24 letters of the alphabet, so there are 24 choices. The fourth letter can be any of the remaining 23 letters of the alphabet, so there are 23 choices. Therefore, by the fundamental principle of counting, the number of possible 4-letter radio-station call letters is $2 \cdot 25 \cdot 24 \cdot 23 = 27,600$.

(b) There are two choices for the first letter. Because repeats are allowed, there are 26 choices for each of the remaining 3 letters. Therefore, by the fundamental principle of counting, the number of possible 4-letter radio-station call letters is $2 \cdot 26 \cdot 26 \cdot 26 = 35,152$.

(c) There are two choices for the first letter. There are 24 choices for the second letter since it cannot repeat the first letter and cannot be R. There are 23 choices for the third letter since it cannot repeat any of the first two letters and cannot be R. There is only once choice for the last letter since it must be R. Therefore, by the fundamental principle of counting, the number of possible 4-letter radio-station call letters is $2 \cdot 24 \cdot 23 \cdot 1 = 1104$.

Copyright © 2013 Pearson Education, Inc.

35. Since each ordering of the blocks is considered a different arrangement, we have the permutation, $P(7,7)$.

$$P(7,7) = \frac{7!}{(7-7)!} = \frac{7!}{0!}$$
$$= \frac{7 \cdot 6 \cdot 5 \cdot 4 \cdot 3 \cdot 2 \cdot 1}{1} = 5040$$

There are 5040 different arrangements of the 7 blocks. Note: This problem could also be solved by using the fundamental principle of counting. There are 7 choices for the first block, 6 for the second, 5 for the third, etc. The number of arrangements would be $7 \cdot 6 \cdot 5 \cdot 4 \cdot 3 \cdot 2 \cdot 1 = 5040$.

37. (a) The first three positions could each be any one of 26 letters, and the second three positions could each be any one of 10 digits. Using the fundamental principle of counting, we have
$26 \cdot 26 \cdot 26 \cdot 10 \cdot 10 \cdot 10 = 17,576,000.$
Thus, 17,576,000 license plates were possible.

(b) Using the fundamental principle of counting, we have
$10 \cdot 10 \cdot 10 \cdot 26 \cdot 26 \cdot 26 = 17,576,000.$
17,576,000 additional license plates were made possible by the reversal.

(c) Using the fundamental principle of counting, we have
$26 \cdot 10 \cdot 10 \cdot 10 \cdot 26 \cdot 26 \cdot 26 = 456,976,000.$
456,976,000 plates were provided by prefixing the previous pattern with an additional letter.

39. Since order does matter, we have the permutation,

$$P(9,9) = \frac{9!}{(9-9)!} = \frac{9!}{0!} = \frac{9 \cdot 8 \cdot 7 \cdot 6 \cdot 5 \cdot 4 \cdot 3 \cdot 2 \cdot 1}{1}$$
$$= 362,880$$

There are 362,880 ways to seat 9 people in 9 seats in a row.

41. He has 6 choices for the first course, 5 choices for the second, and 4 choices for the third. Thus, the student has $6 \cdot 5 \cdot 4 = 120$ possible schedules according to the fundamental principle of counting.

43. The number of ways in which the 3 officers can be chosen from the 15 members is given by the following permutation.

$$P(15,3) = \frac{15!}{(15-3)!} = \frac{15!}{12!} = \frac{15 \cdot 14 \cdot 13 \cdot 12!}{12!}$$
$$= 15 \cdot 14 \cdot 13 = 2730$$

45. (a) $P(5,5) = \dfrac{5!}{(5-5)!} = \dfrac{5!}{0!}$
$$= \frac{5 \cdot 4 \cdot 3 \cdot 2 \cdot 1}{1} = 120$$
The letters can be arranged in 120 ways.

(b) $P(3,3) = \dfrac{3!}{(3-3)!} = \dfrac{3!}{0!} = \dfrac{3 \cdot 2 \cdot 1}{1} = 6$

The first three letters can be arranged in 6 ways.

47. We want to choose 6 group members out of 40 and the order is not important. The number of possible groups is

$$C(40,6) = \binom{40}{6} = \frac{40!}{(40-6)!6!} = \frac{40!}{34!6!}$$
$$= \frac{40 \cdot 39 \cdot 38 \cdot 37 \cdot 36 \cdot 35 \cdot 34!}{34!6!}$$
$$= \frac{40 \cdot 39 \cdot 38 \cdot 37 \cdot 36 \cdot 35}{6 \cdot 5 \cdot 4 \cdot 3 \cdot 2 \cdot 1}$$
$$= 3,838,380$$

49. We want to choose 4 apples from a crate of 25 and order is not important. The number of ways the apples may be sampled is

$$C(25,4) = \binom{25}{4} = \frac{25!}{(25-4)!4!} = \frac{25!}{21!4!}$$
$$= \frac{25 \cdot 24 \cdot 23 \cdot 22 \cdot 21!}{21!4!} = \frac{25 \cdot 24 \cdot 23 \cdot 22}{4 \cdot 3 \cdot 2 \cdot 1}$$
$$= 12,650$$

51. (a) We want to choose 4 extras from a choice of 6 extras and order is not important. The number of different hamburgers that can be made is

$$C(6,4) = \binom{6}{4} = \frac{6!}{(6-4)!4!} = \frac{6!}{2!4!}$$
$$= \frac{6 \cdot 5 \cdot 4!}{2!4!} = \frac{6 \cdot 5}{2 \cdot 1} = 15$$

(b) Since one of the four choices is fixed, we now want to choose 3 extras from a choice of 5 extras and order is not important. The number of different hamburgers that can be made is given by

$$C(5,3) = \binom{5}{3} = \frac{5!}{(5-3)!3!} = \frac{5!}{2!3!}$$
$$= \frac{5 \cdot 4 \cdot 3!}{2!3!} = \frac{5 \cdot 4}{2 \cdot 1} = 10$$

Copyright © 2013 Pearson Education, Inc.

53. The number of 2-marble samples that can be chosen from a total of 15 marbles is $C(15,2)$.

$$C(15,2) = \binom{15}{2} = \frac{15!}{(15-2)!2!} = \frac{15!}{13!2!}$$
$$= \frac{15 \cdot 14 \cdot 13!}{13!2!} = \frac{15 \cdot 14}{2 \cdot 1} = 105$$

The number of 4-marble samples that can be chosen from a total of 15 marbles is $C(15,4)$.

$$C(15,4) = \binom{15}{4} = \frac{15!}{(15-4)!4!} = \frac{15!}{11!4!}$$
$$= \frac{15 \cdot 14 \cdot 13 \cdot 12 \cdot 11!}{11!4!} = \frac{15 \cdot 14 \cdot 13 \cdot 12}{4 \cdot 3 \cdot 2 \cdot 1}$$
$$= 1365$$

55. There are 5 liberal and 4 conservatives, giving a total of 9 members. Three members are chosen as delegates to a convention.

(a) There are $C(9,3)$ ways of doing this.

$$C(9,3) = \binom{9}{3} = \frac{9!}{(9-3)!3!} = \frac{9!}{6!3!}$$
$$= \frac{9 \cdot 8 \cdot 7 \cdot 6!}{6!3!} = \frac{9 \cdot 8 \cdot 7}{3 \cdot 2 \cdot 1} = 84$$

84 delegations are possible.

(b) To get all liberals, we must choose 3 members from a set of 5, which can be done $C(5,3)$ ways.

$$C(5,3) = \binom{5}{3} = \frac{5!}{(5-3)!3!} = \frac{5!}{2!3!}$$
$$= \frac{5 \cdot 4 \cdot 3!}{2!3!} = \frac{5 \cdot 4}{2} = 10$$

10 delegations could have all liberals.

(c) To get 2 liberals and 1 conservative involves two independent events. First select the liberals. The number of ways to do this is

$$C(5,2) = \binom{5}{2} = \frac{5!}{3!2!}$$
$$= \frac{5 \cdot 4 \cdot 3!}{3!2!} = \frac{5 \cdot 4}{2 \cdot 1} = 10$$

Now select the conservative. The number of ways to do this is as follows.

$$C(4,1) = \binom{4}{1} = \frac{4!}{(4-1)!1!} = \frac{4!}{3!1!}$$
$$= \frac{4 \cdot 3!}{3!1!} = \frac{4}{1} = 4$$

To find the number of delegations, use the fundamental principle of counting. The number of delegations with 2 liberals and 1 conservative is $10 \cdot 4 = 40$.

(d) If one particular person must be on the delegation, then there are 2 people left to choose from a set consisting of 8 members. The number of ways to do this is as follows.

$$C(8,2) = \binom{8}{2} = \frac{8!}{(8-2)!2!} = \frac{8!}{6!2!}$$
$$= \frac{8 \cdot 7 \cdot 6!}{6!2!} = \frac{8 \cdot 7}{2 \cdot 1} = 28$$

28 delegations are possible, which includes the mayor.

57. The problem asks how many ways can Dwight arrange his schedule. Therefore, order is important, and this is a permutation problem. There are

$$P(8,4) = \frac{8!}{(8-4)!} = \frac{8!}{4!} = \frac{8 \cdot 7 \cdot 6 \cdot 5 \cdot 4!}{4!}$$
$$= 8 \cdot 7 \cdot 6 \cdot 5 = 1680$$

ways to arrange his schedule.

59. The order of the vegetables in the soup is not important, so this is a combination problem. There are

$$C(6,4) = \binom{6}{4} = \frac{6!}{(6-4)!4!} = \frac{6!}{2!4!}$$
$$= \frac{6 \cdot 5 \cdot 4!}{2!4!} = \frac{6 \cdot 5}{2 \cdot 1} = 15$$

different soups she can make.

61. Order is important in the seating, so this is a permutation problem. All thirteen children will have a specific location; the first twelve will sit down and thirteenth will be left standing.

$$P(13, 13) = \frac{13!}{(13-13)!} = \frac{13!}{0!}$$
$$= \frac{13 \cdot 12 \cdot 11 \cdot 10 \cdot 9 \cdot 8 \cdot 7 \cdot 6 \cdot 5 \cdot 4 \cdot 3 \cdot 2 \cdot 1}{1}$$
$$= 6,227,020,800$$

There are 6,227,020,800 seatings possible.

Copyright © 2013 Pearson Education, Inc.

63. A club has 8 women and 11 men members. There are a total of $8 + 11 = 19$ members, and 5 of them are to be chosen. Order is not important, so this is a combination problem.

(a) Choosing all women implies of the 8 women, choose 5.

$$C(8,5) = \binom{8}{5} = \frac{8!}{(8-5)!5!} = \frac{8!}{3!5!} = \frac{8 \cdot 7 \cdot 6 \cdot 5!}{3!5!} = \frac{8 \cdot 7 \cdot 6}{3 \cdot 2 \cdot 1} = 56$$

56 committees having 5 women can be chosen.

(b) Choose all men implies of the 11 men, choose 5.

$$C(11,5) = \binom{11}{5} = \frac{11!}{(11-5)!5!} = \frac{11!}{6!5!} = \frac{11 \cdot 10 \cdot 9 \cdot 8 \cdot 7 \cdot 6!}{6!5!} = \frac{11 \cdot 10 \cdot 9 \cdot 8 \cdot 7}{5 \cdot 4 \cdot 3 \cdot 2 \cdot 1} = 462$$

462 committees having 5 men can be chosen.

(c) Choosing 3 women and 2 men involves two independent events, each of which involves combinations. First, select the women. There are 8 women in the club, so the number of ways to do this is as follows.

$$C(8,3) = \binom{8}{3} = \frac{8!}{(8-3)!3!} = \frac{8!}{5!3!} = \frac{8 \cdot 7 \cdot 6 \cdot 5!}{5!3!} = \frac{8 \cdot 7 \cdot 6}{3 \cdot 2 \cdot 1} = 56$$

Now, select the men. There are 11 men in the club, so the number of ways to do this is as follows.

$$C(11,2) = \binom{11}{2} = \frac{11!}{(11-2)!2!} = \frac{11!}{9!2!} = \frac{11 \cdot 10 \cdot 9!}{9!2!} = \frac{11 \cdot 10}{2 \cdot 1} = 55$$

To find the number of committees, use the fundamental principle of counting. The number of committees with 3 women and 2 men is $56 \cdot 55 = 3080$.

(d) Choose no more than 3 men.

This means choose 0 men (and 5 women) or choose 1 man (and 4 women) or choose 2 men (and 3 women) or choose 3 men (and 2 women). For each of these choices, we use the fundamental principle of counting. Thus, the number of possible committees with no more than 3 men is

$$C(11,0) \cdot C(8,5) + C(11,1) \cdot C(8,4) + C(11,2) \cdot C(8,3) + C(11,3) \cdot C(8,2).$$

$$C(11,0) \cdot C(8,5) + C(11,1) \cdot C(8,4) + C(11,2) \cdot C(8,3) + C(11,3) \cdot C(8,2)$$

$$= \binom{11}{0}\binom{8}{5} + \binom{11}{1}\binom{8}{4} + \binom{11}{2}\binom{8}{3} + \binom{11}{3}\binom{8}{2}$$

$$= \frac{11!}{(11-0)!0!} \cdot \frac{8!}{(8-5)!5!} + \frac{11!}{(11-1)!1!} \cdot \frac{8!}{(8-4)!4!} + \frac{11!}{(11-2)!2!} \cdot \frac{8!}{(8-3)!3!} + \frac{11!}{(11-3)!3!} \cdot \frac{8!}{(8-2)!2!}$$

$$= \frac{11!}{11!0!} \cdot \frac{8!}{3!5!} + \frac{11!}{10!1!} \cdot \frac{8!}{4!4!} + \frac{11!}{9!2!} \cdot \frac{8!}{5!3!} + \frac{11!}{8!3!} \cdot \frac{8!}{6!2!}$$

$$= 1 \cdot \frac{8 \cdot 7 \cdot 6 \cdot 5!}{3!5!} + \frac{11 \cdot 10!}{10!1!} \cdot \frac{8 \cdot 7 \cdot 6 \cdot 5 \cdot 4!}{4!4!} + \frac{11 \cdot 10 \cdot 9!}{9!2!} \cdot \frac{8 \cdot 7 \cdot 6 \cdot 5!}{5!3!} + \frac{11 \cdot 10 \cdot 9 \cdot 8!}{8!3!} \cdot \frac{8 \cdot 7 \cdot 6!}{6!2!}$$

$$= 1 \cdot \frac{8 \cdot 7 \cdot 6}{3 \cdot 2 \cdot 1} + 11 \cdot \frac{8 \cdot 7 \cdot 6 \cdot 5}{4 \cdot 3 \cdot 2 \cdot 1} + \frac{11 \cdot 10}{2 \cdot 1} \cdot \frac{8 \cdot 7 \cdot 6}{3 \cdot 2 \cdot 1} + \frac{11 \cdot 10 \cdot 9}{3 \cdot 2 \cdot 1} \cdot \frac{8 \cdot 7}{2 \cdot 1}$$

$$= 1 \cdot 56 + 11 \cdot 70 + 55 \cdot 56 + 165 \cdot 28 = 56 + 770 + 3080 + 4620 = 8526$$

8526 committees having no more than 3 men can be chosen.

65. For the first lock, there are 10 choices for the first, second, and third digits. Thus, the number of combinations on the first lock is $10 \cdot 10 \cdot 10 = 1000$. This is the same for the second lock, that is, it also has 1000 different combinations. Since each of the locks is independent from the other, we use the fundamental principle of counting. We have a total of $1000 \cdot 1000 = 1,000,000$ combinations possible.

67. Each of the 12 switches has two possible settings. Since each switch is independent from the others, there is a total of

$$2 \cdot 2 \cdot 2 \cdot 2 \cdot 2 \cdot 2 \cdot 2 \cdot 2 \cdot 2 \cdot 2 \cdot 2 \cdot 2 = 2^{12} = 4096$$

codes possible.

Copyright © 2013 Pearson Education, Inc.

69. Because the keys are arranged in a circle, there is no "first" key. The number of distinguishable arrangements is the number of ways to arrange the other three keys in relation to any one of the keys, which is

$$P(3,3) = \frac{3!}{(3-3)!} = \frac{3!}{0!} = \frac{3!}{1} = 3 \cdot 2 \cdot 1 = 6.$$

In Exercises 71–79, we will be using the formulas

$$P(n,r) = \frac{n!}{(n-r)!} \text{ and } \binom{n}{r} = \frac{n!}{(n-r)!r!}, \text{ where}$$

$r \le n$.

71. Show $P(n, n-1) = P(n, n)$.

$$P(n, n-1) = \frac{n!}{[n-(n-1)]!} = \frac{n!}{(n-n+1)!}$$
$$= \frac{n!}{1!} = \frac{n!}{1} = \frac{n!}{0!} = \frac{n!}{(n-n)!} = P(n, n)$$

73. Show $P(n, 0) = 1$.

$$P(n, 0) = \frac{n!}{(n-0)!} = \frac{n!}{n!} = 1$$

75. Show $\binom{n}{n} = 1$.

$$\binom{n}{n} = \frac{n!}{(n-n)!n!} = \frac{n!}{0!n!} = \frac{n!}{1 \cdot n!} = \frac{n!}{n!} = 1$$

77. Show $\binom{0}{0} = 1$.

$$\binom{0}{0} = \frac{0!}{(0-0)!0!} = \frac{1!}{1!0!} = \frac{1!}{1!1} = 1$$

79. Show $\binom{n}{n-r} = \binom{n}{r}$.

$$\binom{n}{n-r} = \frac{n!}{[n-(n-r)]!(n-r)!}$$
$$= \frac{n!}{(n-n+r)!(n-r)!} = \frac{n!}{r!(n-r)!}$$
$$= \frac{n!}{(n-r)!r!} = \binom{n}{r}$$

Section 7.7 Basics of Probability

1. Let H = heads, T = tails.
The only possible outcome is a head. Hence, the sample space is $S = \{H\}$.

3. Since each coin can be a head or a tail and there are 3 coins, the sample space is
$$S = \{(H,H,H),(H,H,T),(H,T,H),$$
$$(T,H,H),(H,T,T),(T,H,T),$$
$$(T,T,H),(T,T,T)\}$$

5. The sample space is
$$S = \{(1, 1), (1, 2), (1, 3), (2, 1), (2, 2),$$
$$(2, 3), (3, 1), (3, 2), (3, 3)\}$$

7. **(a)** "The result is heads" is the event $E_1 = \{H\}$. This event is certain to occur, $P(E_1) = 1$.

(b) "The result is tails" is the event $E_2 = \varnothing$. This event is an impossible event, so $P(E_2) = 0$.

9. The sample space is $S = \{(H, H, H), (H, H, T), (H, T, H), (H, T, T), (T, H, H), (T, H, T), (T, T, H), (T, T, T)\}$ so $n(S) = 8$.

(a) The event E_1 of all three coins showing the same face is $E_1 = \{(H, H, H), (T, T, T)\}$, so $n(E_1) = 2$. The probability is
$$P(E_1) = \frac{n(E_1)}{n(S)} = \frac{2}{8} = \frac{1}{4}.$$

(b) The event E_2 where at least two coins are tails is $E_2 = \{(H, T, T), (T, H, T), (T, T, H), (T, T, T)\}$, so $n(E_2) = 4$. The probability is
$$P(E_2) = \frac{n(E_2)}{n(S)} = \frac{4}{8} = \frac{1}{2}.$$

11. The sample space is $S = \{(1, 1), (1, 2), (1, 3), (2, 1), (2, 2), (2, 3), (3, 1), (3, 2), (3, 3)\}$, so $n(S) = 9$.

(a) The result is a repeated number" is the event $E_1 = \{(1, 1), (2, 2), (3, 3)\}$, so $n(E_1) = 3$. The probability is
$$P(E_1) = \frac{n(E_1)}{n(S)} = \frac{3}{9} = \frac{1}{3}.$$

(b) "The second number is 1 or 3" is the event
$$E_2 = \{(1, 1), (1, 3), (2, 1), (2, 3), (3, 1), (3, 3)\},$$
so $n(E_2) = 6$. The probability is
$$P(E_2) = \frac{n(E_2)}{n(S)} = \frac{6}{9} = \frac{2}{3}.$$

Copyright © 2013 Pearson Education, Inc.

(c) "The first number is even and the second number is odd" is the event $E_3 = \{(2, 1), (2, 3)\}$, So, $n(E_3) = 2$. The probability is $P(E_3) = \dfrac{n(E_3)}{n(S)} = \dfrac{2}{9}$.

13. Since $P(E) = 0.857$,
$P(E') = 1 - 0.857 = 0.143$.

15. A batting average of .300 means for every 10 times at bat (the sample space), the batter g0t 3 hits. If the event E is "getting a hit," then $P(E) = 0.300$. Thus, we have $P(E') = 1 - 0.300 = 0.700$. The odds in favor of him getting a hit are $\dfrac{P(E)}{P(E')} = \dfrac{0.300}{0.700} = \dfrac{3}{7}$ or 3 to 7.

17. There are 52 cards, so $n(S) = 52$. The deck consists of thirteen spades, thirteen clubs, thirteen hearts, and thirteen diamonds. There are three face cards in each suit, or twelve in total.

(a) E_1 : The card is a spade.
$n(E_1) = 13$. The probability is
$P(E_1) = \frac{13}{52} = \frac{1}{4}$.

(b) E_2 : The card is not a spade.
Since the probability that a card is a spade is $\frac{1}{4}$, $P(E_2) = 1 - \frac{1}{4} = \frac{3}{4}$.

(c) E_3 : The card is a spade or a heart.
There are 13 spades and 13 hearts so $n(E_3) = 13 + 13 = 26$. The probability is
$P(E_3) = \frac{26}{52} = \frac{1}{2}$.

(d) E_4 : The card is a spade or a face card.
Since there are three face cards that are spades,
$$P(E_4) = P(\text{spade}) + P(\text{face card})$$
$$- P(\text{spade and face card})$$
$$= \frac{13}{52} + \frac{12}{52} - \frac{3}{52} = \frac{22}{52} = \frac{11}{26}$$

(e) Using the results from (a) and (b), the odds in favor of drawing a spade are
$\dfrac{P(E_1)}{P(E_2)} = \dfrac{\frac{1}{4}}{\frac{3}{4}} = \dfrac{1}{3}$ or 1 to 3

19. The total of all foreign-born U.S. residents in 2008 is $1436 + 10{,}356 + 4969 + 20{,}150 + 1050 = 37{,}961$ thousand.

(a) $P(\text{born in Asia}) = \dfrac{10{,}356}{37{,}961} \approx 0.273$

(b) $P(\text{not born in Europe})$
$= 1 - P(\text{born in Europe})$
$= 1 - \dfrac{4969}{37{,}961} = \dfrac{32{,}992}{37{,}961} \approx 0.869$

(c) $P(\text{born in Africa or Europe})$
$= \dfrac{1436 + 4969}{37{,}961} = \dfrac{6405}{37{,}961} \approx 0.169$

(d) Let E be "selected resident was born in Latin America."
We have $P(E) = \dfrac{20{,}150}{37{,}961}$ and
$P(E') = 1 - P(E)$
$= 1 - \dfrac{20{,}150}{37{,}961} = \dfrac{17{,}811}{37{,}961}$.
The odds that a randomly selected foreign-born U.S. resident was born in Latin America are
$\dfrac{P(E)}{P(E')} = \dfrac{\frac{20{,}150}{37{,}961}}{\frac{17{,}811}{37{,}961}} = \dfrac{20{,}150}{17{,}811}$
or 20,150 to 17,811.

21. Each suit has thirteen cards, and the probability of choosing the correct card in that suit is $\frac{1}{13}$. Thus, we have P(4 correct choices)
$= \dfrac{1}{13} \cdot \dfrac{1}{13} \cdot \dfrac{1}{13} \cdot \dfrac{1}{13} = \dfrac{1}{28{,}561}$. The probability of getting all four picks correct and winning $5000 is
$= \dfrac{1}{13} \cdot \dfrac{1}{13} \cdot \dfrac{1}{13} \cdot \dfrac{1}{13} = \dfrac{1}{28{,}561} \approx 0.000035$.

23. (a) A 40-year-old man who lives 30 more years would be a 70-year old man. Let E be "selected man will live to be 70." Then $n(E) = 71{,}586$. For this situation, the sample space S is the set of all 40-year-old men, so $n(S) = 95{,}431$. Thus, the probability that a 40-year-old man will live 30 more years is
$$P(E) = \dfrac{n(E)}{n(S)} = \dfrac{71{,}586}{95{,}431} \approx 0.750.$$

Copyright © 2013 Pearson Education, Inc.

(b) Using the notation and results from part (a), the probability that a 40-year-old man will not live 30 more yr is $P(E') = 1 - P(E) = 1 - 0.750 = 0.250.$

(c) Use the notation and results from parts (a) and (b). In this binomial experiment, we call "a 40-year-old man survives to age 70" a success. Then $p = P(E) = 0.750$ and $1 - p = P(E') = 0.250.$

There are 5 independent trials and we need the probability of 3 successes, so $n = 5, r = 3.$ The probability that exactly 3 of the 40-year-old men survive to age 70 is

$$\binom{n}{r} p^r (1-p)^{n-r} = \binom{5}{3}(0.750)^3 (0.250)^{5-3} = \frac{5!}{(5-3)!3!}(0.750)^3 (0.250)^2 = \frac{5!}{2!3!}(0.750)^3 (0.250)^2$$

$$= \frac{5 \cdot 4}{2 \cdot 1}(0.750)^3 (0.250)^2 = 10(0.750)^3 (0.250)^2 \approx 0.264$$

(d) Let F be the event "at least one man survives to age 70." The easiest way to find $P(F)$ is to first find the probability of the complement of the event, $F'.$ F' is the event that "neither man survives to age 70." Since $P(F') = P(E') \cdot P(E') = (0.250)^2 \approx 0.0625,$ we have

$$P(F) = 1 - P(F') \approx 1 - 0.0625 = 0.9375 \approx 0.938.$$

25. The amount of growth would be $11{,}400 - 10{,}000 = 1400,$ so the percent growth would be $\dfrac{1400}{10{,}000} = 0.14.$ Let E be the event "worth at least \$11,400 by the end of the year." This is equivalent to "at least 14 percent growth" which is equivalent to "14 or 18 percent growth." Thus, we have $P(E) = P(14\% \text{ growth or } 18\% \text{ growth}) = P(14\% \text{ growth}) + P(18\% \text{ growth}) = 0.20 + 0.10 = 0.30 = 0.3.$

27. In this binomial experiment, the outcome of interest is "did not get enough sleep." Then $n = 10, r = 4, p = 0.45$ and $q = 1 - 0.45 = 0.55.$

$$P(\text{not enough sleep}) = \binom{n}{r} p^r q^{n-r} = \binom{10}{4}(0.45)^4 (0.55)^{10-4} = \frac{10!}{(10-4)!4!}(0.45)^4 (0.55)^6$$

$$= \frac{10 \cdot 9 \cdot 8 \cdot 7}{4 \cdot 3 \cdot 2 \cdot 1}(0.45)^4 (0.55)^6 = 210(0.45)^4 (0.55)^6 \approx 0.238$$

29. In this binomial experiment, the outcome of interest is "did not get enough sleep on 14 or more days."
$P(\text{not enough sleep on 14 or more days}) = P(\text{not enough sleep on 14-29 days}) + P(\text{not enough sleep on 30 days})$
$$= 0.20 + 0.12 = 0.32$$

Thus, $p = 0.32$ and $q = 1 - p = 0.68.$ "Fewer than 2" means 0 or 1, so $P(\text{fewer than 2})$ is

$$P(0) + P(1) = \binom{10}{0}(0.32)^0 (0.68)^{10} + \binom{10}{1}(0.32)^1 (0.68)^9 = 1(1)(0.68)^{10} + 10(0.32)(0.68)^9$$

$$\approx 0.211392282 + 0.994787209 \approx 0.121$$

31. $P(\text{a student applied to fewer than 4 colleges})$
$= P(\text{a student applied to 1 college}) + P(\text{a student applied to 2 or 3 colleges})$
$= 0.14 + 0.24 = 0.38$

33. $P(\text{a student applied to more than 3 colleges})$
$= P(\text{a student applied to 4-6 colleges}) + P(\text{a student applied to 7 or more colleges})$
$= 0.39 + 0.23 = 0.62$

Copyright © 2013 Pearson Education, Inc.

35. In these binomial experiments, we call "the man is color-blind" a success.

(a) For "exactly 5 are color-blind", we have $n = 53$, $r = 5$, $p = 0.042$, and $q = 1 - p = 0.958$.

$$P(\text{exactly 5 are color-blind}) = \binom{53}{5}(0.042)^5(0.958)^{53-5} = \frac{53!}{(53-5)!5!}(0.042)^5(0.958)^{48}$$

$$= \frac{53!}{48!5!}(0.042)^5(0.958)^{48} = \frac{53 \cdot 52 \cdot 51 \cdot 50 \cdot 49}{5 \cdot 4 \cdot 3 \cdot 2 \cdot 1}(0.042)^5(0.958)^{48}$$

$$= 2,869,685(0.042)^5(0.958)^{48} \approx 0.047822 \approx 0.048$$

(b) "No more than 5 are color-blind" corresponds to "0, 1, 2, 3, 4, or 5 are color-blind." Then $n = 53$, $r = 0$, 1, 2, 3, 4, or 5, $p = 0.042$, and $q = 1 - p = 0.958$.

$P(\text{no more than 5 are color-blind}) = P(0 \text{ are color-blind}) + P(1 \text{ is color-blind}) + P(2 \text{ are color-blind})$
$\qquad\qquad\qquad\qquad\qquad + P(3 \text{ are color-blind}) + P(4 \text{ are color-blind}) + P(5 \text{ are color-blind})$

$$= \binom{53}{0}(0.042)^0(0.958)^{55-0} + \binom{53}{1}(0.042)^1(0.958)^{53-1} + \binom{53}{2}(0.042)^2(0.958)^{53-2}$$

$$+ \binom{53}{3}(0.042)^3(0.958)^{53-3} + \binom{53}{4}(0.042)^4(0.958)^{53-4}$$

$$+ \binom{53}{5}(0.042)^5(0.958)^{53-5}$$

$$= \frac{53!}{(53-0)!0!}(1)(0.958)^{53} + \frac{53!}{(53-1)!1!}(0.042)^1(0.958)^{52} + \frac{53!}{(53-2)!2!}(0.042)^2(0.958)^{51}$$

$$+ \frac{53!}{(53-3)!3!}(0.042)^3(0.958)^{50} + \frac{53!}{(53-4)!4!}(0.042)^4(0.958)^{49}$$

$$+ \frac{53!}{(53-5)!5!}(0.042)^5(0.958)^{48}$$

$$= \frac{53!}{53!0!}(0.958)^{53} + \frac{53 \cdot 52!}{52!1!}(0.042)^1(0.958)^{52} + \frac{53 \cdot 52 \cdot 51!}{51!2!}(0.042)^2(0.958)^{51}$$

$$+ \frac{53 \cdot 52 \cdot 51 \cdot 50!}{50!3!}(0.042)^3(0.958)^{50} + \frac{53 \cdot 52 \cdot 51 \cdot 50 \cdot 49!}{49!4!}(0.042)^4(0.958)^{49}$$

$$+ \frac{53 \cdot 52 \cdot 51 \cdot 50 \cdot 49 \cdot 48!}{48!5!}(0.042)^5(0.958)^{48}$$

$$= 1(0.958)^{53} + 53(0.042)^1(0.958)^{52} + 1378(0.042)^2(0.958)^{51} + 23,426(0.042)^3(0.958)^{50}$$

$$+ 292,825(0.042)^4(0.958)^{49} + 2,869,685(0.042)^5(0.958)^{48}$$

$$\approx 0.102890 + 0.239074 + 0.272514 + 0.203105 + 0.111305 + 0.047822 = 0.976710 \approx 0.977$$

(c) Refer to part (b): $P(\text{none are color-blind}) = \binom{53}{0}(0.042)^0(0.958)^{53-0} \approx 0.102890 \approx 0.103$

(d) Let E be "at least 1 is color-blind." Consider the complementary event E': "0 are color-blind." Thus, we have

$$P(E) = 1 - P(E') = 1 - \binom{53}{0}(0.042)^0(0.958)^{53-0} = 1 - \frac{53!}{(53-0)!0!}(1)(0.958)^{53}$$

$$= 1 - \frac{53!}{53!0!}(1)(0.958)^{53} = 1 - (1)(0.958)^{53} = 1 - (0.958)^{53} = 1 - 0.102890 = 0.897110 \approx 0.897$$

Copyright © 2013 Pearson Education, Inc.

37. (a) First compute

$$q = (1-p)^I = (1-0.1)^2 = 0.9^2 = 0.81.$$

Then, with $S = 4$, $k = 3$, and $q = 0.81$, we have

$$P = \binom{S}{k} q^k (1-q)^{S-k}$$

$$= \binom{4}{3}(0.81)^3 (1-0.81)^{4-3}$$

$$= \frac{4!}{(4-3)!3!}(0.81)^3 (0.19)$$

$$= 4(0.81^3)(0.19) \approx 0.404$$

There is about a 40.4% chance of exactly 3 people not becoming infected.

(b) Compute

$$q = (1-p)^I = (1-0.5)^2 = 0.5^2 = 0.25$$

Then, with $S = 4$, $k = 3$, and $q = 0.25$, we have

$$P = \binom{S}{k} q^k (1-q)^{S-k}$$

$$= \binom{4}{3}(0.25)^3 (1-0.25)^{4-3}$$

$$= \frac{4!}{(4-3)!3!}(0.25)^3 (0.75)$$

$$= 4(0.25^3)(0.75) \approx 0.047$$

There is about 4.7% chance of this occurring when the disease is highly infectious.

(c) Compute $q = (1-p)^I = (1-0.5)^1 = 0.5$.

Then, with $S = 9$, $k = 0$, and $q = 0.5$, we have

$$P = \binom{S}{k} q^k (1-q)^{S-k}$$

$$= \binom{9}{0}(0.5)^0 (1-0.5)^{9-0}$$

$$= \frac{9!}{(9-0)!0!}(1)(0.5)^9$$

$$= 1(1)(0.5^9) \approx 0.002$$

There is about a 0.2% chance of everyone becoming infected. This means that in a large family or group of people, it is highly unlikely that everyone will become sick even though the disease is highly infectious.

Chapter 7   Review Exercises

1. $a_n = \dfrac{n}{n+1}$

$a_1 = \dfrac{1}{1+1} = \dfrac{1}{2}; a_2 = \dfrac{2}{2+1} = \dfrac{2}{3}; a_3 = \dfrac{3}{3+1} = \dfrac{3}{4};$

$a_4 = \dfrac{4}{4+1} = \dfrac{4}{5}; a_5 = \dfrac{5}{5+1} = \dfrac{5}{6}$

The first five terms are $\frac{1}{2}, \frac{2}{3}, \frac{3}{4}, \frac{4}{5}$, and $\frac{5}{6}$.

This sequence does not have a common difference or a common ratio, so the sequence is neither arithmetic nor geometric.

3. $a_n = 2(n+3)$

$a_1 = 2(1+3) = 2(4) = 8;$
$a_2 = 2(2+3) = 2(5) = 10;$
$a_3 = 2(3+3) = 2(6) = 12;$
$a_4 = 2(4+3) = 2(7) = 14;$
$a_5 = 2(5+3) = 2(8) = 16$

The first five terms are 8, 10, 12, 14, and 16. There is a common difference, $d = 2$, so the sequence is arithmetic.

5. $a_1 = 5$; for $n \geq 2$, $a_n = a_{n-1} - 3$

$a_2 = a_{2-1} - 3 = a_1 - 3 = 5 - 3 = 2;$
$a_3 = a_{3-1} - 3 = a_2 - 3 = 2 - 3 = -1;$
$a_4 = a_{4-1} - 3 = a_3 - 3 = -1 - 3 = -4;$
$a_5 = a_{5-1} - 3 = a_4 - 3 = -4 - 3 = -7$

The first five terms are 5, 2, –1, –4, and –7. There is a common difference, $d = -3$, so the sequence is arithmetic.**6.**

$a_1 = 1, a_2 = 3$; for $n \geq 3$, $a_n = a_{n-2} + a_{n-1}$

$a_3 = a_2 + a_1 = 3 + 1 = 4$
$a_4 = a_3 + a_2 = 4 + 3 = 7$
$a_5 = a_4 + a_3 = 7 + 4 = 11$

The first five terms are 1, 3, 4, 7, and 11. The sequence does not have either a common difference or a common ratio, so the sequence is neither arithmetic nor geometric.

7. $a_1 = 4$; $S_5 = 25$

Using the formula $S_n = \frac{n}{2}\left[2a_1 + (n-1)d\right]$ with $n = 5$, we have

$25 = \frac{5}{2}\left[2(4) + (5-1)d\right] \Rightarrow 25 = \frac{5}{2}(8+4d) \Rightarrow$
$25 = 5(4+2d) \Rightarrow 5 = 4 + 2d \Rightarrow d = \frac{1}{2}$

$a_1 = 4; a_2 = 4 + \frac{1}{2} = 4.5; a_3 = 4 + 2\left(\frac{1}{2}\right) = 5;$

$a_4 = 4 + 3\left(\frac{1}{2}\right) = 5.5; a_5 = 4 + 4\left(\frac{1}{2}\right) = 6$

The five terms are 4, 4.5, 5, 5.5, and 6.

Copyright © 2013 Pearson Education, Inc.

9. Arithmetic, $a_3 = \pi$, $a_4 = 1$

$d = a_4 - a_3 = 1 - \pi$

$a_3 = a_1 + 2d \Rightarrow \pi = a_1 + 2(1-\pi) \Rightarrow$

$\pi = a_1 + 2 - 2\pi \Rightarrow 3\pi - 2 = a_1$

$a_2 = a_1 + d \Rightarrow a_2 = (3\pi - 2) + (1-\pi) \Rightarrow$

$a_2 = 2\pi - 1$

$a_5 = a_1 + 4d \Rightarrow a_5 = (3\pi - 2) + 4(1-\pi) \Rightarrow$

$a_5 = 3\pi - 2 + 4 - 4\pi \Rightarrow a_5 = -\pi + 2$

The first five terms are $3\pi - 2$, $2\pi - 1$, π, 1, $-\pi + 2$.

11. Geometric, $a_1 = -5$, $a_2 = -1$

$r = \frac{a_2}{a_1} = \frac{-1}{-5} = \frac{1}{5}$

$a_3 = a_1 r^2 = -5\left(\frac{1}{5}\right)^2 = -5 \cdot \frac{1}{25} = -\frac{1}{5}$

$a_4 = a_1 r^3 = -5\left(\frac{1}{5}\right)^3 = -5 \cdot \frac{1}{125} = -\frac{1}{25}$

$a_5 = a_1 r^4 = -5\left(\frac{1}{5}\right)^4 = -5 \cdot \frac{1}{625} = -\frac{1}{125}$

The first five terms are -5, -1, $-\frac{1}{5}$, $-\frac{1}{25}$, and $-\frac{1}{125}$.

13. Geometric, $a_1 = -8$ and $a_7 = -\frac{1}{8}$

Since $a_n = a_1 r^{n-1}$, we have $a_7 = a_1 r^{7-1} \Rightarrow$

$-\frac{1}{8} = -8r^6 \Rightarrow r^6 = \frac{1}{64} \Rightarrow r = \pm\frac{1}{2}$.

There are two geometric sequences that satisfy the given conditions.

If $r = \frac{1}{2}$, $a_4 = (-8)\left(\frac{1}{2}\right)^3 = (-8)\left(\frac{1}{8}\right) = -1$.

Also, $a_n = -8\left(\frac{1}{2}\right)^{n-1}$ or $a_n = -2^3\left(\frac{1}{2}\right)^{n-1}$

$= -\frac{2^3}{2^{n-1}} = -\frac{1}{2^{(n-1)-3}} = -\frac{1}{2^{n-4}} = -\left(\frac{1}{2}\right)^{n-4}$.

If $r = -\frac{1}{2}$, $a_4 = (-8)\left(-\frac{1}{2}\right)^3 = (-8)\left(-\frac{1}{8}\right) = 1$.

Also, $a_n = -8\left(-\frac{1}{2}\right)^{n-1}$ or $a_n = -2^3\left(-\frac{1}{2}\right)^{n-1}$

$= (-2)^3\left(-\frac{1}{2}\right)^{n-1} = \frac{(-2)^3}{(-2)^{n-1}} = \frac{1}{(-2)^{(n-1)-3}}$

$= \frac{1}{(-2)^{n-4}} = \left(-\frac{1}{2}\right)^{n-4}$.

15. $a_1 = 6x - 9$, $a_2 = 5x + 1$

$d = a_2 - a_1 = (5x+1) - (6x-9)$

$= 5x + 1 - 6x + 9 = -x + 10$

$a_n = a_1 + (n-1)d$

$a_8 = a_1 + (8-1)d \Rightarrow a_8 = a_1 + 7d \Rightarrow$

$a_8 = (6x-9) + 7(-x+10)$

$= 6x - 9 - 7x + 70 = -x + 61$

17. $a_2 = 6$, $d = 10$

To first find a_1, we have

$a_2 = a_1 + d \Rightarrow 6 = a_1 + 10 \Rightarrow a_1 = -4$

Now, to find S_{12}, we use

$S_n = \frac{n}{2}\left[2a_1 + (n-1)d\right]$.

$S_{12} = \frac{12}{2}\left[2(-4) + (12-1)(10)\right]$

$= 6\left[-8 + (11)(10)\right] = 6(-8 + 110)$

$= 6(102) = 612$

19. $a_3 = 4$, $r = \frac{1}{5}$

$a_4 = a_3 r = 4 \cdot \frac{1}{5} = \frac{4}{5} \Rightarrow a_5 = a_4 r = \frac{4}{5} \cdot \frac{1}{5} = \frac{4}{25}$

21. $a_1 = -1$, $r = 3$. $S_n = \frac{a_1\left(1 - r^n\right)}{1-r}$

$S_4 = \frac{a_1\left(1 - r^4\right)}{1-r} = \frac{-1\left(1 - 3^4\right)}{1-3} = \frac{-1(1-81)}{-2}$

$= \frac{1}{2}(-80) = -40$

23. $\displaystyle\sum_{i=1}^{7}(-1)^{i-1}$; $S_n = \frac{a_1\left(1 - r^n\right)}{1-r}$

This is a geometric series with

$a_1 = (-1)^{1-1} = (-1)^0 = 1$ and $r = -1$.

$S_7 = \frac{a_1\left(1 - r^7\right)}{1-r} \Rightarrow$

$S_7 = \frac{1\left[1 - (-1)^7\right]}{1-(-1)} = \frac{1\left[1-(-1)\right]}{2} = \frac{1 \cdot 2}{2} = 1$

25. $\displaystyle\sum_{i=1}^{4}\frac{i+1}{i} = \frac{2}{1} + \frac{3}{2} + \frac{4}{3} + \frac{5}{4} = \frac{24}{12} + \frac{18}{12} + \frac{16}{12} + \frac{15}{12}$

$= \frac{73}{12}$

27. Since

$\displaystyle\sum_{i=1}^{n} i = \frac{n(n+1)}{2}$,

$\displaystyle\sum_{j=1}^{2500} j = \frac{2500(2500+1)}{2}$

$= 1250(2501) = 3{,}126{,}250$

Copyright © 2013 Pearson Education, Inc.

29. $\displaystyle\sum_{i=1}^{\infty}\left(\frac{4}{7}\right)^{i}$

This is an infinite geometric series with

$a_1 = \left(\frac{4}{7}\right)^1 = \frac{4}{7}$ and $r = \frac{4}{7}$. Since $|r| = \left|\frac{4}{7}\right| < 1$,

we can use the formula $S_\infty = \dfrac{a_1}{1-r}$. Thus, we

have

$\displaystyle\sum_{i=1}^{\infty}\left(\frac{4}{7}\right)^{i} = S_\infty = \frac{a_1}{1-r} = \frac{\frac{4}{7}}{1-\frac{4}{7}} = \frac{\frac{4}{7}}{\frac{3}{7}} = \frac{4}{7}\cdot\frac{7}{3} = \frac{4}{3}.$

31. $\displaystyle\sum_{i=1}^{\infty}2\left(-\frac{2}{3}\right)^{i}$

This is an infinite geometric series with

$a_1 = 2\left(-\frac{2}{3}\right)^1 = -\frac{4}{3}$ and $r = -\frac{2}{3}$. Since

$|r| = \left|-\frac{2}{3}\right| < 1$, we can use the formula

$S_\infty = \dfrac{a_1}{1-r}$. Thus, we have

$\displaystyle\sum_{i=1}^{\infty}2\left(-\frac{2}{3}\right)^{i} = S_\infty = \frac{a_1}{1-r} = \frac{-\frac{4}{3}}{1-\left(-\frac{2}{3}\right)} = \frac{-\frac{4}{3}}{\frac{5}{3}} = -\frac{4}{5}.$

33. $24 + 8 + \dfrac{8}{3} + \dfrac{8}{9} + \cdots$

This is an infinite geometric series with

$a_1 = 24$ and $r = \frac{8}{24} = \frac{1}{3}$. Since $|r| = \left|\frac{1}{3}\right| < 1$, we

can use the formula $S_\infty = \dfrac{a_1}{1-r}$. Thus, we

have $S_\infty = \dfrac{a_1}{1-r} = \dfrac{24}{1-\frac{1}{3}} = \dfrac{24}{\frac{2}{3}} = 24\cdot\dfrac{3}{2} = 36.$

35. $\dfrac{1}{12} + \dfrac{1}{6} + \dfrac{1}{3} + \dfrac{2}{3} + \cdots$

This is an infinite geometric series with

$a_1 = \dfrac{1}{12}$ and $r = \dfrac{\frac{1}{6}}{\frac{1}{12}} = 2$. Since $|r| > 1$, the

series diverges.

37. $x_1 = 0$, $x_2 = 1$, $x_3 = 2$, $x_4 = 3$, $x_5 = 4$, and $x_6 = 5$.

$\displaystyle\sum_{i=1}^{4}\left(x_i^{2}-6\right)$

$= \left(x_1^{2}-6\right) + \left(x_2^{2}-6\right) + \left(x_3^{2}-6\right) + \left(x_4^{2}-6\right)$

$= \left(0^2-6\right) + \left(1^2-6\right) + \left(2^2-6\right) + \left(3^2-6\right)$

$= (0-6) + (1-6) + (4-6) + (9-6)$

$= -6 + (-5) + (-2) + 3 = -10$

In Exercises 39–41, other answers are possible.

39. $4 - 1 - 6 - \cdots - 66 = 4 + (-1) + (-6) + \cdots + (-66)$

This series is the sum of an arithmetic
sequence with $a_1 = 4$ and $d = -1-4 = -5$.
Therefore, the *n*th term is

$a_n = a_1 + (n-1)d = 4 + (n-1)(-5)$ or,

$\quad = 4 - 5n + 5 = -5n + 9$

equivalently, $a_i = -5i + 9$.

To find the number of terms in the series, we
realize that the last term of the series is –66.
Since $a_i = -5i + 9$ we solve $-66 = -5i + 9$ for
i: $-66 = -5i + 9 \Rightarrow -75 = -5i \Rightarrow i = 15$
This indicates that the series consists of 15
terms and we have

$4 - 1 - 6 - \cdots - 66 = \displaystyle\sum_{i=1}^{15}(-5i+9).$

41. $4 + 12 + 36 + \cdots + 972$

This series is the sum of a geometric sequence
with $a_1 = 4$ and $r = \frac{12}{4} = 3$.
Therefore, the nth term is

$a_n = a_1 r^{n-1} \Rightarrow a_n = 4(3)^{n-1}$, or, equivalently,

$a_i = 4(3)^{i-1}$. To find the number of terms in
the series, we realize that the last term of the
series is 972. Since $a_i = 4(3)^{i-1}$ we solve

$972 = 4(3)^{i-1}$ for i.

$972 = 4(3)^{i-1} \Rightarrow 243 = 3^{i-1} \Rightarrow$

$3^5 = 3^{i-1} \Rightarrow 5 = i - 1 \Rightarrow i = 6$

This indicates that the series consists of 6
terms and we have

$4 + 12 + 36 + \ldots + 972 = \displaystyle\sum_{i=1}^{6}4(3)^{i-1}.$

43. $(x+2y)^4 = x^4 + \binom{4}{1}x^3(2y)^1 + \binom{4}{2}x^2(2y)^2 + \binom{4}{3}x^1(2y)^3 + (2y)^4$

$$= x^4 + \frac{4!}{1!(4-1)!}x^3(2y) + \frac{4!}{(4-2)!2!}x^2(4y^2) + \frac{4!}{3!(4-3)!}x(8y^3) + 16y^4$$

$$= x^4 + \frac{4!}{1!3!}x^3(2y) + \frac{4!}{2!2!}x^2(4y^2) + \frac{4!}{3!1!}x(8y^3) + 16y^4$$

$$= x^4 + 4x^3(2y) + \frac{4 \cdot 3}{2 \cdot 1}x^2(4y^2) + 4x(8y^3) + 16y^4$$

$$= x^4 + 8x^3y + 24x^2y^2 + 32xy^3 + 16y^4$$

45. $\left(3\sqrt{x} - \dfrac{1}{\sqrt{x}}\right)^5 = \left[3x^{1/2} + \left(-x^{-1/2}\right)\right]^5$

$$= \left(3x^{1/2}\right)^5 + \binom{5}{1}\left(3x^{1/2}\right)^4\left(-x^{-1/2}\right) + \binom{5}{2}\left(3x^{1/2}\right)^3\left(-x^{-1/2}\right)^2$$

$$+ \binom{5}{3}\left(3x^{1/2}\right)^2\left(-x^{-1/2}\right)^3 + \binom{5}{4}\left(3x^{1/2}\right)\left(-x^{-1/2}\right)^4 + \left(-x^{-1/2}\right)^5$$

$$= 243x^{5/2} + \frac{5!}{1!4!}\left(81x^2\right)\left(-x^{-1/2}\right) + \frac{5!}{2!3!}\left(27x^{3/2}\right)^3\left(x^{-1}\right)$$

$$- \frac{5!}{3!2!}(9x)x^{-3/2} + \frac{5!}{4!1!}\left(3x^{1/2}\right)\left(x^{-2}\right) + \left(-x^{-5/2}\right)$$

$$= 243x^{5/2} - 5\left(81x^{2+(-1/2)}\right) + \frac{5 \cdot 4}{2 \cdot 1}\left(27x^{3/2+(-1)}\right) - \frac{5 \cdot 4}{2 \cdot 1}\left(9x^{1+(-3/2)}\right) + 5\left(3x^{1/2+(-2)}\right) - x^{-5/2}$$

$$= 243x^{5/2} - 405x^{3/2} + 270x^{1/2} - 90x^{-1/2} + 15x^{-3/2} - x^{-5/2}$$

47. The sixth term of the expansion of $(4x - y)^8$ is $\binom{8}{5}(4x)^3(-y)^5$. This simplifies as follows.

$$\binom{8}{5}(4x)^3(-y)^5 = \frac{8!}{5!3!}\left(64x^3\right)\left(-y^5\right) = \frac{8 \cdot 7 \cdot 6}{3 \cdot 2 \cdot 1}\left(-64x^3y^5\right) = 56\left(-64x^3y^5\right) = -3584x^3y^5$$

49. The first four terms of $(x+2)^{12}$ are as follows.

$$(x)^{12} + \binom{12}{1}(x)^{11}(2) + \binom{12}{2}(x)^{10}(2)^2 + \binom{12}{3}(x)^9(2)^3 = x^{12} + \frac{12!}{1!11!}\left(x^{11}\right)(2) + \frac{12!}{2!10!}\left(x^{10}\right)(4) + \frac{12!}{3!9!}\left(x^9\right)(8)$$

$$= x^{12} + 12\left(2x^{11}\right) + \frac{12 \cdot 11}{2 \cdot 1}\left(4x^{10}\right) + 220\left(8x^9\right)$$

$$= x^{12} + 24x^{11} + 264x^{10} + 1760x^9$$

51. Let S_n be the statement $1 + 3 + 5 + 7 + \ldots + (2n - 1) = n^2$.

Step 1: Show that the statement is true for $n = 1$.

S_1 is the statement $2(1) - 1 = 1^2 \Rightarrow 2 - 1 = 1 \Rightarrow 1 = 1$, which is true.

Step 2: Show that if S_k is true, S_{k+1} is also true. S_k is the statement $1 + 3 + 5 + 7 + \cdots + (2k - 1) = k^2$. and

S_{k+1} is the statement $1 + 3 + 5 + 7 + \cdots + \left[2(k+1) - 1\right] = (k+1)^2$.

Start with S_k: $1 + 3 + 5 + 7 + \cdots + (2k - 1) = k^2$. and add the $(k+1)$ st term, $\left[2(k+1) - 1\right]$, to both sides of this equation.

$$1 + 3 + 5 + 7 + \ldots + (2k - 1) = k^2$$
$$1 + 3 + 5 + 7 + \cdots + (2k - 1) + \left[2(k+1) - 1\right] = k^2 + \left[2(k+1) - 1\right]$$
$$1 + 3 + 5 + 7 + \cdots + (2k - 1) + \left[2(k+1) - 1\right] = k^2 + (2k + 2 - 1)$$

(continued on next page)

Copyright © 2013 Pearson Education, Inc.

(*continued*)

$$1+3+5+7+\cdots+(2k-1)+\left[2(k+1)-1\right]=k^2+(2k+1)$$
$$1+3+5+7+\cdots+(2k-1)+\left[2(k+1)-1\right]=k^2+2k+1$$
$$1+3+5+7+\cdots+(2k-1)+\left[2(k+1)-1\right]=(k+1)^2$$

It has been shown that S_k implies S_{k+1}. Therefore, by mathematical induction, S_n is true for every positive integer n.

53. Let S_n be the statement $2+2^2+2^3+\cdots+2^n=2\left(2^n-1\right)$.

Step 1: Show that the statement is true for $n = 1$.

S_1 is the statement $2^1=2\left(2^1-1\right)\Rightarrow 2=2(2-1)\Rightarrow 2=2(1)\Rightarrow 2=2$, which is true.

Step 2: Show that if S_k is true, S_{k+1} is also true. S_k is the statement. $2+2^2+2^3+\cdots+2^k=2\left(2^k-1\right)$

S_{k+1} is the statement $2+2^2+2^3+\cdots+2^{k+1}=2\left(2^{k+1}-1\right)$.

Start with S_k: $2+2^2+2^3+\cdots+2^k=2\left(2^k-1\right)$ and add the $(k+1)$ st term, 2^{k+1}, to both sides of this equation.

$$2+2^2+2^3+\cdots+2^k=2\left(2^k-1\right)$$
$$2+2^2+2^3+\cdots+2^k+2^{k+1}=2\left(2^k-1\right)+2^{k+1}$$
$$2+2^2+2^3+\cdots+2^k+2^{k+1}=2^{k+1}-2+2^{k+1}$$
$$2+2^2+2^3+\cdots+2^k+2^{k+1}=2\cdot2^{k+1}-2\cdot1$$
$$2+2^2+2^3+\cdots+2^k+2^{k+1}=2\left(2^{k+1}-1\right)$$

It has been shown that S_k implies S_{k+1}. Therefore, by mathematical induction, S_n is true for every positive integer n.

55. $P(9,\ 2)=\dfrac{9!}{(9-2)!}=\dfrac{9!}{7!}=\dfrac{9\cdot8\cdot7!}{7!}=72$

57. $\dbinom{8}{3}=C(8,3)=\dfrac{8!}{(8-3)!3!}=\dfrac{8!}{5!3!}$

$=\dfrac{8\cdot7\cdot6\cdot5!}{3\cdot2\cdot1\cdot5!}=56$

59. $C(10,\ 5)=\dfrac{10!}{(10-5)!5!}=\dfrac{10!}{5!5!}=\dfrac{10\cdot9\cdot8\cdot7\cdot6\cdot5!}{5!5!}$

$=\dfrac{10\cdot9\cdot8\cdot7\cdot6}{5\cdot4\cdot3\cdot2\cdot1}=252$

61. A person who starts working at age 18 and works until age 65 will work for 47 years. We are seeking the total amount earned during those 47 years. The first year, the person will earn $32,136, the next year, the person will earn $32,136 + $268; the third year, the person will earn $32,136 + 2($268), and so on. We are seeking the sum of the arithmetic sequence defined by $a_n=32,136+(n-1)268$.

$a_1=32,136;\ a_{47}=32,136+46(268)=44,464$

$$S_{47}=\frac{47}{2}(32,136+44,464)=23.5(76,600)$$
$$=1,800,100$$

The person will earn $1,800,100.

63. The person with four years of college will earn $2,996,971 – $1,800,100 = $1,196,871 more than a person with no college attendance. Answers will vary.

65. Three independent events are involved. There are 5 choices of style, 3 choices of fabric, and 6 choices of color. Therefore, using the fundamental principle of counting, there are $5\cdot3\cdot6=90$ different couches.

67. There are nine people on the student body council.

(a) In selecting a 3-member delegation, $C(9,\ 3)$ choices are possible.

$$C(9,\ 3)=\binom{9}{3}=\frac{9!}{(9-3)!3!}=\frac{9!}{6!3!}$$
$$=\frac{9\cdot8\cdot7\cdot6!}{6!3!}=\frac{9\cdot8\cdot7}{3\cdot2\cdot1}=84$$

84 delegations are possible.

Copyright © 2013 Pearson Education, Inc.

(b) If two seniors and one junior must attend, then the two seniors can be chosen in $C(6, 2)$ ways and the junior can be chosen in $C(3, 1)$ ways.

$$C(6,2)=\binom{6}{2}=\frac{6!}{(6-2)!2!}=\frac{6!}{4!2!}$$
$$=\frac{6\cdot5\cdot4!}{4!2!}=\frac{6\cdot5}{2\cdot1}=15$$
$$C(3,1)=\binom{3}{1}=\frac{3!}{(3-1)!1!}=\frac{3!}{2!1!}$$
$$=\frac{3\cdot2!}{2!1!}=\frac{3}{1}=3$$

Thus, there are $15\cdot3=45$ possible delegations.

69. There are 4 spots to be filled by 26 letters and 3 spots by 10 digits. If repeats are allowed, then $26\cdot10\cdot10\cdot10\cdot26\cdot26\cdot26=456,976,000$ different license plates can be formed. If no repeats are allowed, then $26\cdot10\cdot9\cdot8\cdot25\cdot24\cdot23=258,336,000$ different license plates can be formed.

71. (a) Let E be the event "picking a green marble."

$n(E)=4$ and $n(S)=4+5+6=15$ (total number of marbles) Thus,

$$P(E)=\frac{n(E)}{n(S)}=\frac{4}{15}$$

(b) Let E be the event "picking a black marble."

$n(E)=5$ and $n(S)=15$ (from part (a)).

Thus, $P(E)=\frac{n(E)}{n(S)}=\frac{5}{15}=\frac{1}{3}.$

The probability that the marble is not black is given by

$$P(E')=1-P(E)=1-\frac{1}{3}=\frac{2}{3}.$$

(c) Let E be the event "picking a blue marble."

$n(E)=0$ since there are no blue marbles and $n(S)=15$ (from part (a)). Thus, we have $P(E)=\frac{n(E)}{n(S)}=\frac{0}{15}=0.$

(d) Let E be the event "picking a marble that is not white." We have

$$P(E)=\frac{n(E)}{n(S)}=\frac{9}{15}\text{ and}$$
$$P(E')=\frac{n(E')}{n(S)}=\frac{6}{15}.\text{ Thus, the odds in}$$

favor of E are

$$\frac{P(E)}{P(E')}=\frac{\frac{9}{15}}{\frac{6}{15}}=\frac{9}{15}\cdot\frac{15}{6}=\frac{3}{2}\text{ or 3 to 2.}$$

73. (a) Let B be the event "selected student earned a degree in business." Thus, we have $P(B)=\frac{n(B)}{n(S)}=\frac{155,637}{625,023}\approx0.249$

(b) Let H be the event "selected student earned a degree in health professions and related clinical studies" and let V be the event "selected student earned a degree in visual and performing arts." We have

$$P(H\text{ or }V)=\frac{n(H\text{ or }V)}{n(S)}$$
$$=\frac{58,120+14,164}{625,023}\approx0.116$$

(c) Let E be the event "selected student earned a degree in education." Then E' is the event "selected student earned a degree that was not in education."

$$P(E')=\frac{n(E')}{n(S)}=\frac{625,023-175,880}{625,023}$$
$$\approx0.719$$

75. In this binomial experiment, we call rolling a five a success. Then $n=12$, $r=2$, and $p=\frac{1}{6}$.

$$P(2\text{ fives})=\binom{n}{r}p^r(1-p)^{n-r}$$
$$=\binom{12}{2}\left(\frac{1}{6}\right)^2\left(1-\frac{1}{6}\right)^{12-2}$$
$$=\frac{12!}{(12-2)!2!}\left(\frac{1}{6^2}\right)\left(\frac{5}{6}\right)^{10}$$
$$=\frac{12!}{10!2!}\left(\frac{1}{6^2}\right)\left(\frac{5^{10}}{6^{10}}\right)$$
$$=\frac{12\cdot11\cdot10!}{10!2!}\left(\frac{5^{10}}{6^{12}}\right)$$
$$=\frac{12\cdot11}{2\cdot1}\left(\frac{5^{10}}{6^{12}}\right)=66\left(\frac{5^{10}}{6^{12}}\right)\approx0.296$$

Copyright © 2013 Pearson Education, Inc.

Chapter 7   Test

1. $a_n = (-1)^n (n^2 + 2)$

$n = 1: a_1 = (-1)^1 (1^2 + 2) = (-1)(1 + 2) = -3$

$n = 2: a_2 = (-1)^2 (2^2 + 2) = (1)(4 + 2) = 6$

$n = 3: a_3 = (-1)^3 (3^2 + 2) = (-1)(9 + 2) = -11$

$n = 4: a_4 = (-1)^4 (4^2 + 2) = (1)(16 + 2) = 18$

$n = 5: a_5 = (-1)^5 (5^2 + 2) = (-1)(25 + 2) = -27$

The first five terms are –3, 6, –11, 18, and –27. This sequence does not have either a common difference or a common ratio, so the sequence is neither arithmetic nor geometric.

2. $a_n = -3 \cdot \left(\dfrac{1}{2}\right)^n$

$n = 1: a_1 = -3\left(\tfrac{1}{2}\right)^1 = -3\left(\tfrac{1}{2}\right) = -\tfrac{3}{2}$

$n = 2: a_2 = -3\left(\tfrac{1}{2}\right)^2 = -3\left(\tfrac{1}{4}\right) = -\tfrac{3}{4}$

$n = 3: a_3 = -3\left(\tfrac{1}{2}\right)^3 = -3\left(\tfrac{1}{8}\right) = -\tfrac{3}{8}$

$n = 4: a_4 = -3\left(\tfrac{1}{2}\right)^4 = -3\left(\tfrac{1}{16}\right) = -\tfrac{3}{16}$

$n = 5: a_5 = -3\left(\tfrac{1}{2}\right)^5 = -3\left(\tfrac{1}{32}\right) = -\tfrac{3}{32}$

The first five terms are
$-\tfrac{3}{2}, -\tfrac{3}{4}, -\tfrac{3}{8}, -\tfrac{3}{16}$, and $-\tfrac{3}{32}$. This sequence
has a common ratio, $r = \tfrac{1}{2}$, so the sequence is geometric.

3. $a_1 = 2$, $a_2 = 3$, $a_n = a_{n-1} + 2a_{n-2}$, for $n \geq 3$

$n = 3: a_3 = a_2 + 2a_1 = 3 + 2(2) = 3 + 4 = 7$

$n = 4: a_4 = a_3 + 2a_2 = 7 + 2(3) = 7 + 6 = 13$

$n = 5: a_5 = a_4 + 2a_3 = 13 + 2(7) = 13 + 14 = 27$

The first five terms are 2, 3, 7, 13, and 27. There is no common difference or common ratio, so the sequence is neither arithmetic nor geometric.

4. $a_1 = 1$ and $a_3 = 25$

$a_n = a_1 + (n-1)d \Rightarrow a_3 = a_1 + 2d \Rightarrow$
$25 = 1 + 2d \Rightarrow 2d = 24 \Rightarrow d = 12$
$a_5 = a_1 + (5-1)d \Rightarrow$
$a_5 = a_1 + 4d = 1 + 4(12) = 1 + 48 = 49$

5. $a_1 = 81$ and $r = -\dfrac{2}{3}$

$a_n = a_1 r^{n-1} \Rightarrow a_6 = 81\left(-\tfrac{2}{3}\right)^5 = 81\left(-\tfrac{32}{243}\right) = -\tfrac{32}{3}$

6. Arithmetic, with $a_1 = -43$ and $d = 12$

$S_n = \tfrac{n}{2}\left[2a_1 + (n-1)d\right] \Rightarrow$
$S_{10} = \tfrac{10}{2}\left[2a_1 + (10-1)d\right] = 5\left[2(-43) + 9(12)\right]$
$= 5(-86 + 108)$
$= 5(22) = 110$

7. Geometric, with $a_1 = 5$ and $r = -2$

$S_n = \dfrac{a_1\left(1 - r^n\right)}{1 - r} \Rightarrow$

$S_{10} = \dfrac{5\left[1 - (-2)^{10}\right]}{1 - (-2)} = \dfrac{5(1 - 1024)}{3} = \dfrac{5(-1023)}{3}$
$= 5(-341) = -1705$

8. $\displaystyle\sum_{i=1}^{30} (5i + 2)$

This sum represents the sum of the first 30 terms of the arithmetic sequence having $a_1 = 5 \cdot 1 + 2 = 7$ and
$a_n = a_{30} = 5 \cdot 30 + 2 = 150 + 2 = 152$. Thus, we have

$\displaystyle\sum_{i=1}^{30} (5i + 2) = S_{30} = \tfrac{n}{2}(a_1 + a_n) = \tfrac{30}{2}(a_1 + a_{30})$
$= 15(7 + 152) = 15(159) = 2385$

We could have also
$\displaystyle\sum_{i=1}^{30}(5i + 2) = \sum_{i=1}^{30} 5i + \sum_{i=1}^{30} 2 = 5\sum_{i=1}^{30} i + \sum_{i=1}^{30} 2$

Since $\displaystyle\sum_{i=1}^{n} c = cn$ and $\displaystyle\sum_{i=1}^{n} i = \dfrac{n(n+1)}{2}$, we have

$5\displaystyle\sum_{i=1}^{30} i + \sum_{i=1}^{30} 2 = 5 \cdot \dfrac{30(30+1)}{2} + 2(30)$

$= 5 \cdot \dfrac{30(31)}{2} + 60 = 5(15)(31) + 60$
$= 2325 + 60 = 2385$

9. $\displaystyle\sum_{i=1}^{5} (-3 \cdot 2^i)$

This sum represents the sum of the first five terms of the geometric sequence having
$a_1 = -3 \cdot 2^1 = -6$ and $r = 2$. Thus, we have

$\displaystyle\sum_{i=1}^{5} -3 \cdot 2^i = S_5 = \dfrac{a_1\left(1 - r^n\right)}{1 - r} = \dfrac{-6\left(1 - 2^5\right)}{1 - 2}$

$= \dfrac{-6(1 - 32)}{-1} = 6(-31) = -186$

Copyright © 2013 Pearson Education, Inc.

10. $\displaystyle\sum_{i=1}^{\infty}(2^i)\cdot 4$

This is the sum of an infinite geometric sequence with $r = 2$. Since $|r| > 1$, the sum does not exist.

11. $\displaystyle\sum_{i=1}^{\infty}54\left(\frac{2}{9}\right)^i$

This is the sum of the infinite geometric sequence with $a_1 = 54\left(\frac{2}{9}\right)^1 = 54\left(\frac{2}{9}\right) = 12$ and $r = \frac{2}{9}$. Since $\left|\frac{2}{9}\right| < 1$,

we can use the summation formula $S_{\infty} = \dfrac{a_1}{1-r}$.

$$\sum_{i=1}^{\infty}54\left(\frac{2}{9}\right)^i = S_{\infty} = \frac{a_1}{1-r} = \frac{12}{1-\frac{2}{9}} = \frac{12}{\frac{7}{9}} = 12\cdot\frac{9}{7} = \frac{108}{7}$$

12. $(x+y)^6 = x^6 + \dbinom{6}{1}x^5 y + \dbinom{6}{2}x^4 y^2 + \dbinom{6}{3}x^3 y^3 + \dbinom{6}{4}x^2 y^4 + \dbinom{6}{5}xy^5 + y^6$

$ = x^6 + \dfrac{6!}{1!5!}x^5 y + \dfrac{6!}{2!4!}x^4 y^2 + \dfrac{6!}{3!3!}x^3 y^3 + \dfrac{6!}{4!2!}x^2 y^4 + \dfrac{6!}{5!1!}xy^5 + y^6$

$ = x^6 + 6x^5 y + \dfrac{6\cdot 5}{2\cdot 1}x^4 y^2 + \dfrac{6\cdot 5\cdot 4}{3\cdot 2\cdot 1}x^3 y^3 + \dfrac{6\cdot 5}{2\cdot 1}x^2 y^4 + 6xy^5 + y^6$

$ = x^6 + 6x^5 y + 15x^4 y^2 + 20x^3 y^3 + 15x^2 y^4 + 6xy^5 + y^6$

13. $(2x-3y)^4 = (2x)^4 + \dbinom{4}{1}(2x)^3(-3y) + \dbinom{4}{2}(2x)^2(-3y)^2 + \dbinom{4}{3}(2x)(-3y)^3 + (-3y)^4$

$ = 16x^4 + \dfrac{4!}{1!3!}(8x^3)(-3y) + \dfrac{4!}{2!2!}(4x^2)(9y^2) + \dfrac{4!}{1!3!}(2x)(-27y^3) + 81y^4$

$ = 16x^4 - 4(24x^3 y) + \dfrac{4\cdot 3}{2\cdot 1}(36x^2 y^2) - 4(54xy^3) + 81y^4$

$ = 16x^4 - 96x^3 y + 216x^2 y^2 - 216xy^3 + 81y^4$

14. To find the third term in the expansion of $(w-2y)^6$, use the formula $\dbinom{n}{k-1}x^{n-(k-1)}y^{k-1}$ with $n = 6$ and

$k = 3$. Then $k - 1 = 2$ and $n - (k-1) = 4$. Thus, the third term is as follows.

$$\dbinom{n}{k-1}x^{n-(k-1)}y^{k-1} = \dbinom{6}{2}w^{6-2}(-2y)^2 = \frac{6!}{2!4!}w^4(4y^2) = \frac{6\cdot 5\cdot 4!}{2!4!}(4w^4 y^2) = \frac{6\cdot 5}{2\cdot 1}(4w^4 y^2) = 60w^4 y^2$$

15. $8! = 8\cdot 7\cdot 6\cdot 5\cdot 4\cdot 3\cdot 2\cdot 1 = 40{,}320$

16. $C(10,2) = \dfrac{10!}{(10-2)!2!} = \dfrac{10!}{8!2!} = \dfrac{10\cdot 9\cdot 8!}{8!2!} = \dfrac{10\cdot 9}{2\cdot 1} = 45$

17. $\dbinom{7}{3} = C(7,3) = \dfrac{7!}{(7-3)!3!} = \dfrac{7!}{4!3!} = \dfrac{7\cdot 6\cdot 5\cdot 4!}{4!3!} = \dfrac{7\cdot 6\cdot 5}{3\cdot 2\cdot 1} = 35$

18. $P(11,3) = \dfrac{11!}{(11-3)!} = \dfrac{11!}{8!} = \dfrac{11\cdot 10\cdot 9\cdot 8!}{8!} = 11\cdot 10\cdot 9 = 990$

Copyright © 2013 Pearson Education, Inc.

19. Let S_n be the statement $1+7+13+\cdots+(6n-5)=n(3n-2)$.

Step 1: Show that the statement is true for $n=1$.

S_1 is the statement $6(1)-5=1(3(1)-2)\Rightarrow 6-5=1(1)\Rightarrow 1=1$, which is true.

Step 2: Show that if S_k is true, S_{k+1} is also true. S_k is the statement. $1+7+13+\cdots+(6k-5)=k(3k-2)$

S_{k+1} is the statement $1+7+13+\cdots+[6(k+1)-5]=(k+1)[3(k+1)-2]$.

Start with S_k: $1+7+13+\cdots+(6k-5)=k(3k-2)$ and add the $(k+1)$ st term, $[6(k+1)-5]$, to both sides of this equation.

$$1+7+13+\cdots+(6k-5)=k(3k-2)$$
$$1+7+13+\cdots+(6k-5)+[6(k+1)-5]=k(3k-2)+[6(k+1)-5]$$
$$1+7+13+\cdots+(6k-5)+[6(k+1)-5]=3k^2-2k+(6k+6-5)$$
$$1+7+13+\cdots+(6k-5)+[6(k+1)-5]=3k^2+4k+1$$
$$1+7+13+\cdots+(6k-5)+[6(k+1)-5]=(k+1)(3k+1)$$
$$1+7+13+\cdots+(6k-5)+[6(k+1)-5]=(k+1)[(3k+3)-2]$$
$$1+7+13+\cdots+(6k-5)+[6(k+1)-5]=(k+1)[3(k+1)-2]$$

It has been shown that S_k implies S_{k+1}. Therefore, by mathematical induction, S_n is true for every positive integer n.

20. Using the fundamental principle of counting, we have $4\cdot 3\cdot 2=24$.

There are 24 different kinds of shoes if there are 4 styles, 3 different colors, and 2 different shades.

21. We are choosing four people from a group of ten without regard to order, so there are $C(10,4)$ ways to do this.

$$C(10,4)=\frac{10!}{(10-4)!\,4!}=\frac{10!}{6!\,4!}=\frac{10\cdot 9\cdot 8\cdot 7\cdot 6!}{6!\,4!}$$
$$=\frac{10\cdot 9\cdot 8\cdot 7}{4\cdot 3\cdot 2\cdot 1}=210$$

There are 210 ways that four officers can be selected to attend a seminar.

If two women and two men must be included, we have $C(4,2)$ ways to choose the women and $C(6,2)$ ways to choose the men.

$$C(4,2)=\frac{4!}{(4-2)!\,2!}=\frac{4!}{2!\,2!}=\frac{4\cdot 3\cdot 2!}{2!\,2!}$$
$$=\frac{4\cdot 3}{2\cdot 1}=6$$
$$C(6,2)=\frac{6!}{(6-2)!\,2!}=\frac{6!}{4!\,2!}=\frac{6\cdot 5\cdot 4!}{4!\,2!}$$
$$=\frac{6\cdot 5}{2\cdot 1}=15$$

Altogether, then, we have $6\cdot 15=90$ ways to select two women and two men.

22. The student can select the courses in $15\cdot 14\cdot 13\cdot 12=32,760$ ways.

23. **(a)** $P(\text{red three})=\dfrac{n(\text{red three cards})}{n(\text{cards in deck})}$

$$=\frac{2}{52}=\frac{1}{26}$$

(b) Let F be the event "draw a face card." Each suit contains 3 face cards (jack, queen, and king), so the deck contains 12 face cards. Thus $n(F)=12$ and

$P(F)=\dfrac{12}{52}=\dfrac{3}{13}$ The probability of drawing a card that is not a face card is

$$P(F')=1-P(F)=1-\frac{3}{13}=\frac{10}{13}.$$

(c) Let K be the event "king is drawn" and S be the event "spade is drawn." There are

4 kings, so $n(K)=4$ and $P(K)=\dfrac{4}{52}=\dfrac{1}{13}$

There are 13 spades, so $n(S)=13$ and

$P(S)=\dfrac{13}{52}=\dfrac{1}{4}$. There is one card which is both a king and a spade, so

$n(K\cap S)=1$ and $P(K\cap S)=\dfrac{1}{52}$. Thus we have

$$P(K \text{ or } S)=P(K\cup S)$$
$$=P(K)+P(S)-P(K\cap S)$$
$$=\frac{1}{13}+\frac{1}{4}-\frac{1}{52}=\frac{16}{52}=\frac{4}{13}$$

Copyright © 2013 Pearson Education, Inc.

(d) Consider the event F: "draw a face card."
As shown in the solution to part (b),

$$P(F) = \frac{3}{13} \text{ and } P(F') = \frac{10}{13}. \text{ Thus, the}$$

odds in favor of drawing a face card are

$$\frac{P(F)}{P(F')} = \frac{\frac{3}{13}}{\frac{10}{13}} = \frac{3}{13} \cdot \frac{13}{10} = \frac{3}{10} \text{ or 3 to 10}$$

24. "At most 2" means "0 or 1 or 2."

$$\begin{aligned} P(\text{at most 2}) &= P(0 \text{ or } 1 \text{ or } 2) \\ &= P(0) + P(1) + P(2) \\ &= 0.19 + 0.43 + 0.30 = 0.92 \end{aligned}$$

The probability that at most 2 light bulbs are defective is 0.92.

25. In this binomial experiment, we call rolling a five a success. Then $n = 6$, $r = 2$, and $p = \dfrac{1}{6}$.

$$\begin{aligned} P(2 \text{ fives}) &= \binom{n}{r} p^r (1-p)^{n-r} \\ &= \binom{6}{2}\left(\frac{1}{6}\right)^2 \left(1 - \frac{1}{6}\right)^{6-2} \\ &= \frac{6!}{(6-2)!2!}\left(\frac{1}{6^2}\right)\left(\frac{5}{6}\right)^4 \\ &= \frac{6!}{4!2!}\left(\frac{1}{6^2}\right)\left(\frac{5^4}{6^4}\right) = \frac{6 \cdot 5 \cdot 4!}{4!2!}\left(\frac{5^4}{6^6}\right) \\ &= \frac{6 \cdot 5}{2 \cdot 1}\left(\frac{5^4}{6^6}\right) = 15\left(\frac{5^4}{6^6}\right) \approx 0.201 \end{aligned}$$

Copyright © 2013 Pearson Education, Inc.